Advance Praise for Paul Gipe's *Wind Energy for the Rest of Us*

"The ultimate account of how we can take the breeze that washes daily across our planet and use it to power our lives. There's something quietly thrilling about this book!"
—**Bill McKibben**, author of *Deep Economy: The Wealth of Communities and the Durable Future*

"No one knows more about the promise and pitfalls of wind power than Paul Gipe. With stunning photographs, accessible writing, and an eye for fakery and hypocrisy, he has produced one of the finest books on wind power available. Truly a masterwork. We owe him a great debt as we take the first steps toward a more sustainable energy future."
—**Martin J. Pasqualetti**, Professor in the School of Geographical Sciences and Urban Planning, Arizona State University and author of *The Renewable Energy Landscape*

"This is the most comprehensive book on wind power I know of, and its focus on developing wind energy outside the purview of the big utility companies is unique and subversive. Paul Gipe has done us all a great service."
—**Richard Heinberg**, Senior Fellow, Post Carbon Institute and author of *Our Renewable Future: Laying the Path for 100% Clean Energy*

"Germany's Energiewende--its energy revolution--is due to thousands of citizens working together in a grassroots movement. Paul Gipe shows how the energy and enthusiasm of the people created a technology that grew from the bottom-up--not the top down. An important book and an essential message."
—**Hermann Albers**, President, German Wind Turbine Owners Association (BWE)

"Paul Gipe brings over three decades of wind energy experience to bear in this masterful one-stop compendium containing everything we want to know about wind energy."
—**Mark Jacobson**, Professor of Civil and Environmental Engineering, Stanford University

"Over the last 40 years, I've learned more about wind power from Paul Gipe than from any other source. Now he has put it all into one splendid new book. *Wind Energy for the Rest of Us* expertly addresses every question about technology, policy, and economics that is worth answering."
—**Denis Hayes**, Chair of Earth Day 2020 and former director of the National Renewable Energy Laboratory

"Gipe has been on energy in North America . . . effectively straddling the line from technical to political while communicating wind's promise to the general public."
—**Tim Weis**, former Policy Director, Canadian Wind Energy Association (CanWEA)

"Paul Gipe has been involved with wind energy since its revival in the late 1970s. His extensive experience makes him uniquely qualified to be the author of this fascinating and informative book."
—**Tom Gray**, former Executive Director, American Wind Energy Association (AWEA)

"Paul Gipe's monumental new book covers a lot of ground. Big turbines, small turbines, water-pumpers, early developments to the latest advances—whatever your interest—it's all here in this comprehensive—and carefully researched book."
—**Greg Pahl**, author of *Power from the People: How to Organize, Finance, and Launch Local Energy Projects*

"Gipe's call for an ethical energy policy in *Wind Energy for the Rest of Us* is a message that North American politicians should heed. The people deserve nothing less."
—**Glen Estill**, Past President of the Canadian Wind Energy Association, and successful wind entrepreneur

"*Wind Energy for the Rest of Us* is more about sociology than technology; an environmentalist's 'sat nav' for living more lightly on a confused planet. Gipe uses wind as a metaphor for society's efforts to harness the unlimited renewable resources that surround us. Huge technology leaps, cutting the cost of wind power, become portals through which whole energy systems are set to become more accountable, democratic, and sustainable. Gipe's is a book for those who grasp that it's never too late to be all the things we ever wanted to be. Read it. Ride its currents. Then use it to change everything. This is our moment."
—**Alan Simpson**, former British MP, climate campaigner, and energy policy advisor

"This is the most comprehensive book on wind energy you can find. Period. No one has more experience writing about the subject than Paul Gipe. This book is packed with everything you need to know, whether you are thinking about building your own small turbine or investing in the industry. Wonderful photos, illustrations, and charts help the information come alive, as does Gipe's lively writing and a healthy dose of case studies—from wind energy pioneers to a group of nuns who wanted to generate power more in keeping with the Bible. This book is at once a powerful resource and an engaging read. It will help the reader get a solid foundation in this fast-growing, exciting source of clean energy."

—**Brian Howard Clark**, journalist and author of *Build Your Own Small Wind Power System*

"Paul Gipe has done it again--straight talk on wind electricity from a wind journalist without peer. Gipe covers the important ground with clarity and detail, in a no-nonsense style that clears away the clouds of misinformation and hype too common in the popular and social media. Based on science and experience, grounded in a sensible environmental ethic, his newest book will stand the test of time."

—**Ian Woofenden**, Home Power magazine senior editor, and author of *Wind Power for Dummies*

"Few are more qualified to write a book titled *Wind Energy for the Rest of Us* than Paul Gipe. He proves again that he is one of the world's most knowledgeable authorities on wind energy. More importantly, he understands the politics behind energy policy. Following his recommendations would not only create a better environment, but also lead to a fairer and more just world."

—**Stefan Gsanger**, Executive Director, World Wind Energy Association

"Paul Gipe has been involved in the wind industry since the 1970's, wrenching, writing, observing, critiquing, offering kudos and criticism where warranted. His book represents a lifetime of experience that is unmatched by others simply because few have the breadth of involvement that four decades brings. From small wind to megawatt turbines to the latest "thinking out of the box" breakthrough fantasy technologies, Gipe includes them all."

—**Mick Sagrillo**, Sagrillo Power & Light

"Gipe gives us the thrilling story of the social movement that created wind energy and enabled it to grow so rapidly—the movement that took power from the utilities and put into the hands of citizens and local communities. Read it, and become part of the movement!"

—**Tore Wizelius**, author of *Windpower Ownership in Sweden: Business models and motives*

"He must be powered by the wind himself this Paul Gipe. He endlessly shares the stories of those who have made wind energy work. He makes you believe in a sustainable future and feel the wind of change!"

—**Søren Hermansen**, Danish community power advocate on Samsø—the 100% Renewable Energy Island

"Wind Energy for the Rest of Us is the most definitive account of wind energy you are likely to find. Pure and simple. It doesn't matter whether you are an engineer, lobbyist, manager, academic, wind developer, or just somebody who wants to know about wind power, this book is the source for helping you understand the technology, how it developed, where it's going, and how it fits in with the world. Paul Gipe is the master of wind energy, and this book is his masterpiece."

—**David Toke**, Reader in Energy Politics, University of Aberdeen, Scotland

"The wind is like the sun, a gift from the heavens. In all languages the wind is identified with the spirit--and it blows for everyone. This impressive work by Paul Gipe shows that the winds of change are unstoppable and that they are there for all of us. If we want a world free from conflict, we must build as many wind turbines as fast as possible--each turbine a sign of peace."

—**Franz Alt**, journalist, theologian, and author of *Der ökologische Jesus* and *Krieg um Öl oder Frieden durch die Sonne*

"Paul Gipe's written and spoken words have brought many into the world of wind power, both in North America and abroad. He's much more than a journalist, he writes from first-hand knowledge of the wind industry where he's helped define best practices and drive policy initiatives. The North American experience is very different from that of Europe. Paul has seen them both and can share their triumphs as well as their failures."

—**Lisa Daniels**, Founder & Executive Director, Windustry

"A massive expansion of wind power is necessary for the transition to 100% renewable energy if we want to mitigate climate change and avoid conflicts over resources. Paul Gipe's book shows the path forward."

—**Hans-Josef Fell**; former German MP, co-author of Germany's Renewable Energy Resources Act, and President of the Energy Watch Group

"Wind energy has a rich history and a wide array of technology and applications; Gipe covers it all. Anyone interested in wind will have this book on the shelf and refer to it often."

—**Brent Summerville**, Technical Director, Small Wind Certification Council

"Wind energy is more important now than ever and Paul Gipe rises to the occasion with yet another comprehensive but highly readable briefing for all of us."

—**Hugh Piggott**, Scoraig Wind Electric, Scotland

WIND ENERGY
FOR THE REST OF US

*A Comprehensive Guide to
Wind Power and How to Use It*

PAUL GIPE

Introducing Electricity Rebels and
How They Are Changing the
Face of Wind Energy

© Copyright 2016 by Paul Gipe. All rights reserved. No part of this book may be transmitted or reproduced in any form by any means without permission in writing from Paul Gipe.

Designer: Peter Holm, Sterling Hill Productions
Copy Editor: Kate Meuller, Pendragon Productions
Proofreader: Lori Lewis
Indexer: Carol Frenier, The Advantage Group

Printed in the United States

First printing

Library of Congress Control Number: 2016912305
Gipe, Paul B.
Wind Energy for the Rest of Us: A Comprehensive Guide to Wind Power and How to Use It / Paul Gipe.
Includes bibliographical references and index.
ISBN 978-0-9974518-1-8 (pbk.) — ISBN 978-0-9974518-0-1 (ebook)
1.Wind Power. 2. Wind Turbines 3.Wind Turbines—Environmental aspects. 4. Electric power production. 5. Distributed generation of electric power. 6. Community development.
621.31'2136—dc22

wind-works.org
606 Hillcrest Dr.
Bakersfield, CA 93305-1413
(661) 325-9590
www.wind-works.org

All photographs copyright © Paul Gipe, except where noted.

DISCLAIMER: The installation and operation of a wind turbine entails a degree of risk. Always consult the manufacturer and applicable building, electrical, and safety codes before installing or operating your wind power system. Recommendations in this book are not a substitute for the directives of wind turbine manufacturers or regulatory agencies. The author assumes no liability for personal injury, property damage, or loss from using information contained in this book.

DISCLOSURE: Paul Gipe has worked with Aerovironment, ANZSES, An Environmental Trust, APROMA, ASES, AusWEA, AWEA, David Blittersdorf, Jan & David Blittersdorf Foundation, BWEA, BWE, CanWEA, Canadian Co-operative Assoc., CAW, CEERT, Deutsche Bank, DGW, DSF, EECA, ES&T, GEO, GPI Atlantic, IREQ, KWEA, MADE, Microsoft, ManSEA, MSU, NRCan, NRG Systems, NASA, NREL, NZWEA, ORWWG, OSEA, Pembina, PG&E, SeaWest, SEI, TREC, USDOE, WAWWG, WE Energies, the Folkecenter for Renewable Energy, the Izaak Walton League, the Minnesota Project, the Sierra Club, the World Future Council, and Zond Systems and has written for magazines in Australia, New Zealand, Canada, France, Denmark, Germany, and the USA.

TO MY FAMILY AND FRIENDS, WHOSE ENCOURAGEMENT
HAS GIVEN ME THE FREEDOM TO CHOOSE MY OWN PATH.

CONTENTS

Preface, xi
Acknowledgments, xii

1. How We Use the Wind Today — 1

Wind Power Plants, 3: *What's in a Name?*, 4; *Wind Power Plant Arrays*, 4 • Distributed Wind, 7: *Urban and Village Wind*, 7; *End of the Line and Beyond*, 8 • Specialty Applications, 10 • Electric Vehicle Charging, 10 • Heating, 11, **Sacred Winds, 12** • Pumping Water, 13

2. How to Use This Book — 15

How This Book Is Organized, 15 • **Using www.wind-works.org, 16** • Nomenclature: What Are We Talking About?, 17: *Wind Machine and Wind Turbine*, 17; *Power and Energy: There Is a Difference*, 18; *Watt's More*, 19 • Equations: They're Informative, 19 • False Precision—How to Avoid It, 19 • **How *Wind Energy for the Rest of Us* Differs from *Wind Power* (2004), 20** • Units: Yes, Metrics Too, 21 • **Units of Measurement, 21**; Size: It's All Relative, 21, **Wind Energy Workshops, 24, Attention: The Wind Industry is Dynamic, 25**; Some do's and Dont's on Investing in Wind Energy, 26

3. Where It All Began — 27

How Far We've Come, 28 • In the Beginning, 29: *James Blyth*, 29; *Duc de Feltre*, 30; *Brush Dynamo*, 31; *Dead Ends*, 32; *The Danish Edison*, 32 • The Interwar Years, 35: *First Interconnected Wind Turbine*, 35; *Wind Meets Aviation*, 36; *Wind Chargers*, 37 • War Years, 37: *Smith-Putnam*, 38; *Ventimotor and the Third Reich*, 38; *Denmark and F.L. Smidth*, 40; **Wind Technology Known Well by Mid-1950s, 42** • Postwar Years, 42: *Germany's Allgaier*, 42; *Denmark's Johannes Juul*, 44; **Stall Regulation and High Power Ratings, 45**; *Wind Experimentation Elsewhere*, 47

4. The Great Wind Revival — 51

Why the History of Modern Wind Energy is Important, 52 • Large Turbines from the Top Down, 52: *Bringing NASA Down to Earth*, 52; *Germany's Growian*, 57; *Going Beyond Juul—Denmark*, 58 • Denmark's Rebels, 59; **When My Role Began, 60**; *Danish Carpenter*, 60; *Tvindkraft: The Giant That Shook the World*, 61; **Smedemestermølle: The Blacksmith's Turbine, 64**; *Blades That Set the Industry in Motion*, 64; *The Danish Concept*, 68 • The California Wind Rush, 68 • American Designs of the Early 1980s, 69: *Downwind Dominant*, 69; *Lightweights in a Heavyweight Environment*, 70; *US Windpower*, 72; *Enertech's E44*, 74; *Mehrkam*, 74; Wind Energy and the Aerospace Arts, 75; **An Aerospace Success Story: Bergey Windpower, 75** • Bottom-Up Delivered, 77 • Boom and Bust Survivors, 77; **South of the Border (Enercon), 78** • Right Product, Right Place, Right Time, 78 • Wind Turbine Owners' Association, 78 • Design Standards, 80 • Beginning of the Modern Era, 81

5. What Works and What Doesn't 83

Orientation, 84: *Passive Yaw*, 85; **Tail Vanes, 85**; *Active Yaw*, 87 • Lift and Drag, 87 • Aerodynamics, 89: *Apparent Wind and the Angle of Attack*, 90; *Twist and Taper*, 91; *Solidity*, 93; *Betz's Limit*, 94; *Tip-Speed Ratio*, 94; *Blade Number*, 95; **One-Blade Wind Turbines, 96**; *Self-Starting*, 100 • Blade Materials, 101: **Tip Vanes and Winglets, 101**; *Wood*, 102; *Metal*, 102; *Fiberglass*, 104 • Hubs, 105 • Drivetrains, 106: *Small Turbines*, 108; *Medium-size Turbines*, 110; *Large Turbines*, 111; *Other Forms of Transmission*, 113 • Generators, 114: **Permanent Permanent Magnets?, 115**; *Alternators*, 116; **Air-gap or Axial-Flux Generators, 117**; *Variable- or Constant-Speed Operation*, 118; *Induction (Asynchronous) Generators*, 119; *Dual Generators or Dual Windings*, 120 • Overspeed Control, 121; *Horizontal Furling*, 123; *Vertical Furling*, 124; *Coning*, 125; *Changing Blade Pitch*, 126; *Aerodynamic Stall*, 129; *Mechanical Brakes*, 130; *Aerodynamic Brakes*, 131 • Putting It All Together, 134: *Small Turbines*, 134; *Large Turbines*, 135; **Dynamic Breaking: Is it Now Enough?, 135**

6. Vertical-Axis and Darrieus Wind Turbines 137

Lift and Drag, 138; **My Take on VAWTS, 138** • Blade Number, 139 • Towers, 140; **Beware VAWT Resources on the Web, 141** • Φ-Configuration Darrieus Development, 142: *Vestas's Cantilevered Bi-blade Darrieus*, 143; *DAF-Indal*, 143; *Alcoa's VAWT*, 146; *FloWind and The World's Most Successful Darrieus*, 147; *Éole*, 151 • H-Configuration or Straight-Blade Darrieus, 151: *McDonnell Aircraft's Giromill*, 153; *Pinson Cycloturbine*, 154; *Mike Bergey*, 155 • Fixed-Pitch H-Rotor VAWTs, 155: *Musgrove Variable-Geometry VAWT*, 156; *Cleanfield*, 157; *Mariah Windspire*, 159; *Helical Wind Turbines*, 160; **Those Who Don't Build VAWTs, 162** • VAWT Revival: Not Likely to Continue, 163; **Poor Comparison Between Small VAWT and Small HAWT, 163** • Claims and Counterclaims, 164; *Omnidirectional*, 164; *Simpler*, 165; *More Reliable*, 165; *Less Costly*, 165; *More Powerful*, 165; **Monsieur Darrieus and His Wind Turbines, 166**; *More Efficient*, 166; *More Cost Effective*, 166; *Safe for Birds*, 167; *Less Noisy*, 167 • VAWT Design Characteristics, 167: *Efficiency and Performance*, 168; *Stall Control and Overspeed Protection*, 169; *Self-Starting*, 170; *Fatigue*, 170; *Guyed Darrieus*, 170 • VAWTs Now Marginal, 170 • Certification to Minimum Testing Standards, 171 • Conspiracy against VAWTs, 171; **Debunking Pyramidal Power and Magical Mag-Wind, 172**

7. Novel Wind Systems 175

Advice for Inventors of New Wind Turbines and For Everyone Else As Well, 176 • Ducted or Augmented Turbines, 176: *Enflo*, 177; *Eléna 30: Will They Ever Learn?*, 178; *New Zealand's Vortec 7*, 179; **Vortec 7 Promoters on Gipe's Criticism, 179**; *FloDesign (Ogin)*, 180; **Warning: Rebranding DAWTs, 181**; *Better Than Betz?*, 182 • Airborne Wind Energy Systems (Kites), 183; **Mike Barnard on Wind Technology Red Flags, 184** • Wind Ships, 185: *Traction Kites*, 186; *Flettner Rotors*, 187; *Enercon's E-Ship 1*, 188 • The Takeaway, 190

8. Silent Wind Revolution 191

What Is a Wind Turbine?, 192 • Generator Ratings, 192 • Swept Area Trumps Generator Ratings, 192 • Metrics of Productivity, 193 • Measures of Relative Swept Area, 194 • Historical Abuse of Power Ratings, 195 • Wind Turbine Design and Wind Regimes, 196 • Small and Medium-Size Turbines, 197; **Case Study Germany: New Wind Turbines Expand the Wind Resource, 198** • Specific Capacity and Capacity Factor (Full-Load Hours), 199; **Relationship Between Capacity Factor, Yield, and Full-Load, 200** • Why All This Is Important, 201

9. Towers 203

Height, 203; **Stratospheric Heights, 205** • Buckling Strength, 205; **Drag Force and Thrust, 206**; **Rocking and Rolling with Wind, 206** • Tower Types, 207: *Freestanding Towers*, 207; *Guyed Towers*, 212 • Rooftop Mounting, 215; **Rooftop Wind in Action, 217** • Unconventional Towers, 218: *Silos*, 218; *Farm Windmill Towers*, 218; *Steel Pipe*, 218; *Wood Towers*, 218; *Tripod Tower and Platform*, 220 • Other Considerations, 221: *Aesthetics*, 221; *Space*, 221; *Maintenance on Small Wind Turbines*, 222; *Ease of Installation for Small Wind Turbines*, 222; *Access to Large Wind Turbines*, 222

10. Measuring the Wind 225

Wind: What Is It?, 226; **Wind Speed Units, 228** • Wind Speed and Time, 228 • Power in the Wind, 229: **The Beaufort Scale, 230; Wind Speed Notation, 231; Power Density, 231; International Standard Atmosphere, 232;** *Air Density*, 232; **Air Density, 233;** *Swept Area*, 234; *Wind Speed*, 234; *Speed Distributions*, 235; **Frequency Distributions, 237** • Wind Speed, Power, and Height, 238; **Lograithmic Model of Wind Shear, 239; The Wind Shear Exponent α, 241; The Nocturnal Jet, 242** • Published Wind Data, 242; **Calculating the Wind Shear Exponent (α), 243; Online Wind Resources and Wind Calculators, 244** • Surveying the Wind at Your Site, 244: **Estimating the Height of Obstructions, 246;** *Measuring Instruments*, 246; *Anemometer Towers*, 247; *Survey Duration*, 248; **North American Wind Resource Maps, 249;** *Data Analysis*, 250

11. Estimating Performance 251

Swept Area Method, 252: *Small Wind Turbines*, 253; **Calculating Swept Area, 255;** *Large Wind Turbines*, 256; *Annual Yield by IEC Class*, 257; **Swept Area Rules of Thumb, 257; Power Curve Nomenclature, 258** • Power Curve Method, 258: *The Method of Bins*, 259; **Avoid Average Speed Confusion, 259;** *Large Turbine Power Curve*, 260 • Manufacturers' Estimates, 261; **Web-Based Calculators of AEP, 262** • Wind Power Plant Losses, 262 • Estimating Fleet Performance, 262 • Putting It All Together, 263

12. Off-the-Grid Power Systems 265

Hybrids, 266: **Tale of Two Cities, 267** *Reducing Demand*, 267; *AC and DC Systems*, 269; **Cutting Consumption, 269;** *Sizing*, 270; **Micro Hybrid Power Systems, 272;** *Inverters*, 272; *Batteries*, 273; **US Solar and Wind Data, 273; Cabin-Sized Power System, 274;** *Backup Generators*, 275; **Household-Sized Hybred Power System, 275** • Stand-Alone Economics, 276 • Other Stand-Alone Power Systems, 276: *Telecommunications*, 276; *Village Electrification*, 277; **Village Self-Reliance, 278;** *Wind-Diesel Twinning*, 278

13. Interconnection and Grid Integration 281

Models of Interconnection, 282; **Breaking Free From Net Metering, 283** • Interconnection Technology, 283: *Induction or Asynchronous Generators*, 285; *Electronic Inverters*, 285 • Power Quality and Safety, 286: *Power Factor*, 288; *Voltage Flicker*, 289; *Harmonics*, 290 • Net Metering, 290 • Degree of Self-Use, 291 • Dealing with the Utility, 291 • Distributed Generation, 292 • Grid Integration, 293: *Wind's Variability*, 294; *Capacity Credit*, 296; *Balancing Cost*, 297; *Penetration*, 298; *It's All in the Mix*, 298; *Storage*, 299

14. Pumping Water 301

Windmills That Won the West, 301 • Mechanical Wind Pumps, 303: *Pumping Head*, 304 • Estimating Farm Windmill Pumping Capacity, 305 • Counterbalancing for Wind Pumps, 307 • **Farm Windmill Conversion?, 308**; Electrical Wind Pumps, 308 • Storage, 310 • Irrigation and Drainage, 312 • Wind Pump Heritage, 313

15. Siting and Environmental Concerns   315

Antiwind Groups, 316 • Tower Placement, 317: *Exposure and Turbulence*, 318; *Power Cable Routing*, 319; **Wind Turbine Noise: Rumors, Gossip, Lies, and Far-Fetched Stories, 320** • Planning Permission, 322 • Building Permit, 322 • Public Safety, 325: *Falling Blades*, 325; *Falling Ice*, 326; *Attractive Nuisance*, 326; *Height Restrictions on Small Turbines*, 327; *Aviation Obstruction Marking*, 327; *Safety Setbacks*, 328 • Noise, 331: *Decibels*, 332; *Weighting Scales*, 333; *Exceedance Levels*, 333; **Noise Propagation Conspiracy?, 334**; *Noise Propagation*, 334; *Ambient Noise*, 335; **Will It Be Heard?, 335**, *Community Noise Standards*, 336; *Sound Power Levels*, 337; *Wind Turbine Noise*, 338; *Estimating Noise Levels*, 340; *Lowering Wind Turbine Noise*, 341; *Noise Annoyance*, 342; **Source of Small Turbine Noise, 343**; **Noise, Health, and Safety, 344**; *Noise and Public Health*, 344; *Consequences*, 345; *Be Considerate*, 345 • Television and Radio Interference, 346 • Shadow Flicker, 346 • **Shadow Flicker, 347** • Disco Effect, 347 • Birds and Bats, 348: *Pre- and Postconstruction Surveys*, 351; *Bats*, 351; *No Free Lunch*, 352 • Property Values, 352 • Land Area Required, 354: *Land Area Occupied*, 354; *Land Area Used*, 355 • Energy Balance and Energy Return on Energy Invested, 356 • Emissions of CO_2 Equivalent Gases, 357 • Water Consumption, 358 • Removal Bonds, 358; **Replacing the Old with the New, 359** • **Aesthetics Design Summary, 360**; Aesthetics, 360: *Provide Visual Uniformity*, 361; *Remove Headless Horsemen*, 361; *Use Open Spacing*, 361; *Avoid Billboards and Logos*, 361; *Bury Power Lines*, 362; *Always Dress Properly*, 362; *Control Erosion and Promptly Revegetate Sites*, 363; *Harmonize Ancillary Structures*, 363; *Keep Sites Tidy*, 364; *Inform the Public*, 364; *Small Turbines*, 365 • Compatible Land Uses, 365; **Will It Be Seen?, 366**

16. Installation and Dismantling   373

Thoughts on Doing it Yourself, 374 • Parts Control, 375 • Foundations and Anchors, 375; *Anchors*, 376; *Working with Concrete*, 376; *Guyed Towers*, 378; *Freestanding Towers*, 380; *Novel Foundations*, 380 • Assembly and Erection of Guyed Towers, 381; *Guy Cables*, 381; *Using a Crane*, 384; *Using a Gin Pole*, 384 • Freestanding Towers: Assembly and Erection, 385: *Tubular Towers*, 387 • Tilt-Up Towers: Assembly and Erection, 388: *Tilt-Up Guyed Towers*, 390; *Griphoists*, 391 • Wiring, 394: **Up-Tower Block Connectors For Micro Turbines, 395**; *Aboveground and Buried Cable*, 396; *Strain Relief of Tower Conductors*, 396; *Conductors and Conductor Sizing*, 397; **Junction or J-Boxes, 400**; *Conduit Fill*, 400; *Surge Protection*, 400; **Grounding Nets, 401**; *Additional Notes on Wiring*, 401 • Decommissioning and Dismantling, 402 • **Erecting a Micro Turbine with a Griphoist, 403**; Erecting a Household-Size Turbine with Crane, 406; Erecting a Large Turbine, 408; Dismantling a Large Wind Turbine: Windkraft Diemarden, 412

17. Safety   415

Fatal Accidents, 415 • Wind's Mortality Rate, 417; **Deaths in Wind Energy Database, 417** • Hazards, 418: *Falls*, 418; *Spinning Rotors*, 419; **Terry Mehrkam Thrown to His Death, 419**; *Electrical*, 421; *Construction*, 421; *Analysis*, 422 • Tower Safety and Fall Protection, 422; *Positioning Belts and Full-Body Harnesses*, 422; *Lanyards, Lifelines, and Anchorages*, 424; *Snap Hooks, Carabiners, and Slings*, 425; *Fall-Arresting Systems*, 426 • Work Platforms, 427; **Tower Work and Do-It-Yourselfers, 427** • Ladders, 429; **Dynamic Braking or Stop Switches for Small Wind Turbines, 430** • More Tower Tips, 430 • **Steen Aagaard's Crippling Fall, 432** • Blade Root Doors, 433 • Small Turbine Electrical Safety, 433 • Loss Prevention, 435

18. Operation and Maintenance   437

Small Wind Turbine First Rotation, 438: *Interconnected Wind Systems*, 438; *Battery-Charging Wind Systems*, 439 • Monitoring Performance, 439; *Small Wind Turbines*, 440 • Maintenance, 440: *Small Wind Turbines*, 440; *Balance of Remote Systems*, 442; *Large Wind Turbines*, 443 • Cost of Operations and Maintenance, 446: *Small Wind Turbines*, 446; *Large Wind Turbines*, 447

19. Investing in Wind Energy   449

Power Ratings and Cost Effectiveness, 449: **Fantasy Wind Turbines: If It's Too Good to Be True... or How to Spot Scams, Frauds, and Flakes, 451**; *Efficiency or Cost Effectiveness*, 452; *Measures of Cost Effectiveness*, 452 • **Shysters and Bozos, 454**; Small Wind Turbine: Testing and Standards, 454: *Standardized Tests*, 455; *Certification and Labeling*, 456; **Small Wind Turbine Certification, 457** • Buying a Small Wind Turbine for the Home, 458; *Controls*, 459; *Operational History*, 459; **A Small Wind System Is Much More Than a Wind Turbine, 459**; *Product Specifications*, 460; *Evaluating Vendors*, 461; **Ventilators and Squirrels in a Cage, 461**; *Contracts and Warranties*, 462; *What to Expect*, 462 • Financial and Economic Models, 463: *Cost of Energy*, 463; *Payback*, 464; *Cash-Flow Models*, 464; *Profitability Index Method*, 464 • Economic Factors, 465: *Installed Cost*, 465; *Subsidies and Incentives*, 465; **US Federal Tax Credits, 466; Paying for Performance, 466**; *Cost of Capital*, 467; *Annual Reoccurring Costs*, 467; **Wind Turbine Envy and Land Lease Pooling, 469**; *Taxes*, 470; *Revenue*, 470 • Putting It All Together, 474: *Simplified Cash Flow: Large Turbine*, 474; *Tariff Calculation: Large Turbine*, 475 • *Tariff Calculation: Small Commercial Turbine*, 476

20. Community Wind   479

The Third Way, 479 • Community Wind, 480 • What Is Community Wind?, 481 • Why Community Wind?, 482: *Greater Acceptance*, 483; *Greater Economic Benefits*, 483 • Cooperative and Mutual Investment, 484 • Characteristics of Community Wind, 485 • Denmark's *Fællesmølle and Vindmøllelaug*, 486: *Lynetten Vindmøllelaug*, 487; *Middelgrunden Vindmølleaug*, 488; *Hvidovre Vindmøllelaug*, 488 • Dutch Cooperatives, 488 • Germany's Electricity Rebels, 489; **Full Speed Ahead Says Friends of the Earth Germany: Wind Energy Is the Workhorse of the Energy Transition, 490**; *Friedrich-Wilhelm-Lübke-Koog*, 491; **Nordfriesland: Germany's Community Wind Capital and an Electricity Rebel Stronghold, 494**; *Saterland Bürgerbeteiligung*, 496; *German Genossenschaft or Cooperatives*, 496 • Community Wind in Britain, 497: *Baywind*, 497; *Westmill Wind Farm*, 497 • Community Power Down Under, 498 • Community Wind in North America, 499: **Community Wind North American Sources of Information, 500**; *Ontario*, 500; *Nova Scotia*, 503; *Massachusetts*, 504; *Minnesota*, 505; *Nevada, Iowa*, 506 • Who Owns the Wind?, 507 • What's Required to Make Community Wind Happen, 508

21. The Challenge   511

Pitfalls to Avoid, 511: *The Lure of Panaceas*, 511; **Offshore and Near Shore Wind, 512**; *Public Relations Puffery*, 513; *Too Cheap to Meter*, 513 • The North American Challenge, 513: *North American Consumption*, 514; *Swept Area Needed to Meet Consumption*, 514 • The Challenge, 515: *Offsetting Fossil-Fuel-Fired Generation*, 515; *Offsetting Oil in Passenger Vehicle Transport with EVs*, 516; *Manufacturing Capacity*, 517; *Land Area Required*, 518; *100% from Renewables*, 518; **100% Renewable Vision Building: Trend Toward New Targets of 100% Renewable Electricity–And Higher, 519**; **Onshore Wind Returns Three Times More Usable Energy in Transportation Than Investment in Oil, 520**; *Affordable*, 520; *Doable*, 521 • Electricity Feed Laws, 521: *Small Wind Tariff*, 523; *Differentiated Tariffs for Distributed Wind*, 524 • Energy for Life: The Pursuit of an Ethical Energy Policy, 525

Appendix 529

Constants and Conversions, 529 • Scale of Energy Equivalents, 531 • Scale of Equivalent Power, 531 • Battelle Wind Power Density Classes, 532 • American Wire Gauge to Metric Conversion, 532 • Nongovernmental Organizations (NGOs), 533 • Government-Sponsored or Affiliated Laboratories, 534 • Websites, 534 • Electronic Forums on Small Wind Turbines, 534 • Workshops, 534 • Community Wind Organizations, 535 • Community-Owned Projects Mentioned, 535: *Australia*, 535; *Canada*, 535; *Europe*, 535; *USA*, 535 • Prowind Groups, 536 • Historical Sites and Museums, 536; *Museums with Wind Exhibits*, 536; *Open-Air Museums: North America*, 536; *Open-Air Museums: Europe*, 536; *Open-Air Museums: Elsewhere*, 537 • History of Wind Power Additional Sources, 537: *North American Focus*, 537; *International Focus*, 538 • Periodicals, 538

Selected Sources 539

Chapter 3: Where It All Began and Chapter 4: The Great Wind Revival, 539 • Chapter 6: Vertical-Axis and Darrieus Wind Turbines, 540 • Chapter 7: Novel Wind Systems, 541 • Chapter 8: Silent Wind Revolution, 541 • Chapter 9: Towers, 542 • Chapter 10: Measuring the Wind, 542 • Chapter 11: Estimating Performance, 542 • Chapter 12: Off-the-Grid Power Systems, 542 • Chapter 13: Interconnection and Grid Integration, 542 • Chapter 14: Pumping Water, 543 • Chapter 15: Siting and Environmental Concerns, 543 • Chapter 17: Safety, 545 • Chapter 18: Operation and Maintenance, 545 • Chapter 19: Investing in Wind Energy, 545 • Chapter 20: Community Wind, 545 • Chapter 21: The Challenge, 546

Annotated Bibliography 547

Aesthetics and Noise, 547 • Modern Wind Energy History, 547 • Large Wind Turbines, 548 • Small Wind Turbines, 549 • Rigging, 549

PREFACE

Since the mid-1970s I've followed the development of wind energy around the globe. During this time I've been a proponent, participant, observer, and critic of the wind industry. As an observer, I've traveled extensively reporting on the technology and how it's being used. As a participant, I've installed anemometers in Pennsylvania, hunted windchargers in Montana, and measured the performance of small wind turbines in California. As a proponent, I've lectured about the promise of wind energy to groups from Vancouver to New Delhi, from Punta Arenas, Chile, to Husum, Germany. And as a critic, I've called some wind companies to task when their environmental practices were no better than the technologies they intended to supplant.

In the early 1980s, I prepared a daylong seminar on the prospects and pitfalls of wind energy. An early version of this book, published in 1983 under the title *Wind Energy: How to Use It*, grew out of the course notes for these seminars.

At that time there was a chasm between the books written for backyard tinkerers who wanted to build their own wind turbines and those books surveying the entire field of wind energy. There was no book that answered the questions people raised in my seminars about how they could obtain a working wind system and not an experimenter's toy. *Wind Energy* was written to meet that need. The book was unique because it didn't simply look at the technology. It gathered tips and advice from leaders in the field and offered practical guidance on how to select, buy, and install wind turbines—and how to do so safely.

Wind Energy was reissued in 1993 by Chelsea Green Publishing as *Wind Power for Home & Business*. The book became a staple of both homeowners and professionals interested in the subject. In 2004, Chelsea Green again published an extensive revision titled *Wind Power: Renewable Energy for Home, Farm, and Business*.

Today wind energy is a booming worldwide industry, and with the heightened concern about climate change and energy security, this resurgence of interest is here to stay.

Despite wind energy's success—and the plethora of books on the topic—there remains a need for a frank discourse on how to wisely use the technology. For this reason, I have continued to edit and update the book. After a decade on the market, it was time for another extensive revision.

Each new edition has reflected changes in both my view of how best to use wind energy and in the technology available. This version incorporates the lessons I've learned from more than three decades working with wind energy. It also introduces the concept of "community wind" where groups of people invest in large wind turbines that produce commercial quantities of electricity for sale to the utility company. While a seemingly novel concept in North America, it is quite common in Denmark and

Germany. In community wind, farmers, small businesses, and groups of community-minded citizens band together to develop—for profit—"their" wind resources. As Germany's electricity rebels say, "Renewable energy is far too important to be left to the electric utilities alone. We have a responsibility for our own future. We can and will develop our own wind resources for our own benefit and for the benefit of our communities." By proving that it can be done, Germans and Danes have served as models for us in North America as well as for others around the world.

Soon, I hope, we'll see communities across the continent clamoring for the right to connect their wind turbines to the grid—and their solar panels and biogas plants as well—and be paid a fair price for their electricity.

This book is not by any means exhaustive, nor is it intended to be. In the more than three decades I've worked with wind energy, the field has grown so vast that it's no longer possible to confine the technology within the covers of one book.

In 1983, I sought to help newcomers to wind energy avoid the mistakes that I and others had made and to spur development of this renewable resource. *Wind Energy for the Rest of Us* seeks the same end.

Bon vent! (Good wind!)
Paul Gipe
Bakersfield, California

ACKNOWLEDGMENTS

No one can write a book on a subject that crosses so many disciplines as wind energy without the help of numerous contributors. Moreover, after a three-decade career in wind energy, I've had the good fortune to meet many of the visionaries in the field, and their advice and commentary are peppered throughout the text.

My work has been influenced by so many people that in writing these acknowledgements I am certain to forget someone that should be included. Rather than throw up my hands in despair and simply issue a heartfelt thanks to all those who have helped in one way or another, I want to thank as many of those as I can remember. For those I've overlooked, please accept my apologies.

I am especially indebted to Vaughn Nelson at West Texas A&M University's Alternative Energy Institute (AEI). Vaughn first taught me the importance of swept area and how to quickly cut through the hype that often surrounds new wind turbines. AEI's Ken Starcher has been invaluable for his technical expertise as well as his old-fashioned common sense.

My thanks to Mick Sagrillo, Sagrillo Power & Light, and Hugh Piggott, Scoraig Wind Electric, for answering my many questions on battery-charging wind systems. Both Mick and Hugh are fonts of practical, hands-on knowledge of small wind turbine design. I've used their astute observations liberally throughout this book.

Jim Salmon, Zephyr North, and Jack Kline, RAM Associates, were instrumental in the chapter on wind resources, as were Dave Blittersdorf, AllEarth Renewables, and Ken Cohn, Second Wind.

Small wind turbine manufacturers worldwide deserve a note of appreciation for responding to my frequent queries about their products. Over the years, Mike and Karl Bergey, founders of Bergey Windpower, have been notably forthcoming. David Sharman, Ampair, and Brent Summerville, Small Wind Certification Council, have both been a great help in understanding the arcana of certifying small wind turbines.

I again extend my gratitude to Preben Maegaard and Jane Kruse of the Folkecenter for Renewable Energy and to the people of Denmark for a fellowship that allowed me to study the distributed use of wind energy in northwest Jutland. It was at the Folkecenter where I first learned how to use a griphoist to install small turbines.

My appreciation also to Bill Hopwood and Dennis Elliott for their contributions on siting; Mike Barnard, Energy and Policy Institute, on organized antiwind groups; Geoff Leventhall, Consultant in Noise Vibration and Acoustics, on noise propagation; Nolan Clark and Brian Vick, formerly with the US Department of Agriculture; Jim Tangler, formerly with NREL, Henry Dodd, formerly with Sandia, and Peter Jamieson, Garrad Hassan, Peter Musgrove, National Wind Power, on wind turbine design; Peter Schenzle on wind ships; Ken O'Brock and Alan

Wyatt for their help with mechanical wind pumps; Michael Klemen, Eric Eggleston, Claus Nybroe, and Jason Edworthy for their insightful comments on small wind turbine design; Carl Brothers, Frontier Power Systems, on wind-diesel systems.

Capitola Reece, Gene Heisey, Art and Maxine Cook, Phil Littler, Sister Paula Larson, Eli Walter, and Bill Young for sharing their experiences; Gil Morrissey for his tutelage to a sometimes dim-witted apprentice electrician; Ed Butler for advice on how to do the job right; Klaus Kaiser, Christoph Stork, Bernard Saulnier, Charles Dugué, and Charles Rosseel for their help with the lexicon; Heiner Dörner for historical background on FLAIR and ducted wind turbines.

Gottfried Wehr, Windkraft Diemarden, for his series of photos illustrating removal of a large wind turbine; Neal Emmerton for again offering the sequence of photos on the installation of large wind turbines; Martin Ince, M.K. Ince and Associates, and Dr. Ewan O'Sullivan, Harvard-Smithsonian Center for Astrophysics, for their photos of novel wind turbines.

Ed Hale, WindShare, Josef Pesch, Fesa, Kris Stevens, Ontario Sustainable Energy Association, Klaus Rave, Global Wind Energy Council, Hans-Detlef Feddersen, Bürger-Windpark Lübke-Koog, Dave Toke, University of Aberdeen, Grant Taibossigai, M'Chigeeng First Nation, Adam Twine and Liz Rothschild, Westmill Wind Farm Cooperative, David Stevenson, Colchester-Cumberland Wind Field, Henning Holst, Ingenieurburo Henning Holst, Asbjørn Bjerre, Danmarks Vindmølleforening, and Wolfgang Paulsen for the inspiring story of community wind.

Bernard Chabot, BC Consult, and Jens-Peter Molly, DEWI, on the silent wind revolution; Mark Haller, Haller Wind Consulting, Mike Kelly, Mistral Renewable Energy, on operation and maintenance; Martin Hoppe-Kilper, Institut dezentrale Energietechnologien, on the need for feed-in tariffs.

Povl-Otto Nissen and Bjarke Thomassen, Poul la Cour Fonden, historians extraordinaire, Etienne Rogier and Robert Righter, and John Twidell, AMSET Centre, on the dawn of wind-electric generation.

Peter Karnøe, Copenhagen Business School, and Matthias Heymann, Aarhus Universitet, for their work on the great wind revival from the ground up; Henrik Stiesdal, Siemens; Erik Grove-Nielsen, Winds of Change, Birger Madsen and Per Krogsgaard, BTM Consult, Benny Christensen, Danmarks Vinkrafthistoriske Samling, and Britta Jensen and Allan Jensen, Tvind, for the early history of the Danish wind industry; and Herman Drees, Dutch Pacific, on early giromills.

Susan Nelson, Sarah Forth, Joe Maizlish, Glen Estill, Dave Bittersdorf, Bill Hopwood, Mike Brigham, Roger Short, and Malcom Hamilton for their support and faith in the future.

And a special thanks to Nancy Nies, my ever-patient wife, for tolerating me during the long and often arduous process of producing this extensive revision.

1

How We Use the Wind Today

Tout sur terre appartient aux princes, hors le vent.
(Everything in the world belongs to the princes, except the wind.)

—Victor Hugo, *La Rose de l'Infante*

WIND WORKS. IT'S RELIABLE. IT'S ECONOMICAL. IT MAKES environmental sense. And it's here now. Wind turbines are not tomorrow's technology. Whether it's on a giant wind farm in Minnesota, in a small village in Morocco, or in the backyard of a German farmer, wind energy works today in a variety of applications around the world. You too can put this renewable resource to work for yourself, your neighbors, and your community. The following chapters explain how to go about doing just that.

Of course, to use the wind successfully, you must have a good site and select the right machine. You'll also need courage and no small degree of determination. A wind turbine represents a serious investment in the future. And no matter what you choose to do—whether installing a large or small wind turbine, a single unit or dozens—developing a wind project is an undertaking fraught with risk, uncertainty, and often frustration.

Developing wind energy is not simple, nor easy. If it was, it would have been done already by someone else. Fortunately, many have gone before you. Better yet, many have been successful and serve as examples of how to do it right. Like them you must weigh your choices then act. The people who use wind energy are prudent, but they're doers. They are the kind of people who get the things done that need doing.

People use wind energy for many reasons: economic, environmental, and philosophical. The desire to make money—or saving it—is often sufficient for plunging into wind energy. Yet for many people, there's more to it than that. Windmills have fascinated us for centuries and will continue to do so. Like campfires or falling water, they're mesmerizing, indeed, entrancing. People respond almost instinctively. Few escape feeling the excitement at seeing a sleek turbine whirring in the wind.

Working with the wind is more than just a means to low-cost renewable electricity. It becomes a way of life, a way of living in closer harmony with the world around us. Harnessing the wind enables us to regain some sense of responsibility for meeting our own needs and for reducing our impact on the environment. By generating our own electricity cleanly and with a renewable resource, we can reduce the need for distant power plants and their attendant ills. Wind energy can and does make a difference in our

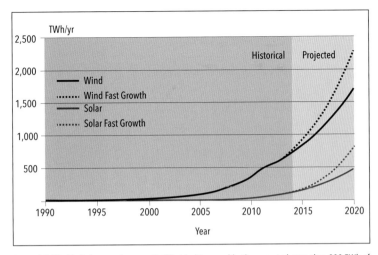

Figure 1-1. World wind generation growth. Wind turbines worldwide generated more than 800 TWh of electricity in 2015. By 2020 the amount of electricity generated by wind turbines could double or triple. Despite the phenomenal growth of solar energy within the last decade, wind turbines will continue to produce more electricity worldwide for the foreseeable future.

We use wind energy in myriad ways. We use the wind to power our appliances, to heat and light our homes, to run our factories. We use the wind in the same way we've used conventional sources of energy in the past. In almost every case, wind turbines can be used to generate electricity to provide the same services as before—but without reliance on fossil or nuclear fuels. Mongolian nomads use modern micro turbines to boil water for tea, the Dutch use wind turbines to power their lock gates, and midwesterners use commercial wind turbines to light their schools. The possibilities are endless—limited only by our imagination.

Not since the wind was used to sail the world's seas and pump water from the lowlands of northern Europe has wind energy been used on such a grand scale as it is today. No longer will wind energy be seen as the domain of a disheveled miller with corn flour in his hair, furling the cloth sails on his wooden windmill. This archaic image has given way to that of trained professionals tending their sleek and powerful aeroelectric generators by computer. From the deserts of California to the rainy shores of the North Sea, from the Mongolian steppes to the Argentinean pampas, wind energy has truly come of age as a commercial generating technology.

From humble beginnings among Danish farmers and community activists in the late 1970s and California's wind developers in the early 1980s, wind energy has grown to become a global industry.

The numbers are telling. In early 2016, there were nearly 250,000 wind turbines in operation worldwide. These wind turbines—some as much as three decades old—were capable of generating more than 400,000 megawatts (MW). Although significant in its own right as an example of the phenomenal pace at which wind energy can grow, far more important is the amount of electricity that they actually produce (see Figure 1-1. World wind generation growth). Wind turbines will generate more than 800 terawatt-hours (TWh or billion kilowatt-hours) of electricity from the wind turbines already in service. Though less than 3% of global consumption today, that's more than enough electricity for the entire country of Germany, or more than twice that consumed in electricity-hungry California.

lives and the lives of our communities—and it's doing so today worldwide.

As evidence of climate change mounts, and the price of fossil fuels edge ever higher, the rocketing demand for renewable energy shows no sign of waning. As a result, wind energy is booming worldwide. Not since the heyday of the American farm windmill has wind energy grown at such a dramatic pace.

While attention in North America has often been on the giant wind farms springing up across the breadth of the continent like giant mushrooms, there's another, often-overlooked side to wind energy: distributed generation—or putting wind power where people live and work.

Distributed wind generation is booming too, mostly in Europe, where the use of wind turbines in or near cities and villages is commonplace. There is also a growing number of small wind turbines finding uses here and abroad on sailboats, at remote cabins, or at new homesteads at the end of the utility's lines, or even off the grid entirely.

The focus of the earlier editions of this book was on small wind turbines used individually at homes, farms, and businesses. However, this version broadens its scope to include wind turbines of all sizes, whether concentrated in large arrays or in applications distributed across the landscape, and whether owned by electric utilities or by groups of landowners and the communities where the turbines are located.

And if success can be measured by the power of one's enemies, wind energy has indeed come of age when the president of the West Virginia Coal Association feels compelled to condemn it on public radio and when former lobbyists for the British nuclear industry find it necessary to form an organization whose sole purpose is to stop wind energy in its tracks.

They have reason to be concerned.

The wind industry continues to grow from 10% to 25% per year. At this rate the amount of electricity generated by wind energy will double every three to seven years. More startlingly, it has grown even faster during the early 2000s, indicating that when the right policies are in place, wind energy can quickly become one of the world's most important sources of electricity. Even now, Denmark generates 40% of its electricity from wind energy, and Danes expect wind turbines to generate one-half of their supply by 2020.

Wind energy has come of age not only for customers of electric utilities but also for those who live beyond the end of utility lines. Wind now works for those living in the Mountain Meadows community near Tehachapi, California, who use small wind turbines instead of extending Southern California Edison's power lines to their homes. And wind works for the women of Ain Tolba, who no longer have to fetch water far from their village in Morocco, now that small wind turbines do the work for them.

No one knows for sure how many small wind turbines are operating worldwide. Data on the growth of the small wind turbine industry is much less reliable than that on large wind turbines. There are likely more than 600,000 small wind turbines altogether, the large majority in China. While they represent only a small amount of generating capacity relative to that of commercial-scale wind turbines, they nevertheless improve the quality of life for people around the globe who often don't have access to electricity.

Though wind energy suffered severe growing pains, and struggled through a stormy adolescence during the 1980s, it has now successfully taken its place as a conventional—albeit renewable—source of energy.

But the way in with which this success was achieved surprised advocates and critics alike. Three decades ago most had envisioned using wind energy either with small wind turbines installed individually on farms and at homes scattered across the countryside, or with large wind turbines erected by electric utilities. Neither approach succeeded. What did work was the installation of tens of thousands of medium-size wind turbines in small clusters and large arrays. In northern Europe most of these turbines were installed by groups of local people or small to medium-size companies with local as well as regional investors. By contrast, in North America most of the wind development has been by unregulated subsidiaries of electric utilities—foreign and domestic—in large wind power plants.

Wind Power Plants

Historically, wind turbines have been used to pump water or provide power at remote sites. This is still an important role even today, especially in the developing world and for those in developed countries who live "off the grid," beyond the reach of power lines. In such applications, wind turbines are typically distributed as single units across the landscape. Wind turbines generating utility-compatible electricity can also be used individually, in small clusters, or concentrated in large arrays to generate bulk electricity much like any other power plant.

Before the 1980s, wind energy development focused on the individual wind turbine. By the late 1970s, this perspective began to change, as attention shifted to maximizing collective generation from an array of many wind turbines. From today's vantage point, this idea seems logically consistent with all prior utility experience: power plants are composed of several generating units.

The idea has historical precedents. At Kinderdyk and elsewhere in the Netherlands, the Dutch bunched clusters of windmills in linear arrays along dikes and canals as needed. However, prior to the 1980s, this concept—as applied to modern

WIND ENERGY HAS TRULY COME OF AGE.

wind turbines—seemed revolutionary. The realization that the wind industry was in the business of building power plants and generating massive amounts of electricity with wind energy and was not simply in the business of installing wind turbines had profound effects.

What's in a Name?

Reflecting the concept's newness are the many terms that arose to describe it: *wind farms, wind parks,* and *wind power plants,* to name a few. Early on, finding the best nomenclature created a dilemma. On the one hand, advocates wanted a term connoting wind's technological success and its coming of age as a conventional source of electricity, conveyed by the term *wind power plant.* On the other hand, proponents also wanted to preserve the association with the enlightened land use—the stewardship—that the term *wind farm* implies.

Wind farms is an expression that still finds adherents. Wind cognoscenti adopted the term in the late 1970s because wind generation and farming depend upon seasonal cycles, the turbines are planted in rows like fields of corn (maize), and there is a literary association between the rural areas where turbines are often sited and with harvesting a renewable crop. Yet the term's agrarian overtones disturbed some. Financiers preferred a more sophisticated term for their well-heeled clientele, thus *wind power electrical generating facility* briefly gained currency in Southern California. Fortunately it died a quick death. But the financiers did point out a need for a more accurate description of wind projects.

The term *wind parks* grew out of the razzle-dazzle world of California real estate development in which groups of commercial buildings become "industrial parks." The term is now pervasive throughout the world. Even the normally sober Danes have adopted it to describe some wind projects using the word *vindmøllepark.* The word *parks,* however, carries with it connotations of sylvan landscapes or natural preserves protected from commercial use. Large assemblages of wind turbines can in no way be construed as *parks.* Critics might even charge that wind energy's proponents deliberately choose to continue using the term *parks* for its positive connotations rather than the term *plants* with its utilitarian overtones.

Eventually it became evident to utility planners and engineers alike that these assemblages of wind turbines were indeed *power plants* differing from conventional plants only in that they were wind driven. For simplicity, *wind power plants* is often shortened to simply *wind plants.*

What then is a wind power plant? Generally, it is any cluster of wind turbines used for the bulk generation of electricity. Wind plants contrasts with a single wind turbine or a small cluster of turbines used to meet on-site needs.

Wind Power Plant Arrays

Wind plants vary widely in size. Today, individual wind plants range in size from three-turbine clusters in Denmark to several hundred machines in one Texas project (see Figure 1-2. Clusters). In Europe, arrays are much smaller than those in North America, averaging in the tens of machines, reflecting Europe's greater population density and smaller land parcel size. Some regions, such as California's Tehachapi Pass, host dozens of wind plants that together represent thousands of wind turbines.

Wind turbines can also be sited to advantage on terrain with strong linear features such as dikes and breakwaters along coastlines, or field boundaries and fence lines on tilled farmland. It has become quite common across Europe to see lines of wind turbines on breakwaters and alongside canals (see Figure 1-3. Linear arrays).

In California's mountain passes, Spain's rugged interior, and parts of France and Great Britain, wind power plants have been built on ridgetops or mountainous plateaus (see Figure 1-4. Ridgetop arrays). In California's mountain passes, thousands of wind turbines have been installed in long, undulating arrays on ridgetops transverse to the near unidirectional wind. These walls of wind turbines act much like a dam in a river, harnessing the winds concentrated by the mountain passes.

On the flat lowlands of northern Europe and across the breadth of North America's Great Plains, thousands of wind turbines have been installed, row upon row, like furrows in a plowed field (see Figure 1-5. Rectilinear arrays). These arrays are the quintessential "wind farms." Some

HOW WE USE THE WIND TODAY

Figure 1-2. Clusters. Inset: Multiple wind turbines in small groups represent a cluster. Main photo: Here are three wind turbines in a cluster near Skibsted Fjord on the west coast of Denmark's Jutland Peninsula. Planning regulations in Denmark define a wind plant as any group of more than three wind turbines. 1998.

Figure 1-3. Linear arrays. Inset: Wind turbines can be aligned with linear features of the landscape such as dikes, breakwaters, fence lines, or canals. Main photo: Modern wind turbines lined up along a canal in the Netherlands' Groningen Province. The wind turbines supply bulk electricity to the utility network served by the giant gas-fired power plant in the background. Mid-1990s.

Figure 1-4. Ridgetop arrays. Inset: Wind plants are also built on ridgetops and mountainous plateaus. Main photo: Some of the thousands of wind turbines installed in undulating rows on ridgetops in California's Tehachapi Pass where they are exposed to the prevailing wind. Mid-1990s.

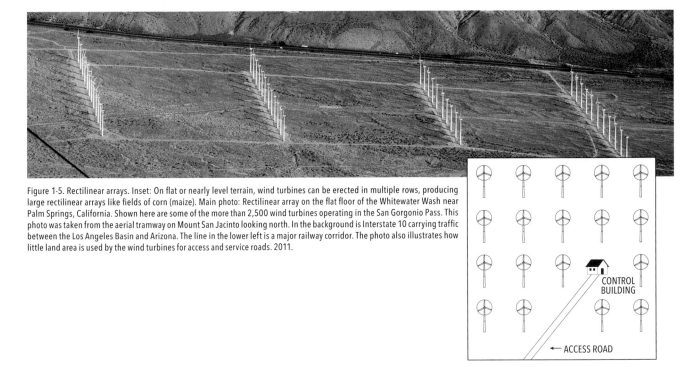

Figure 1-5. Rectilinear arrays. Inset: On flat or nearly level terrain, wind turbines can be erected in multiple rows, producing large rectilinear arrays like fields of corn (maize). Main photo: Rectilinear array on the flat floor of the Whitewater Wash near Palm Springs, California. Shown here are some of the more than 2,500 wind turbines operating in the San Gorgonio Pass. This photo was taken from the aerial tramway on Mount San Jacinto looking north. In the background is Interstate 10 carrying traffic between the Los Angeles Basin and Arizona. The line in the lower left is a major railway corridor. The photo also illustrates how little land area is used by the wind turbines for access and service roads. 2011.

are as small as a few dozen machines; others are so large to an observer on the ground that they to seem to stretch to the horizon.

In contrast to the large arrays of wind turbines now common across the heartland of North America, travelers to Europe quickly note that many of the wind turbines now common there are distributed across the landscape in much smaller groups.

Distributed Wind

What constitutes distributed wind? There's no clear distinction between "distributed wind" and a wind farm. Some define a wind power plant as a large array of turbines connected to the high-voltage system, whereas distributed wind turbines are connected to the medium-voltage and low-voltage system. However, there are small groups of turbines producing bulk electricity that are sometimes connected at transmission voltages. This is not uncommon in Europe.

To small wind turbine advocates in North America, distributed wind is a small to medium-size wind turbine installed to serve a home, farm, or business. Yet in Europe, especially in Germany and Denmark, it's quite common for a farmer, or group of farmers, to install one or a cluster of large wind turbines as a commercial venture, much like building a barn or cooperative dairy. Or for a village to install one to several large wind turbines at a sewage treatment plant, closed landfill, stadium, or other public facility.

Thus, the key characteristic of distributed wind is not the size of the wind turbine, or the voltage at which it is connected, but the number of turbines in a group.

Urban and Village Wind

The distributed use of wind energy need not be limited to rural areas. There are numerous examples both in North America and Europe where wind turbines are used in cities, towns, and villages. Real "urban" wind can be seen throughout the Netherlands, Denmark, and Germany. But it can also be found in the urban core of Toronto, Cleveland, and Boston, as well as in other large North American cities (see Figure 1-6. Urban wind).

Figure 1-6. Urban wind. WindShare's 750-kW Lagerway turbine in downtown Toronto. The cooperatively owned turbine stands prominently on the grounds of Exposition Place between the heavily traveled Gardiner Expressway and Lake Shore Boulevard shown here. The 51-meter (170-foot) diameter wind turbine has been generating nearly one million kWh annually since it began operation in 2003.

Figure 1-7. Playing with the wind. Nordex 600-kW turbine near the playground at the public middle school in Forest City, Iowa. This turbine uses a rotor 43 meters (140 feet) in diameter and generates about 850,000 kWh per year and stands within the small midwestern town of 4,000.

Small towns can take advantage of distributed wind as well. Many midwestern towns and villages are proud of their local wind turbines. Take for example, Forest City, Iowa (see Figure 1-7. Playing with the wind). In January 1999, the school district installed a 600-kW Nordex wind turbine to provide power and light for the city's consolidated school. Through the end of 2012, the turbine has generated nearly 12 million kWh. The 43-meter (140-foot) diameter turbine was still operating in 2014 after 15 years of service, remaining a prominent landmark in the town of 4,000. Forest City Schools maintains a website for their turbine, displaying its annual production as a source of pride in their community.

End of the Line and Beyond
Another common application of distributed wind is at the end of utility lines or beyond the reach of utility service. Wind turbines—most often small and medium-size machines—can prove useful to homeowners, farmers, and small businesses as well as provide a low-cost means for utilities to provide voltage support (see Figure 1-8. End of the line and beyond) at the end of long lines.

Pacific Gas and Electric, for example, found that by reducing the loads at the ends of heavily used lines they can avoid constructing costly new transmission and distribution capacity. They can reduce loads by boosting conservation or by installing modular sources of generation, such as solar photovoltaic panels and small wind turbines, close to the point of demand.

An increasing number of small wind turbines are being put to use by homeowners determined to produce their own power, even though they could just as easily buy their electricity from the local utility (see Figure 1-9. Residential wind). In North America, these small wind turbines offset domestic consumption and are often permitted to run their kilowatt-hour meters backward through net-metering policies. On the other hand, Great Britain and Italy at one time had policies that paid for the electricity generated by small wind turbines, and they paid substantially more than is allowed under net metering in North America.

Next to their reputation for mechanically pumping water, grinding grain, and now powering wind farms, wind turbines are best known for their ability to generate electricity *off the grid* at remote sites. They've distinguished themselves

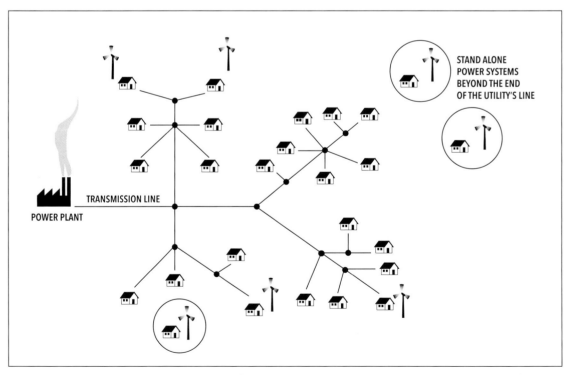

Figure 1-8. End of the line and beyond. Distributed wind turbines, often small, can also be used within a utility's distribution system and at the end of its distribution line. Historically, many small wind turbines have been used in stand-alone or battery-charging applications beyond the reach of a utility.

Figure 1-9. Residential wind. Here a Bergey Excel powers a home near Hinesburg, Vermont. The wind turbine generates electricity that offsets consumption from the local utility. The installation also includes a small rack of solar panels. The Bergey Excel was designed to power an all-electric home in the windy Midwest.

Figure 1-10. Remote telecommunications. Hybrid wind and solar power system powering a telecommunications tower on the frontier between Argentina and Chile. The 1-kW AgroLuz turbine is modeled after the 1930s-era Jacobs wind charger and uses a pitch-regulated rotor 4.1 meters (13 feet) in diameter.

in this role for decades. During the 1930s, when only 10% of North American farms were served by electricity, literally thousands of small wind turbines were in use, primarily on the American Great Plains. These *home light plants* provided the only source of electricity to homesteaders in the days before rural electrification brought electricity to all.

At many remote sites, small wind turbines produce power at less cost than gasoline or diesel generators. Even today, three-fourths of all small wind turbines built are destined for stand-alone power systems at remote sites. Some find their way to remote homesteads in Canada and Alaska, far from the nearest village. Others serve mountaintop telecommunication sites where utility power could seldom be justified (see Figure 1-10. Remote telecommunications).

Stand-alone power systems, however, seldom use wind energy exclusively. In many areas of the world, wind and solar resources complement each other: winter's winds balancing summer sun. Wind and solar hybrids capitalize on each technology's assets, enabling designers to reduce the size—and cost—of each component.

Hybrids contribute even more value in developing countries, where one-third of the world's people live without electricity. Many developing nations are scrambling to expand their power systems to meet the demand for rural electrification. However, extending utility service from the cities to remote villages is a seldom affordable luxury. Thus hybrid systems, though they generate little power in comparison to central power plants, can meet the modest needs of Third World villages.

Low per capita consumption magnifies a hybrid system's benefits because so little electricity is needed to raise the quality of life. One kilowatt-hour of electricity provides 10 times more service in India than it does in Indiana. Two 10-kilowatt turbines, which would supply two

Figure 1-11. Onboard battery charging. Ampair 100 on the deck of a sailboat in Copenhagen's inner harbor. Thousands of micro turbines are used in marine applications worldwide to charge batteries used for communications and navigation. 1998.

Figure 1-12. Electric vehicle charging. Tehachapi, California launched its Ride the Wind program in the late 1990s using three rebuilt 15-meter (50-foot) diameter wind turbines from a nearby wind farm. While the wind turbines didn't charge the city's electric vehicle directly, they offset the additional consumption from vehicle charging as well as offset consumption from the city's sewage treatment plant where the turbines were located.

homes with electric heat in the United States, can pump safe drinking water for 4,000 people in Morocco. Consequently, developing countries are turning to both wind and solar energy as a cost-effective way to meet the electrical needs of rural areas. Because of their relative low cost, wind and solar hybrid power systems enable strapped governments to get power into villages more quickly than by extending lines from the cities. As the central power system expands to these villages, the hybrid systems can then be removed and sent on to even more remote villages.

Specialty Applications

There are also numerous specialty applications for small wind turbines. They include the obscure use of wind turbines for cathodic protection of pipelines, electric fence charging, billboard lighting, and use in recreational boating. Solar photovoltaics has largely superseded small wind turbines in most of these applications, but micro wind turbines are still frequently found on sailboats and yachts to power navigation and communication electronics (see Figure 1-11. Onboard battery charging). In yacht harbors around the world, multiblade micro turbines, such as Ampair and Marlec, are a common sight.

Electric Vehicle Charging

One potentially huge application for wind energy will be powering electric vehicles (EVs). To date, most planners have pictured individual wind turbines installed at homes and businesses for directly charging EVs, much like that of rooftop solar panels. However, this is an unlikely role for wind energy. Instead, it makes more sense to use wind turbines sited to best advantage, delivering their electricity to the grid for use in EV charging where needed (see Figure 1-12. Electric vehicle charging). Conceptually, this is simply an extension of the existing role of wind power plants or clusters of wind turbines. The arrays would simply serve an increase in demand on a utility system caused by the advent of electric vehicles.

But passenger vehicles represent only one form of transportation. The late Don Smith, an engineer on the staff of the California Public Utility Commission, once envisioned thousands of wind turbines powering newly electrified railroads as they cross the Great Plains. Most railroads in North America, unlike those in Europe, are diesel powered. Smith suggested replacing the diesel locomotives with electric locomotives and electrifying the transcontinental routes. He dreamed of installing thousands of wind turbines along the railroads' rights-of-way in a linear array of continental proportions. Fittingly, the High Speed Rail Authority in California depicts the sleek electric trains being powered by wind turbines.

What is clear is that as North America weans itself off fossil fuels in transport, a mix of renewable sources of energy will increasingly drive electric transport of all kinds—rail and passenger vehicles alike. Wind energy will play an ever-growing role in powering the continent's 21st century transport system.

Heating

In temperate climates, heating comprises most of a home's energy demand. In many areas, there is a good correlation between the availability of wind energy and the demand for heat. It's not surprising then that there have been numerous attempts to use wind machines strictly for heating. Advocates of the wind furnace concept believed that the same winter winds that rob a house of its warmth could be used for heating. They argued that generating low-grade energy with either a generator or a mechanical churn would be less costly than producing the same amount of utility-compatible electricity. It's simpler, proponents said, than trying to produce the high-grade electricity demanded by the utility.

Unfortunately, it never worked out that way. Both Danish and American companies tried to commercialize the concept during the 1970s and early 1980s, and again in the 1990s. All were to no avail. Wind turbines are capital intensive, and these researchers thought that by eliminating the need for generating utility-compatible electricity they could build cheap wind turbines. They

Figure 1-13. UMass Wind Furnace. Circa-1980 experimental wind turbine at the University of Massachusetts designed for directly heating the experimental building nearby. Several firms unsuccessfully tried to commercialize the concept.

failed to reckon with low-cost, mass-produced generators. It costs no more to generate utility-compatible electricity with an induction or asynchronous generator than it does to produce low-grade heat.

Experience has shown that in most cases it's cheaper and easier to interconnect the wind system directly with the utility than to generate heat and store the surplus for windless periods. Often, it's more cost effective to produce a high-grade form of electricity that can be used for all purposes, including home heating if desired, than to build a wind turbine that can only be used for one function (see Figure 1-13. UMass Wind Furnace).

There are several examples where household-size turbines have been used to provide supplemental heating. For years, Bergey Windpower has

SACRED WINDS

The wind blows where it will. You hear the sound it makes but you do not know where it comes from or where it goes.
—John 3:8

Imbued with the Holy Spirit and a sense of adventure, a small group of progressive nuns have become a beacon on the wind-swept plains near Richardton, North Dakota. Though the Benedictine sisters live a contemplative life, they are also pragmatic.

The 22 members of the Sacred Heart Monastery wanted to control their utility bill. They also wanted to give something back to the rural community of which they are a part. After learning about the potential, the sisters thought those seemingly ever-present and oftentimes annoying prairie winds could be the answer.

The sisters then bought two used turbines in neighboring Montana. As others have found, you can be in for a rude surprise unless you know what you're buying and who you're buying it from. The sisters learned the hard way, but with the help of Tad Miller, a regional representative for a wind developer, they were able to put both turbines online in the summer of 1997 for $120,000.

"I'd do it differently today," says Sister Paula Larson, the monastery's prioress. "I'd probably get one sized to our specific needs." They would also prefer hiring someone who could do the whole job from start to finish. The nuns found themselves acting as the general contractor on the project, a role that was unfamiliar to them.

The Silver Eagle turbines they bought had a checkered history long before the nuns came across them, and the problems the sisters encountered were nothing new. The turbines were an American copy of a 15-meter (50-foot) diameter Danish Nordtank and were built during the tax credit-driven wind rush of the early 1980s.

The most difficult tasks, says Miller, were negotiating a parallel generation contract with a timid local utility, West Plains Rural Electric Cooperative, and installing a new controller that would meet the utility's unnecessarily stringent requirements. "They [the utility] were terrified of what could happen," he says, and they quite literally "threw the book at us." Ironically, the "book" was interconnection standards for the fossil-fueled plants the utility understood.

The sisters hoped that the turbines would offset one-quarter of the monastery's consumption. If they did, the machines would pay for themselves in 10 years. Fortunately, performance was better than expected. The monastery saved nearly 40% on its electricity bill in the first year.

Sacred Winds. One of two early 15-meter (50-foot) diameter wind turbines operating at the Sacred Heart Monastery near Richardton, North Dakota. Installed in 1998, the turbines provide one-third of the monastery's electricity. (Sacred Heart Monastery, Sister Renée Branigan)

Some three-fourths of the electricity generated was used on-site, offsetting purchases from the utility at $0.092/kWh. The remainder was sold back to the co-op for about half the retail rate. "We should say gave back," at that price, says Sister Paula. "It's appalling, from a spiritual viewpoint, that conservation is penalized."

All told, the turbines operate about one-half the time, spinning out electricity for the monastery. The Silver Eagles have performed well for such early machines. Including outages for major repairs, the turbines generated 2.6 million kWh through the end of 2013 for an average production of 165,000 kWh per year for the 15 years they've been in service. (As is done for other community-owned wind turbines around the world, the sisters proudly display their annual production on the monastery's website.)

"They've been a good investment," says Sister Paula. But there was another equally important reason for the project. The monastery has made a spiritual commitment to help preserve the endangered rural life of the Dakotas, as well as to protect the environment. The wind turbines fulfill both goals.

The twin turbines are visible from I-94, the principal east–west corridor across North Dakota, and have become a local landmark. The interest the turbines have created has allowed the monastery to spread the word that increased use of wind energy would benefit North Dakota's economy.

The sisters' success has dispelled several myths prevalent in a state long known for coal mining and smoke-belching power plants—and now known for its shale-oil boom and open gas flares. Critics said that no one in North Dakota could maintain the "complicated" machines. "We showed that wasn't true," says Sister Paula. The monastery's own staff operates and services the turbines. Others doubted that the turbines would work in the state's extreme cold. Doubting Thomases said they would "freeze up." "They don't," answers Sister Paula bluntly. Some warned that the turbines would be too noisy for the contemplative life. But Sister Paula found that "when the turbines are running, we only hear the howl of the wind. There's been absolutely no disturbance to our life."

The sisters note that using the winds that sweep over the prairie is part of their heritage. A photo from 1916 shows a group of nuns standing in front of their monastery, with an old water-pumping windmill in the background. Their two turbines are now part of their faith in a rural economic revival based on stewardship, not exploitation.

marketed its Excel model to midwesterners in "all-electric" homes, where the wind turbine will provide some or all of the heating load. But the Excel produces utility-compatible electricity that can be used anywhere in the home—or sold back to the utility.

However, renewable energy advocates in Denmark are now examining how they can integrate the country's abundant wind resource for heating on a village and regional scale. Most Danish towns and villages use district heating systems. Each home and business is connected to a central station that distributes hot water. In much of the country, that hot water is produced from the waste heat of a combined heat and power plant that burns natural gas. Because the system uses water to carry the heat, the heat can also be stored in the form of hot water in large insulated storage tanks. Some villages are already storing heat in this way to accommodate peaks in demand.

It is a logical step from storing heat from a gas-fired cogeneration plant to meet peak demand to storing excess electricity as hot water during a windy winter storm. Already wind turbines generate more electricity than can be consumed in Denmark during winter storms. Instead of selling this excess electricity to its neighbors for little profit, Danes could instead store the electricity as heat and offset their consumption of valuable and costly natural gas.

Pumping Water

Wind machines have historically been used to pump water, and pumping water remains an important application of wind energy today, in both the developed and the developing worlds. The American farm windmill, known as the Chicago mill in some parts of the globe, dependably pumps low volumes of water from shallow wells. These multiblade wind pumps are still extensively used for watering remote stock tanks on North America's Great Plains, Argentina's pampas, Australia's outback, and South Africa's veldt. There are probably more than a million of these wind pumps still in use worldwide (see Figure 1-14. Wind pumping).

Researchers at West Texas A&M Univer-

Figure 1-14. Wind pumping. Foreground: American farm windmill on a Colorado ranch near the Wyoming state line. The classic farm windmill was used to pump water for livestock and sometimes domestic consumption. Background: 700-kW NEG-Micon turbine part of the Ponnequin wind power plant.

sity's Alternative Energy Institute and the US Department of Agriculture's Agricultural Research Service have made major advances in water-pumping technology, first with wind-assisted irrigation in the 1980s, and then with wind-electric pumping systems in the 1990s. In cooperation with small wind turbine manufacturers, these researchers have developed pumping systems that couple modern electronics to small windchargers that eliminate the need for cumbersome batteries. Under certain conditions, these wind-electric pumping systems will deliver more water at a lower cost than the traditional farm windmill.

Now that we have a broader picture of the many ways wind energy is used today, we'll turn to how best to use this book and the chapters to come.

2

How to Use This Book

Visquem de l'aire del cel.
(Let us live from the air of the sky.)

—Catalan proverb

IF YOU'RE FORTUNATE ENOUGH AND INSTALL A WIND turbine for yourself or your community, you'll experience sensations few others can share. You'll know what it's like to gaze from the top of your tower at the countryside spread out before you. You'll know the feeling of seeing your wind machine spinning overhead, knowing that you had a part in making it happen. You'll rediscover the sense of accomplishment one gets from a job well done. You and your friends will discover the camaraderie that grows among people after sometimes years of strenuous effort bringing a wind project to fruition. There's nothing quite like it.

Installing a wind machine, however, will never be risk-free. Generating electricity isn't easy; it never has been. Your utility may make it look simple, but it's a tough and sometimes dangerous job. This book is designed to help you minimize the risk, to ensure, as much as possible, that you'll succeed in erecting a wind turbine—or several—that will be reliable, safe, and profitable for you and your community.

Wind Energy for the Rest of Us was written for those who ask questions, who want to know what's going on around them, who want to do what they can for themselves and their community. It's for those who want to make a difference. Yet this isn't another how-to-build-your-own-windmill book, though much of what it contains is essential for building and installing one safely. Instead, this book gives you what you need to make intelligent choices—whether or not to install one for yourself or, instead, install one for you and your neighbors.

This book differs from others on the subject by both describing the technology and explaining how to evaluate what's important and what's not.

Before we plunge into the text, there are a few preliminaries we need to deal with first: the organization of the chapters ahead, the nomenclature that will be used, and some pitfalls to avoid.

How This Book Is Organized

The organization of *Wind Energy for the Rest of Us* differs markedly from the 2004 edition of *Wind Power*. Several chapters have been greatly expanded and several new chapters added.

USING WWW.WIND-WORKS.ORG

Wind Energy for the Rest of Us is designed to use with the World Wide Web, especially my website: www.wind-works.org. Here's a quick guide to what you'll find there.

Books
A list of my other books on wind energy and my reviews of books on wind energy by other authors. Any corrections, changes, updates, or addendums to this book can be found there.

Wind Energy
- Large wind turbines
- Small wind turbines
- Wind energy and the environment
- History of wind power

Articles and commentary for wind professionals and renewable energy advocates on commercial wind development as well as on household-size wind turbines.

Feed Laws
One of the world's most extensive collections of articles, references, and presentations on electricity feed-in tariffs, Advanced Renewable Tariffs, and Renewable Energy Payments. This section includes a country-by-country discussion of feed-in tariff policy developments.

Renewables
- Solar energy
- Geothermal energy
- Community power
- Other articles on renewable energy policy

Although my website was originally developed for my articles on wind energy, it now contains articles on a number of other topics, including nuclear power and renewable energy policy.

About
- Workshops
- Presentations
- Photos
- Biography
- Links

This section includes my presentations, videos, podcasts, and professional background as well as links to organizations and consultants working with wind energy.

Chapters 3 and 4 trace the history of wind energy from the dawn of the first electricity-generating wind turbines in the late 19th century to the great wind revival in the 1970s. These two chapters illustrate that technology does not evolve in a vacuum and that modern wind turbines have surprising roots: wind energy didn't emerge from the drawing boards of the aerospace industry but from the muddy fields of farmers and grassroots activists.

Chapters 5 through 8 explore wind technology: where we've been, where we are today, and where we're headed. Chapter 6 turns a critical eye to the frequent "rediscovery" of vertical-axis wind turbines (VAWTs) and reveals that the inventor of the technology, Georges Darrieus, never built one of the wind turbines himself. Chapter 7 looks at unusual means for using wind energy and debunks one of the common choices of so-called inventors, ducted wind turbines. Chapter 8 introduces an idea that's revolutionizing wind energy. It's not new, it's not sexy, and it's virtually unheard of outside the wind industry, but it makes wind energy more cost effective in more places than ever before. Together, these chapters explain why modern wind turbines look the way they do.

Towers, necessary companions to wind turbines, are the subject of Chapter 9.

The wind itself is the subject of Chapter 10. You'll learn the importance of wind speed, how to find out what you have, and why slight changes in wind speed have such an outsize impact on the energy available. You'll also learn how elevation and temperature affect a wind turbine's performance.

In Chapter 11, you'll find techniques for estimating the amount of energy that a wind turbine may capture and put to work, as well as how to interpret the information published by wind turbine manufacturers. For policy makers and energy analysts, Chapter 11 also explains how to estimate the performance from fleets of wind turbines.

If, instead of generating electricity in parallel with the utility, you prefer to declare your energy independence, Chapter 12 examines the components you'll need for an off-the-grid system. This chapter explains why hybrid power systems using both wind and solar make more sense for remote sites than any one of those technologies by itself.

Chapter 13 explores how to connect a wind turbine with a utility's lines, what's important in the large-scale integration of wind energy into a utility system, and why it's the mix of resources that's more important than any one technology. You'll also learn why storage isn't as critical as once thought and why many of the early fears about wind energy's variability are unfounded.

A modern twist on one of the oldest applications for wind energy, pumping water, is the subject of Chapter 14.

Chapter 15 examines where you can and can't install a wind turbine and why. You'll also learn about potential land-use conflicts, how to avoid them, and—where you can't—how to deal with them. This chapter also debunks several myths about wind energy, explores the risk to public safety, the emissions of noise from wind turbines, the impact wind turbines have on property values, and the amount of land wind turbines require.

Chapter 16 summarizes the steps necessary for installing both small and large wind turbines. This chapter expands the illustrated installation sequence introduced in *Wind Power* (2004) to include the dismantling of a large wind turbine in the repowering of a community-owned wind farm in Germany.

In Chapter 17 we take a close look at a taboo subject not found in other books on wind technology—the safety of those who work with wind energy.

Once your wind machine is installed, you'll need to operate it properly, in order to ensure that it serves you well for many years to come. Chapter 18 reviews some simple start-up procedures for small wind turbines and suggests how to operate, maintain, and monitor the performance of a small wind system. This chapter also summarizes what it costs to operate and maintain both large and small wind turbines.

What's required to minimize your risk and earn a reasonable rate of return on your investment in wind energy is the subject of Chapter 19.

Chapter 20 introduces the third way of developing wind energy—community ownership—and how this model of wind development has powered a renewable energy revolution in Denmark and Germany and can do so in North America.

Chapter 21 challenges North America to pursue an ethical energy policy, weighs how the continent can offset its fossil-fuel-fired generation of electricity, and suggests the policies necessary for doing so.

Nomenclature: What Are We Talking About?

They've been called many things, some unprintable. Most know them as windmills. Whether we call them wind machines, wind turbines, or just windmills, the subjects of this book are kinetic devices intended to capture the wind and put it to work.

Wind Machine and Wind Turbine

The terms *wind machine* and *wind turbine* are simple and unpretentious. They do the job with the least fuss and are used interchangeably in the following pages. *Wind turbine* is used universally by professionals and advocates alike. You'll find no reference to pompous buzzwords like SWECS (for small wind energy conversion systems) or WTG (wind turbine generators) here. Few people who work with wind energy for a living use such jargon. Those who do use such terms do so merely to impress, or worse, to mislead. Don't be fooled. Pretentious terms don't make a wind turbine more reliable, more productive, or more profitable.

The term *wind machine*, as used here, shouldn't be confused with the huge electric fans found in orchards. These fans go by the same name but are used to stir still air during cold winter nights to protect valuable fruit. They move air; they're not moved by the air, as in a wind turbine.

When we're referring to the wind machine, tower, and ancillary equipment as a whole, *wind system* works well. The term *windmill* is reserved for the multiblade, water-pumping wind machine (American farm windmill) and for the European wind machine ("Dutch" windmill) because that's what nearly everyone calls them.

Conventional wind turbines are comprised of three essential components: rotor, nacelle, and tower. The rotor, the spinning part of a wind

DON'T BE FOOLED BY POMPOUS BUZZWORDS.
Pretentious terms don't make a wind turbine more reliable or more productive.

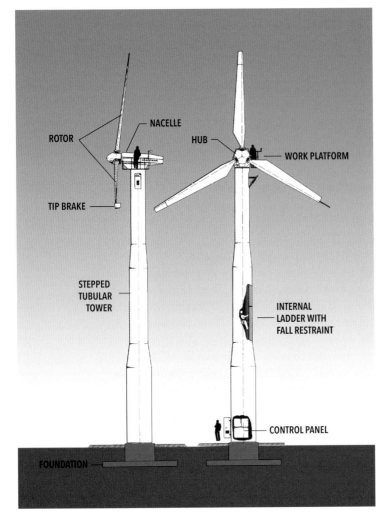

Figure 2-1. Small commercial-scale wind turbine nomenclature. Typical first-generation (55-75 kW) Danish wind turbine, illustrating the rotor, nacelle, tower, and foundation. Wind turbines of this size typically used rotors 15 meters (50 feet) in diameter. Note that even in such early designs, provisions were made to safely service the turbine by the inclusion of a fall-arresting system (on the ladder inside the tower) and a work platform on the nacelle.

turbine, is the most important. Because the rotor determines how much work a wind turbine can perform, the size of a wind turbine will often be described in the text by the diameter of its rotor. The rotor attaches to the wind turbine's drivetrain, which typically sits atop the tower housed inside the nacelle. The tower simply raises the whole works above the ground where it can better catch the wind (see Figure 2-1. Small commercial-scale wind turbine nomenclature).

Power and Energy: There Is a Difference

In casual conversation, we use the terms *power* and *energy* interchangeably. But there is a difference. And that difference is confusing to many people. This confusion allows the unscrupulous to mislead journalists, consumers, and politicians about what wind turbines are capable of and what they actually do. So, it's necessary to delve further into what *power* and *energy* mean. Both terms are related to work.

In the technical sense, work is performed when a force acts through some distance. When you push a stalled car, for example, you are applying a force. For work to be done, something must be accomplished: an object moved, lifted, or turned. If the object does not move, no work is accomplished. If the car did not move, no matter how hard you grunted and groaned to move it, no work was performed.

Energy is defined as the ability to do work or the amount of work actually performed. Both use the same units and are given in the same terms. When the wind strikes the blades of a wind turbine, it imparts a thrust or force that turns the rotor. A finite amount of energy in the flowing wind has been converted to rotational energy in the spinning rotor. When a force does work on an object, energy is transferred from one form to another.

Now couple that spinning rotor to a generator. Work is accomplished when electrons flow from the generator to a load, such as heating a wire in a toaster or turning a motor in a fan. Now take a closer look at the electric motor. The flow of electrons—what we call electricity—transfers the rotational energy imparted by the wind to the wind turbine rotor through wires or power lines to spin the shaft of the motor, giving us, once again, rotational energy. Because the motor spins a fan, work is accomplished. We've gotten something out of it.

The conversion from one form of energy to another is never 100% efficient; that is, you can't get out as much as you put in. There's always some energy lost in the process. Friction is the chief culprit. Though no one has ever built a machine that can convert 100% of the energy in one form to another, people keep trying. Their perpetual motion machines appear periodically as a "startling discovery" in the popular press. Wind turbines are a favorite target of this breed. Such machines never operate as claimed, nor do they usually produce useful work.

As someone who uses energy, you're concerned

about the actual work accomplished and the amount of energy transferred. It's the bottom line of the technical balance sheet. You measure the performance of a wind machine by what it does for you, the work it performs, the energy it transfers and puts to use.

Where does power come in? Power is the rate at which work is performed, the rate at which energy is transferred (changed or released), the rate at which the energy in the wind passes through a unit of area. Power is given in watts (W), in kilowatts (1 kW or 1,000 W), in megawatts (1 MW or 1,000 kW or 1 million W).

When a consumer flips on a light switch, there's an instantaneous demand for power. Wind turbines can provide that power as can all other forms of generation. But what's more important is how long that light switch is left on. The length of time that the light switch is on determines the amount of energy, the electricity, consumed in kilowatt-hours (kWh). As consumers, most of what we pay for electricity is for the consumption of energy in kWh.

In short, power in watts, kilowatts, or megawatts is a measure of what a wind turbine is *capable* of generating—if sufficient wind is present. Energy in kilowatt-hours is a measure of what a wind turbine actually produces. This is the theme in *Wind Energy for the Rest of Us* that's emphasized repeatedly. We want wind turbines that produce electricity in kilowatt-hours. For that the wind turbine needs to sweep a certain area of the wind stream, and it has to work reliably—day in, day out, for years on end. The size of the generator in the nacelle plays a part, but it's not the main part.

Watt's More

In an electrical circuit, *current* is the flow of electrons. It's given in units of amperes or amps. We perceive voltage as the pressure trying to force the electrons to flow through a circuit. But no flow takes place and no work is done unless a load like a toaster is attached to complete the circuit. (No electricity flows out of a receptacle until an appliance is plugged into it.)

Power in watts is the product of voltage times current. This represents the instantaneous rate of work being done when a force (voltage) moves electrons some distance through a wire (current). A toaster in North America, for example, will operate at 120 volts (V) and will draw 10 amps (A), or

$$\text{Power} = 120\text{ V} \times 10\text{ A} = 1{,}200\text{ W} = 1.2\text{ kW}$$

Yet in the average household, a toaster uses little energy (power x time) because it's used so infrequently—only a few minutes every morning. Let's say that the toaster is used six minutes or one-tenth of an hour. How much energy does the toaster use?

$$\text{Toaster: } 1{,}200\text{ W} \times 0.1\text{ hour} = 120\text{ Wh} = 0.12\text{ kWh}$$

The distinction between kilowatts and kilowatt-hours is critically important. Knowing the difference can keep you from being confused by a wind turbine's size, in kilowatts, and how much energy, in kilowatt-hours, it will actually produce.

Equations: They're Informative

There are numerous tables and more than a few equations used in *Wind Energy*. They're there for a reason. They express the relationship between quantities. For example, the equation for the power in the wind succinctly explains why wind speed plays such a critical role. Wherever an equation appears in the text, there'll be an accompanying table to summarize the results. Tables are helpful because they are easier to read than equations, but trends are hard to spot. Where necessary, graphs are used to illustrate trends and general relationships, such as in the distribution of wind speeds over time.

False Precision: How to Avoid It

A long string of digits in a calculation gives a sense of precision that may or may not exist.

HOW *WIND ENERGY FOR THE REST OF US* DIFFERS FROM *WIND POWER* (2004)

- Completely updated from stem to stern with more than 400 illustrations and more than 100 tables.
- New chapter on the history of wind energy, "Where It all Began," and the surprisingly contemporaneous origin of electricity-generating wind turbines in Scotland, France, the United States, and Denmark—and what that means for today.
- New chapter on how modern wind turbines evolved from the ground up, not from the top down, and how a group of activists built a giant that shook the world.
- New chapter devoted to a no-holds-barred critique of vertical-axis wind turbines (VAWTs) and their failed promise.
- New chapter on unusual wind turbines, "Novel Wind Systems," which includes the perennial bane of wind energy, ducted wind turbines (DAWTs), and their strange allure on inventors.
- New chapter on the advent of very large diameter rotors, "Silent Wind Revolution," which make wind energy more economic in more places than ever before.
- Expanded chapter on grid integration and why it's the mix that's all important.
- Expanded chapter on siting and environmental concerns that summarizes research on property values, carbon emissions, and energy balance.
- Expanded section on aesthetic design and compatible land uses.
- Expanded chapter on installation now includes dismantling with a new photo sequence illustrating the removal of a large wind turbine.
- Expanded chapter on how to get the most from an investment in wind energy, introducing the certification system for small wind turbines and the typical royalties paid for leasing land for commercial wind development.
- New chapter on community wind and the electricity rebels who are driving a revolution in who owns—and profits from—wind energy.
- New "back-of-the-envelope" calculation of whether North America has the industrial capacity to offset 100% of its fossil-fuel-fired generation with wind energy, how much electricity would be required to power electric vehicles, and what an ethical energy policy looks like.

Don't be deceived. Estimating the annual energy production of a wind turbine is an inexact science. The art of number crunching for technologies dependent on a natural resource—whether it's farming the wind, mining, or drilling for oil—is finding the best approximation.

In the real world, natural phenomena seldom follow the orderly relationships shown in textbooks. If you plotted a graph of the relationship between wind speed and the instantaneous power generated by a wind turbine, you'd find that the points representing your measurements are scattered. When your job is to interpret this information, you draw a line that best fits the measured data. The resulting line is an approximation of what happened; it doesn't say exactly what happened.

Knowing this, you should be on guard for indicators of false precision. Take the potential performance of a wind turbine as an example. It's absurd to say that a wind turbine will generate 495 kilowatt-hours per month when the data used in the estimate suggests the wind turbine could deliver from 450 to 550 kilowatt-hours per month. Why not simply say 500 kilowatt-hours per month and be done with it? By rounding off the calculation, you indicate uncertainty. Better yet, present the estimate as a range of values from 450 to 550 kilowatt-hours. Then you know that the estimates are only an approximation of what could occur.

Since the advent of pocket calculators in the 1970s, and now the universal availability of personal computer spreadsheets, the results of calculations are often presented in meaningless detail. In the days of the slide rule (yes, this was long ago), every engineering student learned that a calculation was only as accurate as the divisions on the rule—as accurate as the least accurate number in the calculation. You couldn't carry a number out to 10 decimal places, even if you wanted to. Though slide rules have gone the way of the dodo, the concept remains valid: the results of a calculation are only as accurate as the least accurate value used.

Consider average wind speed. It's normally presented to three significant figures. For example, 10.5 mph has two figures to the left and one to the right of the decimal, or 8.25 m/s, which

BE ON GUARD FOR INDICATORS OF FALSE PRECISION.

has two figures to the right of the decimal. If we were to use these average speeds in a calculation to estimate the energy production of a wind machine, the results should likewise be presented to no more than three significant figures. Say the calculation resulted in 22,525.49 kilowatt-hours on your calculator. Only the first three figures are valid, not the seven indicated. The result (22,500 kilowatt-hours) is more realistic considering the accuracy of the numbers used to derive it. Ignoring the concept of significant figures leads to false precision.

Most scientists and engineers are accustomed to approximate arithmetic. They round off numbers every chance they get. It makes their work easier. More importantly, it allows them to get quickly to the heart of a calculation without wasting time on needless detail. The use of approximate arithmetic is how an engineer can take a seemingly complex problem, such as estimating a wind turbine's potential energy production, and solve it mentally or scratch it out on the back of an envelope. Where appropriate, the calculations in *Wind Energy* have been liberally rounded off.

Units: Yes, Metrics Too

The wind industry in North America uses an unholy mix of both the metric and old British Imperial System. Wind turbines are often described by their rotor diameters in meters, the international standard. Yet towers for small wind turbines in North America are often sold in feet. Siting regulations in the United States use distances in feet and miles. We'll use both systems in *Wind Energy*—as needed.

Similarly, the United States still uses miles per hour (mph) for wind speed. However, professional meteorologists use the metric system exclusively. The wind industry in the United States is gradually adopting the metric system for wind speed. For this reason, in *Wind Energy* we'll use the metric system's meters per second (mps). Where we do, the approximate wind speed in mph will also be presented.

Conversion tables for common metric and English units can also be found in the Appendix.

UNITS OF MEASUREMENT

In the United States, we still rely on the old British Imperial System of measurements. Canadians have successfully made the transition to metric. Continental Europeans use metric measurements exclusively. The rotor diameter of wind turbines, because they are sold internationally, is nearly always given in meters. (Some American manufacturers of small wind turbines give rotor diameter in both feet and meters.) Most of those who work with wind energy are accustomed to using the metric system, especially when referring to the size of a wind turbine by its rotor diameter.

When the size of a wind turbine is mentioned in *Wind Energy*, the measurement will be given in meters. If you have an aversion to metric units, don't panic. The approximate conversion to feet will appear in parentheses.

To calculate the power in the wind and to estimate the amount of energy a wind turbine is likely to produce, it will make your life a lot easier to use the metric system. If wind speed is given in mph (miles per hour) or in knots, simply convert the speed to the metric system's m/s (meters per second).

Here are some useful conversions.

1 m/s = 2.24 mph
1 m/s = 1.94 knots
1 mph = 0.447 m/s
1 meter = 3.28 feet

If you have a hard time visualizing the rotor diameter of wind turbines in meters, here are some simple approximations.

1 meter ~ 3 feet (the size of the Marlec 910)
2.5 meters ~ 8 feet (about the size of Bergey's XL.1)
15 meters ~ 50 feet (the size of an old Vestas V15 in California)
30 meters ~ 100 feet
50 meters ~ 150 feet (about the size of WindShare's Lagerwey turbine in Toronto)
100 meters ~ 300 feet (the length of a football field)

Size: It's All Relative

Some wind turbines are so small you can pick them up in your hands. Mongolian nomads carry these micro turbines on horseback from one encampment to the next. Other wind turbines are so large you can see them from commercial airliners as you streak across the sky (see Figure 2-2. Relative size).

Though it's common to use the rated power of a wind turbine as a measure of its size, this is incorrect and can be misleading. As will be explained in the following chapters, it's the area swept by a wind turbine rotor that primarily determines how much electricity it will generate. And this is true regardless of orientation, whether we are describing conventional wind turbines or wind turbines that spin about a vertical axis.

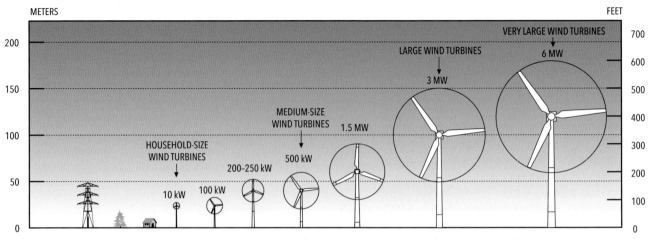

Figure 2-2. Relative size. Wind turbines today span the gamut from micro turbines only 1 meter (3.3 feet) in diameter to very large turbines with rotors greater than 100 meters (328 feet) in diameter. Very large wind turbines intercept 1,000 times more of the wind stream than micro turbines.

For conventional wind turbines—those whose rotor sweeps a circle—the shorthand for the area swept by the wind turbine is its rotor diameter. In wind energy, size—especially rotor diameter—matters. Thus, wind turbine size classes depend primarily upon the diameter of the rotor (see Table 2-1. Wind Turbine Size Classes).

The size classes used in *Wind Energy* are somewhat arbitrary. These classes differ from that used by the International Electrotechnical Commission (IEC). The IEC defines small wind turbines as those that intercept less than 200 square meters (m²) of the wind stream. This is equivalent to a conventional wind turbine with a rotor less than 16 meters (52 feet) in diameter. The IEC defines medium-size turbines as those with a swept area greater than 200 m² but less than 1,000 m² or conventional wind turbines with rotor diameters from 16 meters (52 feet) to 36 meters (117 feet).

In *Wind Energy*, small wind turbines include micro, mini, and household-size turbines. Micro and mini wind turbines are almost solely used in battery-charging applications. Household-size wind turbines are available for both battery charging and for use connected to the grid.

Micro turbines as those from 0.5 to 1.25 meters (2–4 feet) in diameter. These machines include the 200-watt Air Breeze as well as the Ampair 300. Both use rotors 1.2 meters in diameter and intercept about 1 square meter (m²) of the wind stream (see Figure 2-3. Micro wind turbine). Installed on a tall tower at a good site, such a wind turbine should generate up to 500 kWh per year.

Table 2-1. Wind Turbine Size Classes				
	Rotor Diameter		Nominal Swept Area	Standard Power Rating*
	m	~ft	m²	kW
Micro	0.5–1.25	2–4	0.2–1	.04–0.25
Mini	1.25–3	4–10	1–7	.025–1.4
Household	3–10	10–33	7–80	1.4–16
				Typical Manufacturer Power Rating
Small Commercial	10–20	33–66	80–300	10–100
Medium Commercial	20–50	66–164	300–2,000	100–1,000
Large Commercial	50–100	164–328	2,000–8,000	1,000–3,000
Very Large Commercial	100–150	330–500	8,000–18,000	2,000–10,000

*Std. Power Rating for micro, mini, and household-size wind turbines assumes a specific power of 200 W/m².

Figure 2-3. Micro wind turbine. Air Breeze micro wind turbine at the Wulf Test Field in the Tehachapi Pass. The Air Breeze uses a rotor only 46 inches (1.2 meters) in diameter and is rated at 200 watts. It is so small it can be carried in your hands. The turbine intercepts about 1 square meter of the wind stream and at a good site will generate about 500 kWh per year. 2009.

Figure 2-4. Mini wind turbine. Foreground: Bergey 850 in operation at the Wulf Test Field in California's Tehachapi Pass. This cabin-size turbine uses a three-blade rotor 8 feet (2.4 meters) in diameter and was rated at 900 watts. This model has been superseded by Bergey's XL1. Background: Medium-size wind turbines at one of the world's largest wind power plants. At a good site a wind turbine of this size could generate about 2,000 kWh per year. 1998.

Mini wind turbines are slightly larger and span the range between the micro turbines and the bigger household-size machines. They vary in diameter from 1.25 to 3 meters (4–10 feet). One popular turbine in this category is Bergey Windpower's XL.1 (see Figure 2-4 Mini wind turbine). The XL.1 uses a 2.5-meter (8-foot) diameter rotor that sweeps nearly 5 m² of the wind. Thus, the XL.1 is nearly five times larger than the Air Breeze or the Ampair 300.

Household-size wind turbines (a translation of the Danish term *hustandmølle*) are the largest of the small wind turbine family (see Figure 2-5. Household-size wind turbine). As you would expect, wind turbines in this class span a wide spectrum. They include turbines as small as the Skystream with a rotor 3.7 meters (12 feet) in diameter to the Bergey Excel that uses a rotor 7 meters (23 feet) in diameter and weighs in at nearly 500 kilograms (1,000 pounds). The Skystream sweeps slightly more than 10 m²,

Figure 2-5. Household-size wind turbine. Bergey Excel in operation at a home near Tehachapi, California. The Excel uses a rotor 7 meters (23 feet) in diameter and is rated by the Small Wind Certification Council as capable of 8.9 kW at a wind speed of 11 m/s (25 mph). At an average annual wind speed of 5 m/s (11 mph), this turbine is capable of generating nearly 14,000 kilowatt-hours (kWh) per year. Circa 1990s.

IN WIND ENERGY, SIZE, ESPECIALLY ROTOR DIAMETER, MATTERS.

WIND ENERGY WORKSHOPS

Small Wind

Mick Sagrillo, a grizzled veteran of small wind, organizes hands-on workshops annually through the Midwest Renewable Energy Association and Solar Energy International. His multiday program often includes assembly and installation of a small wind turbine. Sagrillo is an energetic and entertaining speaker, regaling his students with amusing anecdotes and homespun advice.

Scoraig Wind Electric's Hugh Piggott, a designer of small wind turbines, offers workshops on how to build your own small wind turbine. Piggott is a frequent contributor to Internet news groups that explain to novices and pros alike the arcane workings of small wind turbines.

Large Wind

Several research organizations, such as ECN (Energie Centrum Nederlands) in the Netherlands, Deutsches Windenergie Institut (DEWI) in Germany, as well as many national trade associations, offer workshops for professionals on the commercial applications of wind energy. See the Appendix or my website for details.

Mick Sagrillo (at right). Sagrillo lectures to a hands-on class in Amherst, Wisconsin, before the start of the Midwest Renewable Energy Fair.

Hugh Piggott (at right). Piggott displays one of the typical rotor blades he and his students make in his workshops.

whereas the Bergey sweeps nearly four times more area than the Skystream and 40 times more than the Air Breeze.

Small commercial turbines range in size from 10 to 20 meters (30–70 feet) in diameter and sweep up to 300 m^2 (see Figure 2-6. Small commercial-scale wind turbine). Turbines in this class, typified by Gaia Wind and Endurance's E3120, are capable of producing from 10 kW to 100 kW. Some of the products in this size class can be used for battery charging; however, most wind turbines in this class are designed for interconnection with the utility.

Medium-size, commercial-scale wind turbines are those used for commercial applications such as for farms, factories, and businesses (see Figure 2-7. Medium-size commercial-scale wind turbine). They can range in size from 20 to 50 meters (70–160 feet) in diameter and sweep as much as 2,000 m^2. Turbines in this class can be rated from 100 kW to more than 1,000 kW. Typical of this size is Northern Power Systems Northwind 100, a wind turbine rated at 100 kW that uses a rotor 21 meters (69 feet) in diameter. The Northwind 100 sweeps nearly 350 m^2 or almost 10 times the area swept by Bergey's Excel.

Large and very large commercial-scale wind turbines are the machines found in modern wind power plants (see Figure 2-8. Very large wind turbine). Though huge on a human scale, they can be found singly or as part of small clusters in or near cities throughout Europe and North America. In Germany, wind turbines of this size are used by farmers, or groups of people, to generate commercial quantities of electricity for sale to the utility.

Figure 2-6. Small commercial-scale wind turbine. While Gaia Wind's 13-meter diameter (43-foot) downwind rotor is only rated at 11 kW, the turbine is capable of generating 40,000 kWh per year at a good site. Note the work platform for scale. The turbine uses a teetered, downwind rotor. 2012.

Figure 2-7. Medium-size commercial scale wind turbine. Northern Power Systems' Northwind 100 uses a rotor 21 meters (69 feet) in diameter and is rated at 100 kW. A turbine of this size is capable of generating about 250,000 kWh per year at a good site such as here at Girvan on the west coast of Scotland. This 100-kW turbine is installed in the parking lot of the regional hospital. 2012.

Figure 2-8. Very large wind turbine. Vestas's V112 uses a rotor 112 meters (~370 feet) in diameter and sweeps nearly 10,000 square meters of the wind stream. The turbine is rated at 3,000 kW (3 MW) and can generate 10 million kWh per year at a good site, such as here on the beach at Hvide Sande, on the west coast of Denmark's Jutland Peninsula. 2012.

For a sense of scale, consider that a wind turbine with a rotor diameter of 113 meters (371 feet) intercepts 10,000 m² of the wind stream. This wind turbine is 10,000 times larger than a micro turbine, and 1,000 times larger than the Skystream household-size wind turbine. A wind turbine in this size class typically stands atop a tower from 100 meters (330 feet) to 140 meters (460 feet) high.

One last word of advice before we begin. Pay attention, ask questions, work safely, and, most important of all, don't believe everything you read.

Let's now turn to how modern wind turbines evolved from the turbulent 1970s and what that means for how we develop policy that will lead to further growth of this critical form of renewable energy.

ATTENTION: THE WIND INDUSTRY IS DYNAMIC

The small and large wind turbine industries are extremely dynamic. Wind turbine manufacturers, models, sizes, and product lines change constantly. Mention of a particular wind turbine or inclusion of a photo of a particular wind turbine in this book is for illustrative purposes only. Books have a long shelf life, often longer than a particular brand of wind turbine. Just because a wind turbine is shown or mentioned here doesn't mean that the turbine is still on the market. For example, the popular household-size wind turbine Skystream was in limbo when *Wind Energy* went to press in 2015. The manufacturer, Southwest Windpower, was defunct, yet some 8,000 turbines had been installed worldwide so someone may resurrect the design. The turbine with its distinctive scimitar-like blades, and its direct-drive, permanent-magnet generator make it a noteworthy design worthy of inclusion in *Wind Energy*.

SOME DO'S AND DON'TS ON INVESTING IN WIND ENERGY

- Do plenty of research. It can save a lot of trouble—and expense—later.
- Do visit the library. Books remain amazing repositories of information.
- Do talk to others who use wind energy. They've been there. You can learn from them what they did right, and what they'd never do again.
- Do read and, equally important, follow directions.
- Do ask for help when you're not sure about something.
- Do build to code. In the end it makes for a tidier, safer, and easier-to-service system.
- Do take your time. Remember, there's no rush. The wind will always be there.
- Do be careful. Wind turbines may look harmless, but they're not.
- Don't skimp, and don't cut corners. Taking short cuts is always a surefire way to ruin an otherwise good installation.
- Don't design your own tower—unless you're a licensed mechanical engineer.
- Don't install your turbine on the roof—despite what some manufacturers may say!
- And, of course, don't believe everything you read in sales brochures, or, for that matter, books about wind energy.

In general, doing it right the first time may take longer and cost slightly more, but you'll be a lot happier in the long run.

3

Where It All Began

> Even now it is not utterly chimerical to think of wind superseding coal in some places for a very important part of its present duty—that of giving light. Indeed, now that we have dynamos and Faure's accumulator, the little want to let the thing be done is cheap windmills.
>
> —William Thomson (Lord Kelvin), in a presentation to the Philosophical Society of Glasgow, 1881

THE HISTORY OF MODERN WIND ENERGY IS A TURBULENT story of ambitious engineers, physicists, and entrepreneurs who went down many technological dead ends before reaching a consensus on what works best. It is an exciting tale of hope and promise, but it is also one of failure and many false starts. In the end, it is a tale of how a small band of activists, dreamers, and entrepreneurs built one of the world's fastest-growing and most dynamic industries.

This chapter briefly examines the history of modern electricity-generating wind turbines. Entire books have been written on the history of wind energy. This chapter and the next only skim the surface. Some books just cover the history of wind in the United States (Righter's *Wind Energy in America*), some just on wind in Germany (Heymann's *Geschichte der Windenergienutzung*), and others just wind in Denmark (*Wind Power—The Danish Way*). Some histories of modern wind energy have looked only at the way the technology developed (van Est's *Winds of Change*), others only one wind turbine (Thorndal's *Gedsermøllen* and Putnam's *Power from the Wind*). Some books on wind energy engineering have included extensive chapters chronicling the development of wind energy (Rapin and Noël's *Energie Eolienne*). These books and other sources on the history of wind-electric generation are listed in the Appendix for those who want more detail.

The focus of this chapter is—of necessity—on Denmark. It was in Denmark that modern commercial-scale wind turbines evolved. Although many people and many nations have played a part, it is to the Danes above all that we owe a debt of gratitude for making wind energy the commercial success it is today.

Technology never develops in a vacuum. It depends on a need, on creative people, and on a nurturing culture. Consequently, wind energy is part of its time and responds to the demands of its day. The history of wind energy illustrates the power of motivation, opportunity, and policy in driving technology.

Today's wind turbines did not come from massive, centrally directed research programs common in the 1970s after the first oil embargo. Wind energy is not like nuclear power. It never needed a Manhattan Project. In some cases, as in the United States and Germany, national research programs hindered rather than helped wind energy development by taking us down the wrong technological paths.

What wind energy has always needed—and needs today—is a reason to use it, a raison d'être. Wind energy grew from the grassroots because the people wanted it—in fact demanded it. The people, in this case the Danish people, demanded to use wind energy instead of nuclear power, and when their government at first didn't listen, they took matters into their own hands.

How Far We've Come

Before we begin, it's useful to see how far we've come. One metric for measuring our progress over the past century is the amount of power a modern wind turbine extracts from the wind, relative to early wind turbines. Modern wind turbines, whether household-size machines or much larger commercial-scale turbines, are typically 10 times more powerful than their forebears (see Table 3-1. Comparison of Modern Wind Turbines with Traditional Windmills).

There are several reasons for this. Most important, today's modern airfoils are derived from a hundred years of development in aviation and aeronautical engineering. We've found that to use these airfoils most effectively, modern wind turbines must operate at higher speeds than traditional windmills. This is reflected in the difference between the tip-speed ratio of the two technologies.

The tip-speed ratio is the speed of the tip of the blade as it moves through the air relative to that of the speed of the wind across the ground. The tip-speed ratio is not a constant; it varies with rotor speed. Consequently, the tip-speed ratio is specified for a particular rotor speed in rpm (revolutions per minute).

Modern wind turbines are "fast runners." They use "high-speed" rotors relative to traditional windmills with tip-speed ratios of about one to two. The development of modern wind turbines has been in part a search for the right combination of airfoils, blades, rotors, and speeds at which they should operate. The terminology becomes somewhat confusing when we talk about some modern wind turbines as "high-speed" turbines relative to other modern wind turbines. These designs may have tip-speed ratios twice those of most other "high-speed" rotors. Today most wind turbines deliver their rated power at a tip-speed ratio of between five and seven.

Commercial-scale wind turbines have grown steadily in size from modest beginnings in the late 1970s and early 1980s. As manufacturers refined the technology, the turbines doubled in size every few years as measured by their rotor swept area. Large wind turbines in the mid-2010s were typically 50 times the size of the commercial turbines installed in the early 1980s (see Table 3-2. Technology Development of Commercial Wind Turbines).

Rated power generally kept pace with increases in swept area, until the mid-2010s when a revo-

Table 3-1. Comparison of Modern Wind Turbines with Traditional Windmills					
	Rotor Diameter		Nominal Swept Area	Effective Capacity	Tip Speed
	m	ft	m²	kW	Ratio
American Farm Windmill	5	16	20	0.5	2
Modern Household-size Wind Turbine	5	16	20	5-6	5-7
Brush Multiblade	17	56	227	12	2
Modern Wind Turbine	17	56	227	75-100	5-7
Dutch or European Windmill	25-30	80-100	500-700	25-30	1
Modern Wind Turbine	25-30	80-100	500-700	250-300	5-7

Table 3-2. Technology Development of Commercial Wind Turbines

Period	Nominal Rotor Diameter		Nominal Swept Area	Nominal Capacity
	m	ft	m²	kW*
Early 1980s	15	50	180	65
Mid 1980s	18	60	250	100
Late 1980s	25	80	500	250
Early 1990s	35	110	1,000	400
Mid 1990s	40	130	1,300	500
Late 1990s	50	160	2,000	750
Early 2000s	70	230	3,800	1,500
Mid 2000s	80	260	5,000	2,000
Late 2000s	90	300	6,400	3,000
Early 2010s**	100	330	8,000	3,000
Mid 2010s	113	370	10,000	3,000

*@60Hz **Development of high specific area, low-specific capacity turbines designed for low and moderate wind sites.

lution swept the industry. Suddenly, rated power no longer increased as wind turbines became larger and larger. That is, the specific power—or power relative to the area swept by a wind turbine's rotor in W/m²—of new wind turbines began to decrease. Why this is a positive development is explained in Chapter 8, "Silent Wind Revolution." What is important here is that one of the pioneers in wind energy recommended this design strategy more than 60 years ago, illustrating why it's useful to know how we came to where we are today.

In the Beginning

It's hard to imagine the 19th-century ferment of invention and discovery that led to so many technologies we now take for granted. This was a time when growth in our understanding of electricity was in full flower. Scientific journals filled their pages with the latest discoveries. Storage batteries were being developed in France. Edison and Westinghouse were fighting the "war of the currents" in the United States. Siemens was building electric trams in Great Britain. Central-station power plants were being built using alternating current generators to power lights and motors. Telegraphy was moving rapidly toward telephony.

Late in the century several inventors on both sides of the Atlantic turned their talents to harnessing the wind to generate electricity. Within a few months of each other, the first wind turbines built for generating electricity were installed in Scotland and France. These were soon followed by one in the United States and a few years later by one in Denmark. It's unlikely these inventors knew about one another's work, though all were current with the latest scientific developments. None are household names now, though they were all major figures of their day. And with the exception of the work in Denmark, none of their experimentation resulted in any lasting technological development.

James Blyth

Our story begins with noted Scottish professor of engineering James Blyth. Late in the summer of 1897, Blyth installed the world's first electricity generating wind turbine. He built a conventional four-blade European windmill using cloth sails. The traditional design required turning the turbine into the wind and furling the cloth sails to control rotor speed. Blyth quickly found that it was difficult to control the rotor in the strong winds Scotland is famous for. He next tried a multiblade farm windmill. He was still dissatisfied. Like a good engineer, Blyth chose to work around his limitations and subsequently in 1891 designed a simple vertical-axis rotor that was inherently self-limiting (see Table 3-3. Wind-Electric Turbines at the Turn of the 19th Century).

Initially, Blyth used cloth sails at the end of long poles, but later he added metal buckets reminiscent of a cup anemometer on which it was patterned. By deliberately choosing a device that became less and less efficient as wind speeds increased, and by keeping his machine within a size he could manage with the materials at hand, Blyth was able to build the world's first vertical-axis wind turbine (VAWT) used to generate electricity. He used it to light his summer home northeast of Edinburgh.

Wind energy then, as now, was not a subject in isolation. It was a part of the great intellectual currents of the day. At the same time as Blyth was teaching at the Royal Institute in Glasgow, William Thomson (commonly known

Table 3-3. Wind-Electric Turbines at the Turn of the 19th Century

Turbine	Country	Diameter (m)	Diameter (ft)	Swept Area (m²)	Power (kW)	Specific Power (W/m²)	No. of Blades	Rotor Orientation	Control	Approx Date Installed
James Blyth	Scotland	8	39	50	3	60	4	upwind	manual	1887
Charles de Goyon	France	12	39	113			multi	upwind	furling	1887
Charles Brush	USA	17	56	227	12	53	multi	upwind	furling	1888
James Blyth	Scotland	10	33	32	3		8	vertical	drag	1891
Poul LaCour	Denmark	12	38	106	18	170	4	upwind	jalousie	1891
James Blyth	Scotland	12	39	38	7.5	195	8	vertical	drag	1895
Poul LaCour	Denmark	23	75	408			4	upwind	jalousie	1901
Lykkegaard	Denmark	14	46	154			4	upwind	jalousie	1905

as Lord Kelvin) was a professor at the University of Glasgow. Thomson was a prolific inventor and scientist famed for his work in thermodynamics, electricity, and other fields. According to Blyth's biographer Trevor Price, both Blyth and Thomson were members of Glasgow's Philosophical Society where in 1881 Thomson gave a famous lecture on using the "sources of energy in nature," specifically mentioning the usefulness of storing electricity generated by wind turbines in rechargeable batteries. It must have been a heady time to be in Glasgow.

Following his experimentation, Blyth became one of the first academics to publish on the topic of using the wind to generate electricity. John Twidell, himself a respected British wind pioneer, notes Blyth made a prescient observation that "any fool can make a wind turbine go round to generate electricity, but the challenge is to make one that can be left unattended without over-speeding to destruction." All wind turbine design could be boiled down to that one observation—essentially the same lesson that Danish wind turbine designers learned the hard way nearly 100 years later.

Blyth filed a patent on his wind turbine in 1891, and in 1892 he presented a scientific paper on the subject of using the wind to generate electricity—a full decade after Thomson had first suggested it. Though Blyth built a third turbine in 1895, no commercial product resulted from his work in an age of abundant coal.

Duc de Feltre

Across the Channel, French noble Charles de Goyon, the Duc de Feltre, and his collaborators were turning their attention to the wind as well. De Goyon's group installed a multiblade farm windmill in early September of 1887 on the coast of France. Historian Etienne Rogier writes that de Goyon coupled a 12-meter (40-foot) diameter Halladay Standard power mill to two dynamos for charging batteries. At a time when most farm windmills were only 2.4 to 3 meters (8–10 feet) in diameter, de Goyon had ambitiously chosen a large wind turbine of the size typically used for mechanical power. Like Brush, de Goyon used his machine for charging batteries.

De Goyon planned to rent the batteries to neighbors at a beachfront community on the Channel. He subsequently modified the design to power the Cap de la Hève lighthouse near the French port of Le Havre for the French government. His efforts mark the first time a wind turbine was used commercially, that is, where the electricity produced would be made for sale to others.

It is noteworthy, says Rogier, that de Goyon used a power mill of the kind that was being installed commercially in North America and in Europe at the time. These windmills were successful because they were automatically self-regulating: they would furl the rotor out of the wind automatically in dangerous winds. They didn't require a miller to furl their sails

BLYTH MADE A PRESCIENT OBSERVATION THAT
"any fool can make a wind turbine go round to generate electricity, but the challenge is to make one that can be left unattended without over-speeding to destruction."

or point them out of the wind. Because of their success, the multiblade windmill, whether for pumping water or providing mechanical power, had become a widespread industrial product.

Brush Dynamo

A few years after Thomson's discourse on the "sources of energy in nature," *Scientific American* posed a similar question on the other side of the Atlantic where wealthy entrepreneur Charles Brush took up the challenge. Brush made his fortune in electric arc lighting long before Edison had developed the incandescent lamp. Like de Goyon, Brush chose to work with a technology that was then commonplace, the American multiblade windmill.

In 1888, Brush installed a massive windmill to drive a 12-kW DC dynamo for powering the lights on his estate outside Cleveland, Ohio (see Figure 3-1. Brush Dynamo). Brush patterned his wind turbine after the water-pumping and power windmills being used across North America's Great Plains. Even so, his wind turbine was large for the day. It stood atop an 18.3-meter (60-foot) tall tower that pivoted with the wind like a traditional post mill common in the lowlands of Europe. The 17-meter (56-foot) diameter rotor was twice as large as de Goyon's (see Figure 3-2. Schematic of early wind-electric generators).

Importantly, Brush's dynamo was automatically self-regulating in high winds. Like multiblade windmills of the day, it used a pilot vane in the plane of the rotor to furl the rotor toward the tail vane in strong winds. The speed of the wind

Figure 3-1. Brush Dynamo. Industrialist Charles F. Brush developed the first electricity-generating wind turbine in North America for his Cleveland, Ohio, estate in 1888. The 17.1 meter (56-foot) diameter rotor used 144 wooden slats in a configuration common among water-pumping farm windmills of the era. The 12-kW wind turbine used a pilot vane to turn the rotor out of the wind. The rotor is furled or out of service in this photo. (Western Reserve Historical Society)

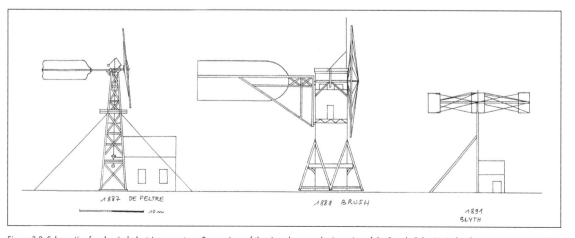

Figure 3-2. Schematic of early wind-electric generators. Comparison of the size, shape, and orientation of the Duc de Feltre's wind turbine near La Havre (1887), Charles Brush's turbine near Cleveland (1888), and James Blyth's later Vertical-Axis Wind Turbine (1891). Blyth's machine was substantially smaller than the other two. (Etienne Rogier)

at which this occurs could be adjusted and the rotor could be manually furled out of the wind as desired. In 1890, the cover story in *Scientific American* was on the Brush Dynamo.

Dead Ends

French historian Rogier argues that de Goyon's experimentation was potentially more important to the development of wind energy than that of either Brush or Blyth. De Goyon used a widely available industrial product of his day rather than a one-off experimental device, says Rogier. If de Goyon had been successful, there was a ready platform of manufacturers who could put the new machine into industrial production.

Not surprisingly, there were many subsequent attempts to commercialize the path chosen by de Goyon well into the 1920s in both North America and Europe. Rogier notes that two French manufacturers, Chêne and Les Etablissements Cyclone, were particularly active in promoting the concept for providing electricity to wealthy clients on their country estates.

Likely there were many other unheralded manufacturers on both sides of the Atlantic who were doing the same thing at about the same time as de Goyon. Certainly for many decades after de Goyon, catalogs of farm windmill manufacturers proposed variations on the concept. Historian Robert Righter notes a commercial attempt in the United States in 1893 and further experimentation with vertical-axis wind turbines in Britain in the mid-1890s.

Nevertheless, the requirements for generating electricity effectively are far different than those for pumping water or providing mechanical power, and the approach of both Brush and de Goyon confirmed that the low-speed, multiblade rotor was a technological dead end. These machines were ideal for pumping water and some mechanical tasks but not for much else. Generating electricity with the wind on a larger scale awaited experimentation with various rotor configurations and results from some of the very first wind tunnel tests ever performed. Enter Poul la Cour.

The Danish Edison

Danes were not newcomers to wind energy, nor were windmills new to Denmark when Poul la Cour began experimenting with the wind to generate electricity around the turn of the 19th century. La Cour, known as the "Danish Edison," was a meteorologist, inventor, and teacher. As a meteorologist, he was familiar with the power of the wind. As an inventor, he rivaled Elisha Gray and Alexander Graham Bell in what today we call telecommunications. And as a teacher, he wrote textbooks on mathematics and geometry. But it's la Cour's work with wind energy that he's best remembered for and where he made his mark.

Significantly for the story of wind energy, la Cour was part of a broader social movement to emancipate Danish peasants from feudal serfdom and raise their standard of living through a system of education for all. In this la Cour was as much the Danish Lincoln as the Danish Edison, where wind was power by the people, for the people. This heritage was seen again much later by leaders in the Danish wind power revival of the 1970s.

La Cour was a follower of Danish theologian N. F. S. Grundtvig, a leading light in the Danish renaissance of the mid-19th century. Grundtvig's teachings had a profound impact on the cultural landscape of Denmark that resonates to this day. Grundtvig's hymns are still sung in Danish churches, and his homilies are found prominently posted on the walls of home and institutions throughout Denmark. His philosophy espoused agrarian reform by liberating Danish peasants spiritually, intellectually, and economically.

As recounted by sociologist Steven Borish, the principal vehicle for Grundtvig's philosophy was through the creation of an extensive network of mostly rural folk high schools (*folkehøjskole*). Through these nontraditional schools, Danes were taught life skills such as faith in oneself and one's community as well as faith in cooperative problem solving—traits that are a thread weaving through the work of la Cour and Danish wind energy pioneers ever since.

La Cour joined the faculty at the *folkehøjskole* in the village of Askov in central Jutland at a time when coal-fired power stations were first being introduced into Denmark's major cities. According to Danish historian Povl-Otto Nissen, la Cour proposed using Denmark's abundant wind instead of the country's limited supply of coal. Thus, it was at Askov that la Cour made history when he installed his first electricity-generating wind turbine in 1891 (see Figure 3-3. Poul la Cour's *klapsejlsmølle*). For the next 17 years, la Cour continued experimenting with wind energy and, more importantly, conducting scientific experiments on what he termed the "ideal" windmill.

The first turbine installed at Askov used cloth sails as was widely used on traditional European windmills at the time, but la Cour soon replaced them with *klapsejls* (clap-sails) or jalousie shutters on a rotor nearly 12 meters (38 feet) in diameter. These self-regulating shutters were not original with la Cour; they were fairly common on windmills of the late 19th century. Nor were the dual fantails on la Cour's turbine unusual for orienting the wind turbine into the wind. They too were found on traditional windmills in northern Europe. These features were all well known to millwrights of the period. What was unusual was la Cour's adaptation of the traditional four-blade windmill to generate electricity by transmitting mechanical power to ground level where the spinning shaft drove an 18-kW dynamo.

In 1897 la Cour installed a much larger turbine atop his new laboratory. This *keglevindfang* or conical wind catcher was the product of a local millwright and represented the intuitive thinking of the day (see Figure 3-4 Askov's conical wind catcher). To the uninitiated, the fanciful wind catcher with its six blades, where each blades was wider at the tip than at the root, appears like it should be much better at, well, catching the wind than la Cour's inelegant four-blade rotor. That wasn't the case, and la Cour wanted to know why.

This question led la Cour to construct one of the world's first wind tunnels a full decade before Gustave Eiffel built his wind tunnel in Paris at the foot of his famed tower. La Cour's results upended the thinking of the day, pointing the way for the later work of physicists who would found the then unknown field of aerodynamics. One controversial conclusion he reached was that the four blades of his original turbine were superior to the six on the newer turbine. La Cour then set about replacing the wind catcher with a

Figure 3-3. Poul la Cour's *klapsejlsmølle*. La Cour's original 1891 electricity-generating wind turbine used *klapsojls* (clap-sails in English), which work like jalousie shutters. The self-regulating rotor was 11.6 meters (38 feet) in diameter and drove an 18-kW dynamo at ground level. The rotor was oriented into the wind with dual fantails. (Poul la Cour Foundation)

Figure 3-4 Askov's conical wind catcher. In 1897, Poul la Cour added a *keglevindfang* or "conical wind catcher" (left) atop Askov's new laboratory. (His earlier turbine is in the background.) The wind catcher used an unconventional six-blade rotor developed by a local millwright. The rotor represented the intuitive thinking of the day but proved unsatisfactory, leading la Cour to his famous wind tunnel experiments on wind turbine rotors. Eventually, the wind catcher rotor was replaced with a more conventional four-blade rotor designed by la Cour. (Poul la Cour Foundation)

BY 1905 THERE WERE 22 WIND POWER PLANTS OPERATING IN DANISH VILLAGES.

traditional four-blade rotor 23 meters (75 feet) in diameter.

Although many historians dwell on la Cour's choice not to charge batteries with his wind turbine as others had, opting instead to electrolyze water into hydrogen and oxygen, this wasn't his most significant contribution. What sets la Cour's work apart from Brush, Blyth, and de Goyon is that his approach built on hard-won mill-building experience aided by the knowledge he gained from his wind tunnel measurements. After all, millwrights had been successfully building windmills for hundreds of years.

Millwrights had learned, for example, that four blades were optimum for the materials at hand. La Cour reached the same conclusion from a different direction—through a methodical scientific approach. Millwrights too had come to the same conclusion as Blyth that the rotor needed to be automatically self-limiting, thus the automatic controls on the jalousie shutters.

La Cour's traditional configuration could also be scaled up to much larger sizes than that using the farm windmill design chosen by de Goyon and Brush. Some traditional European windmills of the day were as much as 25 to 30 meters (80–100 feet) in diameter. Imagine the size of the tail vane that would be needed for a farm windmill of this scale. La Cour's second turbine atop his laboratory was twice the size of that built by Brush, and four times that of de Goyon, both large turbines of their day.

La Cour's contribution didn't end with his scientific discoveries. Because of his disputes with Elisha Gray and Alexander Graham Bell on patents for telephony, la Cour had become disenchanted with the patent system. As a result, la Cour placed his work on the "ideal" wind turbine in the public domain through his scientific publications. And in keeping with Askov's role in the *folkehøjskole* movement, he put his ideas into practice.

As early as 1903, la Cour and some of his students formed the Dansk Vind Elektrisitets Selskab (Danish Wind Power Society) and began publishing the world's first wind energy journal, a quarterly publication that survived for more than a decade. Where the wind turbines of Blyth, Brush, and de Goyon were intended for a single estate or a remote lighthouse, La Cour envisioned adding wind turbines to the then growing number of power stations destined to provide lighting for the villages and small factories dotting the Danish countryside. At the time, only the privileged in the big cities could read by electric lights. La Cour and his students intended to bring light and power to the villages, giving villagers the amenities found in the big cities.

Thus, the Danish Wind Power Society fostered rural electrification in Denmark long before it became a common word in English. Toward this end, the Askov School was turning out the world's first wind energy technicians, and the society was publishing a standardized design of la Cour's "ideal" mill, which the technicians took with them.

Several Danish machine shops subsequently began producing wind turbines to la Cour's design. The most successful was the Lykkegaard Machine Works on the island of Funen. They began building wind turbines in la Cour's time and continued well into the 1950s. These wind turbines were typically 14 meters (46 feet) in diameter driving 6-kW DC dynamos.

By 1905 there were 22 wind power plants operating in Danish villages. All were direct descendants of la Cour's original research unit with four blades using jalousie shutters to regulate power and dual fantails to orient the rotor into the wind. Within three years, another 10 villages added wind turbines to their power stations. In 1906, the plants at Vemb and Oksbøl became the first cooperatively owned wind turbines—another development that would have reverberations in late 20th-century Denmark.

The work of Blyth, Brush, and de Goyon ended with the inventors and their one-off experimentation. In contrast, the results of la Cour's experimentation lived on after his death. By the time la Cour died in 1908, he had built a rich legacy, including a direct line to the development of modern wind turbines. Unlike other wind turbine experimenters of the period, la Cour's systematic investigation of what would soon become the field of aerodynamics led inexorably to modern wind turbines. But it was also the philosophy embodied in *folkehøjskole* movement that la Cour lived that led to the first industrial manufacture of wind turbines and the first extensive use of wind energy for generating electricity to benefit the populace of a country. These are all traits that would be seen again in the 1970s, when Danes once more turned toward wind energy. They had no hesitation in taking wind turbine design into their own hands and solving problems together.

La Cour and the work at Askov were part of something greater—a long-lasting social movement. Moreover, there were few other resources readily available to generate electricity. The British, Americans, and French had abundant coal deposits and not an insubstantial amount of hydropower available. The Danes had none of this, but they had a lot of wind and the desire to liberate their people from drudgery and darkness. Thus, there was a compelling reason to develop and use the wind in Denmark while it languished elsewhere.

For the Danes in the years after la Cour, wind energy would became an important source of power during the privations caused by the Great War. With wartime shortages of oil, and high prices for what was available, local power stations turned to wind turbines—most of them of the la Cour design—to supplement their diesel engines. By the end of the Great War, there were 120 wind turbines generating electricity in Denmark. These were not small machines, each was 15 to 18 meters (50–60 feet) in diameter and capable of generating 25 kW to 35 kW. Historian Jytte Thorndahl notes that although wind energy provided only 1% of the electricity nationwide, it was a significant source in rural Denmark when it was most needed.

The Interwar Years

The Great War brought rapid advances in aeronautical engineering following the pioneering work of British, French, and German physicists. Unlike la Cour's *klapseijlsmølle* wind turbines design, which dominated the prewar period, postwar development soon adopted the airfoils demonstrated so dramatically during the war and applied them to wind turbines. During the 1920s, there was a flowering of experimentation and research into wind energy. Meanwhile, electrification was continuing.

First Interconnected Wind Turbine

Prior to the war, most electrical systems in Denmark were direct current (DC). During the war, the utility serving the area north of Copenhagen introduced alternating current (AC). In 1919 the utility installed a wind turbine at Buddinge and connected it to its lines—a first worldwide, a full two decades before the Smith-Putnam machine in Vermont was connected to the grid (see Table 3-4. Wind-Electric Turbines During the Interwar Years).

The 40-kW Agricco was an odd mix of new and old but surprisingly advanced for its day. The turbine used true airfoils in the blades for the first time, but it used five or six blades instead of the four blades recommend by la Cour. The rotor used struts and stays to brace the blades, much

Table 3-4. Wind-Electric Turbines During the Interwar Years									
		Diameter		Swept Area	Power	Specific Power	No. of Blades	Rotor Orientation	Control	Approx Date Installed
Turbine	Country	m	ft	m²	kW	W/m²				
Agricco	Denmark	13	41	123	40	326	6	upwind	pitch	1919
Flettner	Germany	20	66	314	30	95	4	upwind	N/A	1926
Constantin	France	8	26	50	12	239	2	upwind	pitch	1926
Darrieus	France	8	26	50	2	36	2			1927
Darrieus	France	20	66	314	12	38	2	downwind	stall	1929
Yalta	Soviet Union	30	98	707	100	141	3	upwind	pitch	1931
Jacobs	USA	4.3	14	14	3	210	3	upwind	pitch	1933

like later Danish wind turbines. And the Agricco used fantails to orient the rotor into the wind. Nevertheless, tests in the early 1920s found that the Agricco wind turbine was twice as efficient as la Cour's design.

To produce utility-compatible AC, the Agricco employed an asynchronous or induction generator for the first time. This choice proved significant and would later influence other wind turbine designers in Germany and France. If Agricco's designers had chosen a synchronous generator, they would have had to design a complex, costly, and difficult to maintain pitch mechanism for regulating the speed of the rotor and hence the speed of the generator. Using an induction generator greatly simplified the design and its interconnection with the grid.

As with la Cour and later Danish experimenters, the wind turbine was envisioned as part of the electrical network, providing power to the grid for everyone to benefit. Unfortunately, there was little demand for wind as diesel fuel became more readily available at the end of the war and Agricco reached a dead end. Danish historians describe the Agricco as the right windmill—at the wrong time.

La Cour's design continued to live on into the 1920s and 1930s in the form of the Lykkegaard *klapsejlsmølle*. Although the technology represented a dead end, no other wind turbine was used as extensively in Denmark—or anywhere else for that matter—for power generation. By the mid-1930s, there were 40 Lykkegaard machines at power stations in Denmark.

And unknown by nearly everyone but Danish historians, it was the Lykkegaard Machine Works that pioneered what we now call feed-in tariffs. According to Danish wind historians Jytte Thorndahl and Benny Christensen, Lykkegaard developed a novel marketing model in the 1920s. Lykkegaard didn't sell their wind turbine to power stations. Instead, they negotiated a fixed price that they would be paid for the electricity generated from their wind turbines—and in so doing created the world's first feed-in tariff or purchase-power agreement.

Wind Meets Aviation

By the mid-1920s a veritable who's who of the great names in physics, aviation, fluid dynamics, and the new field of aerodynamics were designing, testing, and experimenting with wind energy—and not a few unsung visionaries who tried to parlay their experience in aviation or ballistics during the war into harnessing the wind.

In France engineer Louis Constantin began experimenting with wind power in the early 1920s. By 1926 he had constructed a wind turbine using a two-blade upwind rotor 8 meters (26 feet) in diameter in France's Massif Centrale. This was the first of several prototypes Constantin would build in central and southern France.

A true visionary, Constantin was one of the founders of modern, high-speed rotors for wind turbines, says French historian Rogier. More important, Constantin anticipated modern wind power plants where multiple wind turbines would be used together in batteries or arrays to produce electricity for the grid. Like la Cour a decade earlier, Constantin recognized the importance of a scientific and commercial association to develop the technology and formed Energie Eolienne, a society to promote wind energy that lasted from 1929 to 1936.

At the same time, Constantin's contemporary, French engineer Georges Darrieus (more on Darrieus in Chapter 6, "Vertical-Axis and Darrieus Wind Turbines"), was installing prototype two-blade turbines at a test field outside Paris and pondering, like Blyth before him, wind turbines that rotated about a vertical axis.

It was during this period that the aerodynamic research laboratory in Göttingen founded by Ludwig Prandtl was testing wind turbine configurations in the wind tunnel. The tests confirmed la Cour's work decades before, but importantly concluded that a three-blade rotor operating at a tip-speed ratio of around 4 would produce the highest overall coefficient of performance—more than 0.4—than wind turbines with fewer or more blades.

In Britain, Oxford University created one of the world's first wind turbine test centers at its agricultural experiment station where the first side-by-side field tests of wind turbines were conducted from 1925 to 1926. The station tested several designs, including Bilau's Aerdynamo and an Agricco from Denmark. The center tested nine turbines in all. There wouldn't be another test center on this scale until nearly half a century later. The results concluded that, though ungainly, Agricco was one of the most cost-effective turbines tested.

But it was the Russians who built the biggest experimental machine of the day. On a bluff overlooking the Black Sea outside Yalta, Russians installed a 100-kW, three-blade wind turbine 30 meters (98 feet) in diameter. Commencing operation in 1931, the Balaklava turbine's upwind rotor drove an induction generator and fed a 6,300-volt line to the 20-MW peat-burning plant at Sevastopol, 20 miles from the site. Palmer Putnam in his *Power from the Wind* reports that at a tip-speed ratio of 4.75, the turbine reached a coefficient of performance of 0.24.

Unlike most wind turbines that yaw the turbine's nacelle about the tower to face the wind, the Balaklava machine turned the entire tower on a circular track. This design feature wasn't to be seen again until the mid-1970s with the Bendix-Schacle turbine in Moses Lake, Washington.

Windchargers

During the 1930s, the Jacobs Wind Electric Company in Minneapolis, Minnesota, was manufacturing thousands of small wind turbines for rural homesteads on North America's Great Plains. Jacobs wasn't the only manufacturer serving this market before rural electrification brought "high-line" power to more and more rural areas, but it was one of the largest and one of the most well known of the period. Paul Jacobs, a son of one of the firm's founders, writes that the company sold 20,000 wind turbines before ceasing business in the mid-1950s. That's a sales record that still stands for wind turbines of this size.

What is most striking about this era in North America was the size of the wind turbines being used. Americans were building very small machines. The largest Jacobs turbine of the era was only 4.3 meters (14 feet) in diameter, powering a 3-kW direct-drive generator. The Danes, Germans, and French were experimenting with wind turbines 8 meters (26 feet) to 20 meters (66 feet) in diameter for connection to the grid. Denmark's Agricco was commercially installing wind turbines 13 meters (41 feet) in diameter in the early 1920s, that is, a wind turbine nearly 10 times larger than the Jacobs turbine.

When American interest in wind energy was rekindled in the 1970s, it was these small turbines to which budding entrepreneurs first turned. Similarly, when Danes reintroduced wind turbines in the 1970s, it was wind turbines similar in size to those of Agricco and Lykkegaard that they turned. This was to have a profound effect on the direction of the modern wind industry. The Danish wind turbines, while large for the day, were of a size manageable for the small to medium-size companies typical of Denmark then—and now.

War Years

When war broke out in Europe, development of wind energy ceased in France and Russia. However, development continued in Denmark—of necessity once again. Surprisingly, work continued in Germany as part of the Nazi war machine. Meanwhile, work was continuing on the greatest North American experiment in wind energy until the great wind revival of the 1970s: the Smith-Putnam turbine (see Table 3-5. Wind-Electric Turbines During the War Years).

Table 3-5. Wind-Electric Turbines During the War Years										
		Diameter		Swept Area	Power	Specific Power	No. of Blades	Rotor Orientation	Control	Approx Date Installed
Turbine	Country	m	ft	m²	kW	W/m²				
Smith-Putnam	USA	53	175	2,231	1,250	560	2	downwind	pitch	1941
F.L. Smidth	Denmark	18	57	241	60	249	2	upwind	stall	1941
F.L. Smidth	Denmark	24	79	452	70	155	3	upwind	stall	1942
Ventimotor	Germany	5.7	19	26	5	196	3	upwind	furling	1942
Lykkegaard	Denmark	18	59	254	30	118	4	upwind	jalousie	1943

Figure 3-5. Smith-Putnam wind turbine. Raising one of the metal-clad blades on the 1,250-kW wind turbine atop Grandpa's Knob near Rutland, Vermont, in 1940. The ungainly turbine used a complex downwind rotor with flapping blades. The blade spar connection to the hub was undersized for the loads even with the flapping hinge, leading one blade to fail after less than 700 hours of operation. No wind turbine rivaled it in size until the early 1980s. (Archive of Carl Wilcox)

Smith-Putnam

In the fall of 1941, an ungainly wind turbine atop Vermont's Grandpa's Knob fed electricity into the lines of Central Vermont Public Service Company. This was a first in North America. Until this time, wind turbines had been used solely to charge batteries at remote homesteads in Canada and the United States.

The story of the turbine's creation—and demise—is told in detail by one of the chief protagonists, Palmer Putnam, in his book *Power from the Wind* (see Figure 3-5. Smith-Putnam wind turbine). The 53-meter (175-foot) diameter, two-blade wind turbine drove a 1,250-kW synchronous generator like that used by electric utilities in conventional power plants. To design, build, and operate the wind turbine from scratch—without any prior experience in wind energy—Putnam and the team organized by S. Morgan Smith Company in York, Pennsylvania, had to overcome numerous technical and logistical problems. That they succeeded at all is a testament to their ingenuity and perseverance.

The Smith-Putnam turbine remains unusual to this day for its choice of a downwind rotor with flapping hinges. The hinges allowed the blades to cone downwind of the tower in gusts, allowing the wind turbine to dynamically balance gusts with the centrifugal forces of the spinning rotor. The large A-frame hinges also allowed the blades to change pitch. The combination of dampened flapping and variable pitch enabled Putnam to control rotor speed accurately enough to drive the synchronous generator at constant speed.

Though it was downwind, Putnam wisely chose to actively yaw the huge machine rather than rely on passive yaw to orient the rotor downwind of the tower. However, as wind engineers Marc Rapin and Jean-Marc Noël note, the flapping rotor and its heavy, complex hinges were not well adapted to the rigors of commercial usage.

In 1943 a blade bearing failed, causing the turbine to be shut down for two years because of wartime shortage of materials. Upon inspection, the blade spars were found to be undersized for the loads, so doubling plates were welded in place. Welding of heavily stressed components is often problematic, and in retrospect this probably was unwise. The rotor was then locked in place for the duration of the war.

The turbine was returned to service in 1945, and cracks were soon found in the blade root at the strengthening welds. Operation continued in hopes of completing the test program. The Smith-Putnam turbine then operated continuously for several weeks until late March 1945 when the repair weld failed and the turbine threw a blade, ending the program. It was three decades before anyone attempted a wind turbine of this size again.

Ventimotor and the Third Reich

After the Nazis seized power in the 1930s, they began a systematic program for assuring energy self-sufficiency in time of war. The development of wind turbines became a part—though never a big part—of this program. Some of the great names in automotive and wind turbine design were associated with the effort. None of the wind turbines themselves left much of a legacy, but the work on them did.

In 1940, the Porsche design bureau installed several small wind turbines at a test field outside Stuttgart. The smallest was a two-blade, 3.4-meter (11-foot) diameter, 130-W machine, and the largest used a three-blade rotor 9.2 meters (30 feet) in diameter powering a 10-kW generator. The sheet metal-covered blades operated at a relatively high tip-speed ratio of 7. The variable-pitch turbine sported an odd twin tail, and the whole assembly resembled a race car of the period, probably an influence of Porche's wind tunnel experience with motor vehicles.

Yet it was the work in Weimar, the one-time capital, that left the biggest mark on wind energy. Though disparaged by Nazi propagandists, Weimar was and is culturally important to Germans. It was a university town, and it was home to two of Germany's most famous writers: Goethe and Schiller. It was here that an ambitious Austrian sailplane designer taught engineering and in 1940 went to work for the Ventimotor Company while he worked on his doctorate. Ulrich Hütter's thesis on wind turbine design principles was not unlike the task undertaken by Poul la Cour 40 years before, defining the "ideal" wind turbine.

One engineer described Ulrich Hütter as the "Werner von Braun of wind energy." Hütter's biographer, Heiner Dörner, called Hütter the *papst* or pope of German wind energy. Whatever the sobriquet, Hütter left a lasting mark on wind turbine technology but not, as is often thought, on the configuration of modern wind turbines.

Like Werner von Braun, Hütter's service to the Third Reich has dogged his legacy. Matthias Heymann in his masterful history of German wind energy (*Die Geschichte der Windenergienutzung*) contains an extensive discourse on Nazi interest in wind turbines. Like Righter in his *Wind Energy in America*, Heymann places the development of wind energy in Nazi Germany in its historical context, that is, within the political currents of the day.

Thus, Heymann describes the Ventimotor Company, its test center near Weimar, and its key personnel. The company was formed in late 1940 by rising stars in the Nazi party to develop wind turbines that could be used to "settle" Germans in the conquered eastern lands. Anyone who has studied the war or Nazi ideology will find that statement chilling. Germans could only be "settled" in the east after the people already there were eliminated.

Into this group stepped Hütter, then a 30-year-old aeronautical engineer. Although Heymann could find no evidence that Hütter himself was a Nazi, a later historian discovered that Hütter had joined the party in 1932 while a student in Vienna.

This is more than just a "footnote of history." Postwar Germany has had to confront its past by asking painful questions about how ordinary Germans could ignore or participate in the crimes of the Third Reich. Future historians may well ask such questions of the engineers who worked for Ventimotor. If nothing else, it calls into question their moral judgment and possibly their technical judgment as well.

What did they know? When did they know it? And, what did they do about it? The questions are relevant because one of Nazi Germany's more notorious concentration camps, Buchenwald, was near Weimar. The camp was located about 5 miles northwest of the city in woodland that is still extant.

According to a 1945 report by Georges Vanier, the Canadian Ambassador to France, Buchenwald was built near Weimar's former zoological garden. The camp was built in 1937 by political prisoners. At the time of its liberation on April 11, 1945, Buchenwald held nearly 60,000 captives. According to Vanier, the number killed at Buchenwald will never be known: "well over 50,000, it may be over 100,000." For comparison, the population of Weimar is 100,000 today.

Hütter lived in Weimar from 1939 to 1943. Again, there is no evidence that he knew of Buchenwald. But it does beg the question, how could he have not known? And, if he did know, what did he do with this knowledge?

In 1943 Ventimotor's activities were sharply curtailed after the 6th Army's defeat at Stalingrad. Hütter was then assigned to the military and subsequently went to work for Graf Zeppelin near Stuttgart designing aircraft, for which he was well known.

Heymann describes Hütter as an artist as well as a talented engineer. Unlike many of those who followed in his footsteps, Hütter emphasized the importance that the aesthetic design of wind turbines would have on subsequent decades. Hütter wrote in his thesis, says Heymann, "These [wind turbines] must for that reason in a deeper sense be of a timeless beauty, so that they do not in three or four decades hence burden a later generation with the heavy task of removing angular skeletons, by our indifference to the imponderable value of our environment."

Reminiscent of Louis Constantine's observation in the 1920s, Hütter also argued that a greater number of medium-size turbines are superior to a small number of large turbines. This lesson was lost on later German engineers who designed the ill-fated Growian.

Hütter's dissertation also concluded that for tip-speed ratios between 4 and 7, three blades were the optimum number of blades in a rotor. Again, this was ignored in the design of Growian and by Hütter himself in his later years, as he devoted most of his research to two-blade turbines. Despite his dissertation, Hütter convinced himself that he could wring more economies from lightweight, two-blade turbines operating at high tip-speed ratios—those above 7—than he could from a three-blade turbine. In this he was proven wrong—time and again.

By the end of the war there were six turbines at Ventimotor's test field. All were forgettable. Even Hütter's 5.7-meter (19-foot) diameter prototype was unremarkable and a throwback to much earlier machines that used a pilot vane to furl the rotor in high winds.

However, in 1942 Ventimotor bought a F.L. Smidth 18-meter (57-foot) diameter Aeromotor turbine from occupied Denmark for testing a directly connected 60-kW induction generator. It is interesting to speculate how the Aeromotor turbine could have influenced Hütter's later work and led German engineers down a path far different—and more fruitful—than that taken by Hütter and the engineers who followed him.

The F.L. Smidth turbines of the war years did influence subsequent Danish designers of the postwar era, including one Johannes Juul.

Denmark and F.L. Smidth

Denmark was occupied by Germany in World War II, and many commodities were in short supply, including electricity. The country again turned to the wind as it had two decades before. By the time the war ended, Danes had installed nearly 90 wind-electric turbines, the most commercial-scale wind turbines of any one country until the great wind revival of the 1970s (see Table 3-6. Early Danish Wind-Electric Turbines).

More than two-thirds of the turbines installed were the 30-kW Lykkegaard wind turbine patterned after la Cour's *klapsejlsmølle* designed at the turn of the 20th century. However, there was a new entrant on the Danish scene—F.L. Smidth's Aeromotors.

F.L. Smidth became one of the world's first firms to marry the rapidly advancing field of aerodynamics to wind turbine manufacturing. The diversified company designed machinery for working with concrete and used this technology to build concrete silos and chimneys. An affiliated company also built small airplanes. They combined these technologies and began manufacturing two-blade upwind turbines that, unlike Lykkegaard, used modern airfoils (see Figure 3-6. F.L. Smidth Aeromotor drivetrain).

It was the second, much larger turbine that would prove most significant. This version used a three-blade, stayed rotor 24 meters (79 feet) in diameter. Though it was rated at only 70 kW, the rotor was nearly twice the size of the two-blade version (see Figure 3-7. Three-blade F.L. Smidth Aeromotor). Like the Danish turbines that had

Table 3-6. Early Danish Wind-Electric Turbines			
	Type	Units	kW	Total MW
During the Great War				
1918	La Cour	120	25	3
During World War II				
1945	F.L. Smidth 17.5 m	12	60	0.7
	F.L. Smidth 24 m	7	70	0.5
	Lykkegaard	67	30	2.0
	Total	86		3.2

come before, the rotor on both models was placed upwind of the tower and used fantails to mechanically point the rotor into the wind.

The F.L. Smidth turbines dwarfed the small machines Hütter had been experimenting with at Ventimotor. They were 10 to 20 times larger. They were designed to serve villages and communities not new individual "settlements." And they were commercial products that entered serial production. Altogether, there were some 19 F.L. Smidth Aeromotors operating by the end of the war. Not many by today's standards, but enough to demonstrate their worth.

One striking feature of these early Danish machines was their longevity, a harbinger of the ruggedness that would make Danish wind turbines famous in the 1980s. F.L. Smidth's two-blade version installed at Ullerslev in the early 1940s operated for 19 years. It was only taken out of service in 1960 when the power station was closed. Their three-blade model at Gedser on the island of Falster south of Copenhagen operated for 16 years. These were durable machines.

Though crude by today's standards, these wind turbines performed well even for the day (see Table 3-7. F.L. Smidth Performance). Each of Smidth's 70-kW models generated about 120,000 kWh at four well-exposed sites in 1944. The better performers delivered specific yields approaching 300 kWh per square meter of rotor swept area (kWh/m²).

Danish wind turbines generated 18 million kWh from 1940 to 1947—when every kilowatt-hour was most needed—reports Danish energy historian Jytte Thorndahl. Though a small part

Figure 3-6. F.L. Smidth Aeromotor drivetrain. A 1941 drivetrain for a two-blade stayed rotor installed by F.L. Smidth at Ullerslev on the island of Funen, Denmark. The original wooden blades were replaced with metal-clad blades in 1954. According to Benny Christensen, board member of the Danish Wind Turbine Historical Collection (Danmarks Vindkrafthistoriske Samling) in Lem, Denmark, this turbine was the last F.L. Smidth in service and operated until 1960. Note the braced rotor, integrated drivetrain, and fantails (without the blades) for orienting the rotor into the wind.

Figure 3-7. Three-blade F.L. Smidth Aeromotor. Later, larger version of F.L. Smidth Aeromotor installed at Skagen, the northernmost tip of the Jutland Peninsula. The three-blade rotor was 24 meters (79 feet) in diameter–nearly twice the size of the two-blade turbine–and was braced with struts and stays, a characteristic of later Danish wind turbines. This unit generated nearly 120,000 kWh in 1944, a remarkable performance for wind turbines of this period. (The FLS-files in Danmarks Vindkrafthistoriske Samling)

Table 3-7. F.L. Smidth Performance			
Dia.	Swept Area	Power	
m	m²	kW	
24	452	70	
	Year	kWh	kWh/m²
Gedser	1944	130,400	290
Frederikshavn II	1944	129,200	290
Skagen	1944	116,610	260
Bogense	1944	102,595	230
		119,701	260
Source: Gedsermøllen: den første moderne vindmølle by Jytte Thorndahl, Elmuseet, 2005, page 36.			

WIND TECHNOLOGY KNOWN WELL BY MID-1950S

It should be clear from this chapter that much of our technical understanding of wind energy and wind turbines was well known by the mid-1920s and certainly their key characteristics by the mid-1950s. Wind turbines look the way they do today–three slender blades rotating about a horizontal axis upwind of the tower–for fundamental reasons. Most of what has followed has been refinements of this basic concept through incremental improvements. Different configurations have been tried–some many times–and have all been found wanting. There may be specific applications, such as pumping low volumes of water, where different configurations are superior–as in the American farm windmill. However, for the large-scale generation of electricity to power a renewable energy transition, wind turbines using a rotor with three blades upwind of the tower has proven to be the configuration of choice.

of Danish electricity supply in percentage terms during the war, these machines were a welcome addition at Danish power stations short of fuel.

Again, wind delivered essential services when fossil fuels were in short supply or simply unavailable—a lesson the world should have taken to heart. Though a few turbines continued operating through the 1950s, the work of F.L. Smidth came to an end with the end of the war.

Postwar Years

Not long after the war, international trade in fossil fuels began to return, lessening the interest in wind energy. However, many remembered wartime scarcity and several nations and the recently created United Nations began programs weighing a future role for wind energy. From the early 1950s to the late 1960s, there were programs in Germany, France, and Denmark. The oil bonanza from Middle Eastern fields wouldn't begin flooding Western markets until well into the 1960s.

Germany's Allgaier

After the war, Ulrich Hütter began working for Allgaier, a German manufacturer of process engineering equipment for the automotive industry. His assignment: Design a wind turbine that could be used in a number of different applications in both the developed and developing world (see Table 3-8. Wind-Electric Turbines During the Postwar Period).

The company, which still exists, built some 200 of the sleek three-blade, downwind design. The turbine was a far cry from the clumsy turbine he had built for Ventimotor (see Figure 3-8. Allgaier) in the early 1940s. In 1949 he erected his first Allgaier at a test field near Holzhausen in central Germany. Subsequently, Allgaier began serial production in the early 1950s.

The first batch of Hütter's WE-10 were nominally 10 meters (33 feet) in diameter and drove a 7.2-kW generator intended for off-grid applications. The turbines operated at a very high tip-speed ratio of 8, reportedly delivering a very high coefficient of performance of 0.47, both features that would come to characterize Hütter's subsequent work. According to German photographer Jan Oelker in his *Windgesichter* (The Face of Wind Energy), Allgaier introduced a version

Table 3-8. Wind-Electric Turbines During the Post War Period										
		Diameter		Swept Area	Power	Specific Power	No. of Blades	Rotor Orientation	Control	Approx Date Installed
Turbine	Country	m	ft	m²	kW	W/m²				
Juul	Denmark	8	25	47	15	322	2	upwind	stall	1950
Allgaier	Germany	11	37	100	10	100	3	downwind	pitch	1952
Juul Bøgo	Denmark	13	43	133	65	490	3	upwind	stall	1952
John Brown	Scotland	15	50	181	100	551	3	downwind	pitch	1955
Juul Gedser	Denmark	24	79	452	200	442	3	upwind	stall	1956
BEST Romani	France	30	99	716	800	1,117	3	downwind	stall	1957
Hütter	Germany	34	112	908	100	110	2	downwind	pitch	1957
Neyrpic	France	21	70	353	132	374	3	downwind	pitch	1958
Neyrpic	France	35	115	962	1,000	1,039	3	downwind	pitch	1963

Figure 3-8. Allgaier. Three-blade downwind 11.3-meter (37-foot) rotor with fantail developed by Urich Hütter in the early 1950s. Several museum pieces remain, including this one in Holzhausen, Germany. Note the work platform, integrated 10-kW drivetrain, blades with full-span pitch control, and a characteristic Hütter flange at the blade root near the hub. Though it was a downwind turbine, it used a fantail (the small rotor) to orient the rotor downwind, a feature overlooked by many later engineers. This unit is one of six that was installed by Klöckner-Moeller, a manufacturer of electrical components. 2005.

with an induction generator for operation with a Schwabian utility in 1952.

In contrast to some of the wind turbine designers before him and the many since, Hütter emphasized the rotor's diameter in designating his designs, not the size of the generator that the rotor powered. For Hütter had early in his research concluded that it was the area swept by the wind turbine that would determine how much energy it could capture.

Thus, the rotor on the WE-10 was eventually enlarged to 11.27 meters (37 feet) in diameter to sweep an even 100 square meters. The larger rotor drove a 10 kW generator, giving the turbine a specific power of an even 100 W/m², one-fifth that of the Smith-Putnam turbine of the early 1940s. Large swept area, small generator size, and the resulting low specific power became another hallmark of Hütter's design approach.

Interestingly, Hütter lived long enough (he died in 1990) that he would go on to design a wind turbine for Sweden that swept exactly one hectare (10,000 m²), illustrating that he found beauty in round numbers and that, again, it was swept area that was important in design, not generator capacity.

Allgaier turbines were sold worldwide throughout the 1950s, making it one of the most widely distributed modern wind turbines before the great wind revival of the 1980s. Oelker also notes that Allgaier installed one group of eight turbines for draining a polder in the lowlands of northern Germany. The turbines, which could lay claim to the title of first modern wind farm, operated for the next decade. Others were used for irrigation pumping in Third World countries.

Klöckner-Moeller, a manufacturer of electrical components, installed six of the turbines at its facilities, including one on the corporate office in Bonn (still visible in the late 1990s) and another in Holzhausen that was still extant in 2005. Allgaier also had installed turbines at a test field near Holzhausen.

There is another Allgaier turbine on the grounds of the University of Stuttgart where Hütter spent the remainder of his life at the Test and Research Institute for Aviation and Space Flight (Deutsche Forschungs- und Versuchsanstalt für Luft- und Raumfahrt).

Hütter's Allgaier turbine remains distinctive in the annals of wind energy. It used three slender airfoils that are bolted to the hub with a flange that would become Hütter's most significant legacy. The rotor incorporated a sophisticated pitch mechanism for closely regulating rotor speed. The integrated drivetrain creates a compact pleasing nacelle. And unlike many later designs that tried to emulate Hütter, the rotor was mechanically oriented downwind rather than relying on passive yaw. And the whole assembly typically rested atop a tower of three tubular legs with a work platform. Until the rise of the Danish wind industry, this was the world's most successful commercial-scale wind turbine.

However, Hütter began a commissioned project to design a wind turbine nearly 10 times larger. In doing so he greatly advanced the fledgling field of fiberglass or composite construction of airfoils and carried his quest for high tip speeds and minimal materials to the extreme. The result was the StGW-34, a 34-meter (112-foot) diameter,

IN CONTRAST TO SOME OF THE WIND TURBINE DESIGNERS
before him and the many since, Hütter emphasized the rotor's diameter in designating his designs, not the size of the generator that the rotor powered.

two-blade, teetering rotor. And, as in the Allgaier, the 100-kW turbine was specifically designed to have a very low specific power.

The StGW-34 was connected to the grid at a test field in the Swabian Alps not far from Stuttgart in December 1957 and subsequently began testing in 1958. It operated intermittently as a test vehicle until it was scrapped a decade later. As an experimental turbine, it suffered extensive outages, eventually throwing a blade in 1961.

It was the design—and the redesign—of these blades in fiberglass that pushed the boundaries of this new technology. The blades were extremely long for the period. This size wouldn't be seen commercially until three decades later. More importantly to the development of wind energy was how Hütter attached the fiberglass filaments around the bolts—or bolt holes—in a wide diameter flange. This became known as the Hütter flange.

Hütter's legacy is not high tip-speed ratios, or lightweight downwind, teetered rotors. From his dissertation to the StGW-34, Hütter had proposed a series of high-performance downwind rotors, including a brief foray to a one-blade prototype with an extremely high tip-speed ratio of 12. These design characteristics have all proven commercially unsuccessful, leading many bright engineers down a dark alley.

Instead, Hütter's legacy is his aesthetic sense that wind turbines must be pleasing to the eye. It is also his insistence on wind turbines designed for low specific power that can be used in areas of low to moderate winds—where most of the world's people live. It is also his contribution to making blades from composite materials and for a practical and durable method of attaching them to the rotor hub: the Hütter flange.

Contrary to popular myth, it was not Hütter that gave us modern wind turbines. That honor goes to a Dane.

> **Contrary to popular myth, it was not Hütter that gave us modern wind turbines. THAT HONOR GOES TO A DANE.**

Denmark's Johannes Juul

Johannes Juul was at the opposite end of the academic and institutional spectrum from Hütter. Juul had little formal education and trained in the Danish craft tradition, graduating from the Askov *folkehøjskole* in la Cour's first class for windmill technicians. But like la Cour before him, and his contemporary Hütter, Juul was inventive and resourceful. Forty years after completing his wind-technician classes, Juul reexamined wind, and for the next 10 years, his work led the world in the development of practical wind energy.

The impetus for Juul's work was continuing shortages of fuel in Denmark and its high cost in the immediate postwar period. In the spring of 1947, Juul began research and development on a new wind turbine for his employer, SEAS, the utility serving the region surrounding Copenhagen.

By 1948, Juul was testing blade designs in a wind tunnel that—like la Cour before him—he built of necessity. In 1950, Juul began testing a prototype 8 meters (25 feet) in diameter for Danish utility SEAS (Sydøstsjællands Elektricitets Aktieselskab) or the Southeast Zealand Electricity Company.

The turbine at Vester Egesborg was a far cry from the turbines that would follow, but it shows that Juul was willing to experiment and not simply take the three-blade F.L. Smidth design and proceed from there. His prototype used two cantilevered blades downwind of the tower. Unlike the F.S. Smidth turbines, only the end of the blades, their roots, were attached to the hub. There were no struts and stays. Juul, like so many wind engineers since, was not immune to the lure of promised simplicity and low cost offered by a two-blade rotor downwind of the tower.

Unlike F.L. Smidth's turbines, which had predominantly driven DC dynamos, Juul went back to Agricco's use of asynchronous (induction) generators for direct connection to the then rapidly expanding AC network. And instead of pitch control requiring a sophisticated hub as on Hütter's Allgaier turbine, Juul used a stall-regulated rotor. According to Danish energy historian Jutte Thorndahl, it was the skilled combination of the asynchronous generator and stall regulation that was Juul's technological breakthrough.

Stall "regulation" is simple, but it has limitations. It only works if the rotor operates at a constant speed and if the generator is large enough to keep the rotor fully loaded in strong, gusty winds. If the generator fails, the grid fails, or if the generator's capacity is exceeded, rotor speed quickly increases, stall is no longer effective, and the rotor can speed up to destruction. It's no surprise then that Juul soon replaced the Vester Egesborg turbine's 10-kW generator with a 15-kW induction generator.

The Vester Egesborg turbine also demonstrated probably one of the most significant developments in wind energy: pitchable blade tips. Despite Juul's belief in stall regulation, he provided a mechanism for protecting the rotor from overspeed in case the generator failed in high winds. This mechanism was simple, yet profound. The outboard section of the blades toward the tip were movable and could change pitch if the rotor went into overspeed, protecting the wind turbine from self-destruction. While not foolproof, this one technology was the principal reason that Danish wind turbines became so successful during the California wind rush of the early 1980s. Relative to their competitors, Danish turbines survived storms in California's windy passes in far greater numbers, thus making them more reliable than technically more complex wind turbines.

The Mill at Bogø

In 1952, SEAS replaced an F.L. Smidth Aeromotor at Bogø with a 13-meter-diameter turbine of Juul's design. The rotor used three blades upwind of the tower—instead of downwind—with Juul's pitchable blade tips. Juul used a prominent bowsprit with struts and stays to brace the thin aerodynamically shaped blades. This was the distinctive configuration that Juul would make famous a few years later at Gedser on the island of Falster.

The Bogø prototype performed surprisingly well for such an early machine. In one nine-month period, the turbine ran unattended. Then the turbine was stopped for inspection and returned to service, said Juul in a report to the UN in 1961. Even by the standards of modern wind turbines during the 1980s wind revival, the Bogø turbine was highly productive, delivering an average specific yield—one of the wind industry's measures of performance—of 600 kWh/m²/yr during an eight-year period (see Table 3-9. Johannes Juul's Bogø Prototype Production). This was double the performance of the F.L. Smidth turbines of the previous decade.

With the experience gained from these two machines, Juul began work on his crowning achievement, the mill at Gedser, the forerunner of later Danish wind turbines.

STALL REGULATION AND HIGH POWER RATINGS

Stall regulation of fixed-pitch rotors require very large generators and hence "rated power" relative to the rotor's swept area. Juul's wind turbines at Bogø and Gedser illustrate this with specific power ratings of 490 W/m² and 442 W/m² respectively. The large generators are necessary to keep a load on the rotor in high winds. For several decades after the great wind revival in the 1970s, wind turbines used stall regulation and fixed-pitch rotors for their simplicity, and consequently most products were designed with high specific power. Once the switch was made to a more complex hub that could change blade pitch in strong winds, high power ratings were no longer necessary. The Allgaier and StGW-34 designs illustrate this approach. Hütter designed both turbines with a specific power of only 100 W/m². Now that all large wind turbines use variable pitch rotors, there are no technical reasons to hold back the introduction of wind turbines with low specific power, as explained in Chapter 8, "Silent Wind Revolution."

Table 3-9. Johannes Juul's Bogø Prototype Production		
Dia.	Swept Area	Power
m	m²	kW
13	133	65
Year	kWh	kWh/m²/yr
1953	87,170	660
1954	90,967	690
1955	68,680	520
1956	91,133	690
1957	78,191	590
1958	78,502	590
1959	73,363	550
1960	72,659	550
Average	80,083	600

Source: Gedsermøllen: den første moderne vindmølle by Jytte Thorndahl, Elmuseet, 2005, page 93.

The Mill at Gedser

Gedser, a small ferry terminal on the island of Falster 150 kilometers (90 miles) south of Copenhagen, is iconic in the field of wind energy. It's at Gedser where the modern wind industry began.

There had been a wind turbine at Gedser before Juul. In 1944, a three-blade F.L. Smidth had generated 130,000 kWh on the well-exposed peninsula. Juul designed a scaled-up version of his Bogø prototype to match the 24-meter (79-foot) diameter of the F.L. Smidth turbine. This was a big leap for Juul. His Gedser design swept three times the area of his earlier machine.

SEAS installed the turbine in 1956 and put the machine into service in 1957. The Gedser mill remained in regular operation from 1959 through 1967, when the turbine was taken out of service after a failure in the chain-drive system.

Like the Bogø turbine before it, Juul's Gedser mill performed well, particularly in the early years, delivering yields of up to 800 kWh/m²/yr. During its lifetime, the turbine generated 2.2 million kWh, producing nearly 370,000 kWh during its best year. In a six-year period, the Gedser turbine generated an average of 275,000 kWh per year (see Table 3-10. Johannes Juul's Gedser Production).

The turbine used an upwind, three-blade rotor braced with struts and stays from a prominent bowsprit (see Figure 3-9. Gedser mill). It used slender, fixed-pitch airfoils and drove a 200-kW induction (asynchronous) generator at relatively constant speed. The rotor's power was limited by stall of the fixed-pitch blades operating at constant speed. For overspeed protection, should stall fail to control rotor speed, the rotor used automatically deployed pitchable blade tips. Unlike the designs of F.L. Smidth and Hütter, the rotor was oriented mechanically with an electric yaw motor rather than a fantail.

Table 3-10. Johannes Juul's Gedser Production

Dia.	Swept Area	Power
m	m²	kW
24	452	200
Year	kWh	kWh/m²/yr
1961	339,020	750
1962	339,210	750
1963	304,450	670
1964	367,140	810
1965	165,140	370
1966	148,890	330
Avg.	277,308	610

Source: Gedsermøllen: den første moderne vindmølle by Jytte Thorndahl, Elmuseet, 2005, page 15.

Figure 3-9. Gedser mill. The grandfather of the Danish design concept. Designed by Johannes Juul, a former student of Poul la Cour, the ungainly but durable machine used struts and stays to brace the rotor as in the F.L. Smidth turbines that preceded it. The Gedser mill operated for a decade beginning in the mid-1950s, generating about 275,000 kWh annually or more than twice that of the same-size F.L. Smidth turbine of the 1940s. The turbine is now on display at the Danish Energy Museum. Note the pitchable blade tips. These were one of Juul's most significant technical innovations and led directly to the success of Danish wind turbines in the 1980s.

Juul, who died in 1967, left a lasting legacy to not just Danish wind engineers but to the world. It was his design of the Bogø and Gedser mills that eventually became known as the Danish concept: three blades upwind of the tower and a stall-regulated rotor with pitchable blade tips for overspeed control. It was Juul's clever combination of stall regulation and pitchable blade tips that were fundamental to subsequent wind turbine design. Other hallmarks were the simple rugged construction that together led to the durability necessary for commercial operation for years on end.

Both the F.L. Smidth turbine and Juul's turbine were still standing at Gedser in 1974 when researchers from NASA and what was to become the US Department of Energy (DOE) visited the site following the first oil crisis. The Danes warned NASA that placing the rotor downwind of the tower as NASA intended for its 100-kW experimental model would create problems—and that it did.

Juul's Gedser turbine was so robust that despite being idle for a decade it was returned to service in 1978 for tests paid for by the United States. When Denmark looked again to wind energy for help in meeting one more energy crisis, the country had a working model still standing.

Wind Experimentation Elsewhere

Juul and Hütter were not alone in developing wind turbines in Europe at the time. There was a small program in Britain and a much larger effort in France as well.

Scottish fabricator John Brown installed an experimental downwind turbine using a complex variable pitch rotor in the Orkney Islands. The Costa Hill turbine saw only limited operation connected with the diesel system on the island of Orkney in 1955. There wouldn't be another wind turbine installed in the Orkneys for three decades.

While the allies were preparing to land on the beaches of Normandy, a French engineer in occupied France was giving a presentation on the role of wind energy after the liberation. Pierre Ailleret would soon become one of the founders of Electricité de France (EDF) when postwar France nationalized its electricity system in 1946. He quickly set about putting his vision

IT WAS JUUL'S CLEVER COMBINATION OF STALL REGULATION AND PITCHABLE BLADE TIPS THAT WERE FUNDAMENTAL TO SUBSEQUENT WIND TURBINE DESIGN.

into action when he became EDF's director of research by creating a division for wind energy. Ailleret would lead this work for the next two decades—the most extensive national undertaking of its kind at the time, presaging the large, and expensive, national programs in the United States and Germany in the 1970s.

Like Constantin, Ailleret envisioned *batteries* of wind turbines supplying the grid. More importantly, he realized that wind energy could play a much greater role in electricity supply if it could be integrated with France's existing hydro system. Ailleret's research found a remarkable match between France's hydro and wind resources. Hydro peaks in the spring and summer. Wind peaks in the fall and winter. What Ailleret needed were wind turbines, and he set out to find companies capable of building them.

Ailleret and EDF settled on two simultaneous programs: one by BEST-Romani, the other by Neyrpic. Etienne Rogier, France's foremost wind energy historian, has written about both efforts, as have Marc Rapin and Jean-Marc Noël in their book *Energie Eolienne*. The experimental French turbines were comparable in size to Hütter's StGW-34 prototype. Thus, they were large turbines for their day and much larger than Juul's Gedser turbine. Unlike Hütter's design, both turbines had very high power ratings. Yet both French programs continued the engineering elite's fascination with downwind, passive yaw turbines.

BEST-Romani

Lucien Romani, a self-taught engineer, was familiar with the role of the wind in aviation, automobiles, and dune formation. In 1946, he and his brother founded BEST (Bureau d'Etudes Scientifiques et Techniques), what we would call today a scientific think tank. Under contract to EDF in 1956, they sought to develop a wind turbine at Nogent-le-Roi, a low-wind site near Paris.

The BEST-Romani turbine was unusual in several aspects. The 30-meter (99-foot) diameter wind turbine was erected on a combination tubular tower atop a tripod truss. The tubular tower section included a fairing to minimize turbulence striking the three-blade downwind rotor. The faring yawed with the wind turbine in response to changes in wind direction. At its rated wind speed of 16.7 m/s (37 mph), the turbine's initial rotor performed well at a fairly low tip-speed ratio of 4.

Surprisingly, the whole of the massive machine—including the truss tower—was hinged and could be tipped to the ground for service and repair. Unlike Juul and Hütter, the BEST-Romani turbine didn't use an induction generator, opting instead for a synchronous alternator, like that in the Smith-Putnam machine two decades before.

Testing began in 1957, and the turbine operated from spring 1958 to spring 1962 during which it generated a total of 220,540 kWh, not a significant amount for a machine of this size even in central France. This was largely due to the turbine being a test bed for EDF where they conducted numerous studies on vibration, pitch regulation, excitation, connection to the grid, yaw, as well as noise. Oddly, they found that the turbine performed better in free yaw than it had in wind tunnel experiments.

Like other stall-regulated turbines to come, the turbine's peak power exceeded its rated power—substantially. During a storm in the fall of 1959, the turbine reached 1,025 kW, or one-fourth more than the turbine's rated power of 800 kW.

All in all, EDF was satisfied with the results, but they and BEST-Romani then took a step too far. In the spring of 1962, they installed a new higher-speed rotor to reduce a stage in the gearbox, despite tests showing a risk of an aeroelastic vibration. The redesigned rotor operated for only 300 hours before losing a blade in the fall of 1963, bringing a sudden end to the program.

Vadot and Neyrpic

The other effort was led by consulting engineer Louis Vadot in conjunction with Neyrpic, a manufacturer—like F. Morgan Smith Co. in the Smith-Putnam project—of hydroelectric turbines. Neyrpic was a major industrial company, and the EDF contract was part of an ambitious program by the company to develop wind energy in France. They would later become famous for the bulb turbines they developed for EDF's *La Rance* tidal barrage—the world's largest tidal power plant that began operation in 1966, despite critics claims it would never work.

Neyrpic eventually installed a series of two experimental turbines at Saint-Rémy des Landes, a windy site on Normandy's Cotentin peninsula. In contrast to BEST-Romani, Vadot chose an asynchronous generator because induction generators were simpler to control and connect to the network than synchronous generators. They were also inexpensive and could become an integral part of industrially produced wind turbines.

In the late 1950s, Vadot had realized that wind energy would more likely become cost effective when it achieved series production than it would by the development of some unforeseen technology. Series production would cut the cost per installed turbine, much like it had for automobiles. (It wasn't until the massive installation of wind turbines in California and subsequently in Denmark and Germany that confirmed Vadot three decades later by dramatically cutting the cost of wind energy.)

Vadot's began testing his first turbine in 1958. Testing continued for the next four years on the turbine with a three-blade rotor 21 meters (70 feet) in diameter. The downwind rotor used variable pitch blades but operated at a constant speed. After losing the first rotor in 1959, Vadot installed a more advanced rotor and tested it from the winter of 1962 to summer 1964. Though originally configured with fantails to mechanically orient the rotor downwind, the machine was operated in passive yaw after replacement of the original rotor. From 1962 to 1966, the turbine generated 700,000 kWh, much less than forecast.

EDF commissioned a larger version, and Vadot developed the 35-meter (115-foot) model.

The second turbine was nearly three times larger than his first turbine and comparable in size to Hütter's StGW-34. Yet Vadot rated his second unit at an incredible 1,000 kW—10 times that of Hütter's prototype—giving the turbine a specific power of 1,000 W/m²!

Installed in mid-1963, the turbine operated until the summer of 1964 when it destroyed the gearbox. Over one seven-month period the turbine generated 500,000 kWh.

As with the BEST-Romani failure, the failure of Neyrpic machine allowed EDF to abandon the development program and close their wind division in 1966. At the time, the cost of electricity from burning oil was cheaper than that from the experimental wind turbines, and EDF was already moving toward a national commitment to nuclear. France had developed a nuclear bomb and, like the United States at the time, was trying to find a civilian use for its weapons program.

As seen a decade later in similar national programs, the participants claimed success but walked away without any lasting technology or products. The centrally directed research by EDF and its contractors' focus on downwind, passive yaw design foreshadowed the failure of top-down technology development seen in the 1970s and 1980s.

But oil wasn't going to remain cheap for long.

4

The Great Wind Revival

Hvad skal væk? Barsebäck!
(What do we want to go away? Barsebäck!)
Hvad skal ind? Sol og vind!
(What do we want in? Sun and Wind!)

—1970s chant by Danish antinuclear demonstrators
against the Swedish reactor at Barsebäck

FOR NEARLY A CENTURY, THE FATE OF WIND ENERGY HAS been inextricably tied to the recurring belief in a never-ending abundance of fossil fuels, particularly oil, and later to the perception that nuclear power would be "too cheap to meter." If oil is available and nuclear is cheap, who needs wind energy? The 1970s shattered this vision.

In 1973 the world awakened to find oil embargoed for the first time. Many nations launched crash programs to develop "alternatives," including wind energy and nuclear power. Western countries panicked. It's hard to imagine today some of the proposals that were taken seriously then. Utilities in California alone considered building nuclear power plants every 50 miles along the state's scenic coastline.

Yet this was only a few years after mass student protests reached a crescendo in the West. Revolution was in the air—along with a whiff of tear gas. Reactor proposals stirred popular uprisings even in France. By the mid-1970s the cost of nuclear had exploded, dashing hopes of it ever being cheap. The development of wind energy grew out of this maelstrom and, despite many false starts, led to the great wind revival—a sustained boom that has survived the vicissitudes of waffling public policy from one country to the next.

After the oil embargo rekindled interest in wind energy, two distinct—and often warring—camps sought to make wind energy work—and to profit by it. These camps reflected the fault lines in society.

One camp represented the establishment thinking of the day. Members of this camp believed in fostering a centrally conceived and centrally directed program acting through aerospace contractors or electric utilities to build large wind turbines. This was the way it had always been done. They would develop wind technology in the same manner as the Manhattan Project developed an atomic bomb—from the top down. They would choose the designs, choose the contractors, and choose the host utilities. By its nature this was an exclusive club, restricted to only a few of the largest firms in the world.

WHY THE HISTORY OF MODERN WIND ENERGY IS IMPORTANT

Wind technology grew from the ground up, not from the top down. That's not apparent today when looking at how sophisticated wind turbines have become and the companies building them. Some histories of wind energy have attempted to portray early R&D programs in a far more favorable light than they deserve. Wind energy is commercially successful today in spite of these programs. Thus, it behooves renewable energy advocates, political leaders, and the public to understand how we came to be where we are today so that we can make informed choices about future energy policy. For more on the failures of early R&D programs see *Wind Energy Comes of Age*, found in most technical libraries.

The second camp, led by an assortment of political activists and entrepreneurs, believed that the development of wind energy should be open to everyone because they wanted to participate. After all, no one owned the wind: it was the "people's" resource.

This camp also didn't trust the "establishment" to get the job done. They argued that incrementally developing wind turbines would be better suited to the skills and knowledge of the time and to the small and medium-size companies that would most likely make wind energy flourish. They thought wind energy should grow from the bottom up, not from the top down.

The "establishment camp" belittled the bottom-up activists as wooly-headed, Birkenstock-wearing, small-is-beautiful followers of E. F. Schumacher—or worse. Their proposals to share the opportunity—and the challenge—of developing wind technology among many different players was reminiscent of Mao Zedong's "let many flowers bloom." Mao's *Little Red Book* was not popular in corporate boardrooms then—or now.

The decisions made during that period still rankle even today. And the fight over who owns the right to develop and use wind energy is still very much alive.

Large Turbines from the Top Down

The top-down strategy focused on multimegawatt wind turbines that would be operated by electric utilities. After all, the thinking went, who makes and sells electricity but utilities and who makes aircraft better than aerospace companies. The names of the companies involved read like a *Who's Who of the Western Aerospace Industry*. In the United States, Hamilton Standard, General Electric, Westinghouse, and Boeing all attempted to develop large wind turbines for the US Department of Energy (DOE). The niche was filled by MBB (Messerschmidt-Bölkow-Blohm) and MAN (Maschinenfabrik-Augsburg-Nürnberg) in Germany and similar companies in Sweden, Britain, and elsewhere.

All failed, some more spectacularly than others.

Bringing NASA Down to Earth

After the moon landings, the space program began winding down, and with it the space agency. NASA (the National Aeronautics and Space Administration) was scrambling to redefine itself, to find new "missions," when opportunity struck in the form of the first oil embargo. What began as mere tinkering by researchers at the agency's Lewis Research Center near Sandusky, Ohio, quickly evolved into the most costly wind energy R&D program in the world.

NASA began by consulting with Hütter and Putnam and studying the operation of Juul's machine at Gedser. In the end they started down a path blazed years before by Putnam. The result, the Mod-0, resembled neither Hütter's lightweight, flexible downwind design nor Juul's rigid, three-blade upwind design. NASA's Mod-0 incorporated none of the lessons from Europe, while abandoning Putnam's most significant design element, his hinged blades. Westinghouse, the contractor on the Mod-0, was subsequently hired to build a more powerful version, the Mod-0A, for extended field tests (see Figure 4-1. Mod-0A).

After the Mod-0A, NASA and US DOE scaled up the configuration and hired General Electric to build the Mod-1, a machine 2.5 times the size of the Mod-0A and only slightly bigger than Putnam's machine. Like its predecessor, the Mod-1 bore two rigid blades downwind of the tower (see Figure 4-2. GE's Mod-1). Installed in 1978 on a ridge of the Appalachian Mountains near Boone, North Carolina, the Mod-1 was the first wind turbine to stir national contro-

THE GREAT WIND REVIVAL

Figure 4-1. Mod-0A. One of four NASA Mod-0A turbines installed in pilot projects across the United States. The Mod-0A used a two-blade rotor 38-meters (125 feet) in diameter to drive a 200-kW generator. This turbine was installed in 1977 for a small municipal utility in Clayton, New Mexico, until it was removed in 1982. Note the lattice tower and how the blades are swept downwind, giving the spinning rotor the shape of a cone oriented downwind. This is characteristic of downwind rotors. The turbine was in operation when this photo was taken in 1979.

Figure 4-2. GE's Mod-1. The US DOE large turbine program got off on the wrong foot with General Electric's Mod-1. The 60-meter (197-foot) diameter, two-megawatt turbine proved disastrous, leaving a record of noise complaints that still dogs the wind industry worldwide three decades after it was removed for scrap. Worse, the turbine logged no recordable operating hours during its brief period of testing, worsening the ignominious failure. (USDOE)

versy. Low-frequency noise from the downwind machine disturbed neighbors in sheltered valleys down slope. The noise was attributed to passage of the blades through the turbulent wind shadow behind the cluttered truss tower, a problem that was noted earlier on the Mod-0 at Sandusky but not corrected. GE's Mod-1 never performed as expected and never logged any operating hours before being quickly removed. It ranks as the world's worst performing large wind turbine ever built.

Despite the problems that plagued the Mod-1, NASA and DOE determinedly proceeded with the next machine in their program, Boeing's Mod-2. Observers at the time questioned the wisdom of proceeding from the failed Mod-1 to an even larger machine. It was as if NASA and DOE were deliberately challenging E. F. Schumacher's observation that "any intelligent fool can make things bigger, and more complex. It takes a touch of genius—and a lot of courage to move in the opposite direction." They clearly were not moving in the opposite direction.

The Mod-2 was nearly three times the size of the Mod-0 and more than twice the size of the Mod-1. NASA "just couldn't think in earthly terms," commented one wit. Four Mod-2 turbines were installed in the DOE program (see Figure 5-20. Mod-2's steel blades). Another was delivered

It was as if NASA and DOE were deliberately challenging E. F. Schumacher's observation that
"ANY INTELLIGENT FOOL CAN MAKE THINGS BIGGER, AND MORE COMPLEX."

to Pacific Gas and Electric (PG&E) in California. After operating sporadically from 1982 to 1988, the PG&E turbine logged only 8,700 hours of operation but generated more than any other Mod-2. In contrast to commercial wind turbines that must be available for operation more than 98% of the time, PG&E's Mod-2 was available for operation only 37% of the time. Worse yet, it generated only 40% of the output possible if the turbine had operated reliably (see Table 4-1. Historical Development of Large Wind Turbines).

All the Mod-2s were eventually sold for scrap, with the PG&E turbine meeting the most spectacular fate. Concluding that it was too costly to dismantle, PG&E felled the tower like a lumberjack cutting a giant redwood. After a series of explosives severed the tower, the last Mod-2 crashed to the ground in front of television news crews helpfully assembled by PG&E—ever the friend of renewable energy.

The Mod-5b contract, the culmination of NASA and DOE's large turbine program, was again awarded to Boeing (see Figure 4-3. Boeing Mod-5b). The Mod-5b performed better than other turbines in DOE's program but after four years of operation delivered only 70% of the revised projection for the site. (Performance relative to the original projection was even worse.)

NASA and DOE had expected to win economies of scale by jumping directly to large turbines rather than take the time to incrementally scale-up smaller turbines after they had become more reliable. Researchers instead found diseconomies of scale due to the costly specialized components required for the limited number of large machines. These one-of-a-kind turbines also needed specialized equipment to service them. Thus, the technology was troublesome from the start, and when it failed, it took a long time to fix.

As French wind turbine designer Constantin had suggested in the 1920s, and Vadot and Hütter had argued in the 1950s, there were clear advantages to operating large numbers of medium-size machines rather than one large turbine. If one of the smaller machines failed, the remaining turbines could continue to operate, generating electricity—and revenues—for their owners until the nonoperating turbine could be brought back on online. However, when one large turbine failed, it removed a significant portion of capacity from the generating mix and contributed to the conventional wisdom that wind would never

Table 4-1. Historical Development of Large Wind Turbines		(Sorted by Operating Hours)								
		Diameter		Swept Area	Power	Specific Power	No. of Blades	Rotor Orientation	Operating Hours	Generation	Approx Date Installed
Turbine	Country	m	ft	m²	kW	W/m²				GWh	
Tvindkraft	Denmark	54	177	2,290	1,000	437	3	downwind	103,000	16	1978
Tjæreborg	Denmark	61	200	2,922	2000	684	3	upwind	44,000	32	1988
Nibe B	Denmark	40	131	1,257	630	501	3	upwind	29,400	8	1980
Maglarp WTS-3	Sweden	79	260	4,927	3,000	609	2	downwind	26,159	34	1982
Mod-5b	USA	98	320	7,466	3,200	429	2	upwind	20,561	27	1987
Mod-0A (NM)	USA	38	125	1,141	200	175	2	downwind	13,045	1	1977
Nasudden	Sweden	75	246	4,418	2,000	453	2	upwind	11,400	13	1983
Mod-2 (PG&E)	USA	91	298	6,504	2,500	384	2	upwind	8,658	15	1982
WEG LS-1	Britain	60	197	2,827	3,000	1,061	2	upwind	8,441	6	1987
Nibe A	Denmark	40	131	1,257	630	501	3	upwind	8,414	2	1979
WTS-4	USA	79	260	4,927	4,000	812	2	downwind	7,200	16	1982
Growian	Germany	100	328	7,854	3,000	382	2	downwind	420		1981
Mod-1	USA	61	200	2,922	2,000	684	2	downwind	0		1979

Figure 4-3. Boeing Mod-5b. The most successful of US DOE's large wind turbines, the Mod-5b marked the end of the line for the program. Though installed nearly one-half decade later, the turbine was as large as Growian with a rotor 98 meters (320 feet) in diameter. Unlike Growian, the Mod-5b used a two-blade rotor upwind of the tower, and instead of the fiberglass used in Growian's blades, the Mod-5b used steel. Yes, steel. It was once considered a blade material.

Figure 4-4. Westinghouse 600. One of the more successful ventures resulting from the US research and development effort was Westinghouse's 600-kW model. Fifteen turbines were installed in Hawaii and operated for a number of years. They were eventually removed, and two surviving turbines now stand at the National Renewable Energy Laboratory at Rocky Flats, Colorado. Late 1980s.

Figure 4-5 Hamilton Standard WTS-4. One of the most poorly performing large wind turbines developed by aerospace contractors during the 1980s. The WTS-4 had an extremely high specific power of 812 W/m² and generated only 16 million kWh during its 7,000 hours of operation before the teetered, downwind rotor struck the tower in 1994. Late 1990s.

work commercially. This was the Achilles' heel of both the US and the German R&D programs.

Unlike the aerospace contractors in the R&D program, who were content to go from one contract to the next, Westinghouse commercialized its work on the Mod-0A during the mid-1980s. In 1985 they installed 15 of their Westinghouse 600 models at Kahuku Point on Oahu. Westinghouse abandoned the downwind orientation of the Mod-0A, putting its teetered rotor upwind of a new tubular tower (See Figure 4-4. Westinghouse 600). It increased the rotor's swept area by 30% and tripled the generator rating for the windy Hawaiian site.

In 1989, the Westinghouse turbines were available for operation 80% of the time, a performance superior to that of most experimental turbines of the era but still well below that of much smaller but fully commercial turbines already operating in California and Denmark. After unsuccessfully attempting to sell more turbines, Westinghouse threw in the towel.

While NASA and DOE were pursuing the Mod program, Hamilton Standard went out of the country to win a portion of a Swedish R&D contract for development of another large turbine. The Connecticut aerospace contractor designed the rotor for both Sweden's WTS-3 and the US Bureau of Reclamation's WTS-4 (see Figure 4-5 Hamilton Standard WTS-4).

THE GREAT WIND REVIVAL

Figure 4-6. Growian. One of German engineering's most spectacular failures was Growian (Grosse Wind Energie Anlage). The downwind, 100-meter (330-foot) diameter turbine operated only 400 hours between the time it was installed in 1983 and when it was taken out of service in 1987. Growian operated even less than the Smith-Putnam turbine in Vermont in 1945. It was more than two decades before German manufacturers again ventured to this size. The two guyed towers are meteorological masts at what was to become the Kaiser-Wilhelm-Koog test center. 1987.

Hampered by bureaucratic and technical problems, the WTS-4 operated intermittently from 1982 until 1994 at its site near Medicine Bow, Wyoming, when it was severely damaged after one of the blades struck the tower.

The demise of the WTS-4 brought to an end the era of large wind turbines in the United States. There was little to show for the effort over nearly two decades, save embarrassment. Sadly, the story of questionable development of large wind turbines by aerospace contractors in Germany was similar.

Germany's Growian

During the late 1970s Germany's ministry for technological development BMFT (Bundesministerium für Bildung und Forschung) called Hütter out of retirement to design a new wind program. He concluded that his 1960s approach

Figure 4-7. MAN WKA 60. Another forgettable MAN design. The boxy nacelle on the 60-meter (197-foot) diameter upwind turbine looks like a Würst stand on a stick as one critic called it. Shown here at the Kaiser-Wilhelm-Koog test center in northern Germany, the same site where MAN's ill-fated Growian had been installed earlier. The turbine was not operating when this photo was taken in the late 1990s.

still represented the state of the art, and that with the technology gained since his turbine had been dismantled, the design could be scaled-up to multimegawatt size. Thus Growian (Grosse Wind Energie Anlage or Large Wind Turbine) was born (see Figure 4-6. Growian).

Hütter remained cautious and, as with his early experimental turbine, recommended loading the rotor lightly. He suggested that a rotor 80 meters (260 feet) in diameter be used to drive only a 1-MW generator. But the contractor had a grander vision; they wanted to claim the title of "world's largest wind turbine" in a race with the United States.

In 1983, MAN completed installation of a 3-MW turbine 100 meters (328 feet) in diameter. The company realized its mistake even before installation was complete. The proposed height of the tower exceeded existing crane capacity, forcing MAN to seek approval for a shorter tower. MAN did succeed in building the world's largest wind turbine—its nacelle alone weighed as much as a jumbo jet—but it also led German engineering to its most spectacular aeronautical failure. In the end, Growian cost nearly twice that estimated and was dismantled in 1987 after operating only 420 hours. To save face after the highly public failure, the project was labeled a "success" by all those involved.

MAN tried to rescue its reputation with two smaller turbines, the WKA 60 for the island of Helgoland off the northwest German coast and the AWEC 60 for Cabo Villano in Spain (see Figure 4-7. MAN WKA 60.) Unlike Growian, the two machines used a three-blade upwind rotor that more closely resembled Danish machines than any of the Hütter-derived designs. Proponents of the turbines argued later that these machines were never intended to compete commercially. They were only demonstrations. What they were demonstrating remains unclear.

Going Beyond Juul–Denmark

Denmark is a small country, and it could never afford the lavish R&D programs seen in the United States and Germany. There was more pressure—or expectation—that what was built had to work. Unlike the United States and Germany, Denmark took the Gedser approach and scaled it up to a size more in keeping with the country's budget and technical capabilities. The result was the Nibe twins, two 630-kW turbines on the Limfjord at Nibe, Denmark (see Figure 4-8. Nibe A and B).

The Nibe turbines were less than three times the size of Juul's Gedser mill. The objective was to compare two different configurations. Nibe A, with its prominent struts and stays, resembled the machine at Gedser. Nibe B, in contrast, used a fully cantilevered blade that could be operated with variable pitch. Like the wing of an airplane, the cantilevered blade was attached at only one

Figure 4-8. Nibe A & B. The twin 630-kW turbines at Nibe, Denmark. Nibe A (top) used a 40-meter (130-foot) diameter upwind rotor braced with struts and stays patterned after those at Juul's Gedser mill; however, the outboard section of each blade changed pitch unlike Gedser's fixed-pitch blades. Nibe B (bottom) used a "modern" rotor with full-span pitch control. Nibe A generated 2 million kWh and Nibe B generated 8 million kWh before the turbines were taken out of service 13 years later. Though functional, the nacelles reminded Danes of a mousetrap. 1980.

point, the hub. However, the blades developed for Nibe were not like those used at Gedser and, more significantly, were not like those used previously by Hütter in Germany (see Figure 4-9. Volund blade).

Nibe A operated briefly before the rotor needed repair. Still, it operated more hours than GE's Mod-1, MAN's Growian, and Hamilton Standard's WTS-4. Nibe B operated more or less successfully, racking up more operating hours than any R&D wind turbine of the early 1980s. The two turbines together cost less than two-thirds the cost of GE's single Mod-1.

THE GREAT WIND REVIVAL

Figure 4-9. Volund blade. Danish manufacturer Voland built this blade using a filament-wound spar. Filament winding was being used for missile casings in the aerospace industry and was an advanced technology in 1980. However, Volund–and this blade technology–were not successful. Successful blade technology came from a much more prosaic source: the Tvind School. 1980.

Figure 4-10. Tjæreborg. The test site at Tjæreborg on the west coast of Denmark's Jutland Peninsula hosted several megawatt-scale turbines in the late 1990s. The oldest turbine, the 2-MW, 61-meter (200-foot) diameter turbine on the concrete tower in the foreground, was installed in 1988. After more than a decade of service, it had operated more than 44,000 hours, becoming the longest-running wind turbine in any national or utility-sponsored R&D program worldwide. The Danes knew how to get their money out of their wind turbines. It was cut down with explosives in 1993 like that of PG&E's Boeing Mod-2. Note that all the turbines in this photo are in operation, a signature of the Danish program. 1998.

On the other side of Denmark, Elsam, the utility serving Jutland, scaled-up the Nibe B design for the Danish R&D program. After several years of fitful operation, the 2-MW turbine at Tjæreborg finally reached 80% availability in 1992 (see Figure 4-10. Tjæreborg). By the time the turbine was removed in 2000—much in the way of PG&E's demolition of its Mod-2—the performance of Elsam's turbine had far exceeded that of any other wind turbine installed in any large wind turbine development program.

Even so, Elsam's Tjæreborg prototype would be outshined by an unlikely turbine with an unlikely origin installed a decade earlier farther north on the west coast of the Jutland Peninsula. Moreover, it wasn't the turbines at Nibe or Tjæreborg that gave us the technology used today, says Danish wind pioneer Preben Maegaard in *Wind Power for the World*. The technology that succeeded came from an unexpected source, one more in keeping with Grundtvig's *folkehøjskole* than either the aerospace or the utility industry.

Denmark's Rebels

Following the first energy crises, Denmark was considering where, not whether, to build its first nuclear power plant. There were two small experimental reactors at Risø National Laboratory, the same Risø that later would become famous for its role testing "small" wind turbines. It was the turmoil around this pending decision that led to the famous Danish logo Atomkraft? Nej Tak! (Nuclear? No Thanks!), recognized worldwide. Meanwhile, across the straits from Copenhagen, Sweden was completing its reactor at Barsebäck, bringing the risks of nuclear power to Denmark regardless of what the Danish people wanted.

The Danish state's move toward nuclear and Sweden's disregard for the safety of the Danish capital led nuclear opponents and renewable energy activists to join together and take matters into their own hands. They decided to develop wind energy—which they knew better than other renewable energy technologies—themselves. Thus, a renewable energy revolution was born.

WHEN MY ROLE BEGAN

In the summer of 1976, I was conducting a geohydrologic study of the groundwater in southeastern Montana to establish a water-quality baseline before massive coal and uranium mining got underway (fortunately, it never did). In my far-flung travels of the region, I came across a number of abandoned windchargers from the 1930s and 1940s. I began collecting them in my Volkswagen bug in hopes of eventually restoring them. I quickly learned how valuable they were and soon found myself joining dozens of other nascent wind energy entrepreneurs in the junk or salvage windmill business. I've been working with wind energy and renewable energy policy ever since.

Figure 4-11. Riisager. One of the pioneering Danish wind turbines during the great wind revival of the 1970s and early 1980s. Laminated wood blades, struts, and stays reminiscent of Juul's Gedser mill and fantails for orientation from the la Cour tradition are all found on this 30-kW version of Christian Riisager's turbine at the Energimuseet, the Danish Museum of Energy near Bjerringbro. Hundreds of derivatives of Riisager's design were installed in California and Denmark. Some were still operating in Californian's Altamont Pass in the mid 2010s. Note the work platform and railing.

End of the hunt. Paul Gipe sitting atop an old Wind King in July 1977 just prior to lowering the turbine to the ground. Teams of entrepreneurs scoured the Great Plains of North America in the mid-1970s looking to salvage windchargers from the 1930s and 1940s. These were the best wind turbines available in North America at the time.

Danish Carpenter

Christian Riisager committed the first act of rebellion when in 1975 he connected his wind turbine to the grid. It was the first interconnected wind turbine in Denmark since Juul's Gedser mill was installed for the regional utility two decades before. Riisager connected the turbine to the distribution system without telling anyone, without authorization, and without utility or planning approval. He just did it—and ran his kilowatt-hour meter backward. When a utility company representative saw it, he didn't believe that it actually worked.

Though the story may be apocryphal, it illustrates the rebellious attitude of Danish wind pioneers at the time. Decades later in the United States, rebellious renewable activists sometimes still found it necessary to connect their small wind turbines and rooftop solar panels to the grid without approval in what was called at the time a "guerrilla" movement.

Subsequently Riisager built two 22-kW turbines in 1976 that he sold commercially. As a carpenter and as a Dane, Riisager built with what he knew best. He made his blades out of wood, and he emulated the mill at Gedser by bracing the rotor with struts and stays (see Figure 4-11. Riisager). He used fantails to orient the 10-meter (33-foot) rotor into the wind. One of Riisager's customers was popular journalist Torgny Møller who went on to found the magazine *Naturlig Energi* and its English-language counterpart *Windpower Monthly*.

Møller's success with his Riisager turbine served to popularize the concept of using wind energy commercially. The turbine generated far more electricity than Møller could consume directly, thus the surplus had to be delivered to the grid. This contrasts with the grassroots movement in the United States at the time.

In North America in the mid-to-late 1970s, most activists and entrepreneurs were either rebuilding used wind turbines from the 1930s or they were designing small, household-size wind turbines. Everyone envisioned a market among "rugged individualists" who wanted to generate electricity for their own home or farm. They didn't imagine building wind turbines large enough to generate electricity to feed the grid for powering the entire community. The United States wouldn't see turbines of the size Riisag-

Table 4-2. Wind Turbines Independently Developed in the Mid-1970 to early 1980s										
		Diameter		Swept Area	Power	Specific Power	No. of Blades	Rotor Orientation	Control	Approx Date Installed
Turbine	Country	m	ft	m²	kW*	W/m²				
Enertech	USA	4.0	13	12	1.5	122	3	downwind	stall	1975
Jacobs**	USA	4.3	14	14	3	210	3	upwind	pitch	1976
Riisager	Denmark	10	33	79	22	280	3	upwind	stall	1976
Kuriant	Denmark	9	30	64	11	173	3	downwind	stall	1977
NIVE	Denmark	10	33	79	22	280	3	upwind	stall	1978
Mehrkam	USA	12.2	40	117	40	342	4	downwind	stall	1978
Carter	USA	10	33	79	25	318	2	downwind	stall	1979
HVK	Denmark	10	33	79	22	280	3	upwind	stall	1979
US Windpower	USA	12.2	40	117	30	257	3	downwind	pitch	1980
HVK-Vestas	Denmark	15	49	177	55	311	3	upwind	stall	1980
Bonus	Denmark	15	49	177	55	311	3	upwind	stall	1981

*Rated at 60 Hz in the USA, 50 Hz in Denmark. **Reused.

er was building for several years (see Table 4-2. Wind Turbines Independently Developed in the Mid-1970s to early 1980s).

Meanwhile, Riisager's turbines and those of competitors were steadily growing in size as well as number. This would have a significant effect on later Danish energy policy because wind turbines were now producing commercial quantities of electricity. By 1979 there were 24 wind turbines operating in Denmark, representing more than 700 kW. Of these, three-fourths were Riisager's.

Although a direct descendant of Juul's Gedser design, the configuration used by Riisager and his imitators presented technical challenges that limited its ability to be scaled-up further. The passive fantails didn't allow direct control of the turbine's yaw. Though simple, they didn't permit the operator to turn the turbine out of the wind when needed, for example, to service the machine. Similarly, the rotor was weak if the nacelle swung around downwind of the tower. Though there were some 200 Riisager and derivative turbines operating in California alone in 1985 (see Figure 5-25. Struts and stays), the design had reached a dead end. Further progress awaited steps being taken elsewhere on Denmark's Jutland Peninsula.

Tvindkraft: The Giant That Shook the World

Those around in the late 1970s may remember seeing magazine photographs of Danish students and volunteers carrying a massive wind turbine blade out of a tent (see Figure 4-12. Tvind people power). That photo captured the world's imagination. It was one of those rare historical moments that became a beacon to citizens everywhere who wanted to develop renewable energy by themselves, for themselves, and for their community's benefit.

Yes, they were not ordinary students. They were on a mission, and they knew at the time they were

Figure 4-12. Tvind people power. The photo seen around the world in 1978 as students at the Tvind School carry one of the wind turbine blades from its assembly hall to the wind turbine. The action sent a political message: Together we are strong. We want wind power and we will build it ourselves. (Tvind School)

THEY MADE ANOTHER MESSAGE CLEAR TOO.
If the Danish government wouldn't act, the people would take the matter into their own hands, as they were doing that historic day, and build their own wind turbines.

Figure 4-13 Tvindkraft. World-famous megawatt-scale wind turbine installed by students at the Tvind School near Ulfborg on the west coast of Jutland in 1978. Like other pioneering Danish wind turbines, the Tvind turbine is still operating after more than three decades–long after other large turbines installed during the period had been removed and sold for scrap. The turbine's striking pop art paint scheme was created in 1999 by architect Jan Utzon to celebrate the turbine's 25th anniversary. Utzon is the son of the architect who designed the Sydney Opera House.

own hands, as they were doing that historic day, and build their own wind turbines.

Tvind was not an ordinary school either. Located near Ulfborg on the windy west coast of Denmark's Jutland Peninsula, the Tvind School was unlike a school in the modern sense and more in the tradition of the Danish *folkehøjskole* movement founded in the mid–19th century by Danish theologian N. F. S. Grundtvig. It was more like the training school founded by Poul la Cour at Askov than a public school. Not surprisingly, Tvind has had a similar influence on the development of wind energy in the contemporary era as the *folkehøjskole* at Askov had at the turn of the 19th century.

In the retelling of the modern wind industry's early history, the construction of the wind turbine by Tvind and its role in pioneering modern wind turbine blades is often overlooked. It's an uncomfortable story for many in positions of power still, because the implications are so profound. How could a group of students, their teachers, and volunteers accomplish what some of the world's most sophisticated aerospace firms with millions in research money could not? How could they build what was then the world's largest wind turbine—a machine that has operated for more than three decades and remains in service to this day—when Boeing, Westinghouse, General Electric, Hamilton Standard, Kaman, Messerschmidt-Bölkow-Blohm, MAN, and others had all failed, their turbines dismantled and sold for scrap?

The work at Tvind was taking place at the same time as NASA was developing its Mod-0A series and GE's subsequent Mod-1. The difference in outcomes couldn't have been starker.

The message delivered by the Tvind School so long ago was that wind energy was too important to be left to aerospace giants, electric utilities, and even to national governments. They demonstrated that unlike nuclear power, which requires massive centralized institutions, wind turbines could be built and owned by common citizens. This is a message that still resonates today.

undertaking a historic task. They had set out to prove to the Danish government that Denmark didn't need nuclear power, that Denmark with its long history of working with the wind could once again do so. They made another message clear too. If the Danish government wouldn't act, the people would take the matter into their

Of course, the Tvind design team had sophisticated engineering knowledge. They and their faculty were not the Luddites some have portrayed them as. The school received valuable technical assistance from Helge Petersen and others from what would become Risø's test station for wind turbines and from the Danish Technical University, for example. This was beneficial to all parties. Tvind was able to deal with some thorny technical problems, while the technical establishment gained valuable experience and hands-on knowledge of a large wind turbine outside the official Danish wind program.

And yes, they built upon a long Danish tradition with wind energy. But they were also willing to depart from that tradition when necessary. After all, they set out to build the first Danish wind turbine using long cantilevered blades instead of a rotor braced with the struts and stays like Juul had used at Gedser. They intended to build what was then considered a "modern" turbine, one that used cantilevered blades mounted downwind of the tower. Just as importantly, they were also willing to borrow good ideas from others, including from their southern neighbor, Germany. It was in this that they made their most significant technical contribution.

Tvind studiously avoided the common affliction that infects most design teams—the not-invented-here syndrome. There's a natural human tendency to want to go it alone, to be the sole inventor of a new idea, and to discount the work of others and ignore the lessons they learned—often at great expense.

To build a long cantilevered blade—one that was connected to the rotor only at the hub—the Tvind design team knew they needed a strong attachment at the blade's root. Only a few decades earlier, Hütter had demonstrated just how to do so. Tvind's development team adopted the concept as its own.

The blades the Tvind team were building were no ordinary blades (see Figure 4-14. Tvind blade). They were big, each was 27 meters (89 feet) long—as long as the blade that failed on the Smith-Putnam turbine in 1945—and massive: each blade weighed 5,200 kg (11,500 lbs). The blades were nearly as big as those being developed at the same time by GE for its unsuccessful Mod-1 turbine.

Figure 4-14. Tvind blade. Preben Maegaard, former managing director of the Nordic Folkecenter for Renewable Energy and one of the pioneers in the Danish wind revival, stands by the root end of one of the original Tvind blades. The blade is part of the Danmarks Vindkrafthistoriske Samling collection of historic Danish wind turbines and components and can be seen at the Folkecenter near Hurup, Denmark. The 27-meter (89-foot) long blade weighs 5,200 kg (11,500 lbs). Note the blade flange where it mounts to the hub. The flange and the technique for attaching the fiberglass in the blade to the flange were originally developed by Ulrich Hütter in the 1950s and 1960s. Tvind adapted the technique to its pioneering wind turbine in 1975.

Hütter's innovation used by Tvind was to carry continuous fiberglass strands down the length of the blade to the mounting flange and wrap them around the bolt holes in the flange. This firmly attached the flange to the blade in what has since become known as the Hütter flange. It was the adoption of this flange by Tvind and subsequent Danish blade designers that revolutionized wind energy.

There were other noteworthy characteristics of the Tvind turbine that embarrassed the aerospace world. Though the design called for controlling power by varying the pitch of the blades, Tvind wisely wanted some means for overspeed protection should the pitch system fail. They opted for parachutes. If the pitch control system failed, parachutes deployed from the end of each blade. Crude as it was it worked, and this technique was later used on some commercial Danish wind turbines.

To our modern eye, the Tvind turbine and its stepped concrete tower look ungainly, and that was largely due to the school's limited budget. Wherever possible, they used surplus components: the turbine's main shaft came from a ship's propeller shaft; the gearbox came from surplus winding gear at a Swedish copper mine; the generator came from a Swedish paper mill. Altogether, the

SMEDEMESTERMØLLE: THE BLACKSMITHS' TURBINE

It was at the famous 1978 *Vindtæf*, or wind meeting, that Smedemester, the Danish association of "master blacksmiths," decided that it wanted to offer its members a way to participate in what looked like a growing industry. A direct translation of the Danish word conjures up an image of a beefy blacksmith in a grimy leather apron pounding out a horseshoe, but this is misleading. The English word *machinist* better captures the skill of these Danish "blacksmiths." In Denmark, blacksmiths are master machinists trained in a long and proud tradition of working with metal.

The association's 2,200 member companies were already accustomed to the local manufacturer of farm machinery, and the representatives at the *Vindtæf* thought, why not wind turbines too. This led to a fruitful collaboration between the blacksmiths' association and NIVE (Nordvestjysk Institut for Vedvarende Energi). The collaboration resulted in a design handbook for a 22-kW turbine driven by Økær's 5-meter blades, the Smedemestermølle, or literally, the Master Blacksmiths Wind Turbine.

NIVE, and later the Folkecenter for Renewable Energy, continued the collaboration by developing ever larger turbines, as each new generation of longer rotor blades was introduced. The handbooks enabled any small machine shop in Denmark to build a working wind turbine. The open, readily available design details led, in part, to the widespread adoption of wind technology across all of Denmark.

Tvind wind turbine cost one-tenth the cost of the Danish government's wind turbine installed later at Tjæreborg.

The huge Tvind project was begun in 1975 when Sweden's Barsebäck reactor went into operation. The turbine was finally completed in 1978. At the time it was the largest wind turbine in the world. It hasn't been all smooth sailing. Out of safety concerns, the original 2-MW design was downgraded to 1 MW, and half of this has been used for heating the Tvind school complex because the local grid wasn't able to take the full 1 MW. The grid has since been upgraded, and in the mid-2010s Tvind was upgrading their inverter to use the full 1-MW permitted capacity of the generator.

One blade failed in 1993 after 15 years of operation, requiring replacement of the rotor. The turbine was later returned to service, and it was still operating in 2015. This is a remarkable accomplishment for any wind turbine, and more so for such an early turbine and for one so large.

At the same time as Tvind was building the big wind turbine, a team of students developed an 11-kW downwind turbine using the same blade-mounting technology they were using on the large turbine. In the spirit of la Cour, Tvind then made the design of these 4.5-meter (15-foot) long blades available to others.

What separates a wind turbine from other power plants is its prime mover: the turbine's rotor. Without rugged and durable blades, a wind turbine is nothing more than a very large lawn decoration. Tvind's blade design—primarily its use of the Hütter flange—and the students' willingness to share the technology they had developed with other experimenters and wind activists was the key element that led to what would become today's wind industry, say Danish wind historians. All that was missing was someone to commercialize the blade technology.

Blades That Set the Industry in Motion
An early experimenter had made a mold set based on the Tvind blade and built another set of 4.5-meter blades. However, one of the blades failed, and they abandoned the project. Wind enthusiasts Erik Grove-Nielsen, Henrik Stiesdal, and Jens Gjerding inspected the failed blade. All would later make their mark in wind energy.

Grove-Nielsen, a young entrepreneur, had correctly identified a problem with the early experimenters in Denmark. None had a good set of working blades. There was a niche that needed to be filled with sets of "off-the-shelf" blades as he explains on his *Winds of Change* website, chronicling the history of Danish wind development. With working blades, experimenters could build working wind turbines.

Shortly after visiting the failed blade, Grove-Nielsen borrowed the Tvind mold himself to build a set of blades. Subsequently, he bought the mold from Tvind and set himself up in business as Økær Vind Energi and began building blades in 1977 before the large Tvind turbine was even completed. His first blades were sold to one of the firms competing with Riisager's Gedser-influenced design. Like Tvind's 11-kW prototype, the Adolfsen turbine placed the three-blade rotor downwind of the tower as Hütter had done with his Allgaier design. Later, Kuriant, another small manufacturer and Adolfsen's successor, used the blades in the same configuration. Both turbines were mounted on guyed lattice masts, as had been the 11-kW turbine installed by Tvind.

Figure 4-15. NIVE prototype. Surprisingly durable 10-meter diameter turbine built by local "blacksmiths" or machinists from off-the-shelf components in 1978. This 22-kW locally built wind turbine had been in service for more than three decades before it was removed. Note that unlike the Riisager design, this turbine uses cantilevered blades and an electrically driven yaw system to point the turbine into the wind. 1998.

Fortunately for Grove-Nielsen and the future of wind energy, a small group of activists, engineers, and millwrights took an interest in the off-the-shelf concept of standardized components and particularly his blades. The group, Nordvestjysk Institut for Vedvarende Energi (Northwest Jutland Institute for Renewable Energy) or NIVE for short, wanted a longer and quieter blade for use on a standardized turbine design they envisioned, one upwind of the tower with active yaw. They didn't want a passively yawing downwind rotor as had Tvind, nor use fantails as Riisager had done. Their design would use a motor to mechanically point the rotor into the wind.

It was at this point that Grove-Nielsen had to make a fateful decision: Which way would the rotor spin: clockwise or counterclockwise? Until then, Danish turbines turned counterclockwise when viewed from upwind. Riisager's turbine and its derivatives turned counterclockwise. Grove-Nielsen chose clockwise, in part to distinguish Økær's blades and the turbines they would be used on from Tvind and from Riisager. Today, the rotors on nearly all—if not all—large wind turbines rotate clockwise as a result.

Grove-Nielsen delivered a set of 5-meter (19-foot) blades for NIVE's prototype 22-kW turbine in 1978. The NIVE turbine used off-the-shelf induction generators, pillow-block bearings, gearboxes, disk brakes, and high-torque motors suitable for yawing the turbine into the wind (see Figure 4-15. NIVE prototype). These were all components that were inexpensively mass produced and available even in the remotest corner of Denmark. With Økær's blades, the NIVE team now had all the essential components for a modern wind turbine, save one critical feature.

The NIVE prototype used a simple welded hub. Thus, there was no pitching mechanism. The fixed-pitch rotor relied on the constant speed of the induction generator to stall the rotor in high winds. Juul knew in the 1950s this wasn't enough. But that crucial detail was lost in the experimenters' enthusiasm.

In the fall of 1978, two wind turbines with Økær blades destroyed themselves when the rotors went into overspeed. This was devastating to Økær. Grove-Nielsen stopped production while he sought a fail-safe air brake.

One of the turbines destroyed was a 22-kW prototype built by machinist Karl Erik Jørgensen with the aid of promising wind turbine designer Henrik Stiesdal. Jørgensen operated a machine shop in Herborg, Denmark.

Fortunately, Jørgensen, Stiesdal, and Grove-Nielsen didn't have far to look for a way to protect the rotor in overspeed. Juul had use pitchable blade tips on his Gedser mill, and it was still standing. While Juul had used hydraulics to pitch the blades on his much larger turbine, the more practical choice on the smaller turbines was a spring-loaded system that was activated centrifugally. In normal operation the spring retained the blade tip in its operating position. When rotor speed increased beyond a certain threshold, centrifugal force threw the blade tip out along a grooved shaft that turned the tip 90 degrees to the direction of motion. They quickly adopted this design.

Grove-Nielsen and Økær then began building blades with the automatic-acting blade tip. By the winter of 1978, he was supplying replacement blades with the new pitchable blade tip, including a set to Jørgensen and Stiesdal at Herborg Vindkraft (see Figure 5-57. Pitchable blade tips). It was the self-acting, pitchable blade tips and the Hütter flange that more than anything else contributed to the durability of Danish wind turbines.

Herborg Vindkraft and Vestas
Armed with fail-safe overspeed protection, Jørgensen decided to go into production with his and Stiesdal's design marketed as the Herborg Vindkraft or HVK in early 1979. Like others of the period, their HVK-10 used off-the-shelf components. They attached Økær's 5-meter blades to a distinctive welded hub and mounted the rotor assembly directly on a Hansen gearbox. The transmission drove two induction generators: one for low wind speeds, the second, much larger 22-kW generator for high wind speeds. The HVK turbine also featured a prominent spinner or nose cone and dual work platforms. Though they sold a few units, the firm didn't prosper.

HVK delivered one HVK-10 turbine with a 30-kW generator to Vestas, a manufacturer of

Figure 4-16. Vestas wind turbine and Økær blades. Background: An HVK wind turbine in Vestas livery at the Vestas plant in Lem. Vestas bought the turbine and design from HVK before launching its successful line of 15-meter (50-foot) diameter wind turbines. Foreground: Økær 7.5-meter (25-foot) blades with pitchable blade tips developed by pioneering Danish blade designer Erik Grove-Nielsen. Note the Hütter flange that made it all work. 1980.

farm equipment and truck cranes in the fall of 1979. For personal as well as financial reasons, HVK sold Vestas the rights to its design. HVK then developed a larger version for Vestas designed to use Økær's new 7.5-meter (25-foot) blades. They installed the HVK-Vestas's V15 in the spring of 1980 at Vestas's plant in Lem on Denmark's windy west coast (see Figure 4-16. Vestas wind turbine and Økær blades).

Vestas substantially redesigned the HVK-Vestas turbine in 1981. They extended the nacelle's bedplate, moved the gearbox rearward, used a long mainshaft supported on two large pillow-block bearings, and placed the disk brake on the low-speed shaft. They retained the hub, the work platforms, and the large conical spinner of the HVK design but housed everything beneath a distinctive fiberglass cover. The Vestas V15, so named for its 15-meter (50-foot) diameter rotor, was born. It became one of the most successful wind turbines of its era. Vestas sold

Figure 4-17 RIGHT: Bonus 55 kW. First Bonus (Danregn Vindkraft) 55-kW wind turbine. Installed in 1981 near the Folkecenter for Renewable Energy in northwest Jutland, the turbine was the forerunner of hundreds of such turbines installed in California in the mid-1980s. Like Vestas's V-15, this Bonus turbine used Økær's 7.5-meter (25-foot) blades. In 2004, Bonus was absorbed by Siemens. LEFT: Bonus 55-kW drivetrain. Note the dual generators and large pillow-block main bearing and welded hub. The drivetrain is part of the Folkecenter for Renewable Energy's collection of historic Danish wind turbines and components. (Preben Maegaard)

thousands of this model and later variants most are still operating three decades later.

Stiesdal and Bonus

With the sale of HVK's design to Vestas, Stiesdal moved on to university where he studied medicine, physics, and biology. Meanwhile, he continued as a consultant to Vestas, working on later Vestas designs, including their first variable-pitch rotor, the V23.

Following a failed attempt to buy Nordtank, another Danish manufacturer of wind turbines, Danregn Vindkraft, began building machines of its own design in 1980. In 1983, Danregn—formerly a manufacturer of irrigation equipment, hence the name, which means Danish rain—began trading under the name of Bonus.

Because they used the same blades as others, Bonus went through the same development cycle as everyone else: 22 kW and then 30 kW. They installed their first 55-kW model in 1981 near the Folkecenter for Renewable Energy in windy northwest Jutland (see Figure 4-17. Bonus 55 kW).

Bonus, like Vestas, opted for putting the brake on the mainshaft. This required a larger and more expensive disk brake than when mounted on the high-speed shaft of the gearbox. However, this arrangement avoided passing braking torque through the gearbox to stop the rotor. This improved reliability and longevity and later served to distinguish the more rugged and durable designs of Vestas and Bonus from other Danish manufacturers, such as Nordtank.

Nordtank, ironically a former manufacturer of tank trucks design to carry liquids such as oil and gasoline, were building similar turbines. Because of the manfacturer's experience fabricating large cylindrical tanks, it pioneered tubular towers. Nordtank's characteristic stepped or "rocket" tower soon became a feature of the Danish landscape.

In contrast to Vestas and Bonus, Nordtank

placed the disk brake on the high-speed side of the gearbox and used sheet metal nacelle covers, as did its competitors Micon and other Danish manufacturers.

Though Bonus joined the rush to the budding California market, the company never overextended itself. Bonus was one of the few Danish manufacturers that survived unmarred by bankruptcy when the California market collapsed in 1986.

In 1987, Stiesdal finished his studies and joined Bonus. He brought his distinctive design style to Bonus and has remained technical director as the company steadily grew, eventually being absorbed by the German electrical conglomerate Siemens.

The Danish Concept
By 1980–1981, blades of the Økær design were being delivered to 20 different manufacturers of wind turbines in Denmark, Germany, the Netherlands, and Belgium. Most were building wind turbines quite similar to one another. Development of wind energy had now passed from the hands of the machinists and activists to those of small and medium-size manufacturers who saw wind energy as a rapidly growing business rather than a political cause.

What was to be labeled the "Danish concept" was in full flower. From a distance, the turbines all looked alike. They used a three-blade rotor upwind of the tower without fantails. The blades all looked alike because of their then common source. To distinguish their products from others, some manufacturers chose distinctive paint schemes. In the field you could identify the different turbines by the color of their pitchable blade tips. Vestas used white tips; Bonus, green; Nordtank, red.

American critics of the Danish concept derided the different designs as mere copies, implying that no real engineer would deign to copy someone else's design. Yet just as Eli Whitney's introduction of interchangeable parts in the manufacture of guns revolutionized manufacturing, the use of standardized parts by Danish wind turbine manufacturers revolutionized the wind industry. Only in this way could small but innovative companies take advantage of two centuries of industrialization by buying low-cost, off-the-shelf components. They could mix and match parts to suit their needs, while focusing on what they did best: designing durable wind turbines.

Using standardized components also enabled the Danish wind industry to scale-up quickly. Instead of building specialized components in house, they could instead rely on what's called today a supply chain of manufacturers who were selling the same product, such as gearboxes, to many different customers. Component suppliers had already won economies of scale by manufacturing products in series to a broadly diversified market. Wind turbines were just one more market to gearbox companies or generator manufacturers.

What looked like a weakness to an American-trained aerospace engineer was a winning combination to a Danish manufacturer of farm equipment. Vestas—in one corporate form or another—went on to become the world's largest manufacturer of wind turbines. Bonus, now Siemens, became another.

Meanwhile, the manufacturers were diversifying the one component that was unique to wind turbines—the blades. Following storm-caused failures of Økær's 7.5-meter blades in the winter of 1980–1981, Vestas decided to design and build its own blades in-house. Elsewhere in Denmark, LM Glasfiber, a manufacturer of fiberglass yachts, began building blades for Windmatic, a Riisager derivative. Later in 1981 Økær entered a licensing agreement with Coronet Boats, another manufacturer of fiberglass yachts, to begin producing blades under the AeroStar trade name. By the mid-1980s, there were three manufacturers of wind turbine blades in Denmark: Vestas, Aero-Star, and LM Glasfiber.

Rotor blades were the linchpin of the Danish concept and with industrial manufacture of blades underway; the Danish wind industry was primed for explosive growth.

The California Wind Rush

The growth of an industry is partly about technology, but also about the market for that technology. As we have seen, when fossil fuels are abundant—or at least are perceived as abundant—the demand for wind energy weakens or disappears altogether.

Thus, the demand for wind energy is due to public policy, often in response to an international crisis. That event was the second oil embargo of 1979.

In response to this event and to pressure from the business community now building wind turbines, Denmark launched its first incentive program. Slowly, the market began to grow as manufacturers established a foothold. Then the California market opened up in the United States, and the stampede was on.

California provided its own tax subsidies on top of subsidies from the federal government, effectively doubling the subsidies for renewable energy available in the state. That alone was insufficient to create a market. Concurrently, California had begun measuring the wind across the state and making the information public. Now everyone knew where the wind was: it was in three principal mountain passes. Still, a piece of the puzzle was missing. There was no place to sell wind-generated electricity. The existing utilities controlled the market.

Fortunately, the utilities blundered badly, antagonizing the progressive governor and the state regulatory commission. The utilities were discovered colluding to violate a commission order to develop renewable energy. The California Public Utility Commission heavily fined the utilities and ordered them to provide connections to independent power producers and to offer generators standard contracts with fixed prices for their electricity.

Now, all the pieces of the puzzle were in place. Developers knew where the wind was. They knew how much they would be paid for their electricity. And they had attractive subsidies that would make it economic to produce and sell wind-generated electricity for a handsome profit. The boom was on. Soon a regular train of shipping containers loaded with Danish wind turbines was snaking its way to the United States, drawn by the great California wind rush. From 1980 through the late 1980s, there were more than 4,600 Danish wind turbines installed in California out of nearly 11,000 wind turbines in the state.

It was a heady time. The stories of that period could fill a book—and have in Peter Asmus's *Reaping the Wind*. What is pertinent for our tale is how various technologies fared in the blistering winds of California's mountain passes. It was a final showdown as the sleek high-speed turbines from the aerospace community faced off against the hulking wind turbines from Denmark.

American Designs of the Early 1980s

California became a battleground between two fundamentally different approaches to wind turbine design. While the Danes initially fell under the "downwind is modern" spell, the activists and the machinists building wind turbines soon learned that Danish tradition led them to upwind turbines like that at Gedser. American designers of medium-size turbines in the early 1980s had no Gedser mill to turn to for inspiration and remained in the thrall of Hütter's downwind turbines to the end.

Downwind Dominant

Many American-designed turbines of the period exhibited most of the characteristics we associate with Hütter's design philosophy. Several, such as the Carter and ESI turbines, used downwind, two-blade, teetered rotors operating at fairly high tip-speed ratios as found on Hütter's StGW-34 turbine (see Figure 4-18. Carter 25). Others, such as Enertech, Storm Master, and US Windpower, opted for three-blade, rigid downwind rotors. Most used integrated drivetrains, as had Hütter, and most placed the brake on the high-speed side of the gearbox. Storm Master, Enertech, and ESI even placed the brake behind rather than in front of the generator, creating even more paths for failure. And similar to Hütter, many of the designers came from the aerospace community either directly or indirectly. Jay Carter Jr., for example, came from Bell helicopter.

The difference in design philosophy between the Danes and the Americans was apparent to any observer in California during the early 1980s. Danish wind turbines appeared far more massive than the US machines: the Danish blades were thicker and their nacelles bulkier than those on the US-designed turbines. Danish turbines operated at lower tip-speed ratios than the US designs—they spun at slower speeds. All in all the Danish designs projected a sense of solidity

Figure 4-18. Carter 25. Developed by the father and son team of Jay Carter Jr. and Jay Cater Sr., the Carter design grew out of the helicopter industry. The nominally 10-meter diameter (33-foot) lightweight, teetered rotor was oriented passively downwind of the tower to drive a 25-kW induction generator. The main spar incorporated a filament winding technology developed for Polaris missile casings. The turbine, created independently of the US wind development program, was introduced in 1979. Downwind two-blade designs such as this generated a characteristic "whop-whop" that annoyed neighbors near Palm Springs, California. Circa 1985.

and durability missing from the frail, frantically flailing US turbines of the period.

Lightweights in a Heavyweight Environment

The difference in appearance betrays an underlying difference in design. First-generation European wind turbines had a much higher specific mass in kg/m² than American wind turbines. That is, there was more mass in the nacelle and rotor of Danish turbines relative to the area swept by their rotors than in American designs.

Few wind turbines epitomize the American lightweight design philosophy better than those designed by the Carter family. They built two models, a 25 kW and a larger turbine variously rated between 250 kW to 300 kW. Both turbines used a flexible downwind, teetered rotor. The Carter 25 had a specific mass of less than 5 kg/m², and the larger model was not much better at slightly more than 7 kg/m² (see Table 4-3. Specific Mass of Rotor and Nacelle Relative to Swept Area for Selected Turbines).

Compare these designs with Bonus's 15-meter (50-foot) diameter turbine and Vestas's V17, a 17-meter (56-foot) diameter machine. (Vestas introduced the V17 into California in 1985.) The Danish turbines contained twice the specific mass of most US-built turbines and three times more than the Carter and Storm Master turbines. In the demanding environment of California's wind farms, specific mass became a surrogate for reliability and durability. Hal Romanowitz, one-time operating officer at Oak Creek Energy Systems in California's Tehachapi Pass, criticized the Carter design simply by noting: "They were not a durable machine."

Other turbines characterizing lightweight designs of the early 1980s include Storm Master, Windtech, and ESI. "Storm Master" was a misnomer. Some Storm Masters failed within a few hours of their installation in the early 1980s. More than 300 of the machines with their whip-like pultruded fiberglass blades were at one time

Table 4-3. Specific Mass of Rotor and Nacelle Relative to Swept Area for Selected Turbines									
		Diameter		Swept Area	Power	No. of Blades	Rotor Orientation	Rotor & Nacelle Mass	Specific Mass
Turbine	Country	m	ft	m²	kW			kg	kg/m²
Carter	USA	10	33	79	25	2	downwind	350	4.5
Carter	USA	23.0	75	415	250	2	downwind	3,020	7.3
Storm Master	USA	12.2	40	117	40	3	downwind	1,050	9.0
US Windpower	USA	17.1	56	229	100	3	downwind	2,539	11.1
ESI-80	USA	24.4	80	467	250	2	downwind	5,221	11.2
Enertech	USA	13.4	44	141	40	3	downwind	1,900	13.4
Bonus	Denmark	15	49	177	65	3	upwind	4,300	24.3
Vestas	Denmark	17	56	227	75	3	upwind	6,200	27.3

Figure 4-19. Storm Master it was not. Lightweight, downwind turbine developed in California during the early 1980s used whip-like pultruded fiberglass blades with pitch weights. Some of these 40-foot diameter (12-meter), 40-kW turbines failed within a few hours of their installation. In contrast, Danish turbines were rugged tanks of durability. 2005.

Figure 4-20. Windtech. One of the most significant failures of US DOE's research program of the 1980s. Note the straps and hinges on the 16-meter (52-foot) torsionally flexible fiberglass blade. Though elegant on paper, the design was notoriously unreliable. Note the technician for scale. 1985.

installed in California (see Figure 4-19. Storm Master it was not).

One of the most spectacularly unsuccessful designs to evolve from the US wind development program was that of United Technologies Research Center (UTRC). Again, the technology came out of the aerospace community and with as much success as elsewhere (see Figure 5-47. Coning).

UTRC was a division of United Technologies, the aerospace giant. Various divisions within the conglomerate build jet engines (Pratt & Whitney) and helicopters (Sikorsky). United Technologies remains one of the major defense contractors in the United States.

The UTRC turbine was a first-generation American design that featured a downwind rotor incorporating passive yaw and passive pitch control of its slender pultruded fiberglass blades. Like Carter, UTRC used a torsionally flexible spar that allowed the blade to change pitch toward stall in high winds. However, unlike the Carter design, the UTRC turbine used an ungainly system of pitch weights and metal straps. Though technically elegant on paper, the flexibility of the blades induced complex and ultimately uncontrollable dynamics between the wind and the rotor.

When United Technologies unceremoniously dumped the program after a failure of the prototype in Boston harbor, the turbine's designer took the technology outside the company. Commercialized under the Windtech trade name, more than 200 of the problem-prone 15.8-meter (52-foot) diameter version were installed in California with disastrous effect (see Figure 4-20. Windtech).

ESI was another commercial spin-off from the US wind R&D program. The ESI turbines were distinctive for their fairly large diameter (comparable to that of Danish turbines), the high speed of their rotors, and the tip brake at the end of each blade. The tip brakes, or flaps, were intended to protect the turbine during high wind emergencies. The design used two fixed-pitch, wood-epoxy blades downwind of its hinged, lattice tower. Two versions were introduced. Nearly 700 of the ESI-54 (16.4-meter) and 50 of the ESI-80 (24-meter) were eventually installed in California (see Figure 4-21. ESI-80).

Figure 4-21. ESI-80. Another US-built series that followed Hütter's design philosophy were the ESI-54 and the later ESI-80 shown here. The ESI turbines also used a downwind, two-blade, teetered rotor driving an induction generator through an integrated gearbox. The brake is mounted on the back-end of the generator (far left). ESI used tip brakes not pitchable-blade tips. The high rotor speed in combination with the downwind design and tip brakes was notoriously noisy. The ESI-80 used an 80-foot (24-meter) rotor made up of laminated wood blades coupled to a 250-kW generator. Circa mid-to-late 1990s.

Whereas Danish manufacturers were, for the most part, already established and accustomed to building heavy-duty products for a demanding rural clientele, US manufacturers were typically small with little or no manufacturing experience. US firms were also undercapitalized and had no staying power. However, there was one exception: US Windpower.

US Windpower

Rebranded in the 1990s as Kenetech, US Windpower (USW) grew out of a successful marriage between the financial and aerospace community around Boston. USW built the first wind farm in the United States when it installed 30 prototype 30-kW downwind turbines on Crotched Mountain, New Hampshire, in 1980. Though the turbines were not successful, the project established the investment vehicle for rapid growth in the hothouse atmosphere created by California's lucrative market.

The company rolled out its first California wind farm in 1981 with its 56-50, a 56-foot (17-meter) diameter, three-blade, downwind rotor with a 50-kW rating. In 1985 USW was operating 600 of this model alone. They also introduced a redesigned, and upgraded version with a 100-kW rating (Figure 4-22. US Windpower 56-100). By the end of 1986, USW had installed 2,700 wind turbines, more than any other manufacturer.

The recipe for USW's success was its ability to raise large amounts of capital quickly and its vertical integration. It both manufactured the turbine and developed its own wind farms. The company's design was more complex than others of the day. It used a more complex hub that allowed full-span pitch control, something that wouldn't be seen on most wind turbines until the late 1990s.

USW didn't cater to customers as other manufacturers did because the company used its wind turbines internally. Consequently, the company was notorious for its secrecy and its heavy-handed treatment of its critics.

By the time the company collapsed in 1996—a victim of its own hubris—USW was operating more than 4,000 wind turbines in California. Unquestionably, USW's turbines were the most successful of US-designed turbines of the era.

Figure 4-22. US Windpower 56-100. One of the most successful of the US-designs from the 1980s, US Windpower's 56-100 used a three-blade, variable pitch rotor, passively oriented downwind of the characteristic three-legged tower. Some 4,000 of these 100-kW turbines were at one time operating in California.

Though most were still in service in 2014, there were plans to remove the turbines and replace them with a fewer number of much larger, modern wind turbines.

Enertech's E44

Though USW installed more wind turbines in California than any other company, some argue that a small New England company, Enertech, designed a more long-lasting configuration. Beginning in the mid-1970s, Enertech introduced a 1,500-watt, three-blade, downwind turbine. In the United States it was innovative for the time because it used an induction or asynchronous generator and could be directly connected to the grid. Initially, the stall-regulated rotor had no overspeed protection. As in Denmark, the company learned the hard way that stall regulation was simply not enough. The turbine was later redesigned with tip brakes and repackaged as the Enertech 1800.

In the early 1980s, Enertech introduced a much larger turbine comparable in size to Danish machines of the time. Enertech's E44, a 13.4-meter (44-foot) diameter downwind turbine, was intended for small commercial and rural applications. The US Department of Agriculture (USDA) installed a 25-kW version at its experiment station west of Amarillo, Texas, in 1982. USDA's unit went through several iterations (see Figure 5-55. Tip brake) and remained in service for most of two decades. Enertech's focus shifted once the California market became apparent.

Like ESI, Enertech used specially designed wood epoxy blades made by the Gougeon Brothers of Bay City, Michigan. Unlike generic Danish blades of the time, the Gougeon blades were designed specifically for each turbine. They were not interchangeable. Nevertheless, the blades were the turbine's most durable component.

By the end of the California wind rush in the mid-1980s, Enertech had sold more than 500 wind turbines to wind farms in the state's three windy passes. After the market collapsed with the expiration of the federal tax subsidies, Enertech, ESI, and most US manufacturers collapsed as well.

However, various enterprises have tried to resurrect the E44 design over the succeeding decades. Atlantic Orient, founded by one of the original partners in Enertech, introduced a slightly larger version as the AOC 15/50 in another US DOE program during the late 1990s (see Figure 5-30. Integrated drivetrain, small turbine). DOE intended its program to spur a revival of US wind turbine manufacturing but, as earlier, to little avail. Several other companies have since marketed derivatives of the AOC 15/50, and the product was still on the market in 2014, the last—and the most long-lived—of the downwind, passive yaw designs of the 1980s.

When well maintained, the E44s have turned in respectable performance and hundreds have been kept in service for nearly three decades. And contemporary versions of the 15/50 have finally remedied many of the defects of the original design. Seaforth Energy's model of the 15/50 now uses a freewheeling rotor, that is, the rotor doesn't need to be motored up to operating speed as in the earlier versions. It's no longer necessary to climb the tower to reset the tip brakes manually. The tip brakes now automatically reclose after restarting the turbine—they didn't before.

Mehrkam

The tragic tale of Terry Mehrkam illustrates the limits to "fly by the seat of your pants" wind turbine design and manufacturing when the necessary skills are lacking. At the same time Riisager was building his turbines in Denmark, Mehrkam was experimenting with wind in his backyard in central Pennsylvania. Mehrkam represented the home-built or craft tradition in the United States, much as his counterpart, Riisager, represented the Danish craft tradition. However, their skills were not comparable. Mehrkam was far less skilled or had far less access to others with the skills he lacked than Riisager, the carpenter, or the other Danish "blacksmiths," that is, professional machinists then building wind turbines.

Additionally, Mehrkham had no wind energy tradition to fall back on. The Danes were steeped in wind energy. Mehrkam was improvising and leapt to a wind turbine that was far too big for his skills. He began building a 12-meter (40-foot) diameter, four-blade, downwind turbine. Like the early Danish turbines, his stall-regulated rotor had no overspeed control. Unlike the Danes who learned the hard way that overspeed controls were essential, Mehrkam didn't adapt his design after he began to encounter serious overspeed incidents. His solution was deadly. When the brakes failed, he would climb the turbine and pry a crowbar into the disk-brake calipers to try and bring his turbine to a halt. His lack of skill and his ambition—or his desperation—eventually cost him his life.

Wind Energy and the Aerospace Arts

The abject failure of the aerospace industry and the national wind development programs that relied on them in the 1970s and 1980s was due to a fundamental misinterpretation. Wind turbines are not airplanes. They were never meant to fly.

Only in superficial ways do wind turbines resemble aircraft. Wind turbines are power plants. They must produce cost-effective electricity. To do so they must perform like other power plant machinery: they must work reliably over long periods with little maintenance.

Whereas a wind turbine must operate many hours on only a few hours of service, an aircraft flies only a few hours relative to many hours of skilled maintenance.

For example, an airline may spend 5.5 hours of maintenance on a relatively low-maintenance, short-haul jet for every one hour in the air. Wind turbines, on the other hand, must operate months on end—trouble-free—if they are to compete with fossil fuels and nuclear power.

The approach by Hütter and his devotees grew out of the design principles embodied in aeronautical engineering under the assumption (an incorrect one, as it turns out) that aerospace was the industrial sector most capable of building wind turbines. After all, Hütter knew sailplanes well. He designed sailplanes before he designed wind turbines.

While US and German engineers were building fighter jets, the Danes were building farm machinery. Fortunately for the future of wind energy, Denmark had no aerospace industry. Had Denmark an aircraft industry, it is probable that its wind turbines would have failed in the same manner as those of their German and US competitors from an overreliance on aerospace technology.

Proponents of developing wind technology from the top down argued that MBB, Boeing, and other aerospace contractors not only had the sophisticated talent needed to master the technology but also had greater financial resources to weather the inevitable downturns in R&D funding or a sluggish market. But large institutions had no more staying power than the poorly financed entrepreneurs and dreamy-eyed inventors who dominated the bottom-up sector. When R&D money evaporated, so did the interest of aerospace contractors.

In Germany by the mid-1990s "all the big companies were kaput," said Uwe Carstensen, a German wind developer. For MAN and MBB, "wind energy was a sideline of little importance." It was the same in the United States. None of the

---

AN AEROSPACE SUCCESS STORY: BERGEY WINDPOWER

There is one noteworthy exception to the disastrous record left by the aerospace industry in modern wind energy's early development: Bergey Windpower.

Karl Bergey and his son Mike founded the eponymous company in the late 1970s. Karl came from the aerospace community. An aeronautical engineer, his résumé includes stints at Piper Aircraft as well as North American Rockwell. He was also a long-time member of the engineering faculty at the University of Oklahoma, where Mike studied mechanical engineering.

Bergey Windpower introduced its first turbine, the Bergey 1000, in 1980 with passively pitched aluminum blades. The aluminum was soon replaced with pultruded fiberglass blades that have been a unique feature of Bergey wind turbines ever since. In 1983 the company introduced its much larger, household-size wind turbine, the Excel, a 7-meter (23-foot) diameter, 10-kW model. Bergey Windpower has periodically updated its product line over the succeeding decades.

One key to the company's enduring success has been keeping it a family-owned business and designing products for maximum simplicity. The Bergeys like to quote Antoine de Saint-Exupéry, the French author and pilot, "A designer knows he has achieved perfection not when there is nothing left to add, but when there is nothing left to take away."

A Bergey wind turbine is about as simple as a wind turbine can be. The Bergeys coupled a special-purpose, direct-drive, permanent-magnet generator with a self-furling rotor, like that used on the successful American farm windmill. This, in combination with durable pultruded blades, is a winning combination in the small wind turbine class.

The company has not only endured but prospered during the past three decades while hundreds of their competitors have fallen by the wayside. It's safe to say that of all the ferment among both large and small wind companies in the United States during the 1970s and 1980s, Bergey Windpower has been the most successful. Its brand is now recognized worldwide.

> "The more sophisticated engineering tools become, the less we realize that engineering is also an art," explains wind turbine designer Peter Jameison. "IF WE HAD WAITED FOR THE RIGHT ANALYTICAL TOOLS, WE WOULD HAVE NEVER BUILT THE FIRST BRIDGE."

contractors in the US wind program, whether they built large machines or medium-size turbines, were active by the late 1980s when Westinghouse finally called it quits.

Fickleness had a price: discontinuity in the development of wind technology. The turbines were designed, built, then destroyed, the teams that created them scattering to other enterprises and carrying with them the knowledge learned at such great public expense. It also gave wind energy a black eye that would take years to overcome.

The late Forrest (Woody) Stoddard argued that sophisticated aeronautical models were no substitute for experience during the industry's early years. Neither aircraft wings nor helicopter blades were directly comparable to wind turbine blades. Their only similarity is that all three use airfoils. "Helicopter design codes have been woefully inadequate" for designing wind turbines," said Stoddard. The wind resource was far more demanding than first thought.

In their treatise *Wind Turbine Engineering Design,* Stoddard and his colleague David Eggleston noted that experience and intuition taught early millwrights their craft. This resulted in a body of knowledge sufficient to build wind turbines up to 28 meters (90 feet) in diameter. Some of these lasted for several hundred years. Saying that "a design life of 350 years isn't bad," Stoddard suggested that millers had a better understanding of the wind than many wind engineers of the 1980s.

Peter Karnøe, a Danish researcher specializing in how technology develops, suggests that success requires the accumulation of knowledge from R&D and—importantly—from field experience. In Denmark, this required blending the values of the Danish craft tradition with the prosaic research needs of fledgling manufacturers aided by Denmark's Risø national laboratory. As the repository of scientific and technical knowledge, Risø had an important role to play, especially later in testing wind turbines.

The Danish industry, says Karnøe, was not handicapped by sophisticated knowledge of aerodynamics. Danes took the lead, he says, by pursuing a bottom-up strategy, because experimenters and small companies had no other choice. Relevant new technological knowledge was acquired by solving problems specific to the open architecture of the "Danish" design.

There was no "ivory tower hierarchy" in the Danish wind industry because of its roots in farm machinery manufacturing, where empirical and hands-on knowledge were highly valued. Danish designers relied on simple, pragmatic principles. They used their intuitive or "seat-of-the-pants" knowledge about materials to reach a first-order approximation of how a wind turbine should be built.

Danes also had the Gedser mill as a model. Like la Cour before him, Juul had developed some simple "rules of thumb" for designing a wind turbine based on his experience with his machine at Gedser.

Danish manufacturers often solved design weaknesses, says German historian Heymann, by oversizing the troublesome component. "Throwing metal at it," as one US aerospace contractor dismissively called such solutions, reflecting a scornful attitude toward Danish wind turbines common among the engineering elite in the United States at the time.

Other engineers were less condescending. "The more sophisticated engineering tools become, the less we realize that engineering is also an art," explains wind turbine designer Peter Jameison. "If we had waited for the right analytical tools, we would have never built the first bridge," he says. And many bridges that were conceived without computer-aided design still serve their function well. "Often analytical tools are used to validate decisions reached intuitively. People expect that an absolute analytical evaluation is possible. But there are no absolutes [in engineering]." Even cost is intrinsically "irrational because it is subject to political decisions."

Bottom-Up Delivered

To Stoddard, an important difference between American and Danish innovation was Danish willingness to start small and increase size incrementally. American designers constantly sought breakthroughs. They wanted to bypass the drudgery of incremental development and bat a home run. They wanted to build a large turbine from the start without gaining the experience necessary to do so successfully. The US R&D program leapt from the Mod-0A to the Mod-1. Ignoring the failure of the Mod-1, DOE continued with the program by scaling-up to Mod-2, a nearly sixfold increase over the Mod-0A. Then they went even further with the 98-meter (320-foot) diameter Mod-5b.

Even though ultimately unsuccessful, Denmark's R&D program offers a sharp contrast with that in the United States and Germany during the 1980s. The Danes built steadily upon the experience of each successive prototype as technical knowledge gradually accumulated. As a small nation, it could ill afford the costly strategy of reaching for the stars, and they certainly could never have justified throwing away all they had learned once they had paid for it by scrapping the turbines after only a few hours of operation.

Danish researcher Karnøe notes that such human factors as being able to lay claim to the world's largest wind turbine played a part in the US program, and German historian Heymann found the same intoxication in the development of Germany's Growian. The wind turbines became as grand as the egos involved.

"We proved we could build them [large machines]," sums up an NREL administrator, "but we didn't get the result everybody hoped for. That's research. You're not always going to be right." That's possibly true, but there were engineers on the outside trying to tell DOE that large machines from the aerospace community were a waste of money from the start. Other critics in the bottom-up camp charged that the program was a time-consuming diversion from the more necessary work of deploying thousands of wind turbines to gain operating experience while delivering useful quantities of electricity. "The Wright brothers didn't start with a [Boeing] 747," says Otech Engineering's John Obermeier. "They [DOE] leap-frogged too far, too fast."

Fortunately, commercial success was found in medium-size wind turbines that gradually and incrementally grew larger and larger as small to medium-size companies worked in markets that wanted wind energy—not just a few showcase projects to temporarily appease a political constituency.

Though learning through failure was the most effective way to gather technological knowledge in the 1980s, it will no longer be as helpful, warns Karnøe, because the machines have grown much larger and the costs of mistakes have risen proportionally. Danish manufacturers have since acquired sophisticated technology and have become more analytical in wind turbine design than they were during the formative years. Today's wind turbines are far most cost effective and reliable as a result.

Modern wind turbines have reached the state of technological sophistication and maturity that the aerospace community in the United States and Germany tried to develop in the 1970s and 1980s. However, success was achieved from the bottom up, not from the top down, and it owes its existence to wind pioneers outside the aerospace and electric utility industry. Many of these pioneers were openly critical of their country's official wind programs and the utilities that were selected to implement them.

Boom and Bust Survivors

When the US tax subsidies expired in 1985, the California market collapsed and with it most wind turbine manufacturers. The exceptions were few: US Windpower in the United States and Bonus in Denmark. The Danes retreated to Denmark where they regrouped, and the industry rebounded serving the Danish domestic market that continued to grow steadily, especially after the introduction of a simple feed-in tariff in 1984.

In the meantime, the German market sputtered to life. Beginning slowly at first, it picked up speed following the introduction of its first feed-in tariff in the early 1990s patterned after that in Denmark. The German market gained

SOUTH OF THE BORDER (ENERCON)

The commonality of Danish blades was of benefit to others outside Denmark as well. Despite the failure of Germany's development of large wind turbines, German entrepreneurs saw a future in wind energy. Like their Danish counterparts, they seized on the availability of standardized rotor blades. In 1981 Enercon's founder Alois Wobben took delivery on his first set of Økær 5-meter blades. By 1984 Enercon was taking delivery of 7.5-meter AeroStar blades.

Wobben's Enercon would go on to become Germany's most important manufacturer of wind turbines, and a pioneer in direct-drive. The distinctive bulb-shaped nacelle of Enercon's direct-drive wind turbines—necessary because of the large-diameter ring generator—is now a common feature of the German landscape today.

momentum after implementation of the precedent-setting Renewable Energy Sources Act (Erneurebare Energien Gesetz or EEG) launched a renewable energy revolution in the world's fourth largest economy. By the early 2000s, Germany was surpassing Denmark as one of the world's largest markets for wind energy.

The United States finally returned to the market in the late 1990s. Though it's been turbulent, the industry has continued expanding worldwide ever since.

Right Product, Right Place, Right Time

There were several other factors that played a decisive role in the growth and ultimate success of the Danish wind industry: geography, timing, access to performance data, and performance standards.

For Danish manufacturers of farm machinery, the second energy crisis in 1979 occurred at an opportune time. During the late 1970s the European market for farm equipment slackened, forcing Danish manufacturers to seek new products for their rural customers. They quickly adapted their surplus capacity to a new market: wind turbines.

The broad distribution of people across the Danish landscape demanding modern wind turbines, a good wind resource, a history of using the wind, and a manufacturing sector accustomed to building heavy machinery for a demanding rural market were all the elements necessary for stimulating a successful domestic wind industry.

Danish success was due in part to "geographical proximity." Because the country is so small, manufacturers could service their own turbines often directly from the factory. More than half of the Vestas turbines in Denmark in the 1980s, for example, were serviced directly by the manufacturer. This enabled companies to learn from their mistakes and quickly make modifications to keep their turbines in operation. These operating turbines served as physical proof to potential buyers that the company's machines were a good investment—that they worked.

The huge geographic size of the United States, as well as the vast difference in terrain and wind speeds from one region to the next, hindered the development of wind technology. As in Denmark, nearly all early marketing efforts in the United States focused on rural applications for homeowners and farmers. In the United States, this resulted in the distribution of single wind turbines across the vast expanse of the continent. Many turbines were installed in low-wind areas, where they produced little electricity. Thus, there was little revenue to pay for repairs that these early turbines inevitably required.

The turbines were so widely dispersed that manufacturers couldn't service them directly and gain hands-on knowledge of what was wrong and how to fix it. When problems arose—as they often did—the turbines could not be repaired quickly, if at all. The machines that failed often remained inoperative for months, if not years, discouraging potential buyers. When these small firms encountered technical problems, they were often unable to correct them and quickly failed as customers ran for the exits.

Wind Turbine Owners' Association

Secrecy among US manufacturers also inhibited the transfer of hard-won field experience, says Karnøe. This cost the US industry dearly. Because Danish companies shared a common platform, they all had a stake in solving common problems. Danish manufacturers also had a different clientele. Prior to the California boom,

Danish manufacturers were competing to win over increasingly sophisticated rural customers.

Buyers of Danish wind turbines were learning more about wind energy than they at first expected due to problems with the early turbines. This led owners to seek answers and remedies as a group.

In the mid-1970s, backyard experimenters, hobbyists, and environmentalists began a series of informal quarterly meetings to discuss wind energy. Soon wind energy activists, designers, consumers, and potential manufacturers began meeting regularly to discuss wind turbine design and the future of renewable energy. These Vindtræf or wind meetings became annual events where common issues were debated publicly and best practices were shared in front of everyone present.

The 1978 Vindtræf was particularly significant. It was at this meeting that the group formed a "safety committee" to establish guidelines for wind turbine design following the failure of several wind turbines. The committee included Preben Maegaard, Henrik Stiesdal, and Erik Gove-Nielsen among others. It was also at this Vindtræf that Flemming Tranæs, journalist Torgny Møller, and others formed Danske Vindkraftværker (literally Danish Wind Power Plants), the Danish Wind Turbine Owners' Association.

The owners' association has been meeting annually ever since. At its peak, says Danish wind historian Benny Christensen, the association represented 150,000 Danish families. That carried clout.

United, the owners began to demand minimum design standards from manufacturers. The most important for the future of wind energy was the requirement for a fail-safe, redundant braking system, such as Juul's pitchable blade tips. This single provision probably did more than any other to further Danish wind technology because it attempted to ensure the survival of the wind turbine when something went wrong—as it occasionally did.

As a part of its self-described role, the association regularly reported both on the performance of installed turbines and on their reliability. It also published an annual survey of owners' satisfaction with the manufacturer of their wind turbines in the association's in-house publication Naturlig Energi.

It was possible for a potential buyer to pick up a copy of the magazine and see exactly how much electricity a particular wind turbine at a specific site was producing; its monthly and annual generation; and what owners thought of the supplier. Such oversight placed constant pressure on manufacturers to deliver on their promises. Eventually, the owners' association compelled manufacturers to fulfill their commitments when the turbines were underperforming. Their magazine provided "market transparency" says German historian Heymann, enabling buyers to insist on reliable performance by selecting their turbines from vendors with the best records.

The exact same structure of the Danish owners' association was replicated in Germany through the Bundesverband WindEnergie (BWE). Though often translated as the German Wind Energy Association, it is more correctly described as the German Wind Turbine Owners' Association. And it has had much the same effect in Germany as the owners' association in Denmark by holding manufacturers accountable for their products.

There was nothing comparable in the United States or Canada. The industry was too fragmented, the suppliers too quarrelsome, the countries too vast for regular meetings of owners. Overall, the North American market was less critical and less well informed about the products it was being sold than that in Denmark.

There was also a penchant for keeping actual turbine performance confidential. US Windpower (USW), for one, elevated corporate secrecy to an art form. Technical questions that would elicit a prompt reply from Danish manufacturers were usually met with cold indifference by USW. Danish manufacturers often publicly provided information in their product literature that USW considered confidential. For example, Danish manufacturers typically published estimates of the annual generation they expected from their turbines in various wind regimes. USW, in contrast, went so far as to withhold data from the state of California on the projected production from its turbines even though required by law to do so.

Design Standards

Defenders of the US wind program and the wind turbine designs that resulted have argued that Danish wind turbines were as much a result of an "official" R&D program as were American designs. "Risø's been there all along," said one US program administrator. "Risø set design standards and they [the manufacturers] built clones."

Karnøe, the Danish researcher, and most other international observers disagree. The pioneers of the modern Danish wind industry were divorced from official research at Risø. They represented a "low-tech" approach dominated by entrepreneurs, machinists, and "hands-on" engineers. For Danish firms, there was no alternative but a bottom-up, incremental development of the technology. The Danish design—an upwind, fixed-pitch, three-blade rotor of heavy construction—resulted largely from a lack of adequate design tools. Early Danish designers knew how to design a turbine, according to Karnøe, even if they did not always know why the design worked.

Risø did not select the Juul approach as many American critics believed, says Karnøe. By 1979 all the principal Danish manufacturers were already using the "Danish" configuration when "system approval" from Risø was first required. It was a design they could manage with in-house skills. These heavy Danish machines were more "forgiving" than American designs and better matched the skills of their proponents.

The Test Station for Small Windmills was established at Risø in 1978—the same year the owners' association was created. Nevertheless, it was the owners' association that established the first design standard when the safety committee issued its guidelines in the spring of 1979. It was the owners' association that was pushing design standards. They demanded turbines that worked.

It was a result of this pressure from the owners' association that the Danish parliament, the Folketing, required that wind turbines had to be approved by Risø as meeting minimum requirements to qualify for the 30% government subsidy introduced that year. Principally, this was for a redundant, fail-safe braking system as demanded by the owners' association.

To qualify for approval, or "certified" as we now say, the turbines had to be tested at Risø. Significantly, the testing was free. This brought about close cooperation between Risø and the manufacturers that benefited the entire Danish industry, consumers and manufacturers alike.

Because the national laboratory at Risø was getting feedback from Danish turbines in the field, Danish designers and Risø researchers were less overconfident than their counterparts in Germany or the United States. Working with the wind was a humbling experience for everyone, including the Danes. But Danish engineers had to accept their limited financial and technical capabilities and work around them. Their solutions may not have been elegant, but overall, they worked—and worked much more reliably than the high-tech approach typical of their competitors from the aerospace community in the United States and Germany.

Risø implemented more comprehensive design standards only after a storm struck Denmark in 1981, laying waste to tens of turbines, particularly to S.J. Windpower's flimsy "wind rose." The following day Risø's Per Lundsager, who acknowledges that Danish wind turbines were "dreadnoughts of iron and steel," resolved to prevent further disasters and required all new turbines to meet minimum design specifications.

American manufacturers steadfastly fought design standards, objecting to the type of approval program popularized by Risø. Preposterously, Americans claimed that standards would lock in designs and their attendant defects. Karnøe argued that this was not the case, not even for Danish machines. It was the Danish market, through the actions of the Danish owners' association, and not Risø's certification program that forced standardization of design. Danish manufacturers had to sell "workable" designs, or the Danish owners' association would warn others to avoid them. This led manufacturers to stay with the Juul's approach or the "Danish concept," which they knew worked and which they knew could be sold to discerning Danish customers.

Beginning in the 1990s, the international market for wind turbines began demanding similar certification. By the mid-to-late 1990s, wind turbine customers as well as their insurance companies were demanding that all large wind turbines be certified to an international standard for much the same reason as the Danish owners' association had demanded nearly two decades before.

Certification for small wind turbines would not arrive in the United States until 2009—three decades after certification was first required in Denmark.

Beginning of the Modern Era

The history of wind energy is often portrayed as a continuous development from one technology to the next. In reality, the history of modern wind energy is one of fits and starts, technological dead ends, and the abrupt collapse of markets as the temporary abundance of fossil fuels, petroleum specifically, lures leaders with promises of inexpensive abundance.

Several times in the past, notably during both world wars and then again in the 1960s, wind was on the verge of breaking out of its niche to become a major source of electric power. Alas, it was not to be until the oil embargoes of the 1970s and the threat of shortages ever since, which have led to its enduring growth.

In the end, it was a group of students, their teachers, and the many volunteers at Tvind that created the initial blade design incorporating Hütter's flange. But it took entrepreneurial wind activists and Denmark's small manufacturers to commercialize the technology through standardized component parts, which led progressively to today's booming worldwide wind industry. The world owes these Danish pioneers a debt of gratitude for the birth of modern wind energy and the spark that started the renewable revolution now sweeping the globe.

Denmark decided not to develop nuclear power in 1985, a year before the world first heard of Chernobyl. Sweden closed the first reactor at Barsebäck in 1999. They closed the second reactor in 2005. Meanwhile, Tvind's turbine continues to spin, generating clean electricity.

Instead of nuclear, Denmark chose to develop wind energy. They've never looked back. In 2014, Denmark was generating nearly 40% of its electricity from wind energy. They fully expect wind energy to provide 50% of their electricity by 2020.

The rebels won.

5

What Works and What Doesn't

"Hey, that's a funny lookin' windmill you got there. What is it?"
"A VAWT."
"Don't get smart with me, son. I asked you a simple question."
"Actually, it's an articulating, straight-blade VAWT."
"What was that? I don't work for the government, you know."
"Some call it a giromill."
"Well, that's better. Why didn't you say so in the first place? For a moment there I thought you were speaking in tongues."

AS WIND TECHNOLOGY HAS GROWN, SO HAS its vocabulary. At times, it may seem as if the wind industry does speak in tongues. Nearly every conceivable wind turbine configuration has been tried at least once—most only once. Designs have run the gamut from the familiar farm windmill to vertical-axis contraptions such as the giromill. Despite the plethora of imaginative designs, only a few approaches have proven successful. These are what we now call "conventional wind turbines" and will be explained in this chapter. Darrieus and vertical-axis wind turbines and giromills will be dealt with in the next chapter. Ducted and "novel" designs will then follow in a third chapter on wind technology.

In this chapter we'll look at where the technology stands today and why certain designs have become commonplace. We'll also look at the important difference between wind machines that use drag to drive their rotors and those that use lift, the reason modern wind machines use only two or three blades, the materials that are used to make these blades, the kinds of controls used to protect the wind turbine, and the types of transmissions and generators now being used.

Again, the single-most important parameter describing a wind turbine—any wind turbine—is the area of the wind stream that the wind turbine intercepts (see Figure 5-1. Medium-size and large wind turbines). It doesn't matter what its orientation, the amount of energy a wind turbine captures is directly related to its swept area.

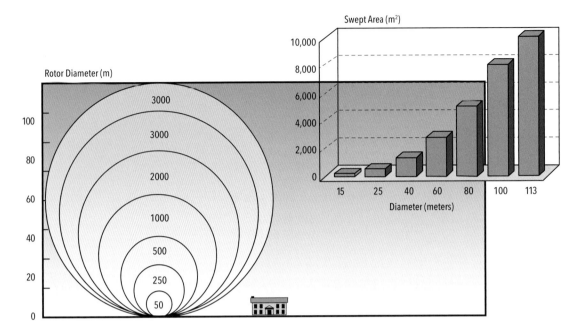

Figure 5-1. Medium-size and large wind turbines. The area swept by large wind turbines has increased steadily since the early 1980s. Today's wind turbines are 50 times larger than those of the early 1980s.

Orientation

There are two great classes of wind turbines: those whose rotors spin about a horizontal axis and those whose rotors spin about a vertical axis (see Figure 5-2. Rotor orientation). Conventional wind turbines, like the Dutch windmill found throughout northern Europe and the American farm windmill, spin about a horizontal axis. As the name implies, vertical-axis wind turbines (VAWTs) operate much like a top or a toy gyroscope. Vertical-axis or Darrieus wind turbines will be discussed in the next chapter. This chapter focuses on conventional wind turbines, but many of the design characteristics discussed apply to both conventional and vertical-axis wind turbines.

Conventional wind turbines must have some means for orienting the rotor with respect to the wind. Traditionally, the rotors of conventional wind turbines have been placed upwind of their towers and have incorporated some device for pointing the rotor into the wind.

With the Dutch windmill, for example, the miller had to constantly monitor the wind. When the wind changed direction, the miller laboriously pushed a long tail pole or turned a crank on the milling platform to face the windmill's massive rotor back into the wind. Later versions liberated millers from this toil by using fantails that mechanically turned the rotor toward the wind (see Figure 5-3. Fantail).

On smaller wind turbines, such as the farm windmill, the task is much easier and a simple tail vane will do. The tail vane keeps the rotor pointed into the wind regardless of changes in wind direction (see Figure 5-4 Tail vane).

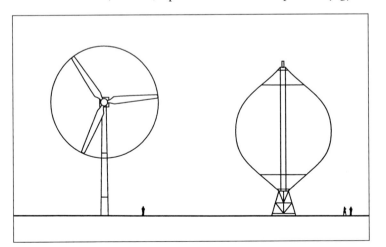

Figure 5-2. Rotor orientation. The rotor on horizontal-axis wind turbines (left) spins about a line parallel with the horizon. The rotor on vertical-axis or Darrieus wind turbines (right) spins about a vertical axis. Wind turbines of the size shown here are capable of generating 200 kW. (Pacific Gas and Electric Co.)

WHAT WORKS AND WHAT DOESN'T 85

Figure 5-3. Fantail. During the 18th and 19th century, European windmills incorporated fantails for automatically orienting the rotor toward the wind. This thatch-covered windmill is located near Pewsum in northwest Germany. Mid-1990s.

TAIL VANES

Tail vanes are deceptively simple. As the many designs on the market attest, as much art as engineering goes into the design of a tail vane for small wind turbines.

Scoraig Wind Electric's Hugh Piggott offers a rule of thumb: the tail vane's surface area should be greater than the square of the rotor diameter divided by 40. He also suggests that the tail boom should be similar in length to that of a rotor blade. The exception is for boating applications where shorter tail vanes are advisable. Experience has also shown that tail vanes with a pronounced vertical shape are more effective than those with horizontal shapes.

Micro turbines often use fixed tail vanes that are part of the generator body. Maritime versions typically have a hole in the tail that enables grappling with a pole hook when it's necessary to take the turbine out of the wind. Britain's National Engineering Laboratory found that the yaw systems on micro turbines, while simple, were overactive in turbulent winds and could cut performance.

Tail vanes for mini and household-size turbines are typically hinged about a near vertical axis. This allows the rotor to fold, or *furl*, toward the tail vane in high winds.

Though some small turbines integrate the tail vane with the nacelle, in most designs, the tail boom carries the tail vane well beyond the nacelle. Manufacturers use booms made from steel pipe, box beams (Bergey), or truss assemblies, or are cable-stayed, as in some Dutch designs.

Dutch, German, and French manufacturers prefer tail booms that elevate the tail vane well above the rotor axis. Others simply extend the tail vane horizontally.

Where furling is used, the furling hinge is sometimes placed at the end of the tail boom, but most designers place the hinge at the nacelle.

Although tail vanes are effective devices for passively controlling yaw, they become unwieldy as turbines increase in size. At the upper end of their range, tail vanes may require dampers or other devices to reduce the rate at which the wind turbine yaws into the wind after furling.

Although tail vanes have been used on American farm windmills up to 17 meters (56 feet) in diameter, they are typically limited to much smaller wind turbines.

Figure 5-4. Tail vane. Dutch LMW turbine uses a tail vane to direct the rotor into the wind at the Folkecenter for Renewable Energy's test field in Denmark. 1997.

Novel twin tail. University of New Castle's 5-kW experimental small wind turbine at Fort Scratchley on Australia's east coast. While unusual, the concept is not unique. During California's wind rush in the early 1980s, a San Diego company (Century) built an unsuccessful 40-foot (12-meter) diameter upwind turbine using dual tail vanes. 1997.

Passive Yaw

Both tail vanes and fantails use the force of the wind itself to orient the rotor upwind of the tower. They passively change the orientation, or yaw, of the wind turbine with respect to changes in wind direction without the use of human or electrical power.

The rotor may also be placed downwind of the tower. Downwind rotors don't necessarily need tail vanes or fantails. Instead, the blades can be swept slightly downwind, giving the spinning

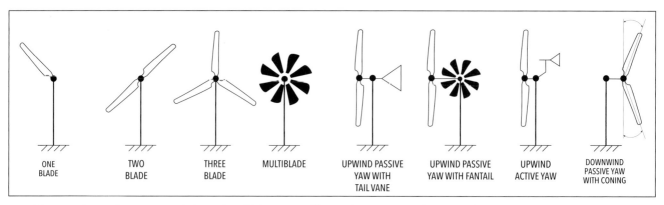

Figure 5-5. Horizontal axis configurations. Upwind, downwind, one blade or two, it's all been tried at one time or another. (Adapted from J. W. Twidell and A. D. Weir, *Renewable Energy Resources*, p. 211. Copyright 1986 by J. W. Twidell and A. D. Weir.)

rotor the shape of a shallow cone with its apex at the tower. This *coning* of the blades causes the rotor to inherently orient itself downwind (see Figure 5-5. Horizontal-axis configurations). On heavy blades the coning angle may be only 1 to 2 degrees, says British aerodynamicist Andrew Garrad, while on lightweight blades it can be as much as 8 to 10 degrees.

Downwind machines are certainly sleeker than wind turbines using either tail vanes or fantails (see Figure 5-6. Downwind rotor). In the 1970s and 1980s, some believed this gave downwind machines a more "modern" look, even though they had been used in the 1940s by Palmer Putnam and in the 1950s by Ulrich Hütter. But these designs pay a price, say proponents of upwind turbines. Downwind machines occasionally get caught upwind when winds are light and variable. Some downwind turbines would occasionally *hunt* the wind. In the early 1980s, Enertech and Storm Master turbines were notorious for "walking" around the tower (precessing) in search of the wind. Critics of downwind turbines add that the tower creates a *shadow* that disrupts the airflow as the blades pass behind the tower. This decreases performance, increases wear because of turbulence, and on two-blade designs emits a characteristic "whop-whop" sound that many find annoying.

One significant disadvantage of passive downwind machines is yaw control in high winds. A common method for protecting upwind turbines in high winds is to orient the rotor perpendicular to the wind. On small turbines with tail vanes, the rotor can be furled, taking the rotor out of the wind. Upwind turbines with active yaw controls can mechanically direct the rotor out of the wind. Unless a downwind machine also has an active yaw drive, the rotor will always—well, nearly always—remain downwind of the tower.

Occasionally a downwind rotor without active yaw control can get caught upwind of the tower. This can prove disastrous if the blades deflect and strike the tower. Storm Master and Carter turbines were plagued by this phenomenon in the early 1980s. Because of these and other problems, no medium-size or large wind turbines today use downwind rotors with passive yaw control. Some small and household-size wind turbines, such as Skystream and Gaia, use downwind rotors with passive yaw. The Endurance 3120, a medium-size turbine, also uses a passively yawed, downwind rotor, but these designs remain the exception. No large wind turbine today uses a downwind rotor.

Figure 5-6. Downwind rotor. The rotor on the Skystream is oriented downwind of the tower (the wind is from the left to the right). The Skystream uses a rotor 3.7 meters (12 feet) in diameter and was rated at 2.4 kW. The scimitar-like shape of the blades serves no functional purpose except to make the wind turbine appear different from competing wind turbines. 2008.

Active Yaw

All large wind turbines today use active yaw control to keep the rotor upwind of the tower. A wind vane mounted on top of the nacelle signals a hydraulic or electric motor to mechanically direct the rotor into the wind. As wind turbines have grown in size, so have the size, number, and sophistication of the yaw drives. Large wind turbines today may incorporate several yaw motors in the nacelle, as well as multiple yaw brakes to reduce the loads on the individual yaw pinions and the large diameter bull gears.

Lift and Drag

Lift is critical to understanding why modern wind turbines look the way they do. All modern medium-size and large wind turbines use aerodynamic lift to drive their blades through the air. These machines use only two or three slender blades to sweep the entire rotor disk, and they operate at several times the speed of the wind that propels them—key characteristics of lift devices. This lift is created by rotor blades that use airfoils or "wings," much like the wings of an airplane. Danes, in fact, call a wind turbine blade a *vinge,* or wing, for this reason. Airfoils typically have a very high lift-to-drag ratio, an important measure of aerodynamic performance.

Drag devices, on the other hand, are simple wind machines that use flat or cup-shaped blades to propel the rotor around its axis (see Figure 5-7. Panemone). The wind merely pushes on the cup or blade, dragging the blade downwind with it at no more than the speed of the wind itself.

Thus, there are inherent drawbacks to drag devices that limit their use for generating electricity. At best only 15% (4/27) of the power in the wind can be captured by such machines. In comparison, the maximum possible for a lift device is 59% (16/27, the Betz limit). Drag devices also require more materials than comparable wind machines using lift.

Backyard tinkerers often turn their attention first to drag devices because these machines are easier to understand and construct (see Figure 5-8. Drag device). The wind pushes on a big wide blade and it moves with the wind. What could be simpler? Though inventors constantly

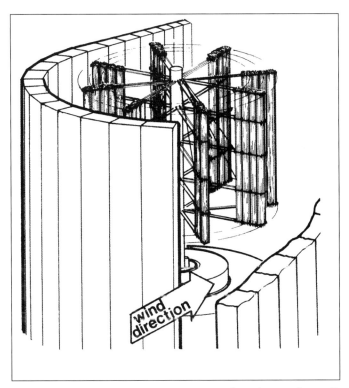

Figure 5-7 Panemone. Simple drag device used in ancient Persia for grinding grain. The vertically mounted blades were made by fastening bundles of reeds onto a wooden frame. The surrounding wall guides the prevailing wind onto the retreating blades. Many backyard inventors unfamiliar with aerodynamics construct similar ducted drag devices. (Sandia National Laboratory)

Figure 5-8. Drag device. Salesmen's model of a 1920s design using drag on the articulating flat panels to drive a water-pumping windmill. Though somewhat more sophisticated than a simple panemone, the aerodynamic performance of this approach is also limited. 1979.

create imaginative ways to use drag to power wind turbines, they always confront the physical limitations of drag propulsion: lift devices are nearly four times more efficient at capturing the energy in the wind, relative to the area of the wind intercepted by the rotor.

When examining the portion of the frontal area occupied by the wind turbine's blades, lift devices easily produce at least 50 times more

> LIFT DEVICES EASILY PRODUCE AT LEAST 50 TIMES MORE POWER PER UNIT OF BLADE AREA THAN THOSE DEPENDENT UPON DRAG.

power per unit of blade area than the same size turbine dependent upon drag, says Oregon State University's Robert Wilson, an authority on aerodynamics. The only way to improve the performance of drag devices is to incorporate some form of lift, as in some Savonius rotors, where there is recirculation of the airflow.

Confusingly, engineers occasionally refer to horizontal-axis wind turbines that use crude blades and operate at low rotor speeds as "drag devices." Examples are the farm windmill and the traditional European windmill. Technically, these machines are not true drag devices because they do use lift. However, drag on the blades as they move through the air is higher relative to the lift produced, the lift-to-drag ratio, than that on modern high-speed airfoils, thus, the description.

Lift has been a driving force in wind turbine development for the past 700 years, writes Robert Gasch, the author of the leading wind turbine engineering textbook in German. He emphasizes that long before there was any theoretical understanding of aerodynamics, builders of traditional European windmills knew what they needed to do to get the most from their machines. For example, British civil engineer John Smeaton in 1759 measured the coefficient performance of traditional smock windmills and found that they were nearly twice as efficient at C_p 0.28 as that of typical drag device of C_p 0.16.

This long history of successfully working with the wind is one reason why experts, such as Gasch, find it so frustrating when modern-day "inventors" overlook or ignore what has been learned at such great expense by others and then go on to reinvent—with much fanfare—simple, yet ineffective, drag devices.

Despite this knowledge, there have been various attempts down through the decades to adapt drag devices for use in wind energy. In 1924, Finnish inventor Sigurd Savonius developed an S-shaped vertical-axis wind machine. Though his was principally a drag device, Savonius improved the rotor's performance by recirculating some of

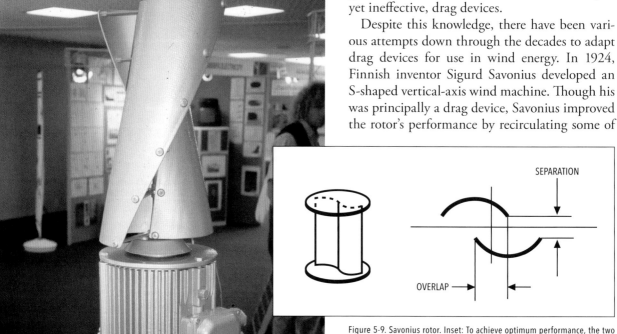

Figure 5-9. Savonius rotor. Inset: To achieve optimum performance, the two blades must be offset to permit some recirculation of flow. Because of its simplicity, a Savonius rotor is often the first choice of do-it-yourselfers and "inventors"; however, the design is material intensive and not as effective as a true lift device. Modern S rotor. Finns remain enamored with Savonius rotors. Windside Oy's S rotor on display at a conference in the late 1990s. Savonious rotors are best suited to very low power, battery-charging applications—and not much more.

Figure 5-10. Taper. Wind turbine blades taper from root to tip. The saber-like shape minimizes solidity but also enables strengthening the root where the blade attaches to the rotor hub. Shown here is a blade for a Siemens 2.6-MW turbine at the Whitelee Windfarm, Scotland. The blade is about 45 meters (150 feet) long. The wind farm near Glasgow is a popular tourist attraction. Note the large diameter Hütter flange at the near or root end. 2012.

the airflow between the rotor's two halves (see Figure 5-9. Savonius rotor). Air striking one blade is directed through the separation between the two halves of the S and onto the other blade. Researchers have measured conversion efficiencies of almost 30% under optimum conditions, considerably higher than those of typical drag devices. In practice, however, S-rotors extract less than 15% of the power in the wind. Because of their poor performance relative to conventional wind turbines, Savonius rotors have never found broad commercial application. Today they're only found sitting in experimenters' garages, powering buoys, or as architectural ornamentation.

Aerodynamics

Lift devices use airfoils with high lift-to-drag ratios to power the rotor. It seems mysterious that a wind machine with only a few blades can operate efficiently. But a modern wind turbine, because it uses lift, can capture the same amount of power with a smaller rotor, using fewer blades than a drag device. Modern wind turbines, those using lift, would make Buckminster Fuller proud: they "do more with less."

Why is this? Why do some wind machines, like the farm windmill, use multiple blades, while others use only a few? Why do some blades taper from the root to the tip, while others taper from the tip to the root? To understand the answers to these questions, we need to delve further into wind turbine aerodynamics.

Wringing the most energy from the wind striking a wind turbine rotor is exceedingly complex. Aerodynamicists Woody Stoddard and Dave Eggleston devoted an entire book, *Wind Turbine Engineering Design*, to the subject. Suffice it to say that modern wind turbine rotors use very few materials to capture the energy in the wind stream.

As contemporary wind energy pioneer Peter Musgrove has written, modern wind turbines present little solidity to the wind: the two or three slender blades occupy only 5% to 10% of the rotor disk. Power densities of 1,000 W/m^2 of blade area are typical. With such little material doing so much work, says Musgrove, the energy used to make the wind turbine is quickly recovered, often in less than one year.

The slender blades on these turbines typically taper and twist from hub to tip (see Figure 5-10. Taper). Why they do so is more difficult to explain.

It's intriguing that a sailboat can travel faster than the wind, and even more so when we learn that the boat sails faster across the wind than with the wind. Mariners discovered this fact centuries ago. Today we explain the paradox by speaking in terms of *lift* and *drag*. The blade of a modern wind turbine is much like the sail of a sailboat—both have good lift to drag, which allows them to travel faster than the wind.

To begin, let's look at the factors affecting the lift from an airfoil like that of a wind turbine blade. Air flowing over the blade causes both lift and drag. When you're driving down the highway and you stick your hand outside the window, lift from the air flowing over your hand (a crude airfoil shape) literally lifts your hand toward the roof. Drag pulls your hand toward the rear of the car. The sum of these two forces generates a thrust that pulls the blade on its journey through the air, much like it pulls a sailboat through the water. This thrust is greatest when the sailboat is sailing across the wind or a blade is slicing through the wind, as on a horizontal-axis wind turbine.

One measure engineers use to rate airfoil performance is the ratio of lift to drag. Designers of high-speed, high-performance wind turbines want a high lift-to-drag ratio. The lift-to-drag ratio is determined by the blade's angle of attack—the blade's angle with respect to the *apparent* wind, the blade's shape, and its aspect ratio.

One characteristic of high-performance airfoils is the shallow angle of attack at which they function best. Slight increases in the angle of attack, for example from 0 to 15 degrees, produce increasing amounts of lift. However, a point is reached where the flow over the blade separates from the airfoil and becomes turbulent. Lift then deteriorates rapidly and drag increases. At this critical juncture, the airfoil is said to *stall*.

Stall is a deadly condition in flight. Airplanes literally fall out of the sky when stall occurs, when there's no longer enough lift to support them. It's one of the leading causes of light plane accidents. In a wind machine, stall can be beneficial. We'll see why in a moment. But first consider the angle of attack: it's a function of the blade's angle to the plane of rotation—its pitch—and the apparent wind. For now, assume the pitch is fixed, which it is on many wind turbines within their operating range.

Apparent Wind and the Angle of Attack

The *apparent wind* is the wind seen by the blade. It is the result of the airflow due to a combination of the blade's own motion and the wind across the ground. If you recall some of your high school physics, you'll note that the resultant is dependent on the relative strength of each. For example, if both were equal in speed and if they were acting at right angles to each other, the apparent wind would be acting at a 45-degree angle between the two (see Figure 5-11. Apparent wind). On a sailboat under sail, the wind you feel on your face is the apparent wind.

If wind speed increases while the blade's speed through the air remains constant, the position of the apparent wind swings toward the wind direction because it has become more influential. As the apparent wind changes position, it also changes the angle of attack. Reversing the process by increasing blade speed relative to wind speed causes the apparent wind to shift toward the direction of the blade's motion, decreasing the angle of attack. Wind turbine designers must decide how best to deal with this relationship for each airfoil because there's an optimum angle of attack, a point where the lift-to-drag ratio is optimum and performance reaches a maximum.

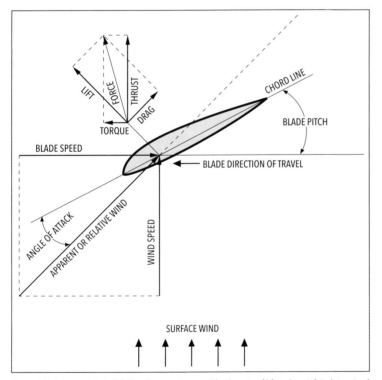

Figure 5-11. Apparent wind. Airfoil performance is gauged by the ratio of lift to drag. Lift is determined by the angle of attack. The pitch of the blade, the speed of the blade through the air, and the speed of the wind control the angle of attack and, consequently, lift.

To maintain an optimum angle of attack as wind speed increases, a fixed-pitch blade must increase its speed proportionally. Thus, to maximize aerodynamic performance, the rotor must spin faster as the wind speed increases. Another way to say it is that the tip-speed ratio—the relationship between blade speed (at the tip of the blade) and wind speed—must remain constant in order to maintain optimum aerodynamic performance. Most designers of small wind turbines try to operate their wind turbines in this manner.

Note that blade pitch on nearly all wind turbines is fixed throughout the turbines' operating range. Blades on variable-pitch turbines are no exception. Contrary to popular belief, these turbines pitch their blades only to start the rotor spinning after the rotor has been at rest and to reduce power when the turbines have reached the generator's rated capacity.

Operators of large wind turbines that use variable-pitch rotors have one significant advantage over users of small wind turbines. The pitch on these turbines can be adjusted to tailor the turbine's performance to the site, say a ridgetop in the Tehachapi Mountains. Today, this can be done by modifying the algorithm that the control system software uses to regulate performance.

On fixed-pitch turbines, the preferred pitch can be set when the blade is first attached to the rotor, or it can, with more difficulty, be adjusted seasonally. One Danish manufacturer modified the hub on its fixed-pitch rotors so that the pitch could be adjusted mechanically just a few degrees to compensate for changes in air pressure. Operators of variable-pitch turbines can change the pitch-control algorithm, letting the rotor's pitch control mechanism do the work.

In contrast, blade pitch is permanently fixed on most small wind turbines. Seldom, for example, can the blade pitch of a small wind turbine rotor be adjusted to compensate for lower air density at higher elevations. The rotor must be designed to work as effectively as possible in a wide range of conditions.

In the past, many medium-size wind turbines operated at a constant rotor speed. As wind speed changed, rotors on these turbines continued to spin at the same speed because of the induction generators they used. Consequently, their rotors operated at varying tip-speed ratios. Designers were willing to sacrifice some performance for the simplicity of fixed-pitch blades driving constant-speed generators. On these stall-regulated turbines, the airfoil began to stall and performance eroded as wind speed increased. This reduced the rotor's power in high winds, making it easier for designers to build protective controls to keep the rotor from destroying itself. On pitch-regulated turbines, the blades change pitch in high winds, dumping excess power, and can ultimately be feathered to take the turbine out of operation.

The amount of thrust driving the rotor is not only a function of the airfoil's lift coefficient, which depends on the blade's angle of attack, but also the area of the blade and its speed through the air.

Twist and Taper

For the sake of simplification, we've been looking at a blade as if the conditions it sees were constant along its entire length. That may be true for airplanes, but it's not so for wind turbines. Even when the pitch of the blade is fixed and rotor speed is constant, the speed through the air of a point on the blade changes with its distance from the hub. On a conventional wind turbine, blade speed is higher at the tip than near the hub because it has more distance to cover in the same amount of time.

Because blade speed increases with distance from the hub, the apparent wind varies as well. The apparent wind increases in strength, and its position shifts toward the plane of rotation as you move out along the blade toward the tip. If the blade designer wants to maintain the angle of attack (to optimize performance) at the same time blade speed is increasing, the angle of attack must decrease toward the tip. As a result, wind turbine blades are twisted from root to tip (Figure 5-12. Twist). The twist reaches a maximum at the root and a minimum at the tip. Glance up at the next wind turbine you see and note that the tip of the blade is parallel with its direction of travel. The blades on the famous mill at Gedser, the distant forerunner of modern Danish turbines, had a twist of 13 degrees at the hub and 0 degree at the tip.

Figure 5-12. Twist. For most wind turbine airfoils, blade angle of attack varies from root to tip. But there's nothing new under the sun. Here the jalousie latticework of Germania Molen's blades twist about the stock or blade spar. Germania Molen, despite its name, is one of the Netherlands nearly 1,000 functioning windmills. Note the airfoil-shaped leading edge. Jalousie slats are controlled by the spider linkage protruding from the main shaft. Yes, that's thatch covering the cap and smock. Inset: Blade taper and twist from root to tip. Note that the blade is much thicker near the hub than the tip. This allows the use of more strengthening materials where the loads are greatest. The blade is about 19 meters (62 feet) long.

Figure 5-13. Multiblade farm windmill. The characteristic curved metal blades of the American farm windmill resulted from the pioneering work of Thomas Perry. The Aermotor brand windmill embodied the principles Perry learned after 5,000 tests on various "wheel" (rotor) designs in the late 19th century. Rotors using Perry's curved blades were almost twice as efficient as those using the wood slats then common. His Aermotor also included back gearing, which allowed the wheel to make several revolutions for each stroke. Perry's design has since been widely copied. Note the work platform for servicing the rotor. This is a common feature of farm windmill towers.

Solidity

Wind machine designers long ago learned that blade area (the number of blades, as well as their length and width) governs the amount of torque, or turning force, that a rotor can produce. The more blades or blade area of a given length for the wind to act on, the more torque it will produce. Greater solidity, the ratio of blade area to the area swept by the rotor, generates greater torque.

The American farm windmill uses multiple "sails," which nearly cover the entire rotor disk (80% solidity). It was designed to deliver high torque for pumping water in the low winds of late summer, and it does its job remarkably well (see Figure 5-13. Multiblade farm windmill).

There's no better example of a high-solidity rotor fitting this description than the American farm windmill. It uses multiple sails that taper from the tip to the root. High torque assures you that the windmill would be able to lift the pump's piston (and the water with it) during late summer when the need is greatest but the winds are lightest.

Yet the demands on electricity-generating wind turbines are different. Wind turbines are not necessary to generate electricity. We are not dependent upon them as the Amish are for water from their farm windmills. For us, there are many other sources of electricity: photovoltaics, fossil fuels, nuclear power, hydro, and so on.

We want power from a source and are willing to pay for it, but only if it's superior in some way to other sources: it's cheaper and cleaner or it provides some combination of benefits. To compete with these technologies, the wind machine must be designed to extract power from the wind in the most cost-effective and environmentally desirable manner possible. The farm windmill is too material-intensive for this task, even though its performance has greatly improved over the past century.

Early farm windmills, for example, used flat wooden slats for blades. In 1888 Aermotor introduced its "mathematical" windmill, which substituted sheet-metal blades for those of wood. Aermotor stamped a broad curve into the metal blade and in doing so directed the air to flow over the backside of the following blade. Much like the slot effect in jib-rigged sailboats, this cascading of air over the following blades improved Aermotor's performance over that of its rivals. Unfortunately, the "new and improved" farm windmills—all of which now use Aermotor's technique—still extract only 15% of the power in the wind.

Intuitively, the multiblade farm windmill looks like it would capture more wind than a modern machine with only two or three blades. We sense that the rotor should have more blades to capture more wind. However, consider what would happen if we carried this belief to its logical extreme. The optimum rotor would cover the entire swept area with blades, in effect producing a solid disk. No air would pass through. The wind would pile up in front of the rotor and flow around rather than through it. The wind speed behind the rotor would be zero. Instead of capturing more wind, we wouldn't capture any. There must be some air moving through the disk, and it must retain enough kinetic energy so it can keep moving to make way for the air behind.

Betz's Limit
We must strike a balance between a rotor that completely stops the wind and one that allows the wind to pass through unimpeded, between the amount of wind striking the rotor and the amount flowing through. German physicist Albert Betz demonstrated mathematically that this optimum is reached when the rotor reduces wind speed by one-third. By conserving the wind's momentum as it passes through the rotor, Betz calculated that the maximum a theoretical wind turbine could capture was 16/27 or 59.3% of the energy in the wind. "You can't beat Betz," says Julian Feuchtwang of the Centre for Renewable Energy Systems Technology in Great Britain, though many have tried.

Real rotors, says the Alternative Energy Institute's Vaughn Nelson, never achieve the Betz limit because of losses due to aerodynamic drag, losses around the blade tip, and losses due to rotation in the wake behind the rotor. As Nelson explains it, the wake spins opposite to the direction of the rotor as a result of conservation of angular momentum. The wind acts on the rotor and in doing so spins the rotor. In the process, the wind itself is deflected and spirals downstream. To optimize rotor performance—to approach Betz's theoretical limit as nearly as possible—designers must opt for high rotor speeds and low torque because high torque increases losses in the wake.

But, you may ask, if we were to lower the rotor's torque, wouldn't we be lowering the power it can produce, even if it's going to be more efficient at producing it? Yes, if we kept everything else the same. We don't. Power is a product of torque and rotor speed. To deliver the same amount of power as we decrease torque, we must increase rotor speed.

This strategy works up to a point, says Scoraig Wind Electric's Hugh Piggott. In high-performance rotors, there are other important losses besides those caused by rotation of the wake, principally drag on the fast-moving blades. There are also tip losses from increased pressure around the end of the blade on rotors using only a few slender airfoils. This causes more air to flow around rather than over the blade—one reason for tip vanes on large aircraft and the appearance of tip vanes on some modern wind turbines.

Tip-Speed Ratio
Blade speed is a function of rotor diameter and rotor speed. Both are described by a single term, the ratio between the speed of the blade through the air at the tip and the wind speed: the tip-speed ratio. Tip speed increases either as rotor speed increases or as the length of the blade increases for a given rpm. Modern, high-performance wind turbines operate at tip-speed ratios of four or more. In contrast, true drag devices operate at tip-speed ratios of less than unity.

Though efficiency improves with increasing rotor speed, there are practical limits. Increasing drag, which reduces the airfoils effectiveness, is one. For small wind turbines, Wisconsin's Mick Sagrillo argues that durability is inversely proportional to tip-speed ratio. That is, small turbines that operate at high tip-speed ratios may wear out faster than those that operate at lower tip-speed ratios.

Scoraig Wind Electric's Piggott suggests that a tip-speed ratio of 5 is aerodynamically opti-

WHAT WORKS AND WHAT DOESN'T

Table 5-1. Tip Speeds and Tip-Speed Ratios of Selected Wind Turbines						
		Dia.		Rated Wind Speed	Tip Speed	Tip Speed
	kW	m	rpm	m/s	m/s	Ratio
Historical						
Farm Windmill	0.5	3.0	78	6.7	12	1.9
Dutch Windmill	25–30	25	25	12	33	2.7
Darrieus HAWT 1927	1.8	8	80	5	34	6.7
Micro Turbines						
Air Breeze	0.2	1.2	1500	10	92	9.2
Household-Size Turbines						
HR3	3	5	308	10.5	81	7.7
Bergey	6	6.2	240	11.9	78	6.5
Bergey	10	7	220	11.6	81	7.0
Gaia	11	13	56	10.5	38	3.6
Medium Turbines						
Endurance	50	19.2	42	9.5	42	4.4
Northern Power 100	100	21	60	14.5	66	4.5
Large Turbines						
Enercon	3,000	101	14.5	12	77	6.4
Siemens	2,300	108	16	12	90	7.5
Vestas	3,000	112	12.8	12	75	6.3
Monopteros 50	640	56	43	11	125	11

At power and rpm noted. Monopteros 50 was an experimental one-blade turbine. Used here for comparison.

mum for small wind turbines with slightly higher tip-speed ratios on rotors used in combination with direct-drive alternators. Large turbines operate at tip-speed ratios between 6:1 to 8:1 at rated power (see Table 5-1. Tip Speeds and Tip-Speed Ratios of Selected Wind Turbines).

Noise is another practical limit to higher tip-speed ratios. Noise is proportional to the speed of the blade tip. At higher tip speeds, vortices are shed from the tip, and it's these vortices that cause the swishing sound as the blades move through the air. For example, the tip speed of large wind turbines, at their rated power, can be from 80 m/s to about 90 m/s (180–200 mph). Higher tip speeds may increase blade noise. Some experimental wind turbines, such as Monopteros 50, operated at unusually high tip-speed ratios near 11:1 and were notoriously noisy as a result.

Blade Number

Wind turbines need only one slender blade to capture the energy in the wind (see Figure 5-14. One blade). To sweep the rotor disk effectively, a one-blade turbine must operate at higher rotor speeds than a two-blade turbine, thereby reducing the gear ratio required for the transmission, and hence the mass and cost of the gearbox. Since one blade should cost less than two or three, proponents argue that one slender blade delivers optimum engineering economy.

The giant German conglomerate Messerschmitt-Bölkow-Blohm built just such a series of one-blade wind turbines. But there are other, equally important design criteria besides lowest initial cost. Two blades are often used for reasons of static balance. Many modern wind turbines use three blades because they give greater dynamic stability than either two blades or one.

Rotors using two and three blades are also more efficient than rotors using only one due to aerodynamic losses at the tip of the blade. British engineer John Armstrong notes that one blade captures 10% less energy than two blades, everything else being equal. And though one blade may be cheaper, engineers say the rotor it is attached to is just as heavy as one with two

ONE-BLADE WIND TURBINES

Although never commercially successful, optimal material economy has for decades been a siren's song luring designers onto the rocks of flimsy, one-blade designs. German engineers have been particularly susceptible because of Ulrich Hütter's 1940s doctoral thesis on the design of inexpensive, high-performance wind turbines. Hütter himself built a one-blade wind turbine prototype in the late 1940s.

Hütter taught for many years at the University of Stuttgart's Institute of Aircraft Design, alongside Professor Franz Wortman, himself well known for airfoil sections at the Institute of Aerodynamics. Wortman and his students pursued Hütter's minimalist design philosophy to its logical conclusion: the one-blade FLAIR or Flexible Autonomous 1-Bladed Rotor.

In the mid-1980s, Wortman installed a FLAIR prototype 8 meters (26 feet) in diameter at the University's test field near Schnittlingen, in southern Germany's Swabian Alps. The novel downwind rotor drove a 5.1-kW, four-pole induction generator—it too was out of the ordinary. The generator on the grid-connected version incorporated high slip (14%), enabling the turbine to regulate rotor speed within a vary narrow range by using its Watt-governor.

Originally developed for the German washing machine company Böwe, FLAIR was subsequently sold to aerospace giant Messerschmitt-Bölkow-Blohm (MBB). From it, MBB developed the Monopteros 15 series: turbines with rotors from 12.5 to 17 meters (40–55 feet) in diameter, driving generators from 15 kW to 30 kW. MBB also introduced a greatly scaled-up version of Wortman's FLAIR: the Monopteros 50 series with rotor diameters from 47 to 56 meters (150–180 feet), rated from 550 kW to 1 megawatt. MBB's work on the turbine ceased in 1986 after Wortmann's death.

Independently of Wortman's group, Riva Calzoni developed its own one-blade design. An Italian heavy-engineering company, Riva Calzoni built 25 of its MP5 and another 25 of its MP7 models by 1992, when it abandoned the small turbine market.

Both MBB and Riva Calzoni found that it was more profitable to build larger turbines for commercial wind farms than small turbines for rural residences. At one time, MBB envisioned building 5-MW versions, dubbed GROWIAN II (Grosse Wind Energie Anlage), along the German coast. But MBB completed only three 640-kW Monopteros 50 models near Wilhelmshaven before abandoning the program. Of the two firms, Riva Calzoni was the most successful, eventually installing more than 100 of its 300-kW version in Italy before discarding the one-blade approach entirely.

Despite these dismal results, one-blade turbines continue to reappear. In the early 2000s, a small German firm introduced a one-blade design with little success. And in 2013 a New Zealand start-up, ThinAir, was trying its hand at bringing a one-blade turbine to market.

Monopteros 20. Smallest entrant in Messerschmitt-Bölkow-Blohm's line of one-blade turbines was a direct descendent of Professor Franz Wortman's FLAIR. Circa 1986. (Messerschmitt-Bölkow-Blohm)

Riva Calzoni MP5. This MP5 was hardly out of production before it was displayed as a novelty in Milan's Museum of Science and Industry. Note counterbalance and pitch weight. 1998. (Nancy Nies)

blades. First, the blade on a one-blade turbine must be stronger than a comparable blade on a two-blade turbine because it must capture twice as much energy in the wind. Second, a one-blade rotor must compensate for the weight of the missing blade by using a massive counterweight. Because of its higher speed and greater aerodynamic loading, one blade will also emit more noise than two.

Ultimately, one-blade rotors provide no cost savings. Some manufacturers claim they can build three simple blades for the cost of one single high-performance blade and the sophisticated teetering hub required for a one-blade rotor.

The advantages of rotors with two blades over those with three are similar to those of rotors with one blade over those with two (see Figure 5-15. Two blades). The rotor is cheaper, lighter, and

Figure 5-14. One blade. Riva Calzoni 33-meter (110-foot) diameter, 300-kW, one-blade wind turbine in the central Apennines in the late 1990s. Nearly 100 of these wind turbines were installed in Italy before Riva Calzoni ceased manufacturing the turbines.

Figure 5-15. Two blades. Wind turbines sporting two blades were much more common in the 1980s and even into the 1990s than they are now. Nedwind 500 kW, 43-meter (140-foot) diameter turbine. Wind pioneer Herman Drees operates 33 of the turbines in the San Gorgonio Pass near Palm Springs, California. Inset: Wind Energy Group's MS-2 at a site in Wales in the early 1990s. Geoff Henderson adapted this British design for his Windflow 500 kW model in New Zealand. Windflow's 33-meter (110-foot) diameter turbine uses a sophisticated teetered hub to relieve dynamic loads. There were 100 such turbines operating worldwide by late 2013. Both turbines use full-span pitch control.

ideally operates at higher speeds, leading to lower-cost transmissions. Two blades are easier to install than three because they can be bolted to the hub on the ground, in the position they will assume on the assembled turbine. The disadvantages of two blades are similar as well. Because of their higher speeds and greater rotor loading, they are often noisier than three blades.

During the 1970s and 1980s, two-blade designs were quite common. At one time there were more than 1,200 two-blade turbines installed in California. Most of the designs grew out of the American aerospace industry and proved problematic in the field. Most if not all of these turbines have been removed for scrap and replaced with larger, more modern and, most important, more reliable three-blade turbines. In the 1990s, a few manufacturers, Nedwind and the Wind Energy Group (WEG), introduced medium-size two-blade turbines in the 500-kW class. In New Zealand, Windflow adapted WEG's turbine for domestic manufacturing. Today, there is a cluster of Nedwinds operating in the San Gorgonio Pass and about 100 Windflow turbines operating in New Zealand. There are probably no more than 250 medium-size, two-blade wind turbines operating worldwide, or no more than 0.1% of all wind turbines globally.

Attempts to develop large, two-blade turbines in the megawatt class have all failed commercially. In the United States, Wind Turbine Company never made it beyond the prototype stage. Nordic Windpower installed a few prototypes of its 1-MW turbine in Sweden and a handful in the United States before collapsing. Only New Zealand's Windflow continues to market a medium-size, two-blade turbine.

Conventional wisdom holds that three-blade machines will deliver more energy and operate more smoothly than either one- or two-blade

Figure 5-16. Three blades. With a few notable exceptions, nearly all wind turbines today use three blades. Inset: Most, if not all, micro turbines, such as this Air Breeze, use three blades. The 200-watt turbine uses a rotor only 1 meter (3.2 feet) in diameter. Nearly all of the world's commercial wind turbines, such as these 3-MW Vestas turbines in the Tehachapi Pass, use three blades. 2013.

Figure 5-17. Multiple blades. Although most wind turbines use two or three slender blades, the ideal rotor would use an infinite number of infinitely thin blades. In the foreground is Windmission's Windflower multiblade rotor at the Folkecenter for Renewable Energy's test field in Denmark. 1998.

turbines (see Figure 5-16. Three blades). They will also incur higher blade and transmission costs as a result of their lower rotor speed. British engineer Armstrong estimates that rotors with three blades can capture 5% more energy than two-blade turbines, while encountering fewer cyclical loads than either one- or two-blade turbines when reorienting the nacelle to changes in wind direction.

The dynamic or gyroscopic imbalance of two-blade turbines with changes in wind direction is a classic engineering challenge. The phenomenon can best be seen in the jerky or wobbly motion of small, two-blade turbines as they yaw with the wind. When the rotor is vertical, there is little resistance to yaw, but as the rotor nears the horizontal, the inertia retarding the rotor from reorientating itself reaches a maximum. This occurs twice every revolution. This causes the turbine to yaw unevenly. On larger machines this effect is dampened with shock absorbers or by allowing the rotor to teeter as on the Windflow design.

Don't be misled by glib talk of a "new" wind turbine that runs in low winds. Anybody can design a rotor to spin in light winds. BUT WHY BOTHER?

Three blades minimize this dynamic problem and are preferred on small machines where yaw dampening or teetering hubs would be too costly. For this reason, Nolan Clark, who for several decades managed the US Department of Agriculture's wind experiments, prefers three blades to two on small turbines.

According to California aerodynamicist Kevin Jackson, a two-blade rigid rotor encounters 10 times the force that a three-blade rotor sees. "This is why two-blade rotors are always teetered" on medium-size wind turbines, explains Jackson. With teetering, the two-blade rotor experiences even fewer cyclic loads than the common three-blade design.

Aerodynamics resembles other branches of engineering where there are always trade-offs. The theoretically ideal rotor, says turbine designer Hugh Piggott, would use an infinite number of infinitely slender blades (see Figure 5-17. Multiple blades). Such a design is impractical.

A less technical reason for preferring three blades is the perception that one- and two-blade turbines are less aesthetically pleasing than those with three.

Self-Starting

Low solidity reduces a rotor's ability to start spinning on its own. Remember that the apparent wind flowing over the blade is partly due to the blade's motion. When the rotor is stopped, the lift on the blades from the wind alone may not be enough to start the rotor into motion. One solution for rotors using fixed-pitch blades is to motor the rotor up to a speed where the aerodynamic forces are sufficient for the blades to drive the rotor.

Many high-performance American designs of the 1970s and 1980s, such as Enertech and ESI, required motoring the rotor up to speed. Similarly, Endurance's 3120, a contemporary 19.2-meter (60-foot) diameter, 50-kW downwind turbine, uses its induction generator to start the rotor spinning in low winds.

Many designers are willing to sacrifice some performance to gain a self-starting capability. With only a few exceptions, nearly all wind turbines today are self-starting. Because large wind turbines all now use full-span pitch control, they can set the pitch to aid self-starting and then, once the rotor is spinning, adjust the pitch to the running position.

Wind turbines typically start spinning at wind speeds from 4 m/s to 5 m/s (8–10 mph). Don't be misled by glib talk of a "new" wind turbine that runs in low winds. Anybody can design a rotor to spin in light winds. But why bother? Because winds below the start-up speed of conventional turbines have so little energy, there's no economic justification for making the effort. The rotor may spin, but it won't produce enough electricity to make it worthwhile.

Blade Materials

Blades are one of the most critical, and visible, components of a wind turbine. Blades can be made from almost any material—and have been. European windmills used blades of wood and canvas, and this tradition survived well into the

TIP VANES AND WINGLETS

When commercial aircraft began sporting winglets to reduce tip vortices, the use of tip vanes at the ends of wind turbine blades came to the fore. Wind turbine engineers freely admit that a portion of the air's flow over the blade, like that over an aircraft wing, is lost at the tip as the wind escapes around the end of the blade.

Eliminating the lost lift by using a tip vane or winglet is nothing new. Aerovironment, the firm that built the Gossamer Condor and other aviation marvels, studied the question in the early 1980s. Due to the difficulties of constructing a tip vane, Aerovironment concluded that it was cheaper to simply extend the length of a conventional wind turbine blade than to add a winglet.

In the early 1990s, designer Henrik Stiesdal employed what he termed a tip torpedo. Instead of boosting performance, his intent was to reduce vorticity-induced noise at the blade tip. By the mid-2000s, winglets had become a common sight on Enercon blades.

Winglet. Blade tip with winglet and protective cover for transport on Enercon E-126, a 6-MW wind turbine with a rotor diameter of 126 meters (410 feet). The winglet increases performance and reduces noise. 2009. (Wikipedia Commons)

19th century. In the 1970s, researchers at Princeton University adapted the technology to build an experimental sail-wing turbine. Philosophy professor Gordon (Corky) Brittan used cloth sails on his Montana Windjammer, an ungainly turbine patterned after a Cretan windmill (see Figure 5-18. Cloth blades).

Wood has always been popular. Early farm windmills used wooden slats, and wind chargers of the 1930s used wood almost exclusively (see Figure 5-19. Wood blades). Wood is still used on small wind machines. It's strong, readily available, easy to work with, relatively inexpensive, and has good fatigue characteristics. "Wood flexes for a living," explains Mick Sagrillo, who

Figure 5-18. Cloth blades. Partially reefed canvas sails on a Cretan sail windmill at the Internationales Muhlenmuseum, Gifhorn, Germany. Reefing allows the miller to control power in the rotor during high winds. Note the prow and stays, a distinguishing feature of Cretan sail windmills. Early Danish wind turbines, such as the mill at Gedser, adapted these features to a rotor using metal clad blades.

Figure 5-19. Wood blades. Mid-19th century farm windmills used wood slats as blades. The slats were cheap, easily made, and easy to replace, though not aerodynamically efficient. One of several historical farm windmills at Nebraska's Windmill State Recreation Area. The rotor on this windmill is furled toward the tail vane, taking it out of the wind.

ran a small turbine repair shop in Wisconsin for many years. "It works well in high-fatigue applications."

Wood

Wood blades for small turbines are built either from single planks of Sitka spruce or from wood laminates. The blades are then machined into the desired shape and coated with a tough weather-resistant finish. The manufacturer then covers the leading edge with polyurethane tape to protect the blades from wind erosion and hail damage. This tape is the same as that used on the leading edges of helicopter blades. It's resistant to ultraviolet light and abrasion. Leading-edge tape is critically important on wood or wood-composite blades. "Without it, you just have driftwood after a few years," says Sagrillo.

Few of those new to wind energy appreciate the wind's erosive force. If you need to be convinced of this, pay a visit to the Texas Panhandle or the Tehachapi Pass during the spring wind season. But don't forget to take your goggles. Sand and blowing grit scour anything in their path. This airborne sandpaper has deeply etched the galvanizing on the windward side of towers in the Tehachapi Pass. In areas prone to blowing sand, wind turbine blades have seen their leading edges damaged after only two years of use.

Though solid wood planks work well for small machines up to 5 meters (16 feet) in diameter, manufacturers prefer laminated wood (like that of a butcher's block) for larger turbines. Laminated wood offers better control over the blade's strength and stiffness, as well as limiting shrinkage and warpage.

In the laminating process, slabs of wood are bonded together with a resin. The resulting block can then be carved into the desired shape. By varying the types of wood, the direction of their grains, and the resin, a material can be produced that is stronger than any one part alone and stronger than a single plank of the same size. Laminated wood blades have been used on small wind turbines of all sizes.

Thinner slices of wood are also used to produce veneers. Layer upon layer of razor-thin slices are sandwiched together with a resin and molded into the airfoil shape. The process is widely used to build the hulls of high-performance sailboats and has been adapted successfully for wind turbine blades both in the United States and Europe.

Wood-composite blades fabricated by Michigan's Gougeon Brothers earned a reputation for strength and reliability in wind turbines up to 43 meters (142 feet) in diameter during the 1980s. Similarly, NEG-Micon used wood composites from its British affiliate on its 1.5-MW turbine. Windflow also uses wood composites on its two-blade, 500-kW turbines.

Metal

In the late nineteenth century, farm windmill manufacturers began replacing wooden blades with stamped, galvanized steel. Thin steel sheets have been used ever since. Steel is strong and well understood. That's why steel was chosen by Boeing engineers for the giant 91-meter (300-foot) diameter Mod-2 and the subsequent 98-meter (320-foot) diameter Mod 5B. The blades were constructed from structural steel—nothing fancy, the same steel used in bridges (see Figure 5-20. Mod-2's steel blades).

But steel is heavy. The hub, drivetrain, and tower must be more massive than on a wind machine with a lighter rotor. Both Boeing and Dutch manufacturers encountered numerous problems with steel rotors and the shafts supporting them.

Figure 5-20. Mod-2's steel blades. The Mod-2 (background) used steel blades welded together in the field. The Mod-2, part of the US DOE's demonstration program, used a two-blade, teetered rotor 90 meters (300 feet) in diameter to drive a 2.5-MW generator. Note the variable pitch blade tips. Foreground Enertech 1500 with tip brakes. Early 1980s, Goldendale, Washington.

Figure 5-21. Aluminum blades. 14-foot (4.3-m) diameter Wincharger on Mormon Row, Grand Teton National Park. Extruded aluminum blades were a unique innovation on this wind turbine when it was introduced in the 1950s. Note work platform. 1997.

Aluminum is lighter and, for its weight, stronger. For this reason, aluminum is used extensively in the aircraft industry. We can fabricate aluminum blades with the same techniques used to build the wings of airplanes: form a rib and then stretch the aluminum skin over it. The blades on NASA's early Mod-OA were built in this way. On smaller machines, a simpler method can be used by stamping a curve into the leading edge, folding the sheet metal over the spars, and then riveting it in place.

Aluminum can also be extruded, eliminating several fabrication steps. Several manufacturers once thought that blades could be mass-produced by extruding blades in the same way we manufacture rain gutters, drain spouts, window moldings, and ladder rails—by squeezing a hot piece of aluminum through a die.

Alcoa and Canadian fabricator DAF-Indal developed extruded aluminum blades for Darrieus turbines. They believed that the Darrieus rotor was ideally suited for their aluminum extrusions because the inertial forces on Darrieus blades are in tension. The blades endure less stress in the Darrieus rotor than they would in a conventional wind machine. They can also use a blade of a constant chord or width, such as those produced by extrusion.

Aluminum, unfortunately, has two weaknesses: it's expensive and it's subject to metal fatigue. Have you ever taken a piece of wire and broken it by flexing it back and forth a few times? That's metal fatigue, and it works the same way in the wing of an airplane or the blade of a wind turbine. Aluminum is a good material—when used within its limits. But on wind turbines, aluminum hasn't been successful.

Most of the problems the hundreds of Darrieus turbines once operating in California encountered were due to metal fatigue where individual blade sections were joined together. The turbines were finally removed in the 1990s and sold for

NO MANUFACTURER TODAY BUILDS WIND TURBINES WITH METAL BLADES.

scrap. Imagine: an aluminum beer can on the shelf in your grocery could have once been on a Darrieus wind turbine in California.

The only remotely successful use of extruded aluminum blades was for home light plants during the 1950s. At the time, Wincharger switched from wooden blades to extruded aluminum. Some Winchargers can still be found with their blades intact (see Figure 5-21. Aluminum blades). How well they performed or how durable they proved to be is unknown.

Metal blades, whether steel or aluminum, may also cause television and radio interference. Metal reflects television signals, and this can cause "ghost" images on nearby TV sets. This has proven to be far less of a problem than first thought, even among the 500 Darrieus turbines that once operated in California. No manufacturer today builds wind turbines with metal blades.

Fiberglass

Fiberglass (glass-reinforced polyester, or GRP to Europeans) or related plastic composites now dominate blade construction (see Figure 5-22. Blade cross section). Like wood, fiberglass is strong, relatively inexpensive, and has good fatigue characteristics. It also lends itself to a variety of designs and manufacturing processes.

Fiberglass can be pultruded, for example. Instead of material being pushed through a die, as in extrusion, fiberglass cloth (like the cloth used in fiberglass auto body repair kits) is pulled through a vat of resin and then through a die. Pultrusion produces the side rails for fiberglass ladders and other consumer products. As with aluminum extrusions, pultruded fiberglass blades are recognizable by their constant chord (width). Bergey Windpower has used fiberglass pultrusions exclusively since abandoning aluminum blades in the early 1980s.

For sailors, fiberglass has become the material of choice. The same techniques used to build fiberglass boats have been successfully adapted by Danish, Dutch, and American companies to assemble wind turbine blades. These manufacturers place layer after layer of fiberglass cloth in half-shell molds of the blades. As they add each additional layer, they coat the cloth with polyester or epoxy resin. When the shells are complete, they are literally glued together to form the complete blade. Blades for nearly all medium-size and large wind turbines incorporate some form of this process.

Filament winding has also been used to

Figure 5-23. Carbon-fiber blade. Handle with care. The trailing edge of Southwest Windpower's Air 403 blade was so sharp, the blade was shipped with a warning label. Note the shark-fin-like planform derived from the Glauert formula for high-performance airfoils. 1998.

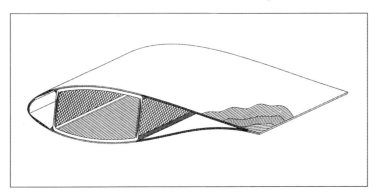

Figure 5-22. Blade cross section. Construction of a fiberglass blade found on many medium-size wind turbines. The central section is the spar, which provides the blade's principal structural support.

Figure 5-24. Blade attachment. Blades can be attached to the hub with stays, or cantilevered (attached at only one point). Inset: Early Danish wind machines used rotors braced with stays. (Danish Ministry of Energy) Most contemporary designs use cantilevered blades, though a few small wind turbines, such as this EasyWind, use cable-stayed rotors. 2012.

produce spars, the main structural members supporting some wind turbine blades. Fiberglass strands are pulled through a vat of resin and wound around a mandrel. The mandrel can be a simple shape like a tube or box beam, or it can be a more complex shape like that of an airfoil. Originally developed for spinning missile cases, filament winding delivers high strength and flexibility. Though some blades have been made entirely from filament winding, the process is often used only for the blade spar. The blade is then assembled in a mold with a smooth fiberglass shell.

Small wind turbines use a variety of materials. Like Bergey Windpower, many use fiberglass. Southwest Windpower used injection-molded blades on its Whisper H40 model with fiber-reinforced composites but used carbon fiber reinforcing in its Air series of micro turbines (see Figure 5-23. Carbon-fiber blade).

Hubs

Like the spokes in a bicycle wheel, the blades become part of the rotor when attached to its hub. The hub holds everything together and transmits the transverse motion of the blades into torque on the wind turbine's drive shaft. Three aspects of the hub are important: how the blades are attached, whether the pitch is fixed or variable, and whether or not the attachment is flexible.

All medium-size and large wind turbines and nearly all small wind turbines today use blades cantilevered from the hub—that is, they're supported only at the hub, just as the wing of a modern airplane is attached only at the fuselage (see Figure 5-24. Blade attachment). During the late 1970s and early 1980s, some European designs used struts and stays to brace the blades, following the pattern of the famous Danish wind machine at Gedser. Struts reduce bending on the root of the blade where it attaches to the hub. Consequently, the spar, the main structural support of the blade, and its attachment to the hub need not be as massive as on a cantilevered blade. Unfortunately, struts and stays increase the drag on the rotor, reducing its performance relative to a cantilevered rotor. In practice, struts and stays worked fine on upwind machines, as long as the turbine remained upwind. They sometimes

Figure 5-25. Struts and stays. Like its Danish predecessor at Gedser, the Windmatic 14S (a derivative of Christian Riisager's original turbine) used struts and stays to brace the laminated wood blades. Note the fantail for mechanically orienting the wind turbine into the wind. The blades used pop-up air brakes to limit rotor speed in emergencies. Though ungainly, these wind turbines have been operating in California's Altamont Pass for three decades.

Figure 5-26. Fixed-pitch rotor hub. Cast-steel hub for a 65-kW three-blade, fixed pitch Danish wind turbine. The slotted bolt holes allow field adjustment of blade pitch, but the blades remain fixed in pitch during operation. This is one of the simplest hub shapes. 1990s.

failed when the rotor was inadvertently downwind of the tower. Early Danish designs, such as the Windmatic 14S, were susceptible to this weakness, which led to an industry-wide abandonment of struts and stays for bracing the rotor in the 1980s (see Figure 5-25. Struts and stays).

Most hubs are rigid: they don't allow the blades to flap back and forth in gusty winds. The blades may change pitch by turning about their long axis, but they don't change from the plane of rotation. For most of the wind turbines installed during the 1980s and 1990s, the blades were bolted directly to a rigid hub (see Figure 5-26. Fixed-pitch rotor hubs).

During the late 1980s and early 1990s, several manufacturers began introducing variable-pitch hubs, enabling more sophisticated control of the rotor in high winds. By the 2000s, all large wind turbines were using variable-pitch hubs (see Figure 5.27. Variable pitch hub).

Several manufacturers have attempted to commercialize wind turbines using two-blade, teetered rotors. The rotor on these machines would *teeter* or rock about the hub. As a unit, the rotor would swing in and out of the plane of rotation like a seesaw. This teetering action relieves stresses on the blade during gusty winds, when the turbine yaws to track changes in wind direction, and when the blade passes through the tower's shadow. Teetering hubs have been used on both upwind (WEG's MS3) and downwind (ESI 54 and 80; Carter 25 and 300) turbines. Though engineers have long stressed teetering's advantages, the only contemporary wind turbines using the technique are Windflow and Gaia.

Even more complex hubs have been used on one-blade wind turbines, with a similar lack of commercial success (see Figure 5-28. Teetering hub).

The hub attaches to the main shaft, which forms part of the turbine's drivetrain: the arrangement of shafts, gearboxes (where used), and generators that convert the motion of the spinning rotor into electricity.

Drivetrains

Just as there is a multiplicity of rotor designs for wind turbines, there is an equally wide range of

Figure 5.27. Variable-pitch hub. Inset: Pinion and bull gear for multimegawatt variable-pitch hub. Each blade has its own pitch motor and gear. Main photo: Hub attaches to the wind turbine's drivetrain on the right. The wind would come from the left. All large wind turbines today use variable-pitch rotors, though whether they use a pitch control linkage or individual pitch motors varies among manufacturers. 2005.

Figure 5-28. Teetering hub. One of MBB's Monopteros wind turbines undergoing blade repair in late 1990 near Wilhelmshaven in northern Germany. Note the massive counterweight, teetering hinge, and work platform. The blade is about 25 meters (80 feet) long.

measures for converting the torque of the spinning rotor into electricity. And over the decades, just about every conceivable means have been used to transfer power from the rotor to the generator. Some turbines have driven the generator directly, many others have used mechanical transmissions, and a few—often short-lived—designs have attempted to use hydraulic or pneumatic transmissions.

Where gearboxes have been used, some designs have integrated them into the structural support of the turbine's rotor. The drivetrains of these designs have a characteristically streamlined appearance as all the drivetrain components—main shaft bearings, gearbox, generator are combined in one unit. The gearbox acts as the main mechanical support for the rotor and all other components.

Conventional drivetrains follow the classic Danish approach and mount the transmission and other components on a bedplate or frame that makes up the nacelle's structural support. Simplest in concept, Danish manufacturers opted for this in the early 1980s because all the components could be ordered off the shelf. Thus the drivetrain's major components were standardized, mass-produced hardware that was cheap, readily available, and could be easily replaced. In the early days, when much went wrong, this was a winning combination in contrast to the integrated drivetrains favored by American manufacturers of the period.

Driving the generator directly with the rotor eliminates the need for a transmission and reduces the complexity of the drivetrain. Direct drive should also offer slightly higher conversion efficiencies since no power is lost in the gearbox. Direct drive, though, requires specially designed slow-speed generators that may be larger and consume greater amounts of expensive materials than conventional transmission-generator combinations.

Small Turbines

Micro and mini wind turbines, because of their small size, limited output, and need for high reliability, use direct drive exclusively (see Figure 5-29. Direct-drive micro turbine). Ampair's 100 and Bergey's XL1 are typical of such turbines.

The choice of designers becomes more varied as turbines increase in size. In the 1930s, household-size wind turbines of the day could be found using simple single-stage speed increasers or direct-drive generators. The most successful of the pre-REA wind chargers, the Jacobs home light plant, used direct drive. It developed a well-deserved reputation for ruggedness and durability largely due to its massive, slow-speed generator.

Jacobs's chief competitor in the 1940s, Wincharger, used a simple, purpose-built gearbox. Wincharger used one large helical gear on the main shaft of the rotor to drive a small gear on the generator mounted directly to the gearbox housing.

During the late 1970s, a number of small-turbine manufacturers flirted with gear-driven machines. Jim Sencenbaugh produced his model 1000 using a similar approach. Sencenbaugh used the transmission to increase the 350-rpm speed of his rotor to the 1,100 rpm needed by the generator. Like Wincharger, Sencenbaugh used a purpose-built gearbox that supported the rotor and the generator as well as the tail vane. Enertech, a popular manufacturer at the time, took a more prosaic approach and mounted all the components on a bedplate.

Other small turbines have used belts or chains. Most home-builts of the 1970s used this approach because belts, pulleys, and sprockets were cheap and readily available. Aeropower, for example, used cogged belts, like the timing belts on auto engines on their small turbine. In practice, belts and chains have proven unreliable.

Most household-size turbines today, such as the Bergey Excel, use direct drive for its simplicity and durability. There are important exceptions. EasyWind, for example uses a simple integrated drivetrain (Figure 5-30. Integrated drivetrain, small turbine). EasyWind is unusual in this size class for many reasons. It uses a four-blade, stayed rotor, and it uses passive pitch control to limit rotor speed. Its drivetrain uses an off-the-shelf speed increaser mated to a flange-mounted generator. Larger turbines have carried this integrated approach much further.

Integrated drivetrains were a popular choice for American designers in the 1970s and 1980s of what we would call small commercial-size wind turbines today. Later, Enertechs, for example, took this path, as did their design successors in

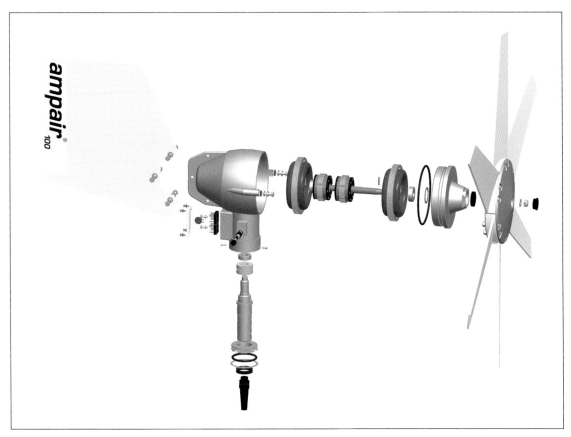

Figure 5-29. Direct-drive micro turbine. In the Ampair 100, the rotor drives the generator directly without a step-up gear. The Ampair 100 uses a multiblade rotor 0.9 meters (3 feet) in diameter. (Ampair)

Figure 5-30. Integrated drivetrain, small turbine. Left: 15-meter (50-foot) diameter, 50-kW wind turbine derived from the Enertech E44. From left to right: hub, main shaft bearing support and gearbox, generator, and brake. The generator hangs on the gearbox flange. The three-blade rotor passively yaws downwind of the tower. Inset: 6-meter (20-foot) diameter, 6-kW household-size turbine, EasyWind, uses an off-the-shelf gearbox and induction generator. As in the Enertech derivative, the EasyWind uses a flange-mounted generator with a disk brake on the end of the generator's shaft (visible here).

Figure 5-31. Conventional drivetrain, small turbine. Most small, commercial-scale turbines use variations on the classic Danish drivetrain developed in the early 1980s: main shaft supported by two independent bearings, off-the-shelf gearbox, bed-mounted, off-the-shelf generator, and disk brake on either the main shaft (best) or high-speed shaft. Left: Gaia 13-meter (42-foot) diameter, 11-kW wind turbine drive train. Generator on left, then disk brake on high-speed shaft, then gearbox. Right: Thy Møllen 7-meter (23-foot), 6-kW drivetrain. Main shaft bearings, disk brake on the main shaft, gearbox, flexible coupling, generator. (Thy Møllen)

the 1990s. Enertech's E44, a 13.5-meter-diameter turbine, epitomized this approach: the bearings supporting the main shaft were housed inside a special casting with a long snout. The casting housed the gear assembly as well as acted as the drivetrain's main frame. The flange-mounted generator hung on the casting.

Most designs using integrated drivetrains in the 1980s, such as ESI, Windtech, and Dynergy, were so problematic that they gave the entire wind industry a reputation for poor reliability that has taken decades to overcome. While Enertech had more than its share of problems, hundreds of the E44 continued to operate in California well into the 2000s. And as late as 2013, there were more than 100 of the turbines still operating in the Altamont Pass—nearly 30 years after they were installed. Because of their relative durability, derivatives of the E44 have seen several reincarnations in Canada and the United States during the late 2000s.

Yet not all household-size wind turbines have opted for either direct drive or integrated drivetrains. Some, such as Gaia and Thy Møllen, have followed their Danish roots and incorporated a conventional Danish drivetrain in a miniaturized version of a Danish wind turbine of the 1980s (see Figure 5-31. Conventional drivetrain, small turbine).

Both Gaia and Thy Møllen mount the drivetrain components on a bedplate. They differ in that Gaia uses a disk brake on the high-speed (output) side of the transmission, while Thy Møllen places the disk brake on the low-speed (input) side. Thy Møllen uses a more traditional approach than Gaia. Its 7-meter (23-foot) diameter, three-blade rotor is mounted upwind of the tower, and its main shaft is supported on two pillow-block bearings. This is the classic Danish design. Gaia, in contrast, uses a 13-meter (42-foot) two-blade rotor downwind of the tower. The main bearings are inside the frame.

Medium-Size Turbines

Most medium-size turbines on the market in the mid-2010s are used machines that have been replaced by repowering old wind farms with newer, larger turbines. These second-hand machines may be 10, 15, or more than 20 years old. As noted above, several manufacturers of medium-size turbines installed in the 1980s followed the integrated path blazed by Ulrich Hütter in Germany during the 1950s. However, most successful companies have mounted critical components independently on a structural frame. In the early 1980s, the difference between integrated drivetrains and traditional ones is what distinguished problem-prone American turbines from what became known as the Danish design.

American designers of integrated drivetrains during the early 1980s also tended to abandon nacelle covers as an unnecessary amenity. This lack of foresight gave US-designed turbines of the period an industrial appearance that certainly did little to further public acceptance of wind energy. And unlike Danish turbines, most US-built machines of the period lacked work platforms for servicing the turbines—another serious oversight.

Rather than integrate the drivetrain components, Danish manufacturers fabricated a metal frame or bedplate to which they independently mounted the main shaft, transmission, generator, and other components, such as yaw drives. The separate main shaft and support bearings on Danish machines allowed the transmission to be readily replaced without requiring removal of the rotor.

While there are few new products in this size class today—turbines greater than 16 meters (52 feet) in diameter but less than 60 meters (200 feet) in diameter—a few stand out: Endurance's E3120 and Northern Power Systems' Northwind 100.

As a passively yawed, downwind turbine, Endurance's E3120 is novel for its size, but otherwise the drivetrain is conventional. The 19.2-meter (63-foot) diameter rotor is supported by two pillow-block bearings and uses an off-the-shelf gearbox and induction generator—all features common of rugged designs of the 1980s (see Figure 5-32. Conventional drivetrain, medium-size turbine).

Northern Power took an entirely different approach for their 21-meter (69-foot) diameter upwind turbine. They developed a special-purpose generator using permanent magnets made from rare earths (see Figure 5-33. Direct-drive, medium-size turbine). In contrast to Endurance's passive yaw system, Northern Power uses an otherwise conventional yaw drive to orient the rotor into or, if need be, out of the wind.

Both turbines have proven popular in niche markets in North America and in Great Britain where there are policies specifically designed for turbines in this size class.

Large Turbines

As wind turbines increase in size, the need for a transmission becomes more pressing. Larger rotors, of necessity, spin at lower speeds, requiring a transmission or speed-increaser to provide the speed necessary to drive the generator. Until the mid-2010s, most manufacturers of large wind turbines, such as Vestas and Siemens, used some variation of the classic Danish design using a gearbox (see Figure 5-34. Conventional drivetrain).

Designers specify gearboxes that use either parallel shafts, the conventional choice for

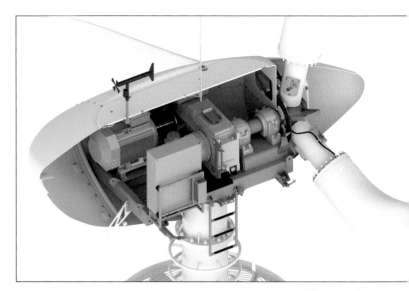

Figure 5-32. Conventional drivetrain, medium-size turbine. Endurance E3120, a 19.2-meter (63-foot), 50-kW downwind turbine. Left: induction generator, disk brake on the high-speed shaft, gearbox, pillow-block bearings, rotor. Note anemometer and wind vane, and lightning rod on nacelle cover. (© Endurance Wind Power Inc.)

Figure 5-33. Direct-drive, medium-size turbine. Northern Power Systems' Northwind 100, a 21-meter (69-foot), 100-kW upwind turbine. Left: rotor main shaft, purpose-built, permanent-magnet alternator, main frame, disk brake on main shaft (not visible). Note yaw drive (blue), yaw bull gear (slewing ring), and aviation warning lights. (Northern Power Systems) Inset: Northwind 100 in operation at Girvan Community Hospital, Ayrshire, Scotland. Note work platform, anemometer mast, and cooling fins on generator casting. Some 300 of the medium-size turbines were installed worldwide through 2013.

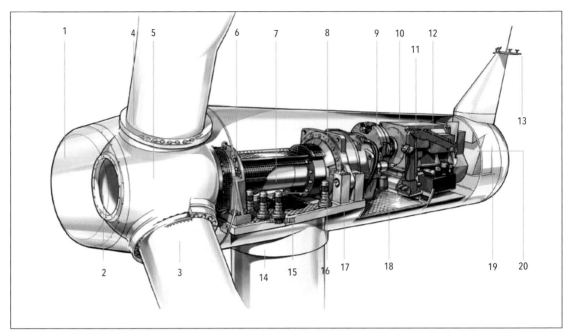

Figure 5-34. Conventional drivetrain: 2.3-MW Siemans' drivetrain for a wind turbine with a rotor 108 meters (350 feet) in diameter. 1. Spinner or nose cone. 2. Spinner bracket. 3. Blade. 4. Pitch bearing. 5. Cast rotor hub. 6. Main bearing. 7. Main shaft. 8. Gearbox. 9. Brake disc. 10. Coupling. 11. Generator. 12. Service crane. 13. Wind vane and anemometer. 14. Tower. 15. Yaw ring. 16. Yaw gear (multiple). 17. Nacelle bedplate. 18. Oil filter. 19. Canopy or nacelle cover. 20. Generator cooling fan. (Siemens Wind Power and Renewables)

medium-size wind turbines or planetary transmissions. Parallel shaft gearboxes occupy more space and weigh more than planetary transmissions but are significantly quieter. Turbines in the multimegawatt class use a combination of planetary and parallel shaft designs. The planetary stage is used on the low-speed shaft from the rotor where noise is less of a problem. The parallel stage is used on the high-speed side of the transmission to keep noise under control.

Transmissions on large turbines are often suspended on the main shaft, rather than mounted directly on the frame of the turbine. Dampening these shaft-mounted gearboxes allows the transmission to absorb fluctuations in torque caused by gusty winds, introducing needed compliance into the drivetrain of big turbines.

Some large European turbines have used a variation of the integrated drivetrain strategy. Although conventional in most respects, these designs hung the rotor directly on the transmission's input shaft without relying on an independent main shaft. This produced a more compact nacelle but at a cost of requiring the transmission housing to withstand thrust loads on the rotor, as well as the torsional forces for which the transmission was designed.

As turbines have grown larger, engineers have sought novel ways to handle the wind's thrust on the rotor, the rotor's weight, and the torque the rotor transmits to the transmission. In one version, much like the drive wheel on heavy vehicles, an extension of the nacelle's frame extends to support the rotor and main bearing, while a small shaft transmits torque to the gearbox. This was the approach Boeing took in its Mod-5 series in the 1980s. Bonus (now Siemens) used the idea as well in the 1990s on its 450-kW turbine.

Gearboxes are not intrinsic to wind turbines. Designers use them only because they need to increase the speed of the slow-running main shaft to the speed required by mass-produced generators. If they choose, designers can opt for special-purpose, slow-speed generators and drive them directly without using a transmission.

Purpose-built, permanent-magnet alternators revolutionized the reliability and performance of small wind turbines by eliminating the need for a gearbox. Since 1994, German manufacturer Enercon has done the same for large wind turbines using conventional electromagnets. After the successful introduction of its 500-kW E40, Enercon grew rapidly to become Germany's largest manufacturer of wind turbines (see Figure

Figure 5-35. Enercon direct drive. By directly driving a large-diameter ring generator, Enercon eliminates the need for a gearbox. Left: This cutaway illustrates an Enercon E82, a 2.3-MW wind turbine 82 meters (270 feet) in diameter. The rotor is coupled to the annular or ring generator, which rotates on the kingpin, or "main carrier" in Enercon's terminology. Note the multiple yaw drives. Right: Technicians wind the coils for the ring generator at one of Enercon's plants in Germany. Enercon turbines account for nearly two-thirds of the German wind turbine market. (Enercon)

5-35. Enercon direct drive), in part based on the high reliability of their direct-drive design.

Enercon pioneered the large-diameter ring generator for wind turbine applications and winds each electromagnet within its own plant. The massive generator, about 10% of the rotor disk, dominates the nacelle, giving the design its characteristic egg shape. By the mid-2010s, Enercon was fielding a range of turbines from 48 meters (160 feet) in diameter and rated at 800 kW to the top of the line at 126 meters (410 feet) in diameter, rated at 7.6 MW.

Dutch manufacturer Lagerway followed Enercon's direct-drive strategy before foundering. Lagerway took a slightly different approach to Enercon, and where Enercon suspends the ring generator on a kingpin, Lagerway spun the rotor on a large ring bearing like the axle on a heavy truck. This unusual design provided direct access to the hub and its pitch motors and pitch gears, making servicing the hub from inside the nacelle possible.

In the 2000s, a number of firms tried to develop hybrid approaches—neither conventional drivetrains, nor fully direct drive. Multibrid (now Areva Wind) developed a hybrid drivetrain by using a large diameter main bearing and special purpose, medium-speed permanent-magnet generator. The generator was coupled with an integrated drive system comprising a single-stage planetary gearbox. Areva has developed a 6-MW version for offshore.

Similarly, Clipper Windpower developed an unusual 2.5-MW integrated drivetrain that featured a large diameter gearbox driving four individual permanent-magnet generators. The design was beset with reliability problems and has been taken off the market.

In the mid-2010s, both Siemens and Vestas were introducing drivetrains on 6-MW models using direct-drive generators using rare-earth, permanent magnets.

Other Forms of Transmission

Hydraulics and pneumatics have been suggested for transmitting the rotor's torque to ground level. In theory this offers some advantages because a hydraulic drive, for example, can more easily be matched to the torque characteristics of a wind turbine rotor than can a mechanical transmission. Again, in principle, they should also be simpler. These advantages are offset, however, by inefficiencies and the complexity of making such a system actually work. The only large-scale test of hydraulic transmissions, the Bendix-Schacle turbine once owned by Southern California Edison Company, ended in ignominious failure. No wind turbine using a hydraulic transmission to generate electricity has ever been either a technical or a commercial success.

Pneumatics has a proud pedigree. Jules Verne's last novel centered on wind turbines powering Paris with pneumatics. Bowjon in the United States and Koender in Canada at one time used wind turbine–driven compressed air to pump water. Yet no one has successfully used pneumatics to produce electricity commercially.

Generators

Generators are not perpetual motion machines. They transfer power but don't create it. Power must be delivered to a generator before you can get power out of it. (In our case the prime mover, as it's called, is the rotor.) Nor are generators 100% efficient at transferring this power. The rotor will deliver more power to the generator than the generator produces as electricity. This leads us to a fundamental principle about the size of wind turbines. The size of a generator indicates only how much power the generator is capable of producing if the wind turbine's rotor is big enough and if there's enough wind to drive the generator at the right speed. Thus, we once again confront the fact that a wind turbine's size is primarily governed by the size of its rotor. (And this has become even more apparent with the introduction of large wind turbines with high specific areas suited for IEC Class III wind conditions.)

The generator converts the mechanical power of the spinning wind turbine rotor into electricity. In its simplest form, a generator is nothing more than a coil of wire spinning within a magnetic field. Consequently, whether generating direct current (DC) or alternating current (AC), a generator must have:

1. coils of wire in which the electricity is generated and through which it flows,
2. a magnetic field, and
3. relative motion between the coils of wire and the magnetic field.

By varying each of these conditions, you can design a generator of any size for any application.

Power in a simple DC circuit is the product of current and voltage. In a generator the armature is the coil of wire where output voltage is generated and through which current flows to the load (see Figure 5-36. DC generator). The portion of the generator where the magnetic field is produced is, simply, the field. Relative motion between the two is obtained either by spinning the armature within the field or by spinning the field within the armature. As you would expect, the stationary part of the generator is the stator; the spinning part is the rotor.

The power produced by a generator depends on the diameter and length of the wires used in the armature, the strength of the magnetic field, and the rate of motion between them. Increase any one of those, and you increase the potential power of the generator. The size of the wire in the armature determines the maximum current that can be drawn from the generator before it overheats, melts its insulation, shorts out, or otherwise destroys itself. The heavier the wire, the more current it can carry. As long as the wind turbine's rotor continues to provide greater and greater amounts of power as wind speed increases, the generator will continue to produce more current until the generator overheats. To prevent such occurrences, generators usually employ a means for limiting current to a safe maximum.

Generators are rated in terms of the maximum current they can supply at a specified voltage and, for AC generators, at a specific frequency. This rating is given on the nameplate in amps and volts (and frequency, where appropriate), as kilowatts and volts, or as kilovolt-amperes (kVA). The generator may be rated for the current it can supply continuously, or the current it can supply for only a short period. If generator size is of concern to you, always check which rating is being used. Reputable manufacturers rate their generators for continuous, rather than intermittent, duty.

Let's turn to voltage, the other half of the power mix. Generated voltage depends on the rate at which magnetic lines of force are crossed by the wire loops in the armature. Designers alter voltage by changing the magnetic field, by changing the rate of motion between them, or both.

The generator's field is provided by magnets. With electromagnets, some power is used to "excite" or "energize" the field around the armature. The strength of this field is a function of the number of wire coils in the field windings and the current flowing through them. For example, if you double the number of coils in the windings, you double the strength of the field, doubling generated voltage.

Many of the wind chargers built during the 1930s produced 32 volts. Resistance losses are proportionally higher when transmitting low-voltage power. Because of this, most reconditioned wind chargers were rewound for 110 volts. The old wire was stripped off the generator and replaced with more turns of thinner wire. Less current could be drawn through the smaller wire than before, but the increased length of wire produced a stronger field, increasing the voltage. Generating capability was not affected, power from the generator remained the same, but the balance between the voltage and the current changed: the voltage increased, and the current decreased by an equivalent amount.

Permanent magnets can also provide the field. They don't require power for excitation because they're inherently magnetic. The principle means for increasing field strength with permanent magnets is to use magnets with greater magnetic density, such as by using neodymium-iron-boron or other rare-earth magnets.

The voltage can be increased by adding more field coils, by adding more permanent magnets, or by increasing the speed at which the armature windings pass through the field. This can be accomplished by increasing the diameter and length of the generator so there's room for more magnets, or by spinning the rotor faster.

Yes, all this does have some bearing on the design of wind-driven generators. To get a feel for how, let's examine two popular 1930s-era wind machines: Jacobs and Wincharger. Both used rotors about 14 feet (4.3 meters) in diameter, thus the power available to the generator and the speed of the rotors were roughly equivalent. Yet Jacobs chose to use a direct-drive generator, whereas Wincharger chose a transmission-generator combination.

To produce the same power and voltage as Wincharger without a transmission, Jacobs had to design a generator that would operate at lower shaft speeds. Jacobs did so by increasing both the diameter and the length of its generator relative to that of Wincharger. This allowed the use of more field coils (six, to Wincharger's four). The coils were also larger.

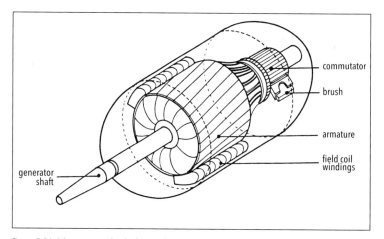

Figure 5-36. DC generator. Sketch of direct-drive generator on 1930s-era Jacobs wind chargers. Power is drawn off the spinning armature through brushes. Some of this power is used to energize the field coils. The commutator of a DC generator is simply a mechanical rectifier, picking off part of the alternator's AC waveform. No commercial wind turbines today use DC generators such as this.

The Jacobs generator's greater diameter also increased the speed at the periphery of the armature where it passed the field coils. Doubling the diameter doubles the rate at which the armature cuts through the field. The effect is the same as that of a 2:1 transmission, where the output shaft spins at twice the speed of the input shaft.

All in all, the Jacobs generator was considerably larger and used much more copper and iron than Wincharger to do the same job. But the Jacobs generator could do that job at a slower speed. Jacobs chose a low-speed generator for long bearing life and simplicity. They believed these advantages offset the generator's greater cost.

Environmentalist Barry Commoner's adage, "There's no such thing as a free lunch," puts it succinctly. Whether it's the design of generators or any other wind machine component, there's always a trade-off. You gain something only by

PERMANENT PERMANENT MAGNETS?

It can come as a surprise to learn that "permanent" magnets may not be permanently magnetized. Permanent magnets may lose some of their magnetization if overheated for extended periods. Southwest Windpower (SWP), for example, attributed some of the poor performance of their early Air series to degaussing, or the weakening of the strength of the turbine's permanent magnets. This occurred, SWP explained, when heat wasn't dissipating fast enough from the Air's alternator. Later versions of the Air were redesigned to better conduct heat from the alternator to the micro turbine's cast-aluminum body. Few other manufacturers have cited such problems.

Figure 5-37. Right-angle drive alternator. This 1980s version of the Jacobs used a right-angle drive to power a vertically mounted alternator. Hundreds of these turbines were at one time installed in California and Hawaii. They have since been removed. This turbine is on display at the Midwest Renewable Energy Fair site in Custer, Wisconsin. 2013.

Table 5-2. Nominal Generator Speed in rpm for Induction Generators		
	Europe	North America
	50 Hz	60 Hz
4-pole	1,500	1,800
6-pole	1,000	1,200

giving up something else. You hope that what you gain is more valuable than what you've lost. That's as true today as it was during the 1930s. Designers of wind turbines who stress long life and low maintenance choose lower generator speeds. The price they pay is a more expensive generator.

Manufacturers of small wind turbines intended for remote sites in harsh environments may opt, as Jacobs did, for building slow-speed generators tailored to their wind turbine. That's what most of today's small wind turbine manufacturers have done. They build special-purpose, direct-drive, slow-speed alternators. Even larger wind turbines destined for remote sites, such as Northern Power's Northwind 100, have followed the same path.

There are some exceptions. French small-turbine manufacturer Vergnet uses off-the-shelf induction generators coupled to gearboxes. Their turbines are intended for both on-grid and off-grid applications. Similarly, Wind Turbine Industries—inheritors of the 1980s Jacobs line—uses a conventional wound-field alternator but, unlike other companies, mounts the alternator vertically at the top of the tower, using a right-angle drive (see Figure 5-37. Right-angle drive alternator).

The trade-offs are also apparent in medium-size wind turbines. During the early 1980s, many American-designed wind turbines that operated at high rotor speeds were installed in California wind farms. These machines were not only noisy, they were also trouble-prone. Danish designs operating at much more modest speeds eventually won more than half the California market. Like Jacobs before them, the rugged Danish designs opted for lower rotor speeds to reduce wear and tear. The Danish turbines typically drove a six-pole generator at 1,200 rpm, while their American competitors used cheaper four-pole generators running at 1,800 rpm. Today, none of the early US designs are still being built (see Table 5-2. Generator Speed in rpm for Induction Generators).

Alternators

1930s-era wind chargers produced DC by spinning the armature within the field. Power was drawn off the rotating armature through a commutator. During the 1960s, the auto industry began replacing DC generators with alternators. Though alternators produce alternating current (AC), they offer several advantages over DC generators. For a given output, alternators cost less than generators. An alternator's slip rings are also more durable because they don't carry the alternator's current, as the brushes do in a generator's commutator.

In today's alternator, the field, rather than the armature, revolves. Power is drawn off the stator from fixed terminals. There are no brushes or commutator to wear out from the passage of high current. There's no arcing at the brushes. Excitation of the alternator's field is provided through slip rings on the rotor. But only enough power passes through the slip rings to excite the field (a small percentage of the alternator's output). Wear is negligible, in comparison to brushes on a DC generator.

There are no slip rings—no moving contacts—in a permanent-magnet alternator because the field is permanently excited.

In a conventional alternator, the field revolves inside the stator. But Bergey Windpower and some other small-turbine manufacturers spin the permanent-magnet field outside the stator. The case to which the magnets are attached rotates outside the stator. In this configuration, the blades can be bolted directly to the case, and they often are. There is also another benefit. Centrifugal force presses the magnets against the wall of the magnet can (see Figure 5-38. Inside out).

The magnets attached to the rotor of a more conventional shaft-driven alternator, such as in Southwest Windpower's Air series, are thrown outward from the spinning shaft. Because of the high rotational speeds found in small wind turbines, especially when unloaded, designers must pay special attention to retaining the magnets. For example, in the Air series, Southwest Windpower strapped the magnets to the rotor with a metal band.

As the name implies, alternators generate alternating current. As the rotor spins, current rises and falls like waves on the ocean (electrons in the armature are first jostled in one direction, then alter course and are jostled in the other). The alternator's frequency is the rate at which current rises and falls; it's given in cycles per second or hertz. The speed of the rotor and the number of poles or coils of wire determine the alternator's frequency. Drive the alternator faster, and frequency increases; slow the rotor, and frequency decreases. This explains why most small wind turbines generate variable-frequency AC. When wind speed rises, the turbine spins faster, increasing frequency (as well as voltage and current). When the wind subsides, frequency decreases.

In a simple alternator, the four coils are wired together in series as a single circuit producing single-phase AC. When three groups of coils are arranged symmetrically around the stator, the alternator produces three-phase AC, each phase one-third out of sync with the next. Most alternators used in wind systems produce three-phase AC. Three-phase alternators do more with less. The designer can more efficiently pack coils within the generator. Voltage is determined by the rate at which lines of force are cut by the armature. Thus we can increase power for the same current flow by increasing the number of coils

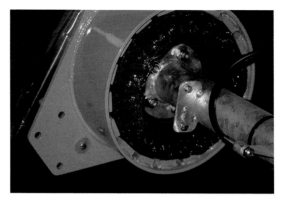

Figure 5-38. Inside out. Bergey Windpower places the permanent magnets (yellow) inside what they call the magnet can. Bergey then mounts the blades directly onto the end plate of the magnet can. Current is drawn off the stator coils at center (red), which are bolted to main-frame flange (left). This Bergey 1500 is no longer produced and was superseded by the Bergey XL1. 1997.

AIR-GAP OR AXIAL-FLUX GENERATORS

Air gap or axial-flux generators are a popular generator design for small wind turbines in Britain such as the Evance 9000. This strategy arrays the magnets and stator coils axially rather than in the more common radial pattern. In one approach, designers use two sets of magnet rings, with stator coils sandwiched between them. To build bigger alternators, a designer can increase the alternator's diameter or add more disks. Jeumont Industrie did both in its 48-meter (160-foot) diameter, 750-kW wind turbines. Jeumont, a manufacturer of discoidal alternators for French nuclear submarines, adapted the technology to wind energy.

Axial-flux alternator. Evance's 9000, a small wind turbine 5.5 meters (18 feet) in diameter drives a permanent-magnet, axial-flux generator. The turbine is distinctive because of the large-diameter, open cowling used to cool the generator. By 2013, more than 1,000 units had been installed in Great Britain.

and taking up all the available space within the generator.

Most small wind turbine alternators produce three-phase AC, to make the best use of the space inside the generator case. Some battery-charging models, such as Southwest Windpower's Air series, rectify the AC to DC at the generator; others, such as Bergey Windpower's XL1, rectify it at a controller, which can be some distance from the generator.

If you've ever spun the shaft of a toy generator in your hand, you remember how it felt when the rotor would stick slightly as the coils in the armature aligned with the magnets in the field. As the coils passed by the magnets, the shaft would turn more easily. This same effect, *cogging*, occurs in large generators and motors. Cogging is of interest in wind machines because it can retard the start-up of the wind turbine in light winds when the poles are aligned. Increasing the number of poles reduces cogging, enabling the turbine to start more easily in light winds. Cogging is also a source of alternator whine, and by reducing cogging, designers can reduce alternator noise. Scoraig Wind Electric's Hugh Piggott notes that skewing the slots in the laminations of the armature reduces cogging, and many small-turbine manufacturers use this or similar techniques.

Variable- or Constant-Speed Operation
Wind machines driving electrical generators operate either at variable speed or at constant speed. In the first case, the speed of the wind turbine rotor varies with the speed of the wind. In the second, the speed of the wind turbine rotor remains relatively constant as wind speed fluctuates.

In small wind turbines, the speed of the rotor varies with wind speed. This simplifies the turbine's controls while improving aerodynamic performance. When such wind machines drive an alternator, both the voltage and frequency vary with wind speed. The electricity they produce isn't compatible with the constant-voltage, constant-frequency AC produced by the utility. If you used the output from these wind turbines directly, your clocks would gain and lose time, and your lights would brighten and dim as wind speeds fluctuated. Eventually, you'd burn up every motor in the house. Unless you have a use for this low-grade electricity (heating, pumping water, and so on), the output from these wind machines must be treated or conditioned first, even if it's simply for charging batteries.

Because batteries can't use AC, the alternator's output must be converted to DC. As in your car alternator, diodes—electrical check valves that permit the current to flow in only one direction—rectify the AC output to DC, which is then used for charging the battery.

To produce utility-grade electricity, either the alternator's AC or rectified DC can be treated with power electronics to produce constant-voltage, constant-frequency AC like that from the utility (see Figure 5-39. Utility-compatible wind turbines). With the improved performance

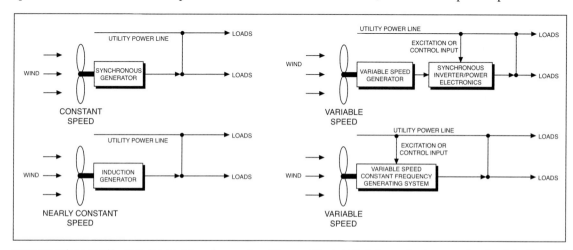

Figure 5-39. Utility-compatible wind turbines. Techniques for generating utility-compatible electricity. In the recent past, many medium-size wind turbines used constant-speed induction or asynchronous generators. Today, all large wind turbines use variable-speed generators with power electronics that convert the generator's AC to DC and then back to AC.

and reliability of power electronics in the past two decades, they have become ubiquitous in the wind industry. Power electronics that convert the generators variable-frequency, variable-voltage AC to DC and then back again to constant-frequency, constant-voltage AC are now an essential component of wind turbine design.

It wasn't always so. In the past, nearly all medium-size wind turbines, such as the thousands of machines installed in California during the early 1980s, operated at constant speed by driving standard, off-the-shelf induction generators.

As in small turbines, operating the rotor of medium-size and larger turbines at variable speed theoretically improves aerodynamic performance. More importantly, variable-speed operation allows the rotor to increase speed as a gust strikes the turbine, reducing potentially damaging loads on the drivetrain. By absorbing these loads in the inertia of the spinning rotor, designers can either cut maintenance costs and extend the life of the turbine, or they can reduce the size of components and lower the turbine's initial cost. One big advantage of variable speed is the reduction of the rotor's aerodynamic noise at low wind speeds relative to that of constant-speed machines.

Induction (Asynchronous) Generators

Induction or, as they are known in Europe, asynchronous generators became popular in medium-size and some household-size wind turbines during the 1970s and 1980s for several reasons: they're readily available, inexpensive, and come in a variety of sizes, and they can supply utility-compatible electricity without sophisticated electronic inverters. At the time, power electronics were costly and less reliable than today.

Induction generators are simply induction motors (like the motor in your refrigerator) in disguise. An induction motor becomes a generator when driven above its synchronous speed. Plug a four-pole induction motor into an outlet in North America, and the motor will turn slightly less than 1,800 rpm, consuming power. Leave it plugged in, but now drive the motor slightly faster. The motor will no longer consume power from the outlet. You're now supplying it. Spin the rotor just a little faster, and the motor not only won't be consuming electricity, it will now be generating it. As you try to spin the motor faster, it gets harder to turn. The utility consumes the additional power as you produce it, without rotor speed appreciably increasing.

In a wind turbine driving an induction generator, as wind speed increases, the load on the generator automatically increases, as more torque (power) is delivered by the wind turbine's rotor. This continues until the generator reaches its limit and either breaks away from the grip of the utility or overheats and catches fire.

Induction generators are not true constant-speed or *synchronous* machines. The speed of induction generators varies slightly and is thus *asynchronous*. As the torque available from the wind turbine rotor increases, the generator speed *slips* by 2% to 5% or 36 to 90 rpm on a 1,800-rpm generator. In an operating wind turbine, this slip is imperceptible. But in a cluster of turbines, the variation in slip from one to the next is detectable. In one moment, the rotors of a small group of turbines will all be in sync—all spinning at exactly the same speed. But they will gradually fall out of sync, becoming more and more out of sync until the cycle is repeated.

Danish manufacturer Vestas exploited slip to its advantage on its V47 series during the mid-1990s. Vestas used electronic controls on the generator rotor to vary the slip by as much as 10% of nominal speed. This enabled Vestas to enjoy some of the benefits of variable speed while avoiding the costs associated with true variable-speed operation.

So-called squirrel-cage induction generators have proved extremely popular for wind turbines because they're widely available in a range of sizes from numerous manufacturers worldwide, and their interconnection with the utility is straightforward. You can literally plug it in and go. Early promotions for the defunct manufacturer Enertech depicted its wind machines being plugged into a wall socket. Interconnection is a little more sophisticated today, but the principle remains the same.

A more costly version of the induction generator uses a wound rotor. These doubly fed induction generators enable operating the wind turbine at variable speed but require an AC-DC-AC inverter for producing utility-compatible

Figure 5-40. Doubly-fed induction generator. Flange-mounted, variable-speed generator for Zond 750-kW integrated drivetrain in the background. 1997.

electricity. Beginning in the late 1990s, most wind turbine manufacturers moved to this technology (see Figure 5-40. Doubly-fed induction generator).

When you're looking at a wind machine's generator, there's no need to be dazzled by the technology employed. Your primary concern is what kind of power it produces. If you want a wind machine for charging batteries at a remote hunting cabin, it's unlikely you'll be able to use an induction generator. If you want utility-compatible power, then you can't use a permanent-magnet alternator that doesn't also include some kind of power electronics.

Don't be swayed by the size of the generator, either. It's only an indication of how much power the generator is capable of producing, not how much it will generate. In the 1980s, if you had asked Danish manufacturers what size generator they used in a given wind turbine, they would have looked at you quizzically and asked, "Which generator?" Danish wind machines of the period often used two induction generators, one for low winds and another, much larger generator for higher winds.

Dual Generators or Dual Windings

Induction generators operate inefficiently at partial loads. For a wind machine with a generator designed to reach its rated output in an 11 m/s to 15 m/s (24–33 mph) wind, the generator would operate at partial load much of the time. To improve performance in the industry's early days, Danish designers would bring a smaller generator online in low winds so that it would operate efficiently at nearly full load. As wind speed increased, they would disengage the smaller generator while energizing the larger or main generator. Thus, both generators operated more efficiently than either did alone, and overall performance of the wind machine improved.

Essentially, Danish designers gained the performance advantage of operating the rotor at variable speed by using two generators. The use of dual generators permitted the wind turbine rotor to operate at two speeds. This enabled them to operate not only the generator at higher efficiencies but the rotor as well. Though they were not true variable-speed machines and couldn't take full advantage of the rotor's optimum tip-speed ratio, these turbines could bracket the optimum range. This was particularly useful in low winds where efficiency was most crucial.

In many parts of the world, wind turbines would operate most of the time at partial load or on the small generator of a typical Danish design from the 1980s (see Table 5-3. Typical Generating Hours for Rayleigh Distribution). For example, in a windy region with a 7 m/s (16 mph) average wind speed, the wind turbine would be producing electricity more than 7,000 hours per year, nearly all of that below the turbine's "rated" capacity. More than half the time the turbine

DON'T BE SWAYED BY THE SIZE OF THE GENERATOR, EITHER.
It's only an indication of how much power the generator is capable of producing, not how much it will generate.

| Table 5-3. Typical Generating Hours for Raleigh Distribution ||||||||
| Annual Average Wind Speed || Generating Hours/yr | % of Total Hours | Hours Below Rated Capacity | % of Generating Hours | Hours on Small Generator* | % of Generating Hours |
m/s	mph						
5	11	5,970	68%	5,960	100%	4,480	75%
6	13	6,710	77%	6,670	99%	4,150	62%
7	16	7,200	82%	7,020	98%	3,650	51%
8	18	7,540	86%	7,090	94%	3,150	42%

Note: *or low-speed windings. Assumptions: cut in, 4 m/s; cut out, 28 m/s; small generator, 4-7 m/s; rated, 16 m/s.

would be operating on the small generator. At Great Britain's first wind plant near Delabole, Cornwall, Peter Edwards found that his Danish turbines operated 77% of the time on the small generator.

The two generators were placed either in tandem and driven by the same shaft, or side by side, with the small generator being driven by belts from the main generator. Usually, both generators were spun at the same time and were not brought online mechanically but by energizing the field electrically.

During the early 1990s, Danish designers combined both generators into one by using generators with dual windings and by switching the number of poles. Typically, capacity of the small generator or low-power windings were about 20% to 25% of the main generator's capacity. For example, in Vestas's V27, the generator's low-wind windings were rated at 50 kW of the generator's full 225 kW. During low winds, the turbine's controller energized six poles. Under these conditions, the generator reached its nominal synchronous speed at 1,200 rpm in the Americas (60 Hz) and 1,000 rpm in Europe (50 Hz). In higher winds, the controller switched to four poles, and the generator reached its synchronous speed at 1,800 rpm in the Americas, 1,500 rpm in Europe.

Today, nearly all small turbines operate at full variable speed, most medium-size turbines operate at full variable speed, and all large wind turbines use some form of variable speed generators coupled with power electronics. The exceptions, which use induction generators, are noteworthy: EasyWind, Gaia, and Endurance's 3120. Used turbines from the 1980s and 1990s will all feature constant-speed induction generators.

Overspeed Control

The rotor is the single most critical element of any wind turbine. It's what confronts the elements and harnesses the wind. Because the blades of the rotor must be relatively large and operate at relatively high speed to capture the energy in the wind, they're the most prone to catastrophic failure. How a wind turbine controls the forces acting on the rotor, particularly in high winds, is of the utmost importance to the long-term, reliable functioning of any wind turbine. It is the critical design condition. All other design features pale in comparison to the survivability of a wind turbine—with little or no damage—under worst-case conditions.

All wind turbines must have some means of controlling the rotor in high winds. This is fundamental to wind turbine design. Gale-force winds are highly destructive, and a wind turbine must be designed to operate safely under such conditions or else in some manner turn itself either off or out of the wind. If not, there is the danger that the rotor will go into "overspeed," which could eventually destroy the wind turbine. Overspeed control is one of the characteristics that set different wind turbines apart.

Danes learned the hard way more than 30 years ago that all wind turbines must have an aerodynamic means of limiting the rotor's speed in an emergency. The Danish wind turbine owners association (Danmarks Vindmølleforening) was formed in part to force Danish manufacturers to

Danes learned the hard way more than 30 years ago that all wind turbines must have an aerodynamic means of limiting the rotor's speed in an emergency.

Figure 5-41. Decreasing frontal area by furling. Bergey 850 furling toward the tail vane in high winds at the Wulf Test Field in California's Tehachapi Pass. Bergey wind turbines furl the rotor out of the wind, reducing the rotor's frontal area. This protects the turbine in high winds. 1998.

abide by this requirement. How did they do this? They simply refused to buy any wind turbine that didn't have it. They did so because they had already lost too many wind turbines to winter storms. Their action was one of the reasons that Danish manufacturers grew to dominate the world wind industry.

Danish consumers literally saved manufacturers' bacon by insisting on this provision—and in so doing saved wind energy. This simple requirement is the reason that there are so many 30-year-old Danish wind turbines still operating in California, while only a few American wind turbines from that period remain in use. Danish wind turbines, although far from perfect, were designed to be fail-safe and, for the most part, were.

Wind turbine technology has not progressed so far that we can now safely eliminate this requirement. Even with mechanical or electrical braking and aerodynamic controls, there is still no certainty that a wind turbine will not destroy itself. However, the likelihood increases that a wind turbine will not survive an emergency intact if there are no aerodynamic overspeed controls.

This lesson has been continually repeated over the decades of modern wind power.

In this section, we look at the various means to limit rotor speed in high winds. Though there is some overlap, the technology chosen for small wind machines differs importantly from that of large turbines.

For the smallest micro turbines, those about 1 meter (3 feet) in diameter or smaller, the absence of controls may be acceptable under certain conditions—where extreme wind speeds may be unlikely. Multiblade turbines such as Marlec's Rutland rely on their relatively low rotor speed and rugged construction to survive high winds without any controls whatsoever. These turbines were designed for use on sailboats, to keep the batteries charged when the boat was moored in a protected harbor where wind speeds are unlikely to be excessive.

In contrast, Southwest Windpower initially relied on the controversial use of electronics and blade flutter to protect its lightweight but innovative Air series of micro turbines. Most wind turbine designers do everything they can to avoid blade flutter, because flutter can destroy both the blades and the wind turbine. "Flutter is like driving downhill with badly balanced wheels," says Scoraig Wind Electric's Hugh Piggott. Southwest Windpower, which originally designed its wind turbine for use on land, not at sea, used flutter to limit the aerodynamic performance of its sleek turbine. Early models were noisy and plagued with failures. Later models used an electronic controller built into the turbine to short the alternator's phases together to dynamically control rotor speed, vastly improving reliability. Eventually, Southwest Windpower wisely abandoned flutter entirely as a control technology.

For micro turbines installed on land, even Marlec recommends its Rutland 913F, the furling version of its rugged micro turbine.

Furling remains the simplest and most foolproof method for controlling a small wind turbine rotor. Furling, in its various forms, decreases the frontal area of the turbine intercepting the wind as wind speeds exceed the turbine's operating range (see Figure 5-41. Decreasing frontal area by furling). As frontal area decreases, less wind acts on the blades. This reduces the rotor's torque, power, and speed. The thrust on the blades (the

Figure 5-43. Pilot vane. In this classic design from the mid-to-late 19th century, the small wind vane below and to the right of the tail vane pushed the rotor toward the tail vane in high winds. This reduced the rotor's frontal area, protecting it in high winds. The farmer could control the wind speed at which the rotor furled by adjusting weights (not shown). Windmill State Recreation Area near Gibbon, Nebraska. Late 1990s.

Figure 5-42. Halladay rosette or umbrella mill. Segments of the rotor furl in high winds by swinging out of the wind's path. The rotor also passively orients itself downwind of the tower–an unusual feature for wind pumps in the mid-to-late 19th century. The large weight (left) counterbalances the weight of the downwind rotor. 1979.

force trying to break the blades off the hub) and the thrust on the tower (the force trying to knock the tower over) are also reduced. This method of rotor control permits the use of lighter weight and less expensive towers than on small wind machines where the rotor remains facing into the wind under all conditions.

Halladay's umbrella mill exemplifies the concept of reducing frontal area (see Figure 5-42. Halladay rosette or umbrella mill). These 19th-century water-pumping wind machines automatically opened their segmented rotor into a hollow cylinder in high winds, letting the wind pass through unimpeded. Each segment was composed of several blades mounted on a shaft, allowing the segment to swing into and out of the wind. When the segments are closed, Halladay's windmill looked like any other water-pumping windmill from the period. But in high winds, thrust on the segments would force them to flip open. This action was balanced by counterweights so the farmer could adjust the speed at which the windmill would open and close.

Horizontal Furling

Later inventors, such as the Reverend Leonard Wheeler, chose to use the same control concept (changing the area of the rotor intercepting the wind) but in a different manner than that used by Halladay. Rather than swinging segments of the rotor parallel to the wind, Wheeler thought it simpler to swing the entire rotor out of the wind. He couldn't do this with the downwind rotor used by Halladay. Instead, he placed his rotor upwind of the tower and used a tail vane to keep the rotor pointed into the wind.

Wheeler hinged the tail vane and rotor—or *windwheel* to old-timers—to permit the rotor to swing or *furl* toward the tail. As it did, the rotor disk took the shape of a narrower and narrower ellipse, gradually decreasing the area exposed to the wind. The mechanism for executing this was the *pilot vane*, which extended just beyond and parallel to the rotor disk (see Figure 5-43. Pilot vane). Unlike the tail vane, the pilot vane was fixed in position relative to the rotor. Wind striking the pilot vane pushed the rotor toward the tail and out of the wind. In the folded position, the rotor and pilot vane were parallel to the wind like the segments of the Halladay rotor. The thrust on the pilot vane was counterbalanced with weights. As in the Halliday design, a farmer could determine the wind speed at which the rotor would begin to furl by adjusting the weights.

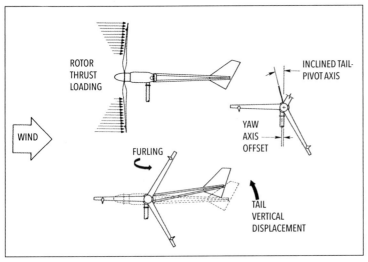

Figure 5-44. Simplified horizontal furling. Bergey wind turbines passively furl in high winds by swinging the rotor toward the tail. Note that the yaw tube (the tube connecting the wind machine to the tower) pierces the nacelle off center. The rotor axis is offset from the yaw axis, causing the rotor to fold toward the tail in strong winds. The nacelle pivots about an inclined axis, enabling the weight of the raised tail vane to push the nacelle and rotor back into the wind. Bergey has been successfully using this control strategy on its wind turbines for the past 30 years. (Bergey Windpower)

The pilot vane went the way of hand cranks on cars as the American farm windmill evolved. Offsetting the axis of the rotor slightly from the axis about which the wind machine yaws around the top of the tower produced the same results: self-furling in high winds. When the rotor axis is offset from the yaw axis, the wind's thrust on the rotor creates a force acting on a small moment arm (lever), represented by the distance between the two axes. The wind's thrust pushes the rotor out of the wind toward the hinged tail vane (see Figure 5-44. Simplified horizontal furling).

On contemporary farm windmills, there are no weights and levers to counteract the furling thrust. Instead, they use springs. By adjusting the tension in the spring, the farmer controls the wind speed at which the rotor furls. To see this for yourself, find an operating farm windmill and watch it in high winds. It will constantly fold toward the tail and unfold without any intervention.

Self-furling is a marvel of simplicity. Millions of farm windmills using Wheeler's approach to overspeed control have been put into service around the world. It's what you might call a proven concept. And if it worked reliably for all those machines for all those years, it should still work today. It does. Many small wind turbines use furling of one form or another.

The Bergey series of small wind turbines carries simplicity even further than does the farm windmill. Rather than using springs to control furling, Bergey uses gravity to return the rotor to its running position.

Bergey skews the hinge pin between the nacelle and the tail vane a few degrees from the vertical. As the rotor furls in high winds, the nacelle lifts the tail vane slightly. When the wind subsides, the weight of the tail pushes the nacelle back into the wind.

Wind machines that use furling, like the water-pumping windmills before them, can be controlled manually by furling the rotor with a winch and cable on models where such a feature is included (Bergey's Excel, for example). When furled, the rotor doesn't come to a complete stop—it will continue to spin—but this can make the difference between survival and failure in an emergency.

Furling design is as much art as science and when done poorly can lead to wildly fluctuating power from the rotor, as well as generate considerable noise. Passive furling is "crude but effective," says Scoraig Wind Electric's Hugh Piggott, "and simplicity pays off in a small wind turbine."

Vertical Furling

If you can furl the rotor horizontally toward the tail vane, you can just as well furl it vertically and equally reduce the rotor's frontal area. During the 1930s, Parris-Dunn built a wind charger that did just that. Rather than turning the rotor parallel to the tail vane, the company chose to tip the rotor up out of the wind. In high winds, the turbine would take on the appearance of a helicopter (see Figure 5-45. Parris-Dunn vertical furling). As the winds subsided, the rotor would rock back toward the horizontal.

Northern Power Systems developed a modern version of the vertical furling wind turbine, the HR3—for high reliability 3 kW—in the 1980s. Unlike most wind turbines—where the hype exceeds performance—the HR3 design has lived up to its billing. HR3s have survived winds of nearly 90 m/s (200 mph) in Antarctica and have been in operation for nearly three decades (see Figure 5-46. Vertically furled HR3). Like the original Paris-Dunn, Northern Power used a spring to control the wind speed at which the

Figure 5-45. Parris-Dunn vertical furling. This 1930s-era wind charger furled the rotor toward the tail vane vertically to protect the rotor in high winds. The wind speed at which the rotor furled could be varied by changing tension in the spring. This rebuilt unit was operating in Montana in 1976.

Figure 5-46. Vertically furled HR3. In high winds, the HR3 model wind turbine furls by tilting the rotor skyward, following the example of a 1930s-era wind charger. A shock absorber dampens the rate at which the rotor returns to the running position. The 3-kW turbine uses a 5-meter (16.4 foot) diameter rotor. This design includes a winch and cable for manually furling the turbine as shown here at the Centro de Estudio de los Recursos Energeticos in Punta Arenas, Chile. 1998.

rotor begins to furl. By varying the tension in the spring, you can govern the wind speed at which the rotor begins to pitch back. A shock absorber dampens the return of the rotor to the running position as winds subside.

Spanish small wind turbine manufacturer Juan Bornay uses a variation on vertical furling. In Bornay's version of vertical furling, the rotor, nacelle, and tail vane act as one unit. As the rotor moves up out of the wind, the tail vane drops toward the tower. Unfortunately, as the tail vane tips down toward the tower, it reduces directional stability. This can lead the wind turbine to erratically yaw about the tower.

Coning

Theoretically, rotor blades can be hinged at the hub allowing them to "cone" downwind from the tower in high winds on downwind turbines. In the 1970s, there were several attempts to adapt helicopter rotor technology to wind turbines. The most notable were those by Carter Wind Systems and by United Technologies Research Center (UTRC).

Carter introduced a 10-meter (33-foot) diameter, two-blade, downwind turbine that used a flexible, filament-wound spar in the blades that allowed them to cone downwind like "the fronds on a palm tree in a hurricane," claimed Jay Carter Jr. Although the blades were not truly hinged, the flexible spar allowed the blades to increasingly cone with increasing wind speeds. Hundreds of these turbines were installed in the United States during the 1970s and early 1980s. Few, if any, exist today.

Similarly, UTRC developed a highly flexible downwind rotor in the 1970s. Like the Carter design, ultimate high wind protection was presumably provided in UTRC's design by allowing the rotor blades to increasing sweep downwind of the tower (see Figure 5-47. Coning). This rotor control strategy proved problematic. Despite repeated efforts by various firms in the early 1980s to perfect the technology, the design was abandoned, and the 200 turbines installed in California were ultimately removed.

Figure 5-47. Coning. UTRC prototype wind turbine at Rockwell International's Rocky Flats Test Center. On January 14, 1982, the 9.8-meter (32-foot) diameter, two-blade, 8-kW prototype survived winds of 90 mph (40 m/s) with the blades fully flexed downwind. Windworks 10-m (33-foot) diameter turbine in the background. (Jonathan Hodgkin, Rockwell)

Some small wind turbine manufacturers have used coning for overspeed protection as well. Proven Engineering (now Kingspan) developed a line of flexible, three-blade downwind turbines in the 1990s. As with earlier ventures, Proven reduced the intercept area in strong winds by allowing the blades to increasingly cone downwind of the tower, shedding loads as they did so. The glass-fiber-reinforced polypropylene blades were attached to a flexible hinge at the hub. The cleverly designed hinge pitched the blades as they coned downwind.

Despite the varied attempts over the years to use coning successfully, the approach has not found favor with users of either small or large wind turbines.

Changing Blade Pitch

When most people first consider the problem of controlling a rotor in high winds, they think immediately of changing blade pitch. This probably results from our exposure to propeller-driven airplanes. Indeed, a wind turbine rotor can be controlled much like the propeller of a commuter plane. Like changing intercept area, changing blade pitch affects the power available to the rotor. By increasing or decreasing blade pitch, we can control the amount of lift that the blade produces.

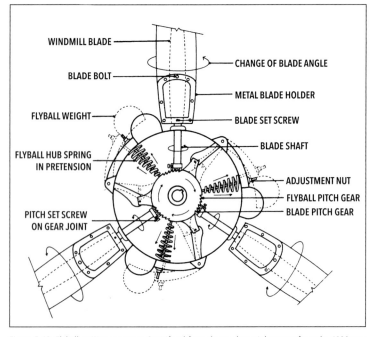

Figure 5-48. Flyball or Watt governor. Centrifugal force throws the weights away from the 1930s-era Jacobs governor, changing the pitch of the blades via a mechanical linkage.

There are two directions in which the blades can be pitched: toward stall or toward feather. When the blade is turned until it's nearly parallel with its direction of travel (perpendicular to the wind), it stalls. Thus, to stall the blade, we need turn it only a few degrees. To feather a blade, on the other hand, it's necessary to turn it at right angles to its direction of travel (90-degree pitch), or parallel to the wind. Feathering a blade requires it to be rotated farther about its long axis than stalling it does, causing the pitch mechanism to act through a much greater distance.

Stall destroys the blades' lift, limiting the power and speed of the rotor, but it does nothing to reduce the thrust on the rotor or the tower. Though it is simpler to build a mechanism for stalling the blade than it is to build a feathering governor, the stalling technique is less reliable. On upwind machines, thrust on the blades in high winds bends them toward the tower.

Small wind turbine designs dependent on blade stall as the sole means of overspeed protection have a poor survival record. Historically, the blades have had the nasty habit of striking the tower. Downwind turbines using stall regulation have had fewer problems because the blades are forced to cone farther downwind and away from the tower. Still, they too have had a poor reliability record, overall.

During the late 1990s, several European manufacturers of medium-size turbines introduced variable-pitch rotors that pitched the blades to stall. These machines and their active controls were far more sophisticated than the passive pitch-to-stall systems sometimes used on small turbines.

Where changing blade pitch is the primary means of control on an upwind rotor of a small wind turbine, experience has shown that the blade should rotate toward full feather. When it does, the drag on the blade is reduced to one-fifth that of a blade flatwise to the wind.

Governors for passively pitching the blades of small turbines appear in a variety of forms. During the 1930s, the Jacobs brothers popularized the *flyball* or *Watt* governor (see Figure 5-48. Flyball or Watt governor). Above normal rotor speeds, the weights would feather all three blades simultaneously via a mechanical linkage.

It's relatively easy to design a mechanism on a small turbine that pitches each blade independently. Jacobs's innovation was to link blade pitching together; thus, all blades changed pitch at the same rate. This avoids an aerodynamic imbalance that can damage the turbine.

When carefully adjusted, Jacobs's massive governor reliably protected the 4.3-meter (14-foot) rotor. Jacobs introduced an even more resourceful solution on later models, called the Allied (after a wind charger on which it first appeared) or blade-actuated governor.

Why use weights when you don't have to? The blade-actuated governor uses the weight of the blades themselves to change pitch (see Figure 5-49. Blade-actuated governor). Unlike the blades on the flyball governor, the blades not only turn on a shaft in the hub, they also slide along the shaft. Each blade is connected to the hub through a knuckle and springs. The knuckle, in turn, is attached to a triangular spider. As the rotor spins, the blades are thrown away from the hub, causing them to slide along the blade shaft. When they do, the blades pull on the spider, which rotates all three blades together. The springs govern the rotor speed at which this occurs. Like the flyball governor, the blade-actuated governor works reliably when properly adjusted and built to the highest material standards.

In the late 1970s, Marcellus Jacobs, the sole surviving founder of the original Jacobs Wind Electric Company, reentered the wind business. (The original firm ceased activity during the 1950s.) Along with his son Paul, Marcellus began manufacturing wind turbines patterned after his earlier models. His company briefly built wind turbines 7 to 8 meters (21–26 feet) in diameter. Jacobs's redesigned machine didn't depend solely on blade feathering to control rotor speed, since the blade-actuated governor was inadequate for a machine of this size. The new Jacobs turbine was also self-furling. The governor feathered the blades to limit power output to the alternator; overspeed protection was provided by furling the rotor toward the tail.

For many manufacturers of small wind turbines, mechanical governors have proven too costly and unreliable. Critics also charge that they are too maintenance intensive for the

Figure 5-49. Blade-actuated governor. The forces acting on the blades at high rotor speeds causes them to collectively change pitch. Here is a 1980s-era Jacobs household-size wind turbine on display at the Midwest Renewable Energy Association's Sustainable Living Center near Custer, Wisconsin. The 23-foot (7-meter) diameter turbine is unusual because of its right-angle drive. The 10-kW generator on this model is mounted vertically. Hundreds of these machines were used in California and Hawaii during the wind rush of the early 1980s. 2013.

Figure 5-50. Pitch weights. The rod protruding from the rotor is one of two pitch weights on this wind turbine undergoing tests at SEPEN (Site Expérimental pour le Petit Eolien de Narbonne) in the south of France. Designs using two blades regulated with pitch weights has characterized a number of small French wind turbines, beginning with Aerowatt, then Vergnet, and subsequently other manufacturers. 2005.

modern small wind turbine market. Still, some French manufacturers uses pitch weights to govern their two-blade rotors (see Figure 5-50. Pitch weights). And there are hundreds of small wind chargers operating in North America from the 1930s and from the Jacobs's revival in the 1970s that rely on mechanical governors. With proper maintenance and a supply of spare parts,

Figure 5-51. Full-span pitch control. In the late 1980s, several manufacturers of commercial medium-size wind turbines, such as Britain's Wind Energy Group (WEG), introduced variable-pitch rotors. WEG's MS2 used wood-composite blades on a variable-pitch rotor 25 meters (82 feet) in diameter. Shown here are a cluster of the 250-kW turbines in California's Altamont Pass. The blades on the turbine in the foreground are feathered; the blades on the turbines in the background are in the operating position. The turbines were still in operation in the late 2000s.

these machines will last for several more decades. Owners of these turbines argue that mechanical governors provide better power regulation in high winds than does furling. In winds above the rated speed, power output drops sharply on some furling turbines, while small wind turbines using mechanical governors are able to maintain near constant peak power. Though "there are substantially more moving parts in a pitching governor [than in a furling governor], I'd take a pitching governor every time," says Wisconsin's Mick Sagrillo.

It's possible to change blade pitch without using a mechanical governor. In the bearingless rotor concept, the blades are attached to the hub with a torsionally flexible spar. At high speeds, the blades twist the spar toward zero pitch, stalling the rotor. Weights attached to the blades are sometimes used to provide the necessary force. There are no moving parts in the hub: no bearings, knuckles, or sliding shafts. Several manufacturers in the 1970s and 1980s, such as Carter and UTRC, attempted to commercialize this technology on what are now considered small wind turbines. None succeeded because they failed to master the complex rotor dynamics involved.

Carter wind turbines typified this overspeed control strategy. The filament-wound, fiberglass spar permitted the blade to twist torsionally. During high winds, small weights inside each blade would rotate the blade toward stall. As noted above, the flexible spar also permitted the blade to cone progressively downwind of the tower in high winds. This design, though elegantly simple, was unreliable and fared poorly over time.

In medium-size turbines, the cost and complexity of variable-pitch hubs precluded their commercial use until the late 1980s. Most wind turbines of the period used rigid, fixed-pitch hubs (see Figure 5-26. Fixed-pitched rotor hub). Part of the problem at the time was how to handle the large loads placed on a variable-pitch hub of a fully cantilevered blade. In the Danish experimental program at Nibe, Risø National Laboratory built two 600-kW turbines in the late 1970s. One Nibe turbine used struts and stays to reduce these cantilevered loads, allowing the turbine to vary the pitch of the outboard blade section only. The second turbine varied the pitch of the blade along its entire length (see Figure 5-24. Blade attachment). The latter proved more successful. But it wasn't until the late 1980s that manufacturers of medium-size wind turbines, such as Vestas and Wind Energy Group (WEG), began introducing full-span pitch control (see Figure 5-51. Full-span pitch control).

Early designs of full-span pitch control split between those that used a kingpin and those that used a large turntable bearing. Both Mitsubishi and Vestas fielded turbines in the late 1980s that used cantilevered blades rotating about a kingpin bolted to the hub. This approach became quickly outmoded as turbines rapidly increased in size. Instead, manufacturers switched to large turntable bearings not unlike the giant yaw bearings.

As wind turbines continued to grow larger, the trend toward full-span pitch control became universal. In large wind turbines, the higher costs and complexity of pitch control is justified by the increase in rotor control, reliability, and longevity.

Aerodynamic Stall

Almost all wind machines without blade pitch control use aerodynamic stall to some extent for limiting the power of the rotor. This is particularly true of medium-size wind turbines using fixed-pitch rotors to drive induction generators, such as Endurance's 3120. In winds above the rated speed, the tip-speed ratio for these turbines declines because the speed of the rotor remains constant. For wind turbines operating at constant speed, the angle of attack increases with increasing wind speed, reducing the aerodynamic performance of the blades.

Designers seldom rely on blade stall as the sole means of overspeed protection on wind turbines driving induction generators. Stall control is dependent upon the generator keeping the rotor at a constant speed. Induction generators, in turn, are dependent on the utility for controlling the load. During a power outage, the generator immediately loses its magnetic field. The rotor, no longer restrained to run at constant speed, immediately accelerates. Stall now becomes ineffectual for regulating rotor speed until a new equilibrium is reached. Unfortunately, this occurs at extremely high rotor speeds.

On an upwind machine with a tail vane, the rotor can protect itself from destruction by furling out of the wind. Since tail vanes are limited to small turbines, medium-size upwind machines—and all fixed-pitch downwind machines-—must use a different strategy. Brakes are the most popular.

Once brakes have been selected as the means of limiting rotor speed during a loss-of-load emergency, they're also frequently used during normal operation. In a typical fixed-pitch wind machine, the brake is applied at its cut-out speed to stop the rotor. Wind turbines using this approach require extremely strong blades, in case the brake should fail and the rotor accelerates to destructive speeds.

Consider the case of a small downwind turbine driving an induction generator that was designed to brake to a halt at the cut-out speed. When the manufacturer, Enertech, first introduced the machine, the company stressed that the rotor was stall regulated and that it could operate safely, if necessary, above the cut-out speed, without the brake. The rotor was braked, asserted Bob Sherwin, then an Enertech vice president, only to minimize wear on the drivetrain at high wind speeds. (One of many claims about this and later turbines that were proven untrue.) The amount of energy in the wind at these higher speeds, he said, did not warrant the cost of capturing it. Mother Nature soon gave Sherwin's Enertechs ample opportunities to prove their mettle. The brake failed on several occasions. Rotors went into overspeed, and several Enertechs destroyed themselves to Sherwin's chagrin. Stall alone wasn't enough to protect the rotor. Enertech later added tip brakes for such emergencies.

Mechanical Brakes

Brakes can be placed on either the main (slow-speed) shaft or on the high-speed shaft between the gearbox and the generator (see Figure 5-52. Mechanical brake on main shaft). Brakes on the high-speed shaft are the most common because the brakes can be smaller and less expensive than the large disks and multiple calipers of those on the main shaft. When on the high-speed shaft, the brakes can be found between the transmission and the generator or on the tail end of the generator. In either arrangement, braking torque places heavy loads on the transmission and couplings between the transmission and generator. Moreover, should the transmission or high-speed shaft fail, the brake can no longer stop the rotor.

In general, brakes on fixed-pitch machines should be located on the main shaft, where they provide direct control over the rotor. (There's always a greater likelihood of a transmission failure than a failure of the main shaft.) But the lower shaft speeds require more braking pressure and greater braking area. As a result, the brakes are larger and more costly than those on the high-speed shaft.

Brakes can be applied mechanically, electrically, or hydraulically. Most operate in a fail-safe manner. In other words, it takes power to release the brake. The brake automatically engages when the wind machine loses power. Springs provide the force in a mechanical brake, such as those on the defunct Enertech and ESI brands; a reservoir provides the pressure in the hydraulic brakes on early Vestas turbines, for example.

Figure 5-52. Mechanical brake on main shaft. Massive disk-brake calipers on a Tacke 600-kW drivetrain. Note that the brake disk is part of the rotor hub. Mechanical brakes are ideally placed on the main shaft in front of the transmission, rather than on the output side of the transmission. Note also the integrated drivetrain and the flange-mounted generator. Tacke's 600-kW model was one of the largest wind turbines ever commercially produced that didn't use an aerodynamic means of overspeed protection, relying entirely on the massive disk brake. The company failed in the mid-1990s. Tacke's technology for its 1.5-MW turbine was absorbed by Enron and subsequently General Electric.

The problem with brakes of any kind is that they can fail—not often, it's true, but once is enough. Brake pads require periodic replacement or adjustment. After extensive use, the calipers have to travel farther to reach the disk. If the brakes are spring applied, pressure from the springs decreases, as does braking torque, as pad travel increases.

In one 12-meter (40-foot) diameter model built briefly during the late 1970s, the undersized brake just wasn't up to the job. And tragically there were no backup devices to protect the turbine. Several machines ran to destruction, but some were brought under control when the designer, Terry Mehrkam, climbed the tower, wedged a crowbar into the brake, and manually forced the pads against the disk. This dangerous practice eventually cost Mehrkam his life.

Experience has taught wind turbine owners and designers alike that wherever a brake is used to control the rotor, there must be an aerodynamic means to limit rotor speed should the brake fail. There are three common choices for aerodynamic overspeed protection on wind machines without furling tail vanes and blade pitch controls: tip brakes, spoilers, and pitchable blade tips (see Figure 5-53. Overspeed controls). These devices were frequently found on medium-size wind turbines using fixed-pitch rotors to drive induction generators in the 1980s. They are much less common today because all large turbines now use full-span pitch control. Large turbines use brakes today only to park the rotor—not to control it.

Aerodynamic Brakes

Aerodynamic brakes are not a new idea. In the early part of the 20th century, during the heyday of pioneering work in aviation, some traditional Dutch windmills began to sport new-fangled shapes on their wooden spars: jalousies, airfoil-shaped leading edges, and air brakes (see Figure 5-54. Air brake).

Tip brakes are similar to air brakes. These plates attached to the end of each blade also use drag to slow the rotor (see Figure 5-55. Tip brake). They're activated by centrifugal force once the rotor reaches excessive speed. They deploy a flat plate transverse to the direction of travel. They're simple and effective and have

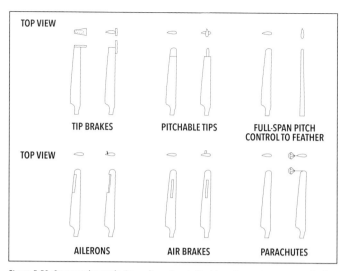

Figure 5-53. Overspeed controls. On medium-size wind turbines, the most common mechanism for protecting the rotor from overspeed is pitchable blade tips. Some turbines, Windmatic 15S for example, have used spoilers; others, such as Enertech, have used tip brakes; and a few have used, yes, parachutes. All large turbines today, and now many medium-size designs as well, use full-span pitch control.

Figure 5-54. Air brake. Miller Cornelis Gerkes releases the air brake on Zilvermeeuw or the herring gull drainage mill in the Netherlands' Groningen province. Note the airfoil-shaped leading edge and the jalousie shutters. The air brake was used to protect the rotor from overspeed.

Figure 5-55. Tip brake. Enertech popularized the use of tip brakes for overspeed protection on downwind, induction wind machines in the late 1970s and early 1980s. Tip brakes were troublesome and noisy and robbed power. Modern versions may be less problematic. The Enertech E44, shown here, was the model for contemporary turbines that use electrically released and reclosable tip brakes.

saved many a fixed-pitch rotor from destruction. Tip brakes, however, have been likened to keeping your foot on the accelerator at the same time you're stepping on the brake. They keep the rotor from reaching destructive speeds but do nothing to reduce the lift of the blade or the thrust on the wind turbine and tower. In the past, tip brakes were noisy and reduced the performance of the rotor under operating conditions by increasing drag at the tip where blade speed is greatest.

Enertech eventually adopted tip brakes on its turbines after discovering that stall wasn't sufficient to protect the machine in high winds. When Bob Sherwin reintroduced the Enertech E44 design as the AOC 15/50 in the 1990s, he continued Enertech's use of tip brakes. The contemporary version of this design uses electrically released and reclosed aerodynamically streamlined tip brakes.

Somewhat more sophisticated was Northern Power Systems' experimental use of what were called *ailerons* on the blade's trailing edge (see Figure 5-56. Ailerons). Tehachapi sailplane pilot Kevin Cousineau says these devices should more correctly be labeled *flaps*. Regardless of what they're called, they've proven noisy and unreliable for use in wind turbines.

Most of the power in the wind is captured by the outer third of the rotor. Consequently, it's not necessary to change the pitch of the entire blade to limit the rotor's power and speed. The performance of the blade in this region can be reduced by using spoilers or movable blade tips. In the Windmatic 15S, if there was a loss of load or the brake failed, centrifugal force activated spoilers along the backside of each blade. The spoilers popped out of the blade's surface and changed the shape of the airfoil, destroying its lift and protecting the rotor from destruction.

The widespread success of wind energy in the 1980s is attributable to the Danish concept of rugged, three-blade upwind turbines driving induction generators. One of the principal reasons for their success was that nearly all of the stall-regulated Danish turbines of the period used pitchable blade tips for overspeed control (see Figure 5-57. Pitchable blade tips). These passively activated tips were more reliable and less noisy than the tip brakes used on Enertechs, for example.

Figure 5-56. Ailerons. Northern Power Systems' experimental turbine unsuccessfully used ailerons (flaps) for overspeed control. The design was unusual for several reasons. It used only two teetered blades, an integrated drivetrain, and a flange-mounted 250-kW induction generator. Note person on platform for scale. This rotor control approach was eventually abandoned. Early 1990s.

Figure 5-57. Pitchable blade tips. Danish wind turbines, like the thousands installed in California during the early 1980s, used centrifugally activated blade tips to passively protect the rotor from overspeed. Later versions actively pitched the blades to stop the rotor under normal operation. The blades shown here are a static sculpture at the Folkecenter for Renewable Energy in Denmark. The tip of the blade on the right is in the deployed position. Fail-safe, pitchable blade tips may have been the single most important technological factor in the success of Danish wind turbine designs and, ultimately, the development of the modern wind industry.

During a loss-of-load emergency, such as a power outage, higher than normal rotor speeds would cause the blade tips to slide along a grooved shaft. As they move along the shaft, the tips would pitch toward feather. This action decreased lift where it's greatest, while dramatically increasing drag. Pitchable blade tips have proven highly successful, though refinements have been necessary. Frequently, for example, only one or two of the blade tips would activate, rather than all three at the same time. Although this was usually sufficient to protect the turbine, it caused many windsmiths sleepless nights.

As the turbines grew larger, it was no longer sufficient to rely on passive activation of the blade tips in emergencies. For large turbines with pitchable blade tips, it was necessary to use active pitching. For example, both Boeing's experimental 2.5-MW Mod-2 and its 3.2-MW Mod-5B actively controlled the pitch of the outboard section of each blade to limit power to the rotor. Though Boeing failed to commercialize its turbines, some Danish and German manufacturers introduced partial-span pitch control, or active blade tip control, in multimegawatt-size turbines during the 1990s. However, pitchable blade tips have been superseded by rotors using full-span pitch control on large wind turbines.

Putting It All Together

We'll now take a look at two classes of wind machines and how their manufacturers put all the pieces together. We'll also look at how they operate under normal and emergency conditions. The first group are the advanced small wind turbines designed specifically for high-reliability and low-maintenance applications often off the grid. The second group are large or commercial-scale wind turbines like those found in wind power plants around the world.

Small Turbines

Small wind turbines should be designed for simplicity, ruggedness, and low maintenance. Most, but unfortunately, not all, are.

Most of today's small wind turbines typically employ an upwind rotor and are passively directed into the wind by tail vanes, though some (Gaia) are passively oriented downwind of the tower. Most use three blades, though some (Gaia) use only two. Nearly all micro and mini turbines drive special-purpose generators directly, without the use of step-up gearing. Among the larger household-size turbines, there are some, such as those of Gaia and Thy Møllen, that use off-the-shelf generators coupled to transmissions.

Nearly all micro and mini wind turbines are designed for stand-alone, battery-charging applications. Some household-size small turbines, such as Bergey Windpower's Excel, also work with special-purpose inverters for interconnection with the electric utility. A few household-size turbines, like those of Gaia and Thy Møllen, drive induction generators and can be interconnected to the utility without an inverter.

With the exception of those driving induction generators, these turbines operate at variable speed. From start-up to rated wind speed, rotor rpm increases with increasing wind speed. Similarly, voltage and frequency increase as wind speed increases. Near the rated wind speed, the blades begin to stall, reducing performance. As wind speed continues to increase, most small turbines begin to furl, swinging the rotor toward the tail vane, while in some designs, the blades change pitch. Regardless of the technique, the turbines spill excess power to protect the rotor. When high winds subside, the turbine's rotor returns to its operating position automatically.

Most small turbines lack a mechanical brake, and many lack a furling cable for manually bringing the turbine under control during emergencies. Practically none include a parking brake for servicing the rotor, or a yaw brake to prevent the wind machine from yawing about the top of the tower. Both are important considerations if you plan to service the turbine atop the tower.

For the most part, the designers of these machines stressed simplicity and ruggedness over greater control. Bergey Windpower carries simplicity one step further by integrating the hub and rotor housing into one assembly. The turbine is designed for little or no maintenance and with good reason. At windy sites, it's not uncommon for a wind machine to be in operation two-thirds of the time, or about 6,000 hours per year. At that rate, a wind machine operates as many hours in the first six months of the year as an automobile driven 100,000 miles (170,000 km). Over a 20-year lifetime, a wind machine 3 meters in diameter will accumulate nearly 3,000 million revolutions, eight times that of a crankshaft in an automobile driven 100,000 miles!

In programs where performance has been monitored, wind machines such as the HR3 and Bergey Windpower's Excel have chalked up impressive records of reliability. After two decades of development, these designs have proven more dependable in remote power systems than the conventional engine generators they were originally designed to supplement. The turbines do require regular service as well as occasional repairs, but they perform the job they were designed to do—day in, day out.

Large Turbines

In contrast to small wind machines, commercial wind turbines found in wind power plants use, for the most part, off-the-shelf generators, transmissions, brakes, yaw drives, electrical sensors, and controls, though this is changing as the turbines grow ever larger.

The numerous components on a large wind turbine require regular maintenance, which can be simplified by clustering the turbines together in one location, as in a wind farm, or within easy reach of the manufacturer's nearest service center.

Like small wind turbines, large wind turbines are self-starting, that is, they do not need to be motored up to operating speed as were some early American-designed wind turbines. From start-up to the cut-in wind speed, the rotor speed varies with wind speed until it reaches the speed at which the generator can be synchronized with the grid.

The supervisory controls—or control algorithms—on large turbines may start the turbine by pitching the blades from the parked or feathered position to a start-up position. Once the rotor reaches a certain speed, the blades can then be pitched to the running position.

From the cut-in to the rated wind speed, the pitch of the blades will stay constant while the rotor speed will vary on fully variable-speed turbines. As wind speed increases, the turbine will deliver more and more power.

Above the rated wind speed, the control system will begin to pitch the blades to dump excess power. When wind speeds exceed the turbine's cut-out speed, or with any abnormal occurrence such as excessive vibration, the controller will pitch the blades to feather. Once the rotor has slowed sufficiently, a parking brake is applied, bringing the rotor to a stop.

When high winds subside below the cut-out wind speed threshold, the controller pitches the blades back to the running position, and the turbine resumes operation. Some control algorithms incorporate hysteresis for when the turbine is pitched to feather and when it returns to normal operation. For example, if the wind turbine is designed to operate up to a cut-out speed of 25 m/s (60 mph), it may not resume operation until the winds have fallen below 20 m/s (45 mph) to prevent the turbine from frequently cycling on and off.

DYNAMIC BRAKING: IS IT NOW ENOUGH?

I've always been a strong proponent of the Danish design rule that all wind turbines must have some aerodynamic means of limiting rotor speeds in a high wind emergency. This simple provision has been the backbone of the wind industry's commercial success. However, the advent of special-purpose, direct-drive, permanent-magnet generators using powerful rare-earth magnets may now require that this proscription be qualified.

Field experience appears to confirm that dynamic braking of well-designed rotor generator combinations on small and medium-size turbines may be sufficient to bring the rotor under control in all conditions. There are some 8,000 Southwest Windpower Skystreams installed worldwide. To my knowledge, only one of these 3.7-meter (12-foot) diameter turbines destroyed itself due to a runaway rotor.

Similarly, a much larger medium-size turbine, Northern Power's Northwind 100 has fared equally well. Northern Power has installed nearly 500 of the 21-meter (70-foot) diameter, 100-kW turbines worldwide without catastrophic failure. The Northwind 100 relies on both dynamic braking and dual mechanical brakes on the permanent-magnet generator's main shaft. Either system is designed to bring the rotor to a halt. Typically, the dynamic brake brings the rotor under control, slowing it down. The mechanical brake then stops the rotor.

Although I continue to emphasize the need for aerodynamic controls, there may be exceptions to this rule.

6

Vertical-Axis and Darrieus Wind Turbines

Vendre du vent et de la fumée.
(Sale of wind and smoke.)

—French expression for not delivering on promises made

ALTHOUGH VERTICAL-AXIS WIND TURBINES (VAWTS) ARE one of the two great classes of wind turbines, they remain a marginal—and marginalized—part of today's wind industry, and for good reason. In the 1970s and 1980s, Darrieus turbines promised lower costs and greater reliability than conventional wind turbines. They didn't deliver. Darrieus designs failed first in the demanding small wind turbine market. They couldn't compete. Then they failed in the commercial or wind farm market. Harsh, yes, but it is a conclusion based on bitter disappointment in the Darrieus wind turbines of the day.

The 2004 version of this book concluded that "work on Darrieus (or VAWT) technology has practically ceased." Since then there has been an explosion of interest in a new crop of small vertical-axis wind turbines for the residential and "architectural" market. And in France, the government is funding development of a large Darrieus turbine for offshore use. Some have called this a VAWT revival.

Have times—and technology—changed? Do these inventors and manufacturers know something that the rest of the wind industry doesn't? That interest and experimentation in new Darrieus technology is high can be clearly seen in the dozens of websites devoted to various vertical-axis designs. The concept is constantly being "rediscovered" by inventors whose "revolutionary" new devices are then fodder for countless news articles touting the technology as a panacea for our energy ills because these machines are not like those "other" wind turbines—the ones that actually work.

This chapter examines VAWTs in much more depth than in previous editions—some may argue in far more detail than VAWTs warrant. The following chapter takes a critical look at VAWTs near cousins on the margins of wind energy: ducted wind turbines. Together, these chapters will serve as a reference for students of the technology, journalists, and renewable energy advocates alike in the hope that so informed they will be better able to cut through the hype about the technology's promise.

We'll first look at various VAWT configurations and rotor designs, then explore the historical development of Darrieus wind turbines, examine some modern VAWTs, and—finally—weigh their advantages and disadvantages. Along the way, we'll discover that Georges Darrieus,

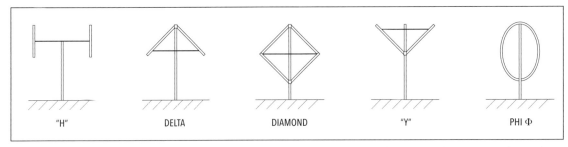

Figure 6-1. Darrieus configurations. There are several configurations of VAWTs besides the common Φ-configuration or eggbeater Darrieus. The second most common configuration is the H-rotor.

the French engineer for which this entire class of wind turbines are named—tellingly—never built one of the devices himself.

Although Darrieus is most famous for the phi (Φ) or *eggbeater* configuration, his patent covers other configurations as well, including the H-rotor. Delta (symbol), Diamond (symbol), and "Y" have all been tried at one time or another (see Figure 6-1. Darrieus configurations).

Lift and Drag

As with conventional wind turbines, VAWTs can use either lift or drag to drive the rotor about its axis. Most VAWTs use lift because a lift device is so much more cost effective than a drag device. Regardless, inventors keep revisiting drag devices often by using curved "blades" or buckets rather than the flat plates of the very simplest wind turbines (see Figure 6-2. Vertica "salad spinner"). Even though these curved plates are somehow intended to "capture" more of the wind, they remain at heart a simple drag device.

MY TAKE ON VAWTS

My view is that we've been there, done that, and moved on. That's why I chronicle the rise–and the fall–of VAWT technology so extensively in this chapter. As a proponent of the rapid expansion of renewable energy, I support what works–what makes economic and environmental sense now–not something that might or might not work in the distant future. If a tinkerer wants to experiment with a VAWT, fine, go ahead. If a budding manufacturer wants to develop a small VAWT for a "niche market," do so. Just don't ask public institutions–or ratepayers–for any support. But of above all else don't get in the way or otherwise obstruct the energy transition that we so desperately need now.

Figure 6-2. Vertica "salad spinner." This "novel" vertical-axis wind turbine is installed on a rooftop within Montreal's Biosphere. The manufacturer said this turbine was capable of 3 kW, about three times what could be reasonably expected. 2007.

Another attempt to improve on the limitations of drag devices is through what can only be called squirrel-cage rotors. In these machines, inventors try to direct the wind through a series of slats onto an inner rotor (see Figure 6-3. Squirrels in a cage). Squirrel-cage designs are not unlike Francis hydroelectric turbines from which the concept may have evolved. However, the wind in a free-stream environment is far different from water confined in a penstock under high pressure.

No squirrel-cage rotor or any of the various forms of VAWTs using curved plates and buckets has ever succeeded in delivering the amount of electricity promised by their promoters.

More successful, or at least longer lasting in the marketplace, are simple VAWTs using Savonius rotors (see Figure 6-4. VAWT demonstration, Crissy Field Center). Though these designs have been around for more than a decade, there was a flurry of new "helical" products introduced in the late 2000s. They have begun appearing atop light poles and in other "archi-

Figure 6-3. Squirrels in a cage. This example of a squirrel-cage, vertical-axis wind turbine has stood—idle—along Interstate 210 near San Fernando, California, since the mid-1980s. The outer ring of slats supposedly directs the wind over the rotor. 2008.

Figure 6-4. VAWT demonstration, Crissy Field Center. Twisted or helical Savonious. Tangarie's "Gale" turbine is typical of the many helical Savonious rotors introduced with much fanfare during the late 2000s. This turbine is part of a group of VAWTs in a so-called demonstration project at Crissy Field in San Francisco. There has been no need for such "demonstrations" for decades. 2012.

tectural" uses. Although the helical design may appear novel to the uninitiated, these turbines remain materially intensive, and most versions suffer from very poor performance in the field. They've been most successful in off-the-grid or remote applications where their poor performance and high cost is acceptable.

Blade Number

As in conventional wind turbines, only one blade is necessary to drive a VAWT's rotor about its axis, and during the heyday of VAWT development in the late 1970s, there were proposals to do just that. These designs typically used a counterweight to provide dynamic balance. Though Sandia National Laboratories (Sandia) designed its final experimental Darrieus wind turbine—the 34-meter test bed—so that it could be operated with only one blade without a counterweight, it was never operated in that mode.

Nearly all commercial Darrieus turbines used two blades, though there were notable exceptions. Two blades provide dynamic and aerodynamic balance, and rotors using two blades are much easier to assemble on the ground before being hoisted into place than rotors with three and four blades.

Alcoa developed a line of Darrieus turbines using three blades in the early 1980s based on Sandia's work (see Figure 6-5. Darrieus wind turbine with three blades). Three blades reduced—but did not eliminate—the torque ripple characteristically produced by the drivetrain of fixed-pitch VAWTs.

Figure 6-5. Darrieus wind turbine with three blades. Alcoa's 500-kW prototype commercial wind turbine at Agate Beach near Newport, Oregon, circa 1982. The turbine operated briefly before Alcoa abandoned its Darrieus program. Alcoa was interested in Darrieus technology because it provided a new application for extruded aluminum, an Alcoa product.

Figure 6-6. Darrieus wind turbine with four blades. One of a cluster of five Adecon wind turbines installed near Pincher Creek, Alberta. Two units had failed, and the remaining turbines were not operating when this photo was taken, circa 1995. The Adecon turbines were also unusual because they used a space-frame construction instead of guy cables to keep the tower erect. The space frame, while a clever alternative to guy cables, increases both visual and aerodynamic clutter.

Canadian developer Adecon introduced a Darrieus turbine in the late 1980s and early 1990s using four extruded aluminum blades (see Figure 6-6. Darrieus wind turbine with four blades).

Towers

Nearly all commercial Φ-configuration Darrieus wind turbines in the 1980s and 1990s used guy cables attached to the top of the turbine to keep the wind turbine erect. The exception was Adecon, which introduced a space frame of lightweight lattice masts instead of guy cables to hold the rotor upright. The idea wasn't original to the Canadian company. In 1978, Canada's National Research Council (NRC) installed a small DAF-Indal turbine in a space frame atop a "model home" near Ottawa, the capital. The turbine was removed in 1983. (It is now in the collection of the Canada Science and Technology Museum.)

One of the signature lessons of conventional wind development during the 1980s was that the benefits of tall towers far outweighed their added cost. VAWT designers were aware of this as well, but it's very difficult to install a guyed Φ-configuration Darrieus on a tall tower. This is a fundamental limitation of the Φ-configuration.

There were various attempts to circumvent this limitation. Some tried to use cantilevered towers. Alcoa, the giant aluminum company, installed a small 8-kW cantilevered Darrieus in the late 1970s in western Pennsylvania and installed another 8 kW on a cantilevered tower strapped to the side of a farm silo at Clarkson University in upstate New York (see Figure 6-7. Cantilevered Darrieus wind turbine).

Because of the severe bending loads at the base

Figure 6-7. Cantilevered Darrieus wind turbine. Installation of an Alcoa 8-kW cantilevered Darrieus turbine in western Pennsylvania's Somerset County. The site wasn't far from Alcoa's research center outside Pittsburgh. 1981.

of the rotor, these cantilevered Darrieus towers were limited to very small turbines. The Swedes proposed giant cantilevered Darrieus turbines for offshore (Poseidon), but they—wisely—never attempted to build one.

Most of the commercial Darrieus turbines used very short truss towers to raise the entire assembly off the ground to allow for access to the drivetrain. Realizing that one of the problems inherent in the Φ-configuration Darrieus is getting the rotor off the ground, Canadian manufacturer DAF-Indal erected their turbines atop truss towers that were much taller than those of its competitors.

Small and medium-size H-rotors of the period were installed on conventional lattice or truss towers. Though Pinson Cycloturbines were occasionally installed on octahedral towers, this wasn't the norm.

BEWARE VAWT RESOURCES ON THE WEB

As in most things on the web, it's caveat emptor (buyer beware). Entries on VAWTs and Darrieus range from thinly veiled promotions to an extensive visual catalog of VAWT experimentation, including a few photos from my own collection.

The English *Wikipedia*'s entry Vertical axis wind turbine is of dubious value. Although its introduction is useful in describing a VAWT, the remainder is not helpful if not deliberately misleading. Parts of the entry read like a sales pitch. For example, here's a paragraph that caught my eye.

> VAWTs are rugged, quiet, omni-directional, and they do not create as much stress on the support structure. They do not require as much wind to generate power, thus allowing them to be closer to the ground. By being closer to the ground they are easily maintained and can be installed on chimneys and similar tall structures.

This is the kind of over-the-top hype that gives VAWTs—of any kind—a bad name. With a little digging you can find that the author of that passage is a promoter of his own H-rotor turbine, which he wanted to mount on . . . chimneys. He issued a few press releases, published some articles, tested some blades, and then disappeared—an all too familiar story.

That paragraph is then followed by an equally misleading statement: "Research at CalTech has also shown that carefully designing wind farms using VAWTs can result in power output ten times as great as a HAWT wind farm the same size."

This statement is true as far as it goes. A professor at the California Institute of Technology (CalTech) installed 24 small VAWTs (Windspires) at a site in California. All told the "wind farm" represented 24 kW of capacity. Yes, that's kilowatts, not megawatts. Until he installs 24 MW of VAWTs in such a dense array, the claims in this statement are unsubstantiated. It's unlikely that such an array can generate 10 times the electricity of an equivalent size conventional array. It would be quite an accomplishment if they could just equal the performance of a conventional array.

On the other hand, the German *Wikipedia* entry for Darrieus-Rotor is extensive and factual. There are several useful and informative links that lead to websites not found via English *Wikipedia*.

One of the German references is a veritable gold mine of detailed information, photos, and graphics on nearly every imaginable form of vertical axis wind turbine ever conceived. The entry is encyclopedic, and the fact that it's in German shouldn't deter English speakers from perusing page after page of VAWTs, new and old. The Buch der Synergie's Chapter on Vertikalachsen-Rotoren is one of the most extensive, if not *the* most extensive, discussion of VAWTs found anywhere on the web. While the text is in German, anyone who wants to see if a "new" invention has been seen before should look at this site first.

Most of the text just lists details on the prototypes or promotional concepts and doesn't delve into whether they've worked in the field or not. Some of the text simply repeats the claims made by the promoters of various designs. However, the extensive section on Anton Flettner alone is worth reviewing.

Another link leads to Heiner Dörner, a retired professor at the aerospace institute at the University of Stuttgart. His web pages under the title Darrieus-Rotor is a trip down memory lane with comments on the status of various ventures and what happened to all those one-time clever designs. Dörner is someone who knows what he's talking about. He wrote the book on Ulrich Hütter, the father of German wind energy.

Φ-Configuration Darrieus Development

British wind turbine designer Peter Musgrove notes that the simplest VAWT configuration is that of the H-rotor. It's effectively the approach that the Chinese and later the central Asians took in building their panemones (see Figure 5-7. Panemone). It's where Musgrove himself started, and it's where many of the newly reinvented Darrieus designs of the "VAWT revival" in the late 2000s begin as well.

Not surprisingly, that's where Darrieus started too. Darrieus, as others before and since, quickly learned that this configuration permits centrifugal forces to induce severe bending stresses in the blades at their point of attachment. His famous patent cleverly dealt with this limitation.

Instead of using straight blades, Darrieus proposed attaching curved blades to the rotor at both the top and bottom of the rotor's torque tube or central mast (see Figure 6-8. Φ-Configuration Darrieus nomenclature). When the turbine was operating, the curved blades would take on the form of a spinning rope held at both ends, Darrieus explained. This *troposkein* shape directs centrifugal forces through the blade's length toward the points of attachment, thus creating tension in the blades rather than bending. Because materials are stronger in tension than in bending, the blades can be lighter for the same overall strength and operate at higher speeds than straight blades for the same loads. Some have likened the Φ-configuration Darrieus turbine to a slice of onion spinning about a vertical axis.

Darrieus's patent faded into obscurity until Canadian engineers Peter South, Raj Ranji, and Jack Templin at NRC reinvented the design in 1966. The "invention" was an innovative way to develop wind energy by using a Canadian commodity: aluminum. The blades of a Φ-configuration Darrieus can be extruded from aluminum, and most Φ-configuration Darrieus turbines have used this metal. Canada was, and remains, a major aluminum producer. Subsequently, Canadian research and development focused solely on Darrieus turbines.

Meanwhile, south of the border, first the National Aeronautics and Space Administration (NASA) and later Sandia began developing the US-version of the Darrieus design. NASA soon focused on conventional wind turbines, and Sandia became the national laboratory responsible for pursuing Φ-configuration Darrieus technology.

From the mid-1970s through the mid-1990s, most commercial VAWT development focused on the Φ-configuration or eggbeater Darrieus, and that effort was predominantly in North America. Some European firms also attempted to commercialize Darrieus technology. For example, Dornier in Germany and Vestas in Denmark (yes, that Vestas), but none built more than a prototype or two.

In Canada, DAF-Indal became the principal NRC contractor and began development of commercial turbines in the mid-1970s. The Toronto area manufacturer of structural aluminum developed a line of Darrieus turbines. Alcoa, the Pittsburgh-based aluminum manufacturer, followed suit with its own line of Darrieus turbines, and FloWind, another US manufacturer, focused solely on the commercial market. Although Alcoa was a much larger company, FloWind was far more successful commercially than Alcoa.

Now, a little about Vestas's brief foray into Darrieus turbines.

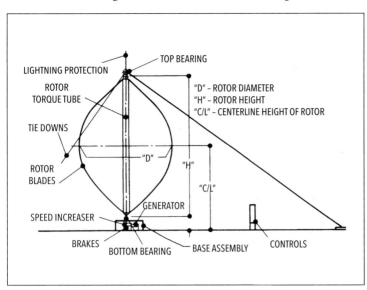

Figure 6-8. Φ-Configuration Darrieus nomenclature. Most Darrieus turbines of the 1970s through the mid-1990s used a guyed rotor. The blades were attached at the top and bottom of the central mast, or so-called torque tube because it transmitted the rotor's torque to the transmission at ground level. Key characteristics of the eggbeater shape are the diameter or width at midrotor, the height of the rotor, and the height of the rotor above the ground.

Vestas's Cantilevered Bi-Blade Darrieus

Vestas, the world's largest wind turbine manufacturer, is famous for its three-blade, upwind turbines. Today, this configuration is so universal that it's known as "conventional" in contrast to Darrieus turbines, which are now "unconventional." But it wasn't always so, and the history of the Danish manufacturer's entry into wind energy illustrates that in the early days no one was sure what design would work best.

Vestas was one of the companies experimenting with Darrieus turbines in the late 1970s. At the time, Vestas manufactured portable cranes and other machinery on the west coast of Jutland near the city of Ringkøbing.

In 1978, Vestas tested a sleek and unusual Darrieus turbine designed by Danish wind turbine pioneer and political activist Leon Bjervig. His Darrieus was unusual because it used two bi-blades that were attached to a cantilevered tapered and faceted torque tube (see Figure 6-9. Vestas's cantilevered bi-blade Darrieus). The 9-kW turbine was also unusual in that the rotor drove three 3-kW generators instead of one single generator. Despite the misuse of the term today, Bjervig's wind turbine was truly innovative—and a work of art. Bjervig later installed a three-blade, cantilevered Darrieus for Vestas.

At about the same, Vestas installed a three-blade upwind HAWT, built by a local machine shop, Herborg Vind Kraft or HVK. Both turbines were at Vestas's plant in 1980, so Vestas, and their chief engineer Birger Madsen, had experience with both types of turbine. Vestas chose to pursue the HVK model, which, for all practical purposes resembled what was to become Vestas's famous V15. The rest is history.

This ended Bjervig's work with Vestas. But according to the Danish Wind Turbine Manufacturer's Association (Danmarks Vindmølleindustrien), he didn't give up the dream of building bigger VAWTs. In 1986 he interested a local distribution utility in a 600-kW version, and by 1988 he had developed a detailed design. Nevertheless, the turbine was never built, and Danish development of Φ-configuration Darrieus turbines came to an end.

Figure 6-9. Vestas's cantilevered bi-blade Darrieus. Vestas experimented with Darrieus turbines before they chose to tie their fortunes to a three-blade upwind rotor configuration. The bi-blades were probably intended to stiffen the blades without the need of struts. The rotor was 8 meters (26 feet) in diameter and the turbine was 14 meters (46 feet) tall.1980.

DAF-Indal

DAF-Indal built about a half-dozen small prototype Darrieus turbines less than 5 meters in diameter in the mid-to-late 1970s. One was still standing—inoperative—outside Toronto in 2007 (see Figure 6-10. Small DAF-Indal Φ-configuration Darrieus turbine). The Canadian company also developed a series of larger, commercial-scale turbines.

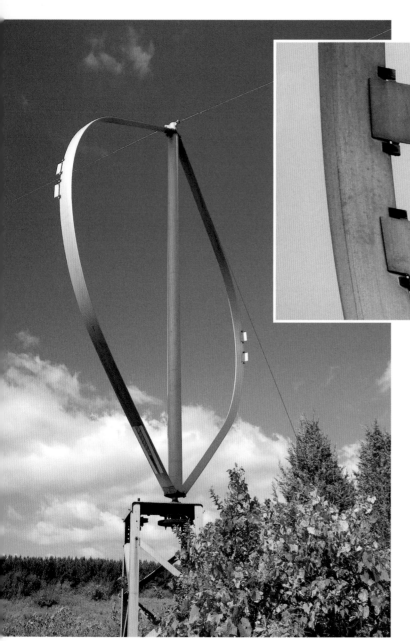

Figure 6-10. Small DAF-Indal Φ-configuration Darrieus turbine. The turbine at the Kortright Centre outside Toronto is a museum piece and hasn't operated for decades. Note the air brakes (inset) at the equator (midpoint) of the blade. These air brakes were a characteristic feature of DAF-Indal turbines. 2007.

In 1977, DAF-Indal installed a 230-kW Darrieus prototype for Hydro-Québec on the Îles de la Madeleine in the Gulf of Saint Lawrence. The experimental turbine soon became well known for an infamous episode disproving the adage that Darrieus turbines are not self-starting.

Darrieus turbines are typically not self-starting, though it is now known that Darrieus turbines can self-start under the right wind conditions. These conditions—though infrequent—do occur. When the Darrieus rotor is at a standstill, only the wind across the ground acts on the blade. Because the pitch of the blade is fixed, the blades stall and nothing happens. Normally, the rotor must be motored up to speed. But the event on the Îles de la Madeleine changed all that and added a new corollary to Murphy's Law: "Wind turbines that won't self-start will."

On July 6, 1978, the turbine was disengaged from its braking system in light winds during routine maintenance. The turbine was left unattended overnight. During the night, the wind picked up. To everyone's surprise the next morning, the rotor was turning, slowly at first, but eventually it began to pick up speed.

The DAF-Indal design was unique for the period. Designers had correctly identified one of the fundamental flaws of the Φ-configuration rotor: there is no way to change the pitch of the blades to reduce lift and bring the rotor under control during emergency conditions. Thus, DAF-Indal's designers provided air brakes at the equator (the widest part) of the rotor. These metal flaps would deploy at high rotor speeds and were intended to keep the rotor from self-destruction.

The air brakes deployed as expected. Unfortunately, as the rotor slowed down the metal plates returned to the normal position, allowing the rotor to again speed up. At some point while the air brakes were cycling from one position to another, one of them became stuck in the running or normal position. The rotor sped up and eventually leapt off the tower, corkscrewing itself into the ground.

It was an embarrassing failure, but DAF-Indal and Hydro-Québec replaced the destroyed turbine with another—improved—version in 1980. One of the lessons learned—a lesson repeated many times in the wind industry since—was the necessity to include a brake on the low-speed shaft as opposed to the less expensive approach of placing the mechanical brake on the high-speed of the gearbox. The second turbine wasn't "retired" by Hydro-Québec until 1986. What is little known in the English-speaking world is that this turbine was still standing as a forlorn tourist attraction on the remote windswept island as late as 2005.

The penchant for VAWT promoters to install their turbines on buildings isn't a modern phenomenon. DAF-Indal had already installed one of their small 4-kW turbines on a residence in 1978. In another noteworthy—if shockingly unwise—effort to find markets for its products, they installed a 17-meter-tall version atop a school in St. Eleanors on Prince Edward Island (PEI) in 1979. This remains the granddaddy of all rooftop VAWT installations to this day. After Malcolm Lodge, one of Canada's early wind pioneers, convinced NRC and DAF-Indal that this was an accident waiting to happen, they moved the turbine to the Atlantic Wind Test Site on the windy north end of the island where Lodge was program manager.

At the time, many were trying to find the right application for Φ-configuration Darrieus turbines in North America. West of Amarillo, Nolan Clark's team at the US Department of Agriculture's experiment station in Bushland, Texas, had installed one of DAF-Indal's 4 kW turbines in a wind-pumping system. Later the USDA installed the 4-kW turbine's big brother, a 40-kW machine in a wind-diesel demonstration project (see Figure 6-11. DAF-Indal 40-kW Φ-configuration Darrieus).

DAF-Indal developed a much more powerful version of their 24-meter, 230-kW turbine with a larger blade operated at a higher rotor speed. They installed one of the 500-kW versions at the Atlantic Wind Test Site and another at Southern California Edison's test site in the San Gorgonio Pass in the early 1980s. When a guy-cable shackle

Figure 6-11. DAF-Indal 40-kW Φ-configuration Darrieus. Demonstration of a Darrieus wind turbine in a wind-assisted pumping application at the USDA's experiment station in Bushland, Texas, to attendees of the American Wind Energy Association's conference, circa 1982. Note the truss tower and the struts at the top and bottom of the rotor.

failed on the latter turbine during precommissioning, the rotor collapsed, taking with it Eric Wright, a young DAF-Indal engineer. The second 500-kW turbine ran for a few years on Prince Edward Island, says Lodge, before one of the blades failed at the blade-strut joint, flinging the blade into one of the guy cables and bringing the whole turbine tumbling down. DAF-Indal replaced the turbine, and testing continued for a few years until the company walked away from Darrieus turbines for good.

In 1997, meteorologist Jim Salmon reported that the 50 kW DAF-Indal turbine at Christopher Point, on the southern tip of Vancouver Island, west of Victoria, British Columbia, had been in continuous service for 16 years. During that time, the turbine had operated 35,000 hours and generated nearly 600,000 kWh. This is likely a world record for a single Darrieus turbine in this size class.

Figure 6-12. Darrieus turbines at the Atlantic Wind Test Site. Two museum pieces still standing at North Cape in 2004. The turbine using a lattice mast for a torque tube was designed by the staff of the test site. The DAF-Indal turbine was finally removed in 2006.

In 2004, there were two derelict DAF-Indal turbines still standing at PEI's Atlantic Wind Test Site, one with a conventional tubular torque tube and another with an unusual lattice mast for the torque tube (see Figure 6-12. Darrieus turbines at the Atlantic Wind Test Site).

Alcoa's VAWT

Alongside DAF-Indal, Alcoa, the aluminum company, developed a line of Φ-configuration Darrieus turbines during the late 1970s and early 1980s. Like their Canadian competitor, Alcoa was looking for new aluminum products, and the Darrieus design used a lot of extruded aluminum in the blades.

The Alcoa turbines were a spin-off from research by Sandia near Albuquerque, New Mexico. Alcoa was the design-build contractor for Sandia's "low-cost 17-meter" Darrieus, a two-blade turbine rated at 100 kW. Paul Vosburgh, Alcoa's program manager, called this turbine the most successful design in the entire US wind program—and this was at a time when there was a flood of federal research and development money flowing into wind energy in the United States. At least three companies spun off commercial products from this design: Alcoa, FloWind, and Forecast Industries.

But unlike Sandia's prototypes, Alcoa's design featured three blades, distinguishing it from DAF-Indal's turbines as well. The tall ellipse that characterized Alcoa's rotor, or its greater height to diameter, also marked a departure from Sandia's more circular two-blade prototype rotor.

Like DAF-Indal, Alcoa grappled with how best to raise their Φ-configuration Darrieus rotors off the ground. In 1978, Alcoa teamed up with a silo manufacturer and mounted their 8-kW Darrieus on the side of a farm silo at Clarkson College in New York state.

This small cantilevered Darrieus anchored Alcoa's budding product line, and the Pittsburgh-based manufacturer installed another unit mounted on the ground at their research center and another outside Somerset in western Pennsylvania (see http://www.wind-works.org/cms/index.php?id=499 for a sequence of installation photos of Alcoa's 8-kW household-size turbine).

Between 1979 and 1981, Alcoa erected its preproduction 25-meter, 500-kW design for the Eugene Water & Electric Board on a recently logged site near Newport, Oregon (see Figure 6-5. Darrieus wind turbine with three blades). About the same time, Alcoa installed a similar 25-meter turbine in the San Gorgonio Pass at the Southern California Edison test site near Palm Springs. On April 3, 1981, the 500-kW Alcoa Darrieus destroyed itself while test engineers were on-site just prior to the American Wind Energy Association's annual conference.

In a famous case of aplomb, Alcoa's Vosburgh told the shocked conference audience that Alcoa had cancelled the planned tour of the test site. He said, "I have bad news and some good news. The bad news is that our turbine has failed and been destroyed. The good news is that we have not had to evacuate Los Angeles." The 1979 nuclear accident at Three Mile Island in Pennsylvania was still fresh in everyone's mind. The message that wind, despite its problems, was still much safer than nuclear power was not lost on the audience or the attending media. It's a message that's even more true today after not only Chernobyl but now also Fukushima.

Following the previous destruction of a smaller turbine a year earlier in Pennsylvania, Alcoa developed cold feet and cancelled its wind program. Vosburgh left Alcoa shortly afterward and joined Forecast Industries.

Forecast had taken the Alcoa-Sandia 17-meter turbine and developed the Vawtpower 185, so named for the rated power of the turbine. In the heyday of the California wind rush, Forecast installed a small farm of 40 Darrieus turbines in the San Gorgonio pass. In 1987, they added another 14 turbines of a slightly larger version. Altogether, Vawtpower installed nearly 9 MW, alone more VAWT generating capacity than DAF-Indal and Alcoa combined and probably more than all the Darrieus wind turbines installed worldwide since then.

Like Alcoa before them, Forecast's fortunes were star-crossed from the start with poorly performing and trouble-prone turbines. By 1988, none of its machines were operational, according to the California Energy Commission's Performance Reporting System.

FloWind and the World's Most Successful Darrieus

The most successful VAWT in history was that developed by FloWind in the early 1980s. Using what had become by then a rather conventional two-blade Φ-configuration or eggbeater Darrieus, FloWind installed more than 500 turbines in California's Altamont and Tehachapi passes (see Figure 6-13. Installation of FloWind Darrieus wind turbine in the Tehachapi Pass). By the end of 1985, FloWind had installed 95 MW of its signature product. For comparison, that's equivalent to more than 15,000 of Quiet Revolution's sleek 5-meter diameter turbine of the 2010s.

Figure 6-13. Installation of FloWind Darrieus wind turbine in the Tehachapi Pass. By the end of 1985, FloWind was operating some 500 Darrieus wind turbines in California.1985.

At their most productive in 1987, FloWind's fleet generated 100 million kilowatt-hours—enough electricity for nearly 20,000 California homes (see Figure 6-14. FloWind Darrieus wind turbine performance in California). No VAWT manufacturer to this day has ever come close to rivaling that accomplishment. On any windy day in Tehachapi during the late 1980s, the sun could be seen glinting from hundreds of FloWind's turbines spinning atop Cameron Ridge. However, generation soon began collapsing as serial failures in the joints

For whatever reason, many VAWT promoters ARE PRONE TO MORE HYPERBOLE than most other wind turbine designers.

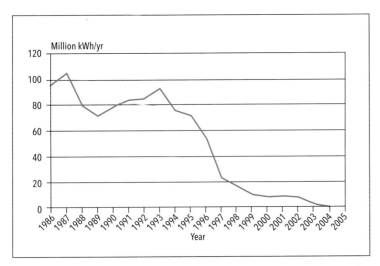

Figure 6-14. FloWind Darrieus wind turbine performance in California. FloWind generated more electricity with Darrieus wind turbines than any other company in the world. By the time the turbines were removed, FloWind's fleet had produced nearly 1 TWh over more than a decade, a record that is unlikely to ever be rivaled.

between the sections of extruded aluminum blades overwhelmed the American company.

Alas, in the late 1990s, FloWind was generating one-tenth the electricity it had in 1987, and by the end of 2004, nearly all of FloWind's Tehachapi turbines had been removed and sold for scrap. Some relics were still standing in the Altamont pass in 2003, though they too were slated for removal.

Nevertheless, when the last of FloWind's turbines were taken out of service, the fleet of Darrieus turbines had generated a lifetime total of nearly 1 billion kWh (1 TWh). It is unlikely that any VAWT will ever challenge that record.

The wind industry worldwide learned a lot from the FloWind experience. FloWind proved without a doubt that VAWTs could reliably generate commercial quantities of electricity—at least for a decade. But FloWind's aggressive marketing and high power ratings have tainted vertical-axis technology ever since. Much of today's cynicism about new VAWTs derives from FloWind's hype about its turbines and its manipulation of power ratings (see Table 6-1. FloWind Darrieus (VAWTs) Characteristics).

FloWind's turbines were the large commercial wind turbines of the day. They built two models: a 17-meter and 19-meter version. The 17-meter turbine, for example, was about 17 times the size of Quiet Revolution's architecturally dramatic QR5.

Characteristic of FloWind's marketing, and that of other Darrieus turbines of the day, including DAF-Indal and Vawtpower, was the turbines' high power ratings. FloWind's 17-meter model was rated at 142 kW at a wind speed of 17 m/s (38 mph), their 19-meter model was rated at 250 kW at a wind speed of 20 m/s (44 mph). That's a Force 8 or "fresh gale" on the Beaufort scale, meaning it's so windy few but the hardiest souls would want to be outside in the wind. (Note that this is the power rating officially reported to the state of California. This may not be the power rating found in product literature, which was often higher still.)

To understand these power ratings, it's necessary to look at the area swept by the eggbeater-shaped rotor. FloWind's 17-meter model swept 260 m^2, equivalent to a conventional wind turbine 18 meters (60 feet) in diameter. The 19-meter model swept 340 m^2, equivalent to a conventional wind turbine 21 meters (70 feet) in diameter.

For comparison, a conventional wind turbine 18 meters in diameter would typically be rated at 100 kW, and a 21-meter turbine would be rated at 150 kW. Thus, the FloWind turbines were overrated in comparison to their peers by at least 50%. Today, those ratings are unfathomable. Consider that the Endurance E3120 is 19 meters (62 feet) in diameter and rated at 50 kW or that Northern Power's 21-meter turbine is rated at 100 kW. FloWind's turbines of the early 1980s were rated more than 2.5 times greater than wind turbines of similar size today!

Table 6-1. FloWind Darrieus (VAWTs) Characteristics						
	Diameter	Height	Swept Area	Rated	Specific Power	Specific Area	Approx. Date
	m	m	m^2	kW	W/m2	m^2/kW	
FloWind	17	23	260	142	546	1.8	1983
FloWind	19	25	340	250	735	1.4	1985
FloWind	25	unkown	515	381	740	1.4	1986
FloWind EHD	18	50	536	300	560	1.8	1992

Figure 6-15. Hiking among FloWind's Darrieus turbines. The Sierra Club's annual hike among the wind turbines during FloWind's heyday in the mid-1980s crosses Cameron Ridge in the Tehachapi Pass. The orange flower blooming is California poppy (*Eschscholzia californica*), the state flower.

The high ratings of the FloWind machines translate into a specific power of 546 W/m² for the 17-meter model and an incredible 735 W/m² for the 19-meter model.

Another way of saying this is that the 17-meter model had a specific area of 1.8 m²/kW and the 19-meter model a specific area of 1.4 m²/kW. Conventional turbines of the day swept 2.5 m² for every kW of generator capacity.

That FloWind was greatly overstating the potential performance of its turbine was reflected in its average capacity factor, a measure of performance relative to the size of the turbine's generator. The capacity factor of FloWind's turbines never exceeded 12% on average and was often less than 10% at a time when conventional wind turbines were delivering twice that.

Why were FloWind's power ratings so high? Yes, fixed-pitch Darrieus turbines were stall regulated, requiring large generators. But conventional wind turbines of the day also were stall regulated. No, there's more to why FloWind's power ratings were inflated.

Wind turbines of that era were often sold to uninformed investors who compared wind turbine prices based on the cost per kilowatt of installed capacity. FloWind's aggressive power ratings enabled it to charge far more for its turbines than they were worth. FloWind's turbines were never truly in the 150-kW or 250-kW size class, but that's what the company charged its investors.

Despite these outlandish power ratings, FloWind's Darrieus designs turned in a respectable performance relative to the area of the wind stream swept by the two-blade rotor (see Figure 6-15. Hiking among FloWind's Darrieus turbines). During good years, Flowind's machines would generate 500 to 600 kWh/m²/year, competitive with conventional wind turbines of the day.

As FloWind struggled to remain relevant in the fast-changing industry, it experimented with two larger designs. The first was a version of its 19-meter turbine scaled up to 25 meters (82 feet) in diameter (see Figure 6-16. FloWind 25-meter Darrieus VAWT). The turbine swept twice the area of FloWind's original 17-meter product.

Figure 6-16. FloWind 25-meter Darrieus VAWT. Installation of a 25-meter (82-foot) diameter Darrieus prototype atop Cameron Ridge in 1986. The turbine swept 515 m² according to the California Energy Commission. This was a large wind turbine for its day and one of the largest Darrieus wind turbines of the era.

The second approach was developed for Sandia to overcome a number of problems with the Darrieus turbines of the day.

FloWind's EHD (extended height-to-diameter) prototype attempted to initially solve the fatigue problem found in the extruded aluminum blades on Darrieus turbines by using pultruded fiberglass. FloWind successfully extruded very large sections, but the stiff fiberglass could not be bent into the typical shape of its other Darrieus turbines. Thus the height of the tower needed to be extended relative to its diameter. This had the additional benefit of raising the midpoint of the rotor higher off the ground relative to its other turbines without the need for an unwieldy tower, thus solving another problem with Φ-configuration Darrieus.

Unfortunately, the EHD design also gave up swept area and increased blade length per unit of swept area—all factors increasing relative costs and against the tall-slim shape. All in all, it was too little too late to save FloWind.

To summarize: FloWind's Darrieus turbines operated for about a decade generating millions of kilowatt-hours and in doing so delivered respectable performance until fatigue and design flaws led to increasing unreliability and eventually the company's bankruptcy. FloWind's turbines, when in regular service, delivered about the same performance as nearby conventional wind turbines relative to their swept area but performed poorly in comparison to their inflated power ratings.

Because of scale effects, it is unlikely that the small wind turbines of the modern VAWT revival will approach the historical performance of FloWind's large (for the day) wind turbines.

Éole

Éole represents the culmination of more than a decade of research in Canada and the United States on Φ-configuration Darrieus turbines. Éole was—and remains—the largest VAWT ever built. The rotor swept 4,000 m² of the wind stream. That Éole is still standing on the south shore of the Saint Lawrence near Cap Chat on Québec's Gaspé Peninsula is in itself remarkable (see Figure 6-17. Eole).

The turbine was not only unique for its sheer size but also because it used a large-diameter, direct-drive ring generator rated at 4 MW. Until then, most wind turbines had used off-the-shelf asynchronous or induction generators.

Éole was a giant for its day, with a rotor 64 meters (210 feet) in diameter and 96 meters (315 feet) tall. Erected in early 1984, Éole went into operation in 1987. The experimental turbine operated until spring 1993 when it was stopped due to damage to its large and expensive lower bearing. During its six years of operation, the turbine generated 12 million kWh.

The specific power of this machine even exceeded those of FloWind with an incredible 1,000 W/m². Once it had begun operating, the turbine was derated, but even then Éole must rank as one of the most overrated wind turbines ever built.

For two decades the turbine has stood idle. However, plans to remove the turbine have been scuttled by locals who use the turbine as a tourist attraction. Thus, Éole is again unique. It is the world's first wind turbine that has been turned into a static sculpture.

Cap Chat's late mayor, Jean-Yves Bérubé, was a dynamic proponent of his community and a great advocate of bringing tourists to see Éole. It was at his insistence that a visitors' center was developed at the base of the tower to meet the needs of tourists interested in wind energy. His efforts live on in the continuing operation of the visitors' center and Cap Chat's regular tours during the summer months. The turbine is also the site of a summer festival featuring electronic music.

Éole effectively represents the end of the line for Φ-configuration Darrieus development in North America.

Figure 6-17. Eole. The world's largest Darrieus wind turbine represented the culmination of more than a decade of development. Installed in the mid-1980s on Cap Chat on the south shore of the Saint Lawrence River, Eole operated for about six years. The 96-meter-tall turbine, now a static sculpture, acts as a tourist attraction. 1995.

H-Configuration or Straight-Blade Darrieus

As noted above, Darrieus in his original patent proposed a VAWT that used straight blades, where the blades were attached to the torque tube or rotating mast with a crossarm, giving the rotor the appearance of an H when viewed from the side.

The H-rotor gives designers more flexibility than that of a Φ-configuration Darrieus. The blades of an H-rotor can be fixed in pitch like that on the Φ-configuration Darrieus or the blades can swivel—articulate—about their vertical axis, changing pitch as they move around the carousel

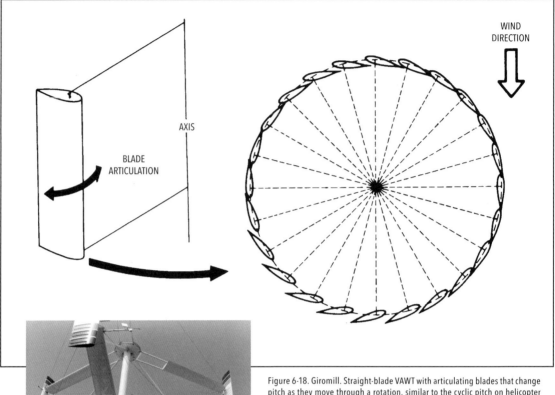

Figure 6-18. Giromill. Straight-blade VAWT with articulating blades that change pitch as they move through a rotation, similar to the cyclic pitch on helicopter rotors. This image (above) was used by Mike Bergey to illustrate the concept of an articulating, straight-blade wind turbine he developed at the University of Oklahoma in the 1970s. The 7-meter (23-foot) diameter, 8-kW VAWT he (below) and other students built won a national design competition in 1977. However, he and his father, Karl Bergey, judged it to be too complex and expensive to put into production when they formed Bergey Windpower later that year. Bergey has been building conventional three-blade, horizontal-axis wind turbines ever since. (Bergey Windpower)

path, much like a helicopter blade changes pitch in flight (see Figure 6-18. Giromill).

Though some make a distinction between the two terms, both design approaches have been called giromills or Cycloturbines by their developers. For example, McDonnell Aircraft called their design a giromill. Herman Drees registered Cycloturbine as a trademark of Pinson Energy Corp. Both used articulating blades.

Articulating blades offer several important advantages over VAWTs with fixed-pitch rotors. They can be designed to start the rotor from its parked position without motoring the rotor up to speed. They can provide a smoother torque profile because they provide lift throughout their path. And most important of all, they can control the rotor in high winds or overspeed conditions by feathering the blades (pointing them into the wind).

In an articulating, straight-blade rotor, a wind vane on the turbine senses the wind direction and changes the pitch of the blades either with a mechanical linkage or electrically through servo motors (see Figure 6-19. Dansk Vindkraft giromill).

During wind energy's renaissance in the 1970s, there were several attempts to build wind turbines using both configurations in the United States, Denmark, Britain, and Germany. Some famous names in aviation and wind power have tried their hand at H-rotor Darrieus turbines. All failed.

To the uninitiated, the H-rotor design may appear novel simply because no one has used it for nearly three decades, but this wasn't always so. There was a significant effort in North America in the late 1970s by Herman Drees and McDonnell Aircraft.

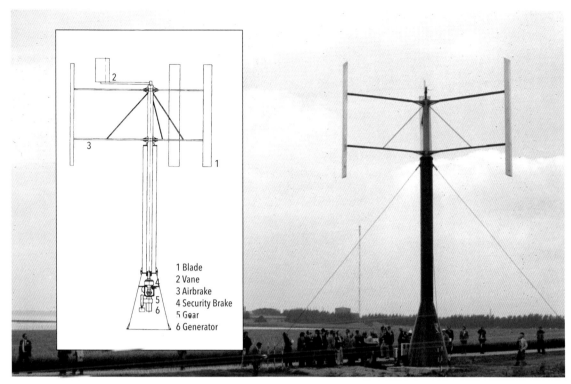

Figure 6-19. Dansk Vindkraft giromill. Danish attempt to use an articulating, straight-blade wind turbine at the Test Station for Small Wind Turbines, Risø National Laboratory, Roskilde, Denmark. As with McDonnell Aircraft's Giromill, this turbine transmitted power to ground level with a shaft inside the tubular tower. Note the prominent wind vane for orientating the blades as they move about the rotor path, the guy cables to support the tower, and the stays on the rotor crossarms. 1980.

McDonnell Aircraft's Giromill

In the mid-1970s, the defense contractor McDonnell Aircraft won a bid to design, build, and test a prototype giromill from the forerunner of the US Department of Energy. The contract—and McDonnell Aircraft's participation—were part of the infamous top-down development model favored by DOE and the aerospace community. After all, this model had worked successfully to put an American on the moon, why couldn't it work with something as prosaic as a wind turbine?

McDonnell's prototype wasn't small (see Figure 6-20. McDonnell Aircraft giromill). With a 17.7-meter (58-foot) diameter rotor and a rotor height of 12.8 meters (42 feet), the rotor swept 227 m². It was a big machine for the day, equivalent to that of a conventional wind turbine with a rotor 17 meters (56 feet) in diameter. Commercial turbines wouldn't reach that size for nearly another decade.

Because this machine was envisioned for agricultural uses, it was designed for moderately windy sites on farms across the Midwest. Thus, McDonnell's turbine was conservatively rated to deliver 40 kW at 11 m/s. The turbine was unusual too in that it was designed to deliver mechanical power at ground level where it could be used in a wind-assisted pumping application or used to generate electricity.

McDonnell installed its contracted turbine at the Rocky Flats test center west of Denver in the late 1970s. It was an ungainly machine on a truss tower. The rotor was cluttered with crossarms and cable stays. It used a massive torque tube, necessary to transmit the low rpm, high torque of the rotor to ground level. One didn't need to be an engineer to see that it was materially intensive.

However, what stood out in McDonnell's attempt to use an articulating, straight-blade VAWT was its hype and chutzpah, a pattern that has been followed by many latter-day "inventors" of the concept. According to McDonnell's statements of the day, it had "proven in wind tunnel tests" that its giromill had approached the Betz limit. In the 1970s, as now, this was big news. In theory this meant that it would perform better and, hence, be more cost effective than conventional wind turbines.

Figure 6-20. McDonnell Aircraft giromill. The wind vane at the top (inset) instructs the blades to change pitch relative to the wind. Only one machine of this design was ever built. This design uses the large diameter shaft to transmit the rotor's torque to ground level for wind pumping applications or for generating electricity. Note that the blades are feathered, that is, pitched so the blades don't provide lift and the rotor can be parked. The fairings, crossarms, and cable stays all add significant aerodynamic clutter that affected the turbine's performance. Unlike Φ-configuration Darrieus, giromills can use conventional towers. Circa late 1970s.

Based on their wind tunnel tests of a scale model, McDonnell engineers estimated that their design would achieve a coefficient of performance (C_p) of 0.54 or nearly the Betz limit of 0.593 from a full-size giromill. These results were widely criticized at the time as being extremely optimistic. Such performance had never been achieved in the field.

It was rumored in the industry that the tests didn't accurately account for the confined space of the wind tunnel. McDonnell engineers acknowledged some unexplained discrepancies in the tests, but their enthusiasm for their design won out over prudence, and they issued their report with the questionable estimate of performance in the very last line. Many have gone down this path since then.

This story may be apocryphal, but for whatever reasons the prototype—along with all the others in the Rocky Flats program—was never commercialized. Ultimately, McDonnell and the other aerospace contractors pulled out of DOE's wind program, finding it much more challenging than they had first thought.

Pinson Cycloturbine

Herman Drees, a young American engineer of Dutch decent, developed his own version of the giromill, the Pinson Cycloturbine, in the mid-1970s. Although his company won some grants from the Rocky Flats program, most of the turbine's development was privately financed, and thus, the product was named Pinson for one of his backers. Drees installed a surprising 140 turbines before his company closed shop (see Figure 6-21. Pinson Cycloturbine).

Drees did find success in wind energy but not with his Cycloturbine. Many years later, Drees developed a wind farm of 20 NedWind, two-blade Dutch wind turbines in the San Gorgonio Pass. The group of 500-kW turbines has been operating continuously for two decades. In 2009, Drees installed 13 more of the turbines at the site after he'd found them in a Dutch warehouse. Drees actively manages the 16.5-MW wind plant and still works in the wind industry.

Mike Bergey

Mike Bergey and a team of engineering students at the University of Oklahoma tried their hand at beating Betz with an experimental giromill in 1977. For a student-led project, they built an impressive 7-meter-diameter, 8-kW turbine. This was far larger than anything Pinson had built—and much larger than many of the small straight-blade VAWTs on the market in 2013.

Bergey, like Drees and like the powerful aerospace team of McDonnell Aircraft, abandoned straight-blade VAWTs because of the cost, complexity, and poor performance relative to conventional wind turbines. Bergey, of course, went on to manufacture small and household-size wind turbines using a conventional horizontal-axis design. The company he and his father, Karl Bergey, founded still exists today.

Fixed-Pitch H-Rotor VAWTs

Most of the small wind turbines in the VAWT revival have been H-rotors using fixed-pitch blades. Three examples that illustrate this approach were turbines made by the Canadian company, Cleanfield, California's PacWind, and Mariah. In contrast to earlier articulating VAWTs, these turbines are much simpler. Of course with this simplicity comes the difficulty of controlling the rotor in high winds and the likelihood that they will perform more poorly than their more complex predecessors. With the exception of Mariah's Windspire, there is little to

Figure 6-21. Pinson Cycloturbine. Household-size giromill near Las Cruces, New Mexico. One hundred forty of these turbines were installed across North America in the late 1970s and early 1980s. Note the wind vane for cyclically pitching the blades and the generator at the top of this octahedron tower. The rotor used collective pitch control and a Watt governor for limiting rotor speed. 1979.

no information on how these turbines performed in the field. This is often a sure sign that the results were not encouraging.

Before we move on to these modern examples, we should examine the world's most extensive development of a fixed-pitch, straight-blade VAWT that took place in Great Britain in the 1980s—and where that led.

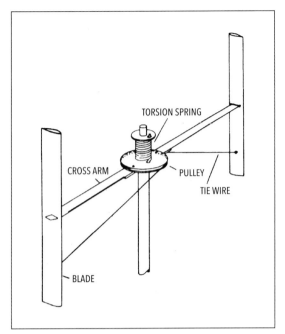

Figure 6-22. Musgrove variable-geometry, straight-blade VAWT. Creative solution by Peter Musgrove to limit the power available to a fixed-pitch, straight-blade VAWT by varying its intercept area.

Musgrove Variable-Geometry VAWT

Peter Musgrove is one of the great names in modern wind energy. In the 1970s, he was a professor of engineering at Reading University where he invented an ingenious means for controlling the rotor of a nonarticulating, straight-blade VAWT. He found that by attaching the blades at something other than the midpoint to the H-rotor crossarm, the centrifugal force on the blades as the rotor spun would fling them toward the horizontal. As the vertical blades approached the horizontal, they would reduce the rotor's intercept area, limiting the amount of power the rotor captured (see Figure 6-22. Musgrove variable-geometry, straight-blade VAWT).

Musgrove's doctoral student, Ian Mays, pursued the concept first with a small household size machine and later with what were then large-scale prototype turbines for Britain's wind energy program. Mays, who later became a central figure in Britain's wind energy development for Sir Robert McAlpine, found that by increasing the rotor's solidity—by making the blades wider relative to their length, for example—the rotor could start itself reliably in low winds. Until then, it was common to motor the rotors on Φ-configuration and H-configuration turbines up to operating speed, that is, they would use the generator as a motor to start the rotor spinning (see Figure 6-23. Variable-geometry H-rotor).

Mays went on to lead Sir Robert McAlpine's foray into VAWTs. McAlpine, one of Britain's storied engineering companies, pursued a series of Variable-Geometry VAWTs in the 1980s through a subsidiary VAWT Ltd. (see Table 6-2. Giromills' Early VAWT Characteristics).

These were not small machines, even by today's standards. The first version, the VAWT 450, was twice the size of McDonnell Aircraft's giromill (see Figure 6-24. Sir Robert McAlpine's VAWT 450). It was equivalent to a conventional wind turbine, 24 meters in diameter. At the time, wind turbines of that size were typically rated at 200 kW.

Noteworthy for then and even now, the turbine designation signaled the swept area. Many European turbines use rotor diameter in their designations as a surrogate for swept area. Rotor diameter alone isn't as meaningful for straight-blade VAWTs because swept area also depends on blade length. McAlpine simply used swept area directly. It was a breath of fresh air at a time in the industry when many suppliers were playing games with rated power to take advantage of tax subsidies in the United States. Even today in a retrospective on Sandia's Darrieus VAWT program, there was little reference to

Table 6-2. Giromills' Early VAWT Characterstics							
	Diameter	Height	Swept Area	Rated	Specific Power	Specific Area	Approx. Date
	m	m	m²	kW	W/m²	m²/kW	
McDonnell Aircraft	17.7	12.8	227	40	177	5.7	1977
Dansk Vinkdraft	9.0	5.5	50	15	303	3.3	1980
VAWT 450	25.0	18.0	450	130	289	3.5	1986
VAWT 260	19.5	13.3	259	100	386	2.6	1987
VAWT 850	38	22.5	855	500	585	1.7	1990

Figure 6-23. Variable-geometry H-rotor. In Peter Musgrove's ingenious design, the straight blades of the H-rotor are hinged so that they tilt toward the horizontal at high rotor speeds, reducing the rotor's intercept area. Shown here is an early attempt by a British firm to commercialize the concept. PI Specialist Engineers 6-meter-diameter variable-geometry VAWT, built in 1979. The figure at the bottom of the machine is Barry Holmes, their Technical Director. (PI Specialists)

Figure 6-24. Sir Robert McAlpine's VAWT 450. This straight-blade, H-configuration VAWT was the largest wind turbine built using Peter Musrgove's variable-geometry system for reefing the blades to limit rotor speed. Tests on this experimental turbine at Carmarthen Bay, Wales, revealed that the struts and blade hinges for the reefing system caused significant drag. (Bill Hopwood)

rotor swept area except when referring to the British program's development of straight-blade VAWTs.

The prototype with Musgrove's variable geometry was installed at Britain's Carmarthen Bay test site in South Wales. Subsequently, May's team developed a smaller 17-meter (56-foot) diameter turbine for isolated grids. They installed one on the Isles of Scilly and another in Sardinia.

The field tests revealed that the reefing system used to vary the rotor geometry was unduly complex and that the turbine could use blade stall to regulate power, the approach then being used across the wind industry. Thus, a later, much larger turbine, the VAWT 850, abandoned the Musgrove control system. This brought to an end development of the Musgrove rotor.

The VAWT 850 was installed at Carmarthen Bay in mid-1990. On February 22, 1991, the VAWT 850 lost a blade. Shortly thereafter, Sir Robert McAlpine dropped the program, bringing an end to straight-blade VAWTs in Britain until the advent of Quiet Revolution more than two decades later.

Investigations into Britain's VAWT program concluded that any benefits from the claimed advantages for VAWTs were offset by generally increased manufacturing costs. In the meantime, Peter Musgrove had left the university and joined the Wind Energy Group, a consortium developing conventional two- and three-blade upwind turbines.

Cleanfield

Cleanfield made a big media splash in Ontario, Canada, during the mid-2000s with its claim of producing a new wind turbine suited for mounting on buildings. (More about rooftop and building mounted wind turbines in Chapter 9, "Towers.") At the time, the Toronto Renewable Energy Cooperative was installing a conventional wind turbine, a 600-kW Lagerwey, at Toronto's Exhibition Place (Ex-Place), and the province was weighing an aggressive renewable energy policy.

Figure 6-25. Nonarticulating, straight-blade VAWT. Cleanfield fixed-pitch VAWT on the top of a building at a technical college near Hamilton, Ontario, Canada. 2007. (Martin Ince)

Opponents of wind energy were quick to seize on the fact that Cleanfield claimed its turbine could be installed on buildings. Thus, there was no need to install the commercial-scale turbine at Ex-Place in downtown Toronto, or any commercial-scale turbines anywhere in Ontario for that matter. If the province wanted wind energy, it could simply just put wind turbines, like solar panels, on rooftops. If Toronto wanted wind energy, it could just put the wind turbines on Toronto's skyscrapers.

The media, politicians, and overenthusiastic renewable energy advocates never bothered to verify the claims of Cleanfield (see Figure 6-25. Nonarticulating, straight-blade VAWT). This was easy enough to do. The design was not new. (There was a lot of historical precedent as explained in this chapter.) The design had never been tested, so there was no way that Cleanfield could honestly make pronouncements on its presumed advantages over conventional wind turbines.

In 2007, Cleanfield told the Region of Pecl, Ontario, that it hadn't completed testing. By 2013 when its website went dark, Cleanfield had still "not completed testing," even though it had installed a unit at the Atlantic Wind Test Site on Prince Edward Island (now known as the Wind Energy Institute Canada). After more than a decade of "development," and after receiving funds from the province of Ontario for its "innovative" technology, Cleanfield never produced any performance data.

Some of the "testing" Cleanfield often pointed to were aerodynamic measurements made in 2007 by McMaster University in NRC's wind tunnel outside Ottawa. Cleanfield crowed about the test results, implying that there was academic endorsement of its product. (This sounds strangely similar to McDonnell Aircraft's wind tunnel tests from an earlier era.) What Cleanfield's promoters wouldn't say is that wind tunnels are useful for testing airfoils but not much else. They are not a substitute for measuring how much electricity a wind turbine will actually produce in the field under real-world conditions.

Worse, in an embarrassing attempt to appease a critic of its design, Cleanfield sent "test results" as proof of its claims. The "test results" were merely a chart of the generator's voltage output relative to wind speed. Whether Cleanfield knew—or cared—that this was meaningless information is unknown. It's simple enough to measure electrical energy; every home in North America has a kWh meter that does just that. When a promoter of a wind turbine presumably designed to produce electricity doesn't know this and can't provide this kind of information, it calls into question the promoter's competency in every other aspect of the wind turbine and its intended use—rooftops in this case.

There also was the odd case of Cleanfield's product literature. Product literature for many newly introduced small VAWTs is often long on hype and short on reliable data. One Cleanfield brochure listed the rotor's diameter, the length of the blades, and the swept area. What was odd was the swept area, 13 m², didn't match the other parameters. Because Cleanfield's turbine used an H-rotor, the swept area is simply the length of the blades times the rotor's diameter. The rotor was 2.5 to 2.75 meters in diameter, depending upon the source, and the blades were 3.1 meters tall. The area swept by the rotor was—at most—9 m². It's hard to imagine how this got by. This measurement is something so fundamental and easy to check with an H-configuration VAWT that not getting it right is inexcusable.

The implication in these three examples is that either Cleanfield didn't understand wind energy, or worse, it was being intentionally misleading. Everyone makes mistakes. It's conceivable that Cleanfield might get something wrong. But in the case of Cleanfield, there was a pattern of mistakes, blunders, and promises that were never fulfilled.

In the spring of 2013, the company defaulted on its debt, according to news reports, and by the fall, their website was dark, nearly a decade after the product was launched in Ontario. Cleanfield lasted longer than most other small VAWT products, and their example is far from the worst case.

Fortunately, the commercial-scale turbine at Ex-Place went ahead, and it has been generating nearly 1 million kWh per year for the past decade. This turbine has since been enormously successful at demonstrating the use of wind energy in an urban setting—and it's not on a building.

Mariah Windspire

Hollywood is justifiably mocked for its "high-concept" movie themes: duds that are long on what at first glance appears as a clever idea but short on plot. Wind energy has its high-concept wind turbines too: clever ideas that on closer examination are short on substance.

Mariah and its Windspire will go down in history as another novelty Vertical Axis Wind Turbine that was doomed from the start. Tall and slender, it was—if nothing else—distinctive. Introduced with much media fanfare in the mid-2000s, the fixed-pitch straight-blade VAWT's claim to fame was a rotor with an extended height to diameter (EHD). Mariah pitched the turbines EHD as a

Figure 6-26. Mariah Windspire. Tall, slender, fixed-pitch blades—a Darrieus rotor on a stick. Unlike its VAWT competitors, Windspire was responsibly rated at only 1.2 kW, but with a limited swept area of only 7 m², short towers, poor siting, and poor performance, the turbines simply didn't produce a lot of electricity. Shown here are two Windspire turbines in downtown Indianapolis, Indiana. 2013.

way to overcome one of VAWTs major limitations, raising the rotor off the ground. Unfortunately, a tall and slender wind turbine mounted on a short tower is still a wind turbine on a short tower (see Figure 6-26. Mariah Windspire).

Unlike Cleanfield, PacWind, and many of the other small VAWTs introduced in the 2000s, Mariah did attempt to test, refine, and document the turbine's performance. Thus, there are a number of reports on the performance, or the lack thereof, of Mariah's Windspire. We know, for example, that the turbine suffered mechanical and electrical failures that required the turbine's redesign. However, even under ideal conditions, with the turbine operating as advertised, the small turbine simply couldn't produce more than a few thousand kWh per year, and often, where they were monitored, the turbines consumed more electricity than they generated.

Through an aggressive marketing program, Mariah installed more of its Windspire turbines than possibly any other small VAWT in North America. At one time the company claimed it was manufacturing 1,000 turbines per year.

The company eventually realized that the design was really only suitable in "architectural" or "urban" applications that were looking for visual bling. For example, two of the wind turbines were installed at Crissy Field in urban San Francisco in a "demonstration" project. (What was being "demonstrated" was never made clear.) Nearby, San Francisco's Public Utility Commission installed four of the turbines on the side of its new "energy-efficient" building.

Architectural ornamentation alone is not enough of a market to support a wind turbine manufacturer, as Mariah learned to its regret. In early 2012, Mariah became a victim of its hubris and sought bankruptcy. By mid-2013, some communities were already ordering removal of the turbines from their urban areas.

Helical Wind Turbines

In the early 2000s, a series of unusual VAWTs began appearing on the European market. The rotors appeared to be transitional between a Φ-configuration Darrieus and a typical H-configuration rotor. The turbines appeared similar to helical rotors developed a decade earlier by Northwestern University's Alexander Gorlov for tidal and in-stream hydro turbines, though they were developed independently.

These turbines used three blades twisted in a helix about the vertical axis. The three-blade, helical design reduces the torque ripple encountered in traditional two-blade, Φ-configuration Darrieus or eggbeater turbines. The same effect, of course, can be achieved with articulating blades but at the cost of greater complexity.

Turby, a Dutch turbine, was the first on the scene, and its promoter immediately began touting "promising" wind tunnel tests at TU-Delft. Like many companies before (see Cleanfield), the promoter used the university's tests as a form of academic imprimatur. Turby and TU-Delft quickly became the bane of the wind industry for their promotion of urban wind turbines installed on building rooftops as a vast new market. It was not to be, and Dutch field measurements of Turby's performance in the late 2000s were disastrous.

However, it was the British company Quiet Revolution that first popularized helical VAWTs in the English-speaking world with its architecturally striking QR5 (see Figure 6-27. Quiet Revolution's helical fixed-pitch VAWT). Since the QR5's introduction, the helical, fixed-pitch rotor has been widely copied by numerous VAWT manufacturers, though often with much less elegance (see Figure 6-28. Venco Twister helical VAWT). Like Turby in the Netherlands, Quiet Revolution sought to remake the market for small wind turbines by promoting its turbine for rooftop, urban installations.

Because of their dramatic aesthetic appeal, helical VAWTs quickly became the darling of architects looking to adorn their buildings with some "green" ornamentation. In North America, the addition of a wind turbine to a building or as part of a building's development gives the building design extra points toward an energy-efficiency certification. Architects seeking top honors became notorious for adding small wind turbines—often helical—as an architectural element, often only kinetic sculptures. Quiet Revolution pioneered this market in Great Britain and was probably the world's most successful company at integrating its wind turbines as part of an architectural design.

Company press releases emphasized the "compact design" of the turbine, making it more suitable, they claimed, for rooftop installation. Of course, what they failed to say is that, by definition, a compact design—or one that has little swept area—will generate very little electricity. Nor did the company explain that rooftop siting of wind turbines was problematic at best.

In the fall of 2008, RWE, one of Germany's big four electric utilities, invested £7 million ($11 million) into the company. RWE is not a significant developer of renewable energy in its home market and is notorious for aggressively opposing expanded renewable energy in Germany.

VERTICAL-AXIS AND DARRIEUS WIND TURBINES

Figure 6-27. Quiet Revolution's helical fixed-pitch VAWT. The strikingly beautiful QR5 has been widely imitated. Despite its appearance, it remains a simple H-rotor that intercepts less than 16 m² of the wind stream. Although the manufacturer rated the turbine at 6.5 kW, the QR5 is equivalent to a conventional wind turbine 4.4 meters (14 feet) in diameter rated at 3.2 kW to 4 kW. Like many other VAWTs, the QR5 used inflated power ratings. 2010. (Andy Dingley, Wikimedia Commons)

Figure 6-28. Venco Twister helical VAWT. This is a German take on the fixed-pitch helical rotor. This is one of two turbines installed in a "demonstration" project at Crissy Field Center, part of San Francisco's Golden Gate National Parks. Neither turbine was operating when this photo was taken. 2012.

Critics quickly suggested that RWE was trying to "greenwash" its image in Britain by promoting unproductive wind turbines. Past experience with utility investment in questionable new wind turbines was usually the kiss of death for the product.

Quiet Revolution launched in 2005, and by 2008 had installed only 30 turbines throughout Britain. In early 2011, the company was hoping to double the number of installed turbines from the 150 then installed. In late spring 2011, the QR5 became the first small VAWT—of any configuration—to become certified to international standards.

Now, after several years of operation, tales of the turbine's performance in the field were becoming available. It was not unexpected, then, to hear the company announce in late 2012 that they were redesigning and recertifying the turbine to double its output and to increase the turbine's return on investment.

Clearly, the QR5 and its marketing had a way to go. The BBC reported in late 2013 that a QR5 installed at a Welsh government office cost £48,000 ($75,000) (architectural bling doesn't come cheap) and had proven unreliable. The BBC estimated that at the rate it was generating

THOSE WHO DON'T BUILD VAWTS

This chapter is in part a litany of failure, a saga of those who have tried to build VAWTs of various shapes and sizes and who no longer do so. Some of those discussed are illustrious names in the field of wind energy, such as Vestas and Bergey, whose success today is built on conventional three-blade wind turbines. There's a message in this. Many have tried. All have failed or have chosen a more fruitful path, and this list doesn't include the numerous developers of small VAWTs of the modern VAWT revival—or the many others worldwide who have tried.

Adecon	DAF-Indal	Heidelberger
Alcoa	Herman Drees	Sir Robert McAlpine
Bergey	Dornier	McDonnell Aircraft
Bristol Aerospace	FloWind	Vestas

electricity, the turbine would require 450 years to pay for itself.

By late 2013, the new turbine had yet to be certified, and Quiet Revolution was promoting a conventional Chinese-built wind turbine on its website. The Chinese wind turbine achieved certification to the British standard. Meanwhile, the certification for the original QR5 was removed. Though it was expected that the redesigned turbine would eventually become recertified, Quiet Revolution failed before this occurred.

Like all its predecessors, what began with much fanfare failed to remake the market for small wind turbines. After nearly a decade of promotion, less than 200 of the QR5 had been installed worldwide, only marginally better than Herman Drees's Pinson Cycloturbine in the early 1980s.

VAWT Revival: Not Likely to Continue

The flurry of small VAWTs that came on the market in the mid-2000s led some to suggest a revival of VAWT technology was underway. Not everyone was thrilled at the idea of a VAWT renaissance. "The driving force [behind the revival of VAWTs]," says Ian Woofenden, "is ignorance of past failures and arrogance about overcoming the problems" inherent in the designs. Woofenden, the author of a popular book on small wind turbines, points out (as noted at length in this chapter) that VAWTs are not new. Tellingly, he admonishes "anyone designing a new VAWT should do their homework first to find out what past designers learned."

As Woofenden notes, few of the "new" designs are in fact new at all. Many of the designs touted as new have been around for decades if not generations. The only exception is the new class of helical VAWTs. These designs had not been seen before. Whether they are revolutionary or not, or whether they are better than preceding Darrieus designs, only time and field experience will tell. The track record so far is not encouraging. For now, helical turbines have followed the same route to commercial failure as their predecessors.

Another of the "show me it works" skeptics is small wind guru Mick Sagrillo. And when he says "show me it works," he means show him that it produces electricity in kilowatt-hours—not simply "spinning" in YouTube videos. After all, it is wind energy in kilowatt-hours that we're after. Sagrillo grumbles about one famous YouTube video showing men installing a purported wind turbine on the roof of a building to Beethoven's *Ode to Joy*, and immediately at the climax, the turbine starts spinning. Indeed, spinning is the operative word because the rotor of this "turbine" wasn't connected to anything, no generator, no gearbox, no drivetrain at all. People are "dazzled by the motion," says Sagrillo in despair. Most viewers never notice that the "turbines" are not doing anything productive, only spinning. As Sagrillo pithily notes, "If that's all you want, buy a whirligig for $12 and be done with it."

VAWT skeptics, such as Woofenden and Sagrillo, hold inventors of conventional wind turbines to the same standards as they do vertical-axis designers. They don't single out vertical-axis designs solely because they use a vertical axis, as some VAWT proponents charge. As Jon Powers, a Tehachapi windsmith with decades of operational experience on conventional wind turbines notes, if inventors develop a nontraditional wind turbine, they should prove it, commercialize it, and deploy it in the thousands. Then it, too, will become the new conventional thinking. And, he could add, everyone, skeptics included, would applaud their success.

POOR COMPARISON BETWEEN SMALL VAWT AND SMALL HAWT

In the spring of 2014, Urban Green Energy (UGE) announced with much fanfare that they had completed third-party performance testing on a new product, the VisionAIR wind turbine. As press releases for new wind turbines go, it wasn't as over the top as many. Nevertheless UGE, an importer of small helical VAWTs, crowed that the tests "confirmed that the turbine... is twice as efficient as several of its competitors." Of course, UGE cleverly avoids listing these competitors so no one can challenge whether the statement is true or not.

UGE's VisionAIR is a relatively small household-size wind turbine that sweeps 16.6 m². Competing turbines in this size class include Bergey Windpower's Excel 6 a conventional upwind turbine intercepting nearly twice the area (30 m²) of UGE's VAWT (see Comparison Between Small VAWT and Small HAWT).

So, how does the VisionAIR VAWT stack up to the Bergey 6? Not so well. Despite UGE's hype, Bergey's turbine is twice as efficient as the VisionAIR at extracting the energy in the wind at the standard wind speed used to rate small wind turbines: 22% for the Bergey and only 11% for the VisionAIR.

Worse, UGE's data confirms the fundamental disadvantage of most VAWTs: they are much more material intensive than conventional wind turbines. The relative mass of a wind turbine—a measure of its material intensity, which includes the rotor and nacelle—is often used as shorthand for its expected cost. The more the relative mass of a turbine, the more it's likely to cost. The specific mass of UGE's VisionAIR is four times more than Bergey's Excel 6.

But the poor comparison doesn't end there. The test report also confirms that UGE's VAWT—like so many other fixed-pitch VAWTs—"has no passive braking means." That is, the only way to limit the rotor's power under emergency conditions is to stop it with a brake, whether that brake is mechanical or by using the generator. The VAWT has no passive or aerodynamic means—other than stall—to limit the power of the rotor in high winds. And stall is not a reliable means for overspeed protection. In contrast, the rotor on the Bergey can passively furl toward the tail vane in strong winds, regulating the rotor's power.

Future UGE press releases may well note that though the VisionAIR only produces 1.5 kW at 11 m/s (25 mph), it will produce a peak power of 2.5 kW at 14 m/s (31 mph) or two-thirds more than at 11 m/s. Yet this only illustrates one of the design's principal weaknesses compared to conventional wind turbines: there is no passive means to control the rotor's power. Compare this with Bergey's Excel 6. Bergey's turbine will produce up to 6.5 kW at an equivalent wind speed–an increase of only one-fifth—indicating that the Bergey has an effective means of limiting the rotor's power in strong winds. Passive furling of the rotor in high winds is a significant design advantage of the Bergey relative to the limited stall control in the VisionAIR VAWT.

To summarize, UGE's VAWT is half as efficient, is four times more material intensive, and has no effective means for limiting the rotor's power in strong winds relative to a competitive conventional wind turbine.

UGE is the same company that installed two of its helical VAWTs inside–yes, inside–the Eiffel Tower. Gustave Eiffel must be turning in his grave.

Comparison Between Small VAWT and Small HAWT

Manufacturer	Model	Rotor Height m	Rotor Height ft	Rotor Dia. m	Rotor Dia. ft	Swept Area m²	Tested Power kW	Tested Speed m/s	Tested Speed mph	Perf. at Tested Power %	Loading at Tested Power W/m²	Tower Top Mass kg	Specific Mass kg/m²
Urban Green Energy	VisionAir	5.2	17.1	3.2	10.5	16.6	1.5	11	25	0.11	90	756	45.4
Bergey Windpower	Excel 6			6.2	20.3	30	5.5	11	25	0.22	182	350	11.6

Tested to IEC 61400-12-1.

For whatever reason, many VAWT promoters are prone to more hyperbole than most other wind turbine designers. Some claim their turbines will produce at less cost and with less impact on the environment than conventional wind turbines. Maybe such claims are due to widespread ignorance of VAWT technology or its long history, as Woofenden charges. Maybe vertical-axis designers imagine that their turbines don't suffer the limitations of "all those other ordinary wind turbines." Certainly, it's easier to visualize how drag devices, like cup anemometers, work than how those thin spindly blades on conventional wind turbines can be so efficient at extracting the energy in the wind.

Do VAWTs work? Of course they do. This isn't in question. Can they produce useable quantities of electricity? Yes, the record in California is clear: FloWind's Darrieus wind turbines generated millions of kilowatt-hours for nearly a decade. Can they compete with conventional wind turbines? Perhaps, but the record suggests it's unlikely. Are there specialized markets for which VAWTs may be ideally suited? Possibly, but by definition a specialized market is a limited market. Do we need wind turbines for specialized markets? Yes, we'll need all the renewable energy we can find in the coming decades. Do VAWTs offer a panacea of limitless renewable energy without the drawbacks of conventional wind turbines? No. Many small VAWTs sweep

Figure 6-29. Small Savonious rotors for remote sites. Savonious rotors have found a limited niche in very low power applications at remote sites such as on marker buoys. Unfortunately, these Savonious rotors were installed in another "urban demonstration project" where they serve more as kinetic sculpture than as wind turbines. 2013.

only enough area of the wind stream to "light an exit sign," warns Sagrillo.

VAWT proponents sometimes claim there's little or no evidence that their designs won't do what they promise. (See the next chapter for more on this logic.) That may be true for a particular product, but the record on VAWTs in general is damning. None have proven commercially successful for more than a decade except in the most limited of niche markets (see Figure 6-29. Small Savonius rotors for remote sites). The burden of proof always lies with the inventors to prove that their claims are true.

Claims and Counterclaims

Here, then, are some of the arguments VAWT proponents use to contrast their technology with that of conventional wind turbines. Note that the claims and the following counterclaims can be applied equally well to any proposed wind turbine that is advertised as a technological breakthrough (see the next chapter, "Novel Wind Systems," for more on "breakthrough" designs).

- They can accept the wind from any direction.
- They are simpler.
- They are more reliable.
- They are less costly.
- They are more powerful.
- They are more efficient.
- They are more cost effective.
- They don't kill as many birds.
- They are less noisy.

Omnidirectional

The most often heard justification for Darrieus wind turbines is that they are omnidirectional, that is, they can accept the wind from any direction. As Ian Woofenden is quick to point out, conventional wind turbines also accept the wind from any direction. Conventional wind turbines must yaw or turn toward the wind, requiring some means to do so. Thus, the principal technological advantage of VAWTs over conventional wind turbines is reduced to "they can accept the wind from any direction without a yaw mechanism."

The question then becomes, how important is this advantage? The answer is, not very much. Worse, what appears as an advantage may in fact be fixed-pitch VAWTs fatal flaw. By accepting the wind from any direction, VAWTs can't be turned—yawed—out of the wind like a conventional wind turbine to limit their intercept area.

Simpler

Yes, some VAWTs are simpler than conventional wind turbines because they don't use a yaw mechanism to orient the turbine relative to the wind. However, some VAWTs are more complex. For example, an articulating, straight-blade VAWT is more complex than a conventional wind turbine that uses fixed-pitch blades. In the end, VAWT designs often trade one form of complexity for another and one advantage for a severely limiting disadvantage.

For guyed, Φ-configuration Darrieus, the generator and gear train can be located on the ground. The presumed advantage is the simplicity of servicing the components on the ground as opposed to servicing them at the top of a tall tower. Although this may be true at first glance, it illustrates that what appears as an advantage hides a fundamental design flaw. We use tall towers—even on flat terrain—because that's where the wind is. Guyed Darrieus turbines are earth bound, limiting the wind they intercept and, hence, the amount of wind energy they can capture.

Further, in a 2003 study for Sandia that looked at both large and small Darrieus wind turbines, consultants concluded that "the savings that a VAWT may enjoy due to lower drivetrain and maintenance costs are unlikely to balance the lower energy capture and higher initial rotor costs" found in Darrieus wind turbines. And this was from a study commissioned by the federal agency assigned to make VAWTs work.

More Reliable

In nearly all cases, proponents have no field experience to support such a claim. Often the claim is based only on a wish, not on real performance. Very few manufacturers of VAWTs provide performance data in the public domain where independent analysts can gauge the reliability of their design. The data that is available is convincing. As chronicled in this chapter, most VAWTs to date have proven less reliable in the field than conventional wind turbines.

Could VAWTs be made more reliable than they have been? Yes, says wind engineer Robert Preus. He should know. Preus was an engineering consultant on the FloWind Darrieus turbines in the 1990s. It's not that we can't do it, argues Preus, it's just more difficult and challenging than for a conventional wind turbine.

Who then is going to pay for all the development work necessary to do so? FloWind failed because it didn't have the money to carry the turbines through another retrofit. Would FloWind have succeeded if it had had the money? We'll never know. It didn't receive the money in time to fix the serial defects in the turbines. Was this unfair? Likely, but the reality is FloWind failed and with it went the dreams for Φ-configuration Darrieus turbines. There have been many manufacturers of conventional wind turbines that have failed too, but the fundamental design lives on. Not so with commercial-scale Darrieus turbines.

Less Costly

Maybe some day in the future, VAWTs could be cheaper than conventional wind turbines, but that hasn't been the case to date. The record for modern VAWTs is clear. They have been more expensive than conventional wind turbines. Not only have they been overhyped, they've also been overpriced. Quiet Revolutions QR5 was being installed in Britain at a startling cost of £40,000 ($62,000) for what is the equivalent of $4,000/m^2 of rotor swept area—four times more than the cost of a commercial-scale, conventional wind turbine and two times more than a conventional small turbine of similar size.

Mariah's Windspire was priced competitively with other small wind turbines at about $2000/m^2, but it wasn't cheaper. And it could be argued that Windspire didn't earn enough on each turbine to fix the defects it found among those it'd already installed.

More Powerful

Fixed-pitch VAWTs, as explained below, must use oversize generators because they depend on stall to limit the power of the rotor in strong winds. Again, what appears to the uninitiated as an advantage is in fact another serious disadvantage. Consequently, the large generators for fixed-pitch VAWTs—relative to their swept area—lead to grossly exaggerated power ratings. And as in the 1980s, ill-informed

MONSIEUR DARRIEUS AND HIS WIND TURBINES

Largely forgotten today, Georges Jean Marie Darrieus was one of France's great engineers. While he is mostly known in the English-speaking world for his patent on vertical-axis wind turbines (VAWTs), he was a prolific inventor in a number of fields from ballistics to turbo-alternators.

When considering Darrieus's role in wind energy, it is useful to remember the time in which he lived. In 1925, Darrieus filed for a patent on his "turbine having its rotating shaft transverse to the flow of the current" in France and a year later in the United States.

Following the Great War, the field of aeronautics was in full flower, and Darrieus was a contemporary of Frederick Lanchester, Ludwig Prandtl, and Albert Betz (of the Betz limit fame). Darrieus's patent specifically makes the link to the theories of all three. Thus, Darrieus was well grounded in what was then modern aerodynamic thought.

Although the patent illustrates the Φ-configuration VAWT that has come to be most closely associated with his name, it wasn't the first design he chose to describe. No, he first describes an articulating, straight-blade, vertical-axis rotor that we know of as a giromill or Cycloturbine. His lengthy patent goes on to describe other forms of H-rotors in what he aptly calls a "squirrel cage," which can be used not only in the wind but also in rivers and tidal currents.

Most telling, however, is that Darrieus did install experimental wind turbines for his employer, Compagnie Electromécanique (CEM) at le Bourget outside Paris. These wind turbines were what we now call conventional wind turbines. Darrieus installed these turbines in the late 1920s–after he filed his famous patent on VAWTs (see Darrieus's Wind Turbines). To reiterate, Darrieus built wind turbines following his patent, but he did not build "Darrieus" wind turbines. He built horizontal-axis wind turbines.

That Darrieus had reached the same conclusion as later designers about the need for "fast-running" rotors was evident is his first turbine in 1927. The tip speed ratio for this four-blade turbine was 5.6, and his second turbine ran at an even higher tip-speed ratio of 7.9.

French historian Etienne Rogier notes that there is no evidence that Darrieus ever built a Darrieus wind turbine. We do know, writes Rogier, that during World War II Darrieus conducted experimental wind tunnel tests on a scale-model, three-blade, H-configuration rotor in Toulouse. Unlike many latter-day VAWT inventors, academics, and engineers who have been lured by the promise of Darrieus turbines, Darrieus himself concluded that the performance of his model was too poor to justify further work–though he remained interested in it the rest of his life.

Darrieus was clearly a man ahead of his time but not necessarily for the invention that bears his name: the vertical-axis wind turbine. The invention of the eggbeater and straight-blade, vertical-axis wind turbine were a minor part of his vast body of work over a long lifetime of achievement. Like many other wind turbine designers in the decades since, Darrieus–in the end–opted for a conventional, horizontal-axis wind turbine.

Darrieus's Wind Turbines				
Diameter		Blade No.	Tip Speed Ratio	Year
m	kW			
8	2	4	5.6	1927
10	4.5	3	7.9	1930
20	15	3	5.2	1929
Installed by Compagnie Electrmécanique (CEM) at le Bourget. Source: Etienne Rogier, Cahiers D'Eole, November, 2000, No.2.				

consumers—and not a few utility executives as well—are then easily misled into believing that a wind turbine with a bigger generator will necessarily generate more electricity than one with a smaller generator but a larger rotor when the opposite is true.

More Efficient

VAWT developers McDonnell Aircraft, Clearfield, and Turby touted their products based on limited wind tunnel tests. Calculation of efficiency from wind tunnel or truck tests says very little about how a wind turbine will operate in real winds. Even where a wind turbine is markedly more efficient than another wind turbine, reliability is a far more critical parameter. A wind turbine may be efficient, but if it is not working reliably, it will produce little or no electricity.

This was a lesson demonstrated time and again in California during the 1980s.

Contrast the results of supposed superior performance obtained by VAWT promoters with that from the inventor himself, Monsieur Darrieus. He found his wind tunnel tests on a scale model VAWT disappointing and the performance so poor he didn't pursue it further. That should say volumes about the prospect for greater efficiency from VAWTs relative to conventional wind turbines.

More Cost Effective

Whether a wind turbine is a good buy or not is a function of its installed cost, the amount of energy it generates, and the cost of operating and maintaining it. Modern VAWTs have fallen far short on each of these parameters.

Safe for Birds

Like so much else about VAWTs, this is simply an unfounded claim. Very few studies have been done on birds and modern VAWTs because there are so few VAWTs in operation. Nearly all studies to date have been conducted on large commercial wind turbines because that is what is being used and deployed on a large scale. These studies find that the number of birds killed by wind turbines is primarily a function of turbine size. A big wind turbine will kill proportionally more birds than a small wind turbine—of any configuration. Nearly all modern VAWTs are extremely small, and the likelihood of a small wind turbine—of any configuration—killing any bird is, therefore, very small.

Proponents further argue that modern VAWTs won't kill birds because their rotor speed is lower than that of conventional turbines. Again, this is simply a claim without substantiation. Whether a structure is moving in the wind or its specific speed of movement is slower than another is not critical to whether it will kill birds or not. Birds fly into stationary windows and glass-walled buildings all the time.

Less Noisy

This is one claim that may have real merit. The blades on modern VAWTs may move through the air at much lower speeds than blades on conventional wind turbines. The lower blade speeds often translate into lower noise emissions than those from conventional wind turbines. Unfortunately, there is very little field experience, and even less publicly available data, to verify this assertion.

VAWT Design Characteristics

Many of the claims by VAWT proponents touch on specific design characteristics, so it may be helpful to look at these in more detail. VAWT design, like the design of any other wind turbine, is a series of trade-offs. For each plus there is a minus. We can't evaluate design elements in isolation. We must consider the wind turbine as a complete package. What is the historical record of this particular design? How many of this design have been installed in the past. Have they been tested? What have been the results?

For example, the lower blade speed of modern

Table 6-3. Selected Vertical Axis Wind Turbine Specific Mass								
	Dia.	Height	Swept Area	Rated	Specific Power	Specific Area	Specific Mass	Approx. Date
	m	m	m²	kW	W/m²	m²/kW	kg/m²	
Φ-configuration								
FloWind	19.2	25.2	340	250	735	1.4	32	1985
DAF-Indal	24.4	36.7	595	500	840	1.2	34	1980
Sandia	34.0	42.5	955	500	524	1.9	76	1987
Eole	64	96	4,000	4,000	1,000	1.0	86	1987
H-configuration								
McDonnell Aircraft	17.7	12.8	227	40	177	5.7	31	1977
Pinson	3.7	2.4	8.9	2	224	4.5	30	1979
Cleanfield	2.8	3.1	9	3	352	2.8	29	2002
Quiet Revolution	3.1	5.0	16	6.5	419	2.4	29	2008
Conventional								
Bergey	7.0		38	10	260	3.8	12	1982
Micon	16.0		201	60	298	3.4	23	1984
USW 56-100	17.8		249	100	402	2.5	26	1984
Vestas	25.0		491	200	407	2.5	21	1987
Southwest Windpower	3.7		11	2.4	223	4.5	7	2008

VAWTs is a design element, a by-product of which is lower noise emissions. However, notes Jim Tangler, a retired aeronautical engineer, VAWTs must derive more of their power from torque than conventional wind turbines because the rotor spins at lower speeds. To handle the greater torque, the blades and their supports must be stronger. This results in greater mass and, hence, often greater cost than for similar components of a conventional wind turbine.

A common measure of materials used in a wind turbine relative to the area of the wind the turbine intercepts is tower head (or nacelle) specific mass in kilograms per square meter (see Table 6-3. Selected Vertical Axis Wind Turbine Specific Mass). This measure includes the rotor, generator, and drivetrain. Darrieus wind turbines of the 1970s and 1980s were more material intensive than conventional wind turbines of the period. This was particularly the case for the largest VAWTs of the day, such as Eole and Sandia's 34-meter test bed.

Yet, this isn't the whole story. Tangler's comment is directed at the rotor alone, and here the difference is far more substantial (see Figure 6-30. Relative rotor mass of VAWTs and HAWTs). Consultants to Sandia found that the specific rotor mass of Darrieus wind turbines was an order of magnitude greater than that for conventional wind turbines. The consultants concluded that "for a given swept area, the mass of the rotor and support structure of a VAWT will be greater than that of an equivalent HAWT. This mass difference is likely to translate into a cost difference." That is, the Darrieus rotor will cost more—not less—than that for a conventional turbine.

Further, says Tangler, the blades of fixed-pitch VAWTs, like the Φ-configuration Darrieus, only operate at optimum aerodynamic performance over a small portion of their carousel path, and they typically use less efficient symmetrical airfoils than the cambered (asymmetrical) airfoils used on conventional wind turbines. All in all, says Tangler, VAWTs of the 1980s weighed more and were less efficient than conventional turbines of the period.

Efficiency and Performance

It is unlikely that any "modern" reinvention of Darrieus designs can perform any better than the experimental turbines Sandia and its contractors developed during the late 1970s and early 1980s. Sandia measured the performance of its 5-meter-diameter, two-blade Darrieus and found a coefficient of performance (C_p) of more than 0.3.

In testing the performance of their 17-meter-diameter, two-blade Darrieus Sandia obtained a C_p of more than 0.4 at a tip-speed ratio of 6 without struts and with struts the C_p was, as you would expect, substantially less, but still well above a C_p 0.30. Thus, Φ-configuration Darrieus wind turbines can technically perform nearly as well as those of conventional machines but certainly not better (see Figure 6-31. Vertical-axis rotor performance).

This performance is probably the reason that Alcoa's Paul Vosburgh was fond of saying that Darrieus wind turbines delivered as high performance in these terms as conventional wind turbines. That doesn't mean the turbines would be more cost effective, more reliable, more bird friendly, or more attractive than any other wind turbine, only that the rotor's aerodynamic performance was as good as conventional machines.

However, aerodynamic efficiency, while critical, is not the sole measure of performance. In a study for Sandia in 2003, consultants compared

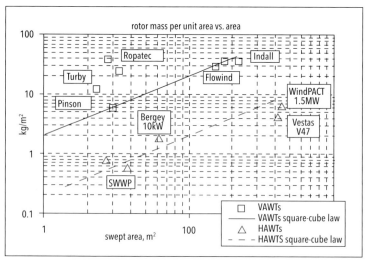

Figure 6-30. Relative rotor mass of VAWTs and HAWTS. In a study for Sandia, Global Energy Concepts found that the relative rotor mass of VAWTs was an order of magnitude greater than that for conventional wind turbines. While components of a VAWT rotor may be cheaper to manufacture than a conventional wind turbine rotor, it would be difficult to overcome such a disadvantage. Pinson, FloWind, Turby, and Indal are all VAWTs. SWWP, Bergey, and Vestas are conventional wind turbines. (Sandia National Laboratory)

VERTICAL-AXIS AND DARRIEUS WIND TURBINES

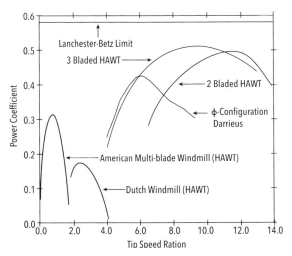

Figure 6-31. Vertical-axis rotor performance. Rotor coefficient of performance, C_p, for historical windmills, conventional two-blade and three-blade wind turbines, and Φ-configuration Darrieus turbines relative to tip-speed ratio. What is unusual in this Sandia version of a common chart found in wind energy is the portrayal of Darrieus rotor performance by the national laboratory charged with developing the technology. Sandia found that rotor performance at tip-speed ratios common to commercial wind turbines was roughly equivalent, though conventional wind turbines had greater opportunity for performance improvements than Darrieus turbines. (Sandia National Laboratory)

the estimated productivity from two Φ-configuration Darrieus wind turbines (DAF-Indal's 500 kW model and Sandia's 34-meter test bed) with that for a hypothetical 1.5-MW conventional wind turbine (see Table 6-4. Theoretical Annual Specific Yield). They found that at moderate wind sites, a conventional wind turbine would generate 20% to 30% more electricity relative to its swept area than the most sophisticated Darrieus turbine of the day, Sandia's 34-meter test bed. At high-wind sites, the conventional turbine would still produce about 10% greater yield than Sandia's design.

Stall Control and Overspeed Protection

As explained in a previous chapter (see Chapter 4, "The Great Wind Revival"), an early lesson from the development of wind energy in Denmark was the necessity of providing an aerodynamic means for overspeed protection. Wind turbines of the day used fixed-pitch rotors driving induction (asynchronous) generators. As wind speed increased, the aerodynamic performance of the rotor would decrease as it approached stall. The increase in power with wind speed would begin to taper off. This phenomenon, known as stall control, required that the wind turbine had a large enough generator to keep the rotor under control through its full operating range.

Stall control worked reasonably well most of the time, but there were the unfortunate exceptions. The worst-case scenario is the generator suddenly being taken off-line—for whatever reason—in high winds. The brakes must operate instantly and bring the rotor to a halt. If they don't, the rotor can speed up to destruction. The Danes learned this the hard way, and eventually all stall-regulated, conventional turbines began using aerodynamic brakes of one design or another to prevent the rotor from self-destruction.

However, among Φ-configuration Darrieus manufacturers, only DAF-Indal employed air brakes on their rotors. Alcoa, Vawtpower, and FloWind all relied solely on massive brakes. This just wasn't good enough, and no doubt contributed to the demise of their designs. The braking system on fixed-pitch VAWTs has to be "redundant and bullet-proof," says wind engineer Robert Preus. Brakes work reliably—most of the time. It's that one time when they don't that can lead to disaster.

This weakness of fixed-pitch VAWTs has been known for generations. In 1961, Louis Vadot described the limitations of Darrieus turbines in a United Nations report on wind power, "Finally," he said, "it is very difficult to design and build a practical furling device to protect these machines in high winds." For him that was the end of the discussion.

Vadot doesn't say it can't be done. He simply says it's very difficult to do so and notes that this is a major flaw. From this he concludes that

Table 6-4. Theoretical Annual Specific Yield			
	WindPACT	DAF-Indal 6400	Sandia 34 m
Mean wind speed	1.5 HAWT	VAWT	VAWT
m/s	kWh/m²/yr	kWh/m²/yr	kWh/m²/yr
6	950	500	679
6.5	1,125	664	865
7	1,296	846	1,059
7.5	1,461	1,038	1,255
8	1,616	1,235	1,447
8.5	1,761	1,432	1,631

Global Energy Concepts. http://energy.sandia.gov/wp/wp-content/gallery/uploads/SAND2012-0304.pdf. Note: From a study done by Global Energy Concepts on large wind turbines: their aerodynamic efficiency will be considerably greater than that for small wind turbines. What this table shows is that large VAWTs with fixed-pitch blades will have a significantly less yield than large HAWTs.

VAWTs have generally been abandoned, probably referring to his French colleague Darrieus's work in the 1920s, and "there will, in all probability, be no major development" of Darrieus turbines. He then moves on, like Darrieus did himself, to focus on those configurations likely to make a difference to the world, despite noting that VAWTs may find use in some "simple, low-power applications," which has proven true.

The exception to Vadot's concern about an ability to furl the wind turbine out of the wind, of course, is articulating, straight-blade VAWTs. Machines, such as the Pinson Cycloturbine or McDonnell Aircraft's giromill, that control the pitch of the blades can also control rotor speed. Thus, McDonnell Aircraft's giromill could use its pitch servo motors to feather the blades in an emergency or to park the rotor. Herman Drees's solution on the Pinson Cycloturbine was to use a Watt-governor to control pitch in an overspeed condition through a collective pitch-control linkage.

Self-Starting

Φ-configuration Darrieus turbines of the 1970s and 1980s were not reliably self-starting. The low-solidity, fixed-pitch blades then being used couldn't drive the rotor up to operating speed from a standstill unless the blades are parked in just the right position relative to the wind. Although this isn't necessarily a serious limitation, it does require motoring the rotor up to speed.

As noted in the destruction of the Darrieus turbine on the Îles de la Madeleine, the rotor could, under the right conditions, start rotating on its own. However, in commercial operation, the turbines were always started by using the generator as a motor to power the rotor up to speed. As a consequence, Darrieus turbines did not freewheel in light winds, unlike most of the conventional wind turbines installed in California. Because they consumed electricity during start-up, FloWind's Darrieus turbines were programmed to sit out light and variable winds. In light winds, the hundreds of FloWind turbines in California would sit idly by as thousands of Danish wind turbines would be spinning merrily. This led to a perception, partly justified, that Darrieus turbines in California were less reliable and less productive than the turbines from FloWind's Danish competitors.

Increasing rotor solidity by increasing the number of blades and their width can add self-starting capability to Darrieus turbines of all configurations. The VAWTs of the modern VAWT revival incorporate this lesson and are self-starting.

Fatigue

In California, Φ-configuration Darrieus turbines were bedeviled by poor performance and poor reliability. The aluminum blades then used often fatigued and sometimes failed catastrophically. This was due in part because the lift forces, which propel the blades, reverse direction every revolution, flexing their attachment to the *torque tube* or central mast. Another source of frequent flexing of the blades is inherent in the rotor's eggbeater shape. When the rotor is at rest, the blades sag due to their own weight, stressing the connection to the torque tube. Moreover, the presumed advantage of housing the drivetrain at ground level was offset by the large bearings and guy cables at the top of the rotor.

Aluminum can be designed to function reliably in a Darrieus wind turbine application, say wind engineers such as Robert Preus. However, most designers today opt for composite materials such as reinforced polyester (fiberglass) or carbon-fiber blades, as have been developed for conventional wind turbines.

Guyed Darrieus

One dilemma all guyed wind turbines face is the difficulty of installing them in rugged terrain because the guy cables and anchors may be spread over an uneven surface. This challenge particularly affects Φ-configuration Darrieus that use a guyed rotor. Although FloWind installed hundreds of turbines in hilly terrain, the majority of these were installed on a gently rolling plateau.

VAWTs Now Marginal

During the 1970s and 1980s, VAWTs were one of the great currents in wind engineering that stood an equal chance, along with conventional wind turbines, of becoming a mainstream technology. As chronicled in this chapter, that didn't happen. Today, there are more than 350,000 MW of wind-generating capacity worldwide from a fleet of nearly 250,000 conventional wind turbines. Collectively, they generate nearly 800 TWh per year.

VAWTs are now so rare they've become a visual curiosity. Altogether, there are probably no more than a few thousand small VAWTs globally. Darrieus turbines account for such a small percentage of small wind turbines worldwide, and small wind turbines account for less than 0.2% of worldwide wind generating capacity, that they've become truly inconsequential.

Another measure of how far VAWTs have fallen out of favor is the relative paucity of technical information on the subject. One of the premier textbooks on wind engineering, Robert Gasch's *Wind Power Plants*, devotes less than two pages in the 550 page treatise to Darrieus wind turbines and their variations.

Certification to Minimum Testing Standards

For more than three decades, a promoter of a small VAWT could make any wild claim about the supposed superior performance of its wind turbine it wanted. Many claims were simply unfounded, unsubstantiated, or meaningless. Some resorted to celebrity endorsements by late-night TV comedians and green living hustlers in lieu of actually testing what these turbines could deliver in the real world.

Critics, such as Sagrillo, could only grumble that a promoter of small VAWTs should confirm its claims through "independent, third-party testing." While there has been internationally accepted testing standards for commercial-scale wind turbines for some time, there were no testing standards, no readily accessible testing sites, and no program to "test and certify" the performance of small wind turbines. Fortunately, that's no longer true.

Today, there are testing and certification standards for wind turbines of all sizes, including small wind turbines of any orientation. Sagrillo would be the first to note that testing and certification to an internationally agreed-upon standard is the bare minimum any wind turbine should comply with. Testing and certification doesn't assure that a wind turbine will work reliably for a long-enough period to recoup its investment or that it will operate problem free. Testing only certifies that the turbine will do at least what was measured.

To their credit, some Darrieus turbines have undergone equivalent testing before the certification program was launched. One of the heavily hyped giromills of the VAWT renaissance was Mariah's Windspire. Though the company claimed that its turbine was designed for a 20-year life, the reality was far different. The National Renewable Energy Laboratory (NREL) began testing a Mariah Windspire turbine in late spring 2008; by the fall NREL "terminated" the duration test and a few months later terminated the remainder of the tests. The turbine lasted less than 1 year under NREL's conditions. The tests, and NREL's results, are all part of the public domain. And what those tests say is that the turbine didn't last long enough to complete the tests.

As of late 2015, no VAWT—of any design—had been certified by the Small Wind Certification Council in the United States, and no VAWT was registered in the Microgeneration Certification Scheme in Great Britain. (Quiet Revolution's turbine had been certified but the certification lapsed.) Only one small VAWT, an H-rotor turbine, has been certified worldwide—and that was certified under the Japanese system. That's it.

Certification and testing cuts through the hype that has been a hallmark of small VAWT promotion. "We're not asking for these folks to stop *trying*" to build a better wind turbine, says Lisa Difrancisco, a 15-year industry veteran and founding member of the Distributed Wind Energy Association in the United States. "We are demanding that they stop *lying*. Not a single one [VAWT] delivered on their promises to verify their claims [in the United States]."

Conspiracy against VAWTs

Some VAWT proponents stray off into a dark world where "big companies" and conventional wind turbine manufacturers conspire against struggling, upstart VAWT designers. These appeals to support the underdog—VAWTs—particularly resonates

DEBUNKING PYRAMIDAL POWER AND MAGICAL MAG-WIND

The pyramids have always been thought to contain some magical power. Maybe they do. Maybe their magic can be applied to wind turbines. Then again, maybe not.

In the mid-2000s, the news media was abuzz—not just the Internet where new "revolutionary" wind turbines are a plague, but also the mainstream media who should know better—about a new "magnetically levitated vertical-axis wind turbine," the Mag-Wind. Yes, you read that right. Not just a VAWT but a levitating, pyramidal, roof-mounted one at that.

The "invention" (I am using the term loosely here) had all the telltale signs of a fantasy wind turbine: hype high, experience low (actually nonexistent). Other tip-offs that this was a hoax were claims that the device was "much smaller than other wind turbines" and that it was "much more efficient than the old-fashioned windmill."

Most such creations last only a few years. They arrive in a blaze of press then quietly disappear—often with their investors' money. Mag-Wind was different. It kept resurfacing, and as it did, it would garner ever more high-profile celebrity endorsements (another tip-off: celebrity endorsements). The alleged endorsements by Jay Leno and Ed Begley led Canada's politically influential *Globe and Mail* newspaper to fawn over the Canadian promoter as the reincarnation of Alexander Graham Bell (one of Canada's most famous inventors).

This whole episode would be laughable if real money wasn't involved—and in the end it was the search for new money that led the promoter afoul of the law. But first, let's look at some of Mag-Wind's claims and see how these can be manipulated by the unscrupulous.

Could Mag-Wind be more efficient than an "old-fashioned windmill"? This is a common ploy of hucksters—setting up a straw man by comparing their device to something other than what you think it is. What is an old-fashioned windmill after all? They never said. Would Mag-Wind be more efficient than a modern, highly efficient wind turbine? That's very doubtful. Modern wind turbines look the way they do for a good reason. They squeeze a lot energy out of the wind with very few materials.

So, let's check Mag-Wind's numbers. To begin, let's simply forget Mag-Wind's conical shape and assume that it's your garden-variety H-rotor. This is to Mag-Wind's advantage. Mag-Wind asserted that its turbine could generate 13,200 kWh per year in an average wind speed of 13 mph (5.8 m/s). First off, that seems like an awful lot for a windmill that will be setting on your roof as shown in their ads.

Their windmill was 6 feet (1.8 meters) tall by 4 feet (1.2 meters) in diameter at its widest point. Thus, the rotor sweeps a paltry 2.2 m². At an average annual wind speed of 6 m/s (13.4 mph), typical small wind turbines will generate no more than 400 kWh/m²/yr. High-performance turbines using neodymium magnets could generate nearly 600 kWh/m²/yr. Again, let's give Mag-Wind the benefit of the doubt. Their device could—theoretically—produce a maximum 1,300 kWh per year. Mag-Wind claimed that its revolutionary new windmill would produce 10 times more than a high-performance small turbine. Ouch.

But it gets worse. Mag-Wind said its wind turbine would produce 5 kW at a rated speed of 28 mph (12.5 m/s). Mag-Wind was about the size of a conventional wind turbine with a rotor 6 feet (1.8 meters) in diameter. This size turbine would have a standard power rating of only 500 watts. For Mag-Wind to produce 5 kW at that wind speed, it would need to be nearly 200% efficient at doing so. That is, it must produce two times the amount of energy—in the wind. That's some windmill. It produces more energy than there is in the wind striking it!

To summarize, could it do what it claimed? No, not on this planet. Anyone with a calculator could figure this out—even an editor at the *Globe and Mail*.

Why is this important? Because the media and political attention that is directed toward such "revolutionary inventions" distracts us from the real work at hand by offering us panaceas. If we just use this new device, all will be well, and we won't have to make any of the painful—and expensive—decisions that are ultimately necessary to meet our energy needs in a sustainable and environmentally acceptable way.

Mag-Wind's antics were particularly damaging because the province of Ontario, where the *Globe and Mail* is published, was in a heated political debate on renewable energy policy at the time. Some of the *Globe and Mail*'s writers were particularly critical of the proposed policy in general and commercial wind energy in particular. Who needs large wind turbines when you can put a Mag-Wind on your roof for less?

Like most such stories of wondrous new wind inventions, Mag-Wind didn't end well. On October 18, 2013, a federal grand jury in Raleigh, North Carolina, returned a two-count indictment charging James Alan Rowan with securities and wire fraud. The US Federal Bureau of Investigation was seeking his extradition from Canada to face charges. If convicted, Rowan could face up to 40 years in prison.

It took a long time, but the long arm of the law finally caught up with Mag-Wind. When it did, there was embarrassment at the *Globe and Mail*, and a lot of investors were missing their money.

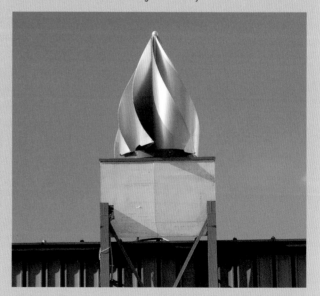

Mag-Wind and levitating pyramidal power. Mag-Wind turbine in Grimsby, Ontario, on July 12, 2007. There was a good stiff wind blowing at the time, and the "wind turbine" was not spinning. Mag-Wind claimed outlandish performance from this VAWT because it presumably was "magnetically levitated." In 2013, the Canadian promoter was indicted for securities and wire fraud in the United States.

with Americans whose foundational myth is our successful revolt against the oppression of the mighty British Empire.

Is there a conspiracy as some VAWT true believers charge? Yes, there is. It's the worst kind of conspiracy, the kind that is almost impossible to overcome. It's a conspiracy of physics, a conspiracy of facts, and a conspiracy of statistics. VAWTs have been tried—and haven't delivered.

Charles Komanoff, an expert at unraveling the optimistic cost projections of nuclear power, warns journalists and politicians to "never confuse promise with performance." A warning that applies equally well to VAWT inventors as it does to proponents of nuclear power. VAWTs promised and their continual reinvention promises cheap, abundant electricity without the disadvantages of conventional wind turbines. They just didn't deliver.

Global Energy Concepts in its study for Lawrence Berkeley National Laboratory sounded an ominous warning: "For all wind turbines, other than those used for more decorative purposes, the cost of energy is important." This one sentence encapsulates what Sagrillo and others have been harping about for years. "New" or "revolutionary" VAWTs have to compete with existing wind turbines, and they must prove their worth. Otherwise, they are just lawn ornaments or "dynamic sculptures." Most often they are not even dynamic and quickly become mere static sculptures—not wind turbines.

Why this is important, why Sagrillo and Difrancisco make such an issue of it, is because the failure of VAWTs—small and large—lends credence to opponents of renewable energy and specifically those who attack wind energy by charging that it doesn't work. Wind energy does work; those quarter million turbines operating today proves that it works. None of that accomplishment is due to Darrieus wind turbines.

Are VAWTs different from those other wind turbines? Yes, of course. They look different, but that's not enough. Wind turbines serve a purpose. They must produce cost-effective electricity in an environmentally sound manner for decades. And if VAWTs don't meet those criteria then being different is simply not good enough. As mathematician and longtime wind advocate Mark Mayhew described them, "VAWTs are the energy source of the future—and they always will be."

7

Novel Wind Systems

> What has been will be again, what has been done will
> be done again; there is nothing new under the sun.
>
> — Ecclesiastes 1:9, New International Version

THIS CHAPTER IS A COMPANION TO THE PROCEEDING chapter on VAWTs. Both chapters are about the technological fringe of wind energy—wind turbines or wind systems that garner a lot of press and political attention but over the decades have failed to deliver on their promises. There are a number of novel wind energy systems that don't fall into the simple dichotomy between HAWTs and VAWTs. We'll discuss them in this chapter. We'll look first at diffuser augmented or shrouded wind turbines, in part because of their promoters' ability to raise money, move briefly to kites, then finish with ship propulsion and how the Magnus effect has been used in wind-assisted shipping.

Before we begin, there are two important concepts we should introduce—concepts that should be used to evaluate the claims of any promoter of a "new" wind turbine. The first deals with the proof necessary that something is indeed new and that it works as intended. The second is the impossible demand by inventors that skeptics prove that their inventions won't work as planned.

Astronomer Carl Sagan popularized the skeptic's credo that "extraordinary claims demand extraordinary proof." That is, anyone asserting that his or her new product or invention is far superior to wind turbines that we know work at reasonable cost has a very high hurdle to overcome before any journalist, politician, or investor should pay any heed.

The second precept is equally important: proving a negative. Promoters of new inventions often complain that their critics can't prove that their invention won't work—or won't work as well as they say. That's true. As skeptics of new-fangled inventions point out, it's impossible to prove that these wondrous new inventions won't work if they've never been installed and tested—but there's no proof they will either. Students in Logic 101 learn that you can't prove a negative. There's always a reason, an excuse really, why many new, earth-saving inventions don't work quite as well as advertised and are then quickly forgotten. Skeptics argue that promoters have to prove their inventions work first, before society should invest scarce political or monetary capital in something without a proven track record.

With these thoughts in mind, let's tackle "novel" wind systems.

ADVICE FOR INVENTORS OF NEW WIND TURBINES AND FOR EVERYONE ELSE AS WELL

James Jarvis, the founder of APRS World, a manufacturer of low-cost data loggers that can be used in the monitoring of small wind turbine performance, has some straightforward advice for inventors of new wind turbines based on his experience developing new technology.

Test early, test often.
Test everything independently and together.
Test more than one unit.

Whether a purported manufacturer of a new wind turbine has met these requirements should be the first question a consumer, journalist, or politician should ask. If they haven't, they don't have a product. They may have a dream. They may have a drawing—or a fancy website—but they don't have a product.

Ducted or Augmented Turbines

Just as some inventors are quick to grasp panemones and their paddles as a mysterious "newly discovered" technology for harnessing the wind, others envision wind turbines encased in shrouds or ducts. As with panemones, shrouded turbines are easy to understand. Much like giant funnels, the shrouds concentrate or *augment* the flow across the wind turbine's rotor.

Unlike panemones, however, ducted turbines are wrapped in mysterious aerodynamic terms such as *diffuser augmentation*, which are beyond the ken of most observers. Even supposedly sophisticated engineers have been snared by what at first appears to be a startling new technology "overlooked" by everyone else. As a result, augmented turbines have been plagued by hucksters and charlatans—the alchemists of wind energy—more than other technologies.

Ducted or shrouded wind turbines have been invented and reinvented throughout the 20th century. One such inventor was Dew Oliver, who installed his "blunderbuss" in California's San Gorgonio Pass in the 1920s. Historian Robert Righter notes that Oliver claimed the 3- to 4-meter (10- to 12-foot) diameter ducted turbine actually worked. Nevertheless, that wasn't enough to protect Oliver from being eventually convicted for fraud. Most inventors today—but not all—are much better at covering their legal tracks than Oliver and escape the consequences of their ventures.

Do diffuser-augmented wind turbines (DAWTs) work? Yes, of course. Panemones and cup anemometers work too. They serve a useful purpose: grinding grain or measuring the wind. Yet, we don't use them to generate commercial quantities of electricity. Do augmented turbines produce the amount of energy promised at the cost promised? Certainly not yet. None have yet to fulfill their often highly touted claims. Modern, high-speed, or "conventional" wind turbines—for all their limitations—reliably deliver quantities of electricity at increasingly competitive costs. DAWTs must somehow be better (cheaper, more reliable, more productive, and so on) than the prosaic wind turbines we already have—or why bother.

Unfortunately, the US Department of Energy (DOE) funded development of the diffuser-augmented concept in the 1970s through a contract with Grumman (of aerospace fame). DOE's contract lent credence to the design just as its funding of McDonnell Aircraft gave credibility to the giromill.

In principle, an airfoil-shaped shroud surrounding a conventional wind turbine draws the wind through the rotor and augments—concentrates—the wind flowing across the blades. By doing so, the diffuser-augmented turbine can achieve conversion efficiencies much higher than conventional wind turbines and even exceed the Betz limit—relative to the rotor. But a DAWT is much more than the rotor because it uses a shroud.

DOE eventually abandoned the idea, along with many other so-called innovative concepts. But the damage was done, and ducted turbines periodically reappear in the press as a promising new technology. Inventors often cite old DOE reports as proving the concept's merits. One effort, New Zealand's disastrous Vortec 7, went so far as to hire Grumman's chief DAWT engineer as a consultant on its project.

Astronomer Carl Sagan popularized the skeptic's credo that
"EXTRAORDINARY CLAIMS DEMAND EXTRAORDINARY PROOF."

One outspoken critic of diffuser augmentation is Heiner Dörner, emeritus professor at the University of Stuttgart's Institute of Aircraft Design. Sure, Dörner says, wind tunnel tests show you can double the wind speed across the rotor, but these turbines haven't performed well in the field. Indeed, this is typical of the results seen by the few ducted turbines that have been built.

Engineers first note in their reports that, yes, they were able to augment the flow; they then go on to say, however, that for various reasons (there's always a host of reasons), the concentrating effect was much less than they had anticipated. Moreover, there were other problems as well. The shroud was more expensive and difficult to construct than they thought, yawing the turbine was more difficult—and a veritable litany of excuses for why the design didn't work as promised.

So let's examine a few of the attempts to commercialize DAWTs, beginning with two small turbines and then moving on to two medium-size turbines. We'll compare the rated power of DAWTs relative to their shroud diameter not their rotor diameter. DAWT promotional materials emphasize only the rotor diameter, but our interest is how the entire wind turbine—and that includes the shroud—stacks up to conventional wind turbines of similar size to the shroud diameter.

Enflo

In the fall of 2005, a Swiss company, Enflo, was displaying a "new" ducted turbine, the Windtec 0071, at the big wind extravaganza in Husum, Germany. The turbine was small and well suited for an indoor display. It was just one of a half-dozen ducted turbines at the show (see Figure 7-1. Enflo Windtec 0071). Enflo's performance claims were the most modest of the DAWTs exhibited.

Enflo's rotor was only 0.71 meters (2.3 feet) in diameter. The shroud, which surrounds the rotor, was 0.87 meters (2.9 feet) in diameter. With the shroud included, the Enflo 0071 was about the size of an Ampair 100. In other words, it was effectively a micro turbine. (For comparison, most turbines at the Husum trade show are so big they build the tents around the nacelles.) However, unlike the Ampair 100, a conventional micro turbine rated at 100 watts, the Enflo 0071 was rated at 500 watts.

Again, let's only consider the area of the wind intercepted by the shroud, about 0.6 m², not the smaller area of the rotor inside the shroud. If a turbine of this sized performed like other turbines in this class, it would be capable of about 120 watts, not 500 watts. By this criterion, the turbine claims it can produce four times more power than other wind turbines of its size. Another way to look at this is the rotor loading at rated power claimed by the manufacturer. The rotor loading for this turbine was 840 W/m² of shroud intercept area compared with the typical 200–300 W/m².

Could the Enflo 0071 do what it claimed? The claims were not far from that theoretically possible. So, it's conceivable. Is it likely? No. The odds were long that this turbine could deliver on its promise of outsize performance. The burden of proof is always on the manufacturer to demonstrate that its turbine can do what it claims, more so in a case like this where the claims are at the theoretical limits. Despite this, Enflo never published any third-party tests of its performance.

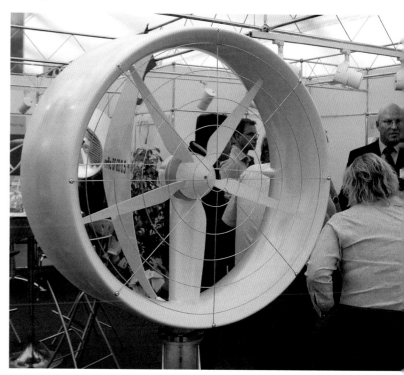

Figure 7-1. Enflo Windtec 0071. Small ducted (shrouded) wind turbine at the Husum wind energy exhibition in 2005. The turbine was rated at 500 W with a shroud 0.87 meters (2.9 feet) in diameter.

In 2014 the website for Enflo no longer contained any information on their ducted turbine. None of the other ducted turbines proffered at Husum in 2005 were still being offered as well.

Eléna 30: Will They Ever Learn?

There's an old Pete Seeger song with the refrain, "When will they ever learn, when will they ever learn?" The unpleasant answer is they never do learn about what works and what doesn't work when it comes to wind energy. There's always another inventor with a "groundbreaking" new wind turbine and a gullible public to believe them. Eléna Energie was another in a long line, but with a French pedigree.

The Maison de l'Air is devoted to all things associated with the atmosphere, including the wind. The building sits atop one of Paris' famed *monts* and commands a spectacular view of the city. (The Eiffel tower can be seen in the distance.) In 2005, there were not just one, but two ducted turbines sitting atop the Maison de l'Air (see Figure 7-2. Eléna Energie).

In some ways it's refreshing to know that it's not just gullible English speakers who fall for the hype of "novel" wind turbine promoters. It feels better somehow to know that the French can fall for them too. France, after all, is the land of the famed TGV (Train à Grande Vitesse), which can take a traveler from Perpignan on the Spanish border to Paris in five comfortable hours. France is also the land of Descartes, Carnot, Pasteur, and famed engineers such as Eiffel and Darrieus.

Alas, the French are human after all, and they succumb to the siren's lure of installing a little windmill on the roof of an urban building, thinking they will produce a lot of electricity without all the bother and displeasure of one those large—but real—wind turbines you see from the TGV as you speed toward Paris.

Eléna Energie, a Grenoble company, installed two of its Eléna 15 ducted turbines on the Maison de l'Air. Placards on the building proclaimed that the turbines would produce 15,000 kWh each at an average annual wind speed of 6 m/s. Now Eléna Energie was very careful not to claim that the turbines would produce this much electricity at this site in Paris. After all, the Maison de l'Air is in one of the world's largest cities and surrounded by taller buildings. Nevertheless, the placards suggested that the turbines could, in fact, produce this somewhere in France, and that's a lot for such a small turbine—ducted or not.

Let's consider the company's bigger model for a brief analysis. The Eléna 30 used a rotor 1.2 meters (4 feet) in diameter with a shroud or duct around it of 1.6 meters (5.2 feet) in diameter. The duct was 2 meters (6.6 feet) long. Eléna Energie rated the Eléna 30 at 6.8 kW! This was nearly 17 times more than that from typical wind turbines of this size. Even granting that there could be some acceleration of flow through the duct, the rating for this turbine was simply outrageous.

Figure 7-2. Eléna Energie. One of two Eléna 15 small ducted turbines atop the Maison de l'Air in Paris 20th *arrondisement*. The two turbines generated no net electricity during their first year of "operation." These "wind turbines" are good examples of the inflated performance claims typical of ducted turbine promoters and the typically poor performance of urban rooftop wind turbines in general. Inset: One of two Eléna Energie turbines on the building in urban Paris.

Developers of many new VAWTs are fond of excessive and unsubstantiated claims of energy generation for their products. But VAWT promoters are pikers in comparison to Eléna Energies. Maybe the altitude in Grenoble at the foot of the Alps or too much time in the *soufflerie* (wind tunnel) affected the inventors' judgment.

The promoters of such machines are quick to say, "Well, we haven't tested it yet, so how can you know that it *can't* produce that much electricity?" Despite having done no testing outside the wind tunnel, Eléna Energie was installing these "wind turbines" in prominent locations throughout France. This was—at the very least—irresponsible.

It was also irresponsible for the assistant mayor of the 20th *arrondisement*, where the turbines were installed, to give these promoters a public venue to showcase their questionable wares. There is a political debate in France—as in many countries—about the future of wind energy. Opponents of renewable energy are quick to seize on installations such as these at the Maison de l'Air to prove one point or another. One can easily imagine them saying, "See we don't need those big ugly monsters. We can use these cute little things instead." Or they could say, "See they don't work. We told you so."

In fact, Denis Baupin, the deputy mayor from the Green Party, said at the dedication of the turbines "*nous ne souhaitons pas détériorer son paysage,*" that is, "we don't want to spoil the [Paris] cityscape with the other kind of wind turbines." If the Greens can be hoodwinked by Eléna Energie and the unfortunate human desire to get "something for nothing," what can we expect of the general public and politicians less supportive of renewable energy?

Sure enough, wind energy's critics quickly seized on the Maison de l'Air installations and gleefully noted that the "wind turbines" had generated no net electricity after a full year of "operation."

Two DAWTs down, but these were small fry in the DAWT sweepstakes.

New Zealand's Vortec 7

Launched two decades ago with much fanfare and attendant hype by a promoter better known for a sea-bed mining venture than wind energy, Vortec 7 became the granddaddy of ducted turbines. Vortec installed its experimental prototype in 1997 south of Auckland, New Zealand. The shrouded turbine with a rotor 7.3 meters (24 feet) in diameter sat on an ungainly and cluttered structure that rotated on a ground-mounted platform. The shroud was probably 12 meters (40 feet) in diameter. Vortec claimed on its website that the turbine was "designed" for 1 MW!

Using the same technique we used with the small ducted turbines, we can compare the claimed performance of the turbine and its shroud to that of conventional wind turbines (See Table 7-1. DAWT Summary of Experimental Products). Vortec's promoters win the prize for most outlandish ducted turbine hype. They claimed that Vortec 7 would extract 20 times more power from the wind than a conventional wind turbine with the diameter of the shroud. It's important to note that they only "claimed" this. They didn't know that it would do this. In fact, they never proved it could do this—or anything close to that. Whether the engineers working on this project believed this, or knew from the beginning that the claim was grossly inflated, will never be known. If the former, it

VORTEC 7 PROMOTERS ON GIPE'S CRITICISM

In June 1997, Vortec–the promoters of a DAWT in New Zealand–posted the following on its home page.

> Paul Gipe is a prolific writer (just search the net). He has an amazing output of stories about wind power. Maybe it is threatening to him that he hasn't foreseen the development of the DAWT. In any case, what he writes about DAWT is a good part speculation and he bases his findings on old data from Grumman. It may turn out to be another case of: "It will never fly, Wilbur."

> Later in July, promoter-in-chief Robin Johannink was quoted by the New Zealand *Star-Times* that he was going to an Australian wind conference where he would counter Gipe's claims. "I'm going for the jugular . . . Paul Gipe hasn't seen the research data we have, he has no involvement with commercial turbines and he's wrong to say we don't represent a quantum leap in wind technology," he said. "Come the end of September when our performance data is released he's going to look very foolish."

> Well, it didn't fly, Wilbur. September came and went. And when Vortec went kaput in 2001, it took 25 million Kiwi dollars with it, sullying the reputation of engineers and politicians alike who bought into the hype–or turned a blind eye to it and didn't speak out.

Table 7-1. DAWT Summary of Experimental Products

Company	Model	Shroud Diameter m	Shroud Diameter ft	Area m²	Mfg. Rated Power kW	Claimed Loading at Rated Power W/m²
Small						
Enflo	71	0.87	2.9	0.6	0.5	841
Eléna Energie	30	1.6	5.2	2.0	6.8	3382
Medium-Size						
Vortec	7	12	39.4	113	1000	8842
FloDesign/Ogin		19.5	64.0	299	100	335
Sample HAWTs						
Small						200–300
Large						200–600

Note: The rotor loading of small turbines can vary from 200 W/m² to 300 W/m². The rotor loading of large turbines has an even wider range: from 200 W/m² to 600 W/m².

calls into question their competence. If the latter, it calls into question their professional ethics.

Developers claimed that the Vortec 7 was the first full-scale DAWT ever built. But like Dew Oliver's blunderbuss, it didn't go well for Vortec 7. Soon, key participants were acknowledging the technical failure of Vortec 7, but they couched their failure in arcane technical language at arcane conferences. Unlike the hype that led up to the "demonstration" of the turbine, the damning conclusions didn't garner the same amount of press—if any. By then it was too late. The money was gone.

They expected power augmentation from 5.5 to as much as 8, based on Grumman's wind tunnel tests in the 1970s. The prospectus said an augmentation of 4 would be necessary for the project to be considered successful. They didn't make it.

The prototype delivered much less than that expected. Technical reports concluded that the simplified models used led to predictions of power augmentations of about 4, but such power levels "were never observed." In layman's terms, Vortec 7 simply didn't deliver anywhere near what was promised. By mid-2001, the company was kaput. The sea-bed mining venture also soon disappeared.

This brief description of the Vortec fiasco doesn't do justice to the disruption that Vortec caused to the development of wind energy in New Zealand. Unsuspecting investors, including the government of New Zealand, poured up to $25 million into this venture, a venture that was questionable—and questioned—from the very beginning.

At the time, wind energy was beginning to flourish in the South Pacific. New Zealand even had one of its own native sons return from the motherland to develop a homegrown version of a conventional wind turbine that had proven successful in Great Britain. Geoff Henderson, a driven entrepreneur and innovator, in the true sense of the word, took an off-the-shelf design and adapted it for local manufacture in New Zealand. Here was a Kiwi company that could have used the money squandered on Vortec.

The Vortec 7 experience soured a lot of Kiwis on wind energy, which is still felt today. Fortunately, the dogged determination of Henderson led to the eventual installation of nearly 100 of his Windflow turbines—50 MW of wind capacity, which has been churning out kilowatt-hours on the South Island for more than decade. It's not an exaggeration to say that Henderson's turbines alone have generated more electricity than all the ducted turbines ever installed. Yet, ducted turbines continue to reappear. It's as though no one examines prior art, including the government of New Zealand.

Vortec 7 remains the common ancestor of all modern versions of ducted turbines. None have fared any better. As disastrous as Vortec was for the people of New Zealand, Vortec's ability to raise money is dwarfed by that of FloDesign.

FloDesign (Ogin)

The Kiwis have an unfortunate relationship with DAWTs. Despite having thousands of working conventional wind turbines across the country as well as a local manufacturer, they are determined to throw their money at new offshore DAWTs instead of investing safely in their own economy. In late 2013, the New Zealand pension fund bought a $55 million stake in Ogin, an erstwhile DAWT start-up. That's two times what Kiwis lost in the Vortec 7 calamity.

Ogin was formerly known as FloDesign, a spin-off from the aerospace community surrounding the Massachusetts Institute of Technology. Their twist on DAWT design was the use of mixer-ejector technology originally developed for jet engines (see Figure 7-3. FloDesign DAWT).

With eerie parallels to Vortec, FloDesign used all the same superlatives to describe its take on an old concept. In 2008, FloDesign was all over the tech press with claims that its turbine would produce three to four times more energy from the wind than conventional turbines. Even DOE's Arpa (Advanced Research Projects Agency) was hyping the device that it had invested $8 million in a "breakthrough." DOE said FloDesign's "innovative" wind turbine "could deliver 300% more power than existing wind turbines of the same rotor diameter by extracting more energy over a larger area." Elsewhere, FloDesign was telling the media it could do all this for 30% less cost than conventional turbines.

Unlike its predecessors, though, FloDesign upped the hype a notch by adding that due to the shroud its turbines were safer and more bird friendly than existing wind turbines. And due to the mixer-ejector technology, the turbines could be sited closer together in a wind farm than conventional turbines. MIT's *Technology Review* raised the ante further by saying the turbine could generate electricity at half the cost of conventional turbines.

Nowhere in this barrage of ballyhoo were there any real numbers on performance. They had none. It was all flash. It was all projections. FloDesign had only a model in a wind tunnel and computer simulations to spin its story to the media. It didn't erect a prototype on Deer Island in Boston Harbor until years later.

Things changed dramatically in late 2013. By then FloDesign/Ogin had raised a total of $80 million from big-name venture capitalists and pension funds with well-honed political connections to former vice president Al Gore. Suddenly,

WARNING: REBRANDING DAWTS

In the mid-2010s, there was a thinly veiled attempt to rebrand diffuser-augmented wind turbines (DAWTs) as compact wind acceleration turbines (CWATs) on Wikipedia.org. It's possible that the disastrous commercial and technical history of DAWTs, shrouded turbines, and concentrators has so poisoned the Internet well that promoters of these designs think the only way to get their message out of a new "revolutionary product" is to coin a new term that's not burdened with such baggage.

The *Wikipedia* entry forthrightly states "CWATs are a new acronym that encompasses the class of machines formerly known as DAWTs (diffuser augmented wind turbines). The technologies mentioned above all use diffuser augmentation that is substantially similar to previous designs as the primary means of acceleration." It then goes on to note that "The concept of these structures has been around for decades but has not gained wide acceptance in the marketplace."

Although the *Wikipedia* entry is not written in a promotional manner, as are some other entries on novel wind turbines, it is clear in the first paragraph that the reason for the entry of a new term is the introduction of several new designs.

Defunct Optiwind is the first mentioned. (Optiwind looked like a modern-day version of TARP for toroidal acceleration rotor platform from the 1970s.) Followed by FloDesign, a company that has been promising a ducted turbine since 2008. Worse, the *Wikipedia* entry also lists Wind Tamer (rebranded Arista) and WindCube—all firms that have a high-hype, low-results factor.

To date, the rebranding hasn't paid off. The companies mentioned all appear to be defunct or in some form of netherworld. They were certainly not commercial in early 2016.

Figure 7-3. FloDesign DAWT. The most recent attempt to develop a commercially viable Diffuser Augmented Wind Turbine. FloDesign has raised more money than most DAWT ventures: $80 million by the end of 2013. Whether they will be more successful than the many others before them is yet to be determined. At the time this photo was taken, this prototype turbine was not operating. Note the lobed diffuser-mixer (the inner ring). The diffuser-mixer is what separates this DAWT from others. 2012. (Dr. Ewan O'Sullivan)

FloDesign was tight lipped. No one was talking except to say it planned to install a cluster of turbines in California's Altamont Pass in 2014. (By the end of 2015 it had failed to do so.)

FloDesign's website and its design changed too. Some of the hype remained, but many of the superlatives used to describe this radically new technology were gone. Gone too were the "lobed mixer-ejector" feature that distinguished its DAWT from all those that had gone before. The digital image looked nothing like the turbine touted by the company in 2008 and nothing like the prototype installed on Deer Island (see Figure 7-4. Ogin California prototype). Gone too were claims about three times the power and half the cost. The company was clearly tamping down expectations. And still no specifications on the turbine or real data on its performance.

What we do know is that the turbine was still rated at 100 kW and that the shroud diameter was likely 19.5 meters (64 feet) by parsing the passages on their website. From that limited data, it appears that the specific capacity or load on the frontal area of the FloDeisgn/Ogin turbine is about 340 W/m².

Conventional wind turbines are rated to produce from 200 W/m² to 400 W/m² of their rotor swept area, though there have been models over the decades with even higher and lower rotor loadings.

In other words, FloDesign/Ogin's much-hyped turbine offers no advantage in specific capacity over conventional wind turbines despite all the rhetoric to the contrary. If this turbine can't be built, installed, and operated for a fraction of the cost of conventional turbines, there's simply no point in doing it. The real world doesn't run on hype and hustle. Business and industry consume real kilowatt-hours of electricity. Utilities and independent power producers generate and sell real electricity. If FloDesign/Ogin can't deliver real kilowatt-hours reliably, then the Kiwis and all the other investors will have to kiss their money good-bye—again.

Better Than Betz?

Wind turbine engineers, manufacturers of commercial wind turbines, and renewable energy analysts throw up their hands in despair at the continuing reinvention of these devices and the rapidity with which the media, the unsuspecting public, and venture capitalists seize on DAWTs as a way to beat the Betz limit. Recall that Betz's theorem explains why conventional wind turbines are limited to extracting no more than 59.3% of the energy in the wind.

Martin Hansen, a wind engineer at the Technical University of Denmark, states unequivocally in his 2008 book, *Aerodynamcis of Wind Turbines*, that "it is possible to exceed the Betz limit" with shrouded rotors relative to the area swept by the rotor. He goes on to explain that "if the cross-section of the diffuser is shaped like an aerofoil, a lift force will be generated by the flow through the diffuser." As noted by Hansen and others, "the results are dependent on the actual diffuser geometry, in other words the amount of lift which can be generated by the diffuser." In short, you can augment wind speed and hence a multiple for the power output of a shrouded rotor, relative to a conventional wind turbine—when designed right.

Few argue with this. Experimentally, you can get multiples of 1.4 to 1.6 increases in power with a shrouded rotor, relative to a bare rotor (conventional wind turbine) of the same diameter—when everything is optimal. Though there have been repeated attempts to develop such technology, "this still has to be demonstrated on a full-size machine," says Hansen. That is, this performance "augmentation" still has to be proven—in the field.

So, DAWT rotors can theoretically beat Betz. What is often missed, however, is that a ducted turbine uses—well—a duct, and it is this big duct, or shroud, that increases the area intercepted by the ducted wind turbine relative to that of a conventional wind turbine of the same rotor diameter. Thus, wind engineers, such as Jim Manwell at the University of Massachusetts, remain unconvinced that a DAWT can beat the Betz limit in the field when the shroud is taken into account. And someone has to pay for that big shroud around the rotor, and pay to turn the whole assembly into the wind, and pay to hang on to it in high winds.

The shroud or duct is large and materially expensive. This causes two, as yet insurmountable problems. You have to pay for the shroud, and so far it has always proven cheaper to simply extend the blade on a conventional wind turbine to make up for whatever multiplier effect the shrouded turbine provides.

More problematic, though, is that the shroud or duct ads to the drag on the wind turbine and tower, requiring the entire assembly and its foundation to be much bigger and heavier than they would be on an equivalent commercial wind turbine. Why? Because all wind turbines have to withstand high winds and storms. This is their design condition. Ducted wind turbines are effectively big blobs of material stuck on a stick in the wind. Conventional wind turbines have long slender blades that can be feathered to ride out storms. The shrouds on ducted turbines can't be feathered—at least no one has yet figured out how.

This objection isn't new. French engineer Louis Vadot made the observation at a United Nations Conference on New Sources of Energy in 1961. He said, "This system has been tried a number of times. It is effective enough in increasing the rotor speed, but is uneconomic, since the entire Venturi-rotor assembly must be kept oriented into the wind." Vadot said this more than 50 years ago when the most advanced calculator they had was a slide rule—and we're still trying to reinvent DAWTs!

Are there ways around these limitations? Possibly, but no one has done it successfully yet. And people have been trying for at least three decades. That should tell us something. In short, conventional wind turbines work, and they work better than the few ducted turbines that have been tested.

Derek Phillips is a one-time researcher on Vortec 7. He designed his own DAWT to surmount these problems and wrote his 400-page doctoral thesis on DAWTs. Yet Phillips concluded that "a rudimentary economic analysis has clearly shown that drag is the dominant economic driver for the DAWT concept" because it determines the cost of the supporting structure and foundation and limits feasible tower height. All of which in turn raise a DAWTs cost of energy. "The economic analysis has shown that due to DAWT cost being determined by the peak design load, DAWTs are uneconomic compared to HAWTs and will remain so until there is a breakthrough in the design to reduce drag at the peak wind speed."

As an inventor of his own DAWT design, Phillips's assessment should be the death knell for DAWT promoters who have a serious interest in generating cost-effective wind energy and not simply in hyping their investment vehicle.

Phillips also made another important observation. He found that as in McDonnell Aircraft's tests of its giromill for DOE, Grumman also erred significantly in the wind tunnel tests of their DAWT design. Grumman's tests were poorly designed, he noted, and didn't account sufficiently for blockage of the wind tunnel. This led inevitably to greatly overestimating the performance improvement likely from a duct or shroud.

All that money. All that time. All that talent. All that hype. All wasted.

And this is only a summary of the DAWTs of various levels of sophistication that have been sold to the public. The list is long, including Wind Cube, Wind Tamer Jetstream, SkyWolf, and SheerWind just to name a few. None have met Carl Sagan's requirement for extraordinary proof that DAWTs offer any significant advantages over conventional wind turbines. Each additional DAWT that fails drives another nail into the technology's coffin. For with each failure, the burden of proof becomes ever greater on the next inventor, as the probability of success grows exponentially smaller.

If investors, politicians, journalists, and the public continue to ignore the science and believe in a dream, then the effort was indeed wasted.

Airborne Wind Energy Systems (Kites)

With the exception of FloDesign/Ogin, shrouded turbines have fallen out of favor with inventors as well as the media. The buzz in 2013 was all about kites, or as their promoters prefer to call them, Airborne Wind Energy Systems (AWES). Likewise, seed capital for new wind energy inventions has shifted from DOE to Silicon Valley with Google's investment in Makani's flying wind turbine.

Kite proponents in North America and Europe exhibit a cult-like fervor, claiming kites are the wave of the future because kites will be more productive and less costly than the wind turbines

MIKE BARNARD ON WIND TECHNOLOGY RED FLAGS

If you want to inoculate yourself against putting money into a bad wind energy product, ask these simple questions, says Mike Barnard, an astute wind energy analyst. If any of the answers are yes, be skeptical.

Technology Red Flags
1. Do they claim to exceed the Betz limit?
2. Is it an old technology masquerading as a new technology?
3. Is the product just a design concept as opposed to at least a working and tested prototype?
4. Are the only test results available from tests that they have performed themselves–as opposed to independent, third-party labs? And do they publish the results?
5. Are patents claimed for devices other than the one they are demonstrating?
6. Are they claiming greater efficiency than existing wind technologies based on anything other than industry standardized tests?

Business Model Red Flags
1. Do the principals have backgrounds in fields unrelated to wind energy?
2. Are they claiming that their product will replace commercial-scale, three-blade wind turbines?
3. Does the product introduce new liabilities, for example downwind throw of solid flying wind turbines such as hard-wing kites, or tether cable, or require flight exclusion zones?

Marketing
1. Do they disparage conventional wind technology to establish their technology's superiority?
2. Do they have a website that is just a façade?

---

we have now. Google's investment has only stoked their certainty. Google couldn't be wrong, could it? In this, kites—as a new technology—are no different than DAWTs or VAWTs or any other wind invention that wins high-profile backing. VAWTs have celebrity endorsers. DAWTs have DOE. Kites have Google.

Most wind engineers don't give kites a second thought. The typical reaction is, "You must be kidding?" As with DAWTs, kites and other flying wind energy systems are not found in any engineering textbooks on wind energy. However, kites and kite promoters have won media attention and with that, political attention—and money—soon follow. For this reason, some analysts have felt it necessary to give kites more than a cursory glance.

No one has done more to analyze the realistic prospect of using kites to generate electricity than wind energy analyst Mike Barnard. His interest in kites and other wind turbine inventions is that of an advocate. He wants wind energy deployed. Anything that stands in the way of expanding the use of wind energy today is ripe for his critical analysis. His approach is cool and level-headed. He's not one to fall for hype.

Barnard segregates kites into several categories based on their wings, where they generate electricity, how they fly, and at what altitude they fly.

- Soft wing, hard wing, lighter than air
- Generation on the ground or in the air
- Single tethers or multiple tethers
- Crosswind flying or static flying
- High-altitude flying or low

When we think of kites, we naturally think of soft wings. EnerKite is just one of many examples of soft-wing kites that use a fabric parasail. Google's Makani, on the other hand, uses a tethered fixed-wing plane that flies to its position. Magenn and its successors propose helium-filled blimps that carry the generator to altitude. Makani and Magenn were designed to generate electricity on board and transmit the electricity to the ground via a tethered power cable. EnerKite, and the others like it, spool and unspool a reel attached to a generator on the ground. Makani flies across the wind as do most soft-wing kites; the Magenn blimp rests at its operating position. Some propose flying their kites at altitudes common to commercial aircraft, others at heights not much different to commercial wind turbines today.

No kite company has built anything more than a prototype. EnerKite, for example, has produced a proof-of-concept 30-kW model that may be suited for niche applications in remote locations, such as for military use. Makani, for all its flash, has only developed a prototype. Although some have made sophisticated measurements under trial conditions, none of the kite companies have published performance results measured under standardized conditions so that their results can be compared to conventional wind turbines. Until kite proponents prove their claims, they remain just that—claims.

Kites, like other novel wind turbines, have

several technical limitations to overcome before they are anything more than a novelty. One of the most obvious problems is the tether whipping around the sky. Flying a kite at anything much more than the height of a conventional wind turbine requires an aircraft flight exclusion zone. This alone is enough to ground most kite concepts.

If the kites can't be flown high enough, because of potential conflicts with aircraft, to take advantage of increased wind speeds at height, then they don't offer any yield advantage over conventional turbines.

For brief use at remote sites, manual operation of the kite may be sufficient. But for anything approaching commercial use, the kite must be automatically flown into the air, automatically flown once the kite becomes airborne, and automatically retrieved when necessary. Automatic flight hasn't been demonstrated on anything more than a prototype scale.

Barnard sums up his take on kites after an exhaustive analysis.

> The potential energy available in the wind flowing high above our heads is alluring, and harvesting it with tethered flying wings has great appeal, but as soon as you start engineering an airborne solution to harvest that energy, the compromises strip away the potential bit-by-bit until it just isn't viable in any incarnation so far attempted. And it's clear that many of the current organizations in the field were started at best with optimistic assessments regarding safety and aviation authority approvals.

By early 2014, the longest that a kite had stayed airborne was a little more than one week. For comparison, conventional wind turbines at good sites operate more than 6,000 hours per year or 35 weeks. While they may not be in continuous operation during that time, conventional wind turbines are ready and available to generate electricity 98% of the full 8,760 hours in a year.

It will take years of steady development if kites are ever to compete with conventional wind turbines—even in niche applications. However, there is one application where kites have already been used commercially: ship propulsion.

Wind Ships

While *Wind Energy* is focused primarily on wind turbines, the role of wind energy is not limited solely to generating electricity. One of the important roles wind has historically served has been to provide motive power for shipping.

As one would expect, shipping cargo under sail in the United States peaked in the mid–19th century about the time of the Civil War and began a long, slow decline with the advent of steam (see Figure 7-5. US operating sail power through 1970).

Figure 7-4. Ogin California prototype. Ogin's experimental DAWT erected on the flanks of the Tehachapi Mountains in mid-2015. The prototype abandons the lobed mixer-ejector feature (the inner ring) that was the hallmark of the original design. This turbine was first announced with much hype in 2008.

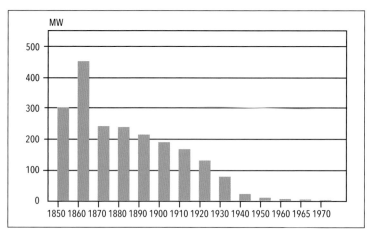

Figure 7-5. US operating sail power through 1970. Shipping cargo by sail peaked in the United States during the mid-19th century and began to decline with the advent of steam. By the early 20th century, sail's days of ruling the seas had passed. (Historical Statistics of the U.S. Colonial Times to 1970, Part 2, Bureau of the Census, Washington DC, 1976, 818)

This pattern of fossil-fuel-fired steam replacing sail was repeated around the globe. Today, sail is reserved for pleasure craft in developed countries and coastal shipping in the developing world.

However, it need not always be so. Just as modern wind turbines are now supplanting fossil-fuel-fired generation as fossil fuels once pushed traditional windmills aside, so too is it possible that modern sailing ships could once again ply the seas? As with wind turbines, today's sailing vessels would use modern technology—especially modern materials—be fully automated, and for the most part, they wouldn't look anything like those of the past.

Initially, modern sails would be used to assist conventional diesel engines—a waterborne version of wind-assisted pumping on land—in shipping bulk cargo, where speed is not critical. Wind-assisted shipping would reduce fossil-fuel consumption and attendant emissions from one of the world's dirtiest fuels—bunker oil.

For the past three decades, naval engineers and sail designers have grappled with what technology will work best (see Figure 7-6. Wind propulsion). Flettner rotors have received the most attention in the past, certainly in Germany. Kite traction, again in Germany, has gained ground in the last decade.

The closest to a traditional sailing rig was the 2006 launch of the *Maltese Falcon*—one of the most expensive yachts ever built. Created for a billionaire venture capitalist, the luxury yacht deployed three masts using 2,400 m² of full-rigged Dyna sails popularized by German mechanical engineer Wilhelm Prölss. The ship also used diesel engines and could reach 24 knots under sail.

That modern materials and aerodynamics can lead to startling performance of what we once simply called "sailboats" is demonstrated by Vestas's Sailrocket. Sponsored by the world's largest manufacturer of wind turbines, the Sailrocket uses carbon fiber throughout and a 22-m² wing and stepped hull. As the boat increases in speed, it rises out of the water to reduce drag. It was designed to have a ship-to-wind-speed ratio of 3:1, that is, the boat's speed across the wind would be three times that of the wind speed. In late 2012, the Sailrocket set the world speed record for boats under sail at 65.5 knots (75 mph, 34 m/s).

The Sailrocket's success followed an equally dramatic run for the record books on land. In spring 2009, the *Greenbird* smashed the land sailing speed record at an incredible 126 mph (56 m/s) on Ivanpah dry lake in California's Mojave Desert.

These speed records may not appear to have any relevance to the prosaic matter of moving goods. Yet it was from the early days of auto racing that many of the technologies were developed that made automobiles the practical conveyance they are today. Such achievements under sail, hardly imaginable just a decade ago, illustrate that there's room for technological improvement in using sail to power ocean-going commerce.

Traction Kites

For more than a decade German kite designer SkySails has been developing an automatic traction kite for wind-assisted shipping. In late 2007, MS *Beluga SkySails* became the first cargo ship to use a kite for traction in commerce. The 160-m² parasail was used to assist the diesel-powered freighter in a transatlantic voyage from Bremerhaven to Venezuela, North America, and Norway before returning to its home port. While under sail, the ship reduced its fuel consumption 10%.

The 2008 financial crisis took its toll on the shipping company, but SkySails is continuing to develop its traction kites (see Figure 7-7. SkySails). By early 2014, SkySails had deployed nearly a dozen traction kites in commercial use and was offering both the original 160-m² kite

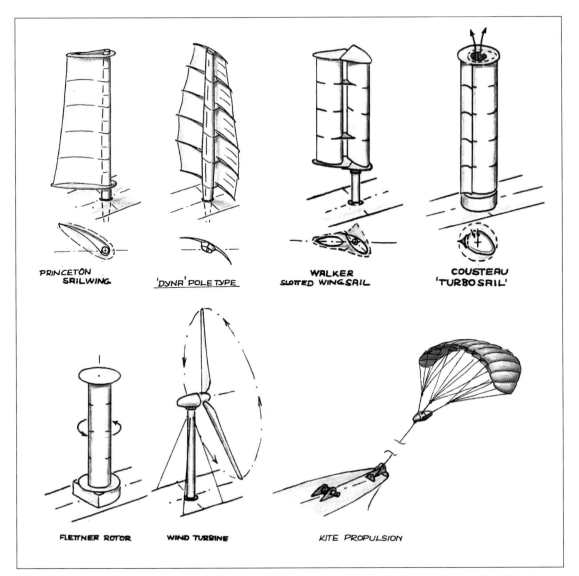

Figure 7-6. Wind propulsion. Modern take on sailing vessels. Here the sails are typically used as an aid to propulsion, not the sole means of propulsion as they were during sail's heyday in the 18th century. Dyna sails, Jacque Cousteau's turbosail, Flettner rotors, and kites have all been used on ocean-going vessels within the past three decades. (Peter Schenzle)

and a newer version with double the surface area of 320 m².

Flettner Rotors

At the same time that German naval engineers were trying to elevate traction kites to commercial status, the founder of Germany's largest wind turbine company, Aloys Wobben, was trying his hand at bringing back one of Germany's long-forgotten inventions: the Flettner rotor. Inventive, some say visionary, Wobben has often turned his attention from his bread and butter of manufacturing direct-drive wind turbines to hydroelectricity, power converters, and now wind-assisted shipping.

Flettner and his rotors hark back to the early days in aviation and fluid dynamics. German aeronautical engineer Anton Flettner was a contemporary of German physicists Albert Betz (of the Betz limit fame) and Ludwig Prandtl and French engineer Georges Darrieus. Flettner made important contributions to aviation, but it's his experimentation with spinning cylinders using the Magnus effect that have come to bear his name.

Though the Magnus effect is named for a

Figure 7-7. SkySails. Commercial kite propulsion on the multipurpose cargo ship MV *Michael A.* 2007. (SkySails)

German physicist, Heinrich Gustav Magnus, who described it in the mid–19th century, the effect of a spinning object in a wind stream had been observed by Isaac Newton in the 17th century and later ballistics engineers (see Figure 7-7. Magnus effect). For North Americans, the Magnus effect is the force that imparts the curve in the curve ball and is due to the lift force that is created by a spinning surface in a wind stream. This force acts at right angles to the wind stream and can be harnessed to propel a wind turbine rotor about its axis or a ship through the water. It can even be used in place of wings to lift an airplane off the ground.

Flettner built several such devices. His rotor ship *Baden-Baden* used two upright, cylinders 15 meters (50 feet) tall by 3 meters (10 feet) in diameter spinning at 250 rpm. In 1926, the vessel sailed across the Atlantic to New York City via South America using the Magnus effect as a means of propulsion. In the same year, Flettner launched an even larger ship, *Barbara*, with three rotors each 17 meters (56 feet) tall and 4 meters (13 feet) in diameter spinning at 150 rpm, which went into commercial service in the Mediterranean Sea. And at about the same time, he built a horizontal-axis wind turbine 20 meters (65 feet) in diameter in which he used four spinning cylinders to drive the rotor. From 1926 through 1932, Flettner toured Germany promoting his rotor ship.

While Flettner was developing his rotor for use at sea, inventor Julius Madaras was attempting to use the Magnus effect on land. In 1933, Madaras constructed a 28-meter tall (90-foot) cylinder 9 meters (28 feet) in diameter at Burlington, New Jersey, in hopes of using the phenomenon to drive cars on an elliptical track. He envisioned an endless train of 18 cars moving around the track with generators on the axles of each car. Madaras rated his massive wind power plant at 18 MW in a wind of 13 m/s (30 mph). Incredibly, as the wagons moved around the track, Madaras's design required bringing each rotor to a stop and reversing its direction to continue the journey.

The onset of the Great Depression brought the dreams of both Flettner and Madaras to a halt. Flettner won far greater commercial success from the development of the Flettner Ventilator—an air extraction fan coupled to a Savonius rotor—than he did with his rotor ships. In the early 1930s, Walter Stern acquired patent rights with the idea of manufacturing the ventilators in Great Britain. Though the Flettner Ventilator Company continues to the present, selling more than 50,000 ventilators per year worldwide, the product uses neither Flettner rotors nor the Magnus effect.

Flettner rotor technology languished until the oil crises of the late 1970s and early 1980s when interest revived due to the high cost of oil. In August 1983, a Flettner-driven wind turbine made the cover of *Popular Science* magazine, and in 1984 a Flettner rotor ship made the cover. By the mid-1980s, naval engineers were holding international conferences on wind-powered shipping. It was at this time that Jacques Cousteau developed the Alcyone wind-assisted research vessel developed by French aerodynamicists. Interest again waned when the price of oil collapsed in the mid-1980s.

Enercon's E-Ship 1

Wobben's Enercon revived the Flettner concept in the mid-2000s, building on the development work done in Germany on rotor ships during the 1980s and 1990s (see Figure 7-9. Flettner rotor ship). Enercon's E-Ship 1 is a specialized cargo vessel using four Flettner rotors 27 meters (90 feet) tall and 4 meters (13 feet) in diameter, two at each end of the cargo deck for a total intercept area of more than 430 m².

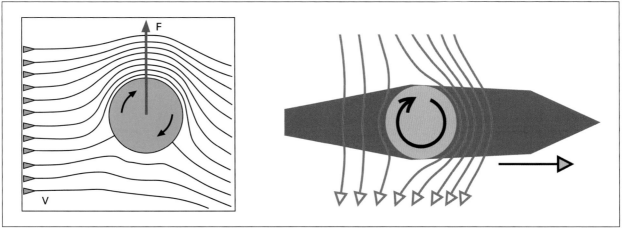

Figure 7-8. Magnus effect. (a) The airflow over a spinning cylinder creates a force, called the Magnus effect, which acts at right angles to the wind stream. (b) The Magnus effect can be used to propel cars on a track, ships on the sea (Flettner rotor ships), drive a wind turbine, or lift a plane into the air. (Wikimedia Commons)

Steam produced by the exhaust gas from the ships diesel engines powers the rotors to provide wind-assisted propulsion. E-Ship 1 is not a pleasure yacht or a plaything of the rich; it's a commercial ship 130 meters (425 feet) long with a beam 23 meters (75 feet) in width, drawing 10,500 DWT, making the E-Ship 1 the largest rotor ship yet built.

After extensive delays, the ship carried its first cargo in mid-2010. Subsequently, it sailed 170,000 nautical miles (300,000 km) in the North and Baltic Seas, North and South Atlantic, and the Mediterranean Sea carrying Enercon wind turbines to projects in Europe and the Americas. While Enercon initially claimed that the technologies used on E-Ship 1 would cut fuel consumption 30% to 50% from both the Flettner rotors and improved hull and propeller design, they later reduced claims of overall fuel savings to 25%.

In mid-2013 the ship was brought into dock for refurbishment. Enercon announced fuel savings of 15% from the Flettner rotors alone. Once the refurbishment was completed in 2014, Enercon returned the ship to the high seas. E-Ship 1 has now become the first Flettner rotor-rigged ship to stay in commercial service for any extended time.

In the mid-1980s, naval architect Peter Schenzle of the Hamburg shipbuilding laboratory explained why there's continuing interest in Flettner rotors for wind-assisted propulsion. According to his analysis, Flettner rotors can provide nearly five times the motive force and average power savings relative to sail surface area of traditional square-rigged sailing vessels.

No one proposes using traditional sailing rigs for cargo ships today, thus, Flettner rotors must compete with other modern wind propulsion systems, such as soft or rigid wings in place of traditional sails. While Flettner rotors may be the champions at low service speeds, says Schenzle, they are less effective at the higher service speeds at which most cargo is transported today (see Figure 7-9. Average driving force of wind propulsion systems).

Unlike modern sail wings, Flettner rotors can't take advantage of increasing service

Figure 7-9. Flettner rotor ship. German wind turbine manufacturer Enercon developed a special-purpose cargo ship for transporting its wind turbines. Enercon's E-Ship 1 used four Flettner rotors 27 meters (90 feet) in height and 4 meters (13 feet) in diameter to provide wind-assisted propulsion for the 10,500-DWT freighter. The ship logged 170,000 nautical miles (300,000 km) in commercial service. (Wikimedia Commons)

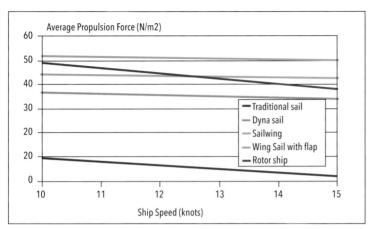

Figure 7-10. Average driving force of wind propulsion systems. Average wind propulsion force in Newtons/m² relative to round-trip ship service speed in knots. Traditional square-rigged sails are shown for comparison. Flettner rotor ships perform well at low service speeds but are hampered by increasing drag at higher service speeds where sail wings and wing sails with flaps perform better. Dyna sails also perform significantly better than traditional sails. (Adapted from Peter Schenzle)

speeds because of increasing drag on the cylinder. The performance of the rotors begins to fall off as ship speed increases. In contrast, as ship speed increases, the relative wind that the ship sees increases as well. Modern sail wings, like that on the Vestas Sailrocket, become ever more powerful as the relative wind increases in strength, while not suffering an equivalent increase in drag.

Perhaps Herr Wobben will next tackle wing sails on the E-Ship 2 and do for wind-assisted shipping what he has done for popularizing direct-drive wind turbines.

The Takeaway

What lessons can you take away from the development of VAWTs in the previous chapter, from DAWTs in this chapter, and other examples of "revolutionary new inventions" that will be mentioned in the pages to come? As with most things in life, don't believe everything you read. There is a deep-seated desire in all of us to believe we can get something for nothing. In wind energy, this trait manifests itself in our willingness to suspend critical judgment, to ask "Does this really work, and if it does where are the results?" Even the technically trained sometimes fall for the latest fad and get swept up in the "madness of crowds." In wind energy, it's caveat emptor. It's "let the buyer beware." More aptly, it should be "let society beware."

This chapter and the previous chapter on VAWTs examined—for the most part—what hasn't worked as touted by their promoters. The next chapter looks at what one expert calls the "silent wind power revolution" because it represents technology that delivers on its promise but isn't sexy enough to garner headlines.

8

Silent Wind Revolution

> The pessimist complains about the wind; the optimist expects it to change; the realist adjusts the sails.
>
> —William Arthur Ward (1921–1994)

THERE IS A TECHNOLOGICAL REVOLUTION UNDERWAY TODAY in wind energy. However, it is not from vertical-axis wind turbines, nor from diffuser-augmented wind turbines. It's not sexy. It's not flashy. It doesn't garner headlines or breathless prose from venture capitalists looking for the next Google. It is, as French renewable energy analyst Bernard Chabot calls it, "a silent revolution."

This revolution is being led by new large-diameter wind turbines with low generator ratings. To the casual observer, these wind turbines look exactly like the wind turbines that they supersede with the exception that maybe the blades are a little longer, a little more slender, and a little more flexible than previously. This is the technology that makes high penetration of wind energy more likely than ever before because it reduces the need for storage and new high-voltage transmission capacity. These are the wind turbines designed for low and moderate wind regimes, whether in central Indiana or the central highlands—the *mittelgebirge*—of Germany.

Some journalists have attempted to describe this phenomenon as a new technology that mysteriously delivers very high generator performance. It's not. We've known how to do this for decades. What's new is that wind turbine manufacturers are now delivering products that meet the need (see Figure 8-1. Silent wind power revolution).

In short, manufacturers are now offering very large diameter wind turbines with relatively low power ratings that are designed for low to moderate wind speed sites. For example, a wind turbine that would have been rated 3 MW or more a few years ago is now being offered as a 2-MW turbine, sometimes even less. These wind turbines will generate much more electricity than earlier turbines rated at 2 MW, delivering very high generator performance. In this regard, they're revolutionary.

Is this a good thing? Yes, absolutely. Here's Chabot's summary of why:

- More generation and higher penetration rates relative to installed capacity,
- Expanded opportunities through use of lower-wind sites,

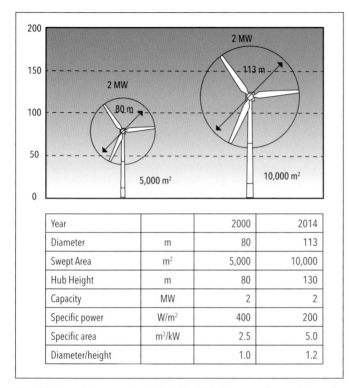

Figure 8-1. Silent wind power revolution. By the late 2010s, there was a silent revolution underway in wind turbine design. Manufacturers began introducing wind turbines with very large rotors relative to their generator ratings, in some cases doubling specific area. They also were installing the turbines on increasingly taller towers relative to their rotor diameter. Together, both factors dramatically increased relative performance in terms of capacity factor and full-load hours. Under moderate wind conditions, the turbine on the right will generate twice the amount of electricity as the turbine on the left, even though they both produce the same peak power.

- Less opposition to wind as less high-wind, high-value sites are now required,
- Less demand on grid operators,
- Less demand for new transmission capacity or capacity upgrades, and
- Wind turbines with large rotors relative to their generator size will allow easier integration of wind energy into the grid, and allow us to put the wind turbines where the people are, that is, near our cities, towns, and villages.

Why this is so requires some explanation.

What Is a Wind Turbine?

In essence, a wind turbine is a rotor to capture the wind and a generator to produce electricity. Physics professor and wind energy authority Vaughan Nelson has emphasized for more than three decades that it is the area of the wind stream intercepted by a wind turbine—the swept area—that largely determines how much energy the wind turbine will capture. Obviously, the generator is a critical component, but it is not the most critical component in what makes a wind turbine—a wind turbine. It is the rotor powered by the wind that separates a wind turbine from a steam generator, for example.

Generator Ratings

Wind turbines are designed with a specific combination of rotor and generator for a specific wind resource. In *Wind Energy*, wind turbines are often described by their rotor diameter—a shorthand for their swept area. However, the media, utility engineers, and even some in the wind industry mistakenly use the generator size in kilowatts (kW) or megawatts (MW) to describe the size of a wind turbine. The wind turbine's generator will produce its "rated power" at a certain wind speed. The wind doesn't always blow at this speed, and this is where descriptions such as this complicate our understanding of what a wind turbine will produce.

Swept Area Trumps Generator Ratings

Let's consider the tale of two wind turbines on the market in the early 2000s: the V82 and the V80. Vestas's V82 is an 82-meter (270-foot) diameter wind turbine capable of generating 1.65 MW. Vestas's V80 is an 80-meter (260-foot) diameter wind turbine variously rated from 1.8 MW to 2.0 MW. Vestas's V82 is a larger—more powerful—wind turbine than the Vestas V80.

How can this be? The V82 intercepts ~ 5% more of the wind stream than the V80. For low and moderate wind sites, the V82 will out produce the V80. At higher wind sites where the V80's larger generator will be used more often, the V80 will generate slightly more than the V82 (see Figure 8-2. Comparison between a V82 and a V80).

To illustrate that this isn't a quirk, here's another case from the early to mid-2000s. Nordex was one of the pioneers in large-diameter turbines with relatively low power ratings. For example, Nordex offered its N80 rated at 2.5 MW while it was also offering its N90 rated at 2.3 MW. Again, the N90 is the larger, more powerful wind turbine. The N90 sweeps ~ 25% more of the wind stream than the N80 and, consequently, would generate considerably more electricity—even though it has a lower generator rating than the N80 (see Figure 8-3. Comparison between a N80 and a N90).

Here's one last example to drive home the point. GE's tried and true 1.5-MW platform was introduced in the early 2000s and has been on the market for more than a decade since. It began with a 71-meter (230-foot) diameter rotor and evolved to the 1.5-MW SL, a 77-meter (250 foot) diameter wind turbine. Though both turbines used the same generator rating, the SL used a rotor that intercepted 18% more of the wind stream. Consequently, the 77-meter turbine would generate more electricity even though it had the same generator rating as the earlier model (see Figure 8-4. GE 1.5-MW platform).

The GE example is noteworthy because GE expanded the platform even farther with its 100-meter (328-foot) diameter model rated at 1.6 MW, effectively doubling the turbine's swept area relative to its earliest model. Wind developers have installed hundreds of this turbine in central Indiana and adjoining states (see Figure 8-5. GE 1.6 MW wind turbine).

While these turbines are, for the most part, no longer on the market, they serve to conclusively illustrate that it is the rotor diameter and, hence, the swept area that determines the amount of electricity a wind turbine will produce.

Metrics of Productivity

There are two principal metrics used to describe the productivity of wind turbines: capacity factor and annual specific yield. Capacity factor is a

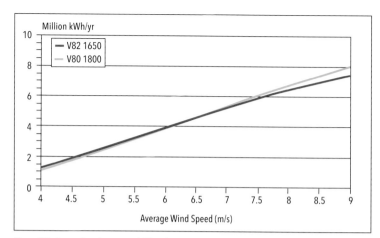

Figure 8-2. Comparison between a V82 and a V80. Vestas's 1.65-MW V82 with a rotor diameter of 82 meters will generate more electricity than Vestas's 1.8-MW V80, a wind turbine with a rotor diameter of 80 meters, at average annual wind speeds of less than 7 m/s (15.7 mph) simply because it uses a larger rotor. The V82's rotor intercepts ~ 5% more area of the wind than the V80.

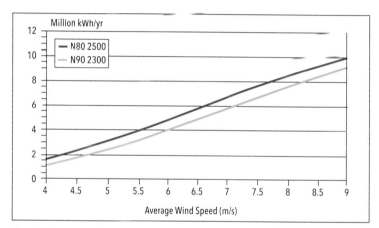

Figure 8-3. Comparison between a N80 and a N90. Nordex's 2.3-MW N90 would generate ~ 25% more electricity than its 2.5-MW N80 even though it has a lower "rated" capacity.

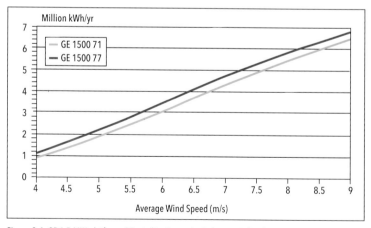

Figure 8-4. GE 1.5-MW platform. GE rated its increasingly larger wind turbines the same as its earliest model: 1.5 MW. The 77-meter diameter, 1.5-MW wind turbine would generate ~ 18% more electricity than its earlier 71-meter, 1.5-MW wind turbine because of its larger diameter and, hence, its greater swept area.

Figure 8-5. GE 1.6-MW wind turbine. German utility E.ON installed 125 of GE's 100-meter (328-foot) diameter turbines in the first phase of its Wildcat Wind Farm near Elwood, Indiana. With a specific area of 4.9 m²/kW–about twice that of wind turbines manufactured only a decade before–GE's 1.6-MW wind turbine has been designed for areas with low to moderate wind speeds. 2013.

measure of how much electricity the wind turbine produces relative to how much it would have produced if the turbine had run all the time at full capacity. It's a common measure in the utility industry in North America, but it's often misused when applied to variable sources of generation such as wind and solar energy. Plant factor is the British expression for the same concept. Continental Europe uses full-load hours, a more direct expression than capacity factor. It is simply how many hours the turbine would have produced at full output.

Capacity factor (CF), because it can be expressed as a percentage, is often confused—sometimes deliberately—with efficiency (η). They are not the same. A wind turbine can have a very low capacity factor and yet be highly efficient at extracting the energy in the wind. Moreover, a wind turbine with a low capacity factor can be cost effective. Capacity factor is simply a measure of the generator's utilization and the effectiveness with which the wind turbine may use the grid. A much more useful measure of wind turbine performance is annual specific yield, or how many kilowatt-hours of electricity a wind turbine generates relative to the area swept by its rotor in kWh/m²/yr.

To determine specific yield or capacity factor, we need actual performance—how much electricity the wind turbine generated over a specific period—or estimate it based on the expected wind resource. When a specific wind resource isn't known, we can compare wind turbines based on their relative swept area.

Measures of Relative Swept Area

There are two measures of relative swept area, that is, how much of the wind stream a wind turbine intercepts relative to its generator capacity: specific power and specific area.

Specific capacity or specific power has been traditionally used and is presented in either watts per square meter of rotor swept area, W/m², or kilowatts per square meter of rotor swept area, kW/m². The new large-diameter wind turbines have very low specific power. GE's 1.6

Table 8-1. Sample Specific Power & Specific Area						
		Rotor Dia.	Swept Area	Rated Power	Specific Power	Specific Area
Manufacturer	Model	m	m²	kW	W/m²	m²/kW
Nordex	N80	80	5,027	2,500	497	2.0
GE	1500	71	3,959	1,500	379	2.6
Nordex	N90	90	6,362	2,300	362	2.8
Vestas	V80	80	5,027	1,800	358	2.8
GE	1500SL	77	4,657	1,500	322	3.1
Vestas	V82	82	5,281	1,650	312	3.2
GE	100-1.6	100	7,854	1,600	204	4.9

MW, 100-meter-diameter turbine has a specific power of 204 W/m². Lower specific power delivers greater capacity factors—or more full-load hours—than turbines with higher specific power for the same wind conditions.

French renewable industry analyst Bernard Chabot principally uses specific area because it is simpler to interpret—a higher number offers better performance for the same wind resource. (Specific area is the inverse of specific power.) Specific area is in units of m²/kW. GE's 1.6-MW, 100-meter-diameter turbine has a specific area of 4.9, almost double that of turbines marketed in the mid-2000s.

The design of the Vestas V80 in our earlier example was typical of its day: 358 W/m², or 2.8 m²/kW. The V82, in contrast, had slightly less rotor loading than the V80, reflecting its greater intercept area. When GE introduced its 1500 in the early 2000s, it had a specific power of more than 400 W/m², or 2.4 m²/kW. While this was typical of the day, GE rated their turbine more aggressively than Vestas did its V82 (see Table 8-1. Sample Specific Power & Specific Area).

Historical Abuse of Power Ratings

The problem with using measures of capacity or power in describing the size of wind turbines arises because it is easy to abuse. Technically, wind turbines can be designed with high power ratings relative to their swept area for very windy sites. Nevertheless, some manufacturers have played on ignorance of how wind turbines work and have marketed wind turbines with very large generator ratings relative to the turbine's swept area (See Table 8-2. Specific Rated Capacity of Selected Wind Turbines in the 1980s). Such abuse was particularly egregious during the California "wind rush" of the early 1980s.

Why? Because unsophisticated buyers often compared wind turbines on their installed cost relative to their installed capacity. By inflating the wind turbine's rating, the manufacturer could charge more money for its product than a competitor and still look cheaper relative to installed capacity.

The most notorious example was Fayette Manufacturing. Its unreliable turbines were the

Table 8-2. Specific Rated Capacity of Selected Wind Turbines in the 1980s					
	Rated Capacity	Swept Area	Specific Power	Specific Area	Approx. Date of Introduction
Manufacturer Model	kW	m²	W/m²	m²/kW	
Fayette 400	400	374	1,070	0.9	1985
Fayette 95	95	95	1,000	1.0	1983
FloWind 19	250	260	962	1.0	1985
Carter 300	300	332	904	1.1	1985
Fayette 75	75	85	882	1.1	1981
Carter 250	250	332	753	1.3	1984
FloWind 25	381	515	740	1.4	1986
Windmaster 200	200	373	536	1.9	1983

bane of the California wind industry during the 1980s. (Fortunately, all but one has been removed.) VAWT manufacturers also overrated their products compared to conventional wind turbines.

Suffice it to say that despite the high specific power of the Fayette wind turbines, they didn't generate any more electricity than other wind turbines with comparable swept area. Because they didn't generate any more electricity than other turbines of similar size, the Fayette wind turbines produced very low capacity factors.

Wind Turbine Design and Wind Regimes

To designate which wind regimes wind turbines are designed to withstand, the International Electrotechnical Commission (IEC) set a design standard. The standard defines what wind conditions the wind turbines must not only endure but remain ready for a return to operation. There are currently three IEC classes and a catchall category. A fifth IEC Class may be added (see Table 8-3. IEC Large Wind Turbine Classes).

IEC Class I is for the windiest sites, those with an average annual wind speed of 10 m/s (22.4 mph) at hub height. Class II is for less windy sites with an average wind speed of 8.5 m/s (19 mph) at hub height. Class III is for even lower wind sites with an average wind speed not to exceed 7.5 m/s (16.8 mph). Class IV is for very low wind speed sites with an average wind speed of 6 m/s (13.4 mph).

Not all wind turbines, then, are created equal. Some turbines are designed for exceptionally windy sites and others for less windy sites. As a result, the specific capacity and specific area will differ between the different IEC classes.

Wind turbines designed for IEC Class I conditions can be used at any site. They will typically have a high specific power and a low specific area. However, wind turbines designed for Class II or Class III sites shouldn't be used at a windy Class I site. Wind turbines intended for Class II or Class III wind regimes will have low specific power and high specific area, reflecting their relatively larger rotor diameters.

When a Class I turbine is used in a low to moderate wind regime, it will deliver a low capacity factor or fewer full-load hours than it would at a windy site because it has a relatively high specific capacity or low specific area. Consequently, specific power and specific area varies with the IEC Class the wind turbine is designed to serve.

For example, a Class I wind turbine will have a specific capacity of 400 W/m^2, or a specific area of 2.5 m^2/kW (see Table 8-4. Specific Capacity and Specific Area of Large Wind Turbines: Example IEC Class I Turbines).

Whereas Class II turbines will have specific capacities of 300 W/m^2 to 400 W/m^2, or a specific area of 2.5 m^2/kW to 3.0 m^2/kW (see Table 8-5. Specific Capacity and Specific Area of Large Wind Turbines: Example IEC Class II Turbines).

And Class III turbines will have specific power of 200 W/m^2 to 300 W/m^2, or a specific area of 3.5 m^2/kW to 5.0 m^2/kW (see Table 8-6. Specific Capacity and Specific Area of Large Wind Turbines: Example IEC Class III Turbines).

The introduction of IEC Class III wind turbines greatly expands the developable wind resource, making entire regions once considered unsuitable for wind energy now attractive.

This trend toward low wind speed turbines is expected to continue. Manufacturers were field-

Table 8-3. IEC Large Wind Turbine Classes					
(Wind speed in m/s)					
		I	II	III	S
Reference Wind Speed	V_{ref}	50	42.5	37.5	Values Specified by the Designer
Annual Average Wind Speed	V_{ave}	10	8.5	7.5	
High Turbulence (A)	I_{ref}	0.16			
Medium Turbulence (B)	I_{ref}	0.14			
Low Turbulence (C)	I_{ref}	0.12			
V_{ref}: 10-minute average wind speed at hub height; I_{ref}: Expected turbulence intensity at 15 m/s; IEC 61400-1, 2005.					

Table 8-4. Specific Capacity and Specific Area of Large Wind Turbines: Example IEC Class I Turbines

Manufacturer	Model	Rotor Dia. m	Swept Area m²	Rated Power kW	Rated Wind Speed m/s	Perf. at Rated Power %	Specific Power W/m²	Specific Area m²/kW	Wind Class
Siemens	2.3-82 VS	82.4	5,333	2,300	13.5	0.29	431	2.3	IA
Siemens	3.6-107	107	8,992	3,600	13	0.30	400	2.5	IA
Vestas	80-2.0	80	5,027	2,000	14	0.24	398	2.5	IA
Nordex	90/2500	90	6,362	2,500	13	0.29	393	2.5	IB
Vestas	112-3.3	112	9,852	3,300	13	0.25	335	3.0	IB

Table 8-5. Specific Capacity and Specific Area of Large Wind Turbines: Example IEC Class II Turbines

Manufacturer	Model	Rotor Dia. m	Swept Area m²	Rated Power kW	Rated Wind Speed m/s	Perf. at Rated Power %	Specific Power W/m²	Specific Area m²/kW	Wind Class
Enercon	E82-2.3	82	5,281	2,300	13.5	0.29	436	2.3	NVN IIA
Enercon	101	101	8,012	3,000	11.5	0.40	374	2.7	NVN IIA
Gamesa	87-2000	87	5,945	2,000	15	0.16	336	3.0	IIA
Vestas	112 3.3	112	9,852	3,300	13	0.25	335	3.0	IIA
GE	1.5-77	77	4,657	1,500	14	0.19	322	3.1	IIA
Vestas	117-3.3	117	10,751	3,300	13	0.23	307	3.3	IIA
Siemens	2.3-101	101	8,012	2,300	12.5	0.24	287	3.5	IIB

Table 8-6. Specific Capacity and Specific Area of Large Wind Turbines: Example IEC Class III Turbines

Manufacturer	Model	Rotor Dia. m	Swept Area m²	Rated Power kW	Rated Wind Speed m/s	Perf. at Rated Power %	Specific Power W/m²	Specific Area m²/kW	Wind Class
Vestas	90-1.8	90	6,362	1,800	12	0.27	283	3.5	IIIA
Siemens	2.3-113	113	10,029	2,300	12.5	0.19	229	4.4	III
Vestas	100-1.8	100	7,854	1,800	12	0.22	229	4.4	IIIA
Nordex	117/2400	116.8	10,715	2,400	12.5	0.19	224	4.5	IIIA
Vestas	110-2.0	110	9,503	2,000	11.5	0.23	210	4.8	IIIA
Gamesa	114-2.0	114	10,207	2,000			196	5.1	IIIA

ing wind turbines for sites with an average annual wind speed less than 6.5 m/s (14.6 mph) in 2014. Turbines in this class have specific power of less than 160 W/m² to 200 W/m² and specific area of from 5.0 m²/kW to more than 6.0 m²/kW. Nearly two-thirds of the wind projects proposed in China, the world's largest market for wind turbines, in 2012 were destined for such sites.

Small and Medium-Size Turbines

Small and medium-size wind turbines are also designed to withstand specific wind conditions represented by an IEC classification similar to that for large wind turbines (see Table 8-7. IEC Small Wind Turbine Classes).

Here too a silent revolution is underway as new products are entering the market with much larger rotor diameters relative to generator size than in the past (see Table 8-8. Specific Capacity and Specific Area of Selected Small & Medium-Size Turbines). Typical wind turbines of just a decade ago, such as Bergey's Excel and Northern Power Systems' Northwind 100, had specific powers from 250 W/m² to 300 W/m², or a specific area of 3.5 m²/kW to 4.5 m²/kW for Class II conditions. Now Northern Power

CASE STUDY GERMANY: NEW WIND TURBINES EXPAND THE WIND RESOURCE

One of the earliest examples of how the new wind turbines entering the market are revolutionizing the wind industry is a study published in mid-2013 for Germany's Environment Agency, the Umweltbundesamt (or UBA). The study by the Fraunhofer Institute for Wind Energy in Kassel reexamined the potential of wind energy on land in Germany.

Several such studies have been done in the past, one as recently as 2010. All have concluded that even in densely populated Germany there's more than enough land area to meet the country's renewable energy targets after excluding national parks, nature reserves, and other sites where development is prohibited.

What was different this time was the scale of the wind resource that the researchers found. They discovered that nearly 14% of Germany's land area or 49,000 km² (~ 19,000 mi²) was suitable for wind energy when newer low-specific power, high-specific area turbines were used. There is the potential, said researchers, to install an astonishing 1.2 million megawatts of wind generating capacity in the country. Such a fleet of wind turbines could generate 2,900 TWh of electricity per year or nearly five times Germany's current consumption (see German Wind Energy Potential on Land).

By including larger diameter turbines installed on towers up to 140 meters (430 feet) in height, researchers could expand the developable wind resource from the North German Plain to the central highlands, the *Mittelgebirge*, and the south of Germany. (Wind speeds typically decrease from north to south in Germany.) They found that the wind resource in Germany doubled when using the new wind turbines designed for low-wind sites.

What is striking is that the Kassel researchers were being conservative. They were using a hypothetical low-wind turbine with a specific power of 314 W/m² and a specific area of 3.2 m²/kW. As noted in this chapter, manufacturers were selling and developers were already installing wind turbines in 2013 that far exceeded these measures, some with a specific power nearly 200 W/m² or nearly 5 m²/kW. These turbines, which exist today, would expand the developable wind resource in Germany even more. That is revolutionary.

German Wind Energy Potential on Land

Region	MW	TWh/yr	%	Full-Load Hours	Capacity Factor
North	526,000	1,378	48%	2,621	30%
Central	287,000	728	25%	2,540	29%
South	375,000	791	27%	2,108	24%
Total		2,897		2,440	29%

Source: Potenzial der Windenergie an Land, Umwetbundesamt, June 2013.

Table 8-7. IEC Small Wind Turbine Classes

(Wind speed in m/s)

		I	II	III	IV	S
Reference Wind Speed	V_{ref}	50	42.5	37.5	30	Values Specified by the Designer
Annual Average Wind Speed	V_{ave}	10	8.5	7.5	6	
Turbulence Intensity at 15 m/s	I_{15}	18%	18%	18%	18%	
Dimensionless slope parameter	a	2	2	2	2	
IEC 61400-2.						

Table 8-8. Specific Capacity and Specific Area of Selected Small & Medium-Size Turbines

Manufacturer	Model	Rotor Dia. m	Swept Area m²	Rated Power kW	Rated Wind Speed m/s	Perf. at Rated Power %	Specific Power W/m²	Specific Area m²/kW	Wind Class
Northern Power Systems	100	21	346	100	15	0.14	289	3.5	IIA
Bergey Windpower	Excel 10	7	38	8.9	11	0.28	231	4.3	II
Northern Power Systems	95	24	452	95	14	0.12	210	4.8	III/S
Evance	R9000	5.5	24	4.7	11	0.24	198	5.1	II
Endurance	3120	19.2	290	50	9.5	0.33	173	5.8	IIIA
Gaia	133-11	13	133	11	9.5	0.16	83	12.1	IIIB

offers a 24-meter (79-foot) diameter version of its turbine for IEC class III sites.

Endurance's E3120 raised the bar further with its 19.2-meter (63-foot) diameter rotor rated at only 50 kW. The specific power for Endurance's turbine is well below 200 W/m² with a specific area of nearly 6 m²/kW. When the turbine was introduced, Nova Scotia regulators were so unaccustomed to such a large rotor on a 50-kW turbine that they excluded it from a program designed specifically for wind turbines up to 50 kW. While the design was being lauded by wind industry analysts for its emphasis on swept area and not its generator rating, Nova Scotia regulators—out of ignorance of what makes a wind turbine work—discriminated against Endurance, who, they thought, was misrepresenting the product.

Scottish manufacturer Gaia takes a large diameter rotor coupled with a small generator to a whole other level. The 13-meter (43-foot) diameter rotor drives an 11-kW induction generator. Before Gaia, this was unheard of in small wind turbines, where many newcomers inflate generator ratings to win attention from the media—and customers. Gaia's two-blade, teetered rotor has a specific power of less than 100 W/m² and a specific area of 12 m²/kW.

While Gaia's large rotor relative to its small generator puts it outside the mainstream of today's wind turbines, the design has good company. In 1958 Ulrich Hütter, the father of German wind turbine design, developed a two-blade, downwind, teetering rotor. At a United Nations Conference on new sources of energy in 1961, one of Hütter's colleagues, Sepp Armbrust, explained why they used such a large 34-meter (112-foot) diameter rotor on their 100-kW turbine.

"The specific power loading of the circular area swept by the wing blades was kept to a low level in order to assure an almost uniform energy output in places with relatively low mean wind speeds. Therefore, contrary to teams in France, Denmark, England, United States, etc., we intentionally chose a design output of only 110 W/m² swept wheel area instead of the usual 300 to 400 W/m²."

Hütter's design represented a specific area of 9.1 m²/kW, or twice that of the Bergey Excel and nearly three times that of the Northwind 100, as well as most of the large wind turbines designed for IEC Class I and Class II sites. The wind industry has now come full circle by building wind turbines that emphasize swept area and not generator size, as Hütter had recommended in the 1960s.

Specific Power and Capacity Factor (Full-Load Hours)

For similar wind conditions, a wind turbine with a low specific capacity or a high specific area will produce a higher capacity factor or more full-load hours because it will simply generate more electricity (see Figure 8-6. Equivalent capacity factor for specific power). In other words, a larger rotor will capture more wind energy than a smaller rotor.

As noted above, this relationship doesn't hold across all wind regimes because some wind turbines are not suitable for all sites. Low wind IEC Class III turbines are not suited for IEC Class I or Class II conditions. Nevertheless, this relationship between capacity factor and specific capacity or specific area explains why manufacturers can advertise high-capacity factors and—

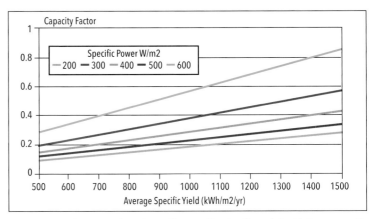

Figure 8-6. Equivalent capacity factor for specific power. There is a direct relationship between specific power (or capacity) and capacity factor relative to the yield of wind turbine at a specific site. For example, a wind turbine with a specific power of 200 W/m², or conversely a specific area of 5.0 m²/kW, at a site with an annual specific yield of 1,100 kWh/m²/yr will produce a capacity factor of more than 60%. Although we've known how to do this for many decades, it is revolutionary that manufacturers are now delivering wind turbines with such low specific power or high specific area to make it possible.

RELATIONSHIP BETWEEN CAPACITY FACTOR, YIELD, AND FULL-LOAD HOURS

We can use the characteristics of the hypothetical turbines in Figure 8-1 to illustrate the relationships between capacity factor, full-load hours, specific power, specific area, and annual specific yield. We'll use the naming conventions of French engineer Bernard Chabot who has done much to popularize this topic.

Capacity Factor

The average annual capacity factor (CF) equals the annual energy production (Ey) divided by the product of a turbine's rated power times the number of hours in a year. It's often presented as a percentage, but it can also appear as a decimal.

$CF = Ey/(8,760*Ps)$

For example, an 80-meter-diameter wind turbine rated at 2 MW that generates 5 million kWh per year will have a capacity factor of

$CF = (5,000,000 \text{ kWh/yr})/(8,760 \text{ hrs/yr}*2,000 \text{ kW}) = 28.5\%$, or 0.285.

Or consider a 113-meter-diameter wind turbine also rated at 2 MW. If this much larger turbine generates 10 million kWh per year, it will have a capacity factor of

$CF = (10,000,000 \text{ kWh/yr})/(8,760 \text{ hrs/yr}*2,000 \text{ kW}) = 57\%$, or 0.57.

Full-Load Hours

Full-load hours (Nh) is another way to express the same idea as capacity factor. Annual full-load hours (Nh) is simply the annual generation divided by a wind turbine's rated power.

$Nh = Ey/Ps$

In our first example, the 80-meter-diameter, 2 MW turbine generating 5 million kWh per year will deliver

$Nh = (5,000,000 \text{ kWh/yr})/2,000 \text{ kW} = 2,500$ full-load hours.

The larger turbine will produce twice the full-load hours as with capacity factor.

Capacity factor and full-load hours are related by the number of hours in a year.

$CF = Nh/8,760$

For example the capacity factor of a turbine with 2,500 full-load hours is

$CF = 2,500 \text{ hrs}/8,760 \text{ hrs} = 28.5\%$.

Annual Specific Yield

Another metric of performance, annual specific yield (Eys), is independent of generator size or generator rating. It's solely a function of annual generation (Ey) and the swept area (S) of the wind turbine in m².

In our example of the 80-meter turbine, annual specific yield is found by dividing annual generation by the rotor swept area.

$Eys = Ey/S$

$Eys = (5,000,000 \text{ kWh/yr})/5,000 \text{ m}^2 = 1,000 \text{ kWh/m}^2/\text{yr}$

Since the large turbine is double the area of the smaller turbine and generates twice the amount of electricity in our example, the annual specific yield of both turbines is the same.

If the conversion efficiency of a class of wind turbines is assumed to be equivalent, then the annual specific yield is a surrogate for the wind resource or annual average wind speed.

Specific Power

Specific power (Ps) in W/m² is found by dividing the swept area (S) of a wind turbine by the turbine's rated power (P). Specific power is one widely used measure of rotor loading.

$Ps = P/S$

Our 80-meter, 2-MW wind turbine sweeps 5,000 m², therefore the specific power is

$Ps = 2,000 \text{ kW}/5,000 \text{ m}^2*1,000 \text{ W/kW} = 400 \text{ W/m}^2$.

Specific Area

Specific area (Su) in m²/kW is simply the wind turbine's swept area divided by the turbine's rated power.

$Su = S/P$

Again, using our example of the 80-meter wind turbine rated at 2 MW, the specific area is

$Su = 5,000 \text{ m}^2/2,000 \text{ kW} = 2.5 \text{ m}^2/\text{kW}$.

Specific area is the inverse of specific power.

$Su = 1/Ps*1000 = (1/400 \text{ W/m}^2)*1,000 \text{ W/kW} = 2.5 \text{ m}^2/\text{kW}$

Relationship Between Capacity Factor and Specific Yield

There is a direct relationship between capacity factor (CF) and annual specific yield (Eys) as a function of specific power (Ps).

$CF = Eys/(Ps*8,760 \text{ hrs/yr}*1,000 \text{ W/kW})$

At a site where a wind turbine with a specific power of 200 W/m² yields 1,000 kWh/m²/yr, the capacity factor is

$CF = (1,000 \text{ kWh/m}^2/\text{yr})/(200 \text{ W/m}^2*8,760 \text{ hrs/yr}*1 \text{ kW}/1000 \text{ W})$
$= 0.57$ or 57%.

There is a similar relationship with full-load hours and annual specific yield.

$Nh = Eys/Ps = (1000 \text{ kWh/m}^2/\text{yr})/(200 \text{ W/m}^2* 1 \text{ kW}/1000 \text{ W}) = 5,000$ hours.

more importantly—deliver high-capacity factors in the field.

Why All This Is Important

What is revolutionary about the new low-specific power, high-specific area turbines is not that they exist—there have always been such turbines—it is that the manufacturers and wind developers have finally embraced them. In the United States, for example, Lawrence Berkeley's Ryan Wiser reported that the average specific power of newly installed wind turbines has dramatically fallen from 400 W/m² in 1998 to 283 W/m² in 2012—a change of 40%.

For many years those who wanted to use wind turbines in lower wind regimes, typically near where people live, were forced to use wind turbines that were designed for high wind sites. While such turbines were adequate for the task, they produced very low capacity factors. This was acceptable as long as wind energy was a small part of the generating mix and there was more than sufficient capacity on the wires and electrical infrastructure to absorb peak power on those occasions when it occurred.

Manufacturers, meanwhile, were selling to commercial wind developers who pick the windiest sites possible to maximize their profits. This was the traditional model of power plant development since the 1940s: power plants were installed where the resource was most abundant often quite distant to where the electricity would be used. It wasn't always so. In the early days of electricity, power plants were built in the cities where the demand was.

All this began to change as more and more of the high-wind sites were developed and the bottlenecks to long-distance transmission of electricity became more problematic.

Countries such as Germany and France went so far as to implement policies—feed-in tariffs differentiated by wind resource—that would enable development at lower wind speed sites. They reasoned that it would be better for the nation if wind development was not solely concentrated on the windiest coastlines or windiest mountaintops but distributed across the breadth of the country. Not only would this simplify integration of wind energy with the transmission and distribution system, it would reduce social conflict by those opposed to wind turbines in scenic, but windy, locales while at the same time spreading economic opportunity to all regions.

France and Germany have been successful in this regard. Wind development is geographically dispersed in both countries. In Germany it's not uncommon to see wind turbines near the great urban agglomerations. For example, numerous wind turbines are visible from the inland harbor of Hamburg and even from the scenic city of Freiburg in southern Germany's Schwarzwald or Black Forest.

Like his French colleague Chabot, Quebec engineer Bernard Saulnier believes the new IEC Class III turbines are not only revolutionary because they allow deploying new wind generating capacity in lower wind speed regions but also because—whether they realize it or not—the manufacturers have declared war on the centralized generation model and the long transmission lines that are an essential part of that model.

This is good news to many environmentalists who have objected to the long-distance transport of electricity. Many environmentalists prefer that generating capacity should be "distributed" among the users of electricity so that new transmission lines are not needed. Distributed generation implies putting wind turbines and solar panels in or near urban areas where consumption is greatest.

> Quebec engineer Bernard Saulnier believes the new IEC Class III turbines are not only revolutionary because they allow deploying new wind generating capacity in lower wind speed regions but also because—whether they realize it or not—the manufacturers have declared war on the centralized generation model and the long transmission lines that are an essential part of that model.

German wind engineer Jens-Peter Molly also points out that low-specific power, high-specific area turbines use the existing network so much more effectively that it drastically reduces the need for storage of a variable resource like wind energy. Thus, argues Molly, we can rethink how best to integrate the high penetration of wind energy into the grid. Incorporating these new wind turbines in the transmission and distribution system will be much more cost effective than adding expensive storage facilities, or expanding transmission capacity with thicker cables on existing lines, or installing controversial new power lines.

Molly explains it this way.

If a wind turbine with a 100-meter-diameter rotor were equipped with a generator of only 1-kW size, it should be clear to everyone that this wind turbine could run throughout the whole year at rated power without requiring expensive storage facilities or overdimensioned grid connections because the capacity factor or the guaranteed power capacity would be almost 100%. The payment for each kilowatt hour generated in this way, however, would have to be very high because with only 8,760 kWh generated, the expenditure for the large rotor, nacelle, tower, foundations, and so on would be spread over so few kilowatt-hours.

On the other hand, the same rotor diameter could be coupled with a 10-MW generator. In this case, the wind turbine would generate the rated power only for a few hours a year, in other words, enormous costs for the mechanical and structural components of the turbine, which are out of all proportion to the increased yield of the wind turbine. The cross sections of the transmission lines would have to be sized to be able to transmit the rated power generated by this turbine during only a few hours per year when it reached its rated capacity. These power lines would then be greatly underutilized and, therefore, much too expensive for the electricity generated.

Obviously, between these two extremes there must be an optimum, says Molly. And that is why there is now a range of wind turbine designs for different wind resources.

IEC Class III turbines are not robust enough for high-wind sites, says Molly, but they are well suited for large areas of countries, such as Germany, where the majority of people live and work—where the load is. He argues that it is much cheaper to pay a little extra for the generation from such a wind turbine than to either pay for storage or increased transmission capacity. Fortunately for Germany, the country's differentiated wind tariffs are easily adapted to this requirement.

Low-specific capacity, high-specific area turbines increase the average power they can deliver for a longer period of time, improving both the predictability of wind energy to grid operators and improving the ability of wind turbines to provide reserve generating capacity for emergencies, such as when a nuclear plant trips off-line. And because the difference between average power and rated power is smaller, there is a much-reduced need for greater transmission capacity.

French engineer Chabot makes a similar observation; wind turbines with low-specific power and high-specific area "represent a strategic advantage for the large-scale integration of wind energy" in the electricity system. Much greater amounts of electricity can now be generated with a lower total installed capacity, he says. And this capacity can be placed nearer the centers of consumption than otherwise, reducing the cost of electrical transmission and distribution. This is a huge advantage, says Chabot, for adapting our existing infrastructure, which was built at such a high cost, to the high penetration of renewables that is coming.

These long-awaited turbines of low-specific capacity and high-specific area are the kind of technology needed to make wind energy an essential low-cost component of supplying 100% of society's electricity with renewable energy. That's revolutionary.

9

Towers

Dieux fournit le vent, mais l'homme doit hisser les voiles.
(God provides the wind, but man must raise the sails.)

—St. Augustine

TOWERS ARE AS INTEGRAL TO THE PERFORMANCE OF A WIND system as the wind turbine itself. Without the proper tower, your wind machine isn't much more than an expensive kinetic sculpture and could even become a hazard to all in the vicinity.

Towers for large wind turbines are designed by the turbine manufacturer specifically for its turbine. There's typically little or no choice, except possibly for optimum height.

However, towers for small wind turbines—as a rule—are one of the few wind-system components in which you have some choice. You may have several types of towers to choose from—at least for household-size and smaller wind turbines.

When considering tower options for small turbines, it's imperative to keep in mind that the tower must be strong enough to withstand the thrust on the wind turbine (the force trying to knock the wind turbine off the tower) and the thrust on the tower (the force trying to knock the tower over). And unless it's a hinged tower that can be lowered to the ground, the tower must have some means for access to the turbine. Otherwise, all service of the turbine will require the use of a crane or person lift.

Foremost among the criteria for the correct tower is whether it's available in the height desired.

Height

As the wind industry has matured—and wind system users as well—selecting a tower of the proper height has become increasingly important. In the early 1970s, anything that would get the wind machine off the ground was acceptable.

Towers for wind turbines used on the Great Plains during the 1930s were never very tall. The flat terrain and the few obstructions present didn't call for towers taller than 60 feet (18 meters). Even so, by the late 1940s Wincharger was installing guyed towers 85 to 105 feet (25–30 meters) in height, and Parris-Dunn was advising its customers that "the higher the tower the greater the power"—an adage that's still true today.

Parris-Dunn was advising its customers that

"THE HIGHER THE TOWER THE GREATER THE POWER"

—an adage that's still true today.

Figure 9-1. Too short a tower. This Bergey 1500 is on too short a tower even for windswept Patagonia. For best performance, a small turbine such as this should be installed on a tower at least 20 meters (60 feet) tall.

Through painful experience, we've learned that economic power generation and good performance are obtained only on a tall tower. This has proven true for both small and commercial-scale wind turbines.

We have known for some time that wind speed and power increase with height, but that didn't begin to sink in until wind systems were being installed in great numbers across the breadth of North America and Europe. We gained far more experience with power-robbing turbulence and what it can do to a wind machine's performance than we ever needed. As a result, recommended tower heights have gradually increased. The commercial success of large wind turbines is, in large part, due to their use of increasingly tall towers.

This is a lesson that has been lost on many small wind turbine aficionados. Mick Sagrillo, never one to mince words when it comes to small wind turbines, is adamant about using as tall a tower as practical (see Figure 9-1. Too short a tower). He tells students in his workshops on how to install small turbines that the three most common mistakes people make with wind energy are using

1. too short a tower,
2. too short a tower, and
3. too short a tower.

Sagrillo likens using too short a tower to "putting solar panels on the north side of the roof so they won't get sunburned." He also advises his students to consider the height of mature trees when sizing towers. "As much as you water them, wind turbine towers don't grow," he warns. Many a wind turbine owner has woken up one morning to find that the trees they planted years before have grown, while their wind turbine has steadfastly remained at the same height.

Manufacturers prefer taller towers because they want their products to perform well and want to minimize turbulence-induced service and warranty claims. Taller towers also allow more flexibility in siting. If buildings and trees are present—and they usually are—a tall tower can redeem an otherwise unusable site. For household-size turbines, a minimum tower height by today's standards is 25 meters (80 feet). And when trees are nearby, 30 to 35 meters (100–120 feet) is the norm.

The height requirements for micro wind turbines are somewhat different. Micro turbines are often used in low-power applications, like weekend cabins, where maximum performance isn't necessary (see Figure 9-2. Micro turbine pipe tower). Because of their relatively low cost, they're often used with inexpensive towers. These turbines are generally not suited for tower heights above 18 meters (60 feet). It doesn't make a lot of sense to install them on taller

Figure 9-2. Micro turbine pipe tower. This relatively short tower made of steel water pipe is adequate for the Marlec 910F in this application, powering a few DC lights in the shed. For anything more demanding, a taller tower is necessary.

STRATOSPHERIC HEIGHTS

The tower heights of large wind turbines are often in the ratio of 1:1 with the turbine's rotor diameter. In the early 1990s, for example, towers for medium-size wind turbines typically reached heights of 30 to 40 meters (100–130 feet) for wind turbines of 25 to 30 meters (80–100 feet) in diameter. By the mid-1990s, towers were reaching heights of 40 to 50 meters (130–160 feet), and by the end of the decade towers, 60 to 70 meters (200–230 feet) tall had become common. In wooded areas of Germany, a tower height to diameter ratio of 1.5 has sometimes been used. It's not uncommon now for wind turbines with rotor diameters of 100 meters (330 feet) or more to be installed on towers 150 meters (~ 500 feet) tall. The use of such tall towers explains why wind turbines have become visible from commercial aircraft flying over Germany's central highlands. Tall towers permit the turbines to stand well above surrounding obstructions–trees and buildings–as well as the terrain. They also make the turbines more visible– even from high-flying aircraft.

towers that cost three or four times as much as the turbine itself, unless you plan to eventually install a bigger turbine.

Andy Kruse, one of the founders of defunct manufacturer Southwest Windpower, contends that what height tower people use with micro turbines depends on what they want from the wind turbine and what they're willing to accept. Kruse argues that Southwest Windpower's Air turbines were so inexpensive that installing them on a taller—and more expensive—tower was seldom justified. The average tower height used for the Air was only 25 feet (7.6 meters). Kruse says, "People felt comfortable with this tower height. Admittedly, this wasn't the optimum, but users shouldn't be so concerned with tower height if they're willing to accept less-than-optimum performance."

There are other times where a nonoptimum tower height may be acceptable. For example, at sites in Scotland, many household-size turbines are installed on freestanding towers only 6.5 meters (20 feet) tall. Scoraig Wind Electric's Hugh Piggott finds that in the open terrain of the Scottish Highlands, these tower heights work acceptably, though they are far from ideal. These short towers were selected because they passed muster with strict local planning codes that severely limited tower height.

Nevertheless, "it's the landscape that determines tower height, not what the manufacturer offers, or the dealer sells, or zoning laws allow," says Sagrillo. "If it doesn't produce enough power to warrant being installed on a suitable tower," he says, "then it's not a serious wind generator. Why bother?"

Buckling Strength

Next to the height, the most important criterion for a tower is its ability to withstand the forces acting on it in high winds. Towers are rated by the thrust load they can endure without buckling. Standards in the United States call on manufacturers to design their wind systems to withstand 120-mph (54-m/s) winds without damage. The thrust on the tower at this wind speed depends on the rotor diameter of the wind turbine and its mode of operation under such conditions— whether it furls the rotor or changes the pitch of the blades, for example.

Two wind turbines of the same rotor area may require entirely different towers because of

DRAG FORCE AND THRUST

The drag on an object in the wind, whether a tall building, tower, or wind turbine is a function of air density, the area intercepting the wind, the speed of the wind, and a dimensionless coefficient that represents the objects shape and its angle to the wind. The drag force in the direction of the wind is the thrust: the force trying to push a tower over or blow a wind turbine off the top of its tower.

$$\text{Drag} = \tfrac{1}{2} \rho A V^2 C_D$$

where ρ is air density in kg/m³, A is the area in m² intercepting the wind, V is wind speed in m/s, and C_D is the coefficient of drag. A flat plate at right angles to the wind has a C_D of 1.1 while that of a closed cylinder is 0.6. It's important to know the maximum amount of thrust you can expect on the wind turbine and tower to determine the needed buckling strength of the tower and the size and depth of the anchors. For small wind turbines and their towers, it's best to assume the coefficient of drag for a flat plate. Some manufacturers include the thrust they expect their turbine to see under worst-case conditions in their list of specifications. Others will provide the data upon request. North American tower manufacturers design their towers to withstand thrust loads at 120 mph (54 m/s) on the top of the tower, with a minimum safety factor of two.

differing approaches to protecting the rotor in high winds. Small wind turbines that furl the rotor reduce thrust loads on the tower substantially compared with those that pitch the blades to stall. For wind turbines that furl the rotor, thrust reaches a maximum at the furling speed and remains fairly constant thereafter. Thrust continues to increase with increasing wind speed on small turbines with mechanical governors that pitch the blades to stall.

For small wind turbines on tall towers, the drag on the tower in high winds adds significantly to the thrust loads the tower must withstand. In contrast, the rotor on a large wind turbine presents far more frontal area to the wind, proportionally, than it does on a small turbine; thus, thrust on the rotor dominates.

All towers flex and sway with the wind—even the massive tubular towers that support commercial wind turbines. In a strong wind, the swaying at the top of a tall tower may be enough to give you the impression of being aboard a storm-tossed ship. One wind turbine dealer discovered this the hard way.

After the dealer finished wiring his newly installed wind turbine to the service panel, he was eager to see it in operation (an affliction that wind pioneer Jack Park diagnoses as "fire-'em-up-itis"). The wind was strong, blowing near the rated speed of his Jacobs wind turbine. To ensure that all was well and to get a bird's-eye view of his new investment, he unwisely climbed up the 33-meter (100-foot), heavy-duty truss tower. Stopping just below the rotor, he decided to check the operation of the feathering governor by unloading the generator and letting the rotor speed increase. When an assistant disconnected the wind turbine, he suddenly found himself hanging on with all his might, as the blades feathered and the tower sprang several feet back into the wind like a giant whip. He was lucky. If he hadn't been strapped to the tower and kept his wits, he could easily have been killed. (This example violates one of the fundamental safety rules of working around wind turbines. Never climb the tower when the rotor is spinning. For more on safety, see Chapter 17.)

Slender tubular towers are far more flexible than truss towers and visibly deflect in strong winds. Deflection isn't a problem unless the turbine and tower are mismatched. When the tower or the blades flex too much or at the wrong time, the blades could strike the tower. This dynamic interaction between the wind turbine and tower is of major concern to manufacturers.

As the rotor and tower deflect in the wind, they begin to oscillate like the swaying spans of a rickety suspension bridge. Should the turbine and tower begin to sway in harmony, the oscilla-

ROCKING AND ROLLING WITH TVIND

I felt the power of the wind in 1980 after riding the elevator up the concrete tower to the nacelle of Tvind's mighty wind turbine on Denmark's windy west coast. There were four of us in the nacelle, and the turbine was operating in a good stiff wind. It felt like a ship at sea once we were in the nacelle. The tower swayed to and fro as each wave of wind hit the turbine. Fortunately, there were plenty of safe handholds. Tools used in the nacelle were securely mounted to the bulkhead like they would be at sea, but one of them clattered to the floor. As the swaying continued to increase, we looked apprehensively at the blond-haired, blue-eyed Danish guide. She smiled and suggested that maybe it was time for us to leave. Yes it was. We quickly agreed.

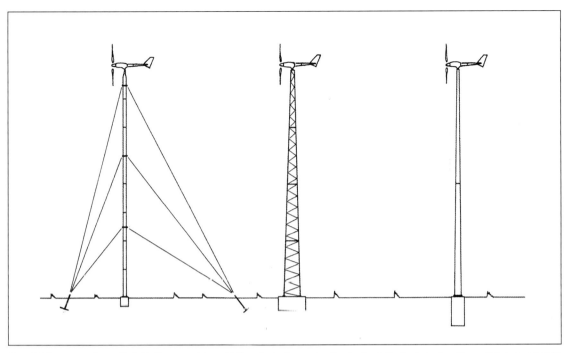

Figure 9-3. Tower types. For small wind turbines, guyed towers (left) are the least expensive. Freestanding lattice (center) and cantilevered tubular towers (right) are more costly but more aesthetically pleasing. (Bergey Windpower)

tions could gradually increase in magnitude until they destroy the wind turbine, tower, or both.

Though guyed towers may appear less secure than truss towers, the reported accounts of tower failures for small wind turbines involve nearly equal numbers of truss towers and guyed towers. In one widely discussed case in the 1980s, a truss tower failed when the bolts holding two 20-foot sections together sheared. The tower manufacturer asserted that the tower was overloaded by the dynamic interaction of the turbine and the tower. Witnesses noted that the tower was vibrating wildly prior to the accident. The turbine manufacturer, Jacobs Wind Electric, countered that the failure was due to "bad steel" in the bolts. Bad steel or not, a wind system vibrating in resonance can exert tremendous force on the tower, creating loads well beyond its design limits.

It's this dynamic interaction between the wind turbine and the tower that leads some manufacturers of small wind turbines to limit the type of tower that can be used for their machines. The pairing becomes increasingly important as size increases. Commercial-scale wind turbines are matched to a limited number of towers that meet the manufacturers' requirements for strength and dynamic response.

Tower Types

Towers fall into two categories: free standing and guyed (see Figure 9-3. Tower types). Freestanding towers, also known as self-supporting towers, are just that—free standing. They depend on a deep or massive foundation to prevent the tower from toppling over in high winds, and they must be strong enough internally to withstand the forces trying to bend the tower to the ground. Guyed towers, in contrast, employ several far-flung anchors and connecting cables to achieve the same ends. For small wind turbines, freestanding towers are more expensive than guyed towers, but they take up less space.

Freestanding Towers

There are two types of freestanding towers: lattice or truss towers and tubular towers. The truss or lattice tower is so called because it resembles the latticework of a trellis. The Eiffel Tower is the best-known example of a truss tower (see Figure 9-4. Truss tower). Truss towers are typically more rigid than tubular towers. The tubular tower is another form of freestanding tower.

Towers can be designed to withstand any load, but as the size of the wind turbine increases, so

Figure 9-4. Truss tower. Gustave Eiffel first used the compound taper seen on these 55-meter (180-foot) tall truss towers on the Cabazon site in California's San Gorgonio Pass. The Z-750 is driven by a rotor 48 meters (157 feet) in diameter. Note power line and buildings for scale. The Z-750 was probably the last commercial wind turbine of this size to use a truss tower in North America. Truss towers for multimegawatt turbines have been used experimentally in Germany but remain rare. Late 1990s.

do the weight and cost of the tower supporting it. The same is true as the tower increases in height. The components become heavier, harder to move, and more costly to ship.

Truss Towers

In North America, truss towers for small wind turbines are assembled from a series of 20-foot (~ 6-meter) sections. For small wind turbines, the sections may be preassembled and welded together prior to delivery. For household-size turbines, the tower is shipped knocked down or in parts and must be assembled on the site. This sounds simple but is labor intensive and requires knowing how to work with heavy awkward components to assemble a tower safely.

Installation of truss towers typically requires a crane. The tower is assembled on the ground, then hoisted into place and bolted to the foundation. Another method is to hinge the tower at its base. The tower is bolted together on the ground, the wind turbine attached, and the whole assembly tipped into place with a gin pole and winch or a small crane. Wind turbines up to 25 meters (80 feet) in diameter have been installed in this manner. This was a popular method for American-designed wind turbines of the early 1980s. With the demise of these manufacturers, tilt-up towers for wind turbines of this size have been abandoned.

Tubular Towers

Nearly all large wind turbines are installed on tubular towers, though there are some notable exceptions. This wasn't always the case. During the great California wind rush of the 1980s, an equal number of turbines were installed on truss and tubular towers. However, the enormous size of today's turbines, as well as aesthetic demands, has led to the almost exclusive use of gently tapered tubular towers (see Figure 9-5. Tubular steel towers).

Small turbines are also installed on tubular towers, though infrequently. These slender towers resemble light standards (see Figure 9-6. Pole tower).

Pole towers for small wind turbines are made from tapered steel tube, steel pipe, wood, concrete, or even fiberglass. Though most are made of steel,

Figure 9-5. Tubular steel towers. Most large wind turbines are installed on tubular steel towers. The towers are composed of rolled steel plate that is then welded in sections. Several sections are subsequently bolted together on site. Shown here are tubular towers supporting a DanWin 160 (left) and a Vestas V29 (right) on Alta Mesa in California's San Gorgonio Pass. An internal ladder provides access to the nacelle. The DanWin 160 is 23 meters (75 feet) in diameter and was rated at 160 kW. The Vestas V29 is 29 meters (95 feet) in diameter and rated at 225 kW. Circa late 1990s.

Figure 9-6. Pole tower. Skystream on a slender cantilevered tubular tower in Alexandria, Indiana. There is no ladder or foot pegs for access to the turbine. To service the machine, a crane is necessary. The Skystream is a household-size turbine 3.7 meters (12-feet) in diameter rated at 2.2 kW. 2005.

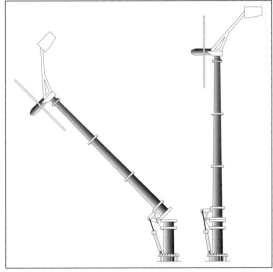

Figure 9-7. Hinged, freestanding tubular tower. This hydraulically operated hinged tower was developed for use at inaccessible Norwegian telecom sites. (Roheico A/S)

pole towers of spun concrete have been used by some manufacturers. For small turbines, pole towers are available only in limited sizes and strengths. For example, the selection of wood and concrete poles suitable for wind machines is limited by the length of pole that can be shipped conveniently. Like truss towers, pole towers are difficult to handle without heavy equipment.

Installation of pole towers usually requires a crane. A pole tower, though, can be hinged at the base and tipped into place with a gin pole. When upright, the tower is then bolted to the foundation. Some European manufacturers have erected hinged tubular towers with powerful hydraulic jacks (see Figure 9-7. Hinged, freestanding tubular tower).

Many observers consider freestanding tubular towers more aesthetically pleasing than truss towers. This is certainly true in foreground views, but it isn't always the case. Surprisingly, tubular towers can be more visible at a distance than lattice towers, especially in silhouette. In arid regions, lattice towers tend to blend into the landscape.

For small wind turbines, pole or tubular towers are significantly more expensive than guyed towers but only modestly more costly than truss towers. However, pole towers require a more substantial foundation than truss towers, which spread the overturning force over a wider base.

Concrete Towers

Yes, concrete. Unlike the wood poles commonly used in North America, utility poles and light standards in Europe are often constructed of reinforced concrete. It was a natural step to use concrete for electricity-generating wind turbines. Since the late 1970s, wind turbines from the multimegawatt turbines in Europe to household-size turbines in the United States have been mounted on concrete towers.

In the 1980s, several large experimental turbines were installed in Britain, Sweden, Germany, and Denmark on concrete towers. These were built in much the same way as concrete chimneys by using slip forms that moved ever higher once the previous pour cured. They were a distinctive feature of large wind turbines of the

Figure 9-8. Concrete tower for small wind turbine. Spun concrete tower for prototype Grumman Windstream 25, a 25-foot (7.6-meter) diameter, 15-kW downwind turbine of the late 1970s. The rotor used an aircraft hub for varying the pitch of the blades.

Figure 9-9. Spun concrete tower for large wind turbine. German manufacturer Enercon popularized spun concrete towers for its 500-kW E-40 turbine. Access to the nacelle is by an external ladder. Note the work platform below the nacelle and the distinctive paint scheme intended to soften the intersection of the tower and the horizon. Mid-1990s.

era (see Figure 4-8. Nibe A & B, and Figure 4-10. Tjæreborg).

Grumman Aerospace installed several of its experimental household-size wind turbines in the United States on spun concrete towers during the late 1970s (Figure 9-8. Concrete tower for small wind turbine). For this type of tower, concrete and reinforcing bar are placed in a mold and then the mold is spun at high speed to remove any air pockets. The towers never caught on for wind turbine applications until Enercon popularized their use in the 1990s.

The German manufacturer built a reputation around the simplicity and the sound-deadening qualities of the spun concrete towers it introduced for its 500 kW E-40 (see Figure 9-9. Spun concrete tower for large wind turbine). Hundreds were installed across Europe with their telltale painting scheme. However, as Enercon's turbine grew in size, the company switched to conventional tubular steel towers and to precast concrete towers.

Enercon's concrete towers are now assembled on-site from sections of precast concrete. Using precast panels of two or three half shells, towers of up to 15 meters (50 feet) in diameter can be assembled like building blocks by connecting and prestressing tendons in the concrete panels. Each concrete panel is made in a factory where dimensions, strength, and quality can be carefully

Figure 9-10. Precast concrete tower. Above: Base of Enercon's E-126 at Dardesheim, Germany. The tower for this 126-meter (410-feet) diameter, 6-MW turbine is 135 meters (440 feet) tall. Left: Reception hall? Coffee, kuchen, and, of course, beer atop picnic tables await guests visiting the E-126 at Dardesheim, Germany. Nancy Nies posing in the base of the precast concrete tower for the 6-MW wind turbine that doubles as a reception hall for visitors to the 100% renewable energy village. Posters on the "wall" recount painting the base of the tower by local students. 2011.

controlled. The joints between the panels are then sealed against the elements when assembled in the field.

This technique allows Enercon to transport and install extremely tall towers without placing (pouring) the concrete on-site. This is useful in Germany's wooded central highlands—the *Mittelgebirge*—where tall towers are necessary to elevate the wind turbine well above the treetops. An advantage of such massive towers is that sound and vibration are absorbed in the concrete and not radiated from the tower.

Enercon pioneered this technology. The distinctive concrete towers are a common feature of the German landscape, but they can also be seen elsewhere in Europe and in Canada. The massive towers for Enercon's E-126, a wind turbine with a rotor 126 meters (410 feet) in diameter are so large you can hold a reception in the base (see Figure 9-10. Precast concrete tower).

Guyed Towers

Guyed towers are—by far—the most common choice for small wind turbines (see Figure 9-11. Guyed lattice towers). They offer a good compromise between strength, cost, ease of installation, and appearance. Unfortunately, they take up more space than freestanding towers, and they suffer from the justifiable fear that a guy cable will fail and the tower will come crashing down.

Guyed towers include a mast, guy cables, and earth anchors. The mast itself may be made from steel lattice, heavy-walled pipe, or thin-walled tube. In North America, most guyed towers for wind machines up to 7 meters (23 feet) in diameter use masts of welded lattice made from steel tube and rod. These masts are popular because they're mass-produced for the telecommunications industry and thus are relatively inexpensive. They're also produced in a convenient range of sizes, from lightweight sections designed for radio antennas to heavy sections for mountain-

Figure 9-11. Guyed lattice towers. Bergey 1500 (foreground) and Bergey Excel (background) on guyed-lattice masts at the US Department of Agriculture's test station near Bushland, Texas. Late 1980s.

Figure 9-12. Guyed pipe tower. Bergey Excel on a guyed pipe tower at the Kortright Centre outside Toronto, Ontario. The household-size Excel uses a 7-meter (23-foot) diameter rotor to drive a 10-kW generator. 2007.

top microwave dishes. Tower height is practically unlimited. A guyed tower using a lattice mast can be assembled by bolting sections together vertically; a section at a time, with a tower-mounted gin pole; or the entire mast can be assembled horizontally, on the ground, and raised upright with a crane.

Masts of steel pipe and tube are also popular (see Figure 9-12. Guyed pipe tower). Masts of the desired height are assembled from several sections bolted or slipped together. Guyed towers using masts of steel pipe or steel tube are usually assembled on the ground and tipped into place with a crane or gin pole (see Figure 9-13. Guyed hinged tower).

The strength of guyed pipe towers is a function of both wall thickness and the fourth power of diameter, says wind systems engineer Dave Blittersdorf. Thus, large-diameter, thin-walled tube towers can be as strong as smaller-diameter, thick-walled steel pipe.

Towers may use three to four guy cables at each level and often require two or more levels. Four

Figure 9-13. Guyed hinged tower. Vergnet's 2-kW prototype at the Chateau de Lastours test field in southern France. The hinged tower is lowered with the attached gin pole (lower left) and an electrically powered grip hoist. Vergnet has built a reputation around installing its turbines on guyed, tilt-up towers for use in hurricane belts around the world. 1990.

Table 9-1. Guy Anchor Radius for a Mini Wind Turbine

Manufacturer	Model	Rotor ft	Rotor m	Mast	Nominal Dia. in	Nominal Dia. mm	Tower Height ft	Tower Height m	Guy Levels	Radius ft	Radius m	Screw Anchor* Dia. ft	Screw Anchor* Dia. m	Screw Anchor* Length ft	Screw Anchor* Length m
Bergey Windpower	XL1	8.2	2.5	tube	4.5	0.114	64	20	3	35	11	0.5	0.15	5.5	1.7
							84	26	4	50	15	0.5	0.15	5.5	1.7

* For normally cohesive soils.

Figure 9-14. Guyed tubular tower. Wind Turbine Company's experimental 46-meter (150-foot) diameter, 500-kW turbine used a stiff, but guyed, tubular steel tower. Access to the nacelle and its unusual flapping downwind rotor was via an internal ladder. This unit was installed at Fairmont Reservoir in California's Antelope Valley in the early 2000s. The experiment was unsuccessful.

guys are used where the site or method of erection requires them. Tilt-up towers, by necessity, use four guys at each level.

Under special circumstances, more guy cables and anchors may be necessary. During the early 1980s Fayette Manufacturing used a novel guying layout on its turbines in the soft soils of California's Altamont Pass. Fayette used guyed towers of steel pipe and anchored the guy cables by driving large screws into the ground. To lessen the risk that a screw anchor would pull out of the ground, they used two anchors and accompanying guy cables at each of three guy points. (The tower was guyed at only one level.) This lessened the loads on each anchor, reducing the chance that any one anchor or cable would fail.

The loads on a guyed tower that might snap a cable or pull an anchor out of the ground are determined by the thrust on the tower and by the guy radius—the distance from the tower to the anchors. The guy radius is a critical aspect of guyed towers and is dependent on the site, the loads imposed on the tower by the wind turbine, and the stiffness of the mast. Because guyed towers need a lot of space, there's a tendency to set the anchors too close to the tower. This can affect the dynamics of the tower and the loads it can endure. Consequently, some manufacturers precisely specify the guy radius (see Table 9-1. Guy Anchor Radius for a Mini Wind Turbine).

Guy radius is limited by the compressive loads the mast can withstand before it buckles and by anchor construction. When the anchors are too close to the tower, for example, the mast may buckle in high winds or the anchors may fail. As a rule of thumb, the guy radius shouldn't be less than one-half the height of the tower. The guy radius can be as great as you like. The tension in the guy cables and the compression of the tower continue to decrease as distance from the tower increases. Usually, there's no reason for going beyond three-fourths of the tower's height.

When the mast is stiff and the tower short, only one guy level may be needed. There are usually two or more guy levels on most guyed towers. One anchor is used to guy all levels. The topmost guy prevents the tower from overturning, and the lower guys prevent the tower from buckling.

Guyed tubular steel towers have been used on experimental downwind turbines up to 500 kW (see Figure 9-14. Guyed tubular tower). Guying the tower reduces the diameter of the tower necessary. Presumably, this cuts cost and limits the tower shadow on the downwind rotor.

Rooftop Mounting

If you can think of it, it's been tried as a substitute for conventional towers: trees, silos, and even rooftops. Unfortunately, rooftop mounting is no substitute for a tower. More than one poor soul has discovered to his dismay that installing a wind turbine on his roof was a good way to destroy a night's sleep.

It seems like a fine idea at first glance: the building gets the turbine above the ground and eliminates the need for a tall tower. Promoters have also pushed the idea as a way for urban dwellers to use wind energy by putting the turbines atop their apartment buildings.

Mounting wind turbines—of any kind—on a building is a very bad idea. In more than three decades of modern wind energy, there is still no installation where this has worked successfully for an extended period. Invariably, rooftop turbines are either "tied off" or otherwise inoperative.

Yes, the old Zenith and Winco wind chargers of the 1930s were sometimes mounted on rooftops on North America's Great Plains. But times and expectations were different then. Ranchers were delighted just to be able to listen to a nightly radio program. They could tolerate the noise of the shiny new turbine on the roof. Today, we expect our machines to work quietly in the background. We also know much more about the wind than we did then.

Much has been written about the small Jacobs wind charger that was installed in the late 1970s on a tenement in the Bronx. True, it was done once, and it can be done again. But what's the point? The Bronx project was intended as a political challenge to Consolidated Edison Company, New York City's utility. It succeeded, and it proved that electricity could be fed back into the utility's network without destroying the city. (Something that was proven by the Danes in 1919!) The turbine was later removed without much fanfare.

Rooftop mounting of small wind turbines shouldn't be controversial. It's simply a bad idea. Yet few topics can stir more heated debate among small wind professionals and newcomers as can rooftop mounting. Many attribute the enduring allure of rooftop mounting to the aggressive marketing of the concept by Southwest Windpower—the one-time manufacturer of the Air series micro turbines. Bizarre as it seems today, Southwest Windpower suggested installing their micro turbine on rooftops as a way to compete with the simplicity of mounting solar photovoltaic panels. Their advertising in the late 1990s generated howls of protest from critics such as Mick Sagrillo.

Dave Calley, one of Southwest Windpower's founders, liked to brag at the time that he lived with four of his company's machines on his roof. Though Southwest Windpower never abandoned rooftop mounting, it began to stress that the building should be unoccupied and eventually the company tempered its advertising by suggesting that buyers install their turbine on "as tall a tower as possible."

All wind turbines vibrate, and they transmit this vibration to the structure on which they're mounted. All rooftops create turbulence that interferes with the wind turbine's operation. Even if Southwest Windpower engineers were able to design a sophisticated dampening system that isolated the wind turbine from the structure, they couldn't eliminate the power-robbing and damaging turbulence created by the building—something they just ignored.

Worse, a rooftop-mounted turbine can provide a nasty surprise, as an owner in upstate New York learned. One stormy night his Air turbine destroyed itself by plunging through his roof. That was the end of his experimentation with rooftop mounting—and nearly the end of his marriage.

Figure 9-15. Rooftop mounting. Four Southwest Windpower Skystreams atop a building in downtown Portland, Oregon. The building was designed by and houses an architectural firm. The towers and their mounting systems were specifically designed for this purpose. Unlike most rooftop installations where the turbines are tied off or otherwise stopped, the wind turbines in this 2012 photo were operating.

Few who consider rooftop mounting ask whether the building can support the loads created by both the wind turbine and tower. The wooden roofs of homes in North America can't support more than a micro turbine at best. A reinforced concrete roof on a commercial or industrial building might be able to withstand a slightly larger turbine. Can the roof, then, handle the dynamic loads—the vibrations—that the tower will transmit to the structure? If the building is an unoccupied warehouse, the vibrations won't bother anyone, but if it's an office building, they may prove annoying.

The greatest hazard of rooftop mounting is caused by turbines that have not been thoroughly tested to international standards. We have no idea how these untested turbines will perform under load in the extremely turbulent conditions found on rooftops. No one wants to see a video on YouTube.com of a rooftop wind turbine falling off a building or destroying itself in an urban setting.

To avoid rooftop turbulence, the wind turbine must be raised well above the roofline. This often negates any potential savings on the tower and increases the complexity of mounting the wind turbine and installing it safely. Such is likely the case at 12 West in Portland, Oregon (see Figure 9-15. Rooftop mounting). In the widely publicized installation, ZGF architects designed their new office and apartment building to include four Skystream turbines on the roof. The company claimed that the building was intended to "serve as a laboratory for cutting-edge, sustainable design strategies." They also promised that the "the turbines will be thoroughly instrumented so that actual performance" could be measured. The implication was that these "cutting-edge" architects would publish the results of their findings.

Five years later the firm still prominently displayed photos of the turbines on its website, but nowhere was there any mention of the actual performance of the turbines. The best that could be gleaned was that the turbines "may" provide enough electricity for the elevators. If ZGF architects learned anything about rooftop mounting, they were not saying publicly. Interestingly, one of the engineering consultants ZGF used on the project, AeroVironment of Gossamer Condor fame, quietly abandoned its own venture into rooftop mounting a few years after the 12 West turbines were installed. As in similar examples elsewhere, it appears the wind turbines were used simply as an expensive form of ornamentation to win the firm new architectural clients who want to "green up" their image.

Great Britain became notorious in the late 2000s for the promotion of small, rooftop-mounted wind turbines and the often over-the-top hype surrounding them. While Britain has certainly not been alone in this—there have been similar outbreaks in Canada, France, the Netherlands, and the United States—British promoters have put more hardware on rooftops than anywhere else. However, widespread complaints of poor performance from rooftop wind turbines installed under Britain's ill-conceived program finally led to a series of studies and actual field trials. The findings were not unexpected.

Britain's Energy Savings Trust published a report in 2009 based on field-testing of several dozen small wind turbines. All together, the trust monitored 38 building-mounted turbines and 19 turbines installed on ground-mounted towers.

Probably most damning was the trust's conclusion that no "building mounted sites generated more than 200 kWh" during 2009. But it gets worse. "In some cases, installations were found to be net consumers of electricity." Rather than "breaking the paradigm," as one rooftop promoter put it, some installations were hard-pressed to break even between the energy produced by the wind turbine and the energy consumed in the inverter. (The turbines in the field trial were all connected to the grid.) In some cases, there was hardly any net energy produced, after accounting for the energy used by the inverter!

The most likely reason for such poor performance, little or no wind, confirms the charge made by critics of rooftop-mounted wind turbines that the obstructions of nearby buildings and the roof itself defeat any presumed benefit of installing a wind turbine on a rooftop. Equally significant, the trust acknowledged in the report that freestanding turbines, that is, those not mounted on rooftops, produced about six times more electricity than those on rooftops.

ROOFTOP WIND IN INACTION

While hiking in the southern Sierra Nevada Mountains, I was surprised to find the controversy over rooftop wind turbines dogging my steps. As we came to the crest of Bald Mountain in the heart of Sequoia National Forest, my eye was drawn immediately to the solar panels—and then to the wind turbine mounted nearby on a railing. The turbine, a Southwest Windpower Air, was not operating. As we came closer, I could tell that it was tied off. "Typical," I thought. Most of the rooftop turbines I've seen are either inoperative or simply tied off (see Rooftop wind in inaction).

The fire warden came out on the platform and hollered hello as we came up the trail. I yelled back a greeting and then asked her why the wind turbine wasn't working, as there was plenty of wind on the summit. She shouted back, "So I don't have to listen to the damn thing."

Enough said. Just another example of improper wind turbine installation and the foolishness of mounting wind turbines on buildings—even a remote watchtower.

Rooftop wind in inaction. Example of the improper installation of a micro wind turbine on a fire watchtower. Note that the turbine is tied off so that it won't operate. Typically, rooftop wind turbines perform poorly or they are tied off—as here—so they won't annoy people in the building. 2005.

The conclusion, says David Sharman of Ampair, a British manufacturer of small wind turbines, is that rooftop wind "can be done in some locations, but it often doesn't make economic sense" to do so.

To paraphrase Shakespeare in *Richard III*, "a tower, a tower, my kingdom for a proper tower."

Unconventional Towers

Silos

Like rooftops, farm silos also seem ideal for a low-cost tower. They're already in place, usually stand well above surrounding farm structures, and are relatively close to where the power will be used. In the late 1970s, Alcoa envisioned strapping its household-size Darrieus turbine to thousands of silos across the vast heartland of North America (see Figure 9-16. Silo tower). They first found that not all silos were structurally suited to the task. Installation and service would also be difficult. Alcoa abandoned the market, and silo mounting hasn't been used since.

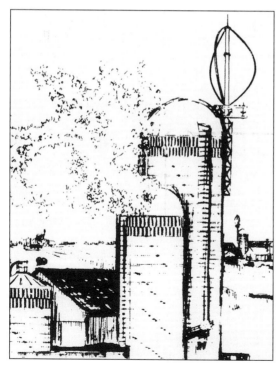

Figure 9-16. Silo tower. Alcoa, in conjunction with a silo manufacturer, mounted a small Darrieus on the side of a farm silo in 1978 at Clarkson College in New York state. The attempt at using farm silos with a wind turbine hasn't been repeated commercially since then. (Alcoa)

Farm Windmill Towers

Farm windmill towers are ubiquitous across North America. Because of their abundance, there's always a temptation to buy a used water-pumping windmill tower and adapt it for a small wind-electric system. Their usefulness is limited, however, unless you plan to use the tower as it was intended.

American farm windmill towers are a special case of the freestanding truss tower. Most are light-duty towers. Farm windmill towers typically have a greater taper than the truss towers used for small wind turbines. (The height of a water-pumping windmill tower is proportionally about five times the width of the base. In contrast, the height of a small wind turbine tower is nine times the base width.) This design enables the water-pumping windmill tower to use less steel in the legs and braces and need a less substantial foundation than required for a tower supporting a similar size modern wind turbine.

Farm windmill towers are also short. Most are no more than 40 to 50 feet (~ 15 meters) tall, particularly in the western United States. Ken O'Brock, who distributes water-pumping windmills along America's East Coast, says farm windmill towers up to 80 feet (24 meters) tall can be found in Amish settlements in Pennsylvania and Ohio. These are exceptions to the rule, and the Amish are not about to part with their towers.

The most commonly used farm windmill is only 8 feet (2.4 meters) in diameter. Towers used with these machines are not suited for larger wind turbines.

Steel Pipe

Well casing or water pipe is a frequent choice for guyed pipe towers supporting micro wind turbines. The strength of water pipe or well casing comes from its thick walls. Steel pipe is readily available and inexpensive. Do-it-yourselfers often choose steel pipe for these reasons. Unfortunately, most micro turbine tower kits using steel pipe use it in 20-foot (~ 6-meter) lengths that are extremely difficult to handle safely.

Wood Towers

Wood is an unlikely material for towers at first glance. However, like concrete, wood has also been used as a tower material for both small and large wind turbines. Sustainably harvested wood is a renewable resource, and if wood could be used to fabricate towers, then more of the wind turbine would be composed of a renewable natural resource. During the mid-1980s, a Danish designer built a sweeping laminated wood tower. Others have used wooden utility poles and even trees to mount small wind turbines.

Wood Truss

Probably the most ambitious attempt to use wood as a tower material was that by a German company. In -2013 it installed a 100-meter (330-foot) tall tower supporting a 1.5-MW wind turbine using proprietary laminated-wood panels. The panels were assembled on-site into a hexagonal-shaped, enclosed tower resembling the more common tubular steel or concrete towers. The company anticipated building a taller tower for even larger wind turbines.

Quebec researchers have explored the concept as well. The Canadian province produces a lot of wood, and with the collapse of the newsprint market throwing thousands out of work, there was hope that the forestry companies could find new markets for their product. The German tower concept offered a path. However, one tower only establishes proof of the concept. Meanwhile, the wind industry has moved to much larger wind turbines than installed on the prototype, and financing for a wind farm project using a novel—and possibly risky—tower concept will be difficult to find.

Trees

If you have to mount your turbine on a tree because your site is wooded and you can't afford a tall enough tower, then wind isn't for you. Trees seldom occur right where you would like your tower. Nor is there usually one lone tree that reaches well above all others. The turbine will also be difficult to install and service. For a tree to be of long-term use, it must remain alive. That's unlikely unless you're a skilled arborist like Ian Woofenden.

Woofenden operates two tree-mounted household-size wind turbines on Guemes Island in Washington state. As a professional, he knows what he's doing. Most won't. Woofenden installed climbing rungs, fall protection systems, and guy cables and ensured he didn't gird either tree's living cambium. Even then, Woofenden found the 3.6-meter (12-foot) sweep of one of his turbines was more than he felt comfortable with on a treetop looming over his house. (After watching his turbines in a fierce squall, I'd agree.)

For these reasons "trees don't make good towers," says Mick Sagrillo. "They're hard to climb safely. They sway too much. Dead trees rot and fall over. Enough said."

Wood Utility Poles

In North America wood poles are as commonplace as farm windmill towers. And, like farm windmill towers, they're frequently considered a choice for a cheap tower. They're strong, rigid, and cheap when bought in quantity. Wood poles can be installed by a crane or utility truck with a special boom.

Wood poles are classified according to their circumference 2 meters (6 feet) from the butt end. Poles of a given class and length are rated to carry approximately the same load. They can handle even greater loads when guyed. Pole lengths suitable for wind systems are found only in Class 4 or better. A Class 4 pole is strong enough for small, self-furling wind turbines up to 5 meters (16 feet) in diameter.

The inexpensive wood poles used by utilities in North America are too short for most wind turbine applications. Utilities often use poles only 40 to 50 feet (~ 15 meters) long. For the minimum 60-foot tower height needed by a small wind turbine, a 70-foot pole is necessary. Longer poles are available, but the cost rises rapidly with lengths beyond the standard sizes used by utilities. Longer poles are also more difficult to transport.

At one time, Northern Power Systems promoted the use of wood utility poles for use with its HR3 turbine. The idea never caught on, and though wood poles are abundant, they've never been widely used for wind machine towers. They're difficult to climb, and the heights available are limited to the length that can be conveniently shipped. They're also unsightly, and even if you don't mind their looks, others might—especially the local zoning officer.

Tripod Tower and Platform

Traditional European smock and post mills used the structure of the windmill itself as the tower. In the smock windmill, only the cap was yawed to face the wind. In the post mill, the entire structure was rotated. Most windmills used a tail pole that the miller would push around on a circular track to reorient the rotor into the wind.

In the early 1930s, the Russians installed an experimental 100-kW wind turbine at Balaklava in the Crimea that used the same principle. The turbine rested on a tripod, and the entire assembly was turned to face the wind. Thrust on the turbine was absorbed by a long tail pole that rested on a wheel that rolled on a rail around a circular track.

The concept resurfaced in the early 1980s when inventor Charles Schacle installed a 100-kW prototype wind turbine at Moses Lake, Washington (see Figure 9-17. Tripod tower and platform).

The venture was unconventional in many ways. The wind turbine used odd-shaped airfoils, drove a generator at ground level via a hydraulic transmission, and oriented the wind turbine into the wind by turning the entire tower and nacelle assembly on a circular track. In a venture with Bendix, Schacle installed a severely overrated 3-MW version at Southern California Edison Co.'s test center near Palm Springs in 1981. The design was unique for several reasons. It was the largest wind turbine that attempted to use a hydraulic transmission. It was the largest wind turbine to use a rotating truss tower on a concrete platform. And SCE's version installed near Palm Springs was the only wind turbine that included a steering wheel, according to longtime wind consultant Mark Haller. Like other experimental turbines installed at the utility's test center, the Bendix-Schacle turbine ended in failure. No one has attempted to use the tripod tower and platform since then.

Figure 9-17. Tripod tower and platform. Following the historical example of the Balaklava wind turbine in the Crimea, Charles Schacle installed an experimental 100-kW wind turbine that rested on a tripod tower at Moses Lake, Washington, in the early 1980s. The tower and nacelle rested on steel wheels (right) that rode on a steel rail embedded in a concrete platform. The entire contraption turned on the platform in response to changes in wind direction. 1977.

Other Considerations

There are a number of other considerations when examining different kinds of towers, such as the space required, how they are installed, access to the tower, and access to the nacelle. One important consideration that is often overlooked is the appearance of the wind turbine on the landscape and how the public will perceive the tower.

Aesthetics

Public perception of the wind turbine and tower combination is critical to acceptance of wind energy. Wind turbines are utilitarian structures, but they are also very prominent features on the landscape. Thus, it behooves engineers, developers, and manufacturers alike to design wind turbines and their towers that are as aesthetically pleasing as possible. We ignore this at our peril. Fortunately, there are some simple rules to follow.

In general, tubular steel and concrete towers are more acceptable to the public than lattice or truss towers. Nacelles and towers should appear as a harmonious whole. There shouldn't be an obvious disjunction between the nacelle and the tower. For example, the width of the nacelle shouldn't be less than the diameter of the tower top (see Figure 9-18. GE's "space frame" tower). Nor should the top of the tower be disproportionally thin where it meets the nacelle.

In the mid-2010s, GE introduced a "new" tower, rebranding an enclosed truss tower as an "aesthetic space frame." Changing the name doesn't change anything. It's still a truss tower. Nor does describing the tower as "aesthetic" make it so. It is anything but aesthetic. Though wrapped with plastic sheathing to prevent birds from roosting on the latticework, the experimental tower is one of the most ungainly towers of the past decade—and GE has stiff competition.

A number of companies manufacture pleasing tubular tower and wind turbine combinations, including GE and its competitors. Cutting corners with an awkward truss tower design is not the way to increase the public's acceptance of wind energy.

It isn't enough to look around and point an accusing finger at other obtrusive objects on the horizon that have found acceptance, such as utility poles and transmission towers. Wind systems

Figure 9-18. GE's "space frame" tower. The experimental truss tower is wrapped in plastic sheathing to prevent birds from roosting on the latticework. Note the angular changes in taper and the mismatch between the top of the tower and the bottom of the nacelle on the GE 100-meter (330-foot) diameter, 1.7-MW turbine. The tower is unlikely to win any architectural awards for "most aesthetically pleasing wind turbine tower of the year." The tower is prominent along a major route through the Tehachapi Pass, engendering lots of comments of the "What were they thinking?" sort. 2014.

shouldn't be an embarrassment to the community. We will all be happier in the long run—and there will be fewer objections from the public—if the tower we select is aesthetically pleasing.

Space

Guyed towers occupy more space than freestanding towers. Normally, this isn't a drawback; where it is, guyed towers can be adapted to small lots by placing the anchors closer to the tower.

GRIPHOIST HAND WINCHES ARE IDEAL
for raising and lowering micro and mini wind turbines.

If you must do this, ask the manufacturer or contact a structural engineer to run through the numbers for you. They may find that the tower has an ample safety margin with the shorter guy radius. On the other hand, if you're approaching the limits of the mast, you may be forced to use a freestanding tower.

Sites with a lot of traffic, either vehicular or pedestrian, will also limit your choice to freestanding towers. Many a guyed Wincharger tower was felled by a rancher on a tractor. Not only is there the danger of losing your wind system; the guy cables themselves may present a hazard to errant snowmobiles, for example. To avoid tragedy, guy cables should be kept out of traveled ways. And they should be well marked.

Maintenance on Small Wind Turbines

Another factor to consider in tower selection is the maintenance of the tower and the wind turbine. Will the tower have to be painted, for example? The steel used in most small wind turbine towers is galvanized. To provide corrosion protection, individual steel members or welded subassemblies are dipped into molten zinc. When parts are shipped from the plant, galvanizing presents a shiny silver finish. It soon oxidizes and takes on a dull gray luster. Galvanized surfaces do not require painting or other treatments.

Access to the turbine shouldn't be overlooked, either. The turbine will require at least occasional inspection, if not maintenance. Some provision must be made for getting to it without hiring a crane. Truss towers can be ordered with climbing lugs or rungs. Guyed towers using lattice masts can be climbed using the cross-girts. Tube towers, whether freestanding or guyed, require the addition of climbing lugs if they are not part of the package.

For small wind turbines, hinged towers are advantageous (see Figure 9-12. Guyed hinged tower). Instead of climbing up to the turbine or working from a boom lift, you bring the machine down to ground level. Hinged, tilt-down towers may be safer to work around once they're on the ground, but safely lowering them can be hair raising if you're inexperienced. More than one installer has found his truck being dragged across the ground as tower and turbine came down with a thump. To work properly and safely, tilt-down towers should be designed by professionals and used with due care. Griphoist hand winches are ideal for raising and lowering micro and mini wind turbines. For larger turbines, power-assisted griphoists or electric winches may be necessary.

Ease of Installation for Small Wind Turbines

Towers for small wind turbines should also be easy to install. You may have found an inexpensive tower that will do the job you want, but if it takes a lot of effort to install, the savings may be offset by greater labor costs. If the tower sections are so large you need a crane or other heavy equipment just to get them off the truck, you'll need an experienced crew with equipment. You pay less for handling and shipping when the sections can be moved around by hand.

When considering how a tower for a small wind turbine is installed, also look at how the wind turbine will be mounted on top of the tower. A tower adapter, sometimes called a stub tower, is needed between the tower and the wind turbine. This isn't a concern with manufacturer-designed tower and turbine kits, but it is when you want to use a nonstandard tower. Also consider how to ensure that your tower is plumb once installed; that is, that the tower is vertical. If it isn't, the wind turbine may not yaw freely.

Access to Large Wind Turbines

Access to the tubular towers commonly used for large wind turbines also needs to be provided as well as access to the nacelle once inside the tower. Safe work practices and consideration for the windsmiths who will service the wind turbine requires steps and railings for access to the tower (see Figure 9-19. Tower exterior access).

The tower interior needs to be well ventilated so the door should have ventilation louvers. The turbine may be serviced after dark, requiring lights near the door. The door should be lockable

Figure 9-19. Tower exterior access. Locked, ventilated steel door provides safe and secure access to the tower. Note security lighting above the door and wide steps with handrails. Not visible is a safety latch on the left handrail (sometimes found on the door itself), which keeps the door secured in the open position. This prevents the heavy door from swinging closed in strong winds and injuring unsuspecting windsmiths as they enter or exit the tower. GE 1.5 MW wind turbine, Gaspé Peninsula, Quebec, Canada. 2006.

Figure 9-20. Tower interior access. On the left, ladder with side rails designed to meet safety standards. Note that windsmiths climb the ladder with their backs to the tower wall. On the right, pendant cables and cable raceway bring power down the tower without the use of slip rings. The turbine's control system counts the number of turns in the cables, and when necessary the control system stops the turbine then yaws the nacelle to unwind the cables.

to prevent unauthorized entry, and there should be some provision for holding the door open in strong winds to prevent it slamming shut, possibly injuring service personnel.

Inside the tower, there should be a ladder with side rails and a fall-arrest system that meets safety regulations (see Figure 9-20. Tower interior access). The turbines and towers have become so large that it is not uncommon for the tower to include an elevator as well as a ladder.

The area around the base of the tower should be kept tidy, and all tower anchor bolts should be treated and capped to prevent corrosion (see Figure 9-21. Capping foundation bolts).

Now that we've examined how we got to where we are today and the technology we use for both small and large wind turbines, it is now time to turn to the wind itself—the force that makes it all work.

Figure 9-21. Capping foundation bolts. As part of a thorough installation, any exposed foundation anchor bolts should be capped to prevent corrosion. Correctly capped anchor bolts on an Epcor project in Ontario, Canada. Inset: uncapped anchor bolts on a Florida Power & Light project in Washington state.

Knowing the wind is essential to working with wind energy. Here a cup anemometer is used to measure wind speed in 1980. In the background is the 640 kW Nibe B turbine in Denmark, one of the world's most successful early research turbines.

10

Measuring the Wind

High winds blow on high hills.
—Thomas Fuller, *Gnomologia*

"HI, I WANT A WINDMILL."

"You sure?"

"Yeah, I can't wait to tell the utility to go you know where."

"Hmm. I'll bet they'll be glad to hear that. How much wind do you have?"

"Wind? Oh, we've got lots of wind. It's always windy here."

More than one investor in wind energy, including some presumably smart Wall Street brokers, have learned an expensive lesson—one that may seem patently obvious. Wind turbines with little wind are like dams with little water. Sure, the rotor may spin, but the turbine may not be as productive as it needs to be to pay off its substantial cost over a reasonable period of time.

There's wind everywhere, but not everywhere has enough. Ample wind is a prerequisite for developing wind energy profitably. The more wind, the better. But just how much is enough? How do you know whether the wind over your site is sufficient? If you're living on the west coast of Denmark or in the Texas Panhandle, you probably have enough wind.

Few people live where the wind has such a well-deserved reputation for being so fierce. Most of us need a better description of the wind than "It's always windy here." Fortunately, today there are a number of online tools and websites with sophisticated and detailed maps of wind resources for most of the developed world—and much of the developing world as well. These are useful for quickly evaluating the wind energy potential of a region and as a "first cut" of whether or not wind energy makes sense. These tools are seldom sufficient for a specific site. For installing anything larger than a household-size wind turbine, the services of a professional meteorologist are needed. Banks often require it.

In this chapter we'll discuss the wind, what it is, how local climate and terrain affect it, and how it changes over time. We'll explore the meaning of wind power and how wind speed and power increase with height. You'll learn where to find wind information for your area and how to determine the winds at your site. As mentioned earlier, our objective is to determine if a site has enough wind to put wind energy to work—profitably.

For many, this chapter provides far more detail than necessary. This is especially true if you employ a professional meteorologist. They will do the work for you. However, it's always wise to understand the fundamentals of any subject, and wind energy is, after all, about the wind. Discovering why the power in the wind changes so dramatically with small changes in wind speed helps you cut through the hype of "new revolutionary" inventions that promise untold riches from wind machines that spin in low winds. Knowing the fundamentals also allows you to check the work of others, to ask informed questions before a project is built, and to monitor its performance after it has gone into operation.

Importantly, there is no simple, straightforward answer to the question of how much wind is necessary to produce wind energy economically. The answer depends on how much you are paid for the electricity you generate. If you sell your electricity for a high price or if you offset high-priced electricity, you need much less wind than if you sell your generation for a pittance. Wind turbines today make economic sense in many more places than they once did because the electricity they generate has become so much more valuable.

Wind: What Is It?

The wind is simply solar energy in another form. The atmosphere is a huge, solar-fired engine that transfers heat from one part of the globe to another. Large-scale convective currents set in motion by the sun's rays carry heat from lower latitudes to northern climes. The rivers of air that pour across the surface of the earth in response to this global circulation are what we call wind, the working fluid in the atmospheric heat engine.

When the sun strikes the earth, it heats the soil near the surface. In turn, the soil warms the air lying above it. Warm air is less dense than cool air and, like a helium-filled balloon, rises. Cool air flows in to take its place and is itself heated. The rising warm air eventually cools and falls back to earth completing the convection cell. This cycle is repeated over and over again, rotating like the crankshaft in a car, as long as the solar engine driving it is in the sky. The cumulus clouds of summer are a sign of the convective circulation that causes winds to strengthen in late afternoon. If you're a pilot, you probably prefer to fly in the early morning hours when winds are light. On the other hand, if you are making a trip to inspect a wind turbine, midafternoon, when you're more likely to find it operating, is better.

Winds are also stronger and more frequent along the shores of large lakes and along the coasts because of differential heating between the land and the water. During the day, the sun warms the land much quicker than it does the surface of the water. (Water has a higher specific heat and can store more energy without a change in temperature than can soil.) The air above the land is once again warmed and rises. Cool air flows landward, replacing the warm air, creating a large convection cell. At night the flow reverses as the land cools more quickly than the water. In late afternoon, when the sea breezes are strongest, winds can reach 5 to 7 m/s (10–15 mph) on an otherwise calm day. Land-sea breezes are most pronounced when winds due to large-scale weather systems are light. The influence of land-sea breezes diminishes rapidly inland and is insignificant more than 3 kilometers (2 miles) from the beach.

The winds along the shore are also higher because of the long unobstructed path (fetch) that the wind travels over the water. Hills, trees, and buildings block the wind on land. The shores of the Great Lakes and the coast of northern Europe have average wind speeds approaching 6 to 7 m/s (12–15 mph), partly due to these effects.

The mountain-valley breeze is another example of local winds caused by differential heating (see Figure 10-1. Mountain-valley winds). Mountain-valley breezes occur when the prevailing wind over a mountainous region is weak and there's marked heating and cooling. These breezes are found principally in the summer months when solar radiation is the strongest. During the day, the sun heats the floor and sides of the valley. The warm air rises up the slopes and moves upstream. Cooler air is drawn up from the plains below, causing a valley breeze. At night the situation reverses, and the mountains cool more quickly than the lowlands. Cool air cascades down the slopes and is channeled through the valley to the plains. Nighttime mountain breezes are generally stronger than valley breezes, with winds reaching speeds of 11 m/s (25 mph) in valleys with steeply sloping floors located between high ridges or mountain passes.

Mountain-valley breezes can be reinforced by the prevailing winds when the two flow in the same direction. One place where the effects of channeling are pronounced is on North America's Pacific coast where onshore winds are funneled into narrow gorges through the mountains. Convective flow can reinforce the funneling effect when heating on interior deserts cause large temperature differences between the coast and the interior. Average wind speeds above 9 m/s (20 mph) are typical, with winds in some seasons averaging well above that.

Many of the windy passes through mountain chains on the West Coast of the United States lie east to west. The tremendous wind potential in the Columbia River Gorge east of Portland, Oregon, is one such example. Another is the Altamont Pass east of San Francisco where cool air off the Pacific Ocean rushes over the low pass into the hot interior of the San Joaquin Valley. The San Gorgonio Pass near Palm Springs, California, is similar. This sea-level pass through the San Bernardino Mountains channels cool coastal air onto the blazing Sonoran Desert. Similar flows occur from the San Joaquin Valley across the Tehachapi Mountains onto the Mojave Desert. Winds through the Tehachapi Pass are more subject to the passage of cold fronts than in California's other passes. The storms associated with the cold fronts push masses of dense air through the Tehachapi Pass sometimes causing winds up to 50 m/s (100 mph).

Long ridges across the path of the wind, like Cameron Ridge in the Tehachapi Mountains, also enhance the flow over the summit (see Figure 10-2. Variation in wind speed over a ridge). Wind speeds may double as the flow accelerates up the gradual slope of a long ridge. As many California wind farmers have learned the hard way, this enhancement occurs only in the last third of the slope near the crest. Turbines lower down the slope perform dismally—unless on very tall towers.

A similar terrain enhancement of the wind has been found on the island of Oahu in Hawaii. There the northeasterly winds sweep around the end of a long ridge that runs the length of the island from north to south. The winds are accelerated as they pass over the ridge, but more so when they pass around the ends. Kahuku Point,

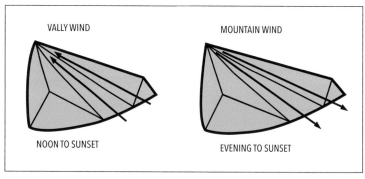

Figure 10-1. Mountain-valley winds. Warm air rises up the valley during the afternoon and descends during the evening. (Battelle PNL)

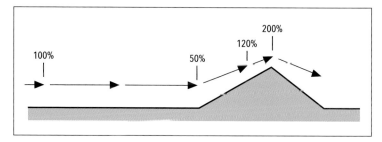

Figure 10-2. Variation in wind speed over a ridge. Wind speed increases near the summit of a long ridge lying across the wind's path. Though there can be some acceleration of the flow on the flanks, wind speeds typically are lower at the foot of the ridge. (Battelle PNL)

at the northernmost end of the ridge, has long been a prime site for wind development because of this local effect.

Mount Washington, in the White Mountains of New Hampshire, is one of the few areas in the eastern United States where terrain enhancement is well known. In fact, this phenomenon was first observed on Mount Washington and the Green Mountains of Vermont. Mount Washington has an average speed of approximately 17 m/s (38 mph), the highest average wind speed on the East Coast. On April 12, 1934, an observatory on the summit recorded the highest wind speed ever measured, 103 m/s (231 mph).

The effect of terrain on wind speeds within the region has been known for some time and has been used to advantage in the past. The Green Mountains of Vermont form a long northeasterly ridge that sits astride the prevailing winds just east of the long unobstructed fetch across Lake Champlain. Because of the potential speed enhancement, Palmer Putnam chose the Green Mountains as the site for his experimental Smith-Putnam wind turbine during the 1940s. Unfortunately, the terrain features that enhance

WIND SPEED UNITS

There are several scales of wind speed in common use. The old Imperial scale of miles per hour (mph) is still used in the United States. Most meteorologists use meters per second (m/s). Sailors use nautical miles per hour (knots). Some national weather services use kilometers per hour (km/h). While conversions from one unit of measurement to another are given in the Appendix, there are some simple rules of thumb that help to keep them straight. There are slightly more than 2 mph for every m/s. Similarly, there is slightly more than 1 mph for every knot, and slightly more than 0.5 mph for every km/h.

1 knot = 1.15 mph
1 m/s = 2.24 mph
1 km/h = 0.621 mph

---

wind speeds also create turbulence that can wreak havoc on wind machines. The 1,500-kilowatt Smith-Putnam turbine eventually succumbed to just such turbulence atop Grandpa's Knob after only limited operation.

Mountains and ridges offer higher winds for reasons other than channeling. Prominent peaks often pierce temperature inversions that can blanket valleys and low-lying plains. Temperature inversions cause a stratification of the atmosphere near the surface. Above the inversion layer, normal airflow prevails, but below the winds are stagnant. The air beneath an inversion layer may be completely cut off from the air circulation of the weather system moving through the region. Temperature inversions are common in hilly or mountainous terrain such as Southern California and western Pennsylvania, both areas notorious for their air pollution episodes. In northern latitudes, inversions are common during the fall and winter.

The inversion layer itself may accelerate the wind. The wind above a temperature inversion, essentially a giant lake of stagnant air, blows across the surface of the inversion virtually unimpeded. Ridgetops may not only possess more frequent winds, they may have stronger winds as well because they may break through this inversion layer. The wind can skip across the inversion layer as though across the surface of a lake until it strikes an exposed mountaintop.

There are numerous regional winds around the globe that can also have a powerful influence on successfully siting a wind turbine. These winds, like the powerful chinook that roars down the east side of the Rocky Mountains, are due to infrequent local meteorological and geographic anomalies. The Santa Ana in Southern California, for example, results from high-pressure systems that occasionally move over the Basin and Range province of the American Southwest. But whether it's the föhn in the Alps, the sirocco sweeping across the Mediterranean from North Africa, the legendary mistral of southern France, or the tramontana howling out of the eastern Pyrenees, these winds are a force to be reckoned with. They can power wind turbines or destroy them. In the early days of the US Department of Energy's wind program, chinooks with winds sometimes above 50 m/s (100 mph) wreaked havoc on the flimsy experimental turbines at the Rocky Flats test center near Denver, Colorado. Today, wind farmers in Tehachapi harness the once-feared Santa Ana winds. And in southern France, operators of wind power plants eagerly look forward to the *tramontana* that sends tourists scurrying for cover at nearby resorts.

Wind Speed and Time

The wind is a variable resource: calm one day, howling the next. Wind speed and direction vary widely over almost all measuring periods. Because wind speed fluctuates, it becomes necessary to average wind speed over a period of time. That most commonly used is the average speed over an entire year.

The average annual wind speed itself is not constant. It varies from year to year. The average speed can change as much as 25% from one year to the next. This can amount to more than 1 m/s (2 mph) in a moderate wind regime where an average of 5 m/s (10 mph) is the norm.

Average wind speeds vary by season and by month. "March roars in like a lion and goes out like a lamb" is a popular adage signifying that early spring is windy while summer is not. For much of North America's interior, winds are light during summer and fall and increase during the winter, reaching their maximum in the spring (see Figure 10-3. Historical monthly average wind speed).

When we looked at the differential heating of the earth's surface and its effects on local winds, we saw that wind speeds often increased during late afternoon after convective circulation had been set in motion. This tells us that wind speeds vary by time of day, not only because of changing weather but also because of convective heating. The effects of local convective winds are greater during the summer when winds are light and solar heating is strong, but they also occur throughout the year. Convective circulation leads to a dramatic difference between wind speeds during daylight hours and those at night (see Figure 10-4. Diurnal hourly wind speed for Wulf Test Field).

The diurnal difference in wind speeds is less marked during winter because there's less convective circulation. During winter and spring, winds are dominated by storm systems. It's the recurring storms of winter that push up the average wind speeds across the Midwest and northeastern United States. Storms formed in the Gulf of Alaska are the source for the winds that funnel through the Tehachapi Pass in California.

Power in the Wind

One of the most important tools in working with the wind, whether designing a wind turbine or using one, is a firm understanding of what factors influence the power in the wind. For the sake of thoroughness, we'll start right at the beginning. As E. W. Golding said so succinctly in his classic *The Generation of Electricity by Wind Power*, "wind is merely air in motion." Good so far. The air about us has mass (think of it as weight, if you're unfamiliar with the term). Though extremely light, it has substance. A bucket of air is similar to a bucket of water, but the bucket of air is lighter. It has less mass than that of water because air is less dense. It's more diffuse. Like any other moving substance, whether it's a bucket of water plummeting over Niagara Falls or a car speeding down the autobahn, this moving air contains kinetic energy. This energy of motion gives the wind its ability to perform work.

When the wind strikes an object, it exerts a force attempting to move the object out of the way. Some of the wind's kinetic energy is given

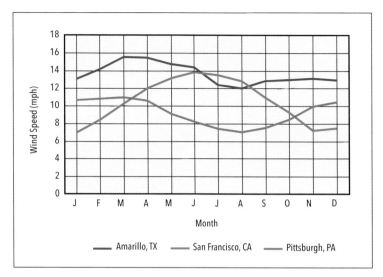

Figure 10-3. Historical monthly average wind speed. For the inland sites of Pittsburgh, Pennsylvania, and Amarillo, Texas, wind speeds increase during the winter and spring. Wind speeds are strongest near San Francisco during the summer when seasonal flows of cool marine air rush toward the blistering San Joaquin Valley. It's this seasonal flow that drives the wind power plants in the Altamont Pass east of San Francisco. The annual average wind speed at Amarillo is 13.7 mph (6.1 m/s), Pittsburgh, 9.10 mph (4.2 m/s), and San Francisco, 10.5 mph (4.7 m/s).

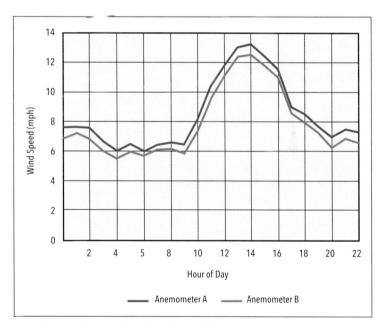

Figure 10-4. Diurnal hourly wind speed for Wulf Test Field. Measurements from two anemometers exhibit a typical diurnal pattern: wind speeds increase in the afternoon due to convective heating. This pattern is evident even at a windy site in California's Tehachapi Pass. The speed at anemometer A is slightly greater than that at anemometer B. Anemometer A is higher on the mast than anemometer B.

up or transferred, causing the object to move. When it does, we say the wind has performed work. We can see this when leaves skitter across the ground, trees sway, or the blades of a wind turbine move through the air.

The amount of energy in the wind is a function of its speed and mass. At higher speeds more

THE BEAUFORT SCALE

One of the more unusual measures of the wind is the Beaufort scale of wind force. Originally developed for use at sea, the scale didn't initially refer to wind speed but to an arbitrary scale of "force." Since its introduction, the scale has been adapted for use on land and to reflect a range of wind speeds in various units. It's a handy tool for sharpening your understanding of the wind. The scale was devised during the Napoleonic wars by British admiral Sir Francis Beaufort. The scale is seldom used by landlubbers in North America, but it is still common today to hear British weather forecasts in terms of wind force, such as a "force 10 gale will strike the coast of Scotland." The beauty of the Beaufort scale is its simplicity. You don't need sophisticated instruments to

Beaufort Scale of Wind Force

Strength or "Force"	Mean Speed		Range of Speed			Wind Pressure		General Description	
	knots	m/s	knots	mph	m/s	lb/ft^2	N/m^2	English	On Land
0	0	0	<1	<1	0-0.2	0.00	0	Calm	Smoke rises vertically, flags hang limp
1	2	1	1-3	1-3	0.3-1.5	0.01	1	Light air	Smoke drift indicates direction, wind vanes don't move, but flags begin to unfurl
2	5	3	4-6	4-7	1.6-3.3	0.08	4	Light breeze	Wind felt on face, leaves rustle, wind vanes begin to move, flags unfurl and begin to extend
3	9	5	7-10	8-12	3.4-5.4	0.27	13	Gentle breeze	Leaves and small twigs in constant motion, light flags extended
4	13	7	11-16	13-18	5.5-7.9	0.57	27	Moderate breeze	Raises dust, leaves, and loose paper; small branches move, dry sand begins to drift
5	19	10	17-21	19-24	8-10.7	1.2	58	Fresh breeze	Small trees in leaf begin to sway, crested wavelets form on inland waters
6	24	12	22-27	25-31	10.8-13.8	2.0	93	Strong breeze	Large branches in motion, overhead wires begin to whistle; umbrellas used with difficulty
7	30	15	28-33	32-38	13.9-17.1	3.0	146	Moderate gale	Whole trees in motion; resistant felt when walking against the wind
8	37	19	34-40	39-46	17.2-20.7	4.6	222	Fresh gale	Twigs break off trees, wind generally impedes progress when walking
9	44	23	41-47	47-54	20.8-24.4	6.6	313	Strong gale	Slight damage to roof and homes, chimney pots and slates damaged
10	52	27	48-55	55-63	24.5-28.4	9.2	438	Whole gale	Trees uprooted, considerable structural damage
11	60	31	56-63	64-72	28.5-32.6	12	583	Storm	Widespread damage
12	68	35	64-71	73-82	32.7-36.9	16	748	Hurricane	Devastation
13	76	39	72-80	83-92	37-41.4	20	935		
14	85	44	81-89	92-103	41.5-46.1	24	1169		
15	94	48	90-99	104-114	46.2-50.9	30	1430		
16	104	53	100-108	115-125	51-56	37	1750		
17	114	59	109-118	126-136	56.1-61.2	44	2103		

Adapted from: *The Generation of Electricity by Wind Power* by E. W. Golding.

gauge the strength or force of the wind because it uses observations of common objects such as smoke rising from a chimney or the swaying of trees. Beaufort devised the scale from 0 to 12, but it was extended in 1955 by the US Weather Bureau to force 17 (see Beaufort Scale of Wind Force).

General Description
At Sea
Sea like a mirror, sails hang limp
Small ripples, no waves
Waves are short and pronounced, crests have a glassy appearance, no whitecaps
Large wavelets, crests begin to break
Long waves, whitecaps
Moderate waves, white foaming crests, chance of some spray
Larger waves, whitecaps everywhere, probably some spray
Very large waves, foam begins to blow in streaks
Moderately high waves, foam blows in well-marked streaks
High waves, dense streaks of foam, crests begin to topple, spray may affect visibility
High waves with long overhanging crests, great foam patches, visibility affected
Ships hidden in troughs of waves, sea completely covered with long white patches of foam
Air is filled with foam and spray, sea completely white with driving spray, visibility seriously affected

WIND SPEED NOTATION

In previous editions of *Wind Energy*, wind speed has been indicated variously as (V) for velocity and as (S) for speed. In this edition (V) is used for wind speed. However, readers will find (U) as well as the lower case (v) in books for professional engineers and meteorologists.

POWER DENSITY

At sea level with a temperature of 15°C (59°F) where air density is 1.225 kg/m³, then power density P/A in watts per area of the wind stream is

$P/A = \frac{1}{2} (1.225) V^3$, where P is in W, A in m², and V is in m/s

$P/A = 0.05472 V^3$, where P is in W, A in m², and V is in mph.

$P/A = 0.00508 V^3$, where P is in W, A is in ft², and V is in mph.

energy is available, in much the same way a car on the highway contains more energy than a car of equal size it passes. It takes more effort—energy—to stop a car driven at 70 mph than it does one at 50 mph. Likewise, heavy cars contain more energy than light cars traveling at the same speed. This relationship between mass, speed, and energy is given by the equation for kinetic energy where (m) represents the air's mass and (V) is its velocity, or speed in common parlance.

$$\text{Kinetic Energy} = 1/2 \, mV^2$$

The air's mass can be derived from the product of its density (ρ) and its volume. Because the air is constantly in motion, the volume must be found by multiplying the wind's speed (V) by the area (A) through which it passes during a given period of time (t).

$$m = \rho AVt$$

When we substitute this value for mass into the earlier equation, we can find the kinetic energy in the wind.

$$\text{Wind Energy} = 1/2 \, \rho AVtV^2 = 1/2 \, \rho AtV^3$$

We've gone through this derivation for a reason. Equations are the language of science, and in this

INTERNATIONAL STANDARD ATMOSPHERE

The following are standard atmospheric conditions used internationally to define the reference conditions for wind turbine performance.

Sea-level pressure (p) = 29.92 in Hg,
= 760 mm Hg,
= 1013.25 mb, or hPa
= 1.01325 × 10^5 N/m² or Pa

Sea-level temperature (T) = 59°F,
= 15°C,
= 288.15 K

Air density of dry air, standard atmosphere (ρ) = 1.225 kg/m³
= 0.07651 lbs/ft³

---

terse, compact script the fundamentals of wind energy are precisely stated. But before we go over each of them, let's complete one more step.

Power, as you may remember, is the rate at which energy is available, or the rate at which energy passes through an area per unit of time.

$$P = 1/2 \rho A V^3$$

Power (P), we've now learned, is dependent upon air density, the area intercepting the wind, and wind speed. Increase any one of these, and you increase the power available from the wind. Most important, slight changes in wind speed produce significant effects on the power available.

Air Density

The wind is a diffuse source of power because air is less dense than most common substances. Water, for example, is 800 times denser than air. Don't be misled, though, the wind can pack quite a punch, as anyone who has survived a hurricane will tell you.

Air density decreases with increasing temperature. Air is less dense in summer than in winter, varying 10% to 15% from one season to the next. On an annual average, seasonal changes in temperature have only a modest influence on the power in the wind. However, changes in elevation can produce substantial changes in air density.

For our performance calculations, we'll assume that we're in an area where conditions approximate those near sea level and the temperature is about 59°F (15°C). Of course, if you plan to install a wind machine at the North Pole or on top of a mountain, these assumptions don't apply.

Although air density is one of the critical factors influencing the power available in the wind, the assumption of standard sea-level conditions is sufficient for most applications. However, overlooking the effect temperature and elevation have on air density can lead to unpleasant surprises.

Air density is inversely related to elevation and temperature: it decreases with increasing elevation or increasing temperature. This simple statement obscures the exponential relationship between air density and elevation. People who live near sea level can often safely ignore elevation. This is true for most of North America east of the Rocky Mountains as well as the lowlands of northern Europe. But once we move into the Rocky Mountains of western North America or, say, the Pyrenees Mountains between France and Spain, elevation has a profound effect on air density and wind turbine performance. For example, the air density at 5,000 feet (1,500 meters) atop Cameron Ridge in California's Tehachapi Pass is about 15% less than that at sea level.

Changes in temperature produce a smaller effect on air density than elevation, yet they are not insignificant. For example, increasing temperature from 59°F (15°C), the standard temperature used when estimating wind turbine performance, to 86°F (30°C) decreases air density 5%, a difference that could be critical in commercial applications. Canadian meteorologist Jim Salmon adds that temperature extremes from −60°C (−76°F) in the arctic north to 40°C (104°F) in the prairie provinces are not unheard of. Changes in air density under such conditions are not merely an academic question. Some generators in wind turbines installed in the far north of Canada have burned out because the air can be as much as 27% more dense than at the standard temperatures for which the turbine was designed.

Thus, it behooves us to examine air density carefully whenever conditions at our site vary much from those of the international standard atmosphere.

AIR DENSITY

For the sake of simplicity, we'll use metric units to find air density in kg/m³. While the following calculations are somewhat involved, they are presented graphically in the accompanying figures.

Air density, as a function of changes in temperature and pressure, can be found using the gas law

$$\rho = p/RT$$

where p is air pressure in N/m² or Pascal, R is the gas constant, 287.04 J/kgK, and T is temperature in Kelvin. For example, the air density at the standard temperature of 15°C (59°F) at sea level is

$$\rho = 101325/[287.04 \times (273.15 + 15)]$$

$$\rho = 1.225 \text{ kg/m}^3 \text{ (see Change in air density with temperature).}$$

Estimating the effect changes in temperature produce on air density are complicated by the normal decrease in temperature with elevation. Temperature typically decreases 6.5°C per 1,000 meters increase in elevation. This is the normal or environmental lapse rate (Γ).

From the hydrostatic equation, we can find the air pressure (p) at a given elevation. Once pressure is known, we can find air density by incorporating the change in temperature caused by the lapse rate.

$$p = p_o [(T_o - \Gamma Z)/T_o]^{g/R\Gamma}$$

where p_o is the sea-level air pressure, T_o is 288.15 K, Z is the elevation above sea level, and g is the acceleration due to gravity of 9.807 m/s².

Air pressure is often given in millibars (mb), 1/100 of a Pascal. Therefore, at an elevation of 1,500 meters in the Tehachapi Pass, the pressure in mb is

$$p = 1013.25 \times \{[288.15 - (6.5/1000 \times 1500)]/288.15\}^{[9.807/(287.04 \times 6.5/1000)]}$$

$$p = 845.55 \text{ mb}$$

The normal temperature at an elevation of 1,500 meters (T_z) is

$$T_z = T_o - \Gamma Z$$

$$T_z = 288.15 - (6.5/1000 \times 1500)$$

$$T_z = 278.40 \text{ K}$$

Thus, air density is

$$\rho = p/RT$$

$$\rho = [845.55/(287.04 \times 278.40)] \times 100$$

$$\rho = 1.058 \text{ kg/m}^3$$

or some 14% less than that at sea level and 15°C (See Change in air density with elevation).

Because the normal lapse rate doesn't always reflect actual temperature changes with elevation, it is sometimes necessary to find the air density for a specific temperature and elevation. For example, if you wanted to know how well your wind turbine would perform at your site in the Sierra Nevada, you would need to find the temperature at the site during your observations.

Again, let's use Cameron Ridge in the Tehachapi Pass, at 1,500 meters above sea level, as an example. We'll assume it's summer and the temperature is 90°F (32°C). To determine air density, we need first to find air pressure. We'll use the hypsometric equation to find pressure changes with changes in elevation where (exp) is the base e exponential and approximately 2.71828:

$$p = p_o \exp(-Zg/RT)$$

$$p = 1013.25 \exp\{(-1500 \times 9.807)/[287.04 \times (273.15 + 32)]\}$$

$$p = 857 \text{ mb}$$

We now substitute this value into the equation for air density.

$$\rho = \{857/[287.04 (273.15+32)]\} \times 100$$

$$\rho = 0.977 \text{ kg/m}^3$$

or some 20% less than air density at sea level and 15°C (59°F). Thus, when conditions vary significantly from those at sea level, we ignore changes in air density at our peril.

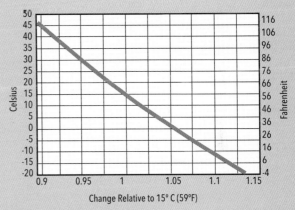

Change in air density with temperature. Percent change relative to conditions at sea level and 15°C (59°F).

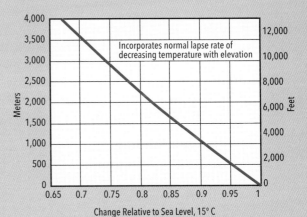

Change in air density with elevation. Percent change relative to conditions at sea level and 15°C (59°F). The chart incorporates the normal lapse rate of 6.5°C per 1000-meter increase in elevation.

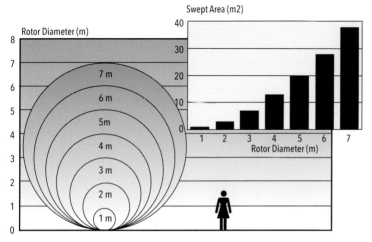

Figure 10-5. Relative size of small wind turbines. The Ampair 100 uses a rotor 0.93 meters (3 feet) in diameter. It intercepts 0.68 m². Bergey Windpower's Excel uses a 7-meter (23-foot) diameter rotor and sweeps nearly 38 m² or 56 times more area of the wind stream than the Ampair 100.

Swept Area

Power is directly related to the area intercepting the wind. Wind turbines with large rotors intercept more wind than those with smaller rotors and, consequently, capture more power. Doubling the area swept by a wind turbine rotor, for example, will double the power available to it. This principle is fundamental to understanding wind turbine design. Knowing this, you can quickly size up any wind machine by noting the dimensions of its rotor (see Figure 10-5. Relative size of small wind turbines). Consider a conventional wind turbine whose blades spin about a horizontal axis. The rotor sweeps a disc the area of a circle

$$A = \pi r^2$$

where (A) is the area and (r) is the radius of the rotor (approximately the length of one blade). This formula gives us the area of the wind stream swept by the rotor of a conventional wind turbine. Swept area is proportional to the square of the rotor's radius (or diameter). Relatively small increases in blade length produce a correspondingly large increase in swept area and, thus, in power. Compare the area swept by one wind turbine with a rotor diameter of 10 and that of another with a rotor diameter of 12. (The units are not important here.)

$$A_2 = (r_2/r_1)^2 A_1$$
$$A_2 = (12/10)^2 A_1 = 1.44 A_1$$

Increasing the rotor diameter by 20% (from 10 to 12) increases the capture area by 44%. Now consider the effect doubling the length of each blade—doubling the rotor diameter—has on the wind turbine's swept area.

$$A_2 = (r_2/r_1)^2 A_1$$
$$A_2 = (2/1)^2 A_1 = 4 A_1$$

Doubling the diameter increases the swept area four times. This exponential relationship between swept area and the power available to a wind turbine explains a crucial wind energy axiom: nothing tells you more about a wind turbine's potential than the area swept by its rotor—nothing. The wind turbine with the bigger rotor will almost invariably generate more electricity or pump more water than a turbine with a smaller rotor. Rotor diameter is the best shorthand for swept area. That's why in this book wind turbines are often referred to by their rotor diameter. It's shorthand for the area of the wind they intercept.

Wind Speed

No other factor is more important to the amount of wind power available to a wind turbine than the speed of the wind. Because the power in the wind is a cubic function of wind speed, changes in speed produce a profound effect on power. Consider the power available at one site with a wind speed of 10 (again, the units are not important) and another site with a wind speed of 12. From an earlier equation, we learned that power is proportional to the cube of wind speed.

$$P_2/P_1 = (V_2/V_1)^3$$
$$P_2 = (12/10)^3 P_1 = 1.73 P_1$$

Although there's only a 20% difference between the wind speeds at the two sites (12/10 = 1.2), there's 73% more power available at the windier location. This is why there's such a fuss concerning the proper siting of a wind turbine: small differences in wind speed caused by bordering trees or buildings can drastically reduce the power a wind turbine can potentially

NOTHING TELLS YOU MORE
about a wind turbine's potential than the area swept by its rotor—nothing.

capture. To grasp the full effect, consider what happens when the wind speed doubles from one site to the next. Doubling wind speed does not simply double the power available. Instead, power increases a whopping eight times.

$$P_2 = (20/10)^3 P_1 = 2^3 P_1 = 8 P_1$$

We can summarize the power equation with these general rules:

- Power can be affected by changes in air density when sites differ markedly from those at standard sea level conditions.
- Power is proportional to the area intercepted by the wind turbine. Double the area intercepting the wind and you double the power available.
- Power is a cubic function of wind speed. Double the speed, and power increases eight times.

At this point an important question arises. What wind speed are we talking about? The average wind speed? If so, what average wind speed? The annual average? Whichever we use determines the results we get.

Using the average annual wind speed alone in the power equation would not give us the right results; our calculation would differ from the actual power in the wind by a factor of two or more. To understand why, remember that wind speeds vary over time. The average speed is composed of winds above and below the average.

To illustrate, let's calculate the power density (P/A), the rate at which energy passes through a unit of area, for an annual average wind speed of 15. *Power density* is a term frequently used by wind energy experts because it's convenient shorthand for how energetic the winds are during a period of time, typically a year. Power density is normally given in units of watts per square meter (W/m^2) but we don't need the units just yet.

$$P/A_1 = 1/2\, \rho V^3$$
$$= 1/2\, \rho \times 15^3$$
$$= 1/2\, \rho \times 3{,}375$$

Now, what happens if the wind blows half the time at 10 and half the time at 20? The average speed still remains 15.

$$(10+20)/2 = 15$$

Yes, but watch what happens to the average power density using these two wind speeds.

$$P/A_2 = 1/2\, \rho \times (10^3 + 20^3)/2$$
$$= 1/2\, \rho \times (1{,}000 + 8{,}000)/2$$
$$= 1/2\, \rho \times 4{,}500$$
$$P/A_2 = 4{,}500/3{,}375\ P/A_1 = 1.33\ P/A_1$$

How can this be? Both have the same average speed. The answer rests with the cubic relationship between power and speed.

Grab a cup of coffee, sit back, and ponder this statement by Jack Park, one of the pioneers in America's wind power revival during the 1970s. *The average of the cube of many different wind speeds will always be greater than the cube of the average speed. Or stated another way, the average of the cubes is greater than the cube of the average.* In this case, the average of the cube for two wind speeds (10 and 20) is 1.33 times the cube of the average.

The reason for this paradox is that the single number representing the average speed ignores the amount of wind above as well as below the average. It's the wind speeds above the average that contribute most of the power.

Speed Distributions

If we plotted a graph of the number of times, or frequency, with which winds occur at various speeds throughout the year, we'd find that there are few occurrences of no wind and few occurrences of winds above hurricane force. Most of the time wind speeds fall somewhere in between these extremes.

The occurrence of winds at various speeds differs from one site to the next but in general follows a bell-shaped curve (see Figure 10-6. Rayleigh wind speed distribution). These distributions of wind speeds can be described mathematically in an attempt to approximate the real world. Meteorologists use the Weibull distribution and its companion, the Rayleigh distribution, to characterize wind resources when the

POWER IS A CUBIC FUNCTION OF WIND SPEED.
Double the speed, and power increases eight times.

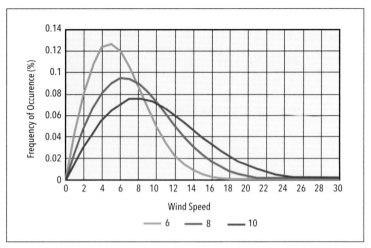

Figure 10-6. Rayleigh wind speed distribution. The average wind speed is all that's needed to define the shape of the Rayleigh distribution. As the average wind speed increases, the curve shifts toward the higher wind speeds on the right of the chart.

actual distribution of wind speeds over time is unavailable.

As we've seen from the previous example, summing the power contributed by a range of wind speeds rather than simply calculating the power from the average wind speed more accurately reflects the power available. For example, the power density calculated from the Rayleigh distribution for a given average wind speed is almost twice that derived from the average wind speed alone. This relationship holds for many sites with moderate to strong average annual wind speeds, but it will underestimate the potential at some and overestimate the potential at others. The real world is never as tidy as the mathematical models portray it.

Sites in the trade winds of the Caribbean often have high average wind speeds. But the winds are steady. They have few occurrences of extremely high winds. At trade wind sites, the Rayleigh distribution overestimates potential generation.

The relationship between the power density derived from the average speed alone, and that from a speed distribution, whether an actual distribution measured at a site or a mathematical distribution, is what Jack Park called the *cube factor* or Golding's *energy pattern factor*. For example, the cube factor for the Rayleigh distribution is 1.91.

The power in the wind at three different sites with exactly the same average wind speed illustrates the importance of Park's cube factor (see Table 10-1. Effect of Speed Distribution on Wind Power Density for Sites with Same Average Speed). Though a site in New York experiences the same average wind speed as one in Puerto Rico, the Caribbean island lies in the trade wind belt and has more constant winds. These steady winds produce less power over time than a temperate wind regime like that of New York. In contrast, the blustery winds that rush through the San Gorgonio Pass contain 66% more power than the gentler winds bathing Puerto Rico.

Meteorologists have characterized the distribution of wind speeds for many of the world's wind regimes. For temperate climates such as that of the continental United States or Europe, the Rayleigh wind speed distribution offers a good approximation.

Because the speed distribution plays such an important role in determining power, it's always preferable to use an actual measured distribution whenever possible. Battelle Pacific Northwest Laboratory notes that the measured distribution for one site near Ellensburg, Washington, produces a power density of 320 W/m^2, twice that from a Rayleigh distribution (160 W/m^2) for the same average speed of 5.3 m/s (12 mph).

Despite its limitations, the Rayleigh distribution remains a useful tool for most sites. Meteorologists often use a more flexible mathematical formula, the Weibull distribution, which can more closely model the wind speeds at a wide range of sites than the Rayleigh distribution. The

Table 10-1. Effect of Speed Distribution on Wind Power Density for Sites with Same Average Speed					
	Annual Average Wind Speed		Wind Power Density	Energy Pattern Factor or Cube Factor	Wind Power Class (at 10 m)
Site	m/s	mph	W/m^2		
Culebra, Puerto Rico	6.3	14	220	1.4	4
Tiana Beach, New York	6.3	14	285	1.9	5
San Gorgonio, Calif.	6.3	14	365	2.4	6
Battelle, PNL Wind Energy Resource Atlas, 1986.					

FREQUENCY DISTRIBUTIONS

While most of those who use wind energy seldom need to employ the following equations for speed distributions, it's important to know when and how they apply. For example, most manufacturer estimates of annual energy output include a brief notation that the estimate derives from a Rayleigh distribution. Some manufacturers substitute the expression "Weibull k = 2"–an arcane way of stating the same condition. In other words, performance of the wind turbine will vary if actual conditions differ from these standard speed distributions.

The Weibull wind speed distribution is a mathematical idealization of the distribution of wind speeds over time. This distribution is determined by two parameters: C, the scale factor that represents wind speed, and k, the shape factor that describes the form of the distribution. A typical shape factor for midlatitude sites is about 2. Sites in the trade wind belt have much higher values for k. The Weibull distribution can be found from

$$f(V) = k/C \, (V/C)^{k-1} \exp[-(V/C)^k]$$

where $f(V)$ is the frequency of occurrence for the wind speed (V) in a frequency distribution, (exp) is the base e exponential function, C is the empirical Weibull scale factor in m/s, and k is the empirical Weibull shape factor.

The Rayleigh distribution is a special case of the Weibull function where k, the shape factor, is 2. Thus, to use the Rayleigh distribution, you need only know the average wind speed. The Rayleigh distribution can be found from

$$f(V_R) = dV \, (\pi/2) \, (V/V_{avg}^2) \exp[-\pi/4 \, (V/V_{avg})^2]$$

where dV is the width of the wind speed bin, V is the speed of the wind speed bin, and V_{avg} is the average wind speed.

Consider two sites where the Weibull parameters are known: Tera Kora and Helgoland (see Comparison between Tera Kora, Curaçao, and Helgoland, Germany).

Tera Kora is on the northeast coast of Curaçao, an island in the Caribbean. Kodela, the island utility, installed a small wind plant there in 1993 of twelve 250-kW turbines (see Figure 10-sb3. Tera Kora). The turbines have been successful partly because the site has a high average wind speed and partly because it is in the trade wind belt. With a shape factor of 4.5, there are few occurrences of very high winds that can damage the turbines. Helgoland is an island off the coast of Germany in the North Sea. The distribution of wind over Helgoland has a shape factor of 2.09. Even though both sites have about the same average wind speed of slightly more than 7 m/s, there is more energy in the wind on Helgoland than at Tera Kora. The shape of the wind distribution on Helgoland indicates that there are more occurrences of high winds there than over Tera Kora. Because of the cubic relationship between wind speed and power, these stronger winds contribute significantly to the overall energy available. This is reflected in the greater power density on Helgoland (400 W/m^2) than that on Curaçao (280 W/m^2) (see Weibull distribution).

The shape of the distribution also affects how well various wind turbines will perform. Most large wind turbines, such as those installed at Tera Kora, are designed for sites with Rayleigh distributions and have high rated wind speeds. However, to improve performance in the Caribbean, where the trade winds dominate, the turbines should be optimized for the wind regime by designing for a lower rated wind speed than elsewhere to better capture the lower speed winds that dominate the energy distribution.

Note that most spreadsheet software contains a function command for the Weibull distribution that greatly simplifies using this equation.

Tera Kora. One of the Caribbean's more successful wind plants, Tera Kora provides about 1% of Curaçao's electricity.

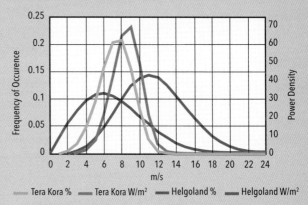

Tera Kora % — Tera Kora W/m^2 — Helgoland % — Helgoland W/m^2

Weibull distribution. Distribution of wind speeds and power density at two sites with Weibull distributions.

Comparison Between Tera Kora, Curaçao and Helgoland, Germany			
		Tera Kora	Helgoland
Avg. Speed, V_{avg}	m/s	7.3	7.1
Shape factor, k	k	4.5	2.09
Scale Factor, C	m/s	8	8
Power Density	W/m^2	280	400
Energy Pattern Factor		1.19	1.83

Rayleigh distribution is a member of the Weibull family of speed distributions.

Where do we stand now? We can calculate power density in two ways. We can sum a series of power density calculations for each wind speed and its frequency of occurrence (the number of hours per year the wind blows at that speed) for the site's distribution of wind speeds. Or we can use the average wind speed and the appropriate cube factor.

At a sea-level site with a temperature 15°C (59°F), air density is 1.225 kg/m³. If the site has a Rayleigh distribution, and thus the Energy Pattern Factor (EPF) is 1.91, then the power density in W/m² is

$$\text{Average Annual P/A} = \tfrac{1}{2}\,(1.225)\,V^3\,\text{EPF W/m}^2,$$
where V is in m/s.

If the site has an average annual wind speed of 4 m/s (9 mph), the annual power density is

$$P/A = \tfrac{1}{2} \times 1.225 \times 4^3 \times 1.91$$
$$= 75\ \text{W/m}^2$$

And once we know the annual power density at a site, we can quickly estimate the *annual* wind energy density in kilowatt-hours per square meter of the wind stream per year (kWh/m²/yr).

$$E/A = P/A\,(8{,}760\ \text{h/yr})\,(1\ \text{kW}/1{,}000\ \text{W})$$
$$= 75\ \text{W/m}^2\,(8{,}760\ \text{h/yr})\,(1\ \text{kW}/1{,}000\ \text{W})$$
$$= 656\ \text{kWh/m}^2/\text{yr}$$

Estimating the amount of energy available annually to the wind turbine becomes simply the product of energy density (E/A) and the turbine's swept area (A) in square meters (see Table 10-2. Annual Wind Power and Energy Density for Rayleigh Distribution).

$$E = E/A \times A$$

For example, Ampair's 100 *intercepts* about 1 m² (12 ft²) of the wind stream. At a site with a 4 m/s average wind speed, it will intercept about 650 kWh per year. It won't *capture* that much, and we'll discuss why in the next chapter. Suffice it to say, it captures much less for a host of reasons.

We're not quite ready to estimate how much of this energy a wind turbine is capable of capturing. There's one step remaining.

Because the wind turbine will be mounted atop a tower that's typically two to three times taller than the tower used to measure wind speed, we need to estimate the wind speed at the top of the tower—the proposed wind turbine's *hub height*. How wind speed and power change with height is the subject of the next section.

Wind Speed, Power, and Height

Wind speed, and hence power, varies with height above the ground. Wind moving across the earth's surface encounters friction caused by the turbulent flow over and around mountains, hills, trees, buildings, and other obstructions in its path. These effects decrease with increasing height above the surface until unhindered airflow is restored. Consequently, as friction and turbulence decrease, wind speed increases.

As you can imagine, frictional effects differ from one surface to another, depending upon its roughness. Friction is higher around trees and buildings than it is over the smooth surface of a lake. In the same manner, the rate at which wind speed increases with height varies with the degree of surface roughness. Wind speeds increase with height at the greatest rate over hilly or mountainous terrain and at the least rate over smooth

Table 10-2. Annual Wind Power and Energy Density for Rayleigh Distribution			
Annual Average Wind Speed		Annual Power Density	Annual Energy Density
m/s	mph	W/m²	kWh/m²
4	9.0	75	656
5	11.2	146	1,281
6	13.4	253	2,214
7	15.7	401	3,515
8	17.9	599	5,247
9	20.2	853	7,471

terrain like that of the Great Plains. Because of this, the benefits of using a tall tower are often greater when siting in hilly terrain than it is on the Llano Estacado, or the Staked Plains, of the Texas Panhandle.

At low wind speeds, the change in speed with height or *wind shear* is less pronounced and more erratic. In light or calm winds, as may be encountered during a temperature inversion, wind speeds may increase slightly between the ground and a certain height, and then begin to decrease. Real-world experience has shown that changes in wind speed with height are not constant. In the Altamont Pass east of San Francisco, Pacific Gas and Electric Company found that above 60 meters (200 feet) wind speeds decreased with increasing height. The utility found this effect after it installed a Boeing Mod-2, the rotor of which was 91 meters (300 feet) in diameter. At times the uppermost part of the rotor would extend above the layer of fast-moving air. On average, however, wind shear is positive, and wind speed increases with height.

This effect is so important that data on wind speeds will always include the height at which the wind was measured. If the height is not specifically mentioned, it is usually assumed to be about 10 meters (33 feet) above the ground, though many measurements at airports in North America were made at 4 to 6 meters (15–20 feet). Most wind turbines will be installed on towers much taller than this to take advantage of the stronger, less turbulent winds aloft. Commercial wind turbines in Europe and North America are now being installed on towers more than 100 meters (330 feet) tall.

The easiest way to calculate the increase in wind speed with height is to use the *power law* method. Another approach using logarithmic extrapolation is common among meteorologists. Logarithmic extrapolation is mathematically derived from a theoretical understanding of how the wind moves across the surface of the earth. In contrast, the power law equation is derived empirically from actual measurements. The power law equation may be less rigorous, but it works well and is more conservative than the logarithmic method.

The following equation illustrates how to use the power law method where V_0 is the wind speed at the original height, V is the wind speed at the new height, H_o is the original height, H is the new height, and α is the wind shear exponent.

$$V/V_0 = (H/H_o)^\alpha$$
$$V = (H/H_o)^\alpha V_0$$

For example, consider the increase in wind speed when doubling tower height from 10 to 20 or from 30 to 60. (The units are unimportant; it's the ratio that counts.)

$$V = (20/10)^{0.14} V_0 = 2^{0.14} V_0 = 1.1 V_0$$

Both systems require the user to estimate surface roughness. Where the rate of increase in wind speed with height is unspecified, it is commonly assumed in North America that the "one-seventh power law" applies, that is, the surface roughness exponent is 0.143 representing open plains. Empirical results indicate that the one-seventh power law fits many, though certainly not all, North American sites (see Table 10-3. Changes in Wind Speed and Power with Height for Selected DOE Wind Turbine Sites in the USA).

On terrain where the *one-seventh power law* applies, doubling tower height increases wind speed by 10%. Increasing tower height five times, say from 10 to 50 or from 30 to 150, may increase wind speed as much as 25% (see Figure 10-7. Increase in wind speed with height). Once again, the one-seventh power law is just a guide.

LOGARITHMIC MODEL OF WIND SHEAR

The logarithmic extrapolation of wind speed with height is

$$V = \ln(H/Z_0)/\ln(H_o/Z_0) V_o$$

where (V_o) is the wind speed at the original height, (V) is the wind speed at the new height, (H_o) is the original height, (H) is the new height, and (Z_0) is the roughness length.

Logarithmic extrapolation has worked well along the coastlines of the North German Plain, particularly in Denmark. In open areas with few windbreaks, such as coastal sites (Roughness Class One in the European system), the logarithmic model produces a result similar to that of the one-seventh power law. Farther inland, results from the two methods diverge. For inland sites, the logarithmic model finds more energy in the wind than that of the one-seventh power law.

The effect of height on wind speed is so great that it often necessitates measuring actual wind speeds at hub height rather than rely on estimates produced by either system. Yet, in the absence of actual measurements, the one-seventh power law is a reasonable, if sometimes conservative, approximation.

Table 10-3. Changes in Wind Speed and Power with Height for Selected DOE Wind Turbine Sites in the USA

	Wind Speed m/s		Speed Increase	Wind Shear Exponent α	Power Increase	Wind Shear Exponent α
Height	9.1 m	45.7 m				
Site	30 ft	150 ft				
Finley, North Dakota	6.1	9.1	1.49	0.25	3.15	0.24
Block Island, Rhode Is.	5	7.4	1.48	0.24	3.06	0.23
Boardman, Oregon	3.8	5.5	1.45	0.23	2.73	0.21
Huron South Dakota	4.7	6.8	1.45	0.23	2.53	0.19
Russel, Kansas	5.3	7.3	1.38	0.20	2.16	0.16
Clayton, New Mexico	5.4	7.3	1.35	0.19	2.06	0.15
Minot, North Dakota	6.5	8.4	1.29	0.16	1.97	0.14
Amarillo, Texas	6.3	8.1	1.29	0.16	2.04	0.15
San Gorgonio Pass	6.2	7.7	1.24	0.13	2.03	0.15
Livingston, Montana	6.8	8.4	1.24	0.13	1.74	0.11
Kingsley Dam, Nebraska	5.3	6.5	1.23	0.13	1.79	0.12
Bridger Butte, Wyoming	7	8.4	1.20	0.11	1.59	0.10

Source: Battelle PNL, Wind Energy Resource Atlas, 1986.

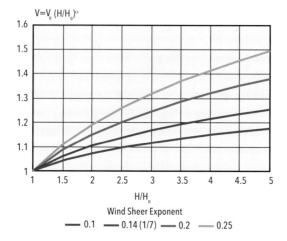

Figure 10-7. Increase in wind speed with height. Simple chart for estimating the increase in wind speed as a function of relative tower height. To use the chart, find the ratio of tower height (H) to anemometer height (H_o), then find the intercept with the line representing the wind shear exponent. The intercept indicates the relative increase in wind speed at the tower height. For example, if the hub height of the wind turbine will be five times taller than the anemometer height and the one-seventh power law applies, find 5 on the horizontal axis, move vertically until you intercept the curve labeled 0.14, then proceed horizontally to the intercept with the vertical axis.

Although wind shear often follows the one-seventh law, it doesn't always. Obstructions significantly reduce wind speeds near the ground. Over row crops such as corn, or over hedges and a few scattered trees, wind speed increases more dramatically with height than that predicted by the one-seventh law: α rises to one-fifth (0.20).

When the surface is rougher still, say with more trees and a few buildings, α increases farther to one-fourth (0.25). The speed profile becomes even steeper over woods and clusters of buildings.

The increase in wind speed with height only holds true for the height above the *effective ground level*. The wind rushing over a field of corn sees the top of the corn not the soil on which it grows as the effective ground level. For a woodlot this is the uppermost point where the branches of the trees are touching, not necessarily the tops of the trees.

Let's see what all this means by examining the effect a tall tower will have on the average wind speed at a site in Kansas where the one-seventh power law applies. In our example the wind speed was measured at 10 meters (33 feet), H_o, and we want to install our wind turbine on a 50-meter (160-foot) tower, H.

$$V = (H/H_o)^\alpha V_o$$
$$V = (50/10)^{1/7} V_o = 5^{1/7} V_o = 1.26 V_o$$

The wind speed at the height of the wind turbine will increase 26% by installing it on a 50-meter tower.

But we can't stop here. Power is a cubic function of speed. Where P_o is the power at the original height of 10 meters and P is the power at the new height, the increase in power on the 50-meter tower in our example is given in the

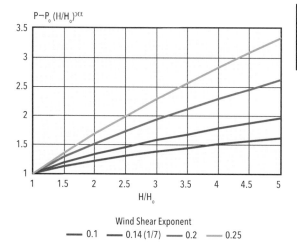

Figure 10-8. Increase in wind power with height. Simple chart for estimating the increase in wind power as a function of relative tower height. To use the chart, find the ratio of tower height (H) to anemometer height (H_o), then find the intercept with the line representing the wind shear exponent. The intercept indicates the relative increase in wind power at the new height. For a fivefold increase in tower height, power almost doubles when the one-seventh power law applies.

THE WIND SHEAR EXPONENT α

The wind shear exponent α varies with the time of day, season, terrain, and the stability of the atmosphere. Shear is low where there is minimum surface roughness and high where there are numerous objects to disturb the flow. German engineer Jens-Peter Molly in his book *Windenergie* presents a simple formula for calculating α from a measure of the surface roughness.

$$\alpha = 1/\ln(z/z_0)$$

where z_0 is the surface roughness length in meters and z is the reference height. For example, when the surface roughness length is 0.01 meter, and the reference height is 10 meters (33 feet), then

$$\alpha = 1/\ln(10\text{ m}/0.01\text{ m})$$
$$\alpha = 1/\ln(1000)$$
$$\alpha = 0.144$$

or about that of the one-seventh power law.

following formula (see Figure 10-8. Increase in wind power with height)

This brings us to another rule of thumb: Increasing tower height fivefold nearly doubles the power available in the wind at sites where the one-seventh power law applies.

$$P = (H/H_o)^{3\alpha} P_o = 5^{3(1/7)} P_o = 1.99 P_o$$

Like all rules of thumb, this is only an approximation of the real world. In Table 10-3 (Changes in Wind Speed and Power with Height for Selected DOE Wind Turbine Sites in the USA), the increase in power with height doesn't exactly follow this rule. The reason? The distribution of wind speeds changes slightly at new heights. For example, the wind shear exponent for wind speed at Clayton, New Mexico, is 0.19, but the shear exponent decreases slightly to 0.15 when considering the increase in wind power density at the new height.

Clearly you can't talk about wind speed without also referring to the height at which it's measured. The two always go together, though sometimes the height is assumed. In general the average wind speed at a specific site refers to the speed at the height of the anemometer. Confusion arises when you start talking to wind turbine manufacturers. The wind speeds they use may be either at the hub height of the wind machine or at some

Terrain	Surface Roughness Length z_0 (m)	Wind Shear Exponent α
Ice	0.00001	0.07
Snow on flat ground	0.0001	0.09
Calm sea	0.0001	0.09
Coast with on-shore winds	0.001	0.11
Snow covered crop stubble	0.002	0.12
Cut grass	0.007	0.14
Short grass prairie	0.02	0.16
Crops, tall grass prairie	0.05	0.19
Hedges	0.085	0.21
Scattered trees & hedges	0.15	0.24
Trees, hedges, a few buildings	0.3	0.29
Suburbs	0.4	0.31
Woodlands	1	0.43

Table sh-10.3. Surface Roughness Lengths and the Wind Shear Exponent α

Relative to a reference height of 10 m (33 ft). Adapted from *Characteristics of the Wind* by Walter Frost and Carl Aspliden in *Wind Turbine Technology*, and *Windenergie: Theorie, Anwendung, Messung* by Jens-Peter Molly.

Increasing tower height fivefold

NEARLY DOUBLES THE POWER AVAILABLE

in the wind at sites where the one-seventh power law applies.

THE NOCTURNAL JET

High wind shear may be a regional phenomenon of North America's Great Plains. It's characteristic of Buffalo Ridge in southwestern Minnesota, according to meteorologist Ron Nierenberg. This contrasts with California's Tehachapi Pass where shear on Cameron Ridge is near zero. At exposed sites in Minnesota, says Nierenberg, wind shear is often double that of the one-seventh power law, from 0.2 to 0.3. It's similar in Iowa and Wisconsin. Trees, he says, are the cause. For comparison, Cameron Ridge in California is virtually treeless.

During summer months when wind speeds are typically low in continental wind regimes, a *nocturnal jet* may occur at a certain height above ground where the wind shear exponent can reach 0.4. This "jet" has nothing to do with the jet stream, Nierenberg explains, it's simply a layer of fast moving air. "There are lots of places in the world where there's a localized zone of high winds, a so-called jet," he says. "Buffalo Ridge, at 600 meters [2,000 feet] above sea level, is possibly just high enough to pierce it." There's a similar jet about 300 meters [1,000 feet] above California's San Joaquin Valley. In part it is this jet that produces the winds in the Tehachapi Pass.

Measurements on Buffalo Ridge near Chandler, Minnesota, found wind shear in the 30- to 50-meter (100 to 160-foot) range typical of the Great Plains. But at 50- to 70-meter (160- to 230-foot) heights, the shear exponent jumped to 0.42. "The resource is very strong," says Rory Artig, an engineer with Minnesota's Department of Public Service. "It's quite different from that in California." Artig found that the 2.5-year average annual wind speed at Chandler was 6.9 m/s (16 mph) at 30 meters (~ 100 feet), 7.6 m/s (17 mph) at 60 meters (~ 200 feet), and 8.2 m/s (18 mph) at 90 meters (~ 300 feet) above ground level. Although the 20% increase in wind speed may seem modest, it produces a 67% increase in the power available.

in valleys than on windswept mountaintops.) Consequently, data from airports, military bases, and weather stations may not reflect the winds that exist at more exposed sites. Using data from these stations alone may lead to underestimating the potential wind power in an area. For example, historical wind data led to an erroneous perception that there was little wind energy in southwestern Minnesota. Later field measurements found just the opposite. The winds across a gentle topographic high called the Buffalo Ridge were more energetic than anywhere else in the state. Now the region is one of the Midwest's largest sources of wind-generated electricity.

The data may be unreliable for other reasons as well: the wind-measuring instruments may have been inaccurate or poorly placed. The instruments at many airports were not properly maintained and frequently were located on or adjacent to the terminal building. At these stations, the data better reflects the turbulence around the building than anything else.

Data from remote sites is even more problematic. In some cases wind data was collected only during daylight hours or for a few hours during the summer months. In either case, the data doesn't represent what could be expected throughout the day or throughout the year. At some stations, so few observations were recorded that the data is useless. The observer was literally noting whether it was *windy* or not.

Historical average speeds are also available from air-quality monitoring stations at both conventional and nuclear plants and from some industries. The limits on data from these sources are the same as those on data from airports. We measure air quality where wind speeds are low and where pollutants concentrate, such as in urban canyons and narrow mountain valleys. These are less than ideal locations for a wind machine, and the wind speeds are unrepresentative of better upland sites. For example, wind speeds measured in the deeply incised valley of the Rhine have little bearing on the Eifel Mountains, a plateau of low hills near Germany's border with Belgium where numerous large wind turbines have been successfully installed.

There are also numerous sources of short-term wind data: government energy offices, universities, and nonprofit organizations.

other height. It makes a big difference. Most rate their products at hub height; some don't. Reputable manufacturers and their dealers will clearly state which method they use.

After three decades of working with modern wind energy, we've learned an important lesson—the hard way. No amount of historical wind data can substitute for knowing the wind at the specific site where you want to put your wind turbine. This includes measuring the wind at the proposed height of the turbine to avoid extrapolations that may or may not reflect actual conditions. Next we will look at what historical wind data is available.

Published Wind Data

In general, wind data in all countries has been gathered near centers of population. People congregate in areas sheltered from storms and severe weather. (Cities are built more often

Wind data is not always presented in the form of annual average wind speeds. Both in the United States and Europe, professional meteorologists have categorized wind resources into a series of wind classes.

In the United States, Battelle Pacific Northwest Laboratories mapped average annual wind power rather than simply wind speed. Battelle devised a numerical rating that corresponds to one of seven wind power classes. Each class represents a range of power densities. For example, Class 4 represents wind power density from 200 to 250 W/m², or wind speeds from 5.5 to 6 m/s (12.3–13.4 mph).

Battelle derived the maps from computer modeling of historical wind data, terrain, and regional weather patterns. The values shown represent only those sites such as hilltops, ridge crests, and mountain summits that are free of obstructions and are well exposed to strong prevailing winds. By giving a range of possible values rather than a single number, Battelle doesn't lure users into the mistaken notion that they can estimate wind speed with precision.

Denmark's Risø National Laboratory has done similar work in Europe. Their European Wind Atlas provides a detailed look at the wind resources across the European Community. Like Battelle, Risø presents the data as a range of values. But Risø goes a step further and suggests the wind speeds likely at sites with differing surface roughness, such as along coastlines.

Large wind turbines in commercial applications are most commonly found in areas with power densities greater than 200 W/m² or average wind speeds above 5.5 m/s (12 mph) at 10 meters above ground level. This corresponds to the wind resource designated Class 4 by Battelle. This resource is equivalent to a power density of 300 to 400 W/m², or an average wind speed greater than 6.5 m/s (15 mph) at 30 meters (100 feet) above the ground, the typical wind turbine height of the late 1980s. At tower heights of 50 meters (164 feet), the same resource is equivalent to a power density of 400–500 W/m², or an average annual wind speed greater than 7 m/s (≥16 mph).

Most of the wind development in California has occurred on windier sites. Wind speeds on the crests of ridges in the Tehachapi Pass average 8 to 8.5 m/s (18–19 mph), while sites atop Altamont Pass' rolling hills are less windy than those in Tehachapi at 6 to 8 m/s (13–18 mph).

Extensive wind development has taken place in Europe along the North Sea coast. At hub height along the coast of the Netherlands, for example, wind speed averages about 7.5 m/s (17 mph). At less well-exposed sites nearby in the German state of Niedersachsen (Lower Saxony), wind speeds at hub height average 6.8 m/s (~ 15 mph).

One of the lessons learned from California's wind rush during the early 1980s was the necessity of understanding the wind resource in complex terrain. (This was a commercial version of the "why it's always windy here, we don't need to measure the wind.") It was common then to monitor the wind with one anemometer for every 150 to 350 proposed turbines. As a result, many operators greatly overestimated the amount of wind their turbines would intercept. In the early 1980s, California wind turbine operators typically projected twice the amount of electricity the turbines actually produced. They were off by 100%! By the late 1990s, California wind devel-

CALCULATING THE WIND SHEAR EXPONENT (α)

If you're measuring wind speeds at different heights, or if you found data for wind speeds at different heights, you can calculate the wind shear exponent (α) for your conditions from

$$\ln(V/V_o)/\ln(H/H_o)$$

where V is wind speed at the upper anemometer, V_o is the wind speed at the lower anemometer, H is the height of the upper anemometer, and H_o is the height of the lower anemometer.

For example, consider the previous example of the diurnal wind speeds measured at the Wulf Test Field. The average wind speed at anemometer A was 8.5 mph, and that at anemometer B was 7.9 mph. Anemometer A is mounted at a height of 56 feet, anemometer B at a height of 36 feet. (The ratio of height A:B is 1.6.)

$$\begin{aligned}\alpha &= \ln(8.5/7.9)/\ln(56/36) \\ &= \ln(1.076)/\ln(1.556) \\ &= 0.073/0.44 \\ &= 0.17\end{aligned}$$

For the period measured, the wind shear approximates that typical for a short grass prairie, which accurately characterizes the site.

ONLINE WIND RESOURCES AND WIND CALCULATORS

There are a number of websites with wind resource data. There are also several websites with online "wind calculators." Use these with caution. Many online calculators don't reveal how they arrive at the result. They simply present a number. I prefer to know the method used to calculate the wind speed at a site so I can gauge the accuracy of the result.

I tested one such calculator on the website of a now-defunct manufacturer. I put my zip code into their "estimator" and was I embarrassed. I've been living in a wind energy gold mine for nearly 30 years and didn't even know it. I'd always thought Bakersfield, California, was a lousy wind area. But no, the estimator told me that the average wind speed was "good" at 11.2 mph (5 m/s). Yes, my neighborhood fell within the recommended zone for one of their wind turbines. The low end, mind you, but still within the "good" category. Suitably chagrined that I didn't know the wind resource where I've lived so long, I decided to look up the wind speed myself from more traditional sources.

I found a NOAA (National Oceanic and Atmospheric Administration) website that had airport data. The airport is 15 minutes from my house. The 50-year average wind speed was only 6.4 mph. Because wind power is a cubic function of wind speed, the estimator told me there was five times more wind power at my site than I could expect in reality!

In the United States, much of the work of collecting wind data from numerous sources was done decades ago by Battelle Pacific Northwest Laboratory for the US Department of Energy. This data, supplemented with numerous wind resource studies for commercial wind development across the continent, has enabled the National Renewable Energy Laboratory (NREL) to generate modern high-resolution digital wind maps derived from sophisticated atmospheric models.

Use these maps (see the sidebar North American Wind Resource Maps). They are scientifically accurate–as much as they can be–and prepared by an organization with the public interest in mind. All the underlying inputs and all the methods used have been published in technical articles in the public domain. They are not trying to sell anyone a wind turbine.

Data collection. Measuring the wind can be as simple as pushing a button and writing down the data or collecting a data chip.

Surveying the Wind at Your Site

To evaluate the potential at your site, begin by asking yourself, what is it that you want? Is it the instantaneous wind speed, the average annual speed, or the distribution of wind speeds? If you want to know when it's too windy to go sailing, instantaneous wind speed will suffice. If you want to estimate the annual or monthly energy output from a wind machine, then at least the average wind speed is necessary. Preferably you'll want the wind speed distribution as well. If you plan to use the wind turbine as your sole source of power at a mountaintop retreat, for example, then more detailed information may be required, such as the number of calm days and the time between them.

In most cases you will be interested in how much energy a wind machine can produce in your area—more specifically, at your site. To estimate the annual energy production, the speed distribution is preferred. The speed distribution gives you the most accurate results, but it isn't always necessary for small wind turbines. Average speed will often suffice, especially if you have some measure of the energy pattern factor too.

opers had seen the light and were monitoring the winds with one anemometer for every 2 to 3 turbines.

Today there are a number of online maps of wind resources that have taken a lot of the guesswork out of finding wind data and evaluating its usefulness. Some of these maps even allow you to "drill down" to an estimate of wind speed at a specific height for a specific location with a resolution of a few square kilometers. Keep in mind that this data is based on models of atmospheric wind speeds that are then brought down to the surface where we mortals live. They seldom take into account local obstructions, such as trees and buildings, and are less accurate in rough terrain.

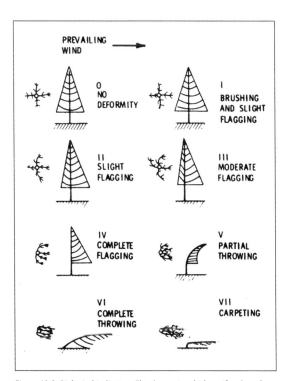

Figure 10-9, Biological indicators. The degree to which conifers have been deformed by the wind can be used as a rough gauge of average annual wind speed (see Table 10-4. Griggs-Putnam Index of Deformity). (Battelle PNL)

Table 10-4. Griggs-Putnam Index of Deformity							
	Index						
Wind Speed	I	II	III	IV	V	VI	VII
mph	7-9	9-11	11-13	13-16	15-18	16-21	22+
m/s	3-4	4-5	5-6	6-7	7-8	8-9	10

Now that you know what's needed, take a look around your area and collect any information you can on your local wind resource. You can start by looking at the vegetation.

Trees and shrubs are frequently touted as a good qualitative indicator of wind speed (see Figure 10-9. Biological indicators). High winds and a harsh environment of ice and snow will deform them. The severity of the deformation, whether the tree is slightly flagged or completely bent to the ground, can be used as a rough gauge of wind speed. The types of deformity are:

- *Brushing:* branches and twigs sweep downwind. This can be observed on both conifers and deciduous trees.
- *Flagging:* branches sweep downwind, upwind branches are cropped short.
- *Throwing:* trunk sweeps downwind.
- *Carpeting:* trunk bends to the ground. Found frequently in Alpine or severe environments where trees grow only a few feet above the ground.

Use the Griggs-Putnam scale of deformity to find the range of wind speeds represented. For example, the divi-divi tree is a popular icon on the island of Curaçao off the coast of Venezuela. The heavily flagged and partially *thrown* tree symbolizes the island as much as their famous orange liqueur. In the Griggs-Putnam index, this suggests average wind speeds of 7 to 8 m/s (15–18 mph). In fact exposed sites on the island experience average wind speeds slightly more than 7 m/s.

There are limitations to this technique. First, don't get excited by one or two odd-shaped trees. One flagged tree is insufficient to make a judgment. There are many other causes for tree deformity besides the wind. You'll need to note several of the same species with an equal amount of deformation to determine if the wind is the cause. (Each species varies in its susceptibility to flagging.) Conifers, especially pine and fir trees, are more reliable indicators of wind strength than are deciduous trees. Moreover, deformation is more obvious where freezing salt spray or ice frequently accompanies high winds. Such conditions are often found along coastlines and on mountaintops.

If you can't find any deformation, don't despair. The absence of flagging doesn't necessarily indicate a low speed. Too often the value of trees as a wind speed indicator is overplayed. At best, it gives only a crude range of possible wind speeds, and even then, it's most useful only where conifers dominate.

Find out if someone else has already done the hard work for you by locating anyone nearby who has installed a wind turbine. Pay them a visit. Or find someone with an anemometer who has measured wind speed. Has this person collected data of the type you're seeking, and has he done so in a systematic manner? What has he found? (Just because someone has stuck a dime-store anemometer on a TV antenna doesn't mean the data collected is useful.) Before you take any of the data to heart, determine if your site and the

ESTIMATING THE HEIGHT OF OBSTRUCTIONS

Remember those comic pictures of a ragged artist thrusting his thumb at the world? He had a good reason for doing that. Artists use the technique to gauge proportions. You can also use it to estimate the height of nearby trees and buildings. A pencil works better, though. Here's how to use it.

Identify an object of known height at about the same distance from you as the tree or building you wish to measure. TV antennas, telephone poles, and houses work well. Hold the pencil at arm's length and sight along it. Line up the top of the pencil with, for example, the top of a tree. Slide your thumb down the pencil until it lines up with the bottom of the tree. Now turn to the object of known height and again sight along the pencil. While keeping the pencil at arm's length, move your arm up or down until your thumb lines up with the bottom of the object. Judge the proportions by noting how much of the pencil extends above the object. Is it twice the height, one-third greater, or the same?

In Wisconsin, Mick Sagrillo uses a similar technique that compares the shadow from an object of known height to the shadow of the object in question. Simply measure the shadow from the object of known height, for example a fence post, and the shadow from the object of unknown height. The relationship between the shadow of the fence post and its height is the same as the relationship between the shadow of the obstruction to its height.

---

others are comparable. Is the data typical of what you could expect?

Next, find sources of wind data for your locale, including that from online resources. Understand the limitations of the data. Your site may experience stronger or weaker winds than that of other sites.

Now, put the pieces together. Estimate what you think is your average wind speed and power. Give yourself room for error. Avoid pinning your hopes on one number alone and instead use a range of values. Be conservative. Most people overestimate the amount of wind available.

With these numbers in hand, use the techniques in the next chapter to estimate the annual energy production from typical wind turbines. Look at their economics.

Next, ask yourself how much risk you're willing to assume. Even though you may have done an admirable job of estimating the wind regime in your area from published sources, obstructions and terrain features can greatly reduce the actual wind energy available. If you don't mind this uncertainty, or if you're on the Great Plains or the steppes of Central Asia and there isn't a tree for miles, then there's little need to go to the trouble of conducting a full-fledged wind speed survey.

For household-size and small, commercial-size wind turbines, measuring the wind isn't always required. Measuring the wind at your site becomes necessary when you're unwilling to take the risk that there's sufficient wind to produce what you expect, or the bank is unwilling to loan you money for your project because of the uncertainty. For commercial wind power plants, the economic risk is too great to proceed without the services of professional meteorologists. This will often entail on-site measurements.

Measurements, if they are to be made at all, should be taken at the intended location of the wind machine and at its proposed height. This is particularly important in rough terrain or where there are obstructions. Reliable measurements can only be made when the anemometer extends well above nearby trees and buildings. Even tall grain crops and low-lying shrubs raise the effective ground level, severely reducing the wind speed measured by anemometers on short towers. If you need to measure the winds at your site, take a survey of the trees and buildings nearby and estimate their heights. You may find that the anemometer, and eventually the wind turbine, should be erected elsewhere.

Assume that the existing wind data you've examined is unconvincing. You want to measure the wind at your site to get a better picture of what's there. What next? Anemometers measure wind speed. Wind vanes indicate direction. That's simple enough. More complex is the kind of recording equipment you'll need. In the next section, we'll go over the equipment that's available and discuss what probably meets your needs best.

Measuring Instruments

To perform a wind resource assessment, you'll need an anemometer, mast, and recorder. A wind-measuring instrument is composed of two parts: the sensor (the anemometer head) and a means for displaying the data it measures. The sensor generates an electrical signal that's proportional to wind speed. Cup anemometers are the most common wind sensor. The spinning cups drive either a DC generator or AC alternator. The least expensive anemometers, widely used even by professionals, produce an electrical pulse, which is then counted over a period of time, often just a fraction of a second.

Whatever system is used, the sensor (the anemometer head) either drives a meter that displays instantaneous wind speed or feeds data to a recorder. Unlike the displays of cheap wind speed meters that indicate only instantaneous wind speed, recorders store information for future retrieval. At one time all meteorological data was recorded on strip charts, but wind prospecting and the boom in electronics have revolutionized wind speed measurement.

Before we go any further, let's clear up a common misconception. Instantaneous wind speed indicators are useless for finding the average wind speed. They're fun to watch, but that's it. To be of value in a site survey, you would have to check them every hour, 24 hours a day, every day for months on end. Instantaneous wind speed meters are only useful for developing a better understanding of the wind.

An electronic odometer or accumulator is an overall better choice for an inexpensive means to log wind data. Similar to the odometer on the dashboard of your car, an accumulator counts the miles or kilometers of wind that pass the anemometer. The average wind speed is found by simply dividing the distance the odometer records by the elapsed time between readings. Today's instruments do that automatically, as well as much more. As Mick Sagrillo emphasizes, you need more than merely the average wind speed. You need some measure of the distribution of wind speeds to accurately assess the potential of a site. Today, many inexpensive instruments can produce both an average wind speed and a measure of the energy pattern factor or, even better, provide a frequency distribution.

As the need for more sophisticated measurements has increased, accumulators have evolved into data loggers. In essence, data loggers consist of multiple accumulators that tally the data falling into each accumulator's domain. For example, each accumulator could represent a given wind speed range. Winds 0–1 would fall into the first accumulator, winds 2–3 would fall into the second, and so on. At the end of the observation period, the contents of each register can be used to plot the speed distribution. This distribution

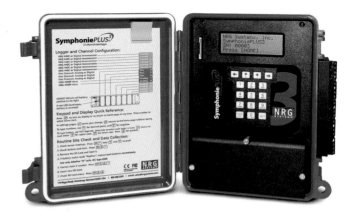

Figure 10-10. Data logger. Electronic data recorders can be used to collect wind and meteorological data or, with the appropriate sensors, monitor wind turbine performance. (Renewable NRG Systems)

can then be used to calculate power density or can be compared with a wind turbine's power curve to project potential performance.

Most data loggers process some of the measured data and record the results. These *smart* recorders significantly reduce the time and expense of analyzing the data later (see Figure 10-10. Data logger). By reducing the amount of data that must be stored, they can record data for months or longer.

Anemometer Towers

From bitter experience, wind prospectors have learned there's absolutely no substitute for measuring the wind at the height where your wind turbine will operate, known in the trade as the hub height. Fortunately, there are hinged anemometer masts of lightweight tubing designed specifically for this purpose (see Figure 10-11. Anemometer mast). These masts can reach heights of 80 meters (260 feet), though 50-meter (160-foot) heights are more common.

As towers for large turbines have grown ever taller, new techniques have become necessary to measure the wind at hub heights, now often exceeding 100 meters (330 feet). Meteorologists use sophisticated LiDAR (Light Detecting and Ranging) and SODAR (Sonic Detection and Ranging) instruments that continuously shoot a beam across the projected height where the rotor will sweep. Developed for atmospheric research, these instruments can measure and record the

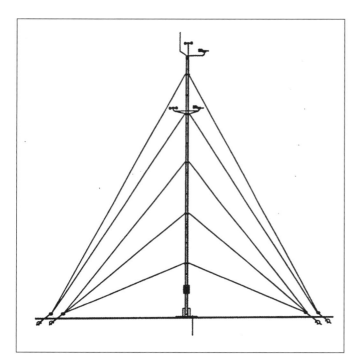

Figure 10-11. Anemometer mast. Temporary, tilt-up masts typically installed with screw-in earth anchors greatly simplify wind measurements at remote sites. (Renewable NRG Systems)

Figure 10-12. SODAR. Sonic detection and ranging instrument for measuring wind speed, direction, and turbulence at heights difficult to achieve with conventional meteorological masts. SODAR and LiDAR devices use the Doppler effect for measuring the wind at the height where modern large wind turbines operate. Shown is a SODAR (foreground) at NREL's Rocky Flats test center. 2012. (Andrew Clifton, NREL)

wind across the entire rotor disk of the hypothetical wind turbine using the Doppler Effect (see Figure 10-12. SODAR).

Some do-it-yourselfers have tried to cut corners by installing an anemometer on the roof of a building. Unfortunately, turbulence around the building affects wind speed dramatically. Avoid this practice. Most of the time, mounting an anemometer (or a wind turbine for that matter) on a building is impractical and a waste of time. Where do you attach the guy cables, for example? Will it be necessary to drill holes in the roof for anchors? If that doesn't stop you, imagine trying to erect a tall slender mast on a steeply pitched roof; it's an accident waiting to happen.

Survey Duration

"Yeah, you got a fine site here," said the dealer as he installed the anemometer George had ordered. "I bet you've got 12 mph."

Two days later the dealer returned. After examining the anemometer he said, "Just as I thought, an easy 12 mph average." The dealer then persuaded George to buy a wind turbine.

A wet finger in the air on the first visit would have been just as accurate. Maybe the dealer didn't know how to measure wind speed. Then again, maybe he was a con artist. It's hard to tell. The site was indeed a good one, and the wind turbine (made by a reputable manufacturer) was installed in a workmanlike manner. The site could have had a 12-mph (5.3-m/s) average wind speed, but the dealer or the buyer wouldn't have known that after two days of measurement.

How long is enough? That's another tough question. Average wind speeds can vary as much as 25% from year to year. But it's obviously impractical to gather 10 years of data from a site before you decide whether or not to install a wind machine. Battelle suggests gathering 1 year of data. Even so, your site's average speed will be dependent on how normal the year has been with respect to the long-term average. Was it a typical year, or was it windier? Was it an El Niño year or a La Niña? To answer that question, you must examine the wind data from the nearest long-term recording station and compare the test year's results with the station's historical average.

Try to establish a correlation between your site and a long-term station. If you're lucky, you may

NORTH AMERICAN WIND RESOURCE MAPS

There are digital maps for much of the developed world–and for many parts of the developing world. These are accessible online (see Digital wind resource atlas for the United States). Here are some online resources for North America. While such maps are a useful tool for a "first cut" at a wind resource assessment, they are no substitute for on-site measurements. The data can be presented in terms of wind power classes in the Battelle system, wind power density (W/m²), or as wind speed in m/s. As the National Renewable Energy Laboratory advises, the average wind speeds "are model-derived estimates that may not represent the true wind resource at any given location. Small terrain features, vegetation, buildings, and atmospheric effects may cause the wind speed to depart from the map estimates."

US DOE's Wind Powering America State Wind Resource Maps:
http://apps2.eere.energy.gov/wind/windexchange/windmaps/

Canada's Wind Energy Atlas:
http://www.windatlas.ca/en/maps.php

Wind Resource Maps of Mexico by Region:
http://www.nrel.gov/wind/international_wind_resources.html#mexico

International Maps:
http://www.nrel.gov/wind/international_wind_resources.html

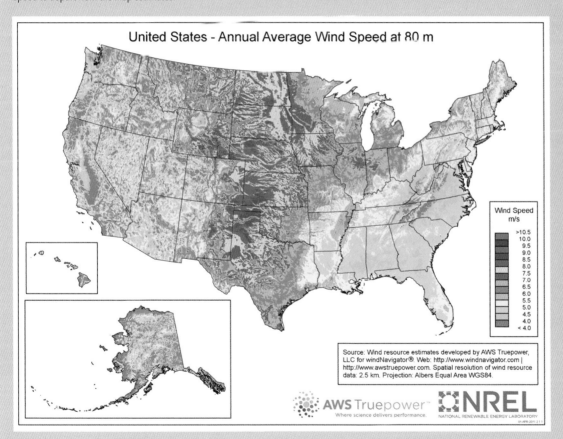

Digital wind resource atlas for the United States. (National Renewable Eneergy Laboratory)

find that a full year of measurements isn't necessary. But four months is a minimum. Anything less is guesswork. If you're not going to gather a full year of data, make sure that you at least capture the wind season, typically winter and spring months for much of the midlatitudes.

Data Analysis

Making sense out of the data from a site survey is more akin to alchemy than science. Much is left to the judgment (or imagination) of the observer. You must determine whether the data is representative of the site and not surrounding obstructions, whether the year is normal, and whether or not there is a direct relationship between the site and, for example, a nearby station with long-term records.

You want to establish the ratio between your site's average speed and the long-term recording station. Then you can adjust the station's historical average by the resulting ratio. Step one establishes whether your site is windier or less windy than the airport. Step two normalizes the results for a typical year.

This approach assumes that a correlation exists between your site and the long-term recording station. There may not be one, particularly in rough or mountainous terrain. Try to use a station in terrain similar to yours and with a similar exposure to the wind. The nearest airport may not always be the best choice.

Another method is to use linear regression analysis. This technique gives a measure of the ratio between the two sites, by testing the degree of correlation, and projects the site's average speed. Linear regression analysis is the same as graphically drawing the best fitting line between two sets of data. The statistical functions in spreadsheet software make the job a cinch.

You'll find that hourly or daily wind speeds are often too erratic to establish a correlation between the two sites. Average weekly speeds are more stable. They tend to smooth out the passage of weather systems and local diurnal variations.

These are just some of the techniques professional meteorologists use. They also test the reliability of the data they've recorded, discarding data that are anomalies and reconstructing lost data when necessary. Meteorologists also provide a measure of the data's reliability by estimating the probability that wind speeds will fall within the range of wind speeds they've determined, a requirement of banks and financiers.

Most potential users of small wind turbines—especially those wanting micro and mini wind turbines—won't take the time or invest the money in a site survey. They will simply turn to online resources. Potential users of commercial-scale wind turbines will employ professional meteorologists who will determine whether an on-site survey is needed.

This doesn't mean you shouldn't study the wind at your site. Monitoring the wind yourself is instructive, says Jason Edworthy, one of Canada's pioneering commercial wind developers. He suggests, for example, holding a simple wind meter to the wind to develop a feel for the wind's strength. Or, says Edworthy, fly a kite. Attach streamers to the line and watch how those streamers near the ground roil and flap and those higher up smooth out. It's easy, he says, and fun too. Those streamers tell you about something that's invisible—the wind. And where those streamers fly smoothly, that's where you want your wind turbine.

Whether you've used online tools, found reliable data from nearby, or surveyed the wind resource at your site, you are now ready to estimate how much electricity a typical wind turbine will generate using the techniques explained in the next chapter.

11

Estimating Performance

Wie wind oogst, oogst de toe komst.
(Who harvests the wind, harvests the future.)

—Dutch proverb

BE FOREWARNED, ESTIMATING HOW MUCH ELECTRICITY A wind turbine might generate is not an exact science. After all, we are dealing with a natural resource that varies from one moment to the next and from one year to another. The characteristics of each site further complicate the picture. However, the task is simpler for large wind turbines than that for small turbines.

Large wind turbines interconnected with the utility feed an infinite sink. The utility system can absorb all the energy the turbine produces. When the wind turbine captures the wind's energy and converts it to electricity, the turbine can then deliver all of its generation to the grid. Thus, the electrical load on the wind turbine is predictable.

Small wind turbines, on the other hand, are notorious for defying expectations, especially in battery-charging systems. Even the experts have trouble getting it right.

The batteries of a small, battery-charging wind turbine may not always be able to take the energy when it's available. When the batteries become fully charged, the turbine must spill or dump the excess energy that's available. Some manufacturers of micro and mini wind turbines offer "dump loads," or circuits for using this excess generation and keeping the wind generator fully loaded. Often, small wind turbines simply spill the energy available in the wind when the batteries become fully charged. Thus, it's difficult to estimate the performance a consumer can expect from a battery-charging wind turbine because the turbine may be spilling the energy in the wind instead of delivering it to a useful load.

Nevertheless with these limitations in mind, there are three methods we can use to estimate performance. The first is a back-of-the-envelope technique using the swept area of the wind turbine. With this method you can quickly evaluate the potential output of any wind machine. You first find the average speed or power as discussed in the previous chapter. Then, you simply calculate the swept area of the wind turbine's rotor to estimate how much electricity it will produce. If you do it often enough, the technique will become so familiar that you'll be able to do it in your head. We'll then elaborate on this technique for fleets of large wind turbines in wind power plants. We'll do so by accounting for wind turbines designed for different IEC wind classes as part of Chapter 8, "Silent Wind Revolution."

The generator tells you very little about the size of a wind turbine.

ROTOR DIAMETER TELLS YOU FAR MORE.

The second method requires a wind speed distribution for your site and a power curve for each wind turbine you would like to evaluate. The third approach simply uses manufacturers' published estimates for typical wind regimes.

Swept Area Method

This is a back-of-the-envelope technique that wind professionals use to quickly size up a wind turbine—and often to size up the people promoting them. If a promoter says its wind turbine will deliver two or three times more energy than that calculated with this method, show him or her the door.

Previously, we learned that rotor diameter, or more correctly the rotor's swept area, is one of the critical factors in determining how much energy a wind turbine can capture. Think about it for a moment. What is it that captures the wind in a wind turbine? Is it the tower, the transmission, the generator? No, of course not; it's the spinning rotor. Yet this concept is difficult for many to grasp—even those working in the wind industry. The tower is important, as we have learned, and so is the generator. But they are not directly responsible for capturing the wind. Inevitably, many people look at the size of the generator first. Yet the generator tells you very little about the size of a wind turbine. Rotor diameter tells you far more.

The other critical factor is wind speed—or the equivalent power density. If we know how typical wind turbines work under ideal conditions, we can use rotor swept area and average annual wind speed to calculate the annual energy production (AEP). (In previous decades, AEP was known as annual energy output or AEO).

Let's assume we want to install a large wind turbine 80 meters (260 feet) in diameter as part of a local wind cooperative. This wind turbine intercepts

$$A = \pi r^2 = \pi(80/2)^2 = \pi \times (40)^2 = \pi \times 1{,}600 \sim 5{,}000 \text{ m}^2 \text{ of the wind stream.}$$

Let's say we plan to install this turbine at a site where the average annual hub-height wind speed is 7 m/s (15.7 mph) and that the distribution of wind speeds mirrors that of a Rayleigh distribution. From the previous chapter, we find that the power density under these conditions is

$$P/A = \tfrac{1}{2} \rho V^3 (EPF)$$

$$P/A = \tfrac{1}{2} \times 1.225 \times (7)^3 \times 1.91 \sim 400 \text{ W/m}^2 \text{ per year.}$$

It's now a simple matter to calculate the annual energy in the wind stream (E) in kilowatt-hours (kWh) striking our hypothetical wind turbine: the product of swept area (A) in m² times the power density (P/A) in W/m² times the hours (h) per year, where t represents time, then

$$E = P/A \times A \times t$$

$$E = \sim 5{,}000 \text{ m}^2 \times \sim 400 \text{ W/m}^2 \times 8{,}760 \text{ h} \sim 17{,}500{,}000 \text{ kWh/yr.}$$

But we don't know yet how much of that the wind turbine will pluck from the wind passing through the rotor because we can't capture all of the energy available.

The maximum aerodynamic efficiency of a wind turbine rotor—its coefficient of performance or C_p—was derived by the German physicist Albert Betz. The Betz limit (sometimes written as the Lanchester-Betz limit) is 59.3% (16/27) of the power in the wind available to the rotor. This is the performance of the rotor alone at a specific wind speed. The rotor is designed to operate near its peak performance over a narrow range of wind speeds, typically from 5 m/s (~ 11 mph) to 11 m/s (~ 25 mph). Outside this range, the rotor is less efficient at capturing the energy in the wind.

In practice, wind turbine rotors deliver less than the Betz limit. Optimally designed rotors can capture nearly 50% of the energy in the wind at selected instantaneous wind speeds. Usable energy is somewhat less because energy is lost in converting the kinetic energy of the rotor to electrical energy.

Well-designed drivetrains operate consistently above 90% efficiency. The efficiency of generators, on the other hand, varies significantly depending on how they are loaded. When running at their rated output, generator efficiency can also be above 90%. But wind turbines infrequently drive their generators at the rated output. Much of the time the generator is partially loaded, and its efficiency suffers as a result. Power conditioning on some wind turbines interconnected with the utility can also contribute to further losses.

Thus, the efficiency (η) of a wind turbine in converting the instantaneous energy in the wind varies with wind speed. More on this in a moment. What we are interested in here is the performance of the wind turbine for the entire year, so we have to account for the distribution of wind speeds over time.

When you put all this together, a large wind turbine can deliver about 30% of the overall annual energy available in the wind at a 7-m/s site with a Rayleigh distribution. This is what you can get out of the wind and actually put to use.

In practice, wind turbines capture from 10% to more than 30% of the annual energy in the wind, depending on the type of wind turbine and the wind regime where it is operating. Wind turbines are designed for use in specific markets under specific wind conditions. Outside these conditions, the wind turbine performs less optimally. That's not to say they're less cost effective, just that the overall system conversion efficiency may be less.

For the sake of simplicity, let's assume that our 80-meter-diameter wind turbine converts 30% of the energy in the wind to electricity. Therefore, the annual energy production is

$$AEP = E \times \eta = 17{,}500{,}000 \text{ kWh/yr} \times 30\% \approx 5{,}250{,}000 \text{ kWh/yr}$$

or about 5 million kWh per year. We'll make this calculation even simpler in a moment.

Small Wind Turbines

Small wind turbines are less efficient at capturing the energy in the wind than large wind turbines. Despite the hype about new airfoils, small wind turbines seldom deliver more than 30% of the energy in the wind over any significant period of time. Most fall short of that.

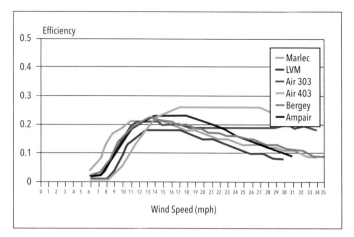

Figure 11-1. Efficiency of selected mini and micro wind turbines. Percentage of the power actually delivered to the batteries relative to the power in the wind at a specific wind speed. The data was derived from power curve measurements at the Wulf Test Field near Tehachapi, California, from 1998 to 2003. The tests included Marlec's Rutland 910F, LVM's 6F, Southwest Windpower's Air 303H and Air 403, Bergey Windpower's 850, and the Ampair 100. The wind turbines were installed to manufacturers' specifications, and the results are typical of what a consumer could expect. For four of the five turbines, peak efficiency of about 22% occurs around 13 mph (5.8 m/s). The Air series of turbines were the only models that used neodymium magnets. In three of the five designs, efficiency gradually declines as wind speed increases. This is typical of most wind turbines, both large and small.

Typical small turbines of the 1980s and 1990s captured less than 20% of the annual energy available in the wind. There are several reasons for this. One reason is that the airfoils on small wind turbines are inherently less efficient than those on large wind turbines, especially those on micro and mini wind turbines. Another reason is that most small turbines of the 1980s and 1990s used generators that were less efficient than those available today.

One manufacturer of small battery-charging wind turbines in the late 1990s crowed that they used "an airfoil so advanced it approached the theoretical limits of efficiency." Although that may have been true in a wind tunnel under ideal conditions, it was extremely unlikely to perform as well in the real world where those wind turbines had to operate. And as we've seen, this claim may refer to the rotor's performance within a very narrow range of wind speeds.

Measurements of small wind turbines at the Wulf Test Field from 1998 to 2003 indicated that of the micro and mini wind turbines tested most delivered less than 25% of the instantaneous power in the wind to the batteries in a stand-alone power system (see Figure 11-1. Efficiency of selected micro and mini wind turbines). They would deliver even less across the spectrum of wind speeds encountered throughout the year.

Table 11-1. Estimated Annual Energy Production for Micro & Mini Wind Turbines

Average Annual Wind Speed		Power Density	Total Efficiency	Average Annual Specific Yield
m/s	~mph	W/m²	η	kWh/m²/yr
4.0	9.0	75	0.190	120
4.5	10.1	107	0.195	180
5.0	11.2	146	0.200	260
5.5	12.3	195	0.190	320
6.0	13.4	253	0.180	400
6.5	14.6	321	0.175	490
7.0	15.7	401	0.170	600
7.5	16.8	494	0.160	690
8	17.9	599	0.150	790
8.5	19.0	718	0.145	910
9	20.2	853	0.140	1,050

Note: Gross generation for a single turbine at hub-height wind speed, based on published manufacturer's data from the 1990s. Actual performance will vary.

Incorporating these and other field measurements and summarizing performance estimates by the manufacturers themselves, we can estimate what typical micro and mini wind turbines of the period would deliver at a range of annual average wind speeds that fit a Rayleigh distribution (See Table 11-1. Estimated Annual Energy Production for Micro & Mini Wind Turbines).

Table 11-1 lists the total conversion efficiency (η) for each average annual wind speed and subsequently lists the *average annual specific yield* in kWh/m²/yr. The table is best applied to micro and mini wind turbines less than ~ 3 meters (~ 10 feet) in diameter. To use the table, find the average annual wind speed at hub height for your site on the left and then move horizontally to the average annual specific yield on the right. The overall conversion efficiency of the turbine you're evaluating could be more or less than assumed here.

Small wind turbines are typically designed to perform best in the low-wind regimes where people live, generally at sites with average wind speeds of 4 m/s to 5 m/s (9–11 mph) at hub height. In locales with higher average annual wind speeds, their performance drops off. This is a normal result of their basic design. Because of the cubic relationship of power with wind speed, there's so much energy available at windy sites that designers can afford to capture only a small part of it.

The objective of this derivation is that once you know the average annual wind speed you can find the average annual specific yield for your conditions in Table 11-1. Now all you need is the rotor swept area to estimate the AEP.

For example, if you're installing a small wind turbine at a site with an average annual wind speed of 6 m/s (13.4 mph), you should be able to expect that the wind turbine will deliver an average annual specific yield of 400 kWh/m²/yr. Let's assume your small wind turbine is 3.6 meters (~ 12 feet) in diameter. Wind turbines in this size class would have a standard power rating of about 2 kilowatts.

$$A = \pi r^2 = \pi \times (D/2)^2 = \pi \times (3.6/2)^2 \sim 10 \text{ m}^2$$

AEP = Average Annual Specific Yield x Swept Area

$$AEP = 400 \text{ kWh/m}^2/\text{yr} \times 10 \text{ m}^2 \sim 4{,}000 \text{ kWh/yr}$$

The beauty of this method is that you can do this quickly in your head, allowing you to quickly cut to the heart of estimating how much electricity a wind turbine will generate. You can always refine the estimate further as needed.

When we look at the product literature describing a wind turbine, the swept area isn't always apparent. What should always be obvious or clearly stated in the literature is rotor diameter. Given rotor diameter, you can calculate the area swept by the rotor. If the product literature doesn't provide the rotor diameter or swept area, the manufacturer is trying to hide something and it's best to move on to other products.

Depending upon the wind turbine, the annual specific yield in Table 11-1 may be optimistic—or too conservative. The table presents an average for each average annual wind speed. For this reason the assumed conversion efficiency is also presented. "Your performance may vary."

Extensive field trials of small wind turbines were undertaken in several countries during the 1990s. The results were disheartening. Many small wind turbines were underperforming their projections. Subsequently, several test sites, including Wulf Test Field, began measuring the performance of small wind turbines to provide feedback to manufacturers and regulators. This had the effect desired. Manufacturers revisited their designs to improve performance and at the

CALCULATING SWEPT AREA

If the swept area of a wind turbine isn't provided, you can easily calculate it with the formulas below. For a conventional wind turbine, use the formula for the area of a circle; for an H-rotor, use the formula for the area of a rectangle; and for Darrieus rotors, use the formula approximating the area of an ellipse.

Conventional rotor	$A = \pi r^2$
H-rotor	$A = DH$
Darrieus rotor	$A = 0.65\,DH$

where r is the radius of the rotor (one-half the diameter), D is the diameter, and H is the height of the blades on a vertical axis wind turbine.

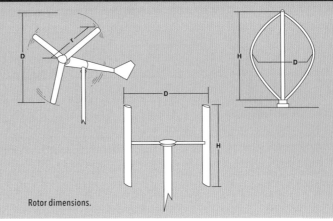

Rotor dimensions.

same time began an industry-wide testing and certification program to weed out the charlatans and underperformers.

Beginning in the early to mid-2000s, manufacturers of small wind turbines began bringing to market household-size wind turbines that used new generator materials. These new wind turbines used direct-drive, permanent-magnet generators like their predecessors. However, these new wind turbines were designed to use high power density neodymium magnets in contrast to the inexpensive but much less powerful ferrite magnets used previously.

Neodymium magnets and the redesign of the wind turbines to use them effectively have greatly enhanced the performance of small wind turbines, dramatically boosting the overall efficiency of the wind turbines (see Figure 11-2. Efficiency of selected household-size wind turbines).

And unlike before, we now have certified measurements by independent third parties of the performance of these new wind turbines, including rigorous estimates of annual energy production under standard conditions (sea-level air density and a Rayleigh wind speed distribution). Though this certification program does not apply to battery-charging wind turbines, it has greatly reduced the uncertainty surrounding the expected performance of household-size wind turbines (see Table 11-2. Estimated Annual Energy Production for Household-Size Wind Turbines).

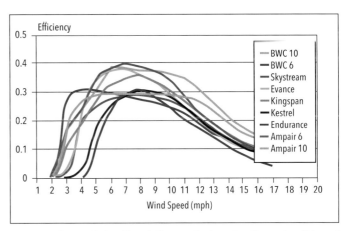

Figure 11-2. Efficiency of selected household-size wind turbines. Overall conversion efficiency of household-size wind turbines derived from SWCC and MCS certified power curves for direct-drive wind turbines using permanent-magnet generators with neodymium magnets.

Table 11-2. Estimated Annual Energy Production for Household-Size Wind Turbines

Average Annual Wind Speed		Power Density	Total Efficiency	Average Annual Specific Yield
m/s	~mph	W/m²	η	kWh/m²/yr
4.0	9.0	75	0.29	190
5.0	11.2	146	0.29	370
6.0	13.4	253	0.26	570
7.0	15.7	401	0.22	770
8	17.9	599	0.18	940
9	20.2	853	0.15	1,080

Note: Gross generation for a single turbine at hub-height wind speed. Derived from SWCC and MCS certified measured performance for direct-drive wind turbines using permanent-magnet generators with neodymium magnets. Actual performance will vary.

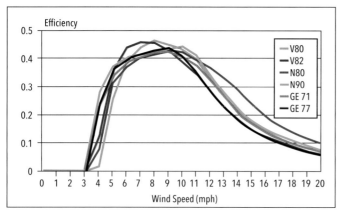

Figure 11-3. Efficiency of selected large wind turbines. Conversion efficiency of large wind turbines relative to the power in wind at the specified instantaneous wind speed. Efficiency was calculated from publicly available power curves assembled by the Idaho National Engineering Laboratory in 2007 of turbines commonly available in the early to mid-2000s. These include Vestas's V80 and V82, Nordex's N80 and N90, and General Electric's 1500 (71 meter) and 1500sl (77 meter) turbines. Note that large wind turbines are nearly twice as efficient as micro and mini wind turbines.

Table 11-3. Estimated Annual Energy Production for Large Wind Turbines

Average Annual Wind Speed		Power Density	Total Efficiency	Average Annual Specific Yield
m/s	~mph	W/m²	η	kWh/m²/yr
4.0	9.0	75	0.37	240
4.5	10.1	107	0.38	350
5.0	11.2	146	0.38	480
5.5	12.3	195	0.37	620
6.0	13.4	253	0.35	770
6.5	14.6	321	0.33	920
7.0	15.7	401	0.30	1,060
7.5	16.8	494	0.28	1,200
8	17.9	599	0.25	1,330
8.5	19.0	718	0.23	1,460
9	20.2	853	0.21	1,570

Note: Gross generation for a single turbine at hub-height wind speed, based on published manufacturer's data for selected models typical of the early to mid-2000s. This table does not take into account IEC classes. Actual performance may vary.

Compare the annual specific yield in Table 11-2 with that in Table 11-1 for an average annual wind speed of 6 m/s: 400 W/m² to 580 W/m². The new, high-performance household-size turbines are nearly 50% more efficient at capturing the energy in the wind than those wind turbines available in the 1990s represented in Table 11-1.

Where certified power curves and AEP estimates are unavailable, it's wise to use the conservative specific yields in Table 11-1. Some small wind turbines deliver even less than the values in Table 11-1. After all, the specific yields in Table 11-1 represent an average. Some small turbines perform better, some worse than the average.

Large Wind Turbines

Large wind turbines are considerably more efficient at extracting the energy in the wind at a specific instantaneous wind speed than small wind turbines (see Figure 11-3. Efficiency of selected large wind turbines). Large wind turbines are nearly twice as efficient as the micro and mini wind turbines of the 1990s and as much as one-third more efficient than the newer, high-performance household-size wind turbines now available (see Table 11-3. Estimated Annual Energy Production for Large Wind Turbines).

We can apply the same method used to estimate the annual energy production for large wind turbines as we did for small wind turbines. In the earlier example of our hypothetical 80-meter-diameter wind turbine, we assumed that the turbine could convert 30% of the energy in the wind throughout the year at an average annual wind speed of 7 m/s (15.7 mph). This is typical of most wind turbines of this size on the market in the early to mid-2000s as seen in Table 11-3.

Using this example, we can estimate the annual energy production from Table 11-3.

$$AEP = Average\ Annual\ Specific\ Yield \times Swept\ Area$$

$$AEP = 1{,}060\ kWh/m^2/yr \times \sim 5{,}000\ m^2 = 5{,}300{,}000\ kWh/yr.$$

This calculation can be even simpler for a quick estimate of what an 80-meter diameter turbine can do at this site. We know, for example, that at an average annual hub-height wind speed of 7 m/s, the typical specific yield is about 1,000 kWh/m²/yr, and we know that an 80-meter diameter wind turbine sweeps about 5,000 m². Therefore,

$$AEP \sim 1{,}000\ kWh/^2/yr \times \sim 5{,}000\ m^2 = \sim 5{,}000{,}000\ kWh/yr.$$

To reiterate, to quickly estimate the AEP of a large wind turbine, find the average annual specific yield in kWh/m²/yr for the conditions at the site and multiply the result by the rotor swept area.

Annual Yield by IEC Classes

As noted in Chapter 8 ("Silent Wind Revolution"), large wind turbines are designed for specific wind regimes. Table 11-3 is based on the average performance calculated from publicly available power curves that were collected by the Idaho National Engineering Laboratory in 2007. Some of these wind turbines were designed for IEC Class I, some for IEC Class II. At the time, few wind turbines were designed for IEC Class III wind regimes. Consequently, Table 11-3 is an oversimplification because it doesn't distinguish between the different classes of large wind turbines. The overall performance of IEC Class II and III wind turbines used only in wind regimes for which they are optimized is superior to IEC Class I turbines used in lower wind speed regimes.

For regions of low to moderate wind resources, the new IEC Class III turbines are revolutionary. The combination of large diameter rotors and relatively small generators gives these turbines greater specific area in m²/kW or lower specific capacity in W/m² than typical wind turbines of the past.

French engineer Bernard Chabot has extensively analyzed the projected performance of these new designs as well as wind turbines designed for windier sites. He's created a chart that maps performance in annual specific yield by IEC Class, specific area, and site average annual wind speed at hub height (see Figure 11-4. Annual specific yield by IEC Class).

Chabot's chart is a sophisticated elaboration on Table 11.3. Because the chart is intended to characterize the performance of a fleet of wind turbines in a wind farm, it assumes that 10% of gross generation is lost for various reasons.

To find the average annual specific yield in kWh/m²/yr, you need both the average annual wind speed at hub height and the specific area of the turbine you're considering. For example, let's consider a site with an average annual wind speed of 7 m/s and a wind turbine with a specific area of 3.1 m²/kW. For the 80-meter-diameter rotor used previously, the wind turbine must be designed to reach a rated power of 1,620 kW.

A fleet of wind turbines under these IEC Class III conditions should produce an annual yield of

SWEPT AREA RULES OF THUMB

There are some simple rules of thumb to remember when comparing the size of different wind turbines in terms of their potential power output. A micro turbine that sweeps 1 m² is roughly equivalent to a power rating of 200 watts. A mini wind turbine that sweeps 10 m² would be rated at about 2 kW. A household-size wind turbine that sweeps 100 m² would be rated from 25 kW to 40 kW. And a large wind turbine that sweeps 10,000 m² could be rated from 2 MW to 4 MW.

Swept Area, Rotor Diameter, and Nominal Power Rules of Thumb

Swept Area	Nominal	Nominal Rotor Diameter		Nominal Power Rating
m²	ft²	m	ft	kW
1	10	1.1	4	0.2
5	50	2.5	8	1
10	110	3.6	12	2
50	540	8	26	10-20
100	1,080	11	37	25-40
1,000	10,800	36	118	300-400
5,000	53,800	80	262	1,500-2,500
10,000	107,600	113	371	2,000-4,000

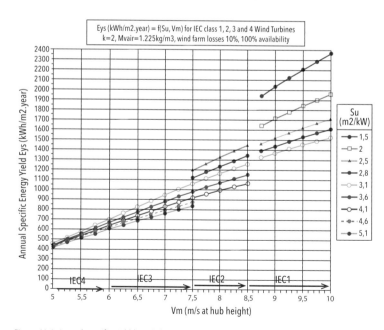

Figure 11-4. Annual specific yield by IEC class. Developed by French renewable energy analyst Bernard Chabot, this chart depicts the average annual specific yield (on the left axis) in kWh/m²/yr by average annual wind speed (on the horizontal axis) in m/s for different IEC classes and specific area in m²/kW. The chart assumes there are 10% losses per turbine in a large wind power plant. (Bernard Chabot)

POWER CURVE NOMENCLATURE

Start-up speed. The wind speed at which the rotor first begins to turn after being at rest. Once spinning, wind turbine rotors can coast to lower wind speeds than those necessary to start the rotor revolving.

Cut-in speed. The wind speed at which the generator first begins to produce power. If the wind turbine in a battery-charging system uses a permanent-magnet alternator, voltage is generated whenever the rotor is turning, but current doesn't begin to flow until the voltage exceeds that in the batteries. For wind turbines interconnected with the utility network, cut-in occurs when the generator is connected to the grid.

Rated Speed and Power. The wind speed at which the generator produces the advertised power. Though frequently used as a reference for the size of a wind turbine, rated speed and power have little utility. Most wind turbines will produce more than their rated power. Small wind turbines certified to an international performance standard are rated at a common 11 m/s (24.6 mph).

Peak power. The maximum power the wind generator is capable of producing. The peak power of many small wind turbines and most older medium-size wind turbines using stall control is substantially greater than the rated power.

Cut-out speed. The wind speed at which the wind generator stops producing power or activates some means of high wind speed protection. This can be accomplished by applying a brake or other mechanism for physically stopping the rotor or by feathering the blades. Most large wind turbines turn themselves off in winds above 25 m/s (56 mph) to reduce wear and tear and protect the turbine from damage. Small wind turbines using permanent-magnet alternators typically have no cut-out speed. They continue producing power in high winds. Most, however, do have a furling speed.

Furling speed. The wind speed at which a small wind turbine using a tail vane begins to fold or furl toward the tail. This reduces the area of the rotor intercepting the wind, protecting the wind turbine.

Wind speed bin. Though wind speed varies over a continuum, measurements used to develop power curves group power measurements into separate discrete registers or bins. For wind speeds in m/s, each bin is typically 0.5 m/s wide. The 5 m/s bin, for example, would represent winds from 4.75 to 5.25 m/s. For wind speeds in mph, each bin is typically 1 mph wide.

electricity supplied by wind in a region—climbs.

While the 80-meter-diameter wind turbine used in this example is hypothetical, there are wind turbines designed with a specific area of 3.1 m²/kW. One example is General Electric's 1500sl. This widely deployed wind turbine uses a 77-meter (250-foot) diameter rotor to drive a 1,500-kW generator.

Although the swept area technique only provides an approximation of what can be expected at a specific site, it is simple and straightforward. It also encompasses all the critical factors affecting how much energy a wind generator will convert to electricity. More importantly, frequent use of this technique reinforces the importance of rotor diameter—or swept area—in wind energy as opposed to generator capacity in gauging the relative size of wind turbines, both large and small.

Power Curve Method

Where you have access to the wind speed distribution for your site and the power curve for the wind turbine you're considering, you can calculate the annual energy production using the power curve. This is the method used by meteorologists when determining the potential generation from a large wind turbine in a commercial wind power plant. Essentially, you match the speed distribution with the power curve to find the number of hours per year the wind turbine will be generating a specific power.

First, a word of caution; the power curves proffered by some manufacturers of small wind turbines that have not been verified by an independent third party can best be characterized as informed guesswork. If the wind turbine's performance has not been certified, take small wind turbine power curves—and the energy

The power curves for large wind turbines have been independently measured by international testing laboratories because their customers— and the banks that finance their projects—demand them.

950 kWh/m²/yr. Each wind turbine in the group should then deliver net generation of

$$AEP = 950 \text{ kWh/m}^2/\text{yr} \times 5{,}000 \text{ m}^2$$
$$= \sim 4{,}800{,}000 \text{ kWh/yr.}$$

Such a wind turbine is desirable from the utility's perspective and from the perspective of public policy. It will deliver very high capacity factors, enabling the much more efficient use of the existing transmission system. Thus, it will greatly enhance the integration of large numbers of such turbines into the existing utility network. This becomes increasingly important as the penetration of wind energy—the percentage of calculations that result—with a good dose of skepticism. The situation is improving, but for three decades, there were no government agencies or independent testing laboratories ensuring the accuracy of published power curves for small wind turbines in North America. It was a jungle out there.

In contrast, the power curves for large wind turbines have been independently measured by international testing laboratories because their customers—and the banks that finance their projects—demand them. The efficiency calculations on a sample of large wind turbines in Figure 11-3 used power curves verified by European test fields—not rosy projections provided by manufacturers.

The Method of Bins

Power curves, like many other aspects of wind energy, are just approximations of what happens in a complex environment. The smooth curve in sales brochures belies a complexity that frequently confuses consumers. Often, the new owner of a small wind turbine will become disenchanted after watching the power being produced at various wind speeds.

In actual use, the instantaneous power from a wind turbine varies for a particular wind speed, depending on whether the rotor was coasting from a previous gust at the time the measurement was made (in which case power would be greater than average) or was coming up to speed and the anemometer registered the gust but the rotor did not (in which case power would be less than the norm). It's very difficult for an observer to simultaneously monitor both fluctuating power measurements and wind speed and make any sense of the resulting data. There's a great deal of scatter in power measurements (see Figure 11-5. Small wind turbine power curve scatter plot).

Because of the wide fluctuations in power and wind speed measurements, manufacturers and testing laboratories have agreed to average the measurements over periods of time sufficiently long for the averages to become repeatable from one test to the next. Professionally produced power curves result from extensive averaging using the method of bins. This technique sorts a series of power measurements into a corre-

AVOID AVERAGE SPEED CONFUSION

Newcomers to wind energy, in their zeal to estimate how much electricity they can generate with a particular wind turbine, sometimes confuse the power curve with graphs of annual energy production. Mike Bergey of Bergey Windpower has found that some customers erroneously apply the *average annual wind speed* at their site to the *wind speed bin* shown on the power curve. Unfortunately, this approach ignores the effect of the speed distribution on a wind turbine's production. If you want to use the average annual wind speed method, you must use tables or charts of the *annual energy production*. If you want to use the power curve method, you must use a wind speed frequency distribution, preferably a distribution for your specific site.

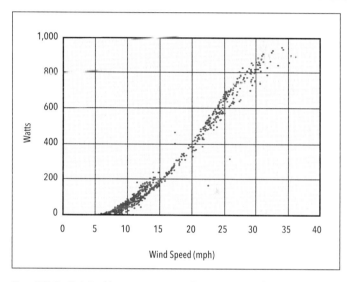

Figure 11-5. Small wind turbine power curve scatter plot. Measurements of power and hub height wind speed collected on an 8-foot (2.4-meter) diameter wind turbine at the Wulf Test Field during a two-week period. Data on wind speed and power was averaged every 10 minutes and recorded. Note that there is a bifurcation in the power curve in winds from 10 to 15 mph. This is typical of the kind of problem that can arise when measuring the power from wind turbines of any size. High winds at this site are consistently from the northwest. However, low winds are both from the northwest and also from the southeast, where they pass over a nearby row of trees. The trees reduce the wind speed seen by the anemometer but have little effect on the turbine. Thus the power curve shifts vertically, giving a misleading indication of low-wind performance.

sponding interval or register of wind speed. For example, a series of power measurements averaged in the wind speed interval from 9.5 to 10.5 m/s represents the average power produced in the 10-m/s bin. For large wind turbines, the international standard is to average a minimum of thirty 10-minute samples (five hours of measurements for each bin) to accurately characterize the power at any one wind speed. Small wind turbine manufacturers may sometimes average a minimum of thirty 1-minute samples to derive

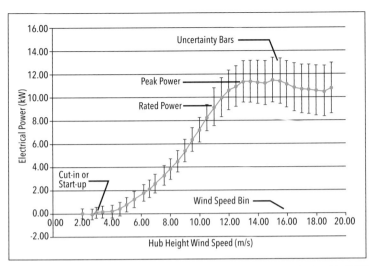

Figure 11-6. Sample power curve nomenclature. Certified measured power curve for Ampair's 10-kW model. Hub height wind speed in m/s bins on the horizontal axis. Power in kW on the vertical axis is the 10-minute average power the turbine will produce within each wind speed bin, not the maximum instantaneous power occasionally reached. The turbine cuts-in or starts producing power at 3 m/s (6.7 mph), reaches its rated power of 9.2 kW at the standard rated wind speed of 11 m/s (24.6 mph), and produces a peak power of 11.5 kW at 13 m/s (29 mph). Published power curves or accompanying technical reports should denote the reference height for the wind speed measurements (usually hub height), whether the data was verified by an independent third party, and what averaging period was used (instantaneous, 1 minute, or 10 minute). Most will also include "error" or "uncertainty" bars to indicate the uncertainty associated with the measurements at a specific wind speed. (Adapted from Ampair)

Table 11-4. Vestas V80 Power Curve and Annual Energy Production for a 7 m/s Site

Diameter: 80m
Swept area: 5,027m^2
Rated power: 1,800 kW
Specific power: 358 W/m^2
Specific area: 2.8 m^2/kW
Specific yield: 1,077 kWh/m^2/yr
Average wind speed: 7 m/s

Wind Speed		Power (kW)	Raleigh Frequency of Occurrence	hrs/yr	Energy (kWh/yr)
m/s	mph				
0	0.0	0	0.0000	0	0
1	2.2	0	0.0315	276	0
2	4.5	0	0.0601	527	0
3	6.7	0	0.0833	729	0
4	9.0	3	0.0992	869	2,608
5	11.2	99	0.1074	941	93,112
6	13.4	260	0.1080	946	246,010
7	15.7	465	0.1023	896	416,759
8	17.9	735	0.0919	805	591,963
9	20.2	1015	0.0788	690	700,305
10	22.4	1345	0.0645	565	760,396
11	24.6	1639	0.0507	444	727,957
12	26.9	1775	0.0383	335	594,851
13	29.1	1797	0.0278	243	437,011
14	31.4	1802	0.0194	170	306,150
15	33.6	1802	0.0131	114	206,075
16	35.8	1802	0.0085	74	133,740
17	38.1	1802	0.0053	46	83,728
18	40.3	1802	0.0032	28	50,589
19	42.6	1802	0.0019	16	29,511
20	44.8	1802	0.0011	9	16,625
21	47.0	1802	0.0006	5	9,048
22	49.3	1802	0.0003	3	4,758
23	51.5	1802	0.0002	1	2,418
24	53.8	1800	0.0001	1	1,187
25	56.0	1800	0.0000	0	564
			Total	8,736	5,415,365

Source: 2007 data from Idaho National Engineering Laboratory.

their power curves (30 minutes of measurements for each bin).

Power curves published by reputable manufacturers depict the average power the turbine will produce within each wind speed interval. These manufacturers will only publish "certified" power curves that have been measured to international standards. Some, shall we politely say, less-than-reputable promoters of small wind turbines still publish power curves produced from measurements of maximum instantaneous power, not the average power. Power curves derived in such a manner are misleading, and seldom will the consumer see the same results as those published by the manufacturer.

Power curves certified by independent third party laboratories will include a measure of the variability or "scatter" in the data. This uncertainty is presented either in tabular form or in a graph depicting "error" bars with the mean value used to plot the power curve (see Figure 11-6. Sample power curve nomenclature).

Large Turbine Power Curve
Let's use a simplified power curve for a Vestas V80 to find the power the turbine will produce at wind speeds from cut-in through cut-out when the wind turbine feathers the blades (Table 11.4 Vestas V80 Power Curve and Annual Energy Production for a 7 m/s Site). We'll use a site with a modest wind resource typical of large land areas in North America and Europe where the average annual wind speed is 7 m/s (15.7 mph) at

hub height and a Rayleigh distribution of wind speeds is likely to be found.

The analysis here has been simplified. There is no measure of uncertainty surrounding the power measured within each wind speed bin, nor is there any measure of the probability that the wind speeds will precisely follow the Rayleigh distribution or that the average wind speed at the site is 7 m/s. Banks and investors in a machine of this size will require a far more sophisticated analysis, and for that a professional meteorologist is required. Nevertheless, this example illustrates the technique used.

The V80 is an 80-meter (~ 260-foot) diameter wind turbine rated at 1,800 kW (1.8 MW). The rotor sweeps nearly 5,027 m², giving the V80 a specific power of 358 W/m² or a specific area of 2.8 m²/kW. The V80 was a fairly typical large wind turbine of the early to mid-2000s.

The turbine cuts-in or starts generating power at a wind speed of 4 m/s (9 mph), though it will coast or "tick over" at lower wind speeds. It reaches its rated power at 13 m/s (29 mph) and cuts-out by feathering the blades at 25 m/s (56 mph).

From the annual distribution of wind speeds, we can calculate the length of time that the winds occur at each speed. For example, the wind will occur at a wind speed of 11 m/s (24.6 mph) 5.07% of the year, or about 440 hours per year. The V80 will produce about 1,600 kW at this speed. The amount of electricity the V80 will generate within this wind speed bin is the product of the power and the hours per year that this power is delivered.

~ 1,600 kW x ~ 440 hours/yr = ~ 700,000 kWh/yr

To calculate the total amount of electricity that the turbine will produce during the year, simply sum the kWh produced within each wind speed bin from 3 m/s to 25 m/s. We find that the V80 will generate about 5.4 million kWh per year in this wind regime across the entire speed range.

Compare the AEP from the power curve calculations for this turbine with that using the swept-area method described earlier. The annual specific yield for the V80 at a 7 m/s site is 1,077 kWh/m²/yr. This is quite similar to the annual average specific yield in Table 11-3 for an annual average wind speed of 7 m/s.

Manufacturers' Estimates

Most manufacturers provide estimates of the amount of energy they expect their turbines will capture under standard conditions: hub-height wind speeds, Rayleigh distribution (Weibull Distribution with a shape factor of 2), air density at sea level (1.225 kg/m³ and 15°C.) The format varies. Some companies provide a chart of AEP at various annual average wind speeds (see Figure 11-7. Chart of annual energy production [AEP]). Others provide the same data in tabular form (see Table 11-5. Table of Estimated AEP for Bergey Windpower's Excel 6).

For example, Bergey Windpower's Excel 6 is a 6.2-meter (20-foot) diameter wind turbine that sweeps 30.2 m² of the wind stream. Searching

Table 11-5. Table of Estimated AEP for Bergey Windpower's Excel 6

Hub Height Annual Average Wind Speed (m/s)	AEP Measured (kWh)	Standard Uncertainty in AEP (kWh)	AEP Extrapolated (kWh)
4	5,522	721	5,522
5	9,919	1,034	9,919
6	14,940	1,287	14,967
7	19,777	1,445	19,944
8	23,749	1,516	24,298
9	26,523	1,521	27,740
10	28,096	1,482	30,200
11	28,657	1,417	31,745

Corrected to sea level air density of 1.225 kg/m². Source: SWCC Summary Report.

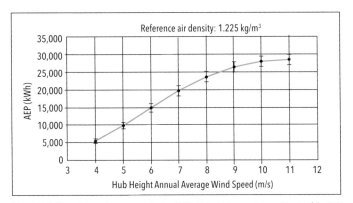

Figure 11-7. Chart of annual energy production (AEP). This chart presents an estimate of the AEP from Bergey Windpower's Excel 6, a small wind turbine with a rotor 6.2 meters (~ 20 feet) in diameter relative to annual average wind speed at hub height. The estimates assume a Rayleigh speed distribution (Weibull shape factor of 2). The same data is also frequently presented in tabular form. Note the error or uncertainty bars. "Your performance may vary." (Bergey Windpower)

WEB-BASED CALCULATORS OF AEP

There are quite a few wind power calculators on the web that will estimate the amount of energy a wind turbine will generate. They are of varying quality. Most operate in the background so you can't look under the hood and see what they're doing. As in most things, it's buyer, or in this case user, beware.

If you're going to use a web-based calculator, what you want is one with an up-to-date library of wind turbines specifications, including the all-important power curve. Some of the most useful calculators are not up to date, and the results from some of the others are spotty to say the least.

Most manufacturers of large wind turbines no longer publish tables of their power curves. They do provide the data to meteorologists but only under confidentiality agreements. In most cases this won't pose a problem. Those wanting to use large wind turbines will nearly always need the services of a meteorologist.

Fraunhofer Institute for Wind Energy and Energy System's Small Wind Turbine Yield Estimator (in English) offers estimates that are slightly more conservative than those in Table 11-1. The estimator also is adaptable, allowing you to key in the power curve for a specific small wind turbine if you have the data. See http://windmonitor.iwes.fraunhofer.de/wind/download/SWT_Yield_Estimator_Eng.xls.

online, we can find a certified estimate of how much the turbine will produce. Despite a minor degree of uncertainty, the chart shows that the turbine's AEP at an annual average wind speed of 6 m/s (13.4 mph) is about 15,000 kWh per year.

We can find similar data in Table 11-4. For an annual average wind speed of 6 m/s, the Bergey Excel 6 should produce 14,967 kWh per year.

Tabular or graphical estimates of AEP are reliable only when they have been certified by an independent testing laboratory. Reputable manufacturers will clearly state this in their product literature. Because such testing is time-consuming and expensive, fly-by-night companies with questionable products won't bother with certification.

Wind Power Plant Losses

It's fairly straightforward to estimate how much any individual wind turbine will produce using the techniques described in this chapter. However, there are numerous losses not accounted for by these simple methods, especially when groups of turbines are clustered together. These include losses for availability, electrical resistance, and array interference. Together, these can be significant.

No wind turbine operates 100% of the time. All wind turbines must be periodically inspected, if not maintained. When a wind turbine is stopped for maintenance, it's no longer "available" for operation, the wind industry's term of art for how much of the time the machine is in service.

Most large turbines are available for operation more than 98% of the time. That is, the wind turbine is stopped for reasons other than a lack of wind less than one week of the year. High availability became such an expected part of wind turbine operations by the mid-1990s that trade publications found newsworthy any hint that a company's availability had fallen to less than 97%.

Electrical losses in large projects with long cable runs can amount to 3%.

More substantial losses can accrue from the interference of one wind turbine with the next in a large array. When wind turbines are clustered together in a wind farm, machines downwind will produce less than those upwind. Interference can cut production 5% to 10%. As seen in some California wind farms, poorly designed arrays, where the turbines are too close together, losses can be even greater.

Accounting for availability, electrical losses, and array interference, actual electricity delivered may be 85% to 90% of that derived from a simple estimate of what a single wind turbine can produce. As seen in Figure 11-4, Chabot estimates overall losses of 10% are a good rule of thumb.

Losses in battery-charging systems can be substantial as well. Batteries that are fully charged may not be able to take more energy when, in a good stiff wind, the turbine is churning it out. Moreover, some of the energy that is eventually stored in the batteries is lost due to the inherent inefficiency of battery storage. Additional losses are incurred when using an inverter to convert the DC stored in the batteries to run AC appliances.

Estimating Fleet Performance

For public policy purposes, regulators, politicians, and industry analysts needed an even simpler approach for how much electricity a fleet of wind turbines would produce. In the 1980s and early 1990s, this was relatively simple. A good rule of thumb then was based on fleets with thousands of wind turbines operating in California. They were producing on average about 2,000 kWh per kilowatt of installed generating capacity per year. Since the industry reports how many megawatts of capacity are installed, it's a simple matter to calculate how much electricity the turbines might produce.

For example, if there were 1,000 MW of wind capacity installed in a region, then the wind turbines should be capable of producing 2 TWh per year.

$$AEP = 2,000 \text{ kWh/kW/yr} \times 1,000 \text{ MW} \times 1,000 \text{ kW/MW} = 2,000,000,000 \text{ kWh/yr}$$

This will give a crude estimate—it will put you in the ballpark. As we've seen, performance is largely determined by the wind resource. Moreover, generator capacity is not a good way to indicate how much a wind turbine will produce. Even in a relatively windy country such as Denmark, performance by generator size class varies widely (see Table 11-6. Danish Wind Turbine Yield for 2011).

In 2011, for example, all the 100-kW wind turbines produced about 2,300 kWh/kW/yr, whereas the 2-MW class turbines were generating more than 3,000 kWh/kW/yr. One reason for this discrepancy is that the 2-MW class turbines are installed on much taller towers than the smaller 100-kW turbines. Consequently, they will perform much better than the turbines on shorter towers. However, another reason for the discrepancy is the introduction of wind turbines with larger rotors relative to their generator size. This will result in more kWh per kW of installed capacity.

Nevertheless, Danish wind turbines monitored by the Danish wind turbine owners association produced about 2,200 kWh/kW of installed capacity in 2011. This is comparable to past performance.

Like hydroelectricity that varies from one year to the next depending upon the amount of rainfall, so too does the average yield of wind turbines vary from one year to the next. Although the 2,000 kWh/kW/yr is a good rule of thumb for the Danish fleet in the late 1990s and early 2000s, the yield from one year to the next varies from 2,000 to 2,500 kWh/kW/yr (see Table 11-7. Fleet-Wide Typical Yields).

In windy countries, such as Ireland, average annual yields of 2,600 kWh/kW/yr have been reached. As the new IEC Class III wind turbines become an ever-larger part of the existing mix of wind turbines, average annual yields will gradually increase as the turbines generate more electricity per kW of installed capacity than older turbines installed in the 1980s and 1990s. For example, in 2014 the average yield in the United States approached 2,800 kWh/kW/yr. This is a revolutionary development that will make wind energy easier to integrate with existing electrical grids.

Putting It All Together

Let's say there's a group of 10 conservative farmers in central Indiana. They've seen hundreds of

| Table 11-6. Danish Wind Turbine Yield for 2011 ||
Size (kW)	Yield (kWh/kW/yr)
15	1,376
55	943
100	2,321
250	1,534
500	2,039
1,000	2,075
2,000	3,241
3,000	2,779
Average	2,196
Selected data from Naturlig Energi, September 2012, page 25.	

Table 11-7. Fleet-Wide Typical Yields (kWh/kW/yr)		
	Range	
	Low	High
Germany	1,500	1,800
France	1,400	1,900
Spain	1,700	2,100
Denmark	2,000	2,500
Britain	1,700	2,600
USA	1,800	2,800
Ireland	2,000	2,600

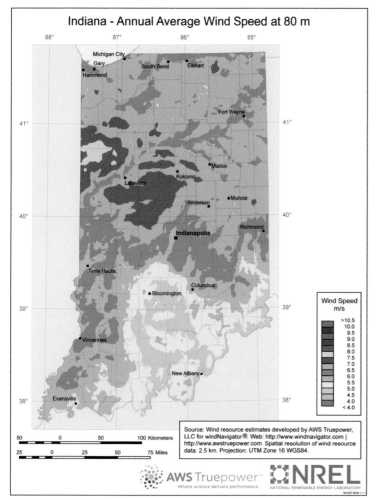

Figure 11-8. Indiana wind resource map. Estimated average annual wind speeds at 80 meters (260 feet) above the ground derived from wind resource data and atmospheric modeling by the National Renewable Energy Laboratory. Once overlooked as a region with little wind, such sophisticated studies have revealed that the state has an abundant wind resource, notably in west central Indiana. (National Renewable Energy Laboratory)

The farmers form a group modeled after the cooperatives that buy their milk. They begin investigating how much wind they have and want to gauge how much electricity they can produce among them. They figure they can install 10 turbines altogether on their farms.

Their farms lie on a low geographic upland between Anderson and Kokomo, Indiana. They look online for wind resource data and find a digital map produced by the National Renewable Energy Laboratory (see Figure 11-8. Indiana wind resource map). The map suggests that there's an average annual wind speed of 7 m/s at 80 meters (260 feet) above the ground.

There are already hundreds of GE wind turbines in the area, so they calculate the specific area of GE's 1500sl. The result is 3.1 m²/kW. They then compare the specific area of this wind turbine with Figure 11-4 for a site with a 7-m/s average annual wind speed and find that on average a group of turbines will generate about 950 kWh/m²/yr.

GE's 1500sl uses a rotor 77 meters in diameter. It intercepts nearly 4,700 m² of the wind stream. Therefore,

$$AEP (10) = 950 \text{ kWh/m}^2/\text{yr} \times 4,600 \text{ m}^2 \times 10 = \sim 45,000,000 \text{ kWh/yr}$$

their small cluster of ten turbines will produce about 45 million kWh per year after intraproject losses are accounted for.

If the farmers were paid $0.10 per kWh, the project would earn $4.5 million per year. Once the turbines were paid for, the project would earn each farmer $450,000 per year. That's enough to intrigue them. They pool some initial capital and begin looking for a professional meteorologist.

Whether a project such as this makes economic sense to the farmers is the subject of a subsequent chapter.

wind turbines installed in the region by large multinational utility companies. The shrewd farmers figure that if wind energy is profitable for a utility company in their area, the turbines could prove profitable for them as well. After all, they own the land—and the wind.

12

Off-the-Grid Power Systems

Go to the end of the road and honk.
(Instructions for finding accommodations in
Chile's Torres del Paine National Park, Patagonia.)

PRIOR TO THE DEVELOPMENT OF INTERCONNECTED WIND turbines, wind generators had historically been used for powering remote sites where utility power was nonexistent (see Figure 12-1. Off-the-grid wind systems). These *home light plants* used small wind turbines and banks of batteries sized to carry the household through winter winds and summer calms. Occasionally, the dealer would throw a backup generator into the mix to charge the batteries during extended calms. The high cost, poor reliability, and maintenance requirements of these early systems discouraged all but the hardiest souls from living beyond the end of the utility's lines.

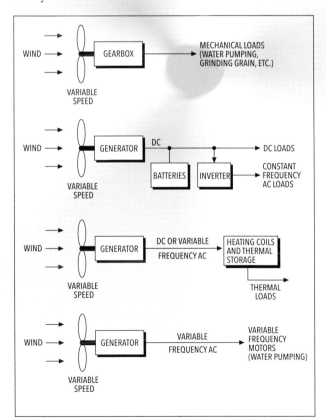

Figure 12-1. Off-the-grid wind systems. Several techniques are available for using wind machines in stand-alone applications. Historically, wind turbines have been used to charge batteries (wind chargers) in remote power systems, to pump water mechanically (farm windmills), or to grind grain (European windmills). Today, wind turbines can also be used to drive AC motors directly in specialized pumping applications and to generate heat at remote sites.

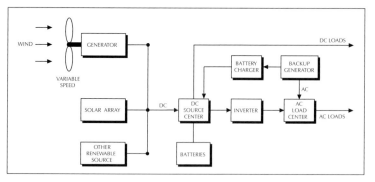

Figure 12-2. Hybrid stand-alone power system. With advances in wind and solar energy, remote power systems are no longer dependent on any single technology. The addition of a backup generator provides even greater flexibility in sizing a system's components.

That's no longer true. Home power systems have become mainstream. Today's improved small wind turbines, new inverters, the widespread availability of low-power appliances, and low-cost photovoltaic panels have revolutionized stand-alone power systems.

Until photovoltaic's entry into the market, remote power systems were solely dependent on small wind turbines (in some special cases, small hydro systems) and backup generators. The modularity of photovoltaics (PVs) has transformed the remote power system market by enabling homeowners to more closely tailor power systems to their needs—and their budgets.

Despite wind's advantages, wind turbines are less modular than PVs. Although the output of some micro turbines is no more than that of most PV modules (50 to 200 watts), each additional turbine requires a separate tower and controls. When scaling up the output from a remote power system, it's easier to add more modules to a PV array than it is to add more wind turbines.

Wind is also far more site-specific than solar. It's safe to assume that nearly everywhere on earth the sun will rise and set every day. Not so with wind. Wind follows daily and seasonal patterns that are less predictable. In the midlatitudes of the Northern Hemisphere, the winds are strongest in the winter and spring and weakest during the summer. Fortunately, this pattern happens to coincide with the attributes of solar energy. Winds are generally strongest when the sun's rays are weakest, and winds are weakest when solar radiation reaches its peak. For this reason, wind and solar are ideally suited for hybrid systems that capitalize on the advantages offered by each technology.

Hybrids

With advances in solar and wind technology, it just doesn't make sense today to design an off-the-grid system using only wind or only solar. Hybrids offer greater reliability than either technology alone because the remote power system isn't dependent on any one source (see Figure 12-2. Hybrid stand-alone power system). For example, on an overcast winter day outside Pittsburgh, Pennsylvania, when PV generation is low, there's more than likely sufficient wind to make up for the loss of solar-generated electricity.

Wind and solar hybrids also permit the use of smaller, less costly components than would otherwise be needed if the system depended on only one power source. This can substantially lower the cost of a remote power system. In a hybrid system, the designer need not size the components for worst-case conditions by specifying a larger wind turbine and battery bank than necessary.

Hybrids often include a fossil-fueled backup generator for similar reasons. In effect, stand-alone systems substitute the fuel and the maintenance of the backup generator for a larger wind turbine or more solar panels. Depending upon the size of the backup generator and the power consumption at the time, the generator can top up discharged batteries and meet loads not being met by the combined wind and solar generation.

Despite the improving cost effectiveness of PVs and small wind turbines, the initial cost of a hybrid system remains high. To keep costs down, it behooves users to reduce demand as much as possible. Fortunately, the development of compact fluorescent and increasingly LED lights as well as energy-efficient appliances now makes this possible with little sacrifice. Today's energy-

IT JUST DOESN'T MAKE SENSE TODAY TO DESIGN AN OFF-THE-GRID SYSTEM TO USE ONLY WIND OR ONLY SOLAR.

TALE OF TWO CITIES

At many midlatitude sites, wind and solar are complementary resources. Wind generation peaks during the winter when solar insolation is at a minimum. Similarly, in the summer a solar array will shine while a wind turbine may sit in the doldrums. Amarillo, Texas, and Pittsburgh, Pennsylvania, illustrate this relationship. No two cities could be more different: Pittsburgh's an old steel town; Amarillo's an old cow town. Their wind resources are equally diverse. Amarillo's windy, Pittsburgh's not. Even so, a wind and solar hybrid power system can provide usable amounts of energy in either location.

Consider a hypothetical hybrid sized so that both the wind and solar systems have approximately the same collector area (see Hybrid wind and solar generation). You wouldn't size a hybrid system this way; this is simply a means to compare the potential contribution of each resource. In this example, six panels of 75-W photovoltaic modules are equivalent to one wind turbine with a rotor diameter of 2.1 meters (7 feet). While the solar resource is 40% greater on the high plains of Texas than in the humid eastern United States, the wind resource is 180% more productive. Sizing an off-the-grid power system must account for such large differences. The Amarillo data illustrates that at a windy site, a wind turbine can contribute far more energy to a hybrid system than a similarly sized PV array. At a low wind site, such as Pittsburgh, wind and PV are equally productive.

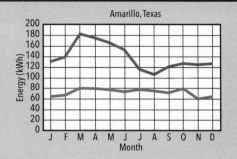

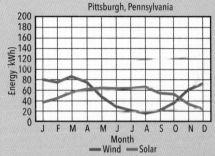

Hybrid wind and solar generation. Wind and solar resource for (a) Amarillo, Texas, and (b) Pittsburgh, Pennsylvania.

efficient appliances permit homeowners to meet their energy needs with smaller, less expensive power systems than were once necessary.

Reducing Demand

The first place to start is by reducing demand. Most North Americans can easily halve their electricity consumption. If one company's advertisement for down comforters "to take the chill off an air-conditioned room" doesn't strike you as absurd, then you need to first examine how much energy you are using before you commit to generating your own electricity. Reducing your consumption by conserving and increasing efficiency improves the services that a renewable power system can provide, by stretching each kilowatt-hour to do as much work as possible.

To reduce demand, find out what you're consuming now by performing an energy audit of your lifestyle. Knowing how, where, and when you use energy is even more important for an off-the-grid system than for an interconnected wind turbine. Determine what appliances you plan to use at your remote site, and estimate how much electricity they will consume (see Table 12-1. Residential Energy Consumption).

Conserve as much energy as possible. It's always cheaper to save energy than to generate it with a hybrid power system. In other words, the return on investment for conserving energy is higher than that for producing it in an off-the-grid power system (see Table 12-2. Return on Investment of Conservation Measures in an Off-the-Grid Power System). To maximize the value of your renewable power system and minimize its cost, carefully pare your electricity consumption to the minimum needed for the services you require.

Turn off all unneeded loads. This should be obvious, but like the example of grabbing a quilt because the air-conditioning has made the room too cold, it isn't. The National Renewable Energy Laboratory found that at one hybrid system in Chile, villagers left lights on 24 hours per day—despite NREL's plea to turn them off!

Decide if there are any electric appliances, such as electric hot-water heaters or electric stoves that

Table 12-1. Residential Energy Consumption		
	Therms/hour	kWh/hour
Heating		
Small gas furnace	0.6	
Large gas furnace	1	
Space heater		1.5
Baseboard heater		3
Electric furnace		10
Heat pump		3–5
Air Conditioning		
110 Volt window unit		1.5
220 Volt window unit		2.6
Central		4.5
Portable fan		0.2
Water Heating		
Electric		300–400
Gas	20–30	
Heat pump		175–225
Refrigeration		
16 cu. ft. frost free refrigerator		100–150
20 cu. ft. frost free refrigerator		115–180
10 cu. ft. manual defrost refrig.		35–60
15 cu. ft. frost free freezer		70–150
Lighting		
General		50–200
Laundry		
Electric clothes dryer		5/load
Gas clothes dryer	22/load	.5/load

Washing machine
  Cold
  Warm wash, cold rinse
  Hot wash, warm rinse
	
	.11/load
	.33/load
	
		.25/load
		.25/load
		.25/load

Appliances		
Stereo		0.03
Color TV		0.23
B&W TV		0.07
Vacuum cleaner		0.75
Microwave/5 minutes		0.1
Toaster/use		0.08
Toaster oven		0.5

Electric range
  Oven
  Surface
  Cleaning/use
		
		1.33
		1.25
		6

Gas range		

Table 12-2. Return on Investment of Conservation Measures in an Off-the-Grid Power System	
	ROI (%)
Compact flourescent lighting	100–140
New refrigerator	10–30
Replace PC with laptop	10–20
Solar photovoltaics	3–12
Source: Rahus Institute	

can be switched to gas or other fuels. It makes no sense to squander your hard-earned electricity on inefficient appliances or on uses where electricity isn't well suited. Heating is one of them. Heating with gas, oil, propane, wood, or better yet the sun is far more economical at a remote site than heating with electricity.

Though cooking consumes little energy overall, electric stoves have high peak power demands that will affect the size of inverters and other hybrid components. Cook with gas or propane, or use a microwave oven instead.

Pacific Gas and Electric (PG&E) found in a study of off-the-grid systems that most remote generation is used for lighting, refrigeration, and water pumping. (Remote sites are seldom served by municipal water sources.) Lighting is the easiest to tackle. Compact fluorescent lamps and the advent of LED lighting can reduce lighting demand significantly. Task lighting (lighting only those areas where light is needed), daylighting, and simply turning off lights when they're not needed can cut lighting consumption two-thirds.

Similar savings can be achieved with refrigeration. Modern refrigerators use as little as 300 kilowatt-hours per year, a fraction of what they used in the 1970s. Sunfrost refrigerators, the efficiency champions, use even less. Generally, if your refrigerator is more than 10 years old, it should be replaced; and whatever you do, don't put that old refrigerator in the garage. Figuratively, drive a stake through its heart. (Don't release the Freon to the atmosphere. Dispose of the refrigerator responsibly.)

Depending on the size of the house and on the climate, air-conditioning can double the consumption of an otherwise energy-efficient home. If you must have air-conditioning, ask yourself whether an evaporative cooler will suffice. Swamp coolers use far less electricity and work well in arid climates, such as the southwestern United States.

The key is to remain flexible. Sacrificing the lifestyle you desire isn't necessary, but some modification of behavior often proves beneficial for optimizing the performance of a hybrid power system. For example, cutting back on energy-intensive discretionary loads on days when the power supply is reduced extends battery life and leaves a little extra in storage available should you need it for more important loads like pumping water. Not unlike our ancestors, learn to synchronize your behavior with the weather. Do the laundry when it's windy or on a bright sunny day. In this way, you take full advantage of the fuel when it's available.

Turning off unneeded appliances isn't much of a burden for those who are energy conscious; it's second nature. But for those going "cold turkey" from a highly consumptive lifestyle where energy's undervalued, it can be a rude awakening. In such cases, it might be wise to gradually reduce your consumption until you're ready to make the transition to producing your own power.

The average North American household should be able to reduce its consumption to about 3,600 kilowatt-hours per year, or about 10 kilowatt-hours per day. This isn't spartan living. Most Europeans live comfortably on this amount or less. How much you're able to reduce your consumption will determine not only what size system you need, but also whether you should wire for DC or AC and at what voltage you should operate your power system.

AC and DC Systems

All stand-alone power systems produce and store direct current (DC). Photovoltaic (PV) arrays produce DC directly. Most wind machines produce alternating current (AC), which must then be rectified to DC, as in an automotive alternator. Direct current is then stored in batteries until needed. The exception to this scenario is the backup generator, which can provide constant-frequency AC for AC loads while also recharging the batteries through a rectifier.

There are three sides to the stand-alone power system: generation, storage, and loads. Since most of the loads will be supplied by the batteries most of the time, the choice becomes whether

CUTTING CONSUMPTION

When California's power crisis struck in 2001, Nancy Nies, my wife, and I carefully examined our domestic consumption. We found that we were already at the *baseline* for our climatic zone: the sunny San Joaquin Valley. Baseline is California's term for what energy planners calculate is the average electricity consumption in a particular region. At the height of the crisis, many wealthy Californians charged that no one could actually live on the baseline. People can. We did. But we found we could do even more–without hardship.

We looked at lighting, refrigeration, laundry, and cooling. We had been using fluorescent task lighting and notebook computers for several years–notebook computers used one-tenth as much electricity as desktop models then available.

We replaced the few remaining incandescent lamps with compact fluorescents. Then we replaced our refrigerator. This alone cut our consumption 600 kWh per year. We also dusted off our "solar clothes dryer" and began using the clothesline instead of the electric dryer.

It's hot in Bakersfield, California, during the summer–very hot. It's not uncommon for temperatures to exceed 110 F° (44 C°). Consequently, our peak consumption occurs in the summer, due to air-conditioning.

During the cooling season, we turned up our thermostat two degrees and found we could live comfortably with the addition of a few strategically placed fans. We also opened our windows at night to cool the interior, and closed them as soon as outside temperatures began rising.

By living more consciously, we cut our consumption 40% to 50%. Our consumption of about 3,000 kWh per year is half that of the typical California household.

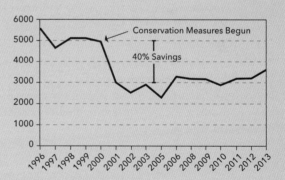

Electricity consumption Gipe-Nies household. The Gipe-Nies household was able to cut their consumption 40% in 2001 with simple conservation measures. Annual consumption fell from an average of 5,100 kWh/yr (slightly below California's base line of 5,200 kWh/yr) to about 3,000 kWh/yr. The years 2004 and 2007 are not included because the house was not occupied.

to feed the loads with DC directly or with AC through an inverter.

For any major off-the-grid application, such as a full-time residence, wire for AC and wire to local building codes. Wiring to code simplifies construction because most electrical fittings readily available are designed only for AC. Wiring

Figure 12-3. On the road and off the grid. Low-voltage DC systems are ideal for motor homes and small vacation cabins. Here two micro turbines charge a motor home's batteries in windy Tehachapi, California. (Nancy Nies)

number of appliances available for the recreational vehicle market.

As demand increases, so too does the optimum system voltage. A rule of thumb is that systems requiring less than 2 kilowatt-hours per day, such as vacation cabins, should stick with 12 volts; those using up to 6 kilowatt-hours per day should go with 24 volts; and those using more than 6 kilowatt-hours per day should opt for 48 volts. Systems that use 24 and 48 volts typically use inverters for powering conventional AC appliances.

Except for the smallest applications, wind turbines, PV arrays, and batteries should be sized for 24 to 48 volts. Long cable runs or heavy consumption may require a 120-volt system. But the large number of batteries needed for 120 volts can be prohibitively expensive except for those systems designed to meet consumption approaching 20 kilowatt-hours per day. High-voltage systems do permit greater flexibility in siting your renewable power sources than do either 24-volt or 48-volt systems. If you need to move your solar array into a more exposed position or move your wind turbine to a hilltop, the 120-volt system may be your only choice to keep resistance losses in the cable, and the cost of the conductors, to a minimum.

Sizing

At extremely small loads, for example a few lights and a radio, a 12-volt PV system comprising one or two 50-watt panels that power DC appliances directly makes economic sense. For such applications, PVs are easier to work with and less expensive than wind. But as loads increase above a few hundred watts, wind and solar hybrids become increasingly more attractive.

Like wind, the output from a solar array isn't constant over time, season, or from one region to another. PV systems in the southwestern United States and Plains states produce at their rated output about five hours per day. For example, a 200-watt system will generate 1 kilowatt-hour per day, or 365 kilowatt-hours per year.

Often after installing a stand-alone power system, a homeowner finds that the noisy backup generator runs more than anticipated. Because PV modules are more modular and far easier to add incrementally than wind turbines, most people simply add more solar panels as needed.

to code also makes mortgage financing possible. Although electronic inverters add complexity and reduce the power system's overall efficiency, modern energy-efficient appliances make up for any losses in the inverter.

Don't discount DC entirely. A small vacation cabin or motor home can function quite satisfactorily with DC appliances (see Figure 12-3. On the road and off the grid). In even larger systems, there may be some major loads, like refrigeration or water pumping, where DC wiring is justified to limit inverter losses. Sunfrost refrigerators are built in both DC and AC versions, for just such situations.

An equally important decision is charging-system DC voltage. The voltage of the charging system determines the size of the cables needed to conduct current from the solar array and wind turbine to the batteries, as well as the number of batteries. For cabin-size systems that use DC directly, 12-volt is preferred because of the large

Many PV arrays are mounted on the roof and, for much of North America, are simply tilted at an angle equal to the latitude. Some users adjust their panels seasonally to optimize performance. To maximize winter production, tilt the panels at the latitude plus 15 degrees. To maximize summer generation, set the tilt angle to the latitude minus 15 degrees.

As wind-electric generation can be increased by installing turbines on taller towers, PV generation can be improved by mounting the array on a tracker (see Figure 12-4. Wind with solar PV tracker). In winter, the tracker boosts performance a modest 10% to 15%. But during the summer, when the winds are most likely to be light, trackers really shine, producing 40% to 50% more energy than a fixed array. For more information on estimating the performance of a PV array, see Joel Davidson's book *The New Solar Electric Home* or *The Real Goods Solar Living Sourcebook*.

How many PV panels and how big a wind turbine will you need? For all systems greater than a few hundred watts, design for both wind and solar. "It's senseless to install a PV-only system," says Mick Sagrillo. If you do, and you undersize the PV array because of the expense, the result is a merely a "PV-assisted generator set," he warns.

In PV-only systems, designers size the arrays to compensate for minimum winter insolation. By adding wind, says Sagrillo, you can size the PV system for maximum summer insolation, reducing the total number of PV modules needed. In a temperate climate, this takes best advantage of each resource by emphasizing solar-electric generation when wind-electric generation is often at a minimum. Sagrillo argues that "in a PV-hybrid system a tracker is critical" because you want to start generating as soon as the sun rises in the morning. A hybrid wind-solar system requires sizing and siting the PV array for summer conditions.

Sagrillo's rule of thumb is to spend two-thirds of your budget on generating sources for wind and one-third for the PV array. Batteries, says Sagrillo, should be sized for six times total capacity. If you follow Sagrillo's advice, your system will overproduce in the spring and fall when the two resources typically overlap.

Figure 12-4. Wind with solar PV tracker. Hybrid wind and solar system near Buena Vista, Colorado, in 2010. The two solar PV trackers follow the sun as it moves across the sky. While more expensive, trackers can improve summertime performance 40% to 50%. The wind turbine is a Skystream 3.7, designating its rotor diameter in meters (12 feet).

Determining the amount of PV panels one would need to meet Sagrillo's criteria is relatively straightforward. Finding the right size wind turbine is less so. In part, this is due to the limited choice of wind turbines available for battery-charging applications. It's also partly due to the confusing "rated power at rated wind speed" nomenclature most novices rely on.

In a study of the economics of hybrid power systems at low wind speed sites—where most such systems are installed—NREL's Peter Lilienthal found that small wind turbines with relatively large diameter rotors relative to their generator capacity offered better returns than those turbines with small rotors and high power ratings. Large-diameter, low-rated power turbines, says Lilienthal, reduce the need for both battery and PV capacity in a hybrid system compared to using a more aggressively rated wind turbine. Lilienthal's study once again confirms

MICRO HYBRID POWER SYSTEM

Remote power systems using a micro turbine and a few solar panels have proven popular for those who want to use renewable energy in a limited application. The low cost of their components also makes them perfect for hobbyists and backyard experimenters. These low-power hybrids can be packaged with small DC-AC inverters to power consumer electronics, but many are used with DC appliances obtainable through specialty houses serving the recreational vehicle (RV) market.

Micro turbines are the wind energy equivalent of solar walk lights: they're inexpensive and easy to install. They're light enough that you can pick one up and carry it home in your arms. Because most micro wind turbines are used for vacation cabins where there's less stringent demand on performance, they're often installed on shorter towers than their big brothers. Ampair and Marlec are two brands often installed on guyed masts using readily available galvanized steel pipe. With due care, micro turbines can be installed by do-it-yourselfers.

Micro Hybrid Power System		Collector Area		AEP
		m²	ft²	kWh/yr
Wind	Ampair 300	1.2	12	480
Solar	4 panels, 300 W	2.4	27	548
Battery storage	4 kWh			
Inverter				
			Total ~	1,000

Assumptions: 6 m/s (13 mph) average wind speed at hub height; 5 sun hours per day.

Micro hybrid. David Nixon displays the "sun flower" at the Kortright Centre for Conservation outside Toronto, Ontario, in 2007. Marlec Engineering's Rutland 910 is a typical micro wind turbine, which could be used in a micro hybrid power system suitable for recreational vehicles.

that what counts in wind energy is rotor swept area and not generator size.

Scoraig Wind Electric's Hugh Piggott agrees. "For practical purposes," he says, "it's much more useful to have a large rotor and a small generator." A big rotor, says Piggott, "gives a gentle charge all the time, which is what batteries like best." In a storm, turbines with high power ratings but small rotors can generate "lots of watts in the middle of the night, usually far more than you need."

Inverters

Whether you're using a PV array, a wind machine, or a hybrid stand-alone system, you will require an inverter to operate conventional AC appliances. Most inverters produce a modified AC sine wave that can serve a wide variety of AC loads—from sensitive electronics, such as computers and stereos, to washing machines.

To determine the size of the inverter needed, add up the demand from all appliances that are likely to operate at the same time. The inverter should be sized to handle both the surge requirements of the induction motors in refrigerators and washing machines and their continuous demand when operating for extended periods. Small appliances often demand 1.5 to 2 times their rated current when they first start. Large appliances such as washing machines and refrigerators can draw 3 to 4 times their rated current when first switched on. For example, an electric motor using 500 watts could require as much as 1,500 to 2,000 watts when starting.

Inverter ratings vary, but all manufacturers list both their continuous and surge capacity. Refer to the *continuous output rating*. This is what the inverter can actually supply over a long period without failure, keeping in mind that few loads operate continuously, and those that do draw little current. Give yourself some room. Sandia National Laboratory recommends sizing the inverter to 125% of the expected load.

A 2-kilowatt inverter should run most minor loads and some major loads, like a washing machine or a microwave, when operated singly. Two 2-kilowatt inverters or one 4-kilowatt inverter may be necessary if there's any chance the refrigerator, washing machine, well pump, and microwave may operate simultaneously. Microwaves, hair dryers, and similar loads draw a lot of power (1–1.5 kilowatts) but are only operated for short periods. They may influence the size of the inverter needed, but they contribute little to total energy consumption.

The inverter should also provide fused protection from the various sources of generation and should offer power factor correction for inductive loads. Ideally, the inverter should also have disconnect switches on both the AC and the DC sides and load-management switches to limit certain loads from exceeding the inverter's capacity.

Never try to operate an electric dryer, electric hot-water heater, or electric stove on a stand-alone power system unless you're using it as a dump load for excess generation. Use bottled gas (propane) instead. These electric appliances consume inordinate amounts of electricity and place an unreasonably high current demand on the inverter.

Some inverters also offer optional battery-charging functions. If you don't use one packaged with the inverter, you'll need to add a battery charger to your system that's capable of handling multiple inputs: those from the renewable sources, as well as the backup generator. Without a battery charger, there's no way to ensure that the batteries stay properly charged.

Batteries

No electrical generator works 100% of the time, even those of the utility. Batteries permit a renewable power system to coast from one spurt of power to the next, from windless night to sunny day. They're integral to a successful home power system that's off the grid.

To illustrate why batteries are important, consider the operation of an Air 403, a high-performance micro turbine, one June at the Wulf Test Field (see Figure 12-5. Daily generation from Air 403 micro turbine). June is one of the windiest months of the year in the windy Tehachapi Pass. Even in a windy month, there are days when the small turbine produces only a fraction of a kilowatt-hour. During June 9, a particularly windy day with an average wind speed of 16 mph (7.1 m/s), this micro turbine generated 1.8 kWh. For the month, the Air 403 generated nearly 17 kWh at an average wind speed of about 11 mph (5 m/s). Batteries enable such a fluctuating source of generation to store production on windy days for use later, when the winds are less productive.

Batteries do add significantly to the cost and complexity of a hybrid system. Though there's a bewildering array of batteries to choose from, conventional lead-acid batteries are the type most commonly used.

US SOLAR AND WIND DATA

Solar radiation and wind data suitable for downloading into a spreadsheet can be found at the Renewable Resource Data Center of the National Renewable Energy Laboratory at www.nrel.gov/rredc/.

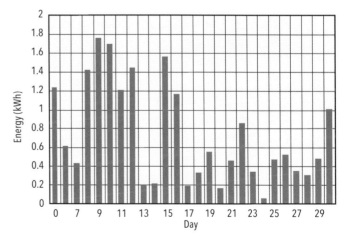

Figure 12-5. Daily generation from Air 403 micro turbine. Batteries are used to store energy produced during windy days, such as during June 8–12 in this example, for windless days, June 12–14. Actual field measurements at the Wulf Test Field in the Tehachapi Pass, June 2000.

CABIN-SIZE HYBRID POWER SYSTEM

These entry level systems are suitable for vacation cabins and are also well adapted for rural second homes that could someday be upgraded into permanent residences. This sample cabin-size hybrid system uses six 75-watt solar panels, a modified-square wave inverter, and sufficient batteries to provide 8 kilowatt-hours of storage on a 24-volt system. Though easily installed with rudimentary tools, this hybrid system will generate nearly 8 kilowatt-hours per day on average—more on windy spring days, less during the summer and fall. The wind turbine provides about two-thirds of the system's total generation. The turbine in this example is Bergey's XL1; it and the tower were designed for owner installation.

Cabin-Size Hybrid Power System				
		Collector Area		AEP
		m²	ft²	kWh/yr
Wind	BWC XL1	4.9	53	1,963
Solar	6 panels, 450 W	3.6	41	821
Battery storage	8 kWh			
Inverter & load center				
			Total ~	2,800

Assumptions: 6 m/s (13 mph) average wind speed at hub height; 5 sun hours per day.

Avoid selecting batteries purely on price. Batteries for stand-alone power systems must be capable of numerous deep discharges. Cheap automotive or truck batteries might be suitable for a tinkerer but not for a remote power system.

Golf-cart batteries, while less than ideal, are popular with do-it-yourselfers and offer good value for entry-level systems. Above all else, use batteries that are designed for the frequent charge-discharge cycles found in remote power systems.

When considering the cost of batteries, don't overlook shipping. Lead and cadmium are among the densest materials known, and shipping batteries made of them any distance at all incurs a hefty freight charge.

Batteries can be finicky. They don't like to be over- or undercharged. To prevent permanent damage, lead-acid batteries shouldn't be discharged to more than 80% of their capacity. Second, lead-acid batteries don't tolerate temperature extremes well.

The capacity of lead-acid batteries decreases markedly with temperature. They're least effective in the dead of winter. They can even freeze, particularly when discharged and the electrolyte becomes more water than acid. One way to avoid damage in cold climates is to store batteries in a cellar below the frost line. But don't bring them indoors without proper ventilation.

Batteries should always be isolated from living areas and from sensitive electronics (see Figure 12-6. Suggested battery and inverter placement). The gases given off by lead-acid batteries are not only highly corrosive, they're also highly explosive.

It's also a good policy to separate batteries from inverters, switches, and service panels. This prolongs the life of the electrical components while guarding against a spark igniting the hydrogen given off during high charge rates (see Figure 12-7. Cabin-size power center).

Size batteries and other fixed hardware to the size of system you eventually want. Avoid undersizing batteries, inverters, and wiring. With a lead-acid battery bank, you're locked in if you find that storage is insufficient after a few years' operation. Unlike solar panels, batteries can't be simply added to the system in small increments. A battery bank is a fixed entity, so additions must

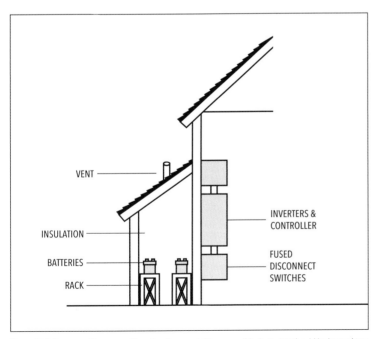

Figure 12-6. Suggested battery and inverter placement. Where possible, batteries should be housed in a well-ventilated room separate from the living quarters.

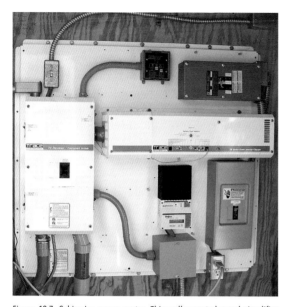

Figure 12-7. Cabin-size power center. This wall-mounted panel simplifies assembly and wiring of a complete DC power system at the Wulf Test Field in the Tehachapi Pass. On the left is the main DC disconnect to the batteries located outside the building in their own ventilated compartment. The disconnect panel includes separate disconnect switches for up to three DC power sources or loads. In this case, the disconnect switch served two separate wind turbine test stands. Center is a 1.5-kW square-wave inverter for AC loads. Below center is a diversion controller for a dump load located outside the building. Lower right is the AC disconnect for the AC loads. Upper right is a charge control panel (unused) for a backup generator.

HOUSEHOLD-SIZE HYBRID POWER SYSTEM

This sample household-size hybrid system should be ample for most off-the-grid homes with a good wind and solar resource (see Household-Size Hybrid Power System). At 3.6 meters (12 feet) in diameter, the Ampair 3-kW turbine is a fairly hefty machine for home owners to install themselves. This system includes almost 1 kilowatt of PVs and the more expensive sine-wave inverters that some find necessary, as well as a large battery bank. Although such a system is capable of producing nearly an average 16 kilowatt-hours per day on Texas's High Plains, for example, most sites will be less productive. North Americans will find it necessary to closely monitor their consumption or provide a backup generator to supplement periods with little wind or sun.

Household-size Hybrid Power System				
		Collector Area		AEP
		m^2	ft^2	kWh/yr
Wind	Ampair 3 kW	10.2	109.5	4,072
Solar	12 panels, 900 W	7.2	41	1,643
Battery storage	16 kWh			
Inverter & load center				
			Total ~	5,700

Assumptions: 6 m/s (13 mph) average wind speed at hub height; 5 sun hours per day.

include a complete new set of batteries of the proper voltage, properly wired.

Battery-charging wind turbines should always include their own charge controller or regulator. (Surprisingly, some don't.) The charge controller will limit battery voltage by either disconnecting the turbine from the load or activating a diversion load when the desired battery voltage is reached. For example, in a 24-volt system, a controller might respond when battery voltage reaches 28 VDC (2.35 volts per cell) and effectively stop charging for a moment until battery voltage falls below 25.2 VDC (2.1 volts per cell), when the controller signals that charging can resume. Like the voltage regulator in an automotive alternator, today's charge controllers use solid-state electronics that perform these tasks many times every second. Some wind turbine charge controllers include a manual switch for using the wind turbine to equalize the batteries during a period of high winds.

Backup Generators

In a properly sized power system with ample battery storage, the backup generator may not be used at all, particularly if you're willing to adapt your usage to the resource. Operate those discretionary loads when the wind is blowing or the sun shining. Do your laundry, for example, only when there's a surplus of power.

Because most remote sites need propane for heating, cooking, and domestic hot water, it's relatively simple to use the same propane to power a backup generator. Propane is modular, portable, and offers good utility.

A backup generator, or gen-set, allows you to design a system with less battery storage than otherwise needed. You can substitute fuel costs for extra batteries. PG&E found in its survey of California stand-alone systems that 80% of the respondents used a backup generator. But the gen-sets provided only 2% of the system's total energy, suggesting that most off-the-grid power systems had enough storage for almost all conditions. The backup generators, even if seldom used, provided a valuable service: peace of mind.

Because generators operate most efficiently near full load, it's best to run them only after lead-acid batteries have fallen to about 20% of their full charge. The generator can then run at full output until the batteries are brought back up to 80% of full charge. The free fuel sources, wind and solar, can then top up the batteries with long-duration charging cycles. This limits the overall running time of the generator, extending its life. It also keeps fuel consumption down. Automatic controls are available for monitoring the batteries' state of charge, which will start and stop the generator as needed.

For long life, look for generators that operate at 1,800 rpm, not 3,600, and that have self-starting capabilities. Avoid the inexpensive portable generators used at construction sites. They're not well suited for remote power systems.

Stand-Alone Economics

Now let's look at the economics of this hybrid system. In general, users of remote power systems don't expect their wind and solar generation to compete with the cost of utility-generated electricity. They typically install a remote power system because the cost of extending utility power to their site is even more costly. California utilities charge new customers about $10 per foot ($33,000 per km) for overhead line extensions (50% more for buried lines). The situation's no different in Europe. Electricité de France charges rural residents in France €20 per meter ($32,000 per mi) or more to extend utility service.

Under these conditions, a stand-alone system can pay for itself in the first year if it's more than ½ mile or 1 kilometer from the utility's lines. If you're considering a stand-alone power system in North America on purely economic grounds, in general it's cheaper to bring in utility power if your home is less than 1,000 feet (300 meters) from existing utility service.

Extending the line may not be the only cost you incur. Many utilities require a minimum purchase of electricity to justify extension of the line. In Pennsylvania, West Penn Power Company at one time required a minimum monthly payment of $100 to $200 for a period of five years from customers requesting line extensions.

Other Stand-Alone Power Systems

The use of wind turbines in hybrid power systems for telecommunications and village electrification are essentially variations of remote power systems for residential use, but each has unique requirements that distinguish it from home light plants.

Telecommunications

Telecommunications demand reliability. Wind machines used in telecommunications encounter more extreme weather, operate more often (sometimes in excess of 7,500 hours per year), and must function unattended for much longer periods of time than home power systems or even commercial wind power plants typically do. Only robust wind machines using fully integrated, direct-drive designs perform satisfactorily in the rugged environments characterized by telecom sites (see Figure 12-8. Wind-powered telecommunications).

Black Island in McMurdo Sound, Antarctica, illustrates the severe conditions that wind turbines must endure to serve telecom applications. It's one of the harshest sites in the world. Shortly after installation, a Northern Power Systems' model HR3 operated for 12 hours in the furled position during a fierce Antarctic storm. The radio station eventually went off the air when the exhaust stack for the backup generator blew away. After the worst of the storm had passed, the HR3 dropped back into its running position, recharged the system's batteries, and brought the station back to life. Twice during the first two years of operation, anemometers at the site blew away, once after recording a wind speed of 126 mph (56 m/s). Since then the site has endured winds up to 197 mph (88 m/s). The project has been so successful that it was expanded. And it has been durable as well. Black Island's HR3 has now been in service for more than a quarter century.

Hybrid systems using small wind turbines have been able to substantially reduce fuel consumption at telecom sites in Canada and the United States. At a 6.3 m/s (14 mph) site on Calvert Island off the coast of British Columbia, two Northern Power Systems HR3 turbines, in conjunction with a 1.2-kilowatt solar array and 84-kilowatt-hour battery bank, were able to cut

Figure 12-8. Wind-powered telecommunications. Bergey Excel powering a Qubectel mobile telephone station at Pointe-au-Père on the south shore of the St. Lawrence River in Quebec, Canada.

The benefits of providing even small amounts of power to remote villages are magnified because so little electricity is needed to raise the quality of life.

generators. But the generators are expensive to operate and often unreliable. Fortunately, more and more developing countries are finding that renewable energy is a less expensive, more reliable, and quicker way to meet the electrical needs of their rural citizens.

Village power systems must meet standards for ruggedness, simplicity, and reliability similar to those demanded by mountaintop telecommunications sites. Though the weather may not be as demanding, Third World villages are distant in both time and space from the technical support and spare parts found in the developed world.

The benefits of providing even small amounts of power to remote villages are magnified because so little electricity is needed to raise the quality of life. A Bergey Excel may supply only one home with electric heat in North America, but it can pump safe drinking water for a village of 4,000 in Morocco.

The typical village system might use two or more wind turbines, batteries, inverters, and a backup generator (see Figure 12-9. Village power). And like hybrid home light plants, village power systems should also include solar PV in a hybrid wind and solar system. The key is to use as much power as possible directly, instead of storing it in batteries and running it through an inverter. This reduces both initial cost and complexity, while delivering more of the power system's energy to do useful work.

the diesel generator's operating time substantially. Overall, the hybrid system reduced fuel use nearly 90%, at half the maintenance cost of a conventional diesel system. At Norway's Hamnjefell telecom station above the Arctic Circle, an HR3 turbine has met 70% of the site's loads since 1985.

Village Electrification

Nearly two billion people live without electricity. Extending utility service from the cities to remote villages in developing countries—where most people live—is costly and difficult to finance and takes years of struggle. To surmount these problems, some villagers have turned to small diesel

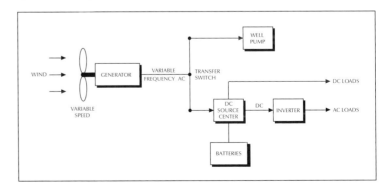

Figure 12-9. Village power. One scheme for using small wind turbines to serve a variety of loads in a village distant from a central station power.

VILLAGE SELF-RELIANCE

Village power systems can meet not only domestic demand but also those of light industrial and commercial uses that spur economic development. Like increasingly popular microcredit programs, villagers can use revenues from the light industrial loads to expand the power system themselves instead of relying on often distant and sometimes indifferent central governments for ongoing aid.

For example, villagers could use electricity from a hybrid power system to refrigerate freshly caught fish for sale at a higher price in nearby markets. Such a system would provide power from five to six hours of light industrial loads and from four to five hours of domestic demand per day. This strategy deals with the mischievous genie unleashed in the past by village power systems: seemingly limitless demand. Soon after installation, everyone wants a personal lamp, television, and other appliances, and the system quickly collapses. The pay-as-you-go approach provides a mechanism for villagers themselves to finance expansion of the power system. In this way they take ownership and responsibility for the success of the system.

If power is used directly to pump water, grind grain, or run other loads not dependent on utility-grade electricity, the need for batteries is diminished. The batteries and inverter then need to be sized only for those loads that must use constant frequency AC. The output from the wind turbine can be manually switched from the direct loads, such as water pumping, to the batteries and inverter as needed. For example, the operator monitors the water level in a storage tank and the batteries' state of charge to determine where the power should be directed. The operator is also responsible for starting the backup generator when the power system can't meet demand. Eliminating automatic switches decreases the likelihood that a minor component could fail and imperil the entire system. It also ensures that one person is always responsible for operation of the power system—a lesson learned from past demonstration projects.

Wind-Diesel Twinning

Inhabited islands share many characteristics with isolated villages that are not connected to a large transmission and distribution system. Since many islands are also windy, integrating wind energy with existing island or remote village power system has long been an attractive option for reducing the high cost of imported diesel fuel. At some Aleut and Inuit villages in North America, diesel fuel is flown in at an extremely high cost.

Power engineers once thought that wind turbines could only off-load a small portion of an island's diesel network without endangering the stability of supply. Because it's also damaging to run the diesel engines at low loads, wind could never be used to meet all or even most of an island's electrical demand. However, electronic controls developed in the 1990s have driven significant advances in marrying the fluctuating wind resource with an island power system's fluctuating load and their diesel generators. Technically, designers *twin* the wind turbines with the diesel generators and the system's various loads.

Longtime wind system designer Lawrence Mott points to St. Paul Island as an example of a hybrid wind-diesel system where wind energy contributes a large portion of electrical demand. Northern Power Systems installed one Vestas V27 on the island in Alaska's Pribilof Islands group in 1999. The 225-kW turbine joined two 150-kW diesel generators. No batteries were provided for storage. As in many household-size hybrid power systems, electronic controls in the St. Paul system dump excess wind generation into resistive heaters producing hot water, a commodity always in demand on the windswept island in the Bering Sea. The electronic controls, coupled with a synchronous condenser, enable the system to operate with the conventional wind turbine—alone. The controls keep voltage and frequency constant without operating the diesel generators. When the wind turbine is insufficient to meet demand, the system starts the diesel.

The project has been so successful that the native-owned and -operated power system subsequently installed two more Vestas V27s at the site in 2007. The wind turbines have cut diesel consumption nearly 50%, reducing the cost of fuel to heat and power the island's airport complex by $250,000 per year.

The installation is also significant because it was one of the first examples of the "high-penetration" of wind into a diesel-electric system. The average penetration of wind, that is the amount of electricity provided by the wind turbines relative to total consumption, is 55%.

Figure 12-10. Newfoundland wind-diesel twinning. Six reconditioned Windmatic 15S in a pilot wind-diesel system on Ramea Island, Newfoundland. The turbines have been operating in this application for the past decade, offsetting expensive diesel fuel for the 700 people living on the island. The 15S uses a rotor 15 meters (50 feet) in diameter to drive a 65 kW induction generator. (Frontier Power Systems)

St. Paul Island isn't an isolated example. In the early 1990s, Electricité de France contracted with French wind turbine manufacturer Vergnet to operate a wind plant on La Désirade, a small island east of Guadeloupe in the French West Indies. The goal: to cut in half the cost of serving the island. Vergnet successfully demonstrated that a cluster of household-size turbines could deliver a significant percentage of the island's electricity. Since then the project has been steadily expanded, harnessing *les alizés*, or the Caribbean's trade winds. Subsequently, an undersea cable was laid to Guadeloupe, and now La Désirade not only is self-sufficient but also exports electricity to the main island. All in all, Vergnet's turbines generate 165% of the island's electrical consumption.

Another example of a remote power system with a high penetration of wind is that at Esperance on the south coast of Western Australia. Esperance, while part of the Australian mainland, is an isolated community with its own grid or *microgrid* in today's parlance.

Like the project on St. Paul Island, Esperance used Vestas's V27, a wind turbine that has seen wide application worldwide both in wind farms and as individual wind turbines or in small clusters. In Esperance, the local power authority installed nine of the turbines at Ten Mile Lagoon in 1993. The 2 MW project operated so successfully that six larger machines, Vestas's V47, were installed in 2003, adding another 3.6 MW to the total.

Though not a wind-diesel system as such, the system works the same way. When the wind is blowing, the turbines offset some of the seven gas-fired turbines at the remote site's power plant. Altogether, the wind turbines generate 9.5 million kWh annually, providing 23% of the system's electricity. The Ten Mile Lagoon turbines have now been in operation in one form or another for more than two decades.

In 2004, Carl Brothers, one of Canada's wind pioneers, installed a small cluster of six reconditioned Windmatic 15S on Ramea Island off the rugged coast of Newfoundland (see Figure 12-10. Newfoundland wind-diesel twinning). The 400 kW of wind turbines are twinned with a diesel-generating system serving an average load of 530 kW, in another example of high penetration in a wind-diesel system. The installed wind capacity is double the village's minimum load of 200 kW. The turbines have operated for more than a decade, serving the isolated population of 700 by offsetting expensive diesel fuel.

As we've seen, wind energy has proven over the past several decades that it can deliver reliable service in high penetration from small household-size systems at remote sites to remote village power systems. Now wind energy is moving toward high penetration in complex utility networks as more and more wind turbines are being integrated into electricity grids worldwide, the subject of the next chapter.

13

Interconnection and Grid Integration

Night has fallen and the sky is clear. December's chill winds whip the Lake Erie shoreline. Drifting snow swirls about the fence posts and outbuildings of George McClain's small farm. The whistling wind rises in crescendo and then dies away in an unpredictable ebb and flow. A faint whirring, rhythmic and ever present, can be heard. Dark, saber-like shapes sweep the starry sky.

"Looks like it's going to be a cold one tonight," George predicts. His two kids, scampering around in their flannel pajamas, flee their mother as she readies them for bed. Darlene, both mother and partner in the McClains' dairy, responds, "George, don't you think we ought to turn the electric heat up? I feel like I'm coming down with something."

"Yeah, Daddy," the kids chime in, "just turn it up to like we used to."

"Now you kids know better than that," he says. "Christmas will be here soon, and we want to buy that new car we've been waiting for so long, don't we," he winks at Darlene. "We only get one more check from Pennelec before the new year, and I want to sell them just as much power as we can. On a night like this, everybody's going to be switching on their electric heaters. The more we save, the more we can feed to Pennelec. Those turbines will really be turning out the juice in winds this high. Just listen to 'em hum."

WHEN THIS PASSAGE WAS FIRST WRITTEN MORE THAN THREE decades ago, it sounded far-fetched: A family that awaits winter's winds and looks to the local electric utility as a source of income? What was far-fetched then has become commonplace today. Farmers such as the fictional McClains have been replaced by pig farmer Niels Mogens Sloth in northwest Denmark, grain farmer Peter Ahmels in northern Germany, or community wind developer David Stevenson in Nova Scotia.

However, the once promising prospect of tens of thousands of household-size wind turbines whirring above farms, ranches, and homes across the breadth of North America feeding electricity into the utility network never fully materialized as envisioned in this vignette. Yes, technically and legally it can be done (there are thousands of small wind turbines distributed across the continent doing just that), but most of the wind-generated electricity produced worldwide is not delivered in the manner described.

The opening vignette is based on an outdated view of how producers of wind energy relate to the electric utility. At the time, it seemed

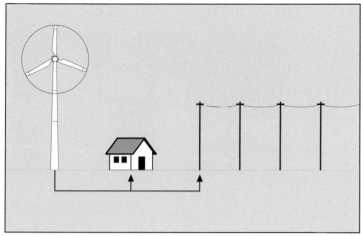

Figure 13-1. Net metering or self-generation. The wind turbine is connected to the electricity distribution system on the customer's side of the kilowatt-hour meter. The wind turbine serves on-site consumption and any "excess" is fed back into the grid. This is seen as "negative load" by the electric utility and as "energy conservation" or energy saving by policy makers.

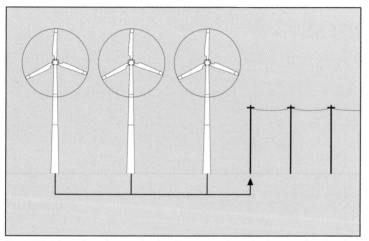

Figure 13-2. Sales to the grid. In true distributed generation, the wind turbine (or turbines) produces all its electricity for sale to the grid. Nearly all the wind turbines distributed across the landscape of Germany and Denmark sell their electricity directly to the local electric utility. Here a cluster of turbines are connected at the voltage of the distribution system.

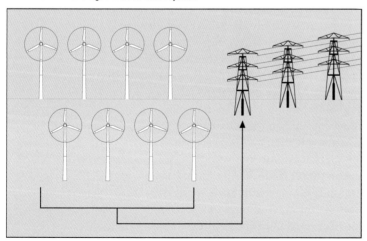

Figure 13-3. Commercial wind power plant. Large arrays of wind turbines are connected to the transmission system like any other power plant.

revolutionary. The wind turbine was used to offset consumption of electricity provided by the utility. Excess generation was sold back to the utility in what's called net metering. When it was windy and consumption was low, the homeowner, or farmer in our fictional family, would sell excess generation to the utility. When the wind wasn't strong enough to match consumption, the electric utility made up the difference. Thus, the father here is encouraging the family to save as much as electricity as possible so they can sell more to the utility.

Today, a far more radical and disruptive model exists. The McClains would install a large wind turbine or a cluster of large wind turbines themselves, or in cooperation with their neighbors, and sell all the electricity generated to the utility. They wouldn't sell "back." They would sell "to" the utility. And they would do so for a profit.

The McClains would become rebels in a revolutionary movement to generate and sell electricity in the same way the utilities once did. They would generate electricity in parallel with the utility. They would buy electricity from the utility as in the past, but unlike net metering, the act of selling electricity to the utility would be separate from the act of buying the electricity. There would be two kilowatt-hour meters: one to measure the amount of electricity consumed, the other the amount of electricity sold to the utility. Probably more than 98% of all the wind-generated electricity worldwide is produced in this manner.

As in most of life, the devil is in the details. In this chapter, we'll look at some issues surrounding the interconnection of wind turbines with the grid as well the question of how to integrate large volumes of wind energy into utility systems.

Models of Interconnection

Small wind turbines are most often connected to the grid on the customer's side of the kilowatt-hour meter in net metering applications (see Figure 13-1. Net metering or self-generation). Most states and provinces in North America have policies specifically designed around this model of wind development.

However, net-metering installations are not

BREAKING FREE FROM NET METERING

For more than two decades, North American renewable energy advocates have pushed net metering—the ability to run your kilowatt-hour meter backward—as the principal means to develop the distributed generation of renewable energy.

Net metering served a useful purpose in the dark days of the post–Jimmy Carter era. Net metering then was a call to arms for hobbyists and guerrilla solar activists out to prove a point: wind and solar works, your meter will run backward, and the lights will stay on.

But net metering was never intended to be a policy for the rapid development of the massive amounts of renewable energy that North America needs. It could not do that alone.

Why? Because retail electricity prices in North America, especially in much of Canada and in the Pacific Northwest, are abysmally low. In most cases, the price offset under net metering is insufficient alone to drive profitable renewable development. Subsidies are needed, and subsidy programs have had a checkered history in North America.

Most subsidy programs have led to widespread abuses that have hurt renewables over the long term. Even today, few subsidy programs for small wind turbines require metering actual generation—one of the fundamental means for monitoring the success or failure of renewable energy programs. Worse, subsidies are dependent upon the appropriation process that is subject to political whim and competing needs for public funds to address the crisis du jour.

On top of that, there's typically a low limit on the amount of renewables that can be installed in net-metering programs. If we really wanted renewables, why set a limit at all? Caps on program size serve only to protect incumbent utilities. Electric utilities tolerate net-metering programs with low caps because the programs pose no serious threat to their markets. Thus, they—and the politicians who listen to them—limit the role renewables can play under net metering. But just to make sure, there's nearly always a limit on the size of any individual installation that can be installed under net-metering programs, often the equivalent of one wind turbine. We certainly wouldn't want renewable energy to rock the utility industry's boat.

Net metering appeals to policy wonks because it rarely threatens entrenched electric utilities, and it gives politicians—as well as the advocates who promote these policies—the perfect cover for appearing to take action on renewable energy, while doing little of substance.

In the end, though, net metering won't get us where we want to go: massive amounts of renewables in the ground quickly. Net metering will never give us "plus energy" houses or "plus energy" buildings because we often literally have to give our surplus electricity to the utility company for free. And where we don't, the payment for our excess generation is insufficient to justify the investment.

Europeans roll their eyes when North Americans speak of net metering. "Was ist das?" or "Qu'est-ce que c'est?" they can be heard saying. They don't bother with net metering.

How then can Europeans be so successful if they don't use net metering? How can they have installed so much generating capacity that the Danes produced 40% of their electricity with wind turbines, and the Germans produced more than 25% of their supply from new renewables in 2014?

The answer is surprisingly simple: they pay for it. Europeans set a tariff or a payment per kilowatt-hour for each renewable energy technology, one sufficient to cover the cost of generation plus a reasonable profit. This ensures that they get the kind of renewables they want, where they want, at the pace they want. The results speak for themselves. No taxpayer subsides are needed.

Net metering was always just a stopgap. A policy that could bide us time until the political climate changed, and we could implement serious renewable energy policies. That time has come.

The time for half-measures—for timid responses like net metering—is past. The public and now some progressive politicians as well are demanding more aggressive policies. For distributed renewable energy to make the substantial market inroads needed in the huge North American market, advocates need to break free from the net-metering straitjacket.

We need new policies, such as electricity feed-in laws, that have a proven record of results.

---

limited to household-size wind turbines; larger wind turbines are used in the same manner (see Figure 1-7. Playing with the wind). Some factories, businesses, and schools install medium-size and large wind turbines to offset a portion of their consumption—as well as green their image. Most jurisdictions limit the size of these installations to the equivalent of one or two turbines. In the United States, net metering has typically been limited to 2 MW of capacity or less.

In Europe it's quite common to see a single large wind turbine or small clusters of large wind turbines connected to the local utility's distribution lines (see Figure 13-2. Sales to the grid). In these commercial applications, the turbines deliver all their electricity for sale to the grid.

In North America, most wind development has taken the form of large wind turbines massed in extensive arrays connected at transmission voltage (see Figure 13-3. Commercial wind power plant). These wind farms are operated in the same manner as any other power plant, producing bulk electricity for transmission to distant cities.

Interconnection Technology

The distributed use of wind energy entails connecting the wind turbine either on the customer's side of the kilowatt-hour meter, or directly to the utility's distribution system

Figure 13-4. Interconnected large wind turbine. Small commercial-scale and large wind turbines may be connected on an industrial or commercial customer's side of the meter or, as with many large wind turbines connected to the utility's distribution lines, via a transformer. This Bonus wind turbine in southwestern Alberta near Pincher Creek is connected to the three-phase distribution system by the transformer on the wooden utility poles. Though this wind turbine was owned by a local farmer, it delivered all its electricity to the grid. Circa mid-1990s. Inset: The transformer steps up the voltage of the wind turbine to that required by the distribution system.

through a dedicated kilowatt-hour meter and transformer (see Figure 13-4. Interconnected large wind turbine).

In the early 1980s, most small wind turbines designed for utility interconnection used induction generators, just like the commercial machines then being installed in California's windy passes. A few, like the Bergey Excel, were a sign of things to come and used permanent-magnet alternators feeding electronic inverters that provided the utility-grade electricity needed in the household. Today, most small turbines destined for the customer's side of the kilowatt-hour meter follow the model pioneered by Bergey.

Induction or Asynchronous Generators

Wind turbines using induction generators are certainly the simplest to connect to the grid. Early promotions by defunct 1970s manufacturer Enertech showed a wind turbine with an electrical cord for plugging into a wall outlet. Most of the medium-size wind turbines operating in wind power plants around the world during the 1980s and early 1990s were nothing more than glorified induction generators with an electrical cord. No mysterious black box was needed to convert the generator's output to the form used by the utility.

The induction generator, or asynchronous generator as it's known technically, uses current in the utility's lines to magnetize its field. Thus, the voltage and current the induction generator produces are always synchronized with that of the utility. In principle, the induction generator is unable to operate without the utility. When the utility's system is down, the induction generator is down too.

Electronic wizards can fool induction generators into thinking the utility's line is present by using capacitors to charge the field, enabling induction generators to be used in stand-alone power systems. Vergnet wind turbines, for example, use induction generators in this manner for application in wind-diesel and battery-charging systems in France and its overseas territories. For nearly all other applications, however, wind turbines using induction generators automatically disconnect from the grid when the utility's lines are down.

Induction wind turbines do use sophisticated electronic controls to tell the wind turbine when to connect and disconnect from the grid. If they didn't, the wind turbine would "motor" in low winds, acting like a giant fan and consuming electricity instead of generating it. These controls also tell the generator to engage the utility's line gradually in small steps rather than suddenly all at once. Such "soft starts" reduce the voltage flicker that once characterized the addition of wind turbines driving induction generators on rural distribution lines.

Electronic Inverters

Years ago, it seemed miraculous when Windworks announced that you could use a synchronous inverter to connect a 1930s-era Jacobs wind turbine with the lines of the local electric utility. Although the technology for doing so had been around for some time, it wasn't widely known among alternative energy enthusiasts. The technology is much less mysterious today. The electronics are now quite commonplace if not passé.

This inverter and the others that quickly followed took DC, in the case of the old Jacobs generator, or rectified the variable-voltage, variable-frequency AC from an alternator to DC, inverted the DC to AC, and synchronized it with the AC from the electric utility. In this way, old synchronous inverters were line synchronized or line commutated. They used SCR (silicon controlled rectifier) switches with analog controls to signal when they would feed bits of current into the utility's lines. Because they were line commutated, they needed the utility's line present to function.

Contemporary inverters are self-commutated. They can produce utility-compatible electricity using their own internal circuitry with IGBT (integrated-gate, bipolar transistors) and digital controls. The new self-commutated inverters greatly improve reliability and power quality over the older line-commutated versions. Advances in inexpensive inverters for solar photovoltaic systems have produced product spin-offs that now benefit small wind turbines.

Some manufacturers of small wind turbines have developed their own, purpose-built inverters. Defunct manufacturer Southwest Windpower designed its inverter to fit into the nacelle of its Skystream (see Figure 13-5. Plug and play?).

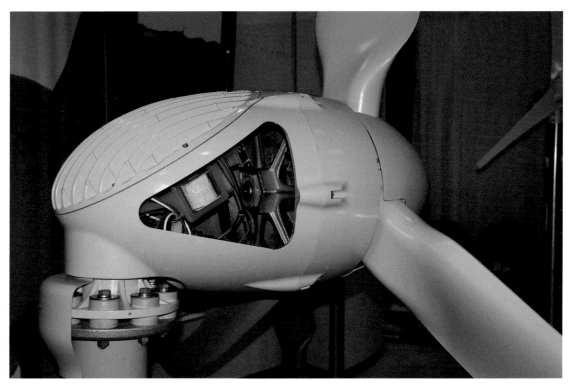

Figure 13-5. Plug and play? Southwest Windpower's Skystream incorporated an inverter built into its nacelle. Note the heat sink and cooling fins on the left of the nacelle. Southwest Windpower's objective was to make the electrical installation as simple as possible. The downside was that when a problem developed with the inverter, the wind turbine had to be lowered to the ground.

Like the Enertech of old, Southwest Windpower hoped its 3.7 model would be as simple to install as plug-and-play computer hardware. The output from the Skystream 3.7 was wired directly into the consumer's electrical service panel.

Advancements in electronics have also made large, direct-drive wind turbines possible. Enercon, the largest supplier to the German wind turbine market, uses electronic inverters in much the same manner as small wind turbines. Similarly, most other large wind turbine manufacturers, those relying on a conventional drivetrain with a gearbox, use doubly fed induction generators coupled to electronic inverters.

With the exception of a few small wind turbine designs and a few medium-size turbines, the wind industry has steadily moved to generator-inverter combinations for producing utility-compatible electricity.

Power Quality and Safety

Utilities on both sides of the Atlantic were initially reluctant to interconnect wind turbines with their lines because of concerns about safety and power quality. To the chagrin of wind energy's critics, there have been few problems with safety, voltage flicker, harmonics, or other technical issues.

Unfortunately, even after more than 30 years of modern wind energy, it may take an attorney and a big bank account to convince the utility to do what they're obligated to do. Yet the data is in. Wind turbines have operated more than 20 billion hours on the lines of electric utilities in Europe, the Americas, and Asia without bringing the world to an end. Wind turbines have generated some 4,000 terawatt-hours (4,000 billion kilowatt-hours) of clean, nonpolluting electricity worldwide. That should be sufficient proof that wind energy can work harmoniously with the existing electricity distribution system.

Wind turbines have operated more than 20 billion hours on the lines of electric utilities in Europe, the Americas, and Asia WITHOUT BRINGING THE WORLD TO AN END.

Wind systems, whether using inverters or induction generators, now have produced line-quality power for more than three decades without endangering utility equipment or personnel (see Figure 13-6. Medium-size, utility-compatible wind systems). Despite this record, some utilities may still have questions that need to be addressed.

The safety of its employees is the utility's principal concern, as it should be. Managers may fear that an interconnected wind system could energize, or deliver power to, a downed line during a power outage and electrocute someone. This fear is not entirely unfounded. Utility workers have been injured by improperly wired emergency generators, and from the utility's perspective, a wind turbine is little different.

Most inverters and all induction generators are line synchronized; that is, without the presence of the utility's line, they can't generate power. Nevertheless, it's possible for induction generators, in rare circumstances, to "self-excite," that is, to provide their own excitation. "Islanding," as it's called, requires a rare match between capacitance, generation, and load. Although the confluence of such conditions is fleeting, it is potentially damaging.

In the past utilities required that all wind machines designed for interconnection with their lines disconnect from the line during an outage, and they must not be able to self-excite the generator. To preclude this, relays or electrical switches were placed on the utility side of the wind turbine's inverter or control panel (see Figure 13-7. Automatic disconnect). When utility power was present, the AC relay, or contactor, was energized, completing the electrical circuit. If utility power was lost for any reason, the spring-loaded relay deenergized (turned off), opening the circuit and disconnecting the wind system from the utility line.

However, within the past decade utilities realized that there were benefits to keeping wind turbines online during brief interruptions, that is, for the wind turbines to "ride through" the outage. The wind turbines would then help the utility recover service more quickly, minimizing disruption to the utilities' other customers. Modern electronics and sophisticated controls make this possible, and this feature has now

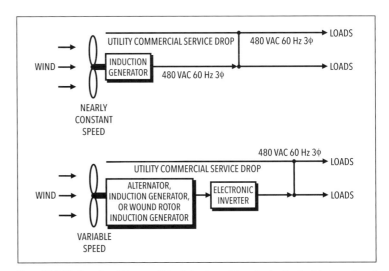

Figure 13-6. Medium-size utility-compatible wind systems. Schematic showing the interconnection of household-size and medium-size wind turbines using induction (asynchronous) generators or alternators and power electronics with the grid. Voltage is typical of farms, ranches, and small businesses in North America. Large wind turbines will use higher voltages.

Figure 13-7. Automatic disconnect. Phil Littler displays the three-phase interconnection and automatic disconnect switch for a 225-kW Vestas V27. That's all there is to it.

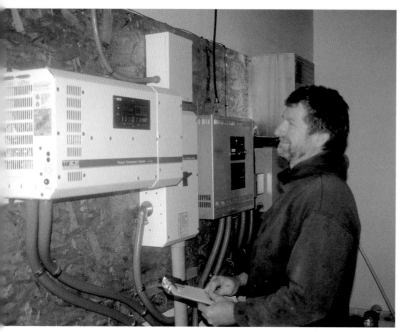

Figure 13-8. Interconnection with battery backup. Greg Nelson logs the number of kilowatt-hours sold back to San Diego Gas & Electric Co. in his net-metering application with a stand-alone inverter. This inverter and a large battery bank (not visible) enable Nelson's wind and solar hybrid system to power his rural home in the event of a power failure.

become a standard requirement for large wind turbines worldwide.

Some inverters are designed for standby or uninterruptible power systems. Coupled with batteries, these inverters are able to provide power to a load when the utility line is down (see Figure 13-8. Interconnection with battery backup). Because of the risk that the batteries could power the inverter and energize the downed utility line, these inverters incorporate electronic switches that prevent the inverter from sending power into the grid.

No matter what protection system is used, utilities require a manual disconnect switch to be visible and accessible to their personnel. When servicing a line nearby, they can "throw" the disconnect switch, severing the connection between the wind turbine and the grid.

Power Factor

Self-excitation becomes a problem when capacitors are used on the wind turbine side of the contactors. Wind turbine manufacturers and utilities frequently use capacitors to correct power factor. Capacitors on the utility side of the AC contactors are disconnected from the wind machine whenever the contactor opens. Any capacitors used with a wind system for power factor correction should be placed on the utility side of the interconnection or should be designed to bleed down (discharge) after power is removed. This prevents self-excitation as well as any electrocution hazard to those servicing the wind turbine.

Why does a utility or wind generator need capacitors to correct power factor? To understand the answer, you need to know the difference between true and apparent power. (Yes, they both exist, and there is an important difference.) Power in watts, as you recall, is the product of voltage and current. This is generally true, but when we're dealing with alternating current, we need to add another parameter to the equation: phase angle.

Don't let phase angle scare you. It merely describes the degree to which rising and falling voltage is in phase with the rising and falling current. If current rises from zero at the same time voltage rises from zero, the two waveforms are said to be in phase.

When this occurs, the cosine of the phase angle (zero degrees) equals unity (one), and true power is the product of voltage and current. In this case, true power and apparent power are equal, and the power factor—the ratio of true power to apparent power—is one. This is the ideal. Unfortunately, the real world seldom looks like this.

Loads on the utility's lines cause the current waveform to shift slightly. Current either leads (starts rising earlier) or lags (starts rising later) voltage. In rural areas, where power must be transmitted over long distances, the length of the line itself is sufficient to cause current to lag behind voltage.

At this point you may cry out, "Who cares?" The utility cares. And it cares a lot. When current and voltage are out of phase, true power decreases (the cosine of the phase angle becomes less than one), yet apparent power remains the same. To deliver the same amount of real or billable power as apparent power increases, the utility must deliver more current. Beyond a certain point, to avoid these resistance losses and to maintain proper voltage, the distribution system must be expanded at a cost to the utility and its ratepayers.

Apparent power, as Bruce Hammett at Wind Energy Conversion Systems in Palm Springs explains, is the power to excite or, as he elegantly puts it, "wake up the iron" in the windings of a motor or generator. Induction, or asynchronous, generators used in many older wind turbines consume apparent power to energize or excite their fields.

The utility's trusty kilowatt-hour meter measures only true power. Therein lies the dilemma. The utility must generate apparent power while it is able to charge only for true power. And true power is always less than apparent power (the power factor is normally either one or less than one). The utility gets short-changed and tries to correct this—to keep the power factor as close to unity as possible—by adding banks of capacitors to its lines.

The foregoing applies to wind systems as well as to electric utilities. It is of importance to you because the power factor of your wind generator may cause the utility to take corrective action by installing capacitors and then charging you for them. Power factor is also a favorite whipping boy for opponents of wind power within electric utilities.

If problems arise, you need to remember that the utility pays only for true power when you sell to it. (You are now in the utility's shoes; that is, trying to maximize the production of true power by getting your power factor as close to unity as you can. Values above 90% are desirable.) Don't think you're selling them horsemeat while charging them for prime beef, as some utilities have implied.

Moreover, your wind system appears to the utility as just another load (albeit a negative load) calling for power factor correction. It's no different than if you added a couple of freezers or new power tools to your home, as far as your impact on the utility is concerned. Utilities don't charge you for power factor correction when you install a new freezer, do they? No, of course not. And they shouldn't charge you for power factor correction when you install a wind generator, either.

Large wind turbines, clusters of turbines, and wind farms do require power factor correction. This is normally part of the wind turbines' electrical system, or it can be added separately. Large wind turbines using variable-speed rotors and power conditioning not only can compensate for the reactive power needed to excite their generator windings but can also deliver reactive power to the grid.

Many large wind turbines today provide reactive power as a service to the utility and the grid. Rather than in days past when the utility charged the wind turbine owner for reactive power, today the utility may pay for the ancillary service that the wind turbine now provides.

Voltage Flicker

Another problem arises with certain wind systems using induction generators. On some early models, the rotor was parked until the cut-in wind speed was reached. The rotor was then motored up to synchronous speed by drawing power from the utility. The effect on the utility was similar to the start-up of a compressor motor in a refrigerator or freezer. There was a momentary surge of current until the rotor came up to speed and the wind began driving the rotor. Because the power rating of wind generators is usually larger than most household appliances, the magnitude of the in-rush current can be large enough to cause a slight voltage drop in the line. The result may be voltage flicker. Your lights may dim briefly whenever the wind generator starts. This isn't a serious problem, but it can be annoying. The utility, though, may claim that an induction wind machine will detract from the level of service it offers other customers and may require a dedicated transformer to mitigate voltage flicker. The transformer isolates the voltage drop to the customer using the wind turbine and often eliminates the problem entirely.

In sparsely populated rural areas, most homes, farms, and small businesses already have a dedicated transformer; that is, only one customer is served by the transformer. In the suburbs, on the other hand, one transformer may be used for a number of customers. You can tell if you have a dedicated transformer by taking a look out your window. The utility primaries, or high-voltage lines, are carried at the top of the utility pole. Leads from the primaries are attached to the top of the transformer. The low-voltage lines, or secondaries, are attached to terminals on the side of the transformer (see the transformer in Figure 13-4. Interconnected large wind turbine). The service drop to your home is always from the secondary (low-voltage) side of the transformer.

If the secondaries are strung directly to your house and to no other, the transformer is dedicated to your service drop. But if the transformer is located several poles away and the low-voltage lines serve several other customers, the transformer is communal, and a household-size wind turbine could affect your neighbors.

The degree of in-rush current or voltage flicker varies from one model wind turbine to the next. Early US designs, for example Enertech, were the most notorious because they motored the rotor up to operating speed. Wind turbines today use freewheeling rotors instead. When there is sufficient wind to turn the rotor at synchronous speed, the AC line contactors are energized and the wind system is brought online. There is still some in-rush current, but it's minor in comparison to what it was with the turbines of the early 1980s. Most medium-size turbines and all large wind turbines use electronic controllers that connect to the grid "softly," minimizing any voltage flicker.

Harmonics

Inverters are not without their power-quality faults. They can produce current harmonics in various degrees. This does not affect most electric appliances but, theoretically, can cause electromagnetic interference with television, radio, and telephones. The degree of interference depends on the inverter, its size and location, and the level of electrical noise on the utility's lines with which it is connected. (Power from the utility is never free from harmonics itself.)

This interference may be noticeable but not necessarily objectionable. For example, one wind machine owner thought his machine was performing normally when his wife walked into the kitchen and wanted to know why he had turned the wind turbine off. Amazed, he asked her what she meant. His wife answered that she had grown accustomed to the faint humming it made on her radio, but she couldn't hear it anymore. He proceeded to check the control panel and found the wind turbine's inverter was indeed down with a tripped circuit breaker.

Current harmonics haven't presented any insurmountable problem to utilities in either North America or Europe. Again, a transformer can be useful in attenuating any disturbance to the customer's immediate vicinity.

Net Metering

If the wind turbine is "behind the meter," as the utility would say—that is, serving on-site loads before any electricity reaches the grid—you would run your kilowatt-hour meter backward, selling any excess energy at the retail rate and buying what you need when you need it. In this way, you use the utility as a battery. The utility stores your energy until you need it. This is the essence of net metering.

To the utility, net metering is losing a customer, and no business likes losing a customer. Consequently, they vigorously fight such policies. North American renewable energy advocates have spent years lobbying—often futilely—to overcome utility opposition to net-metering policies.

Where net metering is permitted, most utilities will replace your kilowatt-hour meter with two ratcheted meters: one to register your purchases, the other to register your sales to the utility. The utility then balances the account on a regular basis.

In true net-metering programs, the account is balanced annually. Beware, there are some faux net-metering programs where the accounts are balanced monthly and some where generation fed into the grid is paid substantially less than the retail rate. (Yes, some utilities call this "net metering.")

There are often onerous restrictions on net metering: who can use it, the size of the system permitted, and the size of the total program. Most jurisdictions limit the size of the wind turbine that can be used under net metering.

Most states and provinces limit the total amount of generating capacity that can be installed under net-metering programs. The limits are often meager, typically just a small percentage—sometimes just a fractional percentage—of the utility's total load.

Net metering can make interconnecting a small wind turbine with the utility more financially attractive than it might otherwise be. With net metering, you can use a larger, more cost-effective wind turbine than you would where net metering was prohibited, but you are still limited in what you can use. If you choose to use a larger wind turbine because it's more cost effective than a smaller one, you may produce more electricity than you consume. Under most net-metering programs, utilities will pay only a token amount for your surplus generation, and in some states, they won't pay anything at all. It's tough to make a good economic case for wind energy under these conditions.

Degree of Self-Use

In states and provinces where consumers are not allowed to run their kilowatt-hour meter backward, it becomes critically important to know how much of the wind generation will be used to offset on-site consumption. This is the degree of self-use, or generation used on-site (see Figure 13-9. Degree of self-use). Output from a wind turbine varies with the wind. Electrical consumption in your home or business varies with the time of day. When you try to match the two together, you get an almost unpredictable mix. Some moments there will be excess generation; other times, there's a net deficit, and you'll need to draw power from the grid.

Homeowners with a wind turbine under these conditions will want to minimize their excess generation. Why sell valuable electricity to the utility for $0.03 per kilowatt-hour, when the utility will turn right around and sell it back to you for $0.10 per kilowatt-hour or more? To avoid selling a surplus, consumers have several choices: they can adjust their consumption by using dump loads as much as possible, they can match their consumption to wind availability as much as possible, or they can use wind turbines smaller than they might otherwise select.

It's a catch-22 situation. Wind turbine cost effectiveness increases with size. But to minimize the amount of energy you sell to the utility at a deep discount during times of surplus, you'll want to buy a wind turbine that produces only a portion of your own consumption. Under German wind conditions and electrical consumption, for example, only one-third of the electricity from a wind turbine sized to meet the annual domestic needs of a home will actually be used in the home. Two-thirds will be sold back to the utility. To use two-thirds of the consumption in the home and only sell one-third back to the utility, you need to use a wind turbine that

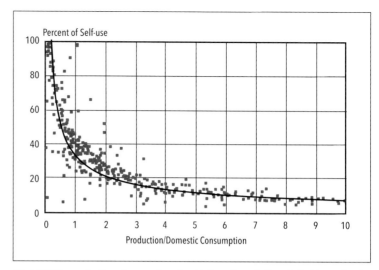

Figure 13-9. Degree of self-use. The percentage of on-site demand supplied by a wind turbine as a function of the wind turbine's annual generation and total consumption. This chart is derived from German data on more than 400 household-size wind turbines—the most extensive survey of its kind in the world. There is some scatter in the data because some owners wisely shift discretionary consumption to periods when the wind is strongest. (Fraunhofer Institut fuer Windenergie und Energiesystemtechnik)

produces half or less of your annual domestic consumption.

Dealing with the Utility

Electric utilities are businesses, and their employees are charged with making them profitable. Utilities are often large institutions. And like any big bureaucracy, the right hand frequently doesn't know what the left is doing.

Corporate policy, as expounded by company managers, may be adamantly opposed to the interconnection of wind turbines of any size. Yet the word may not have drifted down to the lower levels where the work gets done. You may call up and be surprised to find the staff friendly and curious about your project. They may go out of their way to be helpful. This was the case with Metropolitan Edison, the infamous Pennsylvania utility with nuclear reactors at Three Mile Island. In spite of management's public opposition to renewables (they were still enamored of nuclear power even after the accident), Met-Ed's staff were extremely cooperative with the interconnection of a small turbine near Harrisburg.

Of course, there's the other type: utilities that make supportive pronouncements about clean energy sources and brag about all they're doing to improve the environment, but when you contact them, they have a different story. Mike Bergey of Bergey Windpower reports that in some cases it has taken more time to reach agreement with the utility than to build and install the wind system. In a particularly flagrant example, the Los Angeles Department of Water and Power once simply told a wind turbine dealer it didn't want small wind turbines connected to its lines. End of discussion.

Because utilities are bureaucracies, nothing happens quickly. You should notify them of your plans months in advance. Now that wind turbines have become much more commonplace, they may have a clearly defined policy and may be able to promptly give you the answers you seek.

You need to appreciate the utility's point of view because you're dependent on their cooperation. The difference today is that we have almost three decades of actual experience with wind turbines operating on utility networks around the world. Utilities are far less likely to balk at interconnection today than they were 20 years ago. Utility executives dread nothing more than a self-righteous customer strutting into the office and demanding that the utility kneel down and graciously grant a trouble-free interconnection. If you want a fight, there's no better way to find one.

Wind turbine manufacturers have experience dealing with utilities, and they can be helpful in obtaining a fair contract. They can also assist with any technical issues that the utility might raise. Often the manufacturer will talk directly with the utility engineer handling your application. This can facilitate a quick agreement because the two speak the same technical language.

Cooperation is growing between utilities and the wind industry as more utilities gain operating experience with modern wind turbines. Bergey Windpower reports that most utilities they deal with require only one phone call from the manufacturer to grant permission for the interconnection. Some utilities have also allowed net metering, even when they were not required to do so.

Distributed Generation

The small turbine industry in Denmark began much the same way as that in North America. The first wind turbine was connected to the grid without permission, and unbeknownst to the utility, in the mid-1970s. This could have been the world's first example of "guerrilla wind," the counterpart to the 1990s "guerrilla solar" movement countenanced nearly two decades later by *Home Power* magazine. But the Danish industry has since followed a far different trajectory than the small turbine industry has in North America. The once small turbines of Danish manufacturers gradually grew bigger, becoming the core of the world's commercial wind power industry of today.

Early Danish turbines were about the same size as those being installed in the United States at the time: 7 kW, 10 kW, and 15 kW. But they and their manufacturers grew because there was public and political support for them to do so. This support manifested itself in flexible siting policies and in interconnection policies that went well beyond net metering. First Denmark, then later Germany, implemented policies that gave wind turbines priority access to the grid. And, equally as important, Denmark and Germany paid a fair price—a price sufficient to make projects profitable—for wind-generated electricity.

Instead of limiting the interconnection of a wind turbine to a consumer's own property for directly offsetting his or her consumption (as in net metering), Denmark permitted the wind turbine to be located anywhere within the township and eventually within the neighboring township as well. Thus, there was no meter to run backward. There would only be one meter, a new meter where none had existed before, and where it would register delivery—and sales—to the utility.

Danish politicians were responding to a widespread demand for the ability of the public to develop their own renewable resources, especially to install their own wind turbines. But like politicians everywhere, they were cautious. They didn't want to open the floodgates. Denmark, with a long history of cooperative rural economic

The German feed-in law went beyond the Danish program by placing no restrictions on how much electricity could be generated by each owner.

THERE IS NO MENTION OF NET METERING OR OFFSETTING ONE'S OWN CONSUMPTION.

development, targeted their attractive wind policies at local ownership. Each homeowner was allowed to buy shares in the generation from a local wind turbine equivalent to 1.5 times his or her domestic consumption. Eventually, the limit was raised so that a family would not only offset their electricity consumption, but also the equivalent of their total energy consumption as well, including heat.

Denmark, thus, created a program that not only permitted interconnection with the grid, but also encouraged the public's participation in doing so. By the 1990s, 5% of the Danish population owned shares in a neighborhood wind turbine. Today, these cooperatively owned wind turbines can be seen distributed across the Danish landscape from the far northwest of Jutland to the city of Copenhagen.

Similarly, in 1991 Germany responded to public clamor for the ability to generate renewable energy by launching its first electricity feed-in law, the Stromeinspeisungsgesetz: literally the law on feeding electricity into the grid. Germany's feed law required utilities to not just permit but also to enable connections with renewable energy generators, and it spelled out what utilities were to pay for this generation.

The German feed-in law went beyond Danish policies by placing no restrictions on how much electricity could be generated by each owner. There is no mention of net metering or offsetting one's own consumption.

Danish renewable energy policy specifically encouraged distributed local ownership. German policy did not. Distributed development in Germany largely resulted from the complexity of siting wind turbines in a densely populated country.

But in both Denmark and Germany, it was in everyone's financial interest to install the size wind turbine that made the most economic sense at the time. Farmers, homeowners, and wind developers were not forced to choose a wind turbine smaller than they needed because they could only offset their own consumption.

Until recently, most commercial wind development in North America has been the construction of large, central-station wind power plants. In the early 1990s, a large project could be 50 MW to 80 MW. Today, projects of 200 MW to 300 MW are routinely announced. These projects are all connected at transmission voltages. They are far too large to be added to local distribution lines.

However, many wind projects in Denmark, Germany, France, and the Netherlands are connected to the local distribution system. The size of wind project, or the number of turbines it can support, varies widely depending upon the size of the conductors (the lines carrying the electricity), other generators already on the line, and many other factors. Projects could range from 5 MW (two to three turbines) to 20 MW (the equivalent of ten large turbines).

Projects of this scale, the third way of developing wind energy, can be placed close to the loads they will serve: towns, cities, scattered farmsteads, industrial zones, and so on.

The upper Midwest has seen more distributed wind development than any other region of North America, and Minnesota in particular stands out. For more than a decade, Minnesota has encouraged distributed wind development alongside large commercial projects. The state first limited distributed wind projects to no more than 2 MW, but as turbines grew ever larger, this limit was eventually lifted.

Now, there are single large wind turbines or small clusters of turbines throughout the windy parts of the state and even some in less windy areas. Large wind turbines can be found at schools and universities, on farms, and in privately owned, mini–wind farms.

The future promises a rapidly growing contribution from distributed wind in North America as the wind industry, community and renewable energy activists, and political leaders realize the economic and energy benefits. Although large-scale wind farm development will continue to grow rapidly, and some individuals will continue to install small wind turbines, the prospect of locally owned, distributed wind will become increasingly appealing.

Grid Integration

Those opposed to wind energy continually raise the objection that wind energy is difficult and costly to integrate into a utility network. Like a broken record, they've repeated the same charges for three decades despite the obviously successful use of wind energy in increasingly high percentages around the world.

As the name implies, the US Energy Information Administration (EIA) tracks data on energy worldwide. EIA data for 2012 reveals that wind turbines in Denmark generated 36% of the country's electricity; Portugal generated 24% with wind; Spain 18%; and Germany 9%. In every one of those countries, the lights stayed on, the trains ran, and industry continued to hum. In North America, Iowa generated 27% of its supply with wind energy in 2013. Wind was second only to coal, says the EIA. That should be enough data to counter critics' claims.

Wind's Variability

Nevertheless, there remain common myths about wind energy's unreliability. Unlike conventional power plants, wind energy is a variable resource. The amount of generation depends upon the wind available. Although we can turn wind turbines off when we like, we can't call on them to generate electricity if the wind isn't blowing.

Critics like to use the word *intermittent* to describe wind energy as a source of generation. This is deliberately misleading. The generation from wind varies with the wind, certainly, but it's not intermittent. Wind turbines are seldom on, then off. Wind turbines at windy sites will be in operation as much as two-thirds of the hours in a year. That's hardly intermittent. Their output varies widely with the wind, but even this variability is reduced when wind turbines are dispersed geographically.

Wind turbines at windy sites will be in operation as much as two-thirds of the hours in a year. That's hardly intermittent.

On the surface, it appears that a resource that cannot be controlled at will threatens the reliability of the entire grid. Fortunately, a better understanding of wind technology is slowly overcoming the once bedeviling specter of wind energy's seeming Achille's heel—what to do when the wind stops.

While critics are quick to point out the variability of wind and solar generation, they seldom mention the variability of other sources of electricity, such as nuclear, fossil fuels, or large hydro.

Wind is often compared—unfavorably—to large hydro because water can be stored behind a dam, whereas wind cannot. What's apparent from the data—for anyone willing to look—is that generation from hydro is much more highly variable from year to year than that from nonhydro renewables, especially wind.

Utilities must manage variability in both the demand for electricity and in the generation supplying that demand, minute-by-minute and hour-by-hour, as well adjust to seasonal and annual variability. Although wind is a variable resource—that is, the wind doesn't always blow and when it does it doesn't always blow at the same strength—wind is far more reliable than critics charge. In fact, wind is fairly predictable on long time horizons, especially from one year to the next.

The development of wind energy in Portugal is one of the often-overlooked renewable energy success stories. The 11 million Portuguese have made remarkable strides in developing their wind resource. Wind provided less than 1% of supply in 2002, but within one decade, wind generated nearly one-fourth of Portuguese electricity. (Imagine that kind of growth in North America.)

When we examine renewable sources of energy in Portugal, we find large swings in the interannual generation (see Figure 13-10. Variability in hydroelectricity generation in Portugal). This is due to the heavy dependence of Portugal on hydro. Because of its Mediterranean climate, Portugal—like neighboring Spain—is subject to periodic droughts. These play havoc with hydroelectricity. For example, during the drought year of 2005, hydro dams in Portugal generated less than 5 TWh whereas in the rainy year of 2010 hydro generated more than three times as much. Hydro's contribution to Portugal's supply varied from more than 50% to as little as 14% during the past three decades.

In contrast, the generation from wind, solar, and other renewables, while variable throughout the day, is relatively predictable from year to year. The contribution of new renewables, primarily wind, has steadily grown in Portugal. These new sources of generation will in fact increase stability in the Portuguese electricity system from one year to the next.

Likewise, nuclear power is "reliable" until it isn't, as the units at the Fukushima nuclear power plant so dramatically demonstrate. The nuclear disaster still unfolding in Japan isn't the first time the Fukushima plants have been in the news. Earlier they were at the center of a documentation scandal. Several of the reactors were shut down from 2002 to 2005 for safety inspections as a result of the utility's falsification of inspection and repair reports (see Figure 13-11. Interannual variability of generation at Fukushima). Generation didn't return to previous levels until just before the accident brought all the reactors down for good.

The Fukushima 1 plants generated, on average, 30 TWh per year. The key words here are *on average*. Despite nuclear power's reputation as reliable base load generation, the Fukushima plants were anything but reliable over the four decades that the plants were in operation. Annual generation was surprisingly erratic or "lumpy" in the jargon of the trade.

Take Unit 6, the most modern unit, for example. In 2004 generation dropped from 4.6 TWh in 2003 to 1.1 TWh, and both were a far cry from the reported generation in 1997 of more than 9 TWh. That's a lot of generation off-line for even a big system like that in Japan that requires 1,000 TWh per year. (For comparison, the United States consumes 4,000 TWh of electricity annually, Germany 600 TWh.) Similarly, Unit 5's generation fell from 6.2 TWh in 1999 to 1.6 TWh in 2000.

But not just annual generation from individual units varied significantly. Combined generation from Fukushima 1 also fluctuated from one year to the next. The safety shutdown at Fukushima 1 cut generation by two-thirds or nearly 20 TWh from 2002 to 2003.

Compare this with the inexorable rise in generation from wind energy in Germany and Spain. Both countries each produce more electricity with wind than Fukushima 1 did with nuclear power during Fukushima's heyday. Contrast the steady wind generation in Denmark with the wild swings in generation from Fukushima 1. Keep in mind that the Danish fleet is largely comprised of early wind turbines, now more than two decades old, yet there's hardly any annual variability in contrast to that at Fukushima.

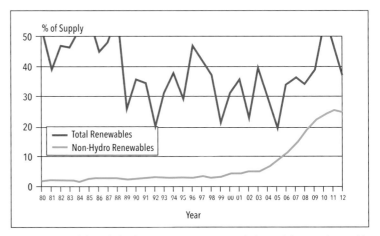

Figure 13-10. Variability in hydroelectricity generation in Portugal. The variability in total renewable generation in Portugal is due to the variability in hydro. Contrary to popular belief, the interannual variability of hydroelectricity is far greater than that of wind energy. If hydro is available behind a dam, for example, it can be dispatched as needed. However, the availability of the resource itself varies dramatically from one year to the next.

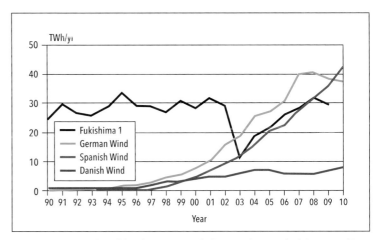

Figure 13-11. Interannual variability of generation at Fukushima. Wind energy in both Germany and Spain now generates more electricity than the reactors at Fukushima 1 did before the nuclear disaster. The interannual wind generation is more predictable, reliable, and stable than that historically from the block of reactors at Fukushima 1.

German wind energy generation has been far more stable from one year to the next than Fukushima 1. Throughout the last two decades, more and more wind generation has been added to the German electrical network. Today, German wind turbines generate as much electricity as the entire Fukushima 1 complex at its peak.

Today, German wind turbines generate as much electricity as the entire Fukushima 1 complex at its peak.

Moreover, it is highly unlikely that an accident at one wind turbine in Germany will affect the more than 20,000 turbines operating across the breadth of the country. The loss of one wind farm with tens or even hundreds of turbines will likewise have little effect on overall wind generation in Germany. And German wind generation is expected to continue growing for the next decade at least.

Unfortunately, all six reactors at Fukushima 1 are down permanently with a loss of 30 TWh per year of generation or nearly 3% of Japan's supply. That's not reliable.

Capacity Credit

These ideas about wind energy's reliability linger because the "technology has surpassed the institution," explained BTM Consults' Per Krogsgaard in the early 1990s. Krogsgaard noted then that utility executives based their perceptions on ideas conceived decades prior. According to conventional wisdom at the time, utilities would still need the same amount of generating capacity with or without wind power plants because it is sometimes unavailable when most needed.

This has led some to argue that wind will never be "more than a supplemental source of power." Disregarding the fact that no technology alone, whether coal, natural gas, or nuclear power, is more than a supplement to a utility with a diversified mix of generating plants, it ignores repeated technical studies that show wind energy does indeed provide capacity benefits.

Although wind turbines may be idle due to a lack of wind at times of a utility's peak demand, there is a statistical probability that they will be available, especially if there are multiple turbines dispersed geographically. In this, wind turbines are no different from conventional power plants. No generating plant operates 100% of the time, and no power plant is 100% dependable during peak loads.

For example, nuclear plants are typically on or off. They may be used as baseload, but when they're off, something else must pick up their lost capacity. In the late summer of 2014, four reactors were taken off-line at the same time in Great Britain. In a reversal of traditional thinking on variability, Britain's fleet of wind turbines picked up the load the reactors had dropped. So much for not being available when needed.

The work of engineers on both sides of the Atlantic, including that of the late Don Smith of California's Public Utility Commission (CPUC), have refuted the notion that wind energy cannot supply secure power. The question then becomes not if there is any capacity value in wind energy, but what its value is in offsetting the construction of conventional power plants. Utilities have traditionally viewed wind energy solely as a fuel saver. Each kilowatt-hour generated by a wind turbine offsets a kilowatt-hour that would have been otherwise generated. Experience now shows that in most cases the capacity value of wind energy to a utility is equal to that of the fuel it offsets.

Smith, for example, found a very close fit between the wind resource in California's Solano County and Pacific Gas and Electric's demand for generation. This justified a capacity credit equivalent to that of a fossil-fuel-fired power plant. Using the same analysis, he found a lesser, but not insignificant, value for the Altamont Pass. Subsequently, the CPUC examined the question in depth during the early 2000s (see Table 13-1. Capacity Credit in California).

As one would expect, a gas-fired power plant can be called on as needed and a geothermal power plant not limited by its steam supply can be counted on for 100% of its capacity. Yet contrary to critics' expectations, wind too could be counted to offset a portion of conventional generation roughly equivalent to its capacity factor.

In a large study on grid integration done in Germany, researchers found that the investment in the grid necessary to integrate the new renewable generation needed to meet German targets was less costly than additional grid expansion if new conventional power plants were built. Investment in the grid to integrate large amounts of renewables would result in making the grid

Table 13-1. Capacity Credit in California	
Resource	Relative Capacity Credit
Medium Gas	100%
Biomass	98%
Geothermal (constrained)	74%
Geothermal (unconstrained)	102%
Solar	57%
Wind (Altamont)	26%
Wind (San Gorgonio)	24%
Wind (Tehachapi)	22%
Source: California RPS Integration Cost Analysis; CEC 500-03-108C, December 2003, page xi.	

more stable and reliable with or without the renewable generation, but with renewables the system would be less costly.

Britain's Energy Research Centre (ERC) issued a lengthy report analyzing the results of some 200 studies on the cost and impacts of integrating wind and other new renewables into electricity networks. Their findings were not surprising. ERC concluded that "wind generation does mean that the output of fossil fuel-plant needs to be adjusted more frequently, to cope with fluctuations in output." Overall, the costs from these effects are much smaller than the savings in fuel and emissions that renewables can deliver.

More significant was ERC's refutation of the antiwind lobby's favorite whipping boy: the need for backup. "There is no need to provide dedicated back-up generators" just for wind, the study found. There is a need to increase the reserve margin on the entire system when wind is added, but the reserve margin is there to cover outages and variability from all generators, not just that of wind. "No electricity system can be 100% reliable, since there will always be a small chance of major failures in power stations or transmission lines when demands are high," the report warned.

Wind generation introduces additional uncertainties in managing the grid, but these can be quantified and the costs determined. Certainly, the introduction of large amounts of renewable generation "will affect the way the electricity system operates" ERC found. One of the costs is for balancing "the relatively rapid short term adjustments needed to manage fluctuations over the time period from minutes to hours," ERC explains.

Balancing Cost

There are costs to compensate for when balancing the need for generation when wind energy is available over short time periods. Specifically there are increased costs to the system for regulating voltage and frequency when wind energy is on the grid as the wind ebbs and flows. Of course, the grid is designed to handle such fluctuations, and operators are trained to manage variations throughout the system in response to wide fluctuations in the load or demand for electricity.

Table 13-2. Regulation Cost of Resources in California 2003

Resource	Regulation Cost $/kWh
Total Load	0.00042
Medium Gas	0.00008
Biomass	0.00000
Geothermal	0.00010
Solar	0.00004
Wind (Altamont)	0.00000
Wind (San Gorgonio)	0.00046
Wind (Tehachapi)	0.00017

Source: California RPS Integration Cost Analysis; CEC 500-03-108C, December 2003, page xii.

The CPUC study found that the cost to manage the fluctuations from wind were negligible. The regulation burden caused by wind energy was less than that imposed by loads the system was designed to serve, said the report. Wind and solar variability were essentially "noise" compared to the variability in the system's load and the variability in other sources of generation (see Table 13-2. Regulation Cost of Resources in California 2003).

Jérôme Guillet and John Evans wrote in an article for *Renewables International* that balancing costs range from €0.002 to €0.004 per kWh ($0.0025–$0.005/kWh). Guillet should know. He arranges financing for offshore wind projects and has access to information from both sides of the negotiating table. The payment for these balancing costs are already collected in some markets, says Guillet.

The regulatory question becomes how the balancing costs are determined, whether they are set on a generator-per-generator basis, or at the scale of the whole system. Guillet points out that "integrating the variability of 1,000 wind farms costs much less than 1,000 times the short-term balancing cost of a single farm." What to charge and who pays is the question that's important.

Guillet's estimate matches that of ERC's of balancing costs for Britain (see Table 13-3. Estimated Cost of Variable Generation in Britain). ERC concluded that the aggregate costs for wind's variability in Britain, including short-run balancing costs and the costs of increasing the system's margin, was in the order of £0.005 to £0.008 per kWh ($0.0080–$0.0129 per kWh). In an article in *The Electricity Journal*, Joseph

Table 13-3. Estimated Cost of Variable Generation in Britain				
	£/kWh		$/kWh	
	Low	High	Low	High
Short run balancing costs	0.002	0.003	0.0032	0.0048
Higher system margin costs	0.003	0.005	0.0048	0.0080
Total	0.005	0.008	0.0080	0.0129
Source: The Costs and Impacts of Intermittency; ERC, March 2006, page 47.				

DeCarolis and David Keith arrived at similar costs ($0.01–$0.02 per kWh) to the system for wind's variability at a penetration of 50%. Although nonnegligible, the costs are not the showstopper critics of wind were searching for—even at high penetrations.

Penetration

Just how much wind can we put on the system? The answer to that question has steadily increased as we've learned more about the technology, the benefits of geographic dispersal, and how best to manage wind's variability. Robust systems that are well interconnected with neighboring regions, such as Denmark, already have higher penetrations of wind than we thought possible as recently as two decades ago.

It's not uncommon at all for Denmark to produce more than 100% of its electricity with wind during windy days. At the same time, Denmark is also producing electricity from its biomass and combined heat and power plants. The surplus generation is fed to its neighbors, including Norway, where it is stored behind the country's numerous dams. At the current pace of development, Denmark will meet its target of generating 50% of its annual electricity consumption with wind energy by 2020.

DeCarolis examined in depth how much wind you could add to the electrical network in the Midwest for his dissertation. The model he devised to analyze how the system would work produced some surprising results. When wind serves upward of 60% of consumption, DeCarolis's model chooses to install more relatively inefficient gas turbines than the expected more efficient but more costly combined cycle plants. The reason was that the simple gas-fired plants were operated so few hours per year that it was more cost effective to install them than the much more costly combined cycle plants. The plants would be used so infrequently that it wouldn't justify their higher costs.

Peter Freere, formerly an engineering professor at Monash University in Australia, notes that fast-reacting conventional generators, including hydro, conventional gas turbines, and diesel generators, are able to compensate for variations in the output of wind generation: "It is correct that in normal electric grids without energy storage, a wind farm would not work well on its own. Some conventional energy sources are also required." But the same requirement applies to conventional sources as well, especially nuclear plants, "whose response time is so slow that they must have a fast responding generation system in parallel. It is also true that due to the large sizes of modern generators in conventional power plants, for example, 500 MW per generator, it is necessary to have a complete spare generator ready to take over when another stops working (for whatever reason)."

"Although there have been numerous studies predicting the maximum penetration level of wind energy in a system, the actual experimental results in the case of diesel systems indicate that 70% penetration" is acceptable, Freer concludes. Some are envisioning even higher penetration levels but not solely from wind.

It's All in the Mix

From the smallest off-the-grid system to giant networks like those in Germany or North America, all work best with a mix of resources. They are more stable, more reliable, and provide overall lower costs—over the long term. No one is seriously proposing a system that uses 100% wind energy, though it can be done. However, many researchers are studying how renewables can be used to generate 100% of the electricity supply—or at least very close to 100%. Of course, this includes wind, and in most cases wind is the "work horse of the energy transition," as the Germans say.

It's not just academics. Politicians—who must be reelected—are staking their careers on a fully renewable path. As mentioned elsewhere in *Wind Energy*, Denmark plans to meet 100% of its electricity and heat with renewable energy by 2035; the north German state of Schleswig-Holstein plans to meet 100% of its electricity consumption with renewables by 2020, as does Scotland.

Evidence continues to mount that it can be done. How it can be done for the least cost with the most stability of the grid has become the focus. Unlike just a few years ago when such ideas were built on dreams, today there are computer banks of data on how wind, solar, and other renewables operate on real systems.

For example, French renewable energy analyst Bernard Chabot has monitored the performance of wind and solar in Germany (see Figure 13-12. Hourly production of wind and solar in Germany). The data from Germany, which has the highest concentration of photovoltaic systems in the world, confirms what has been known for some time: wind and solar work best together. While the generation from both resources is variable, that from solar is far more periodic. Solar generation peaks at noon and disappears entirely at nightfall. If there's sufficient wind on the system, it can dampen these daily peaks and troughs. The more wind relative to solar generation, the better the smoothing effect.

There's also a good match between the two resources seasonally. Solar peaks during the summer when winds are typically light (see Figure 13-13. Average daily productivity of wind and solar in Germany for 2013). Thus, the smoothing effect of the two resources together is more pronounced on an annual average than a daily average. To maximize this smoothing effect of the two resources together, it's critical to optimize the mix. Chabot concludes that for winter-peaking networks, like that in Germany, 60% of the renewable generation capacity on the system should be wind. More specifically, wind generation—the amount of electricity produced—must be two to three times that of solar for system optimization.

Other researchers have reached similar conclusions. It's the mix that's important, and that mix will vary from region to region. In a groundbreaking study, researchers at the University of Delaware found that a mix of renewables could generate from 90% to 99.9% of the electricity needed in 2030 on the Pennsylvania–New Jersey–Maryland (PJM) system at costs comparable to those of today. Diverse sources of generation and the installation of more nameplate generating capacity than seemingly required met

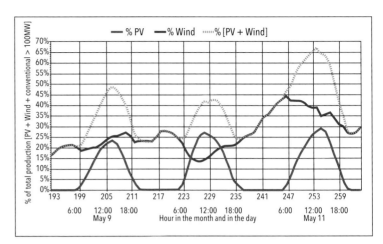

Figure 13-12. Hourly production of wind and solar in Germany. The generation from wind and solar together smoothes out the hourly fluctuations from each technology individually. However, during these three days in May 2014, wind was far less variable than the predictable variation in solar. (Bernard Chabot)

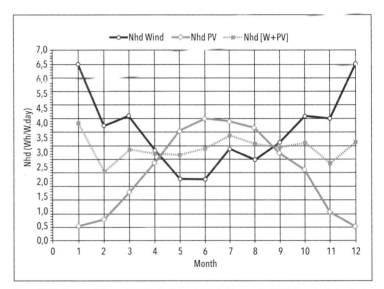

Figure 13-13. Average daily productivity of wind and solar in Germany for 2013. Though wind and solar vary widely throughout the day and through the seasons, combined they offer a much smoother form of generation than either alone. To minimize integration costs, it's necessary to optimize the proportions of each technology on the system. (Bernard Chabot)

the system's need at the least cost with the least amount of storage. For the storage needed, the researchers suggested that the growing fleet of electric vehicles (EVs) could be used as a kind of mobile backup. (For more on EVs and wind, see Chapter 21, "The Challenge.")

Storage

German researchers have reached similar conclusions to those at the University of Delaware: storage isn't really needed until very high penetrations of renewables and even then much less is required than first thought. Thus, another myth—the great bugbear of wind energy—that storage is needed for any but the most insignificant contribution has fallen.

In a study by four academic teams for the German think tank Agora Energiewende, researchers concluded that storage wasn't needed in Germany until renewables reached more than 60% of supply. Germany still has a ways to go. They only get one-fourth of their supply from new renewables today. It could be a decade or more of continued renewable development before Germany needs to deploy storage.

Moreover, the form of storage that may be most useful could be quite different from what we once imagined. Batteries are far from the first choice.

Storage could take the form of hot water, for example. Both Germany and Denmark extensively use district-heating systems. Surplus wind generation during windy winter weather could be stored as hot water and used to heat homes and businesses. Thus, considerably more wind-generating capacity—and that from other renewables as well—could be installed than necessary to meet the demand for electricity alone. The surplus could then be used to meet the need for heat, and by doing so the renewable capacity could take advantage of the simplicity inherent in storing electricity in hot water. Storing energy as hot water is inexpensive. The technology is also readily available.

District heating isn't common in North America today. It once was, and it could be again if we chose to integrate our electricity and urban heating systems to take advantage of what each has to offer.

Another way to store electricity is in the form of "natural" gas. Surplus wind generation could be used to produce hydrogen, which is then "methanated" immediately with carbon dioxide, producing synthetic natural gas. Germany, like North America, has an extensive network of gas pipelines and underground storage. The resulting methane can be used however we currently use methane—to generate electricity, if need be, or to heat our homes and power our industries.

Researchers at Fraunhofer's Institut für Windenergie und Energiesystemtechnik in Kassel estimate that meeting Germany's target of 80% of its electricity from renewable sources will require 60 TWh of storage in addition to regional and cross-border balancing of supply. Currently, Germany has 220 TWh of underground natural gas storage. This is nearly four times the storage needed for electricity in the form of gas, showing that this is well within the realm of possibility.

Today, the electrolysis of water into hydrogen and methanization of hydrogen into synthetic natural gas is too costly, but that could change with more research. In the meantime, we know how to inexpensively store heat as hot water and how better to integrate our energy systems to minimize the need for storage altogether.

In summary then, there are a number of myths about the integration of renewables, especially wind energy, into an electricity network because of the variability of the resource. Though there are technical issues surrounding the grid integration of large amounts of wind energy, they are in no way insurmountable, nor are the costs excessive. Where there has been a desire to do so, as in the utility networks in Denmark, Germany, Spain, and Portugal, there have been the technical and managerial means to do so at modest cost.

We know how to interconnect our wind turbines with the grid. We know how they work and what they do. And we know how to optimize the grid to work best with ever-growing contributions from wind and other sources of renewable energy.

14

Pumping Water

Ech kier die Nuet
On schaff och Bruet
(I turn away need and furnish your bread.)

—Inscription in Low German (Plattdeutsch)
dialect on the beard of a windmill in northern Germany.

FOR THOSE IN INDUSTRIALIZED COUNTRIES SERVED BY community water systems, where water—like electricity—is available on demand, it's hard to imagine the importance of a water pump to the world's rural population. Lifting or carrying water accounts for much of the energy expended in Third World villages. It's also a major load for North American homes beyond the reach of utility lines. In fact, settlement on North America's Great Plains wasn't feasible until experimenters developed a reliable means of pumping water from deep wells. The technology that made settlement possible was the American water-pumping windmill, and it forever changed the face of the landscape.

Windmills That Won the West

Three technological innovations made settlement on the Great Plains possible: the Colt 45, barbed wire, and—the farm windmill. Yes, the windmill—that ubiquitous symbol of rural North America—ranks right alongside the six-shooter, says Walter Prescott Webb, a noted historian.

Webb writes in his seminal book, *The Great Plains,* that the water-pumping windmill was so essential for life in what was then known as the Great American Desert, that settlers warned newcomers that "no women should live in this country that can't climb a windmill or shoot a gun." Promoters extolled the virtues of a land where "the wind pumps the water and the cows chop the wood."

In the semiarid lands west of the Missouri River, the wind did indeed pump the water. Unlike in the eastern United States, few streams coursed across the surface, and seldom were aquifers in reach of simple hand-dug wells. Water was there, but at depths that required pumping by machines—wind machines. (And those poor homesteaders who couldn't find wood for their hearth on the treeless landscape burned cow chips from their bovine lumberjacks.)

The nation's westward migration both caused and was aided by the growth of a great midwestern industry that built windmills. By the late

Figure 14-1. Traditional farm windmill. Dempster's No. 9 at the Windmill State Recreation Area near Gibbon, Nebraska. Thrust on the pilot vane (small vane) furls the rotor toward the tail vane (large vane) in this classic design. Crescent moon weight regulates the wind speed at which the multiblade rotor furls toward the tail vane.

nineteenth century, 77 firms were assembling windmills in one form or another. Catalogs of the day bristled with choices. Sears Roebuck and Wards offered their own house brands.

In 1854 Daniel Halladay invented the first fully self-regulating windmill. Until then, millers had to manually turn the spinning rotor out of the wind or reef (roll up) the sails during storms. Halladay changed all that by constructing a multiblade rotor made up of several movable segments that reefed automatically in strong winds (see Figure 5-42. Halladay rosette or umbrella mill).

Early models of Halladay's rosette or umbrella mill used a tail vane to point the rotor (or wheel to old-timers) windward. Later vaneless versions did away with the tail vane by orienting the wheel downwind of the tower.

Because Halladay's mills could be left unattended, they were ideal for remote pastures where water was scarce. His "patent" mills were immediately popular with farmers for watering livestock. But it wasn't until the post–Civil War boom in railroad construction that the fledgling windmill industry began to flower.

Water was as essential to running a steam locomotive as coal. As the transcontinental railroads pushed westward across the plains, the water-pumping windmill came into its own. Huge windmills (even by today's standards) with rotors up to 60 feet (18 meters) in diameter pumped a steady stream into storage tanks at desolate way stations. Through skillful marketing, Halladay's chief rival, the Eclipse, created a name for itself as the railroad mill.

Invented in 1867 by Rev. Leonard Wheeler, the Eclipse used fewer moving parts and was both cheaper to produce and easier to maintain than Halladay's design. Wheeler built a solid wheel instead of the sectional wheels Halladay popularized. Wheeler protected his windmill in high winds by furling the rotor toward the tail vane. The idea was so successful that even Halladay's U.S. Wind Engine and Pump Co. began producing similar versions under the Standard trade name. Other manufacturers also followed suit (see Figure 14-1. Traditional farm windmill).

The stage was set. The technology existed and an industry was in place. Demand grew rapidly as the railroads poured settlers onto the prairie. The industry blossomed like the homesteaders' gardens that mass-produced windmills soon made possible.

The farm windmill was on its way to becoming an American icon. It quenched many thirsts, provided an occasional bath, and offered salvation through baptism in its associated stock tank. The clanking windmill sang a lullaby to many a sleepless child. The gentry found it useful as well. To their country villas and suburban estates, the windmill brought running water and the convenience of Sir Thomas Crapper's flushable commode.

This was an age of invention and empire building. Steel and the factory system were driving the Industrial Revolution to new heights. Into this milieu stepped LaVerne Noyes, a Chicago industrialist.

In 1883 Noyes hired an engineer, Thomas Perry, to develop a thresher. With his previous employer, U.S. Wind Engine and Pump Co., Perry had conducted over 5,000 tests on different windmill designs. His had been the first scientific attempt to improve wind machines.

Promoters extolled the virtues of a land where

"THE WIND PUMPS THE WATER AND THE COWS CHOP THE WOOD."

By using a steam-driven test stand, Perry had designed a rotor nearly twice as efficient as those then in use. But Halladay's U.S. Wind Engine and Pump Co. wasn't interested. Noyes was. He encouraged Perry's work.

Five years later when Perry's trade secrets were released, Noyes introduced the Aermotor. Derisively tagged the "mathematical windmill" by competitors, the Aermotor incorporated both Perry's design and Noyes's manufacturing sense.

Although it wasn't the first to use metal blades (Mast, Foos and Co.'s Iron Turbine used them in 1872), Aermotor's stamped sheet-metal "sails" revolutionized the farm windmill. Another innovation was Aermotor's method of furling the rotor. Rather than use a pilot vane as on the Eclipse, Perry offset the Aermotor's wheel from the center of the tower. High winds striking the rotor disk forced it to fold toward the tail, eliminating the need for the pilot vane. The furling force was counterbalanced by a spring that held the rotor into the wind. This arrangement worked so reliably that it's still used.

Noyes's ability to mass-produce windmills at Aermotor's Chicago plant reduced its cost to one-sixth that of its competition and led Aermotor to dominate the industry. Eventually, Aermotor captured 80% of the market. Today, if you see an abandoned windmill along a country lane, it's most likely an Aermotor.

The industry peaked in the early part of the twentieth century, collapsing soon after with the extension of electricity to rural areas. Durability played a large role in the industry's demise in the days before planned obsolescence. On many homesteads, the family windmill has been in continuous use for generations.

After a century of refinement, the farm windmill performs its job well. It remains the signature of the Pennsylvania Dutch and Amish settlements in Ohio, Indiana, and Iowa. The Amish, who use them for domestic water, and ranchers of the Southwest, who use them for stock watering, account for the few thousand sold in the United States each year (see Figure 14-2 Chicago or American farm windmill).

Farm windmills, says Vaughn Nelson, founder of West Texas A&M University's Alternative Energy Institute, produce the equivalent of 130 million kilowatt-hours yearly, pumping water for

Figure 14-2. Chicago or American farm windmill. Standard design for multiblade mechanical wind pump used worldwide. Typical stamped, sheet-metal "sails" on Aermotor brand farm windmill seen here in a private collection in California's Central Valley.

livestock on the Great Plains. "They still make sense in a lot of places," he says.

Mechanical Wind Pumps

The Dutch, or European, windmill was well suited to shallow surface sources, but water on the Great Plains was found deep below the surface. The mechanical wind pump, or farm windmill, enabled settlers on North America's Great Plains to pump the relatively low volumes of water needed for domestic uses from deep wells. The farm windmill was a perfect match between the needs of settlers and the abundant winds found on the prairies.

The design proved so successful that it has been widely copied around the world. Today nearly one million remain in use, mostly in Argentina, the United States, Australia, and South Africa (see Table 14-1. Water-pumping Windmills [Wind Pumps] Worldwide). Some have been in quasi-continuous operation for more than 80 years. Because it works, the farm windmill remains popular even today.

Farm windmills, like modern wind turbines, have their own arcane vocabulary and obscure methods for estimating performance. The following section first examines the factors affecting the pumping capacity of farm windmills, and then compares these with those available from modern

Table 14-1. Water-Pumping Windmills (Wind Pumps) Worldwide		
	Units	MW
Argentina	600,000	150
Australia	250,000	63
Brazil	2,000	1
China	1,700	0
Columbia	8,000	2
Curacao, N.A.	3,000	1
Nicaragua	1,000	0
South Africa	100,000	25
US Southern Plains	60,000	15
	1,025,700	256

Sources: Data compiled from multiple sources.

wind turbines. The description of the American or classic, multiblade windmill technology uses English units. However, tables for estimating the pumping capacity of modern wind turbines use both the English and the metric system because the technique is applicable worldwide.

Pumping Head

The energy needed to pump water is a function of the volume and the height, or *head*, it must be lifted (see Figure 12-3. Pumping head nomenclature). It takes as much energy to lift 10 gallons (38 liters) of water 10 feet (3 meters) as it does to lift 1 gallon (3.8 liters) 100 feet (30 meters). Sizing a wind pumping system requires an estimate of not only the volume of water used but also the depth of the well and how far the water must be pumped from the well to its point of use.

The *total dynamic head* includes not only the distance the water must be lifted from the well but also the height the water must be lifted to fill any aboveground storage tank (the *discharge head*). The dynamic head also includes any energy lost due to friction in the pipes (*friction head*). Bear in mind, when estimating head, that the static water level in any well will be lowered, or drawn down, once pumping begins. The depth the water level falls in the well depends on the pumping rate. Thus, the pumping head is always greater than the static head.

Friction in pipes is directly proportional to the rate of flow. You can transfer a given amount of water with large-diameter pipes and a low flow rate, or small-diameter pipes and a high flow rate (see Table 14-2. Friction Head in Iron Pipe). For example, 1-inch pipe may be less costly than 2-inch pipe, but a pump must overcome more than 20 times more friction in the smaller pipe at a flow rate of 10 gallons per minute (38 liters per minute, l/min).

Farm windmill manufacturers advise not to skimp on the size of water pipe. Using a pipe too small for the flow rate can significantly reduce the pumping capacity of the windmill because of friction along the walls of the pipe. They warn that you can add tens of feet to the dynamic head of a well with pipe too small for the job, effectively making a deep well out of a shallow one.

For example, let's assume that you plan to install a farm windmill on a well 100 feet

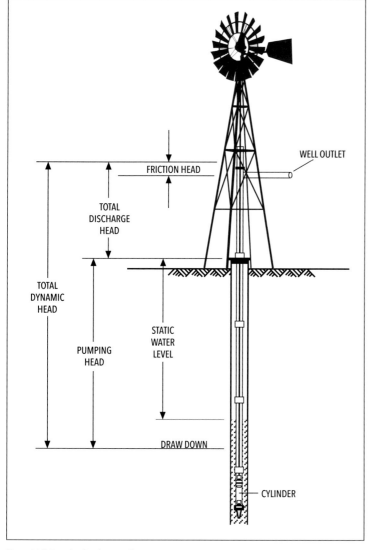

Figure 14-3. Pumping head nomenclature.

Table 14-2. Friction Head in Iron Pipe

Loss of Head in Feet per 100 Feet of Pipe						
Water Flow Rate		Pipe Diameter (inches)				
gallons/minute	l/min	1	1.5	2	2.5	3
5	19	3.25	0.4	-	-	-
10	38	11.7	1.43	0.5	0.17	0.07
15	57	25	3	1.08	0.36	0.15
20	76	42	5.2	1.82	0.61	0.25
Loss of Head in Feet for 90-degree elbows						
		6	7	8	11	15
Adapted from Aermotor Co.						

Table 14-3. Sample Calculation of Total Dynamic Head

	Head	
	ft	m
Static head	100	30.5
Drawdown	5	1.5
Discharge head	10	3.0
Friction head		
pipe	12	3.7
elbows	18	5.5
Total dynamic head	145	44.2

deep. You expect to pump about 10 gallons per minute. At this rate, the pump will draw down the static water level 5 feet (1.5 meters). Let's say you plan to store water in a tank with an inlet 5 feet above the ground, up a small hill with a 5-foot gain in elevation. Now add the final details: a 100-foot run of 1-inch pipe that will need three 90-degree elbows to route the water into the storage tank. Even though the well is only 100 feet deep, the windmill must work against 145 feet (44 meters) of dynamic head (see Table 14-3 Sample Calculation of Total Dynamic Head).

Unless the tank is below the well outlet, you'll need a *packer head* to pump water into a storage tank. The packer head enables the farm windmill to pump water above the height of the well. In the past, farm windmills used for providing domestic water sometimes had storage tanks built right into the towers; others delivered water to a nearby tank on a raised platform. In either case, a packer head is needed when the outlet is above the well opening.

Estimating Farm Windmill Pumping Capacity

Early farm windmills directly coupled the wind wheel (rotor) via a crank to the *sucker rod*. The sucker rod lifted the column of water in the well. Thus, the windmill would lift the sucker rod every revolution of the rotor. In light winds the weight of the water in the well would often stall the rotor, bringing it to a halt. One of the great innovations introduced by Perry and others was *back gearing*. Today most, if not all, farm windmills are back-geared and use a simple gearbox to increase the rotor's mechanical advantage in light winds. This increases the farm windmill's complexity but enables it to pump water more reliably in light winds. Most back-geared windmills lift the piston once every three revolutions. Back gearing works for rotors up to 16 feet (5 meters) in diameter. Farm windmills from 16 feet in diameter to about 30 feet (10 meters) in diameter are crank-driven, or *direct acting*. Mechanical wind pumps greater than 30 feet in diameter are geared up to compensate for the slow rotor speeds and long stroke (see Figure 14-4. Back-geared).

Farm windmills are available in a range of sizes. At one time Australia's Southern Cross

Figure 14-4. Back-geared. Most farm windmills, such as this Dempster model, are back-geared: that is, the rotor turns several revolutions per pump stroke. Rotor hub (far left), rotor drum brake, screw pump to oil rotor bearings, pitman arms, and gearing (right). Unlike many small wind turbines, farm windmills were often provided with a brake for stopping the rotor when service was needed.

Pump Cylinder Diameter	Flow Rate		Maximum Total Pumping Head in Feet Windwheel (Rotor) Diameter (feet)				
inches	gallons/minute	l/min	8	10	12	14	16
2	3	11	140	215	320	460	750
2.5	5	19	95	140	210	300	490
3	8	30	70	100	155	220	360
3.5	11	42	50	75	115	160	265
4	14	53	40	60	85	125	200
4.5	18	68	30	45	70	100	160
5	22	83	25	40	55	80	130

Table 14-4. American Farm Windmill Pumping Capacity

Adapted from Aermotor Co.
Assumes a 15-20 mph (7-9 m/s) wind speed, stroke set for maximum capacity.

windmill was available with windwheels up 25 feet (8 meters) in diameter. The most common size in North America is the 8-foot mill, which is capable of pumping less than 10 gallons (38 liters) per minute from depths of 100 feet (30 meters). How much it can actually pump has often remained a mystery.

Estimating the amount of water that farm windmills might deliver is a dark art. Standard windmill pumping tables are based on instantaneous wind speed, pump cylinder diameter, and the depth of the well. These tables, which were probably derived during the late 19th century—no one knows for sure—give you little idea of how much water can be delivered within a given wind regime. They do illustrate, though, that the performance of the farm windmill is strongly influenced by the relationship between the windmill, the pump, and the pumping head. Too big a pump, and the windmill will stall; too small a pump, and the windmill will operate less efficiently than it might otherwise.

The common 8-foot (2.4-meter) windmill, when matched with a well cylinder 2 inches in diameter, will pump about 3 gallons (11 liters) per minute from a well about 140 feet (43 meters) deep (see Table 14-4. American Farm Windmill Pumping Capacity). According to Aermotor, the pumping capacity remains the same for the same size well cylinders among windwheels (rotors) from 8 to 16 feet (2.4–5 meters) in diameter. By varying the pump's *stroke*—its vertical travel—Aermotor uses the increased power from the larger diameter rotors to increase the height through which the water can be lifted. A 10-foot (3-meter) farm windmill will pump water at the same flow rate as the 8-foot rotor but will be capable of lifting the water half again as much.

For any given rotor, the stroke can be adjusted to vary the proportions between the volume pumped and the head. Adjusting the windmill to a short stroke decreases the volume that can be pumped by one-fourth, but enables the windmill to work against a one-third greater head. For the same head, a shorter stroke will permit the rotor to start pumping in lighter winds.

Nolan Clark and his staff at the US Department of Agriculture's experiment station in Bushland, Texas, undertook a series of tests in the 1990s to determine just how much water the farm windmill could pump under standard conditions. He and his scientists found that on average, the typical 8-foot diameter windmill can pump 1 to 2 gallons (4–8 liters) per minute under the conditions found on the Great Plains.

Pumping tables are the farm windmill equivalent of a wind generator's power curve. Unfortunately, the pumping tables apply for only one wind-speed bin, not for the full range of wind speeds needed to estimate production anywhere the wind isn't blowing constantly at 14 to 20 mph. However, Alan Wyatt at the Research Triangle Institute devised a simple formula for calculating the potential output from mechanical wind pumps operating at various average wind speeds and pumping heads, when the windmill is properly matched to the pump.

The farm windmill is much less efficient at converting the energy in the wind to useful work than modern wind turbines. Although farm windmills can deliver instantaneous efficiencies of up to 15% to 20% in low winds, the average operating efficiency is only 4% to 8%. Assuming an average conversion efficiency of 5% and a Rayleigh wind speed distribution, Wyatt calculates the daily or monthly volume in cubic meters

$$m^3 = (0.4 \times D^2 \times V^3)/H$$

where D is the rotor diameter in meters, V is the average daily or monthly wind speed in m/s, and H is the total pumping head in meters.

For a site with an average wind speed of 6 m/s (13.4 mph), a 10-foot-diameter farm windmill

Table 14-5. Approximate Daily Pumping Volume of American Farm Windmill in m³/day and gallons/day

		Average Annual Wind Speed								
m/s		3		4		5		6		7	
mph		6.7		9.0		11.2		13.4		15.7	
Pumping Head											
m	ft	m³/dy	gals/dy	m³/dy	gals/dy	m³/dy	gals/dy	m³/dy	gals/dy	m³/dy	gals/dy
10	30	10	2,700	24	6,300	47	12,300	80	21,200	128	33,700
20	70	5	1,300	12	3,100	23	6,100	40	10,600	64	16,800
30	100	3	900	8	2,100	16	4,100	27	7,100	43	11,200
40	130	3	700	6	1,600	12	3,100	20	5,300	32	8,400

Rotor Diameter: 3.05 m (10 feet).
Source: Center for International Development, Research Triangle Institute

will pump about 27 cubic meters per day (7,100 gallons per day) from a well 30 meters (100 feet) deep (see Table 14-5. Approximate Daily Pumping Volume of American Farm Windmill in m³/day and gallons/day).

Like the estimates of annual energy production using swept area, these tables of the expected pumping capacities of farm windmills are only crude approximations of what may actually occur.

Counterbalancing for Wind Pumps

The farm windmill no longer has a monopoly on wind pumping. Today there are more options available: from novel mechanical wind pumps, to air-lift pumps, to the improved wind-electric pumping systems now available.

Over the years, experimenters have developed several devices for improving the operation of the traditional farm windmill. The simplest is a counterbalance to the weight of the sucker rod. Farm windmill rotors tend to speed up when the rod begins its downward journey. On the up stroke, the rotor slows down as it lifts both the weight of the rod and the weight of the water in the well. (This is most noticeable in light winds.) The change in speed changes the tip-speed ratio of the rotor and, consequently, its efficiency. To steady the speed of the rotor and maintain the optimum relationship between the rotor and wind speed, some designers have added weights or springs to counterbalance the weight of the sucker rod. Counterbalancing is not unlike the counterweights on oil well pump jacks—the nodding donkeys found in oil fields worldwide—and no doubt where the idea originated.

Another approach tackles a more fundamental problem with wind pumps. As noted before, the power in the wind increases with the cube of wind speed. But the pumping rate of mechanical windmills varies linearly. If the stroke of the windmill is adjusted for optimum production in high winds, the windmill will perform poorly, if at all, in low winds. Ideally, the stroke should vary with wind speed to more closely match the pumping capacity with the power available. This is another strategy that some have attempted.

Though these innovations sound appealing, neither approach became commercially successful. Manufacturers haven't adopted the technology, and they continue building farm windmills the same way they have for the past 100 years.

Dutch researchers, on the other hand, have successfully designed modern versions of the farm windmill. The modern wind pumps developed by CWD in the Netherlands use only 6 to 8 blades of true airfoils, in contrast to the 14 to 18 curved steel plates found on the American farm windmill. As a result, they use fewer materials to do the same job, and are simpler and less costly to manufacture than the American farm windmill. Both aspects are important for Third World countries such as India. Though nearly twice as efficient as the traditional design, modern wind pumps haven't proven as rugged. They may be best suited for regions with light winds.

THE FARM WINDMILL NO LONGER HAS A MONOPOLY ON WIND PUMPING.

FARM WINDMILL CONVERSION?

Because of the prevalence and apparent simplicity of water-pumping windmills, newcomers to wind energy often turn their sights toward adapting farm windmills for generating electricity. Eric Eggleston, a wind engineer, has experimented with farm windmill conversions. His advice: "forget it." While Eggleston acknowledges that the farm windmill's multiblade rotor is superior to that of simple drag devices, it underperforms modern high-speed wind turbines in generating electricity. One measure of this is the tip-speed ratio. For farm windmills, the maximum tip-speed ratio is about 2. In contrast, the tip-speed ratio for high-performance wind turbines is from 4 to 10.

One drawback to the farm windmill is the lack of flexibility in siting. The farm windmill must be located directly over the well. This usually isn't the best location for a wind turbine. In hilly terrain, water is found at the bottom of swales, whereas the wind is often found on hilltops nearby. One solution for low-volume applications has been the introduction of air-lift, or bubble, pumps. The rotors on these wind pumps drive an air compressor, which pumps air through plastic tubing. The use of pliable plastic tubing allows flexibility in locating the wind pump to best advantage (see Figure 14-5. Air-lift wind pump).

Figure 14-5 Air-lift wind pump. Bubbles of air created by this mechanical wind pump can be used to lift water in wells or aerate farm ponds during the winter. Bowjon air-lift pump in operation near Warner Springs, California, in the late 1990s.

Similarly, in a wind-electric pumping system, the wind turbine can be sited where the wind is strongest and the turbine will perform best. The gain in performance more than offsets the cost of the electrical cable to the well.

Electrical Wind Pumps

"Many people are simply unaware that you can use a small wind turbine to pump water," says Brian Vick, formerly an engineer at USDA's Bushland, Texas, station. For the same rotor diameter, wind-electric pumping systems can deliver two or more times as much water as the traditional farm windmill. For example, in the windy Texas Panhandle, a Bergey 850 could pump 2.5 times the volume of water as a much larger farm windmill (see Table 14-6. Approximate Wind-Electric Measured Pumping Rates in Texas Panhandle). However, determining when you want that water available can be critical in choosing a farm windmill or a wind-electric pump.

Previously, the only means of using the wind to pump water without utility power present was the installation of a multiblade farm windmill, or a small wind turbine complete with costly batteries, inverter, and electric pump. Contemporary wind-electric pumping systems, instead, drive well motors directly. The key has been the development of electronic controls that match the pump motor load to the power available from the wind turbine at different wind speeds. This approach frees the wind turbine from the need for batteries and inverter. When wind is available, the wind turbine drives the pump at varying speeds, pumping more in high winds than in low winds. During a period of strong winds, these wind-electric pumping systems bank surplus energy as water in a storage tank, rather than as electricity in batteries.

Direct wind-electric water pumping simplifies matching the aerodynamic performance of the wind turbine to pumping by varying the load electrically, instead of by mechanically changing the stroke of the farm windmill. The control system is the limiting factor, says Nolan Clark, formerly the director of the USDA's Bushland, Texas, station, which pioneered work on the technology. With assistance from West Texas A&M's

PUMPING WATER

Table 14-6. Approximate Wind-Electric Measured Pumping Rates in Texas Panhandle

	Rotor Diameter		Swept Area	Rated	Submersible Motor Rating		Pumping Depth		Pumping rate	
	m	ft	m²	kW	kW	hp	m	ft	liters/min	gal/min
Enertech E44	13.4	44	141.3	40	22.3	30	85	279	946	250
Bergey Excel	7.0	23	38.5	10	5.6	7.5	60	197	227	60
Bergey 1500	3.05	10	7.3	1.5	1.1	1.5	30	98	87	23
SWP Whisper	2.7	9	5.9	1	0.7	1	30	98	47	12.5
AC Solar Pump			18.6	1	0.6	0.75	30	98	42	11.2
Bergey 850	2.4	8	4.7	0.85	0.4	0.5	30	98	38	10
Mechanical wind pump	2.44	8	4.7	0.5	0.4	0.5	30	98	15	4
DC Solar Pump			1.0	0.1	0.05	0.07	30	98	4	1

Source: USDA ARS, Bushland, Texas

Alternative Energy Institute, Clark and his team of researchers developed electronic controls that made wind-electric pumping possible.

As an example of what wind-electric pumping can do Clark notes that two Bergey Excels were used to irrigate cotton in West Texas. The turbines each pumped 20 gallons (76 liters) per minute from 300-foot (~ 100-meter) deep wells—when the wind was blowing. And that remains one of wind-electric pumping's weaknesses: performance in the low winds of late summer, when water demand is greatest. This can be partially compensated for by installing the wind turbine on a much taller tower than the farm windmill.

To test the performance of wind-electric pumping relative to that of mechanical wind pumps, Clark's team evaluated both types at the Agricultural Research Service's wind-pumping test facility. They compared the flow rate and the total volume of water delivered from a 240-foot (73-meter) well by two wind-pumping systems. For the mechanical wind pump, USDA used an 8-foot (2.4-meter) diameter Dempster farm windmill on a 35-foot (10-meter) tower. (The Dempster's performance was adjusted to represent that from a 10-foot [3-meter] diameter rotor.) For the wind-electric system, they used a 10-foot (3-meter) diameter Bergey 1500 on a 60-foot (18.3-meter) tower.

The Bergey, using modern, high-speed airfoils, produced a much higher flow rate than the Dempster (see Figure 14-6. Wind pumping flow rate comparison). However, this isn't the whole story. The Dempster outperforms the modern turbine at low wind speeds: from 3 to 8 m/s (7–18 mph). In the Texas Panhandle, which has a wind resource fairly typical of the American Great Plains, the Dempster will pump considerably more water during the hot summer months than the Bergey. During the critical month of August, the Dempster will pump twice as much water as the Bergey 1500 (see Figure 14-7. Wind pumping volume comparison).

Researchers at Bushland estimated that increasing the Bergey's tower height from 60 to 100 feet (18–30 meters) would allow the wind turbine to match the pumping performance of the farm windmill. Modestly increasing the tower height of a wind-electric turbine doesn't appreciably increase the total system cost, whereas tall towers for farm windmills are costly.

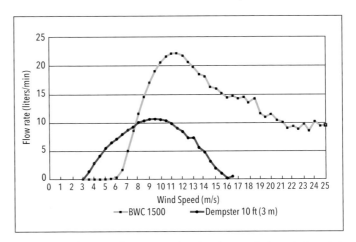

Figure 14-6. Wind pumping flow rate comparison. Pumping tests of a Dempster farm windmill and a Bergey 1500 wind turbine by the USDA's Bushland, Texas, experiment station. The Dempster tested was 8 feet in diameter, but its performance was adjusted to represent that of a 10-foot (3-meter) diameter farm windmill. Note that the farm windmill will pump water in light winds, when the wind-electric system has yet to begin operating. (Adapted from USDA ARS)

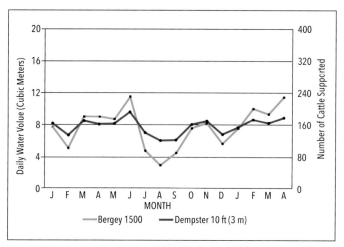

Figure 14-7 Wind pumping volume comparison. While the wind-electric system will pump more water overall, the farm windmill will pump more water when it's needed most: during the summer. (Adapted from USDA ARS)

The tests confirmed an anecdotal report from the livestock manager of the site that for 15 years he never had to haul water. His 10-foot Dempster reliably met the needs for a herd that never exceeded 50 head. But when he switched to a wind-electric system of the same rotor diameter, the wind turbine couldn't meet the herd's needs in the critical summer months; however, the size of the herd did increase to 80 head during the test, which may have been challenging for the Dempster as well.

The farm windmill's high-solidity rotor and its direct conversion of the wind into mechanical power results in a very low cut-in wind speed, explains the USDA's Brian Vick. For example, the cut-in speed for the Bergey 1500 is twice that of the Dempster. The Bergey outperforms the Dempster in the spring but falls short of the Dempster during the summer months. "You don't get any thirstier at 18 mph than at 5 mph," says Wisconsin's Mick Sagrillo, suggesting that high performance is of little benefit if the water isn't available when you need it.

Despite this, there's still a role for wind-electric pumping. Vick himself has operated a Bergey 1500 since 1998 to pump water for his orchard and vegetable garden in the Texas Panhandle. In 2001 there was virtually no rain during the growing season, but the turbine was able to keep Vick's drip irrigation system supplied, saving his 130 fruit and nut trees from drought. Storage of the springtime surplus made the difference. Water storage for wind-powered irrigation systems is just as important as storage for livestock watering or domestic uses, says Vick.

As with so much of wind energy, the application determines which technology will work best. The classic farm windmill "is hard to beat in low-wind regimes," says Eric Eggleston, a wind engineer who has worked on both types of wind pumps. They do their job well, he says.

Like battery-charging home light plants, wind-electric pumping systems are capable of providing power to multiple loads. Specialists in village electrification are finding new uses for the direct coupling of wind-electric turbines to conventional motors. Small wind turbines in these applications can not only pump water, but also power motors for grinding (grains), cooling (vaccine refrigeration), and freezing (fish storage).

When large volumes of water are needed, wind-electric pumping can deliver. For example, Bergey's Excel model can pump upward of 800 gallons (3,000 liters) per minute or pump against heads of 750 feet (225 meters). It's well suited for the high-volume applications that might be found in Third World villages. If one turbine is insufficient to meet the need, several machines can be used at different points in the water distribution system or ganged together in a cluster.

Storage

Storing water in a large tank provides a backup supply when winds are light and, just as importantly, provide an emergency supply for fire protection in rural areas. Storing water in a tank is cheaper and more reliable than storing energy in batteries (see Figure 14-8. Farm windmill and storage tank). A storage tank also provides the fire protection that batteries can't. For rural areas in western North America, fire protection is a critical requirement for any water system. Fire protection will often determine the stor-

> THE CLASSIC FARM WINDMILL "IS HARD TO BEAT [FOR PUMPING WATER] IN LOW-WIND REGIMES." —Eric Eggleston

age needed and the maximum instantaneous demand placed on the water supply.

Proper water system design requires determining both the average water usage and the water demand, or the flow rate at any one instant. The number of faucets that you expect to use at any one time plus other uses, such as a shower or washing machine, governs the maximum flow rate the water system must be capable of meeting. For the stand-alone power systems of the past, pumps were sized accordingly. The pump would then operate whenever water was needed, drawing power from the batteries.

In mechanical systems, the farm windmill alone can't reliably provide the pressure to meet the flow requirement because wind speed and the pumping rate vary. In the past, more or less constant pressure was provided by storing water in a tank on an elevated platform. The tank provided a modest amount of pressure by gravity.

Today nearly all rural water supplies in North America use accumulators, or pressure tanks, to provide the constant pressure demanded by contemporary consumers. Where fire protection isn't necessary, the storage provided by pressure tanks reduces the cycling of the well pump, allowing the pumps to operate more efficiently than if they operated continuously whenever water was required.

Because pressure tanks must be charged by an electric well pump, they're not well suited for the true gravity-fed water supply typically used with mechanical windmills. But a farm windmill can be used to pump water, as available, into a storage tank. The rural water system can then draw water from storage and pump it into a pressure tank with an electric pump driven by a stand-alone power system (see Figure 14-9. Hybrid wind pumping system). Some have adapted electrically powered pump jacks to the reciprocating well pump used by the farm windmill. Thus, if water needs to be drawn from the well and the winds are light; the remote power system switches on and drives the pump jack.

Ample storage shouldn't be avoided. Some building codes and insurance carriers may require it for fire protection. Fire codes may even require the addition of engine-driven pump jacks to ensure an adequate flow during emergencies.

Galvanized steel tanks are a common sight at

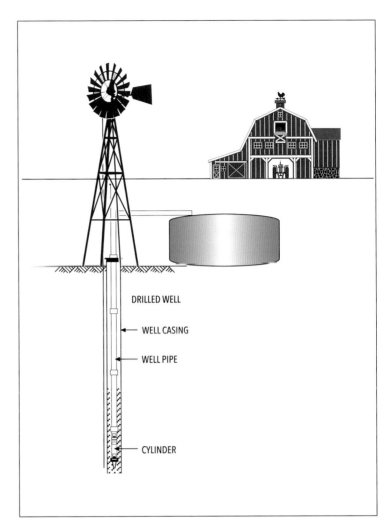

Figure 14-8. Farm windmill and storage tank. Storage is a necessary part of a wind pumping system, whether it's for watering livestock, serving domestic needs, or providing fire protection.

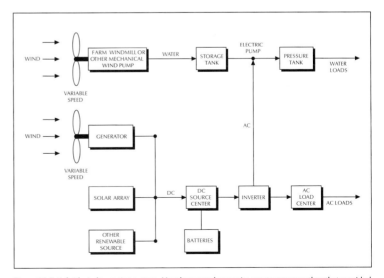

Figure 14-9. Hybrid wind pumping system. Most homes today require greater pressure than that provided by gravity-fed water systems.

rural homes in arid regions of North America. Homesteads on the Mojave Desert of Southern California store a minimum of 2,500 gallons (10,000 liters) in unpressurized aboveground tanks; many prefer 5,000-gallon (20,000-liter or 20 cubic-meter) tanks. In colder climates, cement cisterns can be buried below the frost line to prevent freezing. Ken O'Brock, of O'Brock Windmill Distributors, says this is a common practice among the Amish of Pennsylvania and Ohio.

Irrigation and Drainage

Many applications for water pumping, notably irrigation and drainage, require considerably more capacity than the farm windmill alone can supply. High-discharge irrigation wells in the Texas Panhandle, for example, may pump upward of 1,000 gallons (~ 40,000 liters) per minute from depths approaching 400 feet (120 meters). The American farm windmill isn't big enough to dent such a load.

In the late 1970s, more than 10% of US cropland was irrigated, and at the time irrigation was consuming nearly 90 TWh of electricity nationwide. In the southern Great Plains, irrigation pumping accounted for one-half of the energy used on irrigated farms. In the early 1980s, there were 100,000 irrigation pumps in use. Each pump drew from 20 kW to 100 kW when operating.

Modern medium-size wind turbines can be used for irrigation by either pumping water mechanically or producing electricity to run a large well motor. Because the wind is variable, and irrigation requires such large volumes of water that storage often isn't practical, the wind machine typically drives the well pump in conjunction with a conventional energy source. Coupling wind turbines with conventional sources is a fairly simple task with an electric well pump, but it isn't quite as easy with the engine-driven pumps commonly used on the southern Great Plains.

West Texas A&M University, in cooperation with USDA, developed an ingenious device for mechanically coupling the varying mechanical output of a wind machine with an irriga-

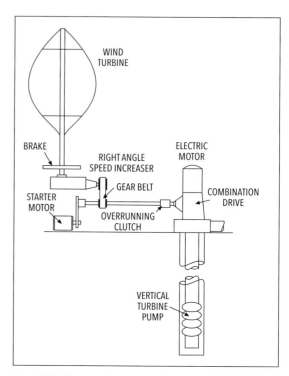

Figure 14-10. Wind-assisted pumping. In this application developed by the USDA's Bushland, Texas, experiment station with West Texas A&M University's Alternative Energy Institute, the vertical-axis wind turbine drives the submersible well pump directly via an overrunning clutch. (USDA ARS)

tion pump: the overrunning clutch (see Figure 14-10. Wind-assisted pumping and Figure 6-11. DAF-Indal 40 kW Φ-configuration Darrieus). When the wind is strong and the wind machine is producing full output, it mechanically drives the well pump entirely on its own via the overrunning clutch. When the winds are weaker, the wind turbine assists in driving the pump. The conventional power source makes up the difference. During a calm spell, the conventional power source operates the pump alone. The overrunning clutch assures that a constant volume of water is pumped regardless of wind speed. At the same time, the overrunning clutch enables using the wind to reduce the consumption of conventional fuels. Tests at USDA's Bushland station in the early 1980s proved that the concept works (see Figure 14-11. Wind-assisted pumping demonstration). In the USDA's tests of DAF-Indal's 40-kW Darrieus, the wind turbine saved 45% of the electricity that would have been normally used.

Using the wind to assist irrigation pumping is even simpler when the pump is driven by an electric motor. Wind machines producing utility-

Figure 14-11. Wind-assisted pumping demonstration. Vaughn Nelson, founder of the Alternative Energy Institute at West Texas A&M University, demonstrates how DAF-Indal's 4-kW Darrieus rotor could be used in wind-assisted pumping for a visiting delegation in 1979. This was one of two DAF-Indal Darrieus wind turbines used to test the application of mechanically coupled wind turbines for irrigating crops in the Texas Panhandle. Both this prototype and the larger 40-kW version used an overrunning clutch to mechanically drive the well pump.

Figure 14-12. Gang of drainage windmills. At Kinderdijk south of Rotterdam, 19 traditional windmills—working in tandem—collectively drained a polder until 1950. This group of windmills, now an open-air museum, is likely the oldest known wind farm in the world.

grade electricity can be connected directly to the well motor in a configuration identical to that for an interconnected wind turbine at a home or business. When wind is available, the wind machine offsets consumption of electricity from the utility. If the wind turbine produced a surplus of energy during periods when pumping loads were light, the excess is sold back to the utility.

The USDA operated an Enertech E-44 at its water-pumping testing laboratory in Bushland, Texas, for two decades. Nolan Clark estimates that wind turbines, like the E-44, will pump water 65% of the time in the Texas wind regime. With the addition of solar-electric pumping, the hybrid could provide water nearly 85% of the time. Overall, the USDA concluded that up to 40% of the energy used for irrigation on the Great Plains could be offset by wind turbines.

Where any single turbine isn't capable of meeting irrigation demand alone, turbines can be ganged together for collectively driving one or several well pumps. Historically, this is how the wind was used to pump large volumes of water. The Dutch used gangs of windmills to drain the polders of the Netherlands (see Figure 14-12. Gang of drainage windmills). At Kinderdijk, for example, 19 Dutch windmills were used in tandem to drain a polder south of Rotterdam. Kinderdijk—most likely the world's oldest wind farm—operated until 1950.

Wind Pump Heritage

American manufacturers of farm windmills no longer have a monopoly on the technology. Today, the few remaining companies must compete head-to-head with other producers worldwide and with newly designed mechanical windmills developed by longtime wind pump designer Peter Fraenkel of IT Power.

Still, the windmill that "won the West" remains enshrined in the American psyche. Farm windmills and their place in rural life continue to intrigue Americans. Much later than the Dutch, who first called for preservation of their windmill heritage in 1923, Americans have slowly begun turning their attention toward preservation. Today, there are nearly a dozen open-air museums in North America that display rare farm windmill collections (see the Appendix for details).

In explaining his fascination for farm windmills, Bryce Black calls them "aeliotropic." To him, the farm windmill follows the wind like a sunflower follows the sun. Black, who has carved out a niche for himself as a farm windmill repairman in western Wisconsin, where some of the machines are still used, says "there's something timeless" about windmills. To Black, the farm windmill was forged from fire, yet provides a living link between sky, earth, and water.

Beyond mere nostalgia for the traditional farm windmill, wind pumping remains an important use of wind energy. Spreading desertification and never-ending population pressure around the globe will, at a minimum, assure a continued demand for wind pumping and could someday create a renaissance among manufacturers of both traditional farm windmills and modern wind-electric pumping systems.

15

Siting and Environmental Concerns

N'allez pas là-bas, disait-il; ces brigands-là, pour faire le pain, se servent de la vapeur, qui est une invention du diable, tandis que moi je travaille avec le mistral et la tramontane, qui sont la respiration du bon Dieu. (Don't go there. Those brigands make bread with steam, an invention of the devil, while I work with the mistral, the breath of the good Lord.)

—The miller advises in Alphonse Daudet's *Lettres de mon Moulin*.

AS ELSEWHERE IN *WIND ENERGY*, THIS CHAPTER MERELY touches the surface of a vast subject: how to best site wind turbines and address both real and fanciful concerns about wind energy. Entire books are devoted to some of the individual topics discussed briefly here, such as the aesthetic design of wind turbines and wind power plants, or the noise from wind turbines.

With care, consideration, and a good measure of patience, you should be able to allay the rational concerns of most neighbors and others in your community. However, problems can arise when some neighbors don't share your enthusiasm for wind energy. It's at such times when you should make an honest appraisal of your site and your community. You may find that the site you envisioned is unsuitable for wind energy because of physical constraints—too many trees and tall buildings, for example—or because of legal restrictions on how you can use the land.

You may also find that the choice of an individual household-size wind turbine is not right for you. Instead, you might realize that the best strategy is for both you and your neighbors—your community in fact—to install one or several large wind turbines that you own together. In this way, everyone shares in the risk but also in the profits of the wind project. When everyone shares in the prospect of profits from wind energy, fewer are prone to object to their placement in the community.

The importance of properly—and promptly—responding to environmental and community concerns about siting wind turbines and wind projects is illustrated by a few examples. Where it exists, criticism of wind energy results largely from fear of the change this new technology may bring to the community. Just as we grew to accept—and now demand—the utility's intrusion on the landscape, gradually we will grow to accept wind turbines, in much the same way and for many of the same reasons.

Though some may unreasonably fear this new technology, most communities try to apply the same standards to wind energy that they do to other, accepted land uses. Proponents of wind turbines need not

ANTIWIND GROUPS

There are organized antiwind groups in most countries (see No to wind turbines and no to parks). These groups are distinct from and should not be confused with environmental organizations that may have genuine concerns about the impact of large wind projects.

Legitimate environmental organizations, such as the Sierra Club or the National Audubon Society in the United States, the David Suzuki Foundation in Canada, or the Royal Society for the Protection of Birds in Great Britain, generally support the use of wind energy, though they may object to specific projects.

In contrast, antiwind groups oppose all wind energy for political or cultural reasons. Worse, many masquerade as environmental groups by adopting names that misleadingly suggest they are only interested in protecting some cherished landscape. You can often identify such groups by their use of negative-sounding code words like "industrial wind turbine." (There is no such thing as an industrial wind turbine. There are large wind turbines, like those used in commercial wind farms, and there are small wind turbines.)

Some of these Astroturf groups are well funded, sophisticated, and utterly ruthless. They are not above intimidation or vandalism. Investigations have uncovered links between their donors and the fossil-fuel and nuclear industries. Often participants in these groups deny anthropogenic climate change and are openly skeptical of the need for any form of renewable energy. These groups have become more vociferous as the phenomenal growth of renewable energy worldwide–predominately wind–threatens the status quo.

While the ire of such groups is generally directed at wind farms, their broadsides don't make distinctions. They paint all wind turbines (large and small), all projects (big and little), and all wind turbine users (individual and corporate) with the same brush. These groups share information electronically. So don't be surprised if someone steps to the podium at a public hearing in Pipestone, Minnesota, and starts talking about wind turbines in Ryd-y-Groes, Wales, or the Causse du Larzac in France.

No to wind turbines and no to parks. Most organized opposition to wind energy has its roots in efforts to protect the status quo of incumbent industries. In France, the powerful nuclear industry dominates the debate about energy choices. Here opponents have defaced a road cut in the Massif Centrale with a sign saying "No to Wind Turbines" (*éoliennes* means "wind turbines"). In the same region is a similar road cut with a fading sign from the 1970s saying "No to a National Park." Despite opposition, the Parc national des Cévennes came into existence and has been a boon to the local economy. Most well-designed wind projects–those sited with care–ultimately receive approval and benefit the community.

ask for special treatment, but they are entitled to equal treatment before the law. Whatever you do, don't bypass the planning and permitting process or informing your neighbors of your plans. You have an obligation to comply with the community's regulations, even if you don't agree with them. And you have a responsibility to your neighbors to keep them abreast of your project.

Here's one example of how not to do it. In New Cumberland, Pennsylvania, an unthinking homeowner impulsively bought a wind turbine to install in his backyard. Then, to his chagrin, his application for planning approval was rejected. Not only was wind energy not permitted in his residential neighborhood, but also his lot was physically too small. He hired an attorney and then engaged in a lengthy and expensive appeal. His neighbors objected vociferously. Then, amid the glare of television lights and a packed hearing room, his appeal was denied. His troubles didn't end there. The dealer then refused to buy back the wind machine, and the homeowner had to sell it at a loss. He didn't do his homework, and it cost him dearly.

This unfortunate homeowner can be excused because of his enthusiasm for wind energy and his ignorance of the planning process. The same can't be said for some so-called wind farm developers who have committed similar blunders. The difference is in the sums of money involved: not thousands, as in the homeowner's case, but hundreds of thousands.

One group of self-styled "professionals" was planning to erect several unreliable wind turbines in a New Jersey residential neighborhood—without planning approval. They were about to begin construction when the local news media broke the story. (There was an exciting mix of New Jersey–style, backroom politics involved.) The scheme was quickly killed in a boisterous public hearing.

These cases illustrate how not to install a wind turbine. There are literally tens of thousands of examples where the appropriate approvals have been obtained in an orderly and businesslike manner and the wind turbine or wind turbines successfully installed with little fuss. Consider the example of an upper-income suburb of Pittsburgh.

Fox Chapel Township has a reputation for strict interpretation of its zoning ordinances. "They'll never let you put one here," some said. Yet the dealer, Bill Hopwood of Springhouse Energy Systems, and the client, the Western Pennsylvania Conservancy, were both respected and thoroughly prepared. (They had to receive approval to erect the anemometer mast, so they were familiar with the process.) They answered all questions forthrightly, allayed the fears of planning officials, and, to the surprise of cynics, won approval. The small wind turbine, an early Bergey, was installed without incident and operated successfully at the site—now administered by the Audubon Society of Western Pennsylvania—for more than two decades.

However, you can do everything right and still encounter determined and politically motivated opposition to your plans. Investigative reporter Wendy Williams dug into one famous—and ongoing—battle off Cape Cod. Her book *Cape Wind* is both a lively and disturbing exposé of how money, class, and politics tried to derail an offshore wind project. The decade-long fight to stop the wind project was led by some of the most powerful names in America, including the Massachusetts' Kennedy clan and Bill Koch, an heir to a multibillion dollar coal-mining and fossil-fuel fortune. Though the project has received all its approvals by late 2014, it now appears the project won't be built. After extensive delays—the preferred tactic of opponents—the utility canceled its contract to buy the energy.

The good news is that for those who do their homework and persevere, the vast majority of projects are approved. And where legally challenged, the courts have overwhelmingly sided with wind energy proponents. Mike Barnard, a senior researcher at the Energy and Policy Institute, found that in 49 cases across five English-speaking countries, courts dismissed all but one appeal of a project approval. Even better, Barnard's detailed analysis discovered that the courts have awarded costs to project proponents. As Barnard notes, while the frivolous court cases are time-consuming and can greatly delay a project, wind turbine proponents will typically not find them costly otherwise.

Figure 15-1. Sidewalk siting. Lagerwey 18/80 next to a fish processing plant along a frontage road in the port of Lauersoog, the Netherlands. The Dutch are accustomed to multiple uses of their limited land area. There are now several more wind turbines on the harbor's jetty. Late 1990s.

There is now a litany of reports debunking the most common fears about wind energy whether it's concerns about health and safety, property values, or energy payback for example.

Tower Placement

One of the most challenging aspects of using wind energy is finding a place to put the wind turbine and tower—a place that's acceptable to you, your family, and the community at large. If you put a household-size wind turbine too

IN GERMANY, IT'S COMMON TO SEE WIND TURBINES LINING THE AUTOBAHN,
while in Denmark rail passengers can watch wind turbines spinning in fields adjacent to the tracks.

close to the house, the wind turbine will suffer from the building's interference with the wind. If you put it too far away, the cost of cabling rises prohibitively, as do wire losses from electrical resistance. Every installation is a balance of these factors. Rarely is there an ideal site.

The first and foremost question is whether or not your site is suitable for a wind turbine. Do you have enough room? There must not only be sufficient space for the tower itself, but also enough space to install it safely. Wind turbines will not work for everyone, everywhere. But wind turbines, both large and small, are used in surprising places. Guyed towers for small turbines have been installed on city lots so small that the anchors have been placed in each corner of the backyard. Freestanding towers have been installed in equally cramped quarters, which on occasion have required a crane to lift the tower over the house to set it on the foundation.

Europeans, accustomed to greater population densities than are common in North America, are more tolerant of placing wind turbines in close proximity to homes, businesses, and public places (see Figure 15-1. Sidewalk siting). Commercial-scale wind turbines have been installed in parks, playgrounds, and parking lots, near soccer fields, and at busy truck stops and lock gates. They can also be found alongside canals, dikes, and breakwaters. In Germany, it's common to see wind turbines lining the autobahn, while in Denmark rail passengers can watch wind turbines spinning in fields adjacent to the tracks.

Exposure and Turbulence

There are two primary rules for siting wind turbines. First, the wind turbine should be as well exposed to the wind as possible. That is, the turbine should not be sheltered behind trees, buildings, or other obstructions. Second, as tall a tower as practicable should be used. Ideally, the turbine and tower should be well clear of any buildings and obstructions in order to minimize the effect of turbulence and maximize exposure to the wind (see Figure 15-2. Zone of disturbed flow).

Turbulence, rapid change in wind speed and direction, is caused by the wake from buildings and trees in the wind's path and resembles the eddies swirling around a rock in a stream. Being buffeted by turbulence can be damaging to modern wind turbines because they use long slender blades traveling at high speeds. Turbulence can wreak havoc on a wind machine, shortening its life.

Buildings and trees also drastically reduce the energy available to a wind turbine. One overriding lesson that has been gleaned from nearly three decades of working with modern wind turbines is that you can't overlook the effect of obstructions, whether buildings or vegetation. Though seemingly less a barrier to the wind than a building, trees, shrubs, and even low hedgerows can rob energy from the wind. It's for this reason that wind turbines are being installed on increasingly tall towers.

Locate the tower far enough either upwind or

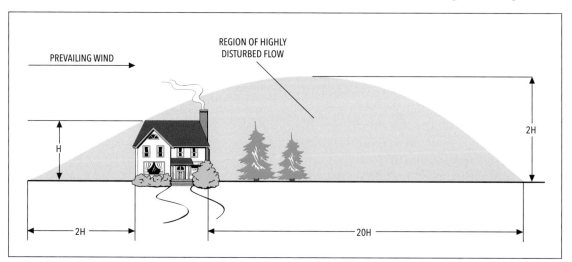

Figure 15-2. Zone of disturbed flow. Wind speeds decrease and turbulence increases in the vicinity of obstructions. The effects are most pronounced downwind but also occur upwind as the air piles up in front of the obstruction. The flow over a hedgerow or group of trees in a shelter belt is disturbed in a similar manner.

downwind to avoid the turbulent zone around nearby obstructions. When this is impractical, use as tall a tower as possible to elevate the wind machine above the turbulence. If neither approach alone is sufficient, use some combination of siting and a taller tower.

From years of experience, small wind turbine manufacturers, consultants, and users have derived a general rule of thumb: the entire rotor disk of the turbine should be least 10 meters (33 feet) above any obstruction within 100 meters (330 feet). If you've determined, for example, that a group of trees along a fence row are 18 meters (60 feet) tall, you'll need at least a 28 meter (90-foot) tower. To ensure the best performance, you should use an even taller tower.

The minimum height tower for large turbines is equal to the turbine's rotor diameter. As mentioned previously, many are installed on taller towers. In forested areas of Germany, it's not rare for the tower to exceed 1.5 times the rotor diameter. Thus, a wind turbine 100 meters (330 feet) in diameter could be installed on a tower 150 meters (~ 500 feet) tall (see Figure 15-3. Clear of obstructions).

If there's a hill on your property with a well-exposed summit, site the wind turbine there instead of lower on the slope, even if the summit is some distance from where you plan to use the electricity.

It's also important that the anchors and guy cables for guyed towers be well outside traveled ways—roads, vehicle tracks, and footpaths. If there are any farm animals roaming the site, guyed towers and their anchors must be fenced or otherwise protected.

Power Cable Routing

Once you've selected the area where the tower will be erected, note how the power will be delivered to your load. At this stage, you need to anticipate any problems that may develop later. They're easier to avoid than to solve. For example, a buried telephone line crossing your path may complicate digging a trench for laying the cable underground. Ideally, the electric service from the wind machine will enter the building near where the utility's lines also enter.

In the past, it was common for installers of small wind turbines to string the power cables

Figure 15-3. Clear of obstructions. These two Vestas V47s stand prominently above the row crops on a hilltop near Montefalcone di Val Fortore, appropriately enough in the *benevento* (good wind) province of Campania in southern Italy.

on poles just like those of the electric utility. The consensus today is to bury all conductors, whether for a small or a large wind turbine. If the service entrance and meter are on the other side of the building from where you are planning to erect the tower, what is the best route for laying the power cables to the service entrance? Are there any sidewalks, driveways, or roads in your path? How will you cross them? These are important questions because the answers affect

WIND TURBINE NOISE: RUMORS, GOSSIP, LIES, AND FAR-FETCHED STORIES

Le bruits de l'éolien, rumeurs, cancans, mensonges et petites histoires is a lovely little book that gives an intriguing glimpse into the French antiwind movement's psyche. As such, it has lessons for English speakers in North America, the British Isles, and the Southern Hemisphere on how a bit of fact can be twisted into a falsehood and quickly become myth–or the more modern term *urban legend*. It was the French, for example, that gave the world the classic urban legend that wind turbines attract sharks. (They don't.)

This is not an academic or technical book. It is intended to debunk urban legends surrounding wind energy. Nevertheless, the book opens by explaining how myths develop before it takes on the various urban legends about wind energy circulating in France.

Based on the work of French sociologists who have specialized in how rumors spread, the book offers an example of a French legend supposedly from the Netherlands to show how rumors develop. Here's the legend.

> After several years of operation the low-frequency noise from wind turbines in the Netherlands destroyed the tissue holding the intestines of not only animals but humans living within 5 kilometers of the turbines, leading to death. For this reason the Dutch were removing the wind turbines and putting them out to sea.

One key to identifying an urban legend, say the researchers, is false precision. In this example, the claim is overly specific: low-frequency noise, 5 km distance from the turbines, the "tissues holding the intestines in place," and so on.

Another identifier is that many of the claims should be relatively easy to verify, but no one bothers. If the wind turbines in the Netherlands had been removed and sent offshore, visitors to the country would have easily seen this. (They haven't.) Similarly, it would be easy to check whether low-frequency sound can liquefy one's intestines. (It can't.) Or, the number of dead from liquefied intestines should be easy to count. (There are none.)

Many other such claims about wind turbines are equally bizarre.

Unfortunately, politicians and journalists are

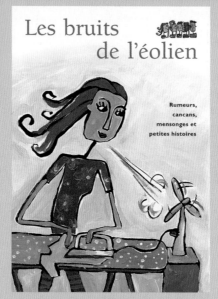

Le bruits de l'éolien (Wind turbine noise). *Le bruits de l'éolien, rumeurs, cancans, mensonges et petites histoires* by Vincent Boulanger, Observ'ER, 2007. (Observ'ER)

not trained in seeing through or taking apart such rumors, or they're too lazy to make the effort. It's as if they have the attitude, "where there's smoke, there must be fire," that is, if enough people repeat the rumor, it must either be true or, even more damaging, "partly" true, therefore, they can act on it without further thought or inquiry.

Rumors can be deadly as in the Russian potato riots of the 1840s. Even when rumors are more benign–no one has been killed by an antiwind protester yet–they can have profound antisocial consequences. Urban legends surrounding vaccinations have led to modern outbreaks of contagious diseases once under control. (In 2014, there was a German measles outbreak in California that began in Disneyland among unvaccinated children.) And if unfounded rumors delay or prevent the growth of renewable energy, the planet and the people on it will be profoundly hurt by the climate change that results.

Effective propaganda, as the Nazis proved during the 1930s, is built around a kernel of truth. This is what makes propaganda believable. And professional opponents of wind energy are adept at manipulating what may seem innocuous facts or observations, for example that wind turbines are big and spin in the wind, to some new, mysteriously dangerous phenomenon.

Le bruits de l'éolien compares the urban legends surrounding wind turbines to the introduction of railroads in France during the 19th century. It was said at the time that the trains would hurt the health of the country's cows. (It didn't.) At the time France was a primarily rural country and anything that affected cows affected the farmers who depended upon them. A modern equivalent, says *Le bruits de l'éolien*, is all the ills that are attributed to high-voltage transmission towers. Today, we would add cell-phone towers or vaccinations to the list of sources causing mysterious diseases, not unlike those attributed to wind turbines or, at one time, railroads.

Nevertheless, *Le bruits de l'éolien* warns that these rumors have a life of their own and, like polio, are very difficult to eradicate, requiring constant rebuttals. *Le bruits de l'éolien* acknowledges that one book won't stop or prevent such rumors from spreading, but it hopes it will at least offer a small countercurrent to combat such misinformation.

Here are a few of the most entertaining rumors.

Effects on Humans

Wind turbines will drive you crazy. (They don't.) Because the wind turbines spin, they can cast shadows (shadow flicker) or reflect light (disco effect). This will put those who watch the wind turbines into a trance and eventually drive them crazy. This legend may have its basis in the mystery surrounding hypnosis.

Another legend related to the effect of the spinning rotor has a particular resonance with the French: decapitation. So it's natural that antiwind activists will associate wind turbines with Dr. Guillotine.

Effects on Animals

Wind turbines lead to a proliferation of mosquitoes. (They don't.) This rumor has its basis in the fact that wind turbines can and do kill some bats. Bats eat mosquitoes; therefore, there are more mosquitoes wherever wind turbines are located.

Wind turbines turn milk sour–in the cows. (They don't.) *Le bruits de l'éolien* isn't sure where this idea came from, but the author relates it to the

19th-century fear that the coming of the railroad would make cows sick. Suffice it to say France has cows, modern trains—and fresh milk.

Wind turbines cause cows to abort. (They don't.) Probably from the fear of the mysterious "infrasound," a rumor has spread that wind turbines cause cows to abort. After all, if infrasound can turn your intestines to mush, it must surely cause cows to abort.

Wind turbines drive wildlife away. (They don't.) I investigated noise effects on wildlife early in my career, specifically the noise from highway traffic. Noise has little or no effect on wildlife, period.

However, hunting is often prohibited around wind turbines to primarily protect employees. This prohibition, or the threat of it, angers hunters, and thus the claim that wind turbines will drive away wildlife—from hunters.

Wind turbines attract sharks. (They don't.) This is one of the best I've heard—and I've heard a lot of outlandish claims. The origin was from a proposed wind farm for La Desirade, an island near Gaudeloupe in the Caribbean. At a public meeting someone made this charge and sent fear racing through the audience. The charge and resulting fear led to a lengthy study of the land-based wind turbines' possible effect on marine life. Despite the conclusion that they wouldn't have any effect, many remained skeptical. (After all, you can't prove a negative.) Only later was it discovered how the rumor began. A hotel operator was looking to cash in on a new marina under development, and he hoped to attract high-powered businessmen to the island—or in the vernacular "the sharks of finance."

Effect on Humans
Wind turbines will create literally the *din of hell*. (They don't.) That is, the wind turbines will roar like jet engines. Wind turbines are audible, and the noise can be heard for some distance, but they surely don't roar like jet engines. This is a classic example of taking a little bit of truth and stretching it out of all proportion to make an exaggerated claim.

Effect on Agriculture
Wind turbines flatten crops downwind. (They don't.) Supposedly, the wind accelerates downwind from the turbines and, therefore, flattens crops. The opposite is true. The wind turbines extract some of the energy in the wind—that's their job, after all—and slow the wind down slightly.

Vibrations from wind turbines will drive earthworms out of the ground. (They don't.) This is related to one of my all-time favorite fables about wind energy. In Mojave, California, a dusty crossroads downwind from one of the world's largest concentration of wind turbines, a rumor was rampant in the early 1980s that vibrations from the wind turbines would drive rattlesnakes from their nesting places toward town, endangering everyone who lived there. In the three decades that the Sierra Club has led hikes through the wind farms of the Tehachapi Pass, we've encountered only one rattlesnake in all that time. Indeed, the town of Mojave still exists, as do its residents. Both the rattlesnake and the human populations seem to have remained in harmony. I suspect this myth began with the ineffective whirligigs sold in hardware stores ostensibly to drive away moles and gophers and keep them from destroying manicured lawns. If vibrations from small whirligigs are sold to drive away gophers, then certainly giant wind turbines would drive subterranean animals even farther from their dens.

The shadow from the wind turbines will reduce the yield of crops nearby. (They don't.) This is another in those seemingly logical series of causation that we were warned about in Logic 101. Wind turbines cause shadows. Shadows reduce crop yields. Therefore, wind turbines reduce crop yields. Although the first two statements are true, the third is not necessarily true. In this case, the third statement is not true.

Wind turbines dry out the land. (They don't.) *Le bruits de l'éolien* suspects this myth began with a misunderstanding of how wind turbines work. Since wind turbines look like giant fans, they must blow the air across the ground, hence drying out the surface. Of course wind turbines are not fans, they are turned by the wind, and in doing so they slow the wind down slightly.

Effect on Climate
Wind turbines ionize the air around them. (They don't.) This myth may be built on the idea that wind turbines are like those static electricity generators we played with in high school physics class, or it could be from the fact that wind turbines generate electricity. In either case, the wind turbines are alleged to produce negative ions, and these cause anxiety among those who live downwind.

In India, wind turbines do not sow the wind, nor do they prevent the whirlwind (the monsoons). After three years of drought in the Indian state of Maharashtra, antiwind activists made the observation that the onset of the drought began with the installation of a large number of wind turbines. Critics argued that the wind turbines chased away the life-giving rains of the monsoons. While laughable, the movement became so powerful that it endangered the operation of 1,700 wind turbines in the state during the monsoon season. Only when the government conducted a study that found—not surprisingly—that the wind turbines didn't cause the drought was the tide turned.

I've often been confronted with a related fear that if we install too many wind turbines we will slow down the rotation of the earth—with untold consequences. (Sometimes all you can do is smile.) Wind turbines are like trees in this regard, and the earth has done quite well in the past when most of the continents were forested.

Economic Impact
Wind turbines will destroy the production of foie gras (goose liver paté). (They won't.) Few things are more serious in France than a charge that something will affect wine or foie gras. Thus, it should be expected that opponents of wind energy would claim wind turbines will endanger geese, and therefore, foie gras farmers would be exposing the animals in their charge to inhumane treatment, threatening them with the loss of their coveted—and highly valuable—appellation. Wind turbines won't have any effect on the geese or the foie gras and, thus, won't endanger the loss of an appellation.

Some French vineyards celebrate wind energy in their *domaine* by adding scenes of modern wind turbines on their wine labels. (Some Italian producers of olive oil also proclaim their green credentials by putting wind turbines on their labels.) I haven't found any similar label for foie gras, but I suspect it's just a matter of time.

While a slim volume, *Le bruits de l'éolien* offers entertaining food for thought for those on the front lines debunking one far-fetched myth about wind energy after another.

the cost of installing the wind system. They also determine how difficult it will be to meet certain institutional restrictions, such as the National Electrical Code in the United States.

Planning Permission

Equally as important as finding the optimal site for the wind system is determining what legal requirements your local community places on structures such as wind turbines. In North America, land-use zoning, building codes, and protective covenants may all apply.

It's always a good idea to talk to your neighbors if you want to develop a wind project, whether it's a micro turbine or something larger. The community and your neighbors do have a say in whether you can install a wind turbine and how you can go about doing it.

If a wind turbine is not a permitted use under the zoning laws in your area, you will need some form of zoning approval. This takes different forms from one region to another. It could be a conditional use permit or a variance from zoning regulations. And if your neighbors are not supportive, the process can be agonizing.

Zoning approval is just one hurdle. The installation must also meet all applicable building and electrical codes. If the turbine is interconnected with the grid, the utility company will have its own approval process for connecting to its lines.

Although your neighbors may have legitimate concerns about wind turbines, their questions can quickly turn into fear if they are not answered promptly. Rational discussions of the benefits and impacts of wind turbines on the community can rapidly degenerate into shouting matches if the myths about wind energy and wind turbines are allowed to spread through the community without rebuttal.

Some myths are quite outlandish—sometimes it's hard not to roll your eyes—but each has to be dealt with. Responding to them all often requires the patience of a saint.

Many who have installed small wind turbines in North America have had few problems, if any, with land-use restrictions. Either their property was not specifically covered by regulations, or where it was, permission was quickly and easily obtained. Many rural areas are not zoned at all, and where they are, there are practically no restrictions on land that is zoned agricultural. The situation changes as you near cities, small towns, and residential neighborhoods. There, the right to swing your fist ends where your neighbor's nose begins.

In some rural locales, wind turbines are specifically permitted unless there is an overriding reason to prohibit their use. In California's Kern County, for example, the use of wind energy is permitted in certain designated agricultural areas—and much of the county is zoned for agriculture.

In most countries, planning approval (or more broadly, the placing of restrictions on how land is used) is a responsibility entrusted to local governments by the public to protect the general health and welfare. Officials will want you to show how your use conforms to the public's general agreement on what can and can't be done on land within a designated area.

Public officials have a moral and oftentimes legal obligation to treat you fairly. Above all, planning officials shouldn't discriminate against you because they're unfamiliar with wind turbines. Treat them cordially. One thing is certain: if you need a building permit, a zoning variance, or other form of planning approval, you want them on your side.

Building Permit

Where planning ordinances apply, you must conform to the law—period. Find out what the requirements are in your area by calling the local building inspector, board of supervisors, or planning office. You want to know how to obtain a building permit (where required) and who is responsible for issuing it (usually the building inspector). Get details. Whoever is responsible should provide a list of what you must do: the forms to fill out, the fees to pay, where and when to file, and any other information that you must supply. Then, methodically deliver what's required.

The intent of this process is to determine conformance with the regulations governing your locale and to alert the public to your project.

Whether you want to install one wind turbine or one hundred, take the initiative and contact anyone who might be affected, especially your neighbors. You have a responsibility to tell them what you're planning and why. Speak to them early in the project so that they feel consulted, rather than pressured into backing you. It's much better to talk with them informally over the back fence than in court or in a shouting match at a public hearing. If you get along well, there should be few problems, but if you've driven over your neighbor's prize rose bush for years, you'd better make amends. Objecting to your building permit is a great opportunity to even the score. You can head off conflict by respecting the needs of your neighbors. Treat them in the same way you would like them to treat you.

At a minimum, you will be required to produce a plan or map showing the dimensions of your site and where the tower will be located. You can prepare this yourself. Drawings of the wind turbine, tower, and foundation with their specifications may also be required. The dealer or manufacturer can supply these.

Planning laws follow one of two approaches. One allows you to do whatever you want, unless specifically prohibited. The other approach prohibits you from erecting any structure unless it is specifically permitted. Where the latter approach is used, your application could be denied simply because no one has ever installed a wind turbine before.

In communities where this is the situation, you can sometimes get permission for a wind machine by bringing it under a permitted category such as radio or cell-phone towers, TV antennas, or chimneys. Building officials may be empowered to make such a determination. If not, formal action before a public board is necessary. They must determine if your use conforms to the intent of the ordinance. Where it doesn't, or where the ordinance specifically excludes wind turbines or similar structures, you must obtain a variance from the regulation.

In the United States, the zoning appeals board or board of adjustment is the final arbiter of permit approval disputes. They're a political body, and if there's a public outcry, they'll respond accordingly within the limits of the law. *Variances*—variations from the law—give the zoning appeals board flexibility in meeting local planning objectives: the protection of the common good without undue restrictions. They'll want to know whether your wind turbine detracts from your neighbors' use of their land, lowers the value of surrounding property, or endangers passersby. The burden of proof is on you, the petitioner.

Frequently, the granting of a variance is little more than a formality. You may not even need to be present. But if the board has questions that you have not answered previously, or if the variance is contested, you'll need to be present and you'll need to be well prepared. On occasion, unfamiliarity with wind turbines—even today—will fuel wild speculation about what they will do to the neighborhood. Often these fears can be quickly dispelled with the facts. Sometimes they can't. When contested, the public hearing can take on the appearance of an expensive courtroom battle, with opponents bringing in their own "expert witnesses" to counter your assertions. It can be rough—even humiliating—if you're unprepared, or if the hearing officers lose control of the meeting.

You have a right to a fair and impartial hearing. You also have a right to argue your case without intimidation—physical or verbal. Public meetings can quickly degenerate into mob rule if public officials and meeting organizers don't limit disruptive behavior. You have an obligation to stem rumors by immediately responding to wild or outlandish claims. Insist on proof or documentation of unsubstantiated charges. The list of real or imaginary problems wind turbines might cause can be endless, limited only by the human imagination. Hearing officers have an obligation to maintain civility. If they can't maintain order, or worse, won't, hold them responsible.

In suburban housing developments or planned communities, deeds may contain restrictions, or *covenants* on how the land can be used. These restrictions are intended to preserve the identity of the neighborhood. Take a look at your deed. Or call your attorney, realtor, or mortgage company for information. If there are any restrictions, they'll know how best to deal with them. Sometimes, the restrictions may be unenforceable or reflect archaic views. For example, some covenants in North America prohibit clotheslines.

Figure 15-4. Stroll through the park. In these unstaged photos, mothers take their children for a stroll around the tower of a Vestas V27 (right), and visitors at the base of the same turbine (left) enjoy the spectacular view of New Zealand's Wellington Harbour on a different occasion. The 225-kW wind turbine, consistently one of the most productive in the world, has become a landmark above the village of Brooklyn. Late 1990s.

California, a presumably progressive state, only removed this restriction in 2015!

Also note the location of any easements on your property for utility rights-of-way. In the United States, *easements* transfer use of the land without transferring outright ownership. Easements are commonly used for a host of public purposes: power lines, underground telephone cables, pipelines, future roads or sidewalks, and so on.

These could all limit your use of the land. You may be unable to encroach on these easements with your wind turbine even though you own the land, there are no restrictive covenants, and you obtained all the proper planning approvals.

Building officials are sometimes bewildered by a request to install a wind turbine. California's San Luis Obispo County officials demanded engineering calculations to assure them that Jim Davis's $1,000 wind system wouldn't pose a hazard to the public. Those calculations would have cost Davis a whopping $5,000 if the wind turbine manufacturer hadn't faxed him an 11-page document that satisfied authorities.

Bergey Windpower's Mike Bergey likens the permit approval process in some states to "medieval torture." Some projects have taken seven months to obtain a permit, says Bergey, far longer than the time needed to build, ship, and install the turbine.

Through experience, other building officials have come to learn what is required to ensure that wind turbines are installed properly and pose fewer challenges. Jonathan Herr, for example, didn't have any problem winning approval to install an Air 403 in California's trend-setting

NO PASSERBY HAS EVER BEEN INJURED OR KILLED BY A WIND TURBINE.

Sonoma County. "I got the building permit over the counter," he says.

For large and complex projects, such as a commercial wind farm, there are consulting companies that specialize in designing, planning, and seeking the approvals necessary to build a project. In Germany, few of the leaders responsible for developing locally owned wind farms of large wind turbines have all the skills necessary to build such a costly project. Often they're independent-minded farmers or community activists, but they know their limits. They know when they need outside help and bring in professionals for the technical studies and design work needed.

Public Safety

The public has a legitimate interest in the safety of wind turbines and the hazards they may pose. There's no point in hiding the fact that dozens of people have been killed while working on or around wind turbines, and some members of the public have died in wind-turbine-related accidents (see Chapter 17, "Safety," for more on this topic). But no passerby has ever been injured or killed by a wind turbine.

Wind turbines, like any large, rotating machinery, should be treated with respect, but there is no reason to fear them unduly. In many parts of the world, wind turbines are part of the community and found in public places (see Figure 15-4. Stroll through the park).

In some North American communities, towers must be set back from the property line a distance equal to their height plus one blade length. Officials reason that if the tower fell over, it would not extend beyond the user's property and present a hazard to neighbors.

However, we think nothing of other man-made and natural hazards that pose a risk similar to if not greater than that of a wind turbine. We've all seen homes sheltered beneath the branches of an old oak tree, where occasionally a storm-weakened limb crashes down onto the roof. We accept this hazard as the price we pay for the benefits the tree provides (shade and visual amenity).

The same is true for radio and television towers. In many ways they are similar to towers for wind machines. They are made of metal and extend visually above the roofline. The public has grown to accept them, and because their failure rate is so low, users often install them adjacent to occupied buildings.

Permitting authorities will be concerned that your tower could collapse. You must show them that the tower meets international standards for wind turbine design and applicable building codes, and that similar towers operate throughout your locale in a host of severe environments without incident. Though towers have failed, the occurrence is rare and far less frequent than that of falling trees or utility poles.

Falling Blades

Authorities will also be concerned that the wind turbine could throw a blade, or worse, fling itself off the tower. While infrequent, neither of these is unknown (see Figure 15-5. A Swiss watch it was not). Once again, you must convince planning officials that the wind machine has been designed and built to accepted international standards and that there's little likelihood that it will throw a blade into the midst of a neighbor's lawn party.

You can best reassure officials that your wind turbine won't become airborne by citing the number of like turbines operating elsewhere, and

Figure 15-5. A Swiss watch it was not. Thirty years ago wind turbines were much less reliable than they are today. While it was uncommon even then for a wind turbine to destroy itself, some did. This Swiss Wenco operated only briefly before failing in California's Tehachapi Pass. This and similar turbines were eventually removed after a complaint by the local Sierra Club chapter that the inoperative turbines were an eyesore along the famed Pacific Crest Trail.

the number of years these turbines have operated without incident. Thus, it behooves you to select a wind turbine with a proven track record: one where a host of units—if not thousands—have operated reliably in a variety of applications for several years. If you plan to install a new, untested, or experimental wind turbine, expect authorities to demand more restrictive setbacks than for wind turbines in widespread use.

Falling Ice

A related question in cold climates is *ice throw*. Under certain conditions, often during cold winter nights when the wind turbine is becalmed, ice can build up on the blades. During the day, sunlight warms the ice, loosening it so that the slightest motion sets the ice moving down the blade (see Figure 15-6. Watch for ice). Occasionally, as at the 600-kW turbine outside the Bruce nuclear power station at Kincardine, Ontario, the turbine will throw the ice some distance. While no one has ever been injured by falling ice, it's prudent to discourage people from walking near the turbine during ice storms or shortly thereafter.

Ice is a common and accepted hazard in cities with severe winters, such as Montreal, Quebec. In such cities, buildings have provisions for breaking up rooftop ice sheets into pieces to minimize the hazards they pose when they eventually slide off. Signs in the parking lots of churches throughout Quebec warn visitors to "watch for ice." In Europe, where wind projects are typically open to the public, operators similarly post signs warning that there may be falling ice and expect visitors to act responsibly.

Large wind turbines destined for northern climes are often constructed with heated blades designed to shed ice as it forms so that it doesn't become a hazard.

Attractive Nuisance

The fear that a wind turbine could become an *attractive nuisance*—that is, attract the attention of vandals or children—is unique to North America. Generally, a property owner is not liable for accidents to trespassers, but a different test is applied to the acts of children.

Swimming pools are thought to entice or attract children to trespass. Because children cannot discern the hazard presented by the pool, the community views it as a public nuisance, and if an accident occurs, a court can hold the owner liable. Permitting regulations allow attractive nuisances when they have met requirements designed to prevent accidents. Swimming pools must be fenced, for example. The same ordinance may require that towers, such as wind turbine towers, be fenced as well.

Fencing isn't the only way to prevent someone from climbing a wind turbine tower. Electric utilities seldom use fencing. On their transmission towers, they simply remove the climbing rungs to a level 10 feet (3 meters) or more above the ground. You can do the same on a freestand-

Figure 15-6. Watch for ice. This sign at the Acqua Spruzza test site in Italy's Apennine Mountains warns, "Wind turbine, watch for ice." Europeans typically urge caution around wind turbines but seldom exclude the public. In this case, falling ice could be a hazard but only during winter storms when few people are likely to visit the windswept site. Similarly, this sign in German (inset) warns "Ice throw possible."

ing truss tower. Or you can wrap the base of a guyed lattice tower in sheet metal or wire mesh. These alternatives should be acceptable to planning officials because they accomplish the same goal as fencing while being less obtrusive (see Chapter 17 for some examples). Utilities seldom erect fences around their utility poles. Imagine the outcry if every utility pole required a fence.

Large wind turbines on tubular towers have no need of a fence to prevent unauthorized entry (see Figure 15-7. Fencing unnecessary). The massive doors to these towers are securely locked. No child or common vandal could climb these towers. Of the thousands of wind turbines operating in Europe, nearly all are fence free. Fencing of tall structures to thwart access by children and vandals is a peculiarly North American phenomenon.

Avoid fencing wherever possible. Fencing increases the aesthetic impact of wind turbines by drawing unwarranted attention to the turbine with the message, "I am dangerous, stay away." Or the equally offensive, "This is my wind turbine. Keep your hands off." In the Tehachapi Pass, unfortunately, wind farm operators are required to do both by Kern County. They must shield their wind turbines behind barbed wire and post signs that say "Keep Out."

Height Restrictions on Small Turbines

The most frequent limitation on the use of small wind turbines in North America is a restriction on the height of the tower. In most residential areas, there's a limit to the height of structures, usually 35 feet (11 meters), a relic of the days when fire brigades had to pump water by hand. Variances to such ordinances can be obtained by pointing out other structures taller than the limit, which have been allowed under the zoning ordinance: radio towers, chimneys, or utility poles. (Note that local officials seldom have control over public utilities. Utilities are regulated at the state and provincial level.)

In Great Britain, some rural residents simply opt for a short tower to avoid the cost and the all-too-frequent controversy surrounding a request to install an appropriately sized tower. Similarly, North Americans also sometimes opt for the path of least resistance. NREL's Jim Green documented one case where the application for a permit to use a tower taller than the

Figure 15-7. Fencing unnecessary. Fencing of this transformer at a wind power plant in Colorado is unnecessary and detracts from the aesthetically pleasing array of wind turbines. Note that the tower is not fenced. The heavy door is locked. It's as though the attorneys representing the wind company were afraid that the cows grazing the site might file suit if they bumped into the transformer.

35-foot limit cost more than the wind turbine. This may explain why Green found in a survey of six small wind installations in Colorado that "every wind turbine I saw could have benefited from being on a taller tower."

Aviation Obstruction Marking

Wind turbines are, by design, tall structures and can interfere with aviation or present a hazard to pilots. Though there are international standards for marking obstructions to aviation, each country controls its own airspace and sets its own regulations. In North America, Transport Canada is responsible in Canada and the Federal Aviation Administration (FAA) is responsible in the United States.

When the height of the wind turbine tower plus one blade length exceeds 200 feet (84 meter), or if you're within one mile (1.7 kilometer) from an airport, the FAA requires that you must register your plans with the agency. This allows the FAA to note an *obstruction* to aviation on maps and alert pilots to the hazard. The FAA also specifies the type of obstruction marking necessary. In the United States, this often requires flashing beacons atop the nacelle: red at night, white during the day.

In Europe, the height threshold is considerably higher, 100 meters (328 feet) before action is required, but most turbines now exceed this height.

Figure 15-8. Aviation obstruction marking. The red stripe on the nacelle, the red band on the tower, and the red and white bands on the blades of this Enercon E126, a 6-MW wind turbine, are designed to warn pilots of the turbine's presence near the village of Dardesheim, Sachsen-Anhalt, Germany, in 2011. Regulations governing how wind turbines must be marked as obstructions to aviation vary from country to country and often include aviation warning beacons atop the nacelle. Inset: Even small turbines may be required to use aviation warning lights as on this Bergey Excel near the Portland, Oregon, international airport.

authorities. However, there are exceptions near airports. In such cases, even a small wind turbine may require aviation warning lights.

While obstruction marking of very tall wind turbines is necessary to warn pilots and prevent a tragic accident, pilots, the wind industry, and regulators are trying to determine just how much lighting is required. The flashing beacons of obstruction marking detract from the night sky, especially in western North America where there are few other light sources but the stars themselves. Some large projects have suffered from regulatory overkill, requiring them to operate beacons on every wind turbine, leading to complaints from neighbors about the light pollution from the wind turbines. For pilots, it may be necessary only to light the outermost turbines in a big project, not all of the turbines. And it may be possible to use lower intensity lighting in some cases or use lighting that is triggered only when an aircraft is nearby.

Safety Setbacks

"Danger-Wind Turbines" proclaim signs prominently posted along the perimeter of California's wind power plants. They make great fodder for opponents of wind energy, who like to point to them and say "See, we told you they were dangerous." The signs imply a degree of risk that is nonexistent and are an outgrowth of California's litigious legal environment. They are a response to county regulations governing what planning officers and lawyers describe as attractive nuisances, anything that attracts trespassers who may subsequently injure themselves. Europeans find the concept hard to grasp. They still have the innocent view that individuals are at least partly responsible for their own actions.

Despite the signs and despite the rare catastrophic destruction of a wind turbine, the operation of wind turbines is "practically risk-free for the public," according to Swiss risk analyst Andrew Fritzsche. In contrast to other energy sources, renewables "have practically no potential for severe accidents" that would endanger the broader public as the reactor accidents at Chernobyl or Fukushima have done.

During tumultuous public hearings in Tehachapi, California, during the early 1980s, some fearful residents charged that the unreliable

Regulations differ from one country to the next. Italy, reacting to the tragic collision of a low-flying US Marine Corps fighter jet with a ski-lift gondola full of skiers in the 1990s, has required obstruction marking of relatively short towers on ridgetops in the Apennines. Very tall turbines in Germany are required to have obstruction markings on the rotor blades, tower, and nacelle (see Figure 15-8. Aviation obstruction marking).

Small wind turbines normally operate well below the height threshold for notifying aviation

Table 15-1. Safety Setbacks for Selected California Counties 2006

County	Property Line	Dwelling	Roads	Reductions in Setbacks
Alameda County	3x/300 ft (91 m), more on slope	3x/500 ft (152 m), more on slope	3x/500 ft (152 m), 6x/500 ft from I-580, more on sloped terrain	maximum 50% reduction from building site or dwelling unit but minimum 1.25x, road setback to no less than 300 ft (91 m)
Contra Costa County	3x/500 ft (152 m)	1000 ft (305 m)	None	exceptions not spelled in ordinance can be filed with county
Kern County	4x/500 ft (152 m) <40 acres or not wind energy zone, 1.5x >40 acres	4x/1000 ft (305 m) off-site	1.5x	With agreement from adjacent owners to no less than 1.5x
Riverside County	1.1x to adjacent Wind Energy Zones	3x/500 ft (152 m) to lot line with dwelling	1.25x for lightly traveled, 1.5x/500 ft (152 m) for highly traveled.	None
Solano County	3x/1000 ft (304 m) adjacent to residential zoning, 3x from other zonings	3x/1000 ft (304 m)	3x	Setback waived with agreement from owners of adjacent parcels with wind turbines

Source: Permitting Setback Requirements for Wind Turbines in California, CEC, November 2006, CEC-500-2005-184. http://www.energy.ca.gov/2005publications/CEC-500-2005-184/CEC-500-2005-184.PDF

wind turbines then in use would throw parts long distances, thus endangering them and their families. Wind turbines, notably during the technology's early days, did indeed destroy themselves, and falling components have damaged trucks, trailers, and even nearby wind turbines. Wind turbines, like other structures, can also catch fire and spectacular videos of burning wind turbines can be found on YouTube.com.

After examining several wind turbine failures in Europe and the United States in the late 1980s, Alexi Clarke concluded that the risk of being hit within 210 meters (700 feet) of a wind turbine is comparable to the risk of dying from a lightning strike; beyond 210 meters, the risk is even lower.

Nevertheless, regulatory authorities have a responsibility to protect the public from the risk of wind turbine accidents even if the probability is remote. They do so by establishing *setbacks* or *safety setbacks* from residences, other land uses, and property lines. (Setbacks to protect the public from the noise impact of wind turbines is often far stricter than that of safety setbacks and will be discussed in the following section.)

Derek Taylor, in a study for Dyfed County, Wales, in the early 1990s, concluded that setbacks of 120 to 170 meters (400–600 feet) were sufficient to protect habitations and that only 50 to 100 meters (150–350 feet) were needed to protect the public using nearby roadways.

Despite the decades since these studies were published, and the tens of thousands of wind turbines installed worldwide since then, the conclusions remain virtually unchanged.

California has more experience with wind turbines than any other jurisdiction in North America. County governments have been regulating commercial wind power plants in the state since the early 1980s. In a 2006 study for the California Energy Commission (CEC), the California Wind Energy Collaborative found that the probability of rotor failure among early 2000s wind turbines then in use was 1 in 1,000 turbine years. Their survey of local regulations concluded that most counties limited large wind turbines to within three times their total height from property lines (see Table 15-1. Safety Setbacks for Selected California Counties 2006).

For example, in Kern County, which administers California's Tehachapi Pass, wind turbines must be set back from roads and trails a distance of 1.5 times the total height of the turbine (hub height plus rotor radius). For a 100-meter-diameter turbine on a 100-meter tower, the setback would 150 meters (~ 500 feet). Wind turbines must be set back from an occupied dwelling 4 times the total height or 600 meters (~ 2,000 feet) for our hypothetical 100-meter-diameter turbine. In rural areas zoned for wind energy, wind turbines in

Table 15-2. Municipal Wind Energy Setbacks in Alberta, Examples			
Municipality	Installed Wind Capacity	Setback from dwellings	Setbacks from property lines
	MW		
Pincher Creek	367	None	Turbine height plus 10%. If noise exceeds 45 dBA, require an easement from affected landowner.
Willow Creek	281	2x turbine height	Turbine height plus 10%
Carston County	37	1.5 x turbine height	Turbine height
Kneehill County	82	4 x turbine height*	Rotor arc >7.6 m from property boundary
Wheatland County	86	2 x turbine height	Rotor arc >10 m from property boundary
Paintearth County	150	2 x turbine height	Rotor arc >10 m from property boundary

*For non-participating landowners.
Source: Survey of Complaints Received by Relevant Authorities Regarding Operating Wind Energy in Alberta by Benjamin Thibault, Tim Weis, and Eli Angen, Pembina Institute, Edmonton, Alberta, July 26, 2013.

Kern County need be only 1.5 times their total height from property lines.

Not surprisingly, safety setbacks are similar in Canada. The province of Alberta has been operating commercial wind farms longer than any other Canadian province. As in California, local municipalities in Alberta regulate safety setbacks (see Table 15-2. Municipal Wind Energy Setbacks in Alberta, Examples). They use the same approach as counties in California. Wind turbines must be set back from dwellings from 1.5 to 4 times the total turbine height, depending upon the municipality.

Safety setbacks are unnecessary in some cases, such as with individually owned wind turbines, and should never impinge on individuals who voluntarily choose to accept the risk: those who want to install wind turbines near their own homes or who grant easements for others to do so.

In the United States, safety buffers from some household-size turbines at suburban locations have been required. Often this is equivalent to the height of the turbine, so that if the tower collapsed, it would not extend across the property line.

Setbacks from roads and certain vistas may be required from commercial wind power plants on aesthetic grounds. However, this has nothing to do with safety setbacks. For example, Palm Springs requires a scenic setback of 500 to 3,500 feet (150–1,000 meters) from designated roads leading into the resort city nestled at the base of Mount San Jacinto. But the impact of hundreds of wind turbines on the landscape differs markedly from that of a single turbine. Most landscapes can easily assimilate a few scattered wind turbines. Thus, setbacks for single wind turbines on aesthetic grounds are inappropriate.

Europeans are more tolerant and less fearful of wind turbines in their midst than North Americans, possibly because of their prior experience with the technology. In at last three officially sanctioned settings in the Netherlands, visitors can stroll unhindered among operating traditional windmills: at Arnheim's open-air museum, at the 17th-century wind farm of Kinderdijk, and at the re-created Dutch village at Zaanse Schans. In each case, the windmills are fully accessible to visitors. Tourists can also visit hundreds of other operating windmills throughout the country.

Modern wind turbines are equally accessible to the public throughout much of Europe. An 80-kW Lagerwey turbine operates prominently at a lock on the northern end of the Afsluitdijk, the dam closing the former Zuider Zee. The turbine stands within 50 meters (150 feet) of the heavily traveled route, in plain view of traffic, which frequently stops for the lock. These and countless examples in Denmark prove that the public need not fear wind turbines and that large safety setbacks are sometimes inappropriate.

Fortunately, modern large wind turbines are now finding their place in towns, villages, and even major cities in North America (see Figure 15-9. Real urban wind). In many of these applications, safety setbacks would have prevented the installation of wind turbines in such locations. Yet, there they are, standing proudly churning out the kilowatt-hours for all to see. Large wind turbines can now be seen in the urban cores of Toronto, Cleveland, and Boston, to name just a few North American cities. One no longer need fly

Figure 15-9. Real urban wind. Remanufactured Vestas V29 at Cleveland's Great Lakes Science Center. The Cleveland Brown's football stadium is off to the left of the 225-kW wind turbine. The V29 is the center of a public art project playing off the shadow from the wind turbine. The turbine is fully accessible to the public.

Table 15-3. Setbacks used in Study of Onshore Wind Energy Potential by Germany's Umweltbundesamt	
	m
Heath resorts	900
Residential areas	600
Industrial & commercial areas	250
National parks	200
Nature conservation areas	200
Special bird protection zones	200
Special bat protection zones	200
Airports	5,000
Airfields	1,760
Funiculars	300
Rail lines	250
Autobahns	100
Roads	80
Power lines	120
Rivers, streams, canals	65
Streams, 2nd and 3rd order	5
Lakes	5
Slopes >30 degrees	0
Source: DEWI Magazin 43, August 2013	

to Germany to see large wind turbines operating peacefully near homes, businesses, and industry.

Of interest to those weighing the safety of wind turbines is a 2013 report by the German Environment Agency, the *Umweltbundesamt* or UBA, on how much wind energy the country could develop on land. Germany has more wind-generating capacity per capita than any other industrial country. (Only Denmark has a higher density of wind turbines than Germany.) Thus, Germans have more experience living with and among wind turbines than anyone else.

The UBA study illustrates the thinking of Germany's federal government about the role of wind energy on the landscape (see Table 15-3. Setbacks Used in Study of Onshore Wind Energy Potential by Germany's *Umweltbundesamt*). They established artificial exclusion zones and setbacks simply to study Germany's ultimate potential. The setbacks were merely limits used by the federal agency in calculating how much land area of Germany could be used by wind turbines. For North Americans, it's noteworthy how close Germans accept wind turbines to roads (80 meters), limited-access highways (100 meters), rail lines (250 meters), and to parks of various classifications (200 meters).

In Ontario, Canada, the provincial government established a minimum setback of 550 meters (1,800 feet) from large wind turbines to allay public concerns about noise. Limits on noise impacts are usually the principal determinant in how far wind turbines must be from occupied buildings or property lines, not safety.

Noise

Noise from operating wind turbines is another frequent community concern. This concern is fueled in part by old reports of noisy wind turbines that were installed in California's San Gorgonio Pass during the early 1980s or by the giant General Electric turbine that operated briefly—very briefly—near Boone, North Carolina. The wind turbines that were the source of the problem are long gone, and since then manufacturers of large wind turbines have made great strides in reducing noise. That's the good news.

The bad news is that manufacturers of small wind turbines began addressing the problem long after manufacturers of large turbines, and only after some customers—and their customers' neighbors—complained. One model, Southwest Windpower's Air 403, was particularly notorious, though other brands were equally at fault.

WIND TURBINE NOISE IS A FIELD WHERE THE TECHNICAL AND THE SUBJECTIVE MEET HEAD-ON.

Fortunately, manufacturers of small turbines have finally heeded customer demand for quieter products.

Noise is especially critical to siting small wind turbines because, as Carl Brothers, one-time manager of Canada's Atlantic Wind Test Site, notes "the smaller they are, the closer they are likely to be placed near someone's house." Mick Sagrillo, one of the founders of the Midwest Renewable Energy Fair, explains, "Noise has a lot to do with acceptability." According to the outspoken Sagrillo, the public's occasional wariness toward small turbines could swiftly shift to outright prohibition if noise hadn't been addressed.

Despite all the technological progress on reigning in noise, no operating wind turbine is or will ever be silent. Wind turbines are audible to people nearby. Whether it's "noisy" or not is far more difficult to determine. Wind turbine noise is a field where the technical and the subjective meet head-on.

Noise, unlike the visual intrusion of wind turbines on the landscape, is measurable. And, because noise is measurable, neighbors will "transfer" their concern about wind energy's aesthetic impact to the increase in background noise attributable to wind turbines. If wind turbines are unwanted for other reasons, such as their impact on the landscape, noise serves as the lightning rod for disaffection.

All wind turbines create unwanted sound, or noise. Some do so to a greater degree than others. And the sounds they produce—the swish of blades through the air, the whir of gears inside the transmission, and the hum of the generator—are typically foreign to the rural settings where wind turbines are most often used. These sounds are not physiologically unhealthful; they do not damage hearing, for example. Nor do they interfere with normal activities, such as quietly talking to one's neighbor. But the sounds are new—and they are different.

Those who live in the rural settings where wind turbines are most often located do so because they prefer the peace and quiet of the country to the noise of the city. Longtime residents are accustomed to the relative quiet of rural life. They are familiar with the noises that exist and have learned to live with them or even to find them desirable: the wind in the trees, the chirping of birds, the creaking of a nearby farm windmill, the hum of the neighbor's tractor. Rather than being nuisances, these sounds reinforce the bucolic sensation of living in the country.

The addition of new sounds, which most residents have had little or no part in creating and from which they receive no direct benefit, can be disturbing. No matter how insignificant they may be in a technical sense, these new sounds signify an outsider's intrusion. The effect is magnified when the source, such as a wind turbine, is also highly visible.

Decibels

First, some background. Noise is measured in decibels (dB). The decibel scale spans the range from the threshold of hearing to the threshold of pain (see Table 15-4. Typical Sound Pres-

Table 15-4. Typical Sound Pressure Levels in dBA			
Source	Distance from the Source		
	ft	m	dBA
Threshold of pain			140
Ship siren	100	30	130
Jet engine	200	61	120
Jack hammer			100
Freight train	100	30	70
Vacuum cleaner	10	3	70
Freeway	100	30	70
Large transformer	200	61	55
Wind in trees	40	12	55
Light traffic	100	30	50
Average home			50
Quiet rural area at night			35
Soft whisper	5	2	30
Sound studio/quiet bedroom			20
Threshold of hearing			0

sure Levels in dBA). Further, the scale is logarithmic, not linear. Doubling the power of the noise source, for example, by installing two wind turbines instead of one, increases the noise level only 3 dB. This alone causes more confusion about noise than any other aspect, because a change of 3 dB is the smallest change most people can detect. Tripling the acoustic energy increases sound level 5 dB, an increase that is clearly noticeable. It takes 10 times the acoustic energy to raise the noise level 10 dB and sound twice as loud.

For most discrete sources, such as wind turbines, the distance to the listener is just as important as the noise level of the source. As in Table 15-4, whenever noise is presented as *sound pressure levels* (SPL), the location is always specified, or implied, because sound levels decrease with increasing distance.

Weighting Scales

The perceived loudness varies not only with the sound level but also with the frequency, or pitch. Human hearing detects high-pitched sounds more readily than those low in pitch. The sound of a complex machine like a wind turbine is composed of sounds from many sources, including the swoosh of the wind over the blades and the whir of the generator. When measuring noise, we try to take into account the way the human ear perceives pitch by using a scale weighted for those frequencies we hear best. The *A-weighting scale* is most commonly used for wind turbine noise. This scale ignores inaudible frequencies—the ones we can't hear—and emphasizes those that are most noticeable.

Some engineers have incorrectly used the *C scale* to measure the noise from wind turbines. The C scale was designed for very loud noises in industrial settings and not for the low sound pressure levels of ambient noise or the emissions from wind turbines. The C scale does not reflect how humans hear noise.

Impulsive sounds, those that rise sharply and fall just as quickly—like a sonic boom, for example—elicit a greater response than sounds that occur at a constant level over time. Wind turbines using two blades spinning downwind of the tower emit a characteristic "whop-whop" as the blades pass through the turbulent wake behind the tower. This impulsive sound and its effect on those nearby may be missed by standard A-weighted measurements. Many of the complaints about wind turbine noise near Palm Springs in the early 1980s had been directed at the impulsive noise from two-blade, downwind turbines. This was also the problem that characterized GE's infamous Mod-1 near Boone, North Carolina.

Noise containing pure tones or impulsive sounds is thus perceived as more annoying than broadband noise. Broadband noise, such as the aerodynamic noise from the wind rushing over a turbine's blades, is composed of sounds across the spectrum of human frequency response. It is less intrusive than either impulsive noise or noise with distinct tonal components.

Exceedance Levels

Another component of noise is time. Noise ordinances specify a noise level that must not be exceeded during a certain percentage of the time. This complicates the task of estimating a wind turbine's noise impact. Unlike trains or airplanes, which emit high levels infrequently

Table 15-5. Selected Noise Limits						
			Urban	Mixed	Residential	Sensitive Receptors
Germany	Day		55	60	50	45
	Night		40	45	35	35
Denmark[1]	6 m/s				42	37
	8 m/s				44	39
Great Britain[2]	Day	L_{90}			35-40	
	Night				43	
Italy	Day		65	60	55	50
	Night		55	50	45	40
Washington State						
	Day	L_{eq}		60		
	Night	L_{eq}		50		
	Day	L_{25}		65		
	Night	L_{25}		55		
	Day	$L_{8.3}$		70		
	Night	$L_{8.3}$		60		
	Day	$L_{2.5}$		75		
	Night	$L_{2.5}$		60		

1. For outdoor living areas within 15 m of a residence.
2. Daytime: For sites with ambient noise of 30-35 dB(A). Add 5 db for sites with >30-35 db(A). Nighttime: For sites with ambient noise <38 dB(A). Add 5 dB for sites with >30 db(A).
Adapted from *Wind Turbine Noise* by by R. Bowdler and G. Leventhall, Multi-Science Publishing, 2012, Chapter 8, Criteria.

NOISE PROPAGATION CONSPIRACY?

In 2014 I was surprised to learn that a passage on noise propagation I'd written was the center of a seething controversy. I mentioned in my 1995 book that as you near a group of turbines, they may "act" as a line source, not a point source, based on work done by NASA. My text on noise propagation was simply to show that the topic is complex and requires a sophisticated understanding of noise emissions, transmission, and adsorption. I didn't elaborate further.

An engineer working for antiwind energy clients charged that my 1995 text proved that the propagation models were wrong. After all, he said, I was an internationally recognized expert on wind energy, and if I wrote that the models were wrong, obviously there was a problem. He stated that not only were the models wrong, exposing the public to more noise than the law allowed, he questioned why this information wasn't made public, darkly hinting that there was a worldwide conspiracy to keep this secret.

While I write and lecture on wind energy, I've never claimed to be an acoustic engineer. I refer to others who are authorities in their fields. If the propagation models consistently got this wrong, we'd know it by now. We'd have measured violations and yanked out the offending turbines. That we haven't done so—worldwide—illustrates that there's nothing seriously amiss. There's certainly no effort to keep all of this—including this highly technical debate—secret.

I've edited the text in this edition to more clearly state what I should have written two decades ago that, as you near multiple turbines, they appear as a line of individual sources.

Noise Propagation

Noise levels decrease with increasing distance as the sound propagates away from the source. Under ideal conditions, sound radiates spherically from a point source, such as a helicopter, and for every doubling of distance, the noise level decreases 6 dB. However, wind turbines seldom hover high above the ground like a balloon. They are earthbound, and their noise emissions spread outward hemispherically.

Over a flat reflective surface such as a lake, noise decays 3 to 6 dB per doubling of distance because a portion of the energy is reflected into the sky. The atmosphere and objects on the landscape absorb some of the noise energy, further attenuating the noise over distance. The International Energy Agency (IEA) assumes hemispherical spreading in its commonly used noise propagation model. This simple model also incorporates a modest amount of atmospheric absorption.

More complex noise propagation models account for ground cover and meteorological effects. Both can greatly influence noise levels. Temperature and wind shear, for example, refract or bend sound waves from those expected, and vegetation can attenuate or absorb more sound than the IEA model assumes.

Noise decays more rapidly with increasing atmospheric absorption. Relatively close to the tower, within 100 to 200 meters (300–600 feet), atmospheric absorption has little effect. As distance increases, absorption increases. Thus the noise attenuated by atmospheric absorption can be important in projecting noise levels surrounding a wind turbine.

Unfortunately, meteorological effects vary with the season, weather patterns, and time of day. Vegetation may vary seasonally as well. Row crops may be tilled in the fall when deciduous trees also lose their leaves, removing much of the vegetation that dampens noise from nearby turbines. Moreover, nighttime temperature inversions refract sound waves, bending them back to earth, increasing the noise level over that estimated by simple models. Valley inversions during the fall and winter produce a similar effect. Anyone living alongside a lake or river has experienced sound carrying great distances during wintertime inversions.

throughout the day, a wind turbine may emit far less noise but do so continuously for days on end. Some find this trait of wind energy more annoying than any other. In windy regions, the sound may appear incessant. The literature of life on the Great Plains is full of references to the ever-present sound of the wind. In the classic 1928 film *The Wind*, the sod-busting pioneer played by silent-screen star Lillian Gish is driven mad by the oppressive wind.

The time-weighting of noise is expressed as the *noise exceedance level*: the amount of time the noise exceeds a specified value. For example, L_{10} is the noise level exceeded 10% of the time; L_{90}, the noise level exceeded 90% of the time; and L_{eq}, the continuous sound pressure level, which gives the same energy as a varying sound level. A noise standard of 45 dBA L_{10} is stricter than a standard of L_{90}, because 90% of the time the noise must be below 45 dBA (see Table 15-5. Selected Noise Limits). Wind turbine noise emissions, that is the noise created by the wind turbine itself, are measured in L_{eq} in order to calculate the sound power generated by the turbine.

SITING AND ENVIRONMENTAL CONCERNS

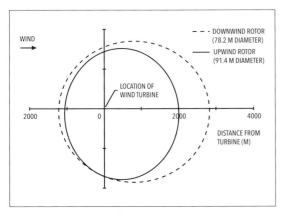

Figure 15-10. Noise footprint. Wind turbine noise is typically greater downwind than upwind of the turbine. Downwind rotors generate more noise downwind than upwind rotors. This image was created by NASA from measurements of two large experimental wind turbines of the 1980s. (USDOE)

There is also little or no atmospheric absorption of extremely low frequency sound. For these reasons, engineers are hesitant to incorporate greater atmospheric absorption into their noise propagation models. Thus, the models remain conservative and this is to the public's benefit.

Multiple wind turbines complicate matters further. From relatively long distances, an array of turbines appears as a point source, and doubling the number of turbines simply doubles the acoustic power, increasing noise levels 3 dB. As you near the turbines, they begin to appear as a line of individual point sources, requiring logarithmic addition of each source. However, Geoff Leventhall explains in his book *Wind Turbine Noise*, a line of point sources is not the same as a line source. The subtle distinction is important because the decay rate for line sources is 3 dB per doubling of distance and not the 6 dB for true spherical propagation, which we have from each turbine in the line.

Even the wind itself will influence how noise propagates. Noise levels are typically higher downwind of turbines, and even higher for downwind turbines (see Figure 15-10. Noise footprint).

Thus, the models used to calculate how noise propagates from a single wind turbine or groups of wind turbines must account for as many of these factors as possible. Estimating the noise at a property line or a nearby home is no simple matter and is fraught with uncertainty. This is why regulators prefer to err on the side of the public and use conservative models. Though noise—unlike aesthetic impact—is quantifiable, interpreting the results of field measurements and mathematical projections of noise levels requires almost as much subjective judgment as it does objective analysis.

Ambient Noise

The total perceived noise is the logarithmic sum of the ambient or background noise and the projected wind turbine noise. Thus, the noise generated by a wind turbine must always be placed within the context of other noises around it. Wind turbines near busy highways will hardly create a problem no matter how noisy they are, though the noise from the wind turbine may still be identifiable above the background noise. Conversely, wind turbines, no matter how quiet, may be heard above ambient noise at great distances in the stillness of a sheltered mountain cove.

The wind itself often masks wind turbine noise by raising the ambient noise level. At exposed locations, there will always be noise from the wind whenever the wind machine is operating, because the wind rustles the leaves in nearby trees or sets power lines whistling. Despite the masking effect of high winds, a wind turbine will still be audible to people nearby, particularly when they are sheltered from the wind.

The sounds emitted by wind turbines are unique and easily distinguishable from those of the wind. The generator or transmission may produce a noticeable whine, for example, or the passage of the blades may generate more discrete sounds. The aerodynamic swish-swish-swish of three-blade rotors is a common wind turbine sound. These sounds may not be objectionable,

WILL IT BE HEARD?

Yes. That's the short answer. If in the heat of a public debate on the noise from a wind turbine or that from a wind farm, resist the temptation to say "It won't be heard." It will. Avoid the equally false statement "You won't hear it over the wind in the trees." They will. The characteristic sounds from a wind turbine–the swishing of the blades, for example–are distinguishable from the background noise of the wind in trees or in our ears. And this distinction can be noted at great distances. Although the noise may not be objectionable, it can be detectable by those who want to hear it–or by those who fear it.

but they are detectable. The whir of the compressor in a refrigerator is audible, for example, but few find the sound objectionable. Some have compared this situation to that of a leaky faucet. Once recognized, the noise is hard to ignore—especially if you objected to installation of the wind turbine in the first place.

Where the background noise level is low, as in a deep valley sheltered from the wind, a new noise may be considered intrusive, particularly at night when few other human-made sounds are present or a nighttime temperature inversion has brought a deathly hush to the valley. Whether or not a noise is intrusive depends upon the nature of the noise; that is, its tonal or impulse character, the perception of the noise source (whether the wind turbines are loved, despised, or merely tolerated), the distance from the source, and the activity (for example, whether one is sleeping inside with the windows closed or conversing with a neighbor in the yard). But no wind turbine, no matter how quiet, can do better than the ambient noise. It is the difference between ambient noise and wind turbine noise that determines how people react.

Community Noise Standards

Local noise ordinances typically state the acceptable sound pressure levels in dBA at the property line or nearest receptor. Many noise ordinances differentiate between acceptable day and nighttime levels and levels for sensitive land uses such as schools and hospitals. The noise levels that wind turbines must meet in Europe and the United States are surprisingly similar.

All community noise standards incorporate a penalty for pure tones, typically 5 dB. If a wind turbine meets a 45-dB noise standard, for example, but produces an annoying whine, planning officers dock the offending turbine 5 dB. The operator must then lower the turbine's overall noise level 5 dB or eliminate the whine.

Turbines emitting characteristic impulsive noise, a loud thumping for example, are also typically docked 5 dB. Thus, they too must meet a more stringent noise standard than a turbine without impulsive noise.

California's Kern County, for example, requires wind turbines not to exceed 45 dBA at $L_{8.3}$ (5 minutes) out of any one-hour period or exceed 50 dBA for any period within 50 feet (15 meters) of a residence or other sensitive land use (school, hospital, church, or public library). The limit is lowered 5 dBA for any pure tones or for any repetitive, impulsive sound like that from early downwind, two-blade turbines.

Despite compliance with community noise standards, operators of wind turbines still run the risk of annoying their neighbors. Whenever wind turbine noise exceeds the threshold of perception, there is the potential for complaints. Fluctuations in ambient noise and variations in the quality or tonal component complicate determining whether wind turbine noise will exceed the perception threshold and stimulate complaints (see Table 15-6. Community Response to Noise from Sources Other Than Wind Turbines).

If there is a noise complaint, public health officers will measure the sound pressure level using a sound level meter and will determine whether the wind turbine complies with the applicable ordinance. This was the situation New Zealand's Graham Chiu found himself in. He received a free noise test courtesy of the Wellington City Council after a neighbor complained about his Air 403. Chiu was found in violation, and the noise control officer ordered the turbine shut off—permanently. Violation of the order could cost Chiu as much as NZ$200,000 in fines.

As in Chiu's case, violating a noise ordinance can result in serious consequences, including removal of the wind turbine. Though not foolproof, noise propagation models are used to project noise levels before a wind turbine is installed. These models use sound power to project noise levels surrounding a wind turbine.

Table 15-6. Community Response to Noise from Sources Other Than Wind Turbines

Amount by Which Noise Exceeds Background Level	Estimated Community Response	
dB	Category	Description
0	None	No observed reaction
5	Little	Sporadic complaints
10	Medium	Widespread complaints
15	Strong	Threats of action
20	Very strong	Vigorous action

Note: This table was derived for noise sources other than wind turbines, and neighbors could be either more or less sensitive to wind turbine noise than that indicated here.
Source: Harvey Hubbard, Kevin Shepherd, NASA, 1990.

Sound Power Levels

The International Energy Agency's model, for example, uses the acoustic energy generated by the wind turbine. Acousticians use field measurements of *sound pressure levels* (SPL), or L_p, to calculate the *sound power levels*, or L_w emitted from the wind turbine. As if the similar-sounding names were not confusing enough, both measures use the same units, dBA. While sound pressure levels will always be specified at some distance from the turbine (or implied), the sound power level will always be presented at the source: the wind turbine itself.

The distinction is important. The sound power level of most wind turbines varies from 90 dBA to more than 105 dBA. For those familiar with sound pressure levels, this appears alarming. Yet a wind turbine emitting a sound power level of 100 dBA can meet a 45-dBA noise limit in sound pressure level, given sufficient distance from the wind turbine. The sound power level can be found by

$$L_w = (L_p - 6 \text{ dB}) + 10 \log(4\pi R^2)$$

where R is the slant distance from the turbine to the sound level meter, L_p is the sound pressure level measured by the meter and -6 dB is a correction to the meter reading to account for using a reflective sound board (see Figure 15-11. Small wind turbine noise measurement).

Sound power data (the emission source strength) on many large wind turbines was publicly available throughout the 1990s. Until recently, there was little comparable data on small wind turbines. Fortunately, within the past decade data on micro wind turbines first became publicly available from the Wulf Test Field and subsequently on household-size turbines from NREL and other independent testing laboratories (Figure 15-12. Noise measurement of micro turbine). Today, the Small Wind Certification Council in the United States publishes summary noise reports on certified small wind turbines designed for interconnection with the utility. Similar data is also available in other countries.

Noise measurements on wind turbines are recorded for two conditions. One condition is the turbine plus ambient; that is, the wind turbine operating as intended (Figure 15-13. Measured Air 403 plus ambient noise). At the Wulf Test Field,

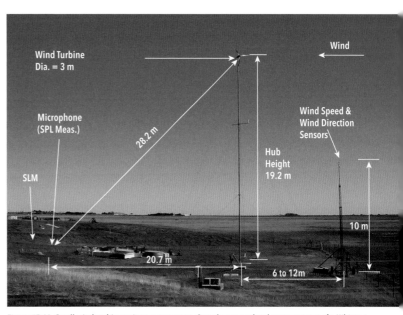

Figure 15-11. Small wind turbine noise measurement. Sound pressure level measurement of a Whisper H80, a 3-meter (10-foot) diameter off-grid, water-pumping, mini wind turbine in 2006 at Bushland, Texas. The sound level meter microphone is placed downwind of the turbine on a reflective soundboard. The distance downwind and the tower height give the slant distance from the turbine to the microphone. This distance is used to calculate the emission source strength–the sound power level–of the turbine at various wind speeds. A similar technique is used for large wind turbines. (Brian Vick, USDA-Agricultural Research Service)

Figure 15-12. Noise measurement of micro turbine. Beginning a sequence of noise measurements downwind from an Ampair 100 at the Wulf Test Field. The recording sound level meter is being inserted into the secondary wind screen mounted on the reflective soundboard. The sound pressure levels measured by the meter are used to calculate the strength of the noise emitted by the wind turbine–the sound power level. In this case, the Ampair wasn't sufficiently noisy at low wind speeds to measure its noise above the background or ambient noise. Late 1990s.

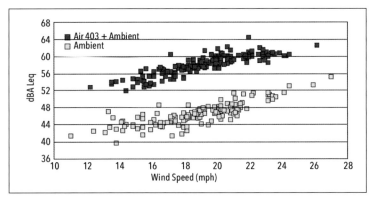

Figure 15-13. Measured Air 403 plus ambient noise. Sound pressure level measurements for an Air 403 charging a constant load at the Wulf Test Field in the Tehachapi Pass in the spring of 2001. This reflects noise from the turbine plus ambient noise–not the turbine noise alone. The lower data set is the ambient or background noise with the turbine off. The slant distance from the turbine rotor is 19.4 meters (64 feet). Note that the Air 403 was an extremely noisy wind turbine. Fortunately, it was eventually superseded by a much quieter design.

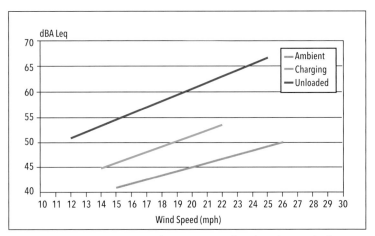

Figure 15-14. Bergey 850 noise summary. A linear regression analysis of sound pressure level measurements at the Wulf Test Field on the Bergey 850 under three conditions: ambient, charging, and unloaded. The Bergey 850 unloaded the generator when the batteries were fully charged, releasing the rotor to spin faster. This generated considerably more noise than when the turbine was charging. Measurements were made at a slant distance of 28 meters (92 feet) from the rotor.

for example, the micro turbines were charging batteries. Another condition is ambient noise alone, that is, with the turbine parked. Once the difference between the turbine plus ambient and the ambient noise is determined, the sound power emitted by the wind turbine can be calculated.

For most small wind turbines, there are only two conditions: operating and parked. For example, the Air series of micro turbines parks the rotor when the batteries are fully charged. Other turbines divert charging to a dump load, keeping a load on the generator and limiting rotor speed. However, Bergey Windpower used a different approach on its 850. The Bergey 850 unloaded the generator when the batteries were charged, releasing the rotor and allowing it spin faster than when charging. (The Bergey 850 is no longer manufactured.) For turbines such as the Bergey 850 then, measurements must reflect all three conditions. The Bergey 850 was noisiest when it operated unloaded (see Figure 15-14. Bergey 850 noise summary).

To compare one wind turbine's noise with another's, you must derive the sound power level, L_{wa}, for a standard wind speed of 8 m/s (17.9 mph), and often 10 m/s (22.4 mph) as well (see Table 15-7. Comparison of Noise from Selected Small Wind Turbines).

The measurement and reporting techniques designed for large wind turbines may not adequately describe the noise characteristics of small wind turbines. Small turbine noise may be most noticeable at wind speeds other than 8 or 10 m/s. For this reason, the SWCC summary reports on certified wind turbines in the United States span a range of wind speeds from 6 m/s to 10 m/s at a height of 10 meters (33 feet) above the ground (see Table 15-8. Sound Power Level for Bergey Excel 6).

When the data is available, sound power levels can be calculated across a full range of wind speeds (see Figure 15-15. Calculated emission source strength). This enables comparisons that otherwise wouldn't be revealed using the standard reporting format. For example, measurements of the Ampair 100 at the Wulf Test Field found that it was too quiet relative to background noise to accurately calculate its sound power level at low wind speeds and that it was significantly quieter than most other turbines tested at wind speeds above 10 m/s (22.4 mph).

Wind Turbine Noise

There are two sources of wind turbine noise: aerodynamic and mechanical. Aerodynamic noise is produced by the flow of the wind over the blades. Mechanical noise results from the meshing of the gears in the transmission, where used, and the whir of the generator.

Unless there is a whistling effect from slots or holes in the blades, aerodynamic noise is principally a function of tip speed and shape. Aerodynamic noise is also influenced by trailing-edge thickness and blade surface finish. The number of blades is also a factor. Neil Kelley, a researcher at the National Renewable Energy Laboratory,

Table 15-7. Comparison of Noise from Selected Small Wind Turbines							
Turbine	Rotor Dia.		Swept Area	Rated Power	Sound Power		Data Source
					@ 8 m/s	@ 10 m/s	
	m	ft	m²	kW	L$_{WA, ref}$	L$_{WA, ref}$	
Late 1990s-Early 2000s							
Ampair 100	0.91	3	0.66	0.1	na[1]	na[1]	Wulf Test Field
Air 403	1.17	4	1.07	0.4	88	91	Wulf Test Field
Air 403	1.17	4	1.07	0.4	81	87	NREL
AirX	1.17	4	1.07	0.2	80	na[1]	Wulf Test Field
AirX	1.17	4	1.07	0.4		85.2	NREL
Whisper H40	2.13	7	3.58	0.9	85	91	NREL
BWC 850	2.44	8	4.67	0.85			
Charging					82	87	Wulf Test Field
Unloaded					92	97	Wulf Test Field
BWC XL 1	2.5	8	5	1		78.7	NREL
Gaia	7	23	38	6.5	88		Risø
BWC Excel-S[2]	7	23	38	6.5	99.5	105.4	NREL
BWC Excel-S[3]	7	23	38	6.5	90.7	93.4	NREL
Mid 2010s							
Skystream	3.7	12	11	2.1	85.5	87.9	SWCC
Bergey Excel 6	6.2	20	30	5.5	89.9	94.1	SWCC
Ampair 10 kW	6.4	21	32	9.2	88.5		MCS
Bergey Excel 10	7	23	38	8.9	84.9	90.2	SWCC
Xzeres 442	7.2	24	41	10.4	87.6	93.7	NREL
Endurance 3120	19.2	63	290	57	87.5	89.3	SWCC

1. Not applicable. Difference between turbine plus ambient and ambient was less than 5 dBA. The turbine was too quiet to measure.
2. BW3 blades (old).
3. SH blades (new).

Table 15-8. Sound Power Level for Bergey Excel 6		
Wind Speed at 10m Height m/s	Apparent Sound Power Level dB(A)	Combined Uncertainty dB(A)
6	84	3
7	86.8	3.3
8	89.9	2.6
9	92.7	2.3
10	94.1	2.8
Source: SWCC Summary Report, Bergey Excel 6.		

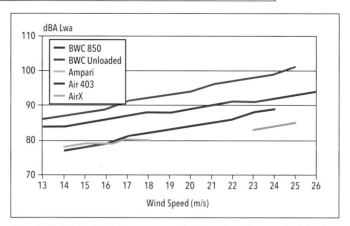

Figure 15-15. Calculated emission source strength. Measured sound pressure level data from the Wulf Test Field on the BWC 850, Air 403, AirX and the Ampair 100 was used to calculate the sound power level or emission source strength (L_{wa}). The sound power level was calculated for each turbine charging a load. It was also calculated for the BWC 850 operating unloaded. The AirX was introduced in part to address complaints about noise from the Air 403. As indicated by the Ampair 100, small wind turbines need not be noisy.

found that the aerodynamic noise of two-blade wind turbines is greater than that of three-blade machines, all else being equal, because the two-blade turbines place higher loads on each blade for an equivalent output. Further, the type of rotor control, whether fixed or variable pitch, affects aerodynamic noise. On rotors with fixed-pitch blades, noise increases when the blades

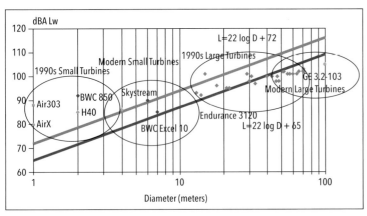

Figure 15-16. Calculated and measured noise emissions. This chart derives from work at ECN by N. C. J. M. van der Borg and W. J. Stam in 1989 on the relationship between source sound power and rotor diameter. Van der Borg and Stam argued that diameter could substitute for tip speed and, hence, determine sound power. The top line was derived from data on experimental large turbines developed in the late 1970s and early 1980s. The bottom line was derived from data on commercial wind turbines being installed in the late 1980s. Published data for commercial turbines in use from the 1990s through 2014 has been added. Noise from small turbines at the Wulf Test Field and other test sites are also included. The chart illustrates that both modern small wind turbines as well as large wind turbines are quieter than would be expected by their rotor diameter. This reflects improvements in noise abatement.

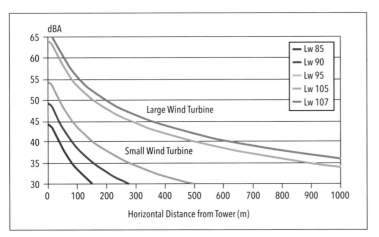

Figure 15-17. Projected noise from a single wind turbine. Using the IEA noise propagation model and the sound power level (L_w), the noise in sound pressure level can be estimated at various distances from the tower. Two examples are shown here: a small wind turbine on a 30.5-meter (100-foot) tower and a large wind turbine on a 100-meter (330-foot) tower. Three sound power levels are shown for the small wind turbine case, two for the large turbine case, and there is no attenuation.

wind turbines designed in the 1970s and early 1980s. Many of these early research turbines operated at very high tip speeds. Van der Borg compared them to commercial turbines available in the 1980s and estimated that the commercial turbines were as much as 7 dB quieter than their predecessors (see Figure 15-16. Calculated and measured noise emissions).

Van der Borg's model can also be used to answer the question of whether small wind turbines are relatively noisier than bigger turbines. In absolute terms, they are not, but relative to their rotor diameter, small wind turbines of the 1990s were noisier. According to van der Borg's model, the Air 403, BWC 850, and Whisper H40 should emit no more than 70 to 80 dBA. Instead, the small turbines were 10 to 15 dBA noisier than would be expected for their size. Fortunately, manufacturers of small wind turbines have addressed the issue, and modern small wind turbines are far quieter than those of a decade ago.

Similarly, modern large turbines are much quieter than would be expected based on experience in the 1980s and even as late as the 1990s. Though data on the sound power level of large wind turbines in 2014 was less readily available than in the past, a quick search of the web will unearth manufacturers touting the reduced noise emissions of their new turbines. For example, GE announced to its clients that its 103-meter (340-foot) diameter 3.2-MW turbine emitted a sound power level of only 105 dBA. GE's competitor Vestas was marketing an even larger turbine, a 112-meter (370-foot) diameter 3-MW model that only produced a sound power level of 106 dBA.

Estimating Noise Levels

Once we know the sound power level, we can use the IEA propagation model to estimate the sound pressure level in dBA at various distances from a wind turbine. In this way, we can estimate the noise level at the property line or at a receptor, such as home or school.

Since we're estimating the noise at a receptor, we don't use the 6 dBA for the reflective sound board.

$$L_p = L_w - 10 \log (4\pi R^2)$$

Again, R, the slant distance or hypotenuse, is a critical parameter in determining the sound level from a small turbine or that from a large

enter stall in high winds. But rotor diameter and speed are the primary determinants of aerodynamic noise. High tip speeds can create greater tip vortices, which are believed to be the dominant emission source.

Dutch researcher Nico van der Borg found that by using rotor diameter as a substitute for tip speed, he could approximate the noise emission of wind turbines. Larger diameter wind turbines generate proportionally more acoustic energy than smaller machines. Van der Borg's model was derived from data on experimental turbine. Compared to large turbines, small wind turbines are installed on relatively short towers. For example, household-size turbines in North America are often installed on towers 100 feet (30.5 meters) tall. Whereas, large turbines being installed in 2014 were often on towers more than 100 meters (330 feet) tall.

Knowing the source strength and the tower height, engineers can estimate the sound pressure level at various horizontal distances from the turbine's tower (see Figure 15-17. Projected noise from a single wind turbine). For example, a household-size wind turbine installed on a 100-foot (30-meter) tower, emitting 90 dBA L_w would produce 45 dBA L_p at 40 meters (130 feet) from the base of the tower. Similarly, a large wind turbine with a source strength of 105 dBA L_w, such as GE's 3.2-MW turbine noted earlier, operating atop a 100-meter tower could meet a 40 dBA noise limit within 500 meters (1,600 feet) of the base.

Because there are no standard tower heights for small wind turbines, the American Wind Energy Association (AWEA) standard reports the sound pressure level 60 meters (~ 200 feet) from the rotor—or the slant distance—to the receptor (see Table 15-9. Estimated Sound Pressure Level from Selected Small Turbines). The AWEA standard reports the noise level for a wind speed of ~ 10 m/s, which corresponds to a 95% exceedance at a site with an average annual wind speed of 5 m/s. However, acoustic data summarized by SWCC on certified small wind turbines includes not only the AWEA sound pressure level but also the sound power level.

Lowering Wind Turbine Noise

Advances in airfoils and reductions in tip speeds have essentially decoupled noise emissions from rotor diameter for large wind turbines. Building quieter turbines not only makes wind energy a better neighbor; it also makes good business sense. In Europe, where competition is fierce, manufacturers find that quieter turbines give them an edge over their rivals. Manufacturers with quieter turbines can site them in areas where planning officials would prohibit their competition, and quieter turbines ensure that there are fewer headaches after installation, as well as less bad press eroding support for wind energy.

Table 15-9. SWCC Estimated Sound Pressure Level from Selected Small Turbines

Turbine	Rotor Diameter (m)	Swept Area (m²)	Rated Power (kW)	AWEA (SPL)*	Source
Skystream	3.7	11	2.1	41.2	SWCC
Bergey Excel 6	6.2	30	5.5	47.2	SWCC
Bergey Excel 10	7	38	8.9	42.9	SWCC
Endurance 3120	19.2	290	57	42.5	SWCC

*AWEA standard reporting format. The AWEA rated sound is the sound pressure level 60 meters from the rotor at 9.8 m/s which is the wind speed not exceeded 95% of the time at a site with an average wind speed of 5 m/s and a Rayleigh distribution.

The most direct route for lowering noise emissions is to reduce rotor speed. One means of lowering rotor speed on a constant-speed turbine is to operate the turbine at dual speeds. This permits operating the turbine at a lower rotor speed in light winds, when there is less wind noise to mask noise from the turbine. Variable-speed operation is also effective, enabling designers to program operation to lower rotor speeds at night, when noise sensitivity is greatest.

Mechanical noise often has tonal components. The gearbox's high-speed shaft is the most critical element, says Henrik Stiesdal, formerly

Figure 15-18. Siemens 6-MW direct-drive prototype. One means to reduce gearbox noise is to eliminate the gearbox. Several manufacturers are following Enercon's lead and developing direct-drive turbines. Siemens began testing its 6-MW prototype in 2012 at the Danish test site for large wind turbines in northwest Jutland. The turbine has a rotor 154 meters (500 feet) in diameter and will sweep 18,000 m². The platform on top of the nacelle is for technicians when the wind turbine is installed offshore. Note the negative coning of the rotor.

the chief designer for Siemens's wind turbines. Mechanical noise can be reduced by redesigning the gearbox and by adding resilient couplings in the drivetrain to isolate vibrations. Acoustic insulation can also be installed inside the nacelle cover to reduce propagation of mechanical noise.

Stiesdal, an early wind pioneer, insists upon totally enclosing the drivetrain and sealing the nacelle canopy. Even ventilation louvers must be carefully designed as sound baffles, he says, or a significant part of the turbine's machinery noise, especially noise at higher frequencies, will escape the nacelle. Stiesdal agrees with NREL acoustician Niel Kelley that noise must be controlled at the source because "once it gets out, you don't know where it will go."

Of course another way to eliminate gearbox noise on large turbines is to eliminate the gearbox. Enercon led the way to gearless wind turbines in the 1990s, and now other manufacturers are beginning to follow suit. Siemens began testing a prototype direct-drive wind turbine in 2012 (Figure 15-18. Siemens 6-MW direct-drive prototype).

Noise Annoyance

In his chapter of the authoritative book *Wind Turbine Noise*, Frits van den Berg examined the technical literature on how people react to wind turbine noise and why some become annoyed. His literature survey found that the thumping, beating, or throbbing character of some wind turbine noise, what's termed *amplitude modulation* in the trade, is not only more perceptible but is also more annoying than the level of the sound alone. Thus, wind turbine noise may meet siting criteria for the noise level—or loudness—but still annoy some of those living nearby, especially if they resent the intrusion of the wind turbines in the first place.

Your Own Pigs Don't Stink

Interestingly, van den Berg's survey found convincing evidence for anecdotal observations about some of the factors affecting annoyance. Dutch farmers have long said, "Your own pigs don't stink." That is, if you have a stake in something, whether it's raising pigs or operating a wind turbine, you're more tolerant of its annoyances.

The scientific literature on noise annoyance confirms this. Dutch researchers concluded that those with no financial benefit from wind turbines found the noise from wind turbines more annoying than those with a direct financial interest. The Dutch observations were reinforced by Health Canada in late 2014 when the federal agency published the results of an extensive survey of wind turbine noise in two Canadian provinces. The study found that "annoyance was significantly lower among the 110 participants who received personal benefit."

Researchers have also found a direct correlation between seeing the wind turbines and noise annoyance, suggesting that if someone had objections to wind turbines on aesthetic grounds, he would find the noise more annoying than someone who did not object to the wind turbines. Similarly, if the observer felt that she had no control over the siting of the wind turbines, she was more likely to object to them on aesthetic and noise annoyance grounds than otherwise.

Van den Berg's literature survey and the results from the Health Canada study substantiates one of the themes coursing through *Wind Energy* and the subject of Chapter 20 on community ownership. Offering the community an opportunity to invest in and profit from wind energy elicits participation from the community, providing a sense of control not otherwise available. Not everyone may want to invest in wind, but if people are given the opportunity, they can make that choice. It is the choice that returns some measure of control over a decision affecting themselves and their community.

The opportunity to participate is one of the reasons for the success of wind energy in Denmark and Germany. Although there are those opposed to wind in both countries, the level of annoyance or complaints about the noise from wind energy is much less prominent than that in North America or in countries where public participation is limited to the most formal and hierarchical or top-down forms of decision making: "They propose, you oppose."

As van den Berg found in his survey of research on acceptance, projects proposed from the ground up require close collaboration at the community level before they can proceed. This cooperation may not eliminate all concerns or eventual complaints about noise because organizers seek consensus—not unanimity—but they can mitigate annoyance of the resulting noise.

Low-Frequency Noise and Infrasound

The low rumbling of noise from wind turbines under certain conditions has also become confused in the popular imagination with low-frequency noise and infrasound. Wind turbines generate noise across a spectrum of frequencies, including low frequencies and frequencies below the audible range—infrasound.

Some have attributed the annoyance of wind turbine noise to this low-frequency component, notably inside buildings. Buildings attenuate higher frequencies more effectively than they do low frequencies. Thus, more low-frequency noise may pass through a building's walls and become perceptible to those inside than the more recognizable wind turbine noise of higher frequencies.

Critics of current noise standards argue that the A-weighting scale misrepresents how humans respond to the lower frequency component of wind turbine noise inside buildings. Measurements in dBA, they charge, misrepresent how humans will respond to wind turbine noise. However, van den Berg noted that the World Health Organization has concluded that any inaccuracy of A weighting is small relative to the much greater normal variability in human sensitivity to noise.

Nevertheless, for all but the simplest noise assessment or when health or code enforcement officers respond to complaints about noise, acousticians will measure the noise across the entire frequency spectrum specifically to determine if there are any frequency components that fall outside norms or exceed permissible levels. These measurements (one-third-octave band analysis) require sophisticated instruments, meticulous technique, and highly sophisticated analysis. For the uninitiated, these measurements and the interpretation of the results may appear as a black art.

The complex nature of low-frequency noise and infrasound—noise that is not directly audible and therefore mysterious—has led to a veritable cottage industry creating urban legends on its deleterious effects on the human body (see the sidebar Wind Turbine Noise: Rumors, Gossip, Lies, and Far-Fetched Stories). Van den Berg, who is with Amsterdam's public health service, concluded that there is no evidence that inaudible (low-frequency) noise can have any effect on the body in part because the body has evolved from constant exposure to normally occurring low-frequency noise, for example, from turbulence in the wind alone. The Health Canada study found, for example, that the noise from wind turbines "was in many cases below background infrasound levels."

Annoyance doesn't in and of itself have an effect on health. However, it can lead to health

SOURCES OF SMALL TURBINE NOISE

Noise from small wind turbines is largely a function of tip speed, blade shape—especially near the tip—and how the turbine regulates power in high winds. Nearly all small turbines operate at variable speeds. As wind speed increases, so does tip speed—and noise.

The Air 403, for example, would reach a tip speed of 90 m/s (200 mph) in winds of 10 m/s (22 mph), nearly twice that of large commercial wind turbines. And the tip speed for the Air 403 would continue to increase until the blades begin to flutter. At that point the noise from the turbine has been described variously as like a hoarse shriek or the buzz of a chain saw. Similarly, when the BWC 850's controller unloaded the generator, the rotor would reach a tip speed of nearly 70 m/s (156 mph). Although this may seem modest in comparison to the Air 403, the Bergey pultruded blade was quite different from the saber-like shape of the Air 403 blade and consequently was noisier. Both turbines are no longer manufactured.

Dave Blittersdorf, himself a one-time wind turbine designer, operated a Bergey Excel in his Vermont backyard for a number of years. A keen observer, Blittersdorf noted that the Excel was noisiest when the controller unloaded the rotor, leading to higher tip speeds. To keep the neighbors happy, he ensured that his turbine always operated under a load.

Wisconsin wind advocate Mick Sagrillo explains that aerodynamic noise can be especially noticeable in small turbines that furl the rotor to limit power in high winds. This behavior differs from one design to another with a resulting difference in noise emissions.

Small turbine manufacturers have heard the message that noise is a subject that won't go away. "Noise is a concern," to us says Bergey Windpower's Mike Bergey. In response, the Oklahoma company introduced new airfoils in the mid-2000s to replace the simple cambered blades that were once the hallmarks of the Bergey design. Dave Calley, Southwest Windpower's one-time chief designer, acknowledged that noise was the "absolute number-one complaint" about the Air series of wind turbines. As a result, the final turbine in the series—the Air Breeze—was significantly quieter than previous models.

Small turbines need not be noisy. "The Marlec is remarkably quiet," says Sagrillo. Among household-size turbines, the 1930s-era Jacobs and the 1980s turbine of the same name were extremely quiet. What's quiet? To Sagrillo, a wind turbine's quiet "when you have to go outside to see if it's running." To Sagrillo, "wind generators should be seen, not heard."

NOISE, HEALTH, AND SAFETY

For links to current popular and scientific articles on the health effects or safety of wind turbines, visit www.wind-works.org and go to the section on Debunking Myths About Wind Energy. There you will find a link to Noise, Health, & Safety. This section and its higher directory also include articles on pseudoscience, chemtrails, scurvy, potato riots, property values, and urban myths about wind energy.

effects if it contributes to anxiety from a fearfulness of mysterious maladies attributed to infrasound.

Noise and Public Health

Recent research now suggests that some members of the public may indeed be developing symptoms of illness—from the fear that they will get sick. This has been attributed to the anxiety created by the scaremongering of antiwind campaigners. Simon Chapman, a professor in public health at the University of Sydney has studied the nocebo effect where people become ill after exposure to an agent they are told is noxious, even though the agent is inert or harmless. Chapman has now identified more than 100 symptoms attributed to the noise or shadow flicker from wind turbines. Chapman concludes that this is more a social phenomenon characteristic of our age than a public health problem.

As reported by Mike Barnard, in nearly every case where courts have been called in to adjudicate a dispute on the noise from wind turbines and their effect on public health, they have dismissed the opponent's objection when those objections were based on suspected health impacts. Bernard notes in his devastating 2014 critique, 21 scientific reviews since 2003 have dismissed claims that wind turbines, notably the noise from wind turbines, have any measurable effect on the health of nearby residents.

The domains of these academic and medical reviews range from Australia to Ontario, Canada, to Oregon and Massachusetts in the United States—areas where claims about health impacts have been particularly virulent.

In 2008, Dr. David Colby, the acting medical officer for the Public Health Unit of the county of Chatham-Kent in southwestern Ontario ruled that the opposition to "wind farms on the basis of potential adverse health consequences is not justified by the evidence." In speaking to a local Rotary Club in 2009, he reiterated his findings by saying that he doesn't doubt some people are bothered by the noise generated by wind turbines, but there was no direct link between wind turbine noise and health problems.

Not satisfied, opponents elevated the debate to the provincial level, where in 2010 Dr. Arlene King, the Chief Medical Officer of Health of the Ontario Ministry of Health concluded

> that while some people living near wind turbines report symptoms such as dizziness, headaches, and sleep disturbance, the scientific evidence available to date does not demonstrate a direct causal link between wind turbine noise and adverse health effects. The sound level from wind turbines at common residential setbacks is not sufficient to cause hearing impairment or other direct health effects, although some people may find it annoying.

And so it has gone in nearly every study since then. Findings by Australia's National Health and Medical Research Council in 2013 led the Australian Medical Association to issue a statement that

> Infrasound levels in the vicinity of wind farms have been measured and compared to a number of urban and rural environments away from wind farms. The results of these measurements have shown that in rural residences both near to and far away from wind turbines, both indoor and outdoor infrasound levels are well below the perception threshold, and no greater than that experienced in other rural and urban environments.

In Oregon the state's public health authority found that "factors unrelated to noise may explain some of the annoyance reported in the few epidemiological studies of wind turbine

> **If there's any doubt as to whether or not your wind turbine might disturb someone nearby, BE A GOOD NEIGHBOR AND CONTACT HIM OR HER IN ADVANCE.**

noise. These factors include being able to see wind turbines."

The Health Canada study of wind turbine noise in the provinces of Ontario and Prince Edward Island—conducted at the insistence of an orchestrated campaign to stop wind energy in Ontario—went beyond the question of health impact alone and also evaluated noise impact based on the World Health Organization's quality-of-life criteria. Their conclusion was not so surprising: "Exposure to wind turbine noise was not found to be associated with any significant changes in reported quality of life for any of the four domains, nor with overall quality of life and satisfaction with health."

Most tellingly is the simple observation that if there were health impacts from modern wind turbines, there would be an epidemic by now. Our clinics and hospitals, especially those in Denmark and Germany, would be bustling with the injured after 30 years of exposure. That they are not should be a sufficiently compelling argument for anyone willing to examine the issue critically.

Nevertheless, public agencies sometimes feel compelled to act if only to allay public concern. This was the situation in 2009 when Ontario's Ministry of the Environment established new guidelines to limit wind turbine noise levels from wind turbines with a source strength of no more than 105 dBA L_w. They set a minimum distance of 550 meters (1,800 feet) from the wind turbine to a receptor to ensure that the noise levels do not exceed 40 dBA. Previously, noise limits alone determined the setback distances—often greater than 550 meters—but the ministry sought to address public perception and opted to establish minimum setbacks. The GE 3.2 MW turbine used in the previous example should meet the intent of Ontario's setback and produce no more than 40 dBA at 500 meters from the base of the tower.

Ontario's setbacks may be greater, depending upon the wind turbine. For example, if a project contained five turbines, each with a sound power level of 107 dBA L_w, the setback is 950 meters (3,000 feet). The province also requires wind projects with more than 26 units of a turbine with sound power level greater than 107 dBA L_w to conduct a full noise assessment.

Consequences

As Graham Chiu found, once noise does get out, the consequences can be costly. Neighborhood reaction to small turbine noise can also affect how or when owners use their turbine. As in the case of Chiu in New Zealand, public authorities can order the turbine removed. Equally damning could be an order not to operate the turbine at night, or in winds above a certain speed. In either case, the operation of the turbines would be so marginalized as to dictate its removal.

The Danish windmill owners' association takes a strong stand on wind turbine noise. The association's members not only are the chief advocates of wind energy in Denmark, but also own most of the wind turbines. Many literally can see wind machines outside their windows. They can speak with authority as people who both want wind energy and demand that it be a good neighbor. Their position is clear: noisy turbines are unacceptable. Noisy machines should either be soundproofed or moved. The goal of the owners' association, one that should be the goal of all wind turbine manufacturers, is to avoid the problem from the start. They have found that once people have been bothered by noise, they remain disturbed—even after the noise has subsequently been abated.

Be Considerate

Our perception of what constitutes noise is affected by many subjective factors. If your neighbors object to your wind machine because you never invite them to dinner, they're more likely than you are to find the sound produced by it objectionable. On the other hand, if your community has fought rate increases with the local utility, the sound of your wind machine whirring overhead may warm their hearts.

Bergey Windpower suggests that if there's any doubt as to whether or not your wind turbine

Figure 15-19. Interference, no; ugly, yes. Vestas V47 with awkward telecom antennas in Germany. While the wind turbine is obviously not interfering with telecommunications here, the antennas and their mounting hardware give this Vestas turbine an undesirable industrial appearance.

might disturb someone nearby, be a good neighbor and contact him or her in advance. Advise them of your plans, and ask for their comments. Answer their questions as forthrightly as you can, and try to incorporate their concerns when designing your installation. Bergey has found that the community's reaction to the noise from a small wind turbine declines after people have had a chance to acclimate to it.

This is equally sound advice for those installing wind turbines of any size. Be considerate. Newer, quieter wind turbines can be good neighbors—when sited with care.

Television and Radio Interference

Neighbors sometimes worry that a new wind turbine will disrupt their radio and television reception. There have been a few cases where large turbines have caused *ghosting* of weak television signals in rural areas. In one case in the early 1980s, Westinghouse's Mod-0A on Rhode Island's Block Island generated complaints as well as electricity. The problem was alleviated by installing cable television on the island. Today, most families across the continent already have cable television.

Interference is a rare phenomenon, and there have been no reported cases due to small wind turbines. Even in the few cases where interference or ghosting has been documented, the effects have been localized.

There are thousands of wind turbines lining the ridges of the Tehachapi Pass, a major corridor for telephone links between Northern and Southern California. The turbines surround the microwave repeater stations but are excluded from the microwave path. This provision is sufficient to prevent any interference.

Small wind turbines are used extensively worldwide to power remote telecommunication stations for both commercial and military uses. The turbines would never have been selected if there had been any hint of interference. Unfortunately, some wind turbine operators have sought additional revenue by renting space on their towers for telecom dishes and antennas, a practice that detracts from the appearance of the wind turbine (see Figure 15-19. Interference, no; ugly, yes).

Shadow Flicker

Shadow flicker occurs when the blades of the rotor cast shadows that move rapidly across the ground and nearby structures. This shadow can disturb some people in certain situations, for example, when the shadow falls across the window of an occupied room.

Small turbines are too small and operate too fast to create a significant shadow. However, large wind turbines can cause shadow flicker, and it can be a nuisance in higher latitude winters, where the low angle of the sun casts long shadows. It can also be more troublesome in areas with high population densities or where neighbors are close enough to be affected by the shadows.

Near Flensburg in Schleswig-Holstein, German researchers examined the effect and found that flicker, under worst-case conditions, would affect neighboring residents a total of 100 minutes per

SITING AND ENVIRONMENTAL CONCERNS 347

Figure 15-20. Shadow flicker avoidance. The device (circled) detects shadow flicker as a function of time of day, season, and light intensity. If conditions are such that shadow flicker will appear on a sensitive site, avoidance software signals the control system to turn the turbine off. Note that this device is distinct from the motion sensor (below left) that turns on the light above the access door when service personnel approach the access stairs in darkness.

SHADOW FLICKER

Many North Americans smile when Europeans begin discussing a phenomenon called shadow flicker. Most Americans have never heard of it and can't imagine what the fuss is about. That was my reaction until one fall when I lived near a 75-kW turbine at the Folkecenter for Renewable Energy in Denmark. One morning while working at my desk I felt uneasy. Something was bothering me. I kept looking up from my work, scanning the room for what was wrong. Finally, I got up from my desk. Then I noticed it: a shadow repeatedly crossing the room. It was still a few moments before I realized I was a victim of shadow flicker. I flipped on the light and went back to work. Often the resolution of shadow flicker is just that simple: turning on a light in a dim room, or adjusting a blind or curtain on the window.

year. Under normal circumstances, the turbine in question would produce a flickering shadow only 20 minutes per year.

There are few recorded occurrences of concern about shadow flicker in North America. Ruth Gerath, however, notes that the flickering shadows from the turbines on Cameron Ridge near Tehachapi have startled her horse and those of others in the local equestrian club. Except for the flickering shadows, she says, the turbines seem to have no effect on the horses. The shadows simply cause the horses to stop briefly, until their riders urge them on.

While few communities have standards regulating shadow flicker, it's wise to be considerate of your neighbors. If there's any question whether nearby residents will be affected, analyze the likely impact before installation. Note, however, fear of shadow flicker triggering seizures is unfounded. A review of the medical literature by the state of Massachusetts in 2012 found that "scientific evidence suggests that shadow flicker does not pose a risk for eliciting seizures as a result of photic stimulation."

Professional wind turbine siting software often includes provisions for calculating shadow flicker. The technique used by these programs is extremely conservative and will project worst-case conditions (bright sun, cloudless sky). Programs for determining shadow flicker on nearby residents can be combined with a wind turbine's control system to shut the wind turbine down during the few hours per year when it might present a problem (see Figure 15-20. Shadow flicker avoidance).

Disco Effect

The *disco effect* is a related phenomenon first noticed in sunny Palm Springs, California. Sunlight glints off the reflective gel coat of the fiberglass blades of the wind turbines in the San Gorgonio Pass. When the blades move, this causes a flash similar to that of a strobe light. As the rotor spins, the flash repeats with a rhythm akin to that of the flashing lights in a discotheque.

To prevent the disco effect from annoying neighbors, Riverside County prohibits reflective blade coatings. The surface finish also dulls after several years in the harsh desert sun, reducing the blades' reflectivity over time.

Birds and Bats

Cynthia Struzik, a special agent of the US Fish & Wildlife Service, strode purposely to the blackboard—her service gun strapped to her hip—where she wrote in bold letters, "Thou shalt not kill," then added as if it was necessary, "acceptable mortality level is zero!" Struzik made her point. And it was not lost on the others in the room who had gathered at Pacific Gas and Electric's research center in San Ramon, California, that December day in 1992 to discuss the conflict between wind turbines and birds in nearby Altamont Pass. A groan of disbelief swept across industry representatives in the room. But it was only the beginning. By the end of the day, Struzik, her colleagues from the California Department of Fish and Game (now Fish and Wildlife), and representatives of the Sierra Club's Mother Lode chapter were openly mocking attempts by the industry and independent biologists to understand what was happening just a few miles away.

One of wind energy's chief attributes is its environmental benefits. When sited with care, wind energy is relatively benign. The key is sensitive siting and a frank acknowledgment that wind turbines do have some environmental impact. Though wind turbines have little or no impact on most plants and animals, they can and do kill some birds and bats. Notably, large arrays of wind turbines have killed significant numbers of birds in the Altamont Pass and near the Straits of Gibraltar. There's no benefit in sugarcoating that fact. Nonetheless, the charge that wind turbines produce more dead birds than electricity is false.

Much has been written about birds and wind turbines. Numerous studies on the topic have been conducted in Europe, North America, and elsewhere. Summaries of this research are available from most national trade associations and environmental groups or regulatory agencies.

No single environmental issue pains wind energy advocates more than the effect wind turbines might have on birds and bats. Clearly, wind turbines should not kill birds, and we should do everything in our power to ensure that they don't. This is the kind of hot-button issue that elicits strong emotional responses that could, if not addressed honestly, derail the use of wind energy.

That some wind turbines kill birds some of the time should come as no surprise. Most tall structures kill birds to some degree, as do most sources of energy. This should never become an excuse for ignoring the issue, but it does help to put it into perspective.

Wind turbines anywhere are capable of killing birds, explains Dick Anderson, a biologist at the California Energy Commission. But nowhere else in the world is the problem as severe as in California's Altamont Pass.

Wind turbines in the Altamont Pass, says Anderson, kill 100 to 300 raptors per year, of which 20 to 50 are golden eagles (*Aquila chrysaetos*). Golden eagles are a protected species in North America but are not rare or endangered. In California, however, they are a "species of special concern." This designation mandates that the California Department of Fish and Game protect them (now called Fish and Wildlife, as of 2013).

Federal law in the United States also prohibits the "taking" of golden eagles under the Bald Eagle Protection Act and the Migratory Bird Treaty Act. Taking is a euphemism for killing, derived from the days when the predecessor of the US Fish & Wildlife Service efficiently managed the extermination of the wolf and other predators. The agency still has teeth. Anyone who "knowingly or with wanton disregard" kills bald or golden eagles commits a felony in the United States, punishable by two years in prison and a fine of up to $250,000. When Struzik wrote her commandments on the board in San Ramon, she was simply stating federal law.

While the death of any bird is unfortunate, biologists prefer to place the death in the context of the total population rather than focus on the number of individual deaths, says Tom Cade, founder of the Peregrine Fund and director of the World Center for Birds of Prey. The deaths, regrettable as they are, "may really have no biological significance," he says.

The number of birds killed in the Altamont

THAT SOME WIND TURBINES KILL BIRDS SOME OF THE TIME should come as no surprise. Most tall structures kill birds to some degree, as do most sources of energy.

Pass could be significant for a species, such as the golden eagle, which has suffered population declines due to urban encroachment throughout its range in California. The state's raptors, or birds of prey, are fast losing their habitat to an exploding human population. In the San Joaquin Valley alone, more than 95% of wildlife habitat has already been converted to other uses. Consequently, wildlife becomes increasingly dependent upon the remaining "islands" of undeveloped land. Some of this land remains undeveloped because high winds make it hostile to human habitation. Thus, there is the potential for increasing competition between raptors and large-scale wind development for the same resource.

The population of golden eagles in the Altamont Pass appears stable, says the CEC's Anderson. The state hasn't yet been able to determine if the number of golden eagles being killed is having a negative effect on the breeding population. Meanwhile, biologists are continuing their fieldwork.

Anderson confirms that wind turbines lining the Tehachapi and San Gorgonio passes in Southern California are also killing raptors, but it's much less of a concern to the state because the numbers killed are significantly lower than those in the Altamont. Fortunately, no rare or endangered birds such as bald eagles (*Haliaeetus leucocephalus*), peregrine falcons (*Falco peregrinus*), or California condors (*Gymnogyps californianus*) are known to have been killed by wind turbines or their power lines anywhere in California.

Despite the problem among the thousands of wind turbines in the Altamont Pass, what little data is available on the impact from single large turbines, small clusters of large turbines, and small wind turbines is that it's unlikely to be a serious issue. It's reasonable to assume that small wind turbines or clusters of larger machines kill birds in proportion to the turbine's size and number. That is, wind turbines with a large swept area will kill more birds than a wind turbine with a smaller swept area.

Toronto has a long history of mass bird kills during the spring and fall migration seasons when tens of thousands of birds cross Lake Ontario, some of whom fly into the brightly lit skyscrapers lining the waterfront. Preventing such disasters is a cause célèbre among bird lovers in the province of Ontario, resulting in the Fatal Light

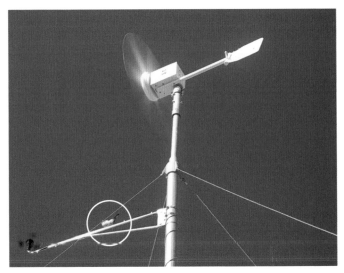

Figure 15-21. Birds and small wind turbines. Bird perching on an anemometer boom (circled) beneath a Marlec 910F at the Wulf Test Field. As this scene illustrates, small wind turbines are not immune to concerns that they pose a hazard to birds.

Awareness Program (FLAP). When the Toronto Renewable Energy Cooperative proposed a wind turbine for a site in downtown Toronto, there was an outcry that the wind turbine would kill birds by the thousands, adding to the slaughter.

The 52-meter (170-foot) diameter wind turbine was eventually installed at a prominent site on the grounds of the Canadian National Exposition (CNE). Following installation, the cooperative began postconstruction monitoring for both the spring and fall migration of 2003. The ornithologists found that the turbine had killed two birds: one an American robin (*Turdus migratorius*), the other a European starling (*Sturnus vulgaris*). They concluded that "the rate of mortality is absolutely insignificant when compared to the thousands that are killed each year by Toronto at tall buildings. The study indicates clearly the wind turbine at the CNE is not going to have any significant impact on bird populations." But the ornithologists went further in trying to put the issue in perspective. "Each of the feral cats seen [at the CNE site] are capable of killing as many as 1,000 small animals per year, including birds and each would have killed more than the wind turbine, probably far more. Every free roaming cat in Toronto probably kills more birds per year than the CNE wind turbine killed."

The question of whether small wind turbines also kill birds does arise as well (see Figure 15-21. Birds and small wind turbines). Consider the case of the Western Pennsylvania Conservancy and Audubon of Western Pennsylvania. They manage a nature center in a Pittsburgh suburb and operate a small wind turbine as part of a display on solar energy. During the mid-1980s, they found a dead duck at the base of the tower. Greatly disturbed, they called the dealer. He was speechless. The next day he inspected the wind turbine for any telltale signs. A bird the size of a duck would have severely damaged the 1-kW Bergey turbine. The dealer found the turbine unscathed.

A few days later, a neighbor called the nature center searching for his pet peacock. Meanwhile, visitors had begun sighting a fox on the grounds. These reports prompted the center's naturalist to reexamine the dead bird, and the mystery was soon solved. He concluded it was the missing peacock and not a duck, after all. And after finding signs of the fox near the tower, the center concluded that the fox, not the wind turbine, was the culprit.

Still, there are anecdotal reports of collisions between birds and the guy cables of small wind turbines. "Birds collide with just about everything," says NREL's Karin Sinclair. Any time a structure, whether a house, a skyscraper, or a fence, is raised above ground level, it will pose a hazard to birds, and that includes small wind turbines.

More damaging to wind energy's reputation than the numbers of birds being killed in the Altamont Pass and elsewhere around the world is the manner in which they die. Two-thirds of the golden eagles were killed after colliding with wind turbines or their towers. This lends itself to blaring headlines and self-styled investigative reports revealing the "true story" behind one green technology.

BioSystems, in a report for the California Energy Commission on the problem with wind turbines in the Altamont Pass, tried to put the issue in perspective by noting that 5 to 80 million birds die annually in the United States from collisions with structures from picture windows on homes to cooling towers on power plants.

In one noteworthy incident in the fall of 2013, a flare at a liquefied natural gas plant in New Brunswick, Canada, drew an estimated 7,500 migrating birds, mostly small songbirds, into its flame. The conditions that night were a perfect storm for migratory birds: foggy with a low cloud cover. How many birds flew past the flare and survived is unknown.

Birds are killed not only in the production of electricity, but also in its transmission and distribution. Birds die by striking overhead power lines (and telephone lines) or by electrocution. While it's difficult to prevent birds from flying into power lines, most deaths by electrocution are avoidable and can be prevented by modifying transmission line towers.

Ornithologists can only speculate on what happens as birds fly near wind turbines. Flying is hazardous, especially for immature birds. "It's a tricky business to be a fast-flying animal at low altitude," said the University of Pittsburgh's Melvin Kreithen. "They make mistakes."

The job of the wind industry should be to make flying around wind turbines less hazardous. But there's no panacea or silver bullet for eliminating the problem. Painting splashy stripes on the blades or adding noisemakers have all been found wanting. The most effective method is avoiding the problem altogether by siting wind turbines where there are no large concentrations of birds that might collide with the turbines and monitoring—postconstruction—that, that is indeed the case.

Wind companies, large and small, must avoid the fortress mentality evoked by the issue of birds crashing into wind turbines. Some companies respond by trying to control the damage instead of trying to solve the problem. As Exxon found with the Valdez, "damage control" may cause as much damage to the company's interests as the disaster itself.

A better approach than damage control is to engage the environmental community before a project is proposed. Environmentalists, including bird lovers, generally support wind energy—when given a chance. In the United States, both the Sierra Club and the Audubon Society support wind energy in principle, though they may object to specific projects in part because of their impact on wildlife.

In Britain, the Royal Society for the Protection of Birds has gone even further than the timid steps of environmental groups in North America who are often satisfied with token installations of solar or small wind turbines as mere architectural adornment. In 2012, the society filed plans to install a large wind turbine at its headquarters north of London. The society issued a very clear statement why they were doing so.

We know that with the right design and location wind turbines have little or no impact on wildlife. We hope that by siting a wind turbine at our UK headquarters, we will demonstrate to others that with a thorough environmental assessment and the right planning and design, renewable energy and a healthy, thriving environment can go hand in hand. . . .

As one of the UK's leading environmental organisations, it is important that we play a pro-active role in leading action towards meeting national carbon reduction targets—particularly given our concern about the threat of climate change to birds and wildlife. . . .

We favour a broad mix of renewables, including solar, wind, and marine power, as long as they are sensitively sited to avoid impacts on wildlife and the wider environment.

Though BUND (Bund für Umwelt und Naturschutz Deutschland) is known for its opposition to fossil fuels and nuclear power, the Friends of the Earth affiliate in Germany proudly notes that its member groups have always been at the forefront, if not the pioneers, of renewable energy in Germany, including the development of wind energy.

In Canada, researchers with the Canadian Wildlife Service estimated the total number of birds that might be killed directly by wind turbines and indirectly by the loss of habitat from wind farm construction. They found that less than 0.2% of the population of any species would be killed by wind turbines in Canada due to direct mortality and habitat loss and concluded that "population level impacts" are unlikely.

Pre- and Postconstruction Surveys

One essential step for projects with large numbers of turbines—and one demanded by environmental groups—is to study how birds use the site beforehand. Ornithologists then determine the level of risk to particular species if the project proceeds. Public authorities and the environmental community must then weigh what risks do exist against the environmental benefits the project provides. Once a large project is in operation, it's also necessary to conduct a postconstruction survey to verify

Figure 15-22. Pre- and postconstruction monitoring of birds and bats. Ornithologist monitoring bird behavior around the wind turbines at Cros de Georand (Ardeche) in France's Massif Centrale. The monitoring was part of a postconstruction study by the Ligue de Protection de Oiseau, the French equivalent to the Audubon Society in the United States. 2005.

that any impacts are within the range expected (see Figure 15-22. Pre- and postconstruction monitoring of birds and bats). And if the impacts are greater than those permitted, operators must propose a plan for taking action or mitigating the impact in some way. Mitigation could include purchase of offsite land for protected wildlife habitat.

In Ontario, mitigation is required when more than 14 birds are killed per turbine per year, or 0.2 raptors are killed per turbine per year. Mitigation may require turning some turbines off during peak migration periods. Some sites in the United States have experimentally used radar systems to detect approaching birds and turn turbines off as needed. However, the "low levels of avian mortality caused by wind turbines suggests this should not normally be necessary," say researchers with the Canadian Wildlife Service.

Bats

Thousands of bats have been killed at some wind projects in the Appalachian Mountains and in southern Alberta. Bats have also been killed in smaller numbers at other sites as well. As with wind energy's impact on birds, more research is underway to both determine the extent of the problem and to examine when bats are most active around wind turbines.

At some sites, collisions between bats and wind turbines are a more serious wildlife problem than that with birds. At some Canadian sites, the number of bats killed per turbine is two or more times greater than that for birds. Unfortunately, less is known about the bat population

THE OVERWHELMING CONCLUSION?
Wind turbines don't have any effect on neighboring property values one way or the other.

in certain areas than that of birds. The number of bats killed by wind turbines, coupled with already declining bat populations due to white-nose syndrome (*Geomyces destructans*) could have an effect on particular species.

Some research has experimented with jamming bat's echolocation so they will avoid tall structures, including wind turbines. Promising work by Bat Conservation International suggests that operators can cut bat deaths dramatically when the turbines are stopped during the low-wind conditions when bats are most active.

No Free Lunch

Though the overall impact on bird and bat populations from wind energy may be slight, the fact that there is an impact at all illustrates, once again, that there are costs to all energy choices. "There's no free lunch," says the CEC's Anderson.

Some birds, including eagles, will fly into wind turbines, regardless of mitigation measures. An unpleasant thought, yes. Yet, to some extent, unavoidable. Those who think otherwise are deluding themselves. "Zero kill?" says Tom Cade of the Peregrine Fund. "That's not ever going to happen."

Property Values

One of the more persistent myths about wind energy is that wind turbines, especially wind farms, will reduce the value of neighboring properties. For most people, their home and land is their most valuable possession. Nothing elicits a fighting response quicker than a threat—real or otherwise—to one's financial nest egg. As an Appalachian coal miner once said to environmental activists trying to regulate strip mining: "Two things you don't touch here: my wife and my wallet." Antiwind energy activists, including those sponsored by coal industry interests such as the Koch brothers, know this, and they ruthlessly play off the fear the potential loss of property value creates.

At one time evidence that wind turbines affected property values one way or the other was hard to come by. That was long ago. Today, there's a veritable mountain of data on property transactions near wind turbines and commercial wind power plants. The overwhelming conclusion? Wind turbines don't have any effect on neighboring property values one way or the other.

The Energy and Policy Institute's Mike Barnard has done more than any other person to weigh the quality and breadth of the many reports on this sensitive topic. He found that 10 major studies in three countries concluded there was no connection between wind farms and property values. He identified a few of the studies coming to the opposite conclusion and found they were based on a limited number of transactions or anecdotal data. Barnard classed them as outliers from the mass of data collected on this subject.

The first extensive study in North America was conducted by the Renewable Energy Policy Project (REPP) in 2003. They examined property transactions near wind turbines at multiple sites in California, Iowa, and Minnesota, as well as sites in several other states. Altogether REPP evaluated more than 25,000 real estate transactions. The study found that property within view of the wind farms generally increased in value faster than property with no view of the wind turbines. REPP defined properties with no view of the wind farms as those beyond 5 miles (8 kilometers) from the turbines. Yet, this report wasn't sufficient to silence critics.

In 2009 the authoritative Lawrence Berkeley National Laboratory (LBNL) collected data on almost 7,500 sales of single-family homes situated within 10 miles (17 kilometers) of 24 existing wind plants in nine different states. The LBNL researchers found that neither the view of the wind turbines nor the distance of the home to those turbines had any "consistent, measurable, and statistically significant effect on home sales" and prices.

The University of Rhode Island weighed in with a 2013 report that examined a total of 48,554 transactions within 5 miles (8 km) of operating wind turbines, including 3,254 transactions as close as 1 mile (1.7 km) to the turbines. The university researchers not only controlled for distance from the turbines, they also controlled for the height and size of the turbines and when the property transactions occurred. They considered transactions before projects were announced, before construction began but after project announcement, and postconstruction. They wrote that "across a wide variety of specifications, the results indicate that wind turbines have no statistically significant impact on house prices. For houses within a half mile of a turbine, the point estimate of price change for properties within ½ mile relative to properties 3 to 5 miles away is -0.2%. . . Our best estimate" is that wind turbines have virtually "no effect on prices of nearby properties."

LBNL followed up its earlier work with an even more extensive study in 2013. The Berkeley Lab analyzed more than 50,000 home sales near 67 wind power plants in 27 counties across nine US states. Again, LBNL evaluated transactions within 10 miles (17 km) of the wind turbines and 1,200 transactions within 1 mile (1.7 km) of the wind farms. They included transactions both before the wind turbines were installed and those after. Again, the LBNL researchers found "no statistical evidence that home values near turbines were affected" either before construction but after the projects had been announced, or after construction.

There was practically a deluge of reports in 2014 on the relationship between wind turbines and property value. An academic paper in an agricultural journal questioned whether empirical evidence matched public perception about the impact of wind turbines on property value. Agricultural economists collected detailed data on 5,414 rural residential sales and 1,590 farmland sales to estimate the impacts of wind turbines in Ontario's Melancthon Township on surrounding property values. They accounted for both the proximity to turbines and their visibility—two factors that they thought could contribute to a "disamenity" effect. The results, the economists concluded, suggest that the wind turbines had no significant impact on nearby property values. "Thus, these results do not corroborate the concerns raised by residents regarding potential negative impacts of turbines on property values," wrote the authors.

RenewableUK (formerly the British Wind Energy Association) contracted with the Centre for Economics and Business Research (CEBR) to analyze property transactions near wind turbines across England and Wales. They examined 82,000 transactions over 18 years within 5 km (3 miles) of a wind farm. Their conclusion doesn't differ from those that preceded it.

> Our analysis of the raw house price data for transactions completed within the vicinity of the wind farms yielded no evidence that prices had been affected by either the announcement, construction or completion of the wind farms for six out of seven sites. . . . One site did see a noticeable downturn following the announcement that a wind farm would be built; however once the turbines were erected, local house price growth returned to the county-wide norm.

But the biggest study yet was that by the University of Connecticut and LBNL for the Massachusetts Clean Energy Center, a state agency. In their 2014 report, the joint study analyzed more than 122,000 home sales in Massachusetts between 1998 and 2012. They found no statistically significant evidence that proximity to a wind turbine affects home values. They considered home sales within 5 miles (8 km) of current or future locations of 41 wind turbines across the Commonwealth. The study found that homes close to turbines sold as frequently as homes farther away. Most surprising of all, they found that property

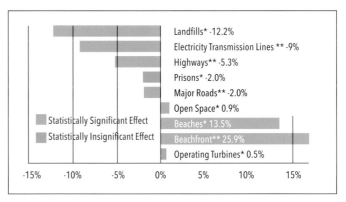

Figure 15-23. Property value near wind turbines. As one would expect, people prefer beachfront property to that near a landfill. What isn't expected is that a measurable number of people prefer living near wind turbines. In the largest study of its kind in North America, the University of Connecticut and Lawrence Berkeley National Laboratory found that in Massachusetts there was a slight (though statistically insignificant) positive benefit of being near operating wind turbines. (Adapted from University of Connecticut, LBNL)

Table 15-10. Property Value Impact Studies Summary				
	Year	Transactions	Distance Limit (km)	Results
Reneable Energy Policy Project	2003	25,000	8	No effect
Lawrence Berkeley National Laboratory (LBNL)	2009	7,500	17	No effect
University of Rhode Island	2013	48,554	8	No effect
Lawrence Berkeley National Laboratory (LBNL)	2013	50,000	17	No effect
University of Connecticut and LBNL	2014	122,000	8	No effect
RenewableUK and Cebr	2014	82,000	5	No effect
Richard J. Vyn and Ryan M. McCullough (Ontario)	2014	5,414		No effect
		340,468		

Table 15-11. Land Area Required NREL 2009						
Land Area Used				Land Area Occupied		
Permanent		Temporary				
ha/MW	ac/MW	ha/MW	ac/MW	ha/MW	ac/MW	
0.3	0.7	0.7	1.7	34.5	85.2	0.9%
Weighted average of 26,000 MW.						
Source: Land-Use Requirements of Modern Wind Power Plants in the United States						
http://www.nrel.gov/docs/fy09osti/45834.pdf						

values increased slightly near operating wind turbines (see Figure 15-23. Property value near wind turbines). The increase, though measurable, was not statistically significant. Still, the finding will surely put a dent in the charge—often without evidence—that wind turbines reduce property values.

After more than a decade of property transaction studies since REPP's first report in 2003, the data is in. Some 340,000 transactions prove that wind turbines have no statistically significant effect on property values (see Table 15-10. Property Value Impact Studies Summary).

Land Area Required

Another frequent concern is how much land commercial wind farms require. The answer depends on the array spacing, that is, the spacing between the turbines, the topography, and what is and is not included in the wind project. Wind turbines in open arrays will occupy more land per MW of installed capacity than a densely packed array. Project boundaries may include only the immediate area where the turbines are located or may include political boundaries that are not actually part of the project.

Land Area Occupied

In the 1990s, California's early wind farms typically occupied from 6 to 7 hectares (ha) per MW (15–18 acres/MW). European projects at the time occupied from 12 to 17 ha/MW (30–40 ac/MW).

The most authoritative analysis of modern wind plants in North America was a 2009 NREL report (see Table 15-11. Land Area Required NREL 2009). NREL examined 172 existing or proposed projects, representing more than 26,000 MW of capacity. The weighted average area occupied was 34.5 ha/MW (85 ac/MW) with a standard deviation of ± 22.4 ha/MW.

Contrary to the implication of earlier studies, the land required for geometric arrays is, paradoxically, independent of size. As turbines become larger, they occupy more land per unit because of the greater spacing required between them. However the relative spacing between turbines, governed by their rotor diameter, remains the same. NREL cited earlier work that found arrays of modern wind turbines spaced 5 rotor diameters (RD) across the wind and 10 RD downwind—a 5 by 10 array—yields an area occupied by the wind plant of 13 to 20 ha/MW (30 to 50 ac/MW).

Energy generation is directly proportional to swept area (rotor diameter) and less so for rated capacity in MW. Thus, the area occupied is a function of the rated capacity relative to the rotor swept area. Modern wind turbines of the silent wind power revolution have a much greater specific area (m²/kW) than wind turbines in the NREL sample. For example, Vestas's V80 was typical of the turbines in the mid-2000s with a rated capacity of 1.8 MW for a specific area of 2.8 m²/kW. Newer wind turbines, such as GE's 100-meter (330-foot) diameter 1.6-MW model, have a specific area of 4.9 m²/kW. These higher specific area turbines require more space between them for the same array spacing (Table 15-12. Summary of Land Area Occupied).

Smaller clusters of turbines typical of the community-owned projects in Germany may require less land area than that found by NREL for the United States. The Deutsches Windenergie Institut (DEWI) found that small clusters

SITING AND ENVIRONMENTAL CONCERNS

Table 15-12. Summary of Land Area Occupied			
	Rotor Specific Area	Area Occupied	
	m²/kW	ha/MW	ac/MW
NREL 2009		34.5	85
5 x 10 Spacing; 2.8 m²/kW	2.8	17.8	44
5 x 10 Spacing; 4.9 m²/kW	4.9	31.3	77
8 x 10 Spacing; 2.8 m²/kW	2.8	28.4	70
8 x 10 Spacing; 4.9 m²/kW	4.9	50.0	124

of turbines on small parcels may require half as much land as larger projects because they can take advantage of edge effects along the property's boundaries. This could be an important argument in favor of small, distributed projects, making them easier to integrate into a densely populated landscape as in Germany. For example, the German state of Saxony-Anhalt has one of the highest concentrations of wind turbines in the country. In a 2014 study, DEWI estimated that there remained sufficient room for growth at a density of 3.7 to 4.5 ha/MW (9–11 ac/MW), one-tenth of that in the NREL survey.

Fortunately, the total land area "occupied" by the wind turbines and all its ancillary structures is not the land actually "used," that is permanently disturbed by the wind plant.

Land Area Used

Though wind energy's land requirement is much less than that first thought, even less land is actually disturbed by the construction and operation of the wind turbines. Development and continued use disturb only a small portion of the land occupied by the wind power plant. The amount of land affected depends upon the terrain, the size and number of roads, mounting pads, buildings, and other structures.

Wind projects disturb more land in hilly terrain than on level ground. Grading a road on steep slopes demands a cut on the uphill side while the loosened soil is pushed down slope. These cut-and-fill slopes extend the disturbance beyond the immediate vicinity of the road or tower pad. On level terrain, grading disturbs less soil because there are no cut and fill slopes.

NREL breaks the land used by a wind farm into two categories: land permanently impacted for roads and the turbine's foundation, and land only temporarily impacted. Of the land permanently used by the wind farm, nearly 80% is accounted for by the roads necessary to install and service the wind turbines.

Figure 15-24. Land area used by a wind turbine. Enercon turbine standing in a tilled field near Oldenburg, Niedersachsen, Germany. Note that the field is tilled to the base of the tower and that the access road to the turbine is limited. This is typical in Germany, where land is at a premium. 2010. (Wikimedia Commons)

The area temporarily cleared averaged 0.7 ha/MW (1.7 ac/MW), mostly for assembling the rotor. Though some projects required more land area permanently disturbed than the average, NREL found that in 80% of the projects land permanently disturbed was less than 0.4 ha/MW (1 ac/MW). Land permanently used by the wind turbines, their service roads, and other structures averaged 0.3 ha/MW (0.7 ac/MW), equivalent to somewhat less than 1% of the total land area occupied by the wind plant. The latter confirms a rule of thumb in the wind industry that a well-designed project should use no more than 1% of the total land area.

European projects may use less land than wind plants in North America. It's not uncommon in Europe to find land tilled to the base of the tower (see Figure 15-24. Land area used by a wind turbine).

Significantly, the amount of land permanently disturbed by the typical large, central-station solar power plant built in the United States, whether a solar photovoltaic plant or a solar thermal plant, is nearly 10 times greater than that for a wind project. Solar One, for example, an early solar power plant using a central receiver, cleared 3.6 ha/MW (9 ac/MW) of desert shrubs when it was constructed. The yield of solar plants is also significantly less than that of a wind plant. Thus, the amount of land cleared for solar energy in the United States is much greater than the number of ha/MW would indicate relative to a comparably sized wind project. Wind turbines are typically twice as productive per MW as solar power plants. Thus, for a comparable amount of generation, solar power plants—at least in the United States—may use as much as 20 times more land area than a wind plant. Of course, this need not be the case. In Germany, the land beneath most solar power plants is not cleared of vegetation. Sheep graze among the solar panels. That this is not practice in the United States is more an indictment of American project developers and their regulators than it is of solar energy.

Energy Balance and Energy Return on Energy Invested

The energy generated by wind turbines pays for the materials used in their construction within a matter of months. Yet the question as to whether they do, thought by industry analysts to have been effectively answered during the 1970s, is continually raised by critics of wind energy. It was the first avenue that California desert activist Howard Wilshire sought in his quest to find a magic bullet that would kill the wind energy monster for all time.

The energy generated by wind turbines pays for the materials used in their construction WITHIN A MATTER OF MONTHS.

The question possibly arises from old reports about the poor energy balance of photovoltaics. Early solar cells—those for the space program—consumed more energy than they produced. That's no longer true for solar and has never been true for wind energy. Wind turbines have always paid for themselves quickly, despite the use of such energy-intensive materials as steel, fiberglass, and concrete.

It seems that every decade there's a flurry of interest in the subject, resulting in a flood of scientific papers all reaching similar conclusions. Then interest wanes until the next crop of doctoral students need a topic for their dissertation.

In the early 1990s, researchers examined the question thoroughly for the time. Two Danish studies considered a typical Danish wind turbine of the period, operating under typical Danish conditions, and a German study examined the energy payback of wind turbines from as small as 10 kW to as large as 3 MW. The results of the three studies were comparable: medium-size wind turbines installed in areas with commercially usable wind resources would pay for themselves easily within one year.

At fairly windy sites, like those on the North Sea coast or in California's mountain passes, wind turbines would return the energy used in their materials in three to five months. At sites typical of North America's Great Plains, wind turbines would pay for the energy they contained in four to six months. Even at low-wind sites, the turbines would pay for themselves in less than one year.

Since the 1990s researchers have introduced a new term related to energy payback: energy return on investment (EROI, or energy return on energy invested). EROI is the ratio of energy generated relative to the primary energy used to build the wind turbines from raw materials to their final dismantling and disposal.

Much of the energy used to manufacture the turbine is contained in the fiberglass blades of the rotor and the steel in the nacelle, but more than one-third of the total energy consumed by the wind turbine was found in the concrete foundation and the steel tower. The early German study found wind turbines produce 4 to 33 times more energy during their 20-year lifetimes than that used in their construction.

SITING AND ENVIRONMENTAL CONCERNS

Table 15-13. Sampling of Wind Turbine Energy Payback Studies

		Turbine		Energy Payback (Years)		EROI/Energy Ratio (kWh$_e$/kWh$_{prim}$)	
		Size (MW)	Lifetime (Years)	Low	High	Low	High
Grum-Schwensen	1990	0.1		0.28			
Gydesen	1990	0.15		0.40	0.68		
Hagedorn & Ilmberger	1991	3		0.32	0.70		
Elsam, onshore	2004	2		0.65			
Elsam, offshore	2004	2		0.75			
Kubiszewski, Cleveland, et al	2009	<1				19.8	25.2
Gupta & Hall	2011					18.1	24.6
IPCC	2011			0.28	0.71	5	40
Lambert, Hall, et al	2012					20	
Guezuraga, Zauner, Pölz	2012	1.8-2.0	20	0.58			
Haapala & Prempreeda	2014	2	20	0.43	0.53		
Palomo & Gaillardon	2014	3	20	1.03		19.6	

During the 2000s, numerous academics and researchers on both sides of the Atlantic have flogged a very dead energy payback horse, with each trying to outdo the other in sophistication. For the most part, they all reach the same conclusion: wind turbines pay for the primary energy used to make and dispose of them easily within one year (see Table 15-13. Sampling of Wind Turbine Energy Payback Studies). Where they looked at EROI, most found that wind turbines returned about 20 times their primary energy.

Probably the most authoritative review of the published work on the topic was that by the Intergovernmental Panel on Climate Change (IPCC) in 2011 when it examined the impact and benefits of renewable energy in mitigating climate change (see Figure 15-25. Energy payback). The IPCC surveyed 20 different studies after screening them for quality. Its conclusion? Wind turbines pay for themselves within 3.4 to 8.5 months (0.28 to 0.71 years) with a median of 5.4 months (0.45 years)—a result not much different than that found in the early 1990s.

Let's hope this is the end of this question, once and for all.

Emissions of CO$_2$ Equivalent Gases

As with energy payback, recent research has confirmed the obvious that the life-cycle emissions of global-warming gases are lower from

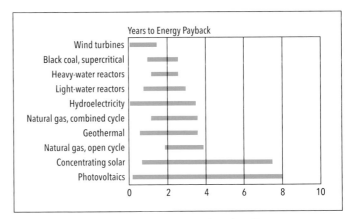

Figure 15-25. Energy payback. The Intergovernmental Panel on Climate Change (IPCC) conducted a meta-analysis of the energy payback and the energy ratio of electricity-generation technologies. Like many other sources, the IPCC found that wind energy has one of the shortest energy paybacks and one of the best energy returns for energy invested of any technology.

wind energy than almost any other technology. For example, one study of 2-MW turbines found that overall the wind turbines emitted 9 grams (g) of CO$_2$ per kWh. A similar study by French researchers of wind turbines in France estimated that wind turbines generated 11.8 g CO$_2$ equivalent per kWh.

The IPCC examined 126 different estimates for the emissions due to wind energy from 49 different studies. They found a range from 8 to 20 g/CO$_2$ equivalent per kWh for the 25th and 75th percentiles with a median of 12 g/CO$_2$ equivalent per kWh. For comparison, the typical coal-fired power plant emits 100 times, or two magnitudes,

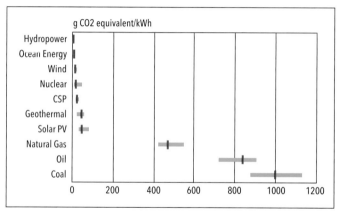

Figure 15-26. Life-cycle greenhouse gas emissions. The Intergovernmental Panel on Climate Change (IPCC) conducted a meta-analysis of the emissions of global warming gases from electricity generation. Their findings surprised no one. Wind energy is one of the lowest sources of greenhouse gas emissions of any technology.

more global-warming gases than wind energy; natural gas nearly 50 times more (see Figure 15-26. Life-cycle greenhouse gas emissions).

The emission of CO_2 equivalent gases is a reasonable surrogate for all other air pollution emissions as well. Wind energy, because it reduces the generation from more polluting sources, is an important contributor to clean air and addressing climate change from burning fossil fuels.

Water Consumption

To someone in a temperate climate, the water required for conventional power plants seems of little significance. But in arid areas of the world, such as in the southwestern United States, such considerations as water, who has it, and how it's used are volatile political issues.

Unlike thermal power plants, wind turbines have no need for cooling water. Conventional power plants use enormous amounts of water for the condensing portion of the thermodynamic cycle. Much of the water required for conventional power plants simply passes through the cooling system, and the heated water is returned to its source. In the case of coal cleaning, the wastewater will be heavily laden with suspended solids and will need treatment. Though wind turbines and photovoltaic cells need no cooling water to produce electricity, they may use slight amounts of water in arid areas for cleaning the blades or the surface of the panels.

In arid regions, the amount of water used is less critical than the amount of water consumed, that is, the water lost to evaporation from the cooling cycle (see Figure 15-27. Water consumption in electricity generation). In an NREL study, researchers found that wind turbines consumed a maximum of 1 gallon (gal) per 1,000 kWh (0.004 l/kWh) in comparison to average consumption for a large solar PV plant of 26 gal per 1,000 kWh (0.1 l/kWh), a nuclear power plant using a hyperbolic cooling tower of 670 gal per 1,000 kWh (2.54 l/kWh), and a coal-fired power plant using a similar cooling tower of 690 gal per 1,000 kWh (2.6 l/kWh). Even concentrating solar power plants consumed an order of magnitude less water than conventional power plants.

Removal Bonds

All wind turbines should be decommissioned and dismantled when they are no longer "used and useful." In most cases the sites where the turbine stood are repowered, that is, one or more turbines are removed and replaced with a new turbine. Regardless, when a wind turbine, of any size, is no longer functioning, it should be removed and not permitted to stand derelict for any extended period of time.

Though individual turbines at wind plants may need replacement after 20 to 30 years, there is no compelling reason for abandoning the site. New, more cost-effective turbines can be added, the infrastructure upgraded if necessary, and

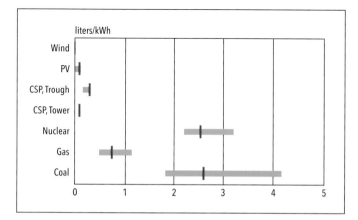

Figure 15-27. Water consumption in electricity generation. In 2011, NREL surveyed the technical literature on the consumption of water in the generation of electricity. Their conclusion was similar to that found two decades ago and should be intuitively obvious: wind energy doesn't consume water in comparison to conventional sources of electricity. This is a critical benefit of wind energy in arid regions of the world.

the plant returned to use. This is one of wind energy's most significant attributes: it is renewable. Because there is no accumulation of wastes (as in a nuclear plant) or exhaustion of the fuel source (as in a fossil-fuel-fired plant), production of wind-generated electricity is sustainable indefinitely.

Nevertheless, it's still useful to look at the costs of dismantling because removal of individual turbines may be necessary from time to time. The public also wants assurance that removal costs are within the financial reach of salvage companies, should a wind turbine or an entire wind power plant become inoperative.

Decommissioning a wind plant, in its simplest form, requires removal of the turbines. Of course decommissioning also includes site restoration as well. This entails removal of the roads, removal of the foundation to plow depth, removal of ancillary structures, and revegetation.

In California, the salvage value of the wind turbines has often offset much if not all of the removal and site restoration costs. In mid-2014, the cost to remove a foundation to plow depth for a typical 1980s-era wind turbine was $1,000, or about one-third of the total cost of $3,500 to remove the turbine and restore the area around the turbine. Most bidders in 2014 to remove old Danish turbines in the Tehachapi Pass bid only for the salvage value, that is, there was no charge to the property owner to remove the turbines and restore the site.

To reassure the public that wind turbines will be promptly removed when they no longer server their function, regulators often require removal and restoration bonds. This is the case in Germany. Gottfried Wehr explains that when their local group began developing Windkraft Diemarden (see Chapter 16, "Installation and Dismantling"), they had to prove to local planners that there was a bank guarantee—or bond—with sufficient funds to pay for dismantling the turbines. Their 1993 decommissioning bond was substantial: the equivalent of €25,000 (~ $30,000). Yet when they removed the turbines in 2012 to replace them with much larger, more modern turbines, they didn't need to use the bond. The salvage value paid for the decommissioning costs. The old turbines were bought by a company in Northern Ireland,

REPLACING THE OLD WITH THE NEW

Unlike a mine-mouth, coal-fired power plant that exhausts its resource and has to be abandoned, wind is inexhaustible. It will always be there. If a wind power plant has proven productive and profitable, there's no reason why the wind plant can't operate indefinitely. However, the wind turbines do age, and as they do, it may become more profitable to replace them with newer machines on the same site.

Repowering occurs when new wind turbines replace older ones. This is not abandonment; it's replacement.

Because modern turbines are much larger and more productive than those installed as recently as a decade ago, repowering usually replaces a number of the original turbines with a fewer number of the newer, larger turbines. In repowering a project of first-generation turbines, for example, one of today's modern turbines can replace 20 to 40 units—and generate even more electricity than before. In one repowering project on the German island of Fehmarn, generating capacity nearly tripled from 40 MW to 136 MW.

When repowering removes earlier turbines, the machines of Danish or German origin are seldom discarded. (Nearly all the early, unreliable wind turbines—mostly of American design—have long since been removed and scrapped.) They have an intrinsic value. The turbines that have remained operating until repowering replaced them have put in long service, but they have proven their mettle. With competent reconditioning and care, they can put in many more decades spinning out wind-generated electricity.

where the turbines have since been returned to service.

Windkraft Diemarden proved to the area's critics that the old turbines would be removed, and that there was no need for explosives to remove the foundation. They broke up the foundation with a hydraulic hammer. They even recycled the old reinforcing bar from the foundation. As Wehr points out, most of the owners of Windkraft Diemarden live in the community. They are proud of their wind turbines and proud to be good neighbors.

ALL WIND TURBINES SHOULD BE DECOMMISSIONED AND DISMANTLED WHEN THEY ARE NO LONGER "USED AND USEFUL."

AESTHETIC DESIGN SUMMARY

- **Provide Visual Uniformity**
  Turbine, Tower, Color, Direction of Rotation
- **Keep Them Spinning**
  Remove Inoperative Turbines
- **Avoid Visual Clutter**
- **Keep Sites Tidy**
  Remove Litter and Scrap
- **Avoid Billboards and Logos**
- **Always Dress Your Wind Turbine Properly**
- **Bury Power Lines**
- **Avoid Steep Slopes and Minimize Roads**
- **Control Erosion and Promptly Revegetate Disturbed Soils**
- **Harmonize Ancillary Structures**
- **Inform Public**
- **Use Proper Proportions**
- **Use Open Array Spacing**
- **Be a Good Neighbor**

Kern County, California, doesn't require a removal and restoration bond, but it does require the landowner to remove any wind turbine that has not operated within a consecutive 12-month period. If the turbine isn't removed, the county will remove the turbine and charge the property owner for the expense of doing so.

Aesthetics

For some the appearance of a wind turbine on the skyline is symbolic of responsible stewardship—a step toward a sustainable future. To others it's industrial blight and a call to arms. Concern about the visual effect wind machines may have

Figure 15-28. Visual uniformity. This pleasing array of Ecotecnia turbines on Spain's Galician coast near Malpica is partly attributable to the visual uniformity of the turbines. 1998.

on a landscape, and the communities of which they are a part, should not be dismissed lightly. How we design our wind turbines and our wind plants determine whether wind energy becomes a respected member of the community—or an unwelcome intruder.

Much has been written about the place of wind turbines in the landscape, how to minimize their visual intrusion, and how to increase their acceptance. What follows are some general guidelines. Most fall under the rubric of "Be a good neighbor."

Provide Visual Uniformity

Although large wind turbines can be installed as single units, like small wind turbines, more often they're installed in clusters or large arrays—wind farms. When there are more than one or two turbines in visual proximity with each other, it is critical to provide visual uniformity of turbine, tower, color, and direction of rotation. This is the single most important step planners can take to successfully integrate wind turbines into the community. The turbines need not be identical, but they must appear similar (see Figure 15-28. Visual uniformity).

Remove Headless Horsemen

As with any business, some wind projects succeed, some fail (see Figure 15-29. Headless horsemen). The community has a right to demand that operators repair or replace any "headless horsemen," that is, towers without turbines on top. If the turbine is not returned to operation, then the turbine, tower, and support equipment should be promptly removed and the site restored to its preproject state.

Use Open Spacing

Avoid visual clutter by designing arrays with open spacing. Don't place the turbines too close together. One Tehachapi wind farm operator placed his turbines so close together that their rotors tangled and the turbines had to be repaired, then moved.

Avoid Billboards and Logos

There are already too many billboards littering the North American countryside. Wind turbines shouldn't contribute to this visual blight (see Figure 15-30. No billboards). Don't paint billboards or corporate logos on the tower or nacelle.

Figure 15-29. Headless horsemen. Dead and dying Windmaster turbines on a wind farm in California's Altamont Pass. When wind turbines are no longer "used and useful," they and their supporting infrastructure should be promptly removed. Mid-1990s.

Figure 15-30. No billboards. The tacky billboards and cell-phone repeaters on the tower of this Tacke turbine earn the owner of this truck stop near Herbolzheim in Baden-Würtemberg Germany a few more Euros at the visual expense of anyone who sees this eyesore. 2005.

Figure 15-31. Logo-free. Corporate logos on the sides of wind turbine nacelles are an unnecessary visual distraction from their clean lines. Riverside County prohibits logos on wind turbines, as here on these Vestas V27s near Palm Springs, California. Mid-1990s.

Regulators in some locales prohibit manufacturers or the project operator from advertising on the side of the nacelles. Keep the nacelle and tower free of corporate logos (see Figure 15-31. Logo-free).

Bury Power Lines

Bury all intraproject power lines and the transmission lines leading to the project site. Aboveground power and transmission lines at large wind projects detract from the otherwise rural character of the landscape, giving such projects an industrial feel (see Figure 15-32. Bury power lines). For sites on hilltops or ridges, all ancillary structures (buildings, transformers, substations, and so on) should be placed on the valley floor or beyond the viewshed of those who live in or travel through the valley below the project.

Always Dress Properly

Always dress the turbine properly. Some manufacturers have been so intent on cutting costs that they will sell and install wind turbines without nose cones (spinners) or nacelle covers. These wind turbines appear angular, mechanical, and, in a word, "industrial." They say to neighbors, "We don't care what you think." Similarly, some

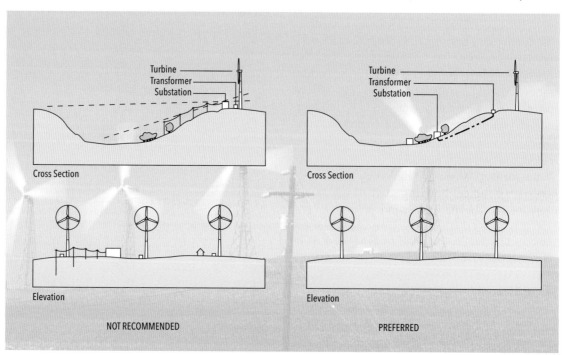

Figure 15-32. Bury power lines. Minimize visual intrusion by burying power lines and removing all ancillary structures from among the wind turbines. Wherever possible, place transformers inside the tubular towers used large wind turbines. (Chris Blandford Associates)

California wind farm operators, in a misguided drive to squeeze every last cent out of aging turbines, remove the nose cones and nacelle covers or fail to replace them when they blow off (see Figure 15-33. Dress turbines properly). Turbines in such projects become "junkyards in the sky," fueling wind energy's detractors who are fond of describing them as "industrial wind turbines."

Control Erosion and Promptly Revegetate Sites

Control erosion by minimizing or eliminating road construction, especially in steep or arid terrain. Too many unnecessarily wide roads can give an otherwise well-designed wind project the appearance of a mining site instead of the pastoral scene wind advocates envision for the technology (see Figure 15-34. Control erosion).

Promptly replant disturbed soil after construction with native vegetation, and ensure that the plants revegetate the site (see Figure 15-35. Revegetate the site). Proper site restoration requires more than hydroseeding—spraying a green-colored mix of fertilizer and seed—over cut-and-fill slopes. Revegetation demands thorough planning to find the right plant mix for the site and thorough execution to plant at the right time of year. Success may also require more than one application and several years of monitoring.

Harmonize Ancillary Structures

Harmonize ancillary structures with other structures on the landscape. Ancillary structures on a wind project should blend in with their surroundings. At the Italian test site of Acqua Spruzza, control and transformer buildings were built to resemble other rural buildings (see Figure 15-6. Watch for ice). When Zilkha Renewable Energy needed an office building and maintenance shop for its Top of Iowa wind plant, it could have chosen a typical slab-sided metal building. Instead, the company took an abandoned barn and adapted it to its needs. The barn was more in keeping with other nearby farm buildings, Zilkha decided, than the industrial structure it would have used. Similarly, transformers should be placed inside the tower, and where that's not feasible, should incorporate a façade that obscures their industrial features (see Figure 15-36. Architectural transformer treatment).

Figure 15-33. Dress turbines properly. Kenetech KVS 33 without a nacelle cover in the San Gorgonio Pass near Palm Springs, California. Kenetech (US Windpower) turbines were notorious for losing their nacelle covers in high winds, which were seldom replaced, giving the turbines and the projects they were a part of the appearance of an aerial junkyard. Fortunately, the turbines have since been removed and sold for scrap metal. 2008.

Figure 15-34. Control erosion. Erosion gullies (circled) caused by excessive road construction in the steep, arid terrain of California's Tehachapi Pass gives wind energy a black eye. Embarrassingly, the company's CEO said, "Erosion, what erosion?" Early 2000s.

Figure 15-35. Revegetate the site. Trees have been planted to revegetate this NEG-Micon project near Oupia (Aude). Revegetation of the rugged and arid terrain of the *maquis* in southern France requires special attention to detail. Here the trees are protected from grazing goats with wire cages and from errant service personnel in their utility vehicles by the line of limestone boulders. The boulders prevent the technicians from driving off the road. 2005.

Figure 15-36. Architectural transformer treatment. Where transformers cannot be placed inside the tower, transformers should be shrouded with an architectural treatment. This can take the form of a façade to harmonize what otherwise would be an industrial structure with other nearby structures on the landscape. Note the concrete façade (left) of the transformer and the open-cell concrete pavers to harden the access track to the NEG-Micon tower. The pavers allow rain to percolate through to the groundwater table. This turbine and the Lagerwey 18/80 in the background were owned and operated by Noud de Schutter, a Dutch farmer in the Wieringemeer polder north of Amsterdam at the time the photo was taken in the late 1990s. The façade treatment (right) of transformers in the Cros de Georand use untreated logs typical of the remote area of France's Massif Centrale. 2005.

Keep Sites Tidy

Operators of wind projects should pick up any litter on their sites and eliminate on-site storage of spare parts, damaged wind turbines, oil drums, and other industrial detritus. Trash and litter quickly make a pastoral array of clean, modern wind turbines into an industrial site that just happens to use wind machines (see Figure 15-37. Remove litter and boneyards).

Inform the Public

Inform the public about your project. They have a right to know. Tell them about the wind turbines and what they do, how much electricity they generate, the pollution they offset, the water they save, and how they benefit the community. Tell them who owns the turbines and who to get in touch with for more information or if there's a problem. Provide space for parking, and include a picnic area if you can. Information kiosks need not be elaborate. Wind farms are not nuclear power plants (see

Figure 15-37. Remove litter and boneyards. Another poor site practice by the United States' one-time largest wind developer, Kenetech (US Windpower). Scrap tires and electronics dumped at the base of a Kenetech KVS 33 wind turbine in southwest Minnesota. The untidy project presented the image of a junkyard or dump—not that of a modern green technology. Late 1990s.

Figure 15-38. Inform the public. Informational kiosks need not be elaborate as seen at this car park explaining what the Bonus 600-kW turbines are doing at Royd Moor in South Yorkshire, England, in the late 1990s.

Figure 15-38. Inform the public). They don't need million-dollar visitor centers to convince the public of their benefit.

Small Turbines

Manufacturers of small wind turbines have paid far less attention to aesthetic design than manufacturers of larger turbines. Some small turbines are crude contraptions that a reasonable person may not want in the neighborhood. One, the 1980s Jacobs, is an ungainly design that looks like it came directly from a 1930s machine shop, which it did. Many manufacturers of small turbines could use a good industrial designer.

Try to incorporate the community's wishes when you're considering the type of tower to use. There's much less objection to the clean lines of a tapered tubular tower than to a wooden utility pole. Likewise, a tubular tower is more pleasing in foreground views than a truss tower. However, truss and lattice towers should be considered. In distant views, guyed lattice towers and truss towers become nearly invisible. From an aesthetic perspective, the type of tower that's most acceptable (whether truss, guyed lattice, or tubular) depends on the viewpoint and the distance between the observer and the tower.

If someone objects on aesthetic grounds, point out similar structures on the horizon that we've learned to tolerate, if not accept. You have as much right to erect a wind machine as the local radio station has to install a tower or the utility to string a transmission line across town. Although it's true that the utility's power lines and the radio's broadcast tower provide community-wide benefits, each person benefits individually. Your installation of a wind system differs little from the utility building a power line to your house. The appearance may differ, but the purpose remains the same.

Don't overlook some obvious ways to adapt the turbine to the community's or your own tastes. Patrick Campbell is a Kern County firefighter who has operated his Bergey Excel since 1998. Campbell is the type of customer who knows what he wants. And what he wanted was a wind turbine that matched the color of his home. Bergey Windpower obliged and painted the turbine to match. "Any customer can request it," says Campbell, but few do.

Avoid garishness. Don't string lights from your turbine for any reason. Be respectful of your neighbors and of the night sky. It's our common heritage to be enjoyed by all. A wind turbine is not a Christmas tree.

Compatible Land Uses

In contrast to most conventional sources of energy, wind energy is compatible with most existing land uses. Farming, recreation, and numerous other activities can and do occur in and around wind farms. Grazing is the most common example of a land use compatible with wind development. In the Altamont Pass, and to some extent in the Tehachapi Pass, land leases for wind development specifically restrict the amount of land taken out of pasturage for cattle. Sheep peacefully graze beneath wind turbines along dikes in the Netherlands and on sites throughout Britain. To a comment about the numerous sheep grazing on a site during construction of a wind farm in Wales, the site manager responded "Of course!" as if to say "Why wouldn't sheep be grazing here during construction? There's no conflict." To the site manager it was the norm, and he never thought anymore about it.

However, grazing is but one example. In northern Europe the land beneath some wind power plants is tilled to the base of each tower. At others, the turbines are planted within the hedgerows and field boundaries. Some of the wind turbines in Denmark and Germany are installed on the owners' farms. Businesses in Great Britain, the Netherlands, Germany, and Denmark install wind turbines in the parking lots of their establishments, as does Taylor University in Upland, Indiana.

It's not uncommon to see wind turbines operating in public places such as schools and sporting centers. On popular strands, mothers can be seen pushing their prams unconcernedly beneath operating wind turbines. Fisherman sit beneath wind turbines on breakwaters. Yachtsmen anchor at harbors graced by wind turbines.

Popular author Helen Colijn suggests biking to the wind turbines on the dike at Urk in her *Backroads of Holland*. She describes how cyclists can find wind farms on Dutch tourist maps, reassuring them that cycling is permitted near the wind turbines, just as it is on bikeways at dikes

WILL IT BE SEEN?

Yes. In all likelihood a wind turbine will be visible on the landscape or at sea–for a great distance. The human eye is sensitive to motion–so we can detect and avoid that predator waiting to eat us. And the whole point of a wind turbine is to be in motion, capturing the energy in the wind. While visualizations and computer simulation are helpful in giving people an idea what wind turbines will look like from various viewpoints, there's no need to overstate the case by saying, "You'll hardly see it." One of the beauties of wind energy is that what you see is what you get. There's no insidious pollution–no black lung, for example. Wind turbines operate out in the open for everyone to see. They are fully transparent. They don't hide behind a massive containment building, shielding them from public scrutiny. When they don't work, everyone knows. When they fail or catch fire–as some do–they make the front page of the local paper.

throughout the country. Once there, cyclists will find sign-posted directions to sights along the route. Tourists cycling the scenic bike path on the east side of Denmark's Ringkøbing Fjord are tempted to detour into the heart of the adjacent wind power plant and visit the information kiosk before continuing on their way. Farther south, across the German border, the Wiedingharde Tourism Association may lead an evening walk along the dikes to the community-owned wind farm at Friedrich-Wilhelm-Lübke-Koog.

In practical terms, nearly all land uses except hunting and wilderness preservation can be compatible with wind development. By definition, wilderness requires the absence of structures, and wind development is, after all, development.

Hunting, though, is not inherently incompatible: California wind turbines have suffered surprisingly little damage from gunfire. And wind turbines do not hinder the hunters. However, hunting endangers those who service the wind turbines. For this reason alone, hunting is prohibited at all California wind plants.

Some ill-informed critics have charged that wind development effectively closes public lands to public use. This is not the case in either North America or Europe. In the San Gorgonio Pass, for example, hikers on the Pacific Crest Trail (PCT) have full access to the trail near hundreds of turbines on a Bureau of Land Management (BLM) lease atop Alta Mesa just as they do on the PCT where it snakes across the Tehachapi Pass on Cameron Ridge.

Coal Clough in England's South Pennines practically sits astride a footpath across the "long causeway" above the historic dale of Calder. A public car park solely for the use of hikers is just across the road from the turbines. The footpath itself crosses the wind farm at several points via gates installed for the purpose. As at Coal Clough, most wind projects in England and Wales can be reached by sign-posted public footpaths that existed long before the wind turbines were in place. Most turbines on public land, and many on private land, can be reached by public walking or cycling paths throughout northern Europe.

Normally, private land in the United States is closed to the public. Wind plants in the Tehachapi Pass have enabled access to previously restricted private land where the PCT crosses BLM and private leases. The trail, which stretches from Canada to Mexico, is the longest footpath in the United States. Through the vehicle of easements, both private and public lands are now open to hikers along the PCT through the heart of Tehachapi's wind development. To publicize the trail, and the public's free access to it, the Kern-Kaweah chapter of the Sierra Club sponsors an annual spring hike among the wind turbines. They have been doing so since 1986. In 1990, nearly 100 people took the hike.

Clearly the experience of hiking among wind turbines is different from that of hiking in a verdant forest. And the mere thought of hiking among wind turbines enrages some purists. A local Audubon activist, Steve Ginsberg, not one to forsake hyperbole when it suits his purpose, proclaimed that "any hiker with the slightest sense of aesthetics or ecology might question such an absurd suggestion" that hiking among the wind turbines in the Tehachapi Pass is a pleasing recreational activity. Yet even he begrudgingly admits that "hundreds have taken the hikes," apparently responding to the "absurd suggestion."

Hiking across Tehachapi's Cameron Ridge on the PCT resembles the kind of hiking found throughout Europe or in the eastern United States, for example on the Appalachian Trail, where the evidence of human activity is often present. Cameron Ridge is not a wilderness, park, or preserve, nor is it wooded. The sense of space, so important to the hiking experience in the western United States, remains. Hikers can still enjoy the spectacular views of the Mojave

SITING AND ENVIRONMENTAL CONCERNS

Desert, the Garlock Fault, and other prominent features. In short, the experience is different from walking through a coniferous forest, but many of the attributes of hiking on the desert remain.

Wind turbines, when sited with care and respect for the environment and the communities where they are located, can be good neighbors. Today, wind turbines can be found in almost every conceivable application (see Figures 15-39 through 15-53).

Figure 15-39. Compatible with grazing. Sheep grazing peacefully beneath Vestas's V82s in the Manawatu Gorge or Te Apiti, Maori for "narrow passage," near Palmerston North on New Zealand's North Island. 2005.

Figure 15-40. Compatible with farming. Next to grazing, wind turbines are most associated with farming operations, including tilled crops, as here in Friedrich-Wilhelm-Lübke-Koog (polder) in Schelswig-Holstein in northern Germany near the Danish border. 2012.

Figure 15-41. Compatible with urban areas. Wind turbines can be found in urban areas on most continents. The WindShare turbine was the first "urban wind turbine" installed in North America. Since then, Cleveland, Boston, and a number of other cities have followed suit. Urban installations are quite common in Europe. WindShare's Lagerwey turbine is at a prominent location on the grounds of the Canadian National Exhibition (Ex-Place) in downtown Toronto. 2007. (Nancy Nies)

Figure 15-42. Compatible with commercial harbors. A cluster of Ampair 10-kW wind turbines at Osprey Quay Marina in Portland, Dorset, a sailing venue for London's 2012 Olympics. Note that the turbines are installed on tubular masts and stand in the midst of the harbor and pedestrian walkways. (Ampair)

Figure 15-43. Compatible with cycling. French cyclist passing one of the wind turbines of Cros-de-Géorand in the Massif Centrale. 2005.

SITING AND ENVIRONMENTAL CONCERNS 369

Figure 15-44. Compatible with hiking. Tony Swan (left) leads a windmill-wildflower hike through a field of grape soda lupine (*Lupinus excubitus*) on a cold spring day in 2005. Swan has been leading the annual hike on the Pacific Crest Trail across Cameron Ridge in the Tehachapi Pass since 1986.

Figure 15-45. Compatible with public footpath. Nancy Nies posing at a gate leading to a public footpath across St. Breock Down in Cornwall. The footpath leads to the village of Wadebridge in the background. 1986.

Figure 15-47. Compatible with vineyards. Nordex wind turbines among the vinyards near Donzère in the Rhône Valley of southern France. Few countries take their wine as seriously as the French, where enterprising vintners use the imagery of modern wind turbines to distinguish their wine from others. Inset. The Nevian wine cooperative's Domaine de l'Eolienne won a gold medal in a national competition in 2006. Near the cooperative are 21 wind turbines, which the label proudly celebrates.

Figure 15-46. Compatible with dikes and breakwaters. Wind turbines line the dike separating the IJsselmeer from the Noordoostpolder in the Netherlands. The informational panel describes the waterfowl that can be seen along the dike. The public has full access, though motor vehicles, camping, and dogs are prohibited—and "Don't bother the sheep." Mid-1990s.

Figure 15-48. Compatible with tourism: paid tours. Paying tourists clamber aboard a four-wheel drive vehicle after a tour of a wind farm at Château de Lastours overlooking the Golfe du Lion. The site, with a gourmet hotel-restaurant and specialty wines in the Corbières foothills near Port-la-Nouvelle, France, has been host to an array of wind turbines for more than two decades. 2000.

Figure 15-49. Compatible with tourism: observation decks. Wind turbines can become tourist attractions in themselves. An Enercon E66 in lower Austria offers an observation platform for intrepid tourists. The turbine is one of five installed in 2000. (Wikimedia Commons) Inset: Leitwind 77-meter (253-foot) diameter, 1.5-MW turbine atop Grouse Mountain, a ski resort near Vancouver, British Columbia. The turbine was inaugurated by the premier of the province just prior to the 2010 Winter Olympics. An elevator takes tourists to the observation deck. (Jonathon Simister, Wikimedia Commons)

SITING AND ENVIRONMENTAL CONCERNS 371

Figure 15-50. Compatible with scenic destinations. Copenhagen's oft-photographed Little Mermaid (of Hans Christian Andersen fame) in the foreground, with a view of several wind turbines in the Lynetten cooperative in the background. The Lynetten cooperative is just one of several wind cooperatives within the Danish capital.

Figure 15-51. Compatible with tourism: destination. The Whitelee Windfarm, Europe's second largest, is only 15 km (9 miles) from the center of Glasgow, Scotland. The sprawling project on an upland moor has become an ecotourist destination with hiking, cycling, and horseback riding. The site is an "official tourist destination" with a visitors' center and café overlooking Glasgow, the Clyde Valley, and, of course, the wind farm. 2012.

Figure 15-52. Compatible with historic sites. Trekroner Fortress (foreground) was designed to protect Copenhagen. The fort was the last line of defense during the Battle of Copenhagen in 1801, when the British navy firebombed and burned the Danish capital, inflicting a devastating defeat on the Danes. Thus, the site has significant historical import for Danes and is a popular tourist destination. (The tactic was later used by the British on Fort McHenry in the harbor of Baltimore, Maryland.) In the background is the Lynetten wind cooperative. 1998.

Figure 15-53. Compatible with sacred ground. An Enercon turbine in Dardesheim, Sachsen-Anhalt. In the background is Brocken (mountain), the highest part of the Harz Mountains and the site of Goethe's witches sabbath in Faust. 2011. Inset: Chapel near the wind turbines above the village of Montefalcone in Italy's Campania province. Late 1990s.

16

Installation and Dismantling

He loosened the last bolt. The generator was now ready to swing free.

"All ready?" he yelled.

"Yeah, let 'er rip," replied the ground crew.

"You sure that pulley's secure?" he asked, his voice less certain now.

"Yeah, it's not going anywhere. Let's get this one down and go for a beer."

The old generator rocked on its saddle. Slowly it rolled off toward the gin pole. Suddenly, there was a loud twang and the squeal of steel cable over pulleys as the 400-pound mass of copper and iron whizzed by, crashing through the platform next to him.

He looked about in dazed silence.

"Are you all right?" they asked from below.

He glanced at his feet. Yep, still there. Then to his hands. They were too, as were all his fingers. *Lucky this time*, he thought.

"I'm OK. What the hell happened anyway?"

"That pulley broke loose from the tower."

THIS INCIDENT ACTUALLY TOOK PLACE. IT HAPPENED TO an experienced crew working professionally. Though it occurred while removing rather than installing a wind turbine, it illustrates what can happen without thorough planning, preparation, and—equally as important in this case—execution.

This chapter focuses primarily on the installation of small wind turbines, mostly micro and mini wind turbines. If you're handy with tools and don't mind hard physical labor, this chapter will offer you guidance on installing a micro or a mini wind turbine yourself. You gain by replacing the skill, time, and expense of the dealer-installer with your own. You will develop a sense of accomplishment in doing it yourself, and you will learn more about your wind system than in any other way. You will know its strong points and also what can go wrong and where and how much effort it will take to fix it. The process will also give you an appreciation for the effort and skills required to reliably produce your own electricity.

However, the installation of any turbine larger than the smallest household-size turbine requires professionally trained construction crews. The installation of most household-size wind turbines and all medium-size and larger wind turbines are beyond the skills of do-it-yourselfers.

At the end of this chapter are a sequence of installation photos for a micro turbine, a household-size turbine, and a modern, large wind

THOUGHTS ON DOING IT YOURSELF

When I wrote the first edition of this book in 1982, I believed most homeowners with a modicum of tool skills and common sense could safely install a household-size wind turbine themselves. I figured, "Heck, if I can do it, anyone can." I've since learned that's not true. I can't say whether this conclusion is due to a decline in our collective knowledge about how to use hand tools or work around machinery, or to my becoming more cautious over the decades. I've certainly made my share of mistakes. Unfortunately, my position in the wind industry does make me aware of the mistakes others–including some professionals– have made and the injuries that have resulted. As a consequence, I now believe that homeowners should only attempt installing wind turbines less than 3 meters (10 feet) in diameter on lightweight tilt-up guyed masts. Products that fit this description are the many micro turbines on the market, as well as Bergey's XL1. Homeowners should avoid installing larger turbines, freestanding truss towers, or heavy-duty guyed towers without hands-on training. Most lack the skills, specialized tools, and safety equipment necessary. The tools can be purchased, and the skills required can be learned. Workshops, such as those that Mick Sagrillo teaches, or installer training programs offered by manufacturers, are worth the money and are the best way to learn how to install wind turbines safely. A book is no substitute for the hands-on learning that's required.

turbine. These installation sequences are followed with a series of photos illustrating how a large wind turbine is dismantled, an essential part of any wind turbine project.

Wind turbines of less than 3 meters (10 feet) in diameter may be installed by the homeowner or hobbyist with basic construction skills. The introduction of lightweight, tilt-up masts for machines of this size has made such installations even easier. The work can still be hazardous but no more so than other projects around the home or farm. With proper respect for the hazards involved and close attention to detail, you can safely install micro and mini wind turbines.

As the Alternative Energy Institute's Ken Starcher points out, risk is proportional to size. Larger turbines entail proportionally more risk. The components are heavier and may require special equipment and techniques unfamiliar to most do-it-yourselfers. Similarly, the installation of large wind turbines requires the skilled use of heavy machinery beyond the ability of even the most resourceful farmer. Leave the installation of these machines to professionals.

The following sections provide general information required by any installer of small wind turbines in North America. Although the materials suggested may differ on other continents, the principles and techniques remain the same.

Always consult the manufacturer's installation manual for more detailed descriptions of anchoring and installation techniques. If you plan to install the turbine yourself, the information in this chapter will help you to select the tower, anchors, and erection methods that best suit your talents and the conditions at the site. After reading this chapter, you may opt to hire a contractor instead of installing it yourself. The information gained, however, will enable you to track the progress and evaluate the performance of the contracted installer.

Whatever route you choose, thorough planning is essential. You must anticipate what will be needed at each step along the way, the problems you may encounter, and how to respond to them. You must coordinate the schedules of your subcontractors, suppliers, and erection crew to keep the project moving smoothly. If you lack any of the required skills, you must find someone who has them.

Pace yourself. Assume it will take twice as long as you expect. A skilled two-person crew can install a micro or mini wind turbine in one day. It may take novices much longer.

Prepare for the installation by collecting the parts, fittings, and tools for the job. Learn how components will be assembled and what tools are needed. Make sure you have met all legal requirements and that you're insured for any accidents that may occur. If you're installing the wind system yourself, it's a good idea to check whether your insurance will cover hospitalization and liability for friends who lend you a hand.

Without proper execution, all your planning and preparation may be for naught. You may know the right way to do a task and have the right tools to do so, but if you don't follow through under the press of time and conditions, the results can be disastrous. Consider the anecdote that opened this chapter. You may be tired, and a trip down the tower to check a pulley may seem unnecessary. It isn't. That's the time to take extra care to do the job right.

One final caveat: Build to local codes even when it's not required. You'll be glad you did.

Doing so makes for a sounder, safer, and more serviceable installation.

Parts Control

Installation can be hindered and proper operation of the wind turbine prevented by components damaged in shipment. Before accepting delivery from the freight carrier, examine the invoice or billing form to determine the number of crates shipped. Make sure all are present, and then carefully examine the crates for external damage. If damage is found, open the crates and look at the contents. The crates are designed to take some abuse while still protecting the product inside. Note any damage as precisely as possible and immediately contact the dispatcher at the freight company. (Digital photographs with a date stamp can be helpful in verifying claims.)

Tower sections and sensitive electronic components are the most easily damaged during shipment. Damage to control boxes and inverters is much harder to determine. The best you can do is identify any loose parts.

Catalog the serial numbers on the wind generator, blades, control panel, inverter, and tower. If you have to make a warranty claim, the numbers are much easier to find in your files than at the top of the tower. Serial numbers will also aid in troubleshooting if problems develop.

Make an inventory of all parts received as soon as possible. Many manufacturers provide a parts checklist for this purpose. Use it. The time to realize that an essential bolt is missing is prior to installation.

For those wind systems where the manufacturer doesn't also build the tower, the wind turbine and tower will be shipped separately. Often they will be delivered by separate carriers. Unless you have a special reason for removing the contents from the crates (for an inventory, possibly), leave them as delivered until you're ready for the installation.

Foundations and Anchors

The tower and guy cables (where used) must be kept clear of vines, trees, and shrubs. It may be necessary to clear the site of any plants that could

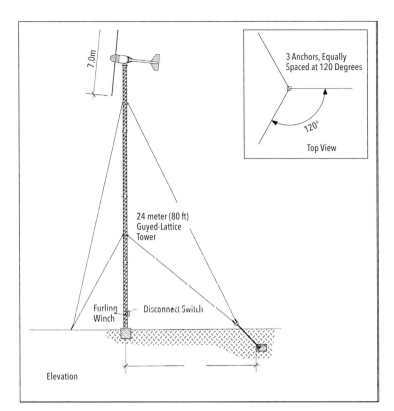

Figure 16-1. Guyed lattice mast. Typical guyed tower for household-size wind turbine in North America. (Bergey Windpower)

eventually interfere with the tower or guy cables. The site doesn't have to be level, so there's no need to grade the site to bare earth.

Anchors are used to prevent guyed towers from overturning (see Figure 16-1. Guyed lattice mast). Anchors resist uplift. Piers, on the other hand, resist loads in compression. On a guyed tower, for example, the anchors hold the tower upright and resist the forces trying to knock the tower over. The pier beneath the central mast supports the weight of the tower and wind turbine and resists the reactive forces from the guy cables trying to drive the mast into the ground. On freestanding towers, the legs act alternately as piers and as anchors, depending on the direction of the wind.

The type of anchor or pier used is contingent on the tower and the site. If you plan to install a guyed tower, there are several anchoring options to choose from: concrete, screw, expanding, and rock anchors. The best choice for your site is determined by the engineering properties of the soil, the depth to bedrock, and the power equipment available in your area. For a freestanding tower, the choice is limited to concrete.

Anchors

Anchors must withstand the static and dynamic loads acting on the wind system, under all weather conditions, for the life of the system. They must do so without appreciable creep toward the surface or settling. The holding power of anchors depends on the area of the anchor, its depth, the soil in which it is embedded, and the soil's moisture content. Weight is a factor as well, but it's not as important as you might think.

Soils vary in their ability to resist creep. Resistance to creep is controlled by the soil's shear strength: the resistance of soil particles to sliding over one another. Shear strength is a function of soil type and whether the soil is wet or dry. Shear strength ranges from a maximum in solid rock to a minimum in mucky or swampy soils.

One anchor manufacturer divides the shear strength of soils into two broad categories: cohesive and noncohesive. Cohesive soils, such as those with high clay content, stick together; the particles cling to each other. These soils have high shear strength. Noncohesive soils are generally those with a high sand content. In such soils the soil particles slide right by each other. Wind system manufacturers specify that their standard anchor designs are intended only for normally cohesive soils, those with high clay content.

Anchor holding capacity also decreases as the moisture content increases. Creep can be troublesome in saturated soils because the soil particles become fluid and tend to flow around the anchor. Water also increases the buoyancy of the anchor. The holding capacity of anchors can be reduced 50% in wet soils. Wherever possible, anchors should be placed below the level of periodic saturation from heavy rains but above the water table.

Frost heave causes similar problems. When soil freezes, it expands slightly, just as ice occupies a greater volume than water. If the anchor is not below the frost line, the cycle of freezing and thawing will heave or jack the anchor toward the surface. This is more of a problem for anchors than for piers because the existing load acts to pull the anchor out of the ground. The forces on piers act counter to frost heave. The frost line varies from year to year and depends on the severity of the winter and the soil cover. Bare soil freezes more quickly and to a greater depth than a soil with a grass cover. (The grass and the organic soil it grows in act as an insulator, slowing the soil's winter heat loss.)

To determine the soil-holding capacity at your site, you can test the soil with a probe or examine nearby road cuts. Better yet, talk to soil scientists. Explain your plans to them. Describe what it is you want to know and why it's important. (You don't want the anchor pulling out of the ground.) They will be able to tell you not only what kind of soil you will be working with but also the depth to the water table and the average frost penetration.

Local excavation companies are another good source. They have a feel for subsurface conditions since they work with them daily. They are in business to make money, though, not to give out free information. If you want their help, you should hire them.

The requirements for piers are less stringent than those for anchors. Most soils are strong in compression. With an adequate bearing surface, concrete piers of standard dimensions are used throughout North America.

Working with Concrete

The most common method for anchoring a tower or constructing a pier is to excavate a hole and partially—sometimes, completely—fill it in with reinforced concrete.

Concrete is literally man-made rock, conglomerate, to be specific. It's strong in compression, weak in tension. Thus, it works well as a pier, or foundation, but poorly as a beam. Tensile strength is improved by reinforcing the concrete with steel rods commonly called rebar (reinforcing bars).

Concrete is rated by its compressive strength. In North America, concrete is rated as obtaining its minimum, ultimate strength after curing for 28 days. Strength is a function of the water-cement ratio and the degree to which curing has taken place. The lower the water-cement ratio (the more cement in the mixture), the stronger the concrete. Strength also increases with curing time.

Curing is rapid in the first few days. (Concrete sets or becomes rigid within an hour of adding water.) Hydration doesn't go forward if too much water evaporates in hot weather or if the concrete

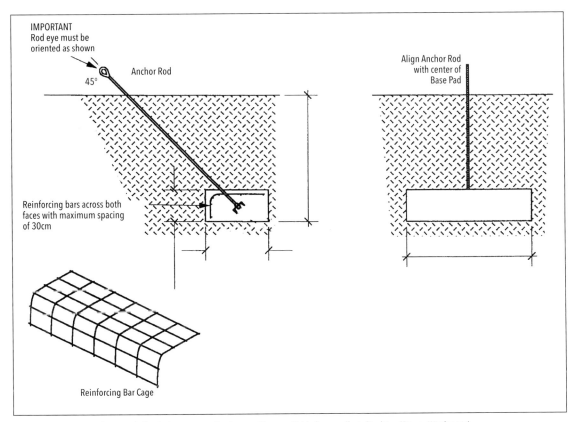

Figure 16-2. Concrete anchor. Detail of typical concrete anchor for guyed tower suitable for a small wind turbine. (Bergey Windpower)

becomes too cool in cold weather. Curing should take place for a minimum of seven days before any load is placed on the concrete. The concrete will continue to gain strength if moisture and temperature conditions remain favorable for complete hydration. If you heed the above precautions, concrete can be placed year round.

Installation drawings invariably show nice neat anchors and piers that look like they were made with a cookie cutter (see Figure 16-2. Concrete anchor). Except where an anchor or pier is exposed at the surface, this precision isn't necessary. Where the soils are stiff and will not collapse into the hole, the concrete can be poured in place. For anchor blocks below the surface, this is superior because the concrete acts directly on undisturbed soil. Forms are necessary where the hole is larger than the anchor or pier desired, where the concrete will extend above the surface, or when the anchor is in sandy soil.

The concrete is placed over a grid or cage of rebar to give the concrete the necessary tensile strength. The rebar is tied together with wire so it won't move when the concrete is placed over it.

All rebar must be covered by at least 3 to 4 inches (about 100 mm) of concrete. When the rebar is closer to the surface of the concrete, acid-laden water can enter the concrete, corroding the steel rebar. As the rebar corrodes, it expands slightly, causing the concrete to spall or chip. For long life, the concrete must seal the rebar from corrosion.

To ensure that the rebar stays where you want it when the concrete is placed, it should be staked down in the excavation or tied to the forms so it won't "swim" around. Pieces of rock or brick can be used to keep the rebar cage off the bottom of the excavation. This keeps the rebar from being too close to the concrete's bottom surface.

Forms can be built from heavy plywood and a wooden frame. The frame, when staked to the side of the excavation, will hold the plywood form in place while the concrete is hardening. Where the soil is stiff and the excavation is no larger than the pier desired, a short form can be made from wooden planks set in the excavation to a depth that gives the finished pier the desired height above the surface. Cylindrical forms can be purchased for placing concrete columns.

Before placing concrete, moisten the forms or the excavation to prevent them from absorbing water from the surface of the concrete mix and reducing its strength. Avoid placing concrete from a height greater than 4 feet (1 meter) or the aggregates will begin to separate. Once in the form, work the concrete to eliminate air pockets by poking a board or shovel into the concrete. Work it around the rebar and along the sides of the forms. Don't overwork, or the aggregates begin to settle. Special gasoline-powered vibrators can be rented for working the concrete after it has been placed.

You can simplify the whole process by hiring a contractor to excavate the hole and pour the concrete. Get firm quotes before you do, and make sure the contractor understands what the concrete will be used for. The fact that a lot is riding on the contractor's work may discourage him from cutting any corners.

Guyed Towers

Most small wind turbines are installed on guyed towers. And many of those are installed on fixed (non-tilt-up) lattice masts. To support a guyed lattice mast, you'll need a pier for the mast and at least three anchors to keep the mast erect.

The pier can be constructed simply by excavating a hole of sufficient size and depth (see Figure 16-3. Concrete pier). In weak soils or with an extremely tall tower, a larger pier than specified by the manufacturer may be need. Similarly, it may also be necessary in some areas to extend the pier deeper than normal to get below the frost line. Strengthen the pier with a rebar cage before placing the concrete.

Guyed towers also need a means to keep the mast in place so it won't scoot off the pier. Before the concrete sets, insert a pin or threaded rod into the center of the pier. Most masts have a base plate that slips over this pin.

Anchors are a little more complex. The type of anchor used depends on several factors: soil strength, the depth to bedrock, and your access to power equipment. In normally cohesive soils, you can choose from concrete, expanding, and screw anchors. Concrete anchors are the most adaptable. Screw anchors are the simplest to install.

Though screw anchors are widely used by electric utilities, building inspectors and contractors are more familiar with concrete. The equipment needed to excavate the hole for a concrete anchor is also more readily available than the utility line truck often used to drive screw anchors. Concrete anchors can also be adapted to weak or soft soils by simply expanding the anchor's bearing surface. Excavators work best at digging trenches and are well suited for making the excavations for concrete anchors. Once the rebar cage has been positioned, the concrete can be placed.

An easier and quicker method than using an excavator is to auger the holes for both pier and anchors. Where soils are not rocky and bedrock is well below the surface, a truck-mounted power auger, like that used by utility-line crews to set wooden utility poles, can drill holes for the pier and anchors in a matter of minutes.

For the pier, auger a hole in the dimensions needed. Set a short form into the top of the hole so that it extends above the surface, place the rebar, and cast the concrete in place.

After the pier, the auger can then be used to drill holes for expanding anchors. Expanding anchors work much like toggle bolts and similar fasteners used in plaster walls. Once the arrow-shaped anchor is inserted into the hole and expanded, the barbs resist being pulled back out.

The strength of expanding anchors is controlled by the soil type, the size of the anchor, and the firmness of the backfill. Expanding anchors hold better in heavy, stiff soils, less well

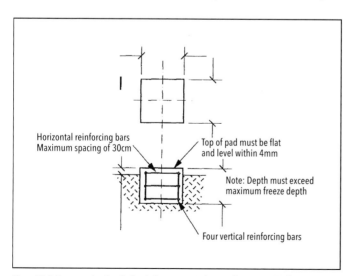

Figure 16-3. Concrete pier. Detail of typical concrete pier for guyed tower suitable for a small wind turbine. (Bergey Windpower)

in sandy or swampy soils. The larger the diameter of the anchor, the greater is its holding power. To reach its rated strength, the anchor must be fully expanded into undisturbed soil and the hole backfilled properly.

After drilling a hole at the correct angle, set the anchor at the bottom. Strike the anchor with a heavy bar to force the leaves into undisturbed soil. Once the anchor is fully expanded, attach the anchor rod. Gradually backfill the hole, and compact the soil with a tamping bar. That's all there is to it.

The development of screw anchors has further simplified the installation of anchors for guyed towers (see Figure 16-4. Screw anchor installation). It's not unusual to install the three anchors needed for a small wind turbine in less than 30 minutes. Many truck-mounted augers have been adapted to drive screw anchors by replacing the auger bit with a special tubular wrench. The hydraulic boom controls both the angle and the rate at which the anchor enters the soil. Unfortunately, screw anchors can't be used everywhere. Rocky soils, in particular, can retard the anchor from advancing.

Screw anchors are sized by the diameter of the screw, the number of helixes on the anchor shaft, and the length of the anchor rod. Their holding strength is once again based on the cohesiveness of the soil (see Table 16-1. Screw anchors).

To test the soil yourself, use a technique that soil scientists use. Take a fist of dry soil in your hand and squeeze it. If it doesn't stick together, it isn't cohesive. Pull-out strength is a function of the soil's cohesiveness or its ability to stick together. Dense, hard clay soils have the highest pull-out strength and sandy soils the least.

If using the anchor in a cold climate, ensure that the anchor helix or plate is below the frost line. If the anchor helix is within the frost zone,

Figure 16-4. Screw anchor installation. Truck-mounted power augers make quick work of driving screw anchors. (A. B. Chance)

frost heave will gradually lift the anchor toward the surface. Eventually, the anchor will pull out, bringing the tower down.

Soil strength greatly affects the holding capacity of any anchor. For a small wind turbine 3 meters (10 feet) in diameter in a normally cohesive soil, the 8-inch (203 mm) screw anchor has a holding capacity over three times greater than the expected maximum load on the anchor. But in weaker soils, the safety factor drops to less than two. In heavy soils, the 8-inch anchors are more than sufficient for this size turbine. In weaker soils, or if there's any doubt about the holding capacity of the soil at your site, check with the wind turbine manufacturer.

Where necessary, you can double the anchors at each guy. This was the approach taken by Fayette Manufacturing on their 1,500 turbines that once operated in the Altamont Pass. Fayette installed their 10-meter turbine on guyed pipe towers using screw anchors. Because of soil conditions in the Altamont Pass, Fayette used two 12-inch

Table 16-1. Screw Anchors					
Wind Turbine Rotor Diameter		Anchor Diameter		Length	
m	ft	mm	in	m	in
1	3	152	6	1.7	66
2.5	8	152	6	1.7	66
3	10	203	8	1.7	66
For normally cohesive soils. For taller towers and less cohesive soils, refer to manufacturer's specifications.					

(305 mm) screw anchors at each guy point. It was one of the few things Fayette did right.

For micro and mini turbines, screw anchors can be installed by hand. It's hard work in dense soils, and it can be challenging to screw the anchor in at the correct angle. Screw anchors are not set like tent pegs at a sharp angle to the tent guy. Instead, screw anchors are driven parallel to the guy cables from the tower.

If you're unfortunate enough to encounter solid rock at or near the surface, none of the preceding anchoring methods can be used. You'll have to drill a hole at the proper angle with an air drill and compressor. A rock anchor and rod are then inserted down the hole and wedged in place. These anchors have a high holding capacity, but installation is time-consuming and expensive.

For small wind turbines, one anchor rod is sufficient for all guy levels in most installations. To minimize bending, the anchor rod must depart the ground at an angle that coincides with the resulting angle of tension in the guy cables. The angle of departure depends on the height of the tower and the guy radius, and it's usually 45 to 70 degrees. In the field, it's easiest to use a 45-degree angle of departure by measuring a rise of one over a run of one, but it's always better to follow the manufacturer's recommendations on the appropriate guy angle.

Freestanding Towers
Freestanding towers—whether truss towers or tapered tubular towers—require an excavation and the placement of concrete. The easiest method, where the depth to bedrock permits, is to use a power auger.

For tubular towers, a large-diameter hole is drilled for a pier to support the entire tower. This is the preferred technique for even the largest wind turbines. Massive augers excavate holes for large wind turbines that are large enough to swallow pickup trucks—and have. At one wind farm site near Tehachapi, a pickup truck disappeared and was later found at the bottom of one of the augered excavations. Fortunately, no one was injured.

On truss towers, holes are augered for a pier at each leg of the tower. In sandy or swampy soils, piers are insufficient for truss towers. Instead, like those of the large turbines on the sandy Whitewater Wash near Palm Springs, California, the tower must rest on a massive concrete pad.

Knowing the soil conditions at a site is just as critical when using freestanding towers as when using guyed towers. In 2001 a number of multimegawatt wind turbines were installed on tubular towers in Texas using a modified pier commonly used on wind farms in North America. After installation, several of the multimillion dollar wind turbines were lurching off the vertical within a few months. The problem? Voids in the limestone bedrock at some of the turbines didn't provide sufficient overturning resistance. Correcting the massive "leaning towers of Texas" was no simple task.

On household-size wind turbines atop truss towers, the piers for each leg may be up to 3 feet (1 meter) in diameter and up to 10 feet (3 meters) deep. Often the excavations are left unfinished, that is, in their circular form. The rebar and anchor bolts for anchoring each tower leg are then added, and the concrete is placed.

You must make certain that the base tower sections or anchor bolts used in the foundation don't swim around when the concrete is placed. They must also accurately fit the foundation template provided with the tower. Otherwise, you could have a rude awakening when you go to set the tower on the base.

Novel Foundations
If power-installed screws work well as anchors, why can't they also be used for the pier supporting the mast of a guyed tower? In theory at least, they can. The foundations for light standards and transformers at substations have been installed this way for years. No one, however, has adapted this technology to wind systems. When they do so, an installer will be able to drive the anchors and pier in just minutes instead of waiting days for the concrete in the pier to cure. The whole wind system could then be erected in one day or less.

Using power-installed screw anchors to secure the legs of a freestanding truss tower is another possibility. There may be engineering limitations, particularly in weaker soils, but the advantage of quick and easy installation justifies a look into whether it's possible.

Kits for tilt-up guyed towers for some micro and mini wind turbines use a base plate instead of a pier. The base plate is simply staked to the surface of the ground to prevent it from sliding out from under the tower.

One major innovation in foundation design was developed during the 1990s for tubular towers installed in California. Rather than construct the pier with a solid mass of concrete and rebar, the new design uses a cylinder of reinforced concrete the diameter of the tower that it supports. The hollow core of the cylinder is subsequently filled with the excavated soil. This foundation substantially reduces the amount of concrete needed over that of the traditional method.

Assembly and Erection of Guyed Towers

Two types of guyed towers are used for small wind turbines: fixed, guyed masts and hinged, tilt-up towers. Fixed, guyed towers use three anchors. Tilt-up towers typically use four anchors.

The guyed lattice masts commonly used for household-size wind machines in North America can be assembled a section at a time with a tower-mounted gin pole, or the entire tower can be assembled on the ground and hoisted into place with a crane. The method you use depends on whether a crane is available and can get to your site. In either case, the guy cables must first be cut to length and attached to the guy brackets that fit around the mast.

Guy Cables

Steel cable is shipped in coils or on reels. Avoid damaging kinks in the cable by rolling the coil along the ground to lay out the cable. If the cable is on a spool, use a spool stand to unreel the cable, or carefully walk the spool along the ground to unreel the cable.

If the manufacturer doesn't specify the length of the guy cables in the installation manual, calculate cable length by using the Pythagorean theorem

$$\text{Guy Length} = \sqrt{(GR^2 + GH^2)}$$

where GR is the guy radius and GH is the guy level or height above ground. Give yourself plenty of extra cable to allow for sag and for slight errors in the position of the anchors. You'll need three lengths of cable for each guy level on a fixed tower, four lengths for a hinged, tilt-up tower.

Mark the length of each guy cable on the ground. Unreel the cable and cut it—carefully—to length with bolt cutters. When doing so, make sure that both ends of the cable can't whip around. Working with wire rope is like working with an unwound spring.

Guy Cable Attachments

How you attach the cable to the guy brackets on the mast may vary. Tower kits provided by manufacturers use swaged connections. Where swaged connections are not provided, wire rope clips or grips will often be used.

North American utility companies sometimes use strand vises and preformed cable grips for this task on guyed utility poles. In a strand vise, the guy cable is passed through the vise and, when tensioned, wedges the cable in the grip of the vise. These are unsuited for use with wind turbine towers. Because the towers sway with the wind, guy cables are not always in full tension. Consequently, strand vises, and other cable attachments like them, can release the guy cable, causing the tower to crash to the ground.

Figure 16-5. Preformed cable grip. Legs or strands of the preformed wire grip are wrapped around the cable. Tension in the legs and a fine grit grips the cable. Note turnbuckle (lower left), thimble through turnbuckle eye, and one leg of the grip before being wrapped around the cable.

Preformed cable grips have been in use by utilities since the 1950s. These attachments use a fine grit embedded in an adhesive to grip the cable. Tension on the grip pulls the strands tighter together. Preformed cable grips allow quick adjustment in the guy cable during installation by being easy to remove and reapply. They were used with guyed towers for Storm Master turbines installed in California and on Bergey wind turbines during the 1980s (see Figure 16-5. Preformed cable grip). However, most installations of small wind turbines will use wire rope clips.

Wire Rope Clips and Thimbles

Wire rope clips are simple to use, but they must be used correctly. The clip's forged saddle must bear on the live end of the guy cable; that is, the portion of the cable connecting the tower to the anchor. The U-bolt must bear on the dead end, or the "tail" of the cable after it passes through the anchor eye (see Figure 16-6. Never saddle a dead horse). As Mick Sagrillo admonishes, "Never saddle a dead horse."

Always use a thimble when attaching wire rope. The thimble distributes the load on the cable over a large radius. A sharp bend in the cable will severely weaken it (see Figure 16-7. Always use a thimble). Some guy brackets, particularly those used in the utility industry, have built-in thimbles. Similarly, the eyes of some anchors have large knuckles, obviating the need for a thimble (see Figure 16-8. Anchor knuckle).

Where thimbles are necessary, the guy cable is wrapped around the thimble and passed through two or more clips. The clips prevent the cable from slipping off the thimble (see Figure 16-9. Cable thimble). Thimbles can be awkward to fit onto the anchor eye. Hugh Piggott, in his helpful

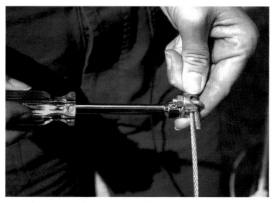

Figure 16-6. Never saddle a dead horse. Forged saddle of wire rope grip must act on live guy cable. U-bolt acts on dead end of guy cable.

Figure 16-8. Anchor knuckle. The guy cable must not pass over a sharp bend. Large-diameter anchor eye used on some installations doesn't require a cable thimble.

Figure 16-7. Always use a thimble. Improper attachment of guy cable at a university test field. The sharp radius of the turnbuckle eye has caused the guy cable to flatten. The weakened cable could fail catastrophically. A metal thimble of the proper size should have been used here.

Figure 16-9. Cable thimble. A thimble must be used to distribute the load from the guy cable over a sufficiently large radius to avoid damage to the cable.

INSTALLATION AND DISMANTLING

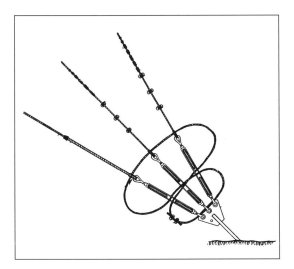

Figure 16-10. Turnbuckle safety cable. Use of a cable through the center of the turnbuckles prevents the turnbuckles from unscrewing and releasing the tower. It's also cheap insurance should a turnbuckle fail. (Bergey Windpower)

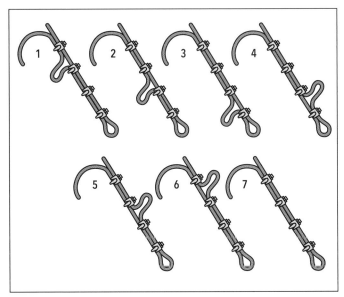

Figure 16-11. Taking up slack. Never release all the wire rope clips when a cable is under tension. Instead, loosen the top-most grip, take up the slack, retighten the grip, and proceed until the slack has passed through all the grips. (Vergnet)

book *Windpower Workshop*, describes the simplest way to open the throat of a stamped thimble. Rather than prying the thimble ends apart, twist the ends past each other. After passing the thimble over the anchor eye, simply twist the ends back together. Pliers on either leg of the thimble provide a little leverage that makes this a snap.

Manufacturers will specify a minimum turnback distance for the tail of the guy cable and the minimum number of clips. Since the wire rope clips are critical to safety, use at least three of them even if the manufacturer says two are sufficient. Apply one clip at the minimum turnback distance. Apply another clip near the thimble sufficient to keep the cable from slipping off the thimble. Place the other clips equidistant between these two. Tighten the nuts on the clips evenly, alternating until the desired torque is reached.

Retighten the clips after the load has been applied. With load, the guy cable will stretch slightly and shrink in diameter, loosening the clip's grip on the cable. Periodically retighten as needed.

On fixed, guyed towers, turnbuckles are used to mechanically tension the cables. It can be just as dangerous to overtension the guy cables, causing the tower to buckle, as to allow too much slack in the cables. There are several methods for measuring the amount of tensioning in a guy cable. It's best to check with the manufacturer for the method they prefer. Where turnbuckles are used, add a safety cable (see Figure 16-10. Turnbuckle safety cable).

Tilt-up towers for micro and mini wind turbines seldom use turnbuckles. The cables are tightened by hand. However, never release all the cable clips at one time when the cable is under tension, or you may find your turbine and tower slamming into the ground. Instead, loosen the top-most grip, pull out the slack, and then retighten the grip. Continue until you've worked the slack through all the clips (see Figure 16-11. Taking up slack).

Figure 16-12. Substandard turnbuckles. Improper use of turnbuckles that were not designed for wind turbine applications at a university test field. The open jaw of the turnbuckle could release the guy cable causing a catastrophic failure. The installer recognized the problem and welded a gate across the open jaw. Welding the forged turnbuckle could reduce the metal's strength, itself contributing to failure of the guy cable. The clevises used here should also include a cotter pin or device to prevent the clevis screw from backing out. All in all, the installer should start over and do it right.

Always use cable and anchor fittings designed for the purpose. Anchors, turnbuckles, and wire rope clips are no place to cut corners (see Figure 16-12. Substandard turnbuckles). Beware of substandard clips, warns Jon Powers, an experienced Tehachapi windsmith. Sometimes there may be telltale excess flashing around the forged saddles. If so, throw them out. They're not worth the risk. Michael Klemen has found such defective clips—after the fact. His Whisper H900 slammed to the ground when he was lowering the tower, and it may have been due to a defective cable clip. It's an expensive lesson, says Klemen.

Using a Crane

Either the tower and wind turbine can be lifted into place together with a crane, or each can be lifted separately. If you're lifting a complete assembly, the crane doesn't have to be taller than the tower. For household-size wind turbines, the combined tower and wind turbine can be lifted at some level below the top of the tower. Because the lattice mast is somewhat frail, a nylon sling should be used to spread the lift over the entire cross section of the tower and not act on one leg alone. Lifting a complete assembly also requires that the wind machine must be fully assembled. This may require lifting the tower off the ground slightly to attach the blades and tail vane (where used).

If a crane is to be used, bolt all tower sections together, making sure that the section with the guy brackets and cable is in the right position. Attach the wiring conduit and thread the power cable through it as discussed in the subsequent section on wiring. While the crane holds the tower upright, position it over the base plate or pier. Rest the tower on the pier and connect the guy cables to the anchors, pulling them taut. Then tighten all connections at the guy anchor. Once this is done—and only then—can the crane be released and final adjustments made to the guy cables.

Using a Gin Pole

For reasons of safety, the trend has been away from using gin poles to raise components on fixed guyed towers. For household-size turbines, installation with a crane is preferred. For micro and mini turbines, manufacturers predominantly use tilt-up guyed towers. However, the following technique may be used in special circumstances for raising a guyed lattice mast of Rohn 25G or 45G sections.

If a gin pole is used, bolt two tower sections together and attach temporary guys. Tip the two sections up onto the pier and tie off the temporary guys to the anchors. They will hold the tower in place until the first guy level is reached. Someone must now climb the tower and bring up the gin pole.

The gin pole is a boom, or davit, that extends above the top of the tower. It permits tower sections or the wind turbine itself to be lifted up the tower and set in place without the use of a crane. Gin poles need be nothing more than a long section of pipe strong enough to handle the expected loads and having some means of being attached to the tower. Some gin poles are a little more sophisticated and incorporate a horizontal arm that allows the load to be centered on the tower. The gin pole used to lift the modular tower sections of Rohn's lattice mast is simply a 12-foot (4-meter) length of aluminum pipe with clamps and a pulley at the top. The clamps are designed for attaching the gin pole to one leg of the lattice mast. This same gin pole has been used numerous times to hoist small wind turbines, though its strength limits it to machines no larger than 3 meters in diameter.

Pulleys are used to direct the hoisting rope over the gin pole to the load. Never use a gin pole without first routing the hoisting rope through a pulley at the base of the tower. This pulley permits the hoisting crew to stand clear of the tower and be well away from any falling objects. It also prevents any unnecessary bending of the gin pole. With a base pulley in place, the hoisting tension acts directly on the gin pole from below. This minimizes the bending forces on the gin pole—and on the tower.

An improperly attached base or down-tower pulley caused the death of one small wind installer in front of his wife in 1993 when the tower he was on buckled due to the lateral loads on an improperly rigged gin pole. Using a tower-mounted gin pole is not something to trifle with.

Similar mishaps have occurred when the attachment of the gin pole to the tower has failed. It's paramount that the gin pole not

ERECTING A WIND TURBINE IS DANGEROUS.
Always use extreme caution. When in doubt, consult the manufacturer.

only be firmly attached to the tower, but also that it not move laterally when the load is applied. The manufacturer's recommendations for the materials used in the gin pole and for its attachment to the tower should be followed scrupulously. Erecting a wind turbine is dangerous. Always use extreme caution. When in doubt, consult the manufacturer.

With the gin pole now in place, the next tower section is brought up. The third section will usually have the lower guy bracket attached. Once the section has been bolted down, the guy cables can be strung to the anchors. These cables are tensioned by hand until the three assembled tower sections are vertical and the tower is straight.

The gin pole is then released and moved to the top of the top-most section. A new section is hoisted up and bolted into place and so on until the tower is completed. After each set of guy cables is attached, the tower should be checked for plumb and twist. If the lower sections have been aligned properly, it is possible to simply sight along the tower to check the alignment of the upper levels. A transit can be used to be certain. The tower must be vertical for proper yawing of the wind turbine.

Normally, the guys on the Rohn towers can be tensioned by simply turning the turnbuckles. For household-size wind turbines, it may be necessary to mechanically tension the guys with a coffing hoist or come-along attached to the guy cable with cable grips. The tension is then measured by a dynamometer in line with the hoist. Follow the manufacturer's directions for attaching the hoist and for the amount of tension required.

Always attach the guy cable directly to the anchor or to a turnbuckle that is attached to the anchor. The coffing hoist or come-along acts in parallel with the guy cable and is used only for tensioning the cable, not for connecting the cable to its anchor. Tensioning hoists and the grips that they use can, and have, failed disastrously. Steen Aagaard was crippled when such a tensioning system inadvertently released and the tower he was on crashed to the ground.

After the guys are in place, the tower is aligned, and the guys are properly tensioned, you're ready to raise the wind turbine. Do so only on a calm day. Any wind at all will make it more difficult to position the turbine once it's atop the tower.

For small wind turbines, it's best to do as much of the assembly on the ground as possible, since it's awkward to work at the top of the tower. Household-size turbines may require the use of a machine stand or cribbing; this permits attachment of the blades and other components.

Attach the hoisting cable and one or two tag lines. The wind turbine may have an eyebolt used for lifting, or it may require a special lifting jig. Whatever's used, it's important that the hoisting line lift at the wind turbine's center of gravity. If not, the wind turbine will be a lot more difficult to handle on its way up the tower, and once you're on top, you'll have a heck of a time mounting it to the tower. The turbine may be easy to move around on the ground, at the end of a long lift line, but it's a lot more difficult at the top of the tower. Everything you do on the tower is more difficult, and more dangerous, than when you do it on the ground.

Use the tag lines to keep the machine from banging into the tower and tangling with the guy cables. The tag lines must be longer than the tower is high. Keep the pull on the tag lines moderate (just enough to prevent the machine from hitting the tower), particularly as the turbine nears the top of the tower. It's easy to buckle a gin pole if too much force is used on the tag line.

Once the turbine is mounted atop the tower, string the conduit and fish the power cable to the generator. Torque all tower fasteners to the specified value and apply locking nuts where required.

Freestanding Towers: Assembly and Erection

Technically, truss towers can also be erected with a tower-mounted gin pole, but no one does it in practice. Only in rare cases, such as remote sites in Alaska or other inaccessible areas, are gin poles used to assemble truss towers. The sections are much heavier and more awkward to work with than those on a guyed tower. Most installers simply call in a crane.

Each individual member on a truss tower is so heavy that by the time the first section is bolted together you're not going to move it anywhere without heavy equipment. Ideally, you'd like the crane to simply drive up to the tower, raise it in one lift, set the tower on its foundation, and leave. You don't want the crane to move sections of the tower around the site because you didn't thoroughly plan the assembly. This is particularly important if the wind turbine has been mounted on the tower. The more moving that's required, the greater the likelihood of damaging the turbine.

The tower can be raised first, and the wind turbine mounted on top in a second lift. When using this method, the boom of the crane must extend above the top of the tower to allow for the crane hook and the lifting jig to clear the top of the tower.

A crane with a shorter boom can do the same job if it raises the wind turbine and tower at one time. When lifting the turbine and tower together, the lift should be made some distance below the top of the tower, yet well above the tower's center of mass. As the tower is raised, its weight keeps the bottom sections on the ground while the upper sections move toward the vertical. If the tower has been positioned correctly during assembly, the bottom will slide across the ground toward the foundation. Because the lift is being made below the top of the tower, the rotor blades are able to clear the boom as the tower nears the vertical.

With the skillful use of hand trucks, dollies, and come-alongs, even the heaviest towers can be fully assembled without power equipment. But it's wise to assemble the tower where it will make the crane operator's job as simple as possible. Begin by bolting the lowest and heaviest section together. Its placement will determine the location of the remaining tower sections, and it won't be moved again until the crane arrives.

Truss towers are like giant erector (meccano) sets and will be puzzling until a few of the cross-girts are bolted into place. Lightly tighten the bolts on the cross-girts but don't overtighten. As you move along the tower section, you'll find that some of the pieces won't fit easily. There are always a few pieces that need some "convincing" before they fall into place.

Erection wrenches (also called *spud wrenches*) and *drift pins* are helpful in these situations (see Figure 16-13. Erection wrenches). An erection wrench has a long tapered shaft that's used by ironworkers to solve stubborn alignment problems between two pieces of metal. The shaft is inserted into the holes, and it's used to lever the pieces into position with the help of a little muscle. A bull pin is a similar device. It too has a long tapered shaft, but instead of a wrench on one end, it has a striking face. The bull pin is dropped into the holes needing a little nudge, and then driven with a hammer until the holes aligned. Both of these tools are well suited for aligning holes on the flange plates between each tower section.

Bolt all sections together until the tower is fully assembled. The wind machine is then mounted on the tower and the conduit for the power cable installed. You'll find that strapping the conduit to the tower while it's on the ground and fishing through the conductors is much easier than trying to do it after the tower is erected. Once you have erected the tower, tighten the bolts to the desired torque.

An old trick from the aerospace industry is to install bolts on truss tower legs upside down. If the nut loosens, the bolt falls free, notifying the user that the bolt needs to be replaced. Although the towers may look massive, constant vibration from the turbine can loosen fasteners. Because of these vibrations, the nuts on all bolts must be prevented from loosening. Use self-locking nuts, or add special locking nuts after the nut and bolt have been tightened. These are easiest to install while the tower is still on the ground. Don't overlook this step. Locking nuts are specified for a reason; towers have failed without them.

The tower is now ready for the lift. Attach the lifting sling so that the stress is distributed

Figure 16-13. Erection wrenches. Two useful tools for working with lattice tower sections: erection wrench or spud wrench (top) and bull pin or drift pin (bottom). (Klein Tools)

onto tower members strong enough to take the load. Don't, for example, wrap the sling around a tower cross-girt. Instead use a tower leg; better yet, use two legs. Note that the sling must not slide along the leg as the tower is being lifted.

The crane will slowly set the tower down on the flanges or threaded bolts in the foundation. On foundations with flanges, align the holes with the drift pin and judicial use of a pry bar. Drop in the bolts once the holes are aligned. Before removing the sling, level the tower.

There are two ways to level the tower. On lighter towers, shims are forced in between the section flanges. The heavier towers use threaded bolts with adjusting nuts between the foundation and the lower tower section. The nuts are used to level the tower. When the tower is level, tighten the mounting bolts and install the locking nuts. The sling can now be safely removed and the crane sent on its way.

Tubular Towers

During the 1980s, some small wind turbines were installed on freestanding tubular towers. The towers relied on a slip fit between sections. For these towers, the first section is placed on the foundation and leveled with adjusting nuts. The next section is slipped onto the first and so on. Gravity does the rest. The wiring run is then strung inside the tower and the wind machine installed. Or the sections are slipped together on the ground then secured with fasteners. Then the turbine is attached, the tower conductors are connected, and the turbine and tower are lifted into place. There are still a few small wind turbines installed on tapered tubular towers (see Figure 16-14. Tapered tubular tower). Unfortunately, most tapered tubular towers for small wind turbines, especially those in Europe, are much too short—often limited to 6 meters (20 feet) in height—for good exposure to the wind.

In contrast, nearly all large wind turbines use large-diameter tubular towers. While a few large turbines have been installed on lattice towers in recent years—some up to 130 meters (426 feet) tall—they are the exception. Tubular towers are now preferred for their clean lines, low maintenance, and ability to shelter service personnel during inclement weather.

For these towers, the first section is raised

Figure 16-14. Tapered tubular tower. Kingspan (formerly Proven) wind turbine on a freestanding tubular tower in northwest Denmark. The Kingspan rotor operates downwind of the tower.

upright, often with two cranes. One crane lifts the top of the section toward vertical, the second lifts the bottom of the section so that it just clears the ground. This prevents the bottom of the first section from scooting across the ground, damaging the flange upon which it will rest. Once the section is vertical, the crane positions it over the bolts protruding from the foundation. Additional sections are added in a similar manner, and the sections are bolted together.

When raised, the interior components of the tubular sections are nearly complete. The ladders, cable raceways, lighting, and, in some systems, even the fall protection rails are in place.

Figure 16-15. Raising assembled rotor. Two cranes are used to lift the assembled rotor and hub on this 100-meter (330-foot) diameter, 1.7-MW turbine. The small crane prevents the lower of the three blades from striking the ground. For larger turbines, the nacelle may be mounted on top of the tower, drivetrain components and hub added, then each blade raised separately. (Neal Emmerton)

The nacelle, sans rotor, is then lifted onto the top of the tower, followed by the rotor. In the recent past, the rotor has typically been fully assembled on the ground and lifted as one piece, again using two cranes in the same manner as for raising the tower sections (see Figure 16-15. Raising assembled rotor). However, the very long blades on some turbines preclude raising the rotor fully assembled, and the blades are individually hoisted into position and bolted to the hub in a delicate aerial ballet using a specialized lifting jig.

Tilt-Up Towers: Assembly and Erection

All towers for small wind turbines, whether guyed or freestanding, can be hinged and tipped upright into the vertical position, eliminating the need for a crane (see Figure 16-16. Tilt-up tower). Hinged towers simplify assembly because all tower and wind turbine components can be added while they are on the ground. They simplify service and repair for the same reason. Rather than climbing the tower and handling awkward components in the air, you can lower the tower and do the job on the ground. Avoiding the need for a crane also reduces the cost of installation and service. Tilt-up towers also allow turbines to be lowered in cyclone-prone regions,

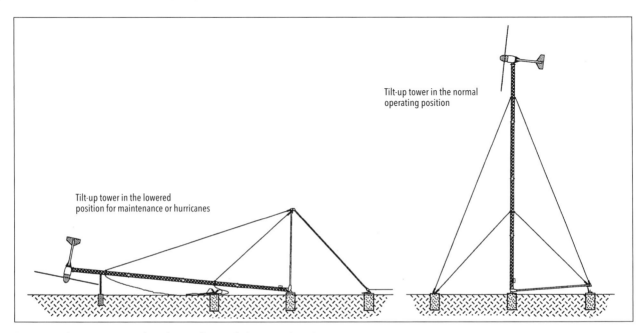

Figure 16-16. Tilt-up tower. Erecting a hinged tower with a gin pole. (Bergey Windpower)

Figure 16-17. Tilt-up tower for medium-size turbine. Vergnet uses hinged, tilt-up towers for wind turbines up to 26 meters (85 feet) in diameter for application in the hurricane belt. Shown here is one of Vergnet's turbines on a tilt-up, guyed lattice mast as tourists visit the Château de Lastours test site in southern France. Inset: Massive two-piece gin pole with block and tackle used by Vergnet to lift its 26-meter-diameter wind turbine. 2005.

such as the Caribbean, for protection during hurricanes. With a gin pole, these towers can be raised by a heavy-duty industrial winch, tow vehicle, griphoist, or small crane.

The hinges do add to the cost and complexity of the installation, and they also introduce a potentially weak link in the tower structure. Worse, many a hinged tower has been dropped by an installation or service crew unfamiliar with the loads involved. Even experienced crews have dropped turbines. Some have watched helplessly while their tow vehicle was dragged across the ground, and a tower with an expensive wind turbine atop it headed inexorably to earth. Despite these limitations, hinged towers make sense when designed and operated properly.

The advantages of hinged towers dictate that they'll continue to be used for small wind turbines and even some larger machines. For example, French manufacturer Vergnet uses guyed tilt-up towers for its entire product line. They offer one model, a 220-kW turbine with a rotor 26 meters (85 feet) in diameter, on tilt-up masts 50 meters (164 feet) tall (see Figure 16-17. Tilt-up tower for medium-size turbine).

Vergnet's towers feature a guyed, tubular mast with accompanying gin pole. The tower is prevented from moving laterally during erection by two guys at right angles to the lift. The free guy cable is routed to the top of the gin pole. The tower is raised by drawing a block and tackle together between the gin pole and the free anchor or in Vergnet's case with a powered—and powerful—griphoist.

Towers with three guy cables may also be used, but there's no lateral restraint on the mast during the lift. This lateral motion can be prevented by using a two-piece gin pole in the shape of an A-frame. The A-frame can be built inexpensively from four sections of lattice mast, but the bases must be hinged to allow the gin pole to move with the tower (see Figure 16-18. A-frame gin pole).

Truss towers have been raised in the same way. All the ESI and most of the Storm Master turbines installed in California during the early 1980s were raised on hinged lattice towers. One manufacturer used the A-frame gin pole; others used a single pole. Either the gin pole is

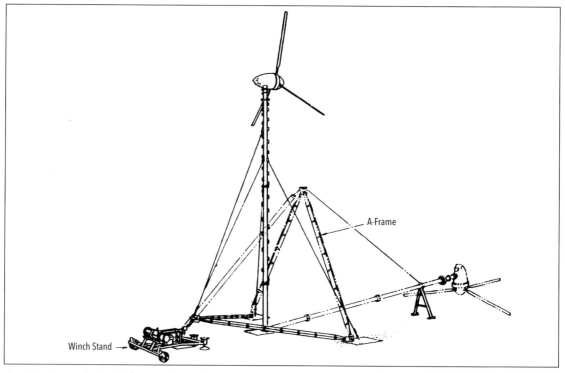

Figure 16-18. A-frame gin pole. An A-frame gin pole is used to stabilize a hinged tower that is only guyed at three anchors.

attached to the tower's foundation with a hinge, or it stands separately. When standing apart from the tower, the gin pole must be guyed to prevent it from moving laterally out from under the load.

Freestanding tubular towers have also been erected with a gin pole mounted at the base of the hinged tower. In the case of both the truss tower and the tubular tower, the hoisting cable is passed over the gin pole to the top of the tower. Once upright, the raising cable can either be removed or left in place for service calls.

In Europe, several small wind turbines have been installed atop hinged tubular towers with the use of a powerful hydraulic cylinder (see Figure 9-7. Hinged, freestanding tubular tower).

Tilt-Up Guyed Towers

For micro and mini wind turbines, a tilt-up guyed tower is both cheaper and easier to install than any other tower choice (see Figure 16-19. Micro wind turbine on tilt-up tower). While installing such a tower isn't a snap, it's much simpler than erecting a freestanding lattice tower with a crane, for example. There's also the advantage of being able to raise and lower the tower whenever you want to

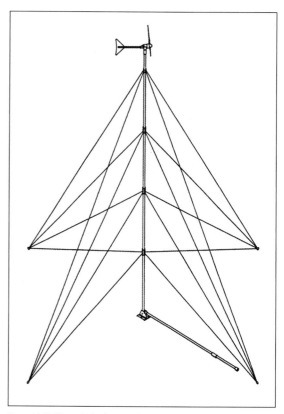

Figure 16-19. Micro wind turbine on tilt-up tower. The gin pole used for raising the tower rests at a right angle to the tower. This turbine and mast are similar to that in the series of photos titled "Erecting a Micro Turbine with a Griphoist."

service the turbine. And some turbines have to be serviced far more frequently than others.

Raising a guyed tilt-up tower for a micro or mini wind turbine can be safely accomplished with as few as two people, but it is not risk-free. Consider the admonition never to walk under the tower when it's being raised. Wind experimenter Michael Klemen explains why this precaution is important after he dropped one of his towers. Klemen calculates that an 80-foot (24-meter) tower would hit the ground within two seconds after the raising cable was released. By the time you realize there's a problem, Klemen says, there's no time to get out of the way. For his part, Mick Sagrillo likens a falling guyed tower to a "Wisconsin cheese slicer." Obviously, not something to trifle with.

Tilt-up tower kits use a gin pole to raise the mast. The gin pole is effectively a second, shorter mast at a right angle to the tower. The gin pole acts as a lever to reduce the lifting loads.

The masts of tilt-up tower kits use thin-walled tubing like that widely used in the wind industry for meteorological masts. Kits have been offered with 3.5-inch (89-mm) diameter tubing for towers up to 42 feet (13 meters) in height. These were being shipped in 7-foot (2.1-meter) lengths, allowing transport by parcel delivery services worldwide. Bundles of three sections in this size are easy to easy to work with. Taller towers, for example for the Bergey XL1, require 4.5-inch (114-mm) diameter tubing. This tubing is shipped in 10-foot (3-meter) lengths in bundles of three pieces each. These require transport by motor freight, and the cartons are hefty. Though the cartons require some effort to move around, the individual sections are easily managed.

For these kits, five anchors are needed: one each for the four guy cables and the fifth for raising the tower. Because the tower sections slip together and the gin pole comprises several sections, the fifth anchor must be directly below the gin pole when the tower is fully upright.

Griphoists

Everyone who has chosen a tilt-up tower to support a small wind turbine has had to face the difficult question of how to raise it. The most common technique among do it yourselfers in North America is to raise tilt-up towers with a truck or tractor. Unless you're skilled with hand signals and with working around heavy equipment, raising a tower with a truck or farm tractor can be a ticklish and dangerous operation. One do-it-yourselfer was killed when the truck driven by his brother pulled the tower over before anyone had time to react.

Properly using a vehicle for raising a tilt-up tower demands a large crew. Sagrillo recommends two on the truck (one to drive, and one to watch for hand signals) and one for each anchor, or a minimum of six people. Gathering together such a large group always presents a challenge. When the inevitable glitches arise, it puts you in the awkward position of either asking everyone to come back another day or forging ahead and taking chances you shouldn't. You can quickly wear out your credit with friends and family if the tower raising doesn't go as planned. You don't want a bunch of your friends standing around asking, "Hey, are we going to install this windmill or not?"

Communal tower raising can be a rewarding experience, like Amish barn-raising, bringing people together for a common purpose. But barns last indefinitely. You put it up, and it stays up. Not so with a wind turbine. Whether we like it or not, small wind turbines do need repairs, and we have to bring them down before we can haul them off to the local repair shop. Some are up and down a lot. Gathering six people together

Figure 16-20. Griphoist. Super pull-all model griphoist, with wire rope, snap hook and keeper, and lever operating handle used to raise mini wind turbines on lightweight tilt-up towers at the Wulf Test Field. The griphoist is also a handy tool around the farm or ranch. Note that safety keepers or gates on lifting hooks ensure that the hooks stay coupled to the load when there's unintended slack in the cable. They're absolutely essential.

every time you want to raise or lower your turbine quickly becomes tiresome.

Fortunately, griphoists are an alternative common where cranes are either too expensive or too difficult to use. Hand-operated griphoists are used throughout Europe and Canada for a variety of applications, which include raising wind turbines and meteorological masts (Figure 16-20. Griphoist).

Scoraig Wind Electric's Hugh Piggott uses a griphoist to install wind turbines in Scotland. A griphoist is "hard to beat for erecting tilt-up towers," says Piggott, "because it's slow and fail-safe." Unlike the driver of a truck or other vehicle being used to raise a tower, the operator of the winch has full control of the operation, and there's no dependence on hand signals or risk of missed cues.

To Piggott, this tool is a *tirfor*. Tractel, the world's largest manufacturer of griphoists, officially calls them a griphoist-tirfor-greifzug product. In English, the word *griphoist* says it all. But the tool was originally sold as a *tirfor*, which in French says much the same. *Greifzug* is the German equivalent.

The griphoist pulls a few inches of cable through the tool's body on each stroke of the operating lever, both on the back stroke and on the forward stroke. Because it's a simple mechanical device, you can actually feel the tension in the cable. This gives the operator a tactile sense of the load. The heavier the load, the more difficult it is to move the lever. The loads in tower raising are greatest when the tower is on the ground and least as the tower nears the vertical. Operating the griphoist takes the most effort when the tower first begins to leave the ground. For loads nearing the limit of a particular size, operating the griphoist will take some effort (see Table 16-2. Griphoists).

Another hoist option, the one used by some American meteorologists to install anemometer masts, is an electric winch (see Figure 16-21. Raising tilt-up tower with electric winch). They typically power the winch from the battery of a truck or haul in special-purpose batteries. But Zephyr North's Jim Salmon prefers a griphoist to raise meteorological masts in Canada. "They [griphoists] are easier to control," than either winches or vehicles, he says, "and in some cases much safer than [electric] winches."

Unlike electric winches, griphoists are readily portable. You can lug a griphoist into areas where you would never consider hauling an electric winch and battery, or even driving a truck, for that matter. Endless Energy's Harley Lee swears by the heavy-duty griphoist he used to raise a 130-foot (40-meter) anemometer mast on a rugged Maine mountaintop. Lee says the griphoist and winch cable, though heavy, were easier to carry up the mountain than the batteries, electric winch, and backup generator that would otherwise have been necessary.

Griphoists are not come-alongs; the latter is a lightweight tool found in North American hardware stores that uses a spool for coiling a short length of wire rope. Ranchers, for example, use come-alongs to tighten fencing, and for that purpose they don't need a long cable.

It's the spool or drum that sets come-alongs, as well as winches in general, apart from griphoists. Technically, griphoists are not winches but hoists. Winches use a drum to spool the hoisting cable, like the large drum on a crane. Griphoists, in

Table 16-2. Griphoists							
				Wire Rope			
Type	Model	Maximum Capacity		Diameter		Weight	
		kg	lbs	mm	inches	kg	lbs
Griphoist	Pull All	300*	700	4.72	3/16	1.8	4
Griphoist	Super Pull All	500*	1,500	6.5	1/4	3.8	8
Griphoist	T-508	800*	2,000	8.3	5/16	6.6	17
Griphoist	T-516	1600*	4,000	11.5	7/16	13.5	35
Griphoist	T-532	3200*	8,000	16.3	5/8	24	58
*Derated for European market.							

Figure 16-21. Raising tilt-up tower with electric winch. Installation crew raising a 4.5-inch (114-mm) diameter lightweight mast with electric winch. Inset: The tower was then lowered and a Bergey 1.5-kW turbine was mounted on top, and the tower was raised again.

contrast, pull the hoisting cable directly through the body of the hoist. Tractel likens the locking cams inside the griphoist with the way we take in a rope "hand over hand." To use a griphoist, you move a lever forward and back. This lever pulls the cable through the tool.

The hoisting cable for a griphoist can be any length, since there is no need to spool the cable on a drum. (Capstan winches can also use cables of any length, but they pass the cable over a drum.) When specifying the size griphoist needed, don't overlook the amount of cable required. The standard lengths shipped with most griphoists are insufficient for raising tilt-up towers. Simply order the correct length required from the griphoist manufacturer. Don't try to skimp by using guy cable in the griphoist.

There are several brands of griphoists on the market. They come in a range of capacities suitable for most small wind turbine applications. The manufacturer of either the wind turbine or the tower will specify the capacity (in pounds or kilograms) needed to raise a tilt-up tower of a given height and the amount of "run" or cable necessary.

If you have to buy any one tool for your off-the-grid wind system, says Hugh Piggott, it should be a griphoist.

Wiring

The bible on wiring in the United States is the National Electrical Code (NEC). It's the rule book for what can and can't be done, but the final say in electrical matters is in the hands of your local electrical inspector, fire underwriter, or code enforcement officer.

Licensed electricians in your area will be familiar with the local application of the codes. They won't be familiar, however, with wind turbines and their specific requirements, especially as they relate to interconnection with the utility. This section will be helpful to your electrician and essential if you plan to wire the wind system yourself. In either case, when unsure of your next step, check with your local electrical inspector before you begin work.

Why is the approval of the electrical inspector or fire underwriter so important? Because your fire insurance may be void without it. And in some communities, you can't sell your home without electrical wiring that's up to code.

The electric utility is responsible for all wiring from the nearest transformer to the service drop (where the conductors are secured to your house above the kilowatt-hour meter). Your responsibility begins at this point.

The conductors from a wind turbine are much like the power lines from a small generating station. The lines can be strung above ground or buried. They enter your home or business in a manner similar to those of the utility. Let's look first at wiring on the tower; then, we'll consider how we're going to deliver the power to your farm, home, or business.

In all cases, the leads from the generator must be connected (*terminated* or spliced, in the jargon of electricians), to the wires (*conductors*) running down the tower. These connections must be permanent and weatherproof. How this is accomplished depends on the kind of wind turbine and on the manufacturer.

When the generator is stationary within the tower, as in a vertical-axis wind turbine or in some horizontal-axis machines with right-angle drives ("modern" Jacobs), the generator leads can be directly connected with the conductors on the tower. On conventional small wind turbines, though, there must be a mechanism for transferring power from the moving platform of the wind machine to the stationary tower. Slip rings and brushes usually perform this task on small wind turbines.

On medium-size and large wind turbines, however, the power cable from the generator hangs freely through the center of the tower. As the machine yaws, or turns to face the wind, the *pendant* conductors twist. The cable is permitted to twist several revolutions before it must be unwound. Large wind turbines monitor how much the nacelle yaws in any direction, and when a set number of twists in the pendant cable are counted, the rotor is braked to a halt. The nacelle then yaws to unwind the cable. Once that is completed, the nacelle yaws back into the wind and the rotor is released, returning the turbine to service.

Good terminations between conductors are those that are mechanically tight, electrically insulated, and corrosion resistant. Terminations may be made in a number of ways: with split bolts, wire nuts, crimp (compression) connectors, or insulated termination blocks. Each has its merits.

Split bolts are easy to use and come in sizes suitable for even the heaviest power cable. The stripped ends of the conductors are inserted into the jaws of the bolt and the faces tightened with wrenches (spanners). These connections need an insulated boot and may loosen under a wind turbine's constant vibration.

Wire nuts are popular connectors for home wiring because of their ready availability and ease of use. Nearly every do-it-yourselfer is familiar with them. They are not well suited to wind turbine applications. They, too, may loosen under vibration and need to be secured with electrical tape. Moreover, they're most practical on solid, rather than stranded, wire and only on the smaller wire sizes. Wire nuts for heavy-gauge conductors are hard to find and are more difficult to use.

Crimp or *compression connectors* are the best all-around termination. They are mechanically sound and will not loosen as a result of vibration. Some include an insulated covering, but all can be insulated with a boot of heat-shrink tubing slipped over the connector. Their chief disadvantage is the need for a special crimping tool, and in the larger wire sizes, this tool is both expensive and awkward to use.

The connection between the leads from the slip rings and tower conductors should be made inside a junction or J box. The conductors are then run down the tower inside metal or plastic conduit.

In the past, some installers have taken unapproved shortcuts, using twist-lock plugs at the top of the tower and then lacing the power cable down a tower leg. They would attach the cable to the tower with electrical tape or with nylon cable ties. This technique doesn't meet building or electrical codes in many communities and presents a potential electrical hazard, to say nothing of creating future maintenance problems.

The turbine's power conductors should be protected from physical damage. Conductors on the exterior of the tower should be protected within conduit. Either electrical metallic tubing

UP-TOWER BLOCK CONNECTORS FOR MICRO TURBINES

One challenge facing installers of micro and mini wind turbines is making good terminations between the conductors on the wind turbine and the conductors running down the tower. Often the space available is cramped, making good connections problematic. Some use wire nuts. These are less than ideal because they can work loose from the wind turbine's vibration. "We never use 'wire nuts' over here," says Hugh Piggott in Scotland.

Some manufacturers suggest split bolts. These are cumbersome to use in the tight space available for a micro turbine. Bergey Windpower recommends NSI brand insulated connectors for up-tower terminations. While costly, these connectors are rated for both copper and aluminum and include an antioxidant compound and a plug for sealing the setscrew, which clamps the conductor in place. Best of all, these connectors are easy to use in the field, can be easily removed and then reused. None of these connectors are designed for supporting the weight of the conductors dangling down the tower. The down tower conductors must be supported with a strain relief.

Up-tower connectors. Block connectors used with a Creative Designs' up-tower junction box simplified installation of this micro turbine at the Wulf Test Field.

(EMT) or plastic tubing can be used. Metal conduit is preferable because it's strong and provides some shielding against voltage transients and lightning. Each 10-foot section of EMT is joined by weatherproof compression couplers and mounted on the tower with conduit hangers. The hangers should be spaced two per section with one at each end near the coupler. If you're using PVC (polyvinyl chloride) conduit, make sure it's rated for electrical use. Because PVC flexes use three or four conduit hangers per section, feeding the conductors down the inside of tubular towers; guyed tubular masts provide the necessary protection without the need for conduit.

Aboveground and Buried Cable
Almost no one runs cable above ground today. In Denmark, the electric utilities bury all distribution lines. Only high-voltage lines remain above ground. It's just simpler to excavate a trench and lay direct-burial cable, and that's how it's done on wind farms around the world. For small wind turbines though, it is better to lay PVC conduit in the trench and pull or fish the conductors through the conduit. This allows replacement of the conductors should that ever be necessary. There's also less likelihood that the conductors will be damaged as the trench is backfilled.

In many cases you'll want to install a disconnect switch in the line between the wind turbine and the wind system's control panel. The disconnect switch is necessary to permit isolation of the wind system from the load center or control panel during emergencies or when maintenance of the wind turbine is required.

Although a disconnect switch may seem like a redundant safety device on interconnected wind systems (because the wind turbine automatically disconnect from the utility line during power outages), it is not. The switch gives utility personnel or windsmiths a positive mechanism for protecting themselves. By throwing the switch off and inserting a lockout (a metal tag warning that someone is working on the line) through the switch handle, service personnel know with certainty at which point the circuit is broken.

From the utility's viewpoint, a lockable disconnect switch for a wind turbine should be located near its service entrance or kilowatt-hour meter. This is where fire crews will want to see the disconnect switch as well. For practical purposes, a second disconnect switch should be located at the base of the wind turbine tower. This is useful for isolating the wind turbine when servicing the machine.

The wind generator's output is then wired to a control panel or inverter, depending on the type of wind system used. In no case should DC output ever be connected directly with the AC service panel. Generators producing DC must incorporate a synchronous inverter before they can be interconnected with utility-supplied AC. Those wind systems producing "wild" AC, that is, AC of varying voltage and varying frequency, must also use an inverter.

For household-size wind turbines intended for interconnection with the utility, conductors from the control panel or inverter are wired to a dedicated circuit in the building's existing service panel. Output from the wind system must not be plugged into a wall outlet or wired to a circuit that's already supplying a load in the building, such as a refrigerator or a series of receptacles. Instead, a circuit breaker must be dedicated to the wind system.

Strain Relief of Tower Conductors
Always support the power cables on the tower with some form of strain relief to carry the weight of the cables. Disregard the suggestion by some manufacturers that the leads from their micro turbines are strong enough to resist being yanked out of the generator. It's always good practice, and often required by electrical codes, to use strain relief on the tower conductors. Otherwise, the leads will eventually pull out of the generator, causing big headaches later.

On guyed tubular towers, thread the power cables down the inside of the mast. Support the weight of the conductors with a strain relief wire net. The net works like a Chinese finger puzzle to grip the cable bundle. Hang the strain relief from an attachment point at the top of the tower. Strain relief can make or break a wind turbine installation.

For the connections inside the tower, you can use compression connectors for making vibration-proof splices between the wind turbine leads and the cables carrying power down the tower.

These connectors are often used for connecting the power supply to submersible well pumps, and can be found in farm supply and plumbing stores. Pros such as Sagrillo and Piggott warn against using split bolts, a common alternative, as they can work loose from vibrations in the tower. Compression terminals can obviate the need for heavy-duty and expensive crimping tools when working with heavy-gauge cable. Barrel connectors are another handy choice for terminating heavy, stiff conductors to terminal blocks or circuit breakers.

Sagrillo argues that manufacturers should ship termination kits and strain reliefs with their turbines to ensure that the job gets done right the first time. France's Vergnet, for example, offers components with its turbines and towers that most other manufacturers leave out. Vergnet includes 35 meters (115 feet) of power cable with all its smaller turbines. They also include junction boxes, grounding, and lightning protection with their tower kits.

Conductors and Conductor Sizing

The electricity produced by a wind generator is seldom of benefit at the top of the tower where the turbine is located. The electricity must be transmitted to the point of use before it provides any benefit. Thus, the cables connecting the wind turbine to the load are as integral to a wind power system as the tower supporting the turbine.

No practical conductor transmits all the electricity that is passed through it. Some is lost due to resistance. The length, diameter, and material of the cables connecting the wind turbine to the load—whether a service panel in your cellar or batteries in the barn—determine the amount of power and energy transmitted and, conversely, the amount lost.

These losses are proportional to the type of material (copper has less resistance than aluminum), diameter (thick cable has less resistance than thin cable), and distance to the batteries (short cables have less resistance than long cables). These resistive losses are reflected in the voltage drop between the wind turbine and the load.

Wind turbine manufacturers specify the cable size and material for a range of wind turbine-to-load distances that will allow their product to

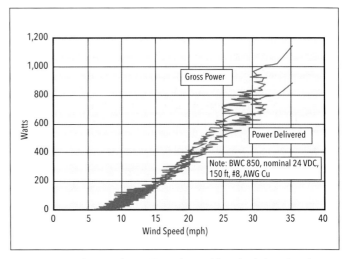

Figure 16-22. Power lost in conductors. Measured power delivered to the batteries and an estimate of the gross power generated by a Bergey 850 at the Wulf Test Field as a function of wind speed. The difference is the amount of power lost in the copper conductors between the wind turbine and the batteries. At the rated wind speed and above, 15% to 20% of the power produced by the wind turbine is lost in the conductors specified by the manufacturer.

perform as designed. For example, a BWC 850 was installed 150 feet (45 meters) from its load at the Wulf Test Field in the Tehachapi Pass. Bergey Windpower recommended using #8 AWG copper wire for the 24-V battery-charging system. These specifications assume that a portion of the electricity produced by the wind turbine will be lost in the conductors (see "American Wire Gauge to Metric Conversion" in the Appendix).

Operators of commercial wind power plants fully understand this phenomenon and account for it. Most consumers do not and are unpleasantly surprised when their shiny new wind turbine doesn't produce quite as much as advertised in the glossy brochures. Although there are many reasons why small wind turbines sometimes underperform, one is undersizing the conductors from the generator to the load (see Figure 16-22. Power lost in conductors).

Resistance in the conductors is seen as a voltage drop between the wind turbine on the tower and the load, for example, at the batteries. The voltage drop not only affects the amount of power transmitted, but, if severe enough, can also affect the operation of electrical equipment.

On interconnected wind systems, the voltage drop between the wind turbine and the load is critical to proper operation. Take the case of a vocational-technical school that installed a 6-meter (20-foot) diameter wind turbine driving an induction generator. The school was responsible for wiring the system into their service panel. Students and their instructor mapped out the conduit run and laid the conductors. When all was finished the installer flipped the switch—and nothing happened. He checked the wind turbine. Everything seemed fine. He then checked the students' wiring: nothing wrong there, either. As they sat and scratched their heads, some wise guy suggested they measure the voltage.

The problem? Low voltage. The wind turbine's control system sensed a voltage below its disconnect value and wouldn't energize the generator. (Manufacturers incorporate this feature to detect a power outage so a wind generator cannot energize a downed line and potentially kill someone.)

The dealer and the students then went back to the books. The wiring was sized according to the installation manual, or so it seemed. Their mistake was failing to take into account that the panel where the wind system was interconnected with the school's service was a long way from the utility's entrance to the school. The wind turbine's conductors were sized properly for the run from the service panel in the outlying classroom. But the long distance from the utility's service entrance to the wind turbine's control panel caused an excessive voltage drop. They remedied the situation by installing heavier gauge conductors.

Acceptable voltage drops range from 1% for interconnected wind systems using power electronics to 3% for those with induction generators.

Using Ohm's law we can calculate the voltage drop in a circuit and, subsequently, estimate the amount of power lost due to resistance.

$$I = V/R$$

where I is current in amps (for *intensité*, as Ampère referred to it), V is voltage in volts, and R is resistance in ohms (Ω). Solving for volts.

$$V = RI$$

The voltage drop increases in direct proportion to resistance and current. An increase in current increases the voltage drop through the conductor. If we wanted to use a 1.2-kilowatt wind generator in a 24-volt battery-charging system, we'd need to pump 50 amps through the conductors. If we sized the conductors so they would handle 5 amps at 24 volts with only a 1% voltage drop (0.24 volts lost), we'd encounter a 10% voltage drop (2.4 volts) at 50 amps.

If we increase the voltage of the 1.2-kilowatt system, say from 24 to 48 volts, the conductors will carry only half the current at rated power as before: 25 amps. Consequently, the voltage drop at rated power would be only 5%.

Now consider the amount of power that would be lost in both systems, where P is power:

$$P = VI$$

Substituting the value for volts from Ohm's law.

$$P = (RI)I$$
$$P = RI^2$$

While doubling system voltage cuts the current in half, it reduces the power lost in the conductors by a factor of four.

For this reason, system voltage increases with the size of the wind turbine. Micro turbines operate at 12 to 24 volts, mini wind turbines at 24 to 48 volts, household-size turbines at 48 to 240 Volts, medium-size wind turbines at 400 to 600 volts, and large wind turbines in the thousands of volts.

The total resistance, R_T, in a circuit is

$$R_T = (\Omega/\text{unit of length}) \times 2L$$

Where L is the wire run or distance from the wind turbine to the load, and 2 is the number of conductors for a DC circuit. For a three-phase circuit

$$R_T = (\Omega/\text{unit of length}) \times 3L$$

Many micro and mini wind turbines rectify AC at the alternator and transmit DC to the load; thus, calculating the voltage drop is straightforward. Other small wind turbines transmit three-phase AC to a rectifier near the batteries. Unfortunately, calculating resistance or power loss due to current flow in three-phase AC-rectified circuits is somewhat less straightforward.

Piggott explains that in rectified three-phase circuits, the AC phase current is about 0.82 of the current on the DC side of the rectifier for low impedance alternators. For high impedance alternators, the AC phase current is nearer 0.74 of the DC current.

Measurements on a BWC 850 at the Wulf

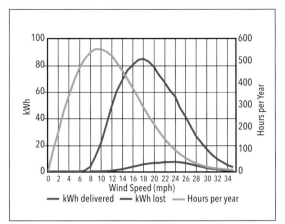

Figure 16-23. Energy lost in conductors. Calculated energy delivered to the batteries and the energy lost in the conductors for a 12-mph (5.4-m/s) annual average wind speed with a Raleigh distribution. The calculation is based on actual power measurements of a Bergey 850 at the Wulf Test Field. Though as much as 20% of the power is lost at higher winds, only 10% of the annual energy generation is lost to resistance in the conductors because the wind occurs at these higher speeds only a few hours per year.

Test Field confirmed this. At high wind speeds and high current, the AC phase current was 0.72 times the DC current. At low wind speeds and low current flows, AC phase current was more than 0.8 DC current.

For the three-phase BWC 850 at 300 feet (90 meters) in the example, the voltage at the load at rated power is

$$V = [(P/V) \times (0.72/\text{phase})] \times R_T$$

$$V = [(850 \text{ W}/24\text{V}) \times (0.72/\text{phase})] \times [(3 \text{ phases}) \times (0.0008090 \, \Omega/\text{ft}) \times (150 \text{ ft})]$$

$$V = 9$$

Or a drop of more than 20%, for the nominal 24-volt system.

The power lost at rated power is:

$$P = I^2R, \text{ and}$$

$$P = [(850 \text{ W}/24\text{V}) \times (0.72/\text{phase})]^2 \times [(3 \text{ phases}) \times (0.0008090 \, \Omega/\text{ft}) \times (150 \text{ ft})]$$

$$P \sim 240 \text{ W}$$

If the turbine is designed to deliver 850 W to the batteries at rated power then it must generate

$$P = 240 \text{ W} + 850 \text{ W} \sim 1{,}090 \text{ W}.$$

Consequently, at rated power, this BWC 850 would lose about 22% of the power produced (240 W/1,090 W).

In the real world, the conductors don't operate at 75°C. Typically, they are buried in the ground, and their starting temperature would be much lower, closer to 20°C. Temperature plays an important role in resistance losses in the conductors. Resistance can be adjusted for temperature of copper conductors where R_1 is the initial resistance, and R_2 is the resistance at the new temperature.

$$R_1 = 0.8090 \, \Omega/1{,}000 \text{ ft (305 meters)}$$

$$T_2 = \text{Ground temperature, 20°C}$$

$$R_2 = R_1[1+\alpha(T_2\text{-}75)], \text{ where } \alpha \text{ for copper} = 0.00323$$

$$R_2 = 0.8090 \times [1+0.00323(20\text{-}75)]$$

$$R_2 = 0.8090 \times 0.822$$

$$R_2 = 0.665 \, \Omega/1{,}000 \text{ ft}$$

Substituting the resistance at ground temperature into the earlier equation for power loss,

$$P \sim 200 \text{ W}.$$

Total power at the turbine then is

$$P_T = 850 \text{ W (delivered)} + 200 \text{ W (lost)} = 1{,}050 \text{ W}.$$

Power lost in the conductors at 20°C is

$$200 \text{ W}/1050 \text{ W} = 19\%.$$

Consequently, possibly as much as 20% of the total power produced by the wind turbine could be lost in the conductors at the rated wind speed.

Fortunately, this isn't the whole story. At many sites, wind turbines operate very little of the time near their rated power (see Figure 16-23. Energy lost in conductors). This diminishes the effect of the power lost in the conductors on the total energy delivered by the wind turbine to the load. For the BWC 850 in this example at a 12 mph (5.5 m/s) site, less than 10% of the energy produced is lost in the conductors.

While a 10% loss in a battery-charging system, such as the BWC 850, isn't critical, losses of this magnitude can make or break commercial wind power plants. For this reason, conductor sizing is an important element of project design.

JUNCTION OR J-BOXES

One challenge facing do-it-yourselfers wiring a small wind turbine is how to work with the heavy-gauge conductors needed on long runs from the turbine to the load. Disconnect switches, for example, are rated by their current-carrying capacity and not by the size of conductor they can accept. If you use a switch of the correct current rating, it will be too small for heavy-gauge conductors. On the other hand, if you buy a switch large enough for heavy conductors, it will be much more costly than necessary. You can solve the problem by using junction boxes at each end of a long cable run. The heavy-gauge wire is joined with a wire sized appropriately for the control panel or disconnect switch (see Junction box). It's important to make good splices when using junction boxes, using insulated split bolts or appropriately rated block connectors. This prevents shorts or grounds and keeps the overall resistance of the wiring run to a minimum.

Junction box. The junction box (bottom) allows connection of heavy gauge conductors used for long wire runs to the terminals in the disconnect switch (top).

Conduit Fill

The size of the wire you use will determine the size of conduit needed for protected cable runs such as on the outside of a tower or in a trench. For practical reasons, there's a limit to the amount of cable that can be stuffed into conduit. Electrical codes also limit the size and number of conductors used inside conduit of a given size.

Electricians generally prefer larger conduit than the size necessary for meeting code requirements. The larger diameter eases the task of pulling the conductors through the conduit. Bergey Windpower, for example, recommends conduit one to three sizes larger than required by electrical codes to simplify long wire runs. When you're pulling heavy, stiff conductors a long distance through conduit, you'll quickly appreciate why. It may cost a few cents more, but it makes your life a lot easier.

For short runs, you can simply push the cables through the conduit. On longer runs, such as through a buried trench from the tower to your service panel, you can push a metal *fish tape* through the conduit first. The fish tape has a woven basket or *cable grip* to which the conductors are attached. The tape is then used to pull the wires through the conduit. Another technique is to feed a nylon rope through each section or "stick" of conduit as you assemble the sections. You then pull the conductors through the assembled conduit with the rope.

Surge Protection

Wind turbines, like most electrical appliances, are susceptible to damaging voltage spikes. Electrical systems are grounded to limit voltage surges due to nearby lightning strikes. Grounding also ensures the prompt operation of fuses, circuit breakers, and protective devices from other electrical faults.

Lightning is only the most obvious of several sources of voltage spikes; passing clouds and the rapid opening and closing of switches on the utility's distribution system are others. Many wind turbine owners have found that voltage surges caused by the utility's lines are more frequent than those caused by lightning.

Lightning can short circuit a wind generator in less than a second. Though it occurs almost instantaneously, lightning can be of sufficient voltage to break down thick insulation and arc over insulators. We try to minimize the effects of lightning by using lightning arresters and ground rods.

Lightning or surge arresters furnish a path to the ground when a greater than normal voltage exists in a conductor. This drains off the excess voltage. After the voltage has returned to normal, the flow to the ground ceases. Though there's no foolproof protection against lightning, arresters do provide some degree of protection to power lines and other electrical components.

Equipment and buildings can also be protected by raising the effective ground level with a static line or lightning rod. The static line on the utility's distribution system and the lightning rods on farm buildings are attempts to do just that. Lightning rods, such as those at utility substations, offer a 45-degree cone of protection beneath them. When lightning strikes a lightning rod, the lightning passes directly to ground without first going through the object shielded below. For this reason, some wind turbines sport lightning rods above the nacelle (see Figure 5-32. Conventional drivetrain, medium-size turbine).

Large turbines incorporate lightning protection in the blades as well as the nacelle. This may not prevent all damage during a direct strike, but it does minimize damage. Manufacturers estimate that large wind turbines in Europe will be struck at least once per year.

Most systems rely on thorough grounding to drain off any static charge. Reducing the buildup of static electricity during a storm minimizes the possibility of a direct strike. Lightning doesn't always strike the tallest object. There are many documented cases where telecommunications towers, because they are thoroughly grounded, have been spared a direct hit, while nearby trees have been incinerated. Proper grounding lessens the possibility of a direct strike and minimizes the damage if one does occur.

To ground the tower, drive several copper-clad ground rods deep into the soil. When bedrock is near the surface, the ground rods can either be driven in at an angle or buried in a trench.

On freestanding towers, use two or more ground rods. On guyed towers, drive a ground rod near the mast and near each concrete anchor. Connect the ground rods to the tower or guy cables with heavy-gauge copper wire with brass or bronze clamps.

In areas of high lightning incidence or where the soil is dry and sandy, Jim Sencenbaugh recommends installing a ground net. On guyed towers the mast and each anchor are wired to their own ground rod. Then all the ground rods are tied together with a buried ground wire. On freestanding towers, there should be a ground rod for each leg of the tower. Sencenbaugh also electrically bonded together each tower section with a jumper wire on his installations. This provided a continuous path down the tower to ground.

On large wind turbines, there's often a bonding conductor between tubular tower sections.

Additional Notes on Wiring

Electrical codes in the United States require that all metal electrical enclosures, disconnect switches, control panels, and inverters be grounded so that any fault (short circuit) between a live or hot conductor and the enclosure will be conducted safely to ground. Grounding causes the circuit's protective devices to function—fuses to blow, circuit breakers to trip. This prevents the metal enclosure from becoming energized and presenting a shock hazard.

Control boxes and switches must be properly mounted, and all holes or cutouts not in use must be sealed. There must also be sufficient clearance in front of any panel or junction box for safely servicing the circuit. Control panels and inverters with ventilation louvers must be located so as to allow for free air circulation.

GROUNDING NETS

The passage on the grounding technique used by Jim Sencenbaugh was originally written in 1982. Unfortunately, arrogance–and ignorance–breeds anew. In the mid-1990s, Zond Systems, at the time the sole North American manufacturer of large wind turbines, installed a small wind farm in a lightning-prone area of Texas. The project, partially subsidized by the US Department of Energy, was plagued by lightning-induced outages. After extensive–and expensive– studies on the cause of the problem, the company installed a ground net much like the one Jim Sencenbaugh had used on his wind turbines in the late 1970s– nearly two decades before. The turbines have since been removed and likely the knowledge gained at such great expense has been lost again.

All terminals, connectors, and conductors must be compatible. Poor connections and the use of dissimilar materials such as copper and aluminum are a major cause of electrical fires. Aluminum is particularly troublesome. It oxidizes when exposed to the atmosphere and forms a highly resistant crust. This increased resistance causes the connection to heat up under heavy loads and is believed to be responsible for numerous fires. Whenever aluminum is used, terminal blocks, split-bolts, and other connectors must be rated for aluminum (Al) or for copper and aluminum (CO/AL, or Cu-Al). Aluminum connections should also be coated liberally with an antioxidizing compound.

Once the installation is complete, you're ready for the final check of your wind system before you begin operation.

Decommissioning and Dismantling

All good things must come to an end, but for wind turbines this often isn't quite true. Though wind turbines are typically designed to operate 20 years, there are many wind turbines still operating—reliably—after more than 30 years in service. Moreover, many of these older turbines can be reused elsewhere when they're eventually removed.

In the 1970s, the wind industry in North America began with entrepreneurs dismantling wind chargers of the 1930s and 1940s and putting them back in operation. Now repowering is again gaining ground in North America and Europe. Older turbines are being decommissioned and dismantled, and new turbines installed on the original sites. The dismantled turbines are then reconditioned and shipped to new sites where they begin a new life.

Dismantling requires the same construction skills as installation, the same attention to detail, and the same attention to safety. Taking a wind turbine down is just as dangerous—if not more so—than putting one up. As noted in the opening of this chapter where a 1930s-era wind charger was being dismantled, mistakes can be deadly.

Nevertheless, dismantling a wind turbine after it has served its purpose is an obligation of all owners and operators. The turbine and tower can be taken down in segments: blades, nacelle, and then the tower sections as shown in the following dismantling sequence.

The concrete foundation can be chipped into pieces with a hydraulic hammer, the rebar separated from the concrete and sent to recycling. The concrete can be broken up for aggregate or sent for disposal. The blades, if not reused, must be broken up and sent for disposal. The metal components—if not reused—can be recycled.

The foundation is usually taken down to below plow depth or about 1 meter (3 feet), seldom more. Then the site is regraded and restored.

ERECTING A MICRO TURBINE WITH A GRIPHOIST

The following sequence illustrates installation of a micro turbine on a 45-foot (13.7-meter) lightweight mast at the Wulf Test Field in the Tehachapi Pass. Using Tractel's Super Pull-All model griphoist, only two people were necessary to safely raise and lower the turbine. Three elements make the lightweight mast system easy to use at remote sites such as the Wulf Test Field: screw anchors that can be driven by hand, thin-walled steel tubing that slips together, and precut lengths and swaged fittings on the lifting cables.

Taking delivery of a micro turbine and lightweight tower kit.

Taking inventory. Making sure all the components needed are on site.

Assembling the mast's base plate. The base plate on this mast kit substitutes for a pier in a fixed-mast, guyed tower.

Carefully aligning anchor positions relative to the base plate. Precision at this step allows raising and lowering the tower with minimal adjustments.

Positioning base plate and staking in place with rebar.

Screw anchor. The holding power of screw anchors is a function of soil conditions, the diameter of the helix, and the depth that the helix is driven into the ground. This tower kit relies on screw anchors.

ERECTING A MICRO TURBINE WITH A GRIPHOIST *continued*

Driving the screw anchors by hand. This can be good exercise, especially in dense soil, as here.

Assembling the mast. Sections of thin-walled tubing slip together.

Unreeling guy cable. This and similar tower kits include all guy cables and attachments that are needed.

Gin pole and rigging. Yellow ropes stay the gin pole while raising the tower. They are later removed. Lifting cables (with red tags) are clipped into snap hook shown.

Raising the tower without the turbine. This is a necessary step to ensure that the guy cables are properly adjusted before raising the wind turbine and tower. Always attach rear guy cable before raising the tower!

Making connections. NSI-brand insulated connectors makes quick and secure work of up-tower connections, requiring only an electrician's screwdriver.

INSTALLATION AND DISMANTLING

Strain relief. Wire net suspends tower conductors from eyebolt in tower wall. Cable connections (terminations) are not rated for supporting the weight of tower conductors.

Mounting the wind turbine.

Raising the turbine.

Raising the turbine and tower with a griphoist. Note tag line on rear guy cable and that the second person stands at a right angle to the tower's fall line, and well outside the fall zone.

Turbine installed.

ERECTING HOUSEHOLD-SIZE TURBINE WITH CRANE

This sequence illustrates the installation of a Bergey Excel on a guyed tower near Tehachapi, California. The tower's location and height were chosen in consultation with the installer. The concrete pier for the mast and the three concrete anchors were placed some weeks prior to scheduling the crane. This allowed ample time for the concrete to cure. Prior to the crane's arrival, the turbine and tower have been assembled and two of the three blades have been bolted to the turbine. The installer uses barrels and wood blocks to support the wind turbine and tower during assembly. Note that this is a professionally trained crew. They do this for a living.

Crane arrives.

Crew attaches nylon sling to tower. Sling attaches to crane hook.

Initial lift to clear the ground for attaching the third blade.

Attaching nosecone or spinner.

Crane raises turbine and tower as ground crew guides the mast toward its pier.

Mast is lowered onto pier. Note disconnect switch mounted on the tower.

Ground crew attaches guy cables to turnbuckles. Note equalizer plate attached to anchor rod and grounding conductor connected to ground rod.

Releasing the crane's boom. Windsmith climbs tower to unhook the crane's boom from the sling, but only after all guy cables are secured to the guy anchors and tensioned. This ensures that the tower is fully rigged and won't fall over.

Turbine installed. Crane prepares to leave site as windsmith descends tower.

ERECTING A LARGE WIND TURBINE

Erecting a large wind turbine is a complex operation requiring highly trained and skilled construction crews. Note that the construction techniques used vary from project to project. This is especially true for the foundation and for attaching the blades. The type of foundation and the soils will determine the system used. In the past, most blades were attached to the hub on the ground and the rotor raised as a unit to the nacelle. In this sequence, each blade on this rotor is attached individually to the hub.

All photos by Neal Emmerton.

Grading pad. Grading the site before excavating the foundation.

Excavating for the foundation. Depending upon the site conditions and the size of the turbine, the foundation may be excavated with a large-diameter auger.

Assembling rebar cage. Note the lower tower section's threaded rebar bolts and mating flange.

Assembled rebar cage. Note electrical conduit is in place for power conductors and communication cables.

Rebar cage ready for concrete placement.

Placed and cured concrete.

Raising lower tower section.

Placing lower tower section. Note control cabinet has been assembled on the foundation so that the tower section is lowered over the cabinet. In some installations, the transformer is also placed inside the tower, and the lower tower section is placed over the transformer and control cabinet.

Final positioning of lower tower section. The moment of truth. Holes in the tower flange must match up with the bolts in the foundation.

Bolting tower flange to foundation. The tower flange does not sit directly on the concrete. It rests on spacers.

Grouting tower flange. Grout is pumped into the space between the tower flange and the foundation ring spacers.

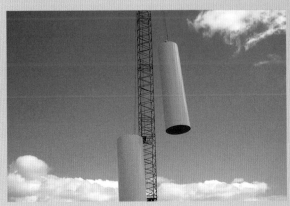

Raising midtower section.

ERECTING A LARGE WIND TURBINE continued

Raising top tower section.

Nacelles arriving by rail. Two Vestas V90 3-MW nacelles on rail flatcar.

Raising nacelle off flatbed trailer.

Rotor hub. Readying the rotor hub for mounting to the nacelle. Note three pitch actuators.

Raising nacelle to tower top.

Nacelle installed on tower.

INSTALLATION AND DISMANTLING

Unloading rotor blades.

Raising blade to nacelle. Note special carrier for attaching blade to the hub on the nacelle. The blades can also be attached to the hub on the ground and the assembled rotor lifted to the nacelle.

Ready to lift third and final blade to the nacelle.

Erection complete. The turbine is now ready for its finishing touches and final check out.

DISMANTLING A LARGE WIND TURBINE: WINDKRAFT DIEMARDEN

Windkraft Diemarden began when a group of determined citizens met unofficially in the summer of 1992 to form a locally owned wind company. By 1993 they had installed one 150-kW Nordex N27 turbine, making Diemarden the first village in the area southeast from Göttingen to generate electricity with wind energy. (Göttingen, of course, is where the famed aerodynamics laboratory employed Ludwig Prandtl and Albert Betz–of the Betz limit fame.) By December 1994 they added a 250-kW N29 turbine. The successful *Bürgerbeteiligung* or community-owned wind project later expanded to include two other sites in the region. In 1996 they added two Bonus 600-kW turbines near Deiderode and from 2000 to 2002 they added three Bonus 1.3-MW turbines at Bischhausen.

As the original turbines were nearing two decades of service, Windkraft Diemarden decided to repower the site. They took the old Nordex turbines down in the winter of 2012 and shortly afterward replaced them with three new Enercon E-101 turbines. The 3-MW Enercon turbines use a rotor 101 meters (330 feet) in diameter. The rotor sweeps 8,000 m², 14 times more area of the wind stream than the Nordex N27.

This sequence of photos illustrates the dismantling of a large wind turbine. Importantly, the photos show how the concrete foundation is broken up with a hydraulic hammer and the rebar is separated from the broken concrete for recycling. The concrete can be used for aggregate or sent to a disposal site.

The wind turbines and towers were also recycled. They found a new home in Northern Ireland where they were reconditioned, repainted, and returned to service for another couple of decades.

In the fall of 2013, 300 guests celebrated the 20th anniversary of the locally owned wind turbines at Diemarden's wind turbine festival with beer, *würst, gemütlichkeit* (good cheer), and tours of the wind turbines.

All photos by Gottfried Wehr, 2012.

Removing the rotor from the nacelle.

Lowering the rotor to the ground for disassembly.

Lowering the nacelle. The turbine and tower were shipped to Northern Ireland where it found a new home. Note image on the tower at left. Many wind turbines in Germany are "adopted" by local groups who paint fanciful patterns on the towers.

Disassembling the tubular tower.

Lowering the top tower section.

Placing the top tower section on a flatbed trailer. The two tower sections, blades, and nacelle were then transported to Northern Ireland where the turbine was reconditioned and is now in operation once again.

Exposing the tower foundation.

Readying the foundation for removal. Gottfried Wehr posing with foundation for scale.

Hydraulic hammer breaking up foundation.

Foundation section and rebar cage exposed. Gottfried Wehr posing for scale prior to separating the rebar from the foundation.

DISMANTLING A LARGE WIND TURBINE continued

Separating rebar from concrete. After breaking up the concrete foundation the rebar is removed for recycling.

Enercon E-101 nacelle. Gottfried Wehr posing in front of the E-101 nacelle before it is installed to replace the two Nordex turbines removed at Diemarden.

New E-101 installed. One of three Enercon E-101 turbines installed by Windkraft Diemarden.

17

Safety

> We obey the law to stay in business,
> but we obey the laws of physics to stay alive.
>
> —Anonymous windsmith

THE CAPTURE AND CONCENTRATION OF ENERGY—IN ANY form is inherently dangerous. Wind energy exposes those who work with it to hazards similar to those in other industries. Of course, there are the hazards that, taken together, are unique to wind energy: high winds, heights, rotating machinery, and the large spinning mass of the wind turbine rotor. Wind energy's hazards, like its appearance on the landscape, are readily apparent. Wind energy hides no latent killers, no black lung, for example. When wind kills, it does so directly and with gruesome effect.

In this chapter, we'll first examine the record and glean what we can from fatal accidents with wind energy. We'll compare wind's record with that of other sources of generation. Then we'll turn to the tools and practices necessary for working safely with the technology. Unpleasant as this topic may be, it emphasizes the need to work safely—because your life quite literally depends on it.

Fatal Accidents

Death in the maw of a wind machine is nothing new. H. C. Harrison recounts in *The Story of Sprowston Mill* how his great-grandfather Robert Robertson was killed in 1842 after becoming entangled in the sack hoist on his English windmill. There are historical accounts of similar deadly accidents in France, and no doubt like tales can be found in other countries where wind energy has been used.

Since its rebirth in the 1970s, wind energy has directly or indirectly killed more than 80 people. The first recorded was Tim McCartney, who fell to his death near Conrad, Montana, in 1980 while trying to salvage a 1930s-era wind charger. There are few details on McCartney's death other than that his broken body was found near the tower. News reports said simply that he fell during high winds. McCartney was soon followed by Terry Mehrkam, a pioneering Pennsylvania designer and manufacturer of wind turbines. Mehrkam was killed in late 1981 near Boulevard, California. Unfortunately, Mehrkam was not the last person to lose his or her life on a wind turbine.

Table 17-1. Summary of Deaths in Wind Energy	
1980 to Early 2014	
Number of Deaths in Construction (Installation or Removal)	45
Number of Deaths in O&M	20
Number of Deaths of the Public	7
Number of Deaths in Manufacturing	2
Number of Deaths in Training	7
Suicides	1
Total	82

Table 17-2. Deaths by Region	
Country	Deaths
Europe	
Denmark	6
Germany	10
Great Britain	7
Ireland	1
Netherlands	5
Portugal	1
Spain	2
Sweden	1
Sub Total	33
North America	
USA	28
Canada	5
Sub Total	33
South America	
Argentina	2
Brazil	1
Sub Total	3
Asia	
China	12
New Zealand	1
Sub Total	13
Total	82

More than half of all fatalities have occurred during construction and the equally dangerous removal of wind turbines for repowering (see Table 17-1. Summary of Deaths in Wind Energy). Twenty people have died during the normal operation of a wind power plant or an individual wind turbine. A surprising number have been killed in training accidents—five in just one incident in China. And there is one documented case of a suicide. There may have been more.

There have been an equal number of deaths on the North American and European continents. In North America, far more have been killed in the United States than in Canada with its much smaller industry. Germany, with the largest concentration of wind turbines in Europe, dominates the mortality statistics, closely followed by Great Britain with a much smaller installed base (see Table 17-2. Deaths by Region).

The fatality rate appears independent of the wind turbine's size. Large turbines are no more deadly than small wind turbines. Three people have died working on or playing in the vicinity of household-size turbines (see Table 17-3. Deaths by Size). Tim McCartney was removing a 1930s-era wind charger, and Robert Skarski was installing a household-size turbine on a used tower when he was killed.

Most of those killed were men, reflecting the male dominance of the construction and service industry. The median age of the fatalities is 34. The oldest was 85 years old and died of a heart attack climbing a tower in Europe. The youngest was 3. The child was crushed to death in Ontario, Canada, while playing on an unsecured gin pole used for a mini wind turbine.

There have been several members of the public who have died around wind turbines, though no passersby have been killed. A teenager climbed a wind turbine tower at his high school in Ohio as a prank and fell off. A snowmobiler drove into a fence surrounding a wind farm in Canada. In a bizarre year 2000 accident, a young parachutist crashed into a wind turbine on the German island of Fehrmarn during her first unassisted jump. (Her instructor landed safely.) And a Texas crop duster pilot was killed when he flew his plane into the guy cables supporting a meteorological mast in 2005. At least two people have been killed in collisions with trucks carrying wind turbine components.

One man died during a training class after climbing a tower and becoming unconscious. In another case, an instructor fell from a tower in Indiana during a tower training exercise at a community college.

There have been a number of cases of multiple fatalities during one accident. The worst to date

SAFETY

Table 17-3. Deaths by Size

	kW	Deaths
Small	<30	4
	<=100	13
	<=1000	12
Large	<2000	11
	<3000	21
	<10000	6
	Total	67

Note: Total is less than total fatalities because the size of the wind turibne is not always reported.

was in China when five were killed, including a Chinese Communist Party official, when a crane collapsed during a demonstration. In 1990, two were killed in Denmark while they were servicing the rotor on a large wind turbine from a basket suspended beneath a crane. The rotor started turning, sending the workers to their deaths. In another service-related accident, two windsmiths died during a fire in the nacelle of a wind turbine in the Netherlands in 2013.

As one of the newest and one of the fastest growing markets, offshore wind has accounted for a relatively high number of fatal accidents. Seven have died working with wind energy offshore just since 2009, with four of those in waters off Germany.

More details on these accidents and others can be found at www.wind-works.org (see the sidebar Deaths in Wind Energy Database). The details are often gruesome and not for the faint of heart.

Wind's Mortality Rate

There's no way to sugarcoat the occupational hazards of working with wind energy. The early days were plagued with a series of accidents that were shocking for an industry so new. For example, eight men died during the period of the California wind rush, predominantly during construction. The mortality rate—in terms of deaths per TWh of electricity generation—was extraordinarily high because so little electricity was being generated at the time. It was the industry's formative period, and wind generation didn't exceed its first TWh worldwide until 1986. However, from 1980 to 1986

DEATHS IN WIND ENERGY DATABASE

I have been tracking the number of deaths in wind energy since Tim McCartney fell from a windcharger in 1980. I make this information available as a public service to researchers and anyone else interested in safe work practices. My database–in the form of a spreadsheet–can be found on my website by looking under "Accidents & Safety" in either the large or small wind turbine directories and searching for "Wind Energy–The Breath of Life or the Kiss of Death: Contemporary Wind Mortality Rates." I update the statistics periodically as new information becomes available, and I include data from countries worldwide. Unfortunately, it is difficult to monitor the industry in countries more noted for their secrecy than their transparency and where the information never appears in English. Note that I only track mortality not accidents in general and that I only track deaths that are directly attributable to wind energy. If you have information on a fatal accident that you would like to share with me, you can reach me through my website: www.wind-works.org.

the death rate dropped from nearly 40 per TWh to 0.

There was a collective sigh of relief when there were no recorded deaths from 1985 through 1988.

Alas, the mortality rate jumped to 0.4 deaths per TWh in 1989 when John Donnelly was killed near Palm Springs servicing a Danish wind turbine. The year 1990 was even deadlier. Leif Thomsen and Kai Vadstrup died in the same accident servicing a wind turbine in Denmark, and Dirk Hozeman died in the Netherlands on his own turbine. These deaths raised the mortal-

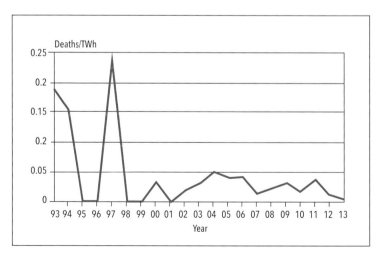

Figure 17-1. Occupational mortality in wind energy. For the two decades from 1993 to 2013, the mortality rate in wind energy worldwide fell from 0.25 per TWh to less than 0.05 per TWh. From 2009 to 2013, the mortality rate averaged 0.020 per TWh or less than that from nuclear generation and two orders of magnitude less than that from coal-fired generation.

ity rate to an alarming 0.93 per TWh. Fortunately, the mortality rate fell dramatically the following year and has been falling ever since.

For the past two decades—1993 to 2013—the death rate has remained below 0.25 per TWh and since 1998 has remained below 0.05 per TWh (see Figure 17-1. Occupational mortality in wind energy). For the decade ending in 2013, the mortality rate has averaged 0.027 per TWh. During the five-year period ending in 2013, the average death rate was 0.020 per TWh.

There are two reasons why the death rate has fallen so steadily. The most important is that the industry—and those who work in it—take safety seriously. Most companies have designated safety officers assigned to stay abreast of safe work practices and the regulations governing occupational safety. Yes, it wasn't always so, and in the early days we performed tasks in ways that were unacceptable in other industries. Sadly, the industry learned some lessons the hard way, but it has learned them.

The second reason for the fall in mortality rates is the ballooning increase in electricity generation. Despite the loss of three men in wind farm accidents during 2013, the industry generated nearly 650 TWh of electricity, cutting the mortality rate to 0.005 per TWh for the year. The goal, of course, is a zero mortality rate, and that has been achieved as recently as 2001 and in 4 of the 10 years during the 1990s.

Although any deaths that can be attributed to wind energy are unacceptable, wind generation is one of the least risky ways to produce electricity. Statistics on the mortality rate in power generation are hard to come by, confusing, and often conflicting. Nevertheless, a number of studies from the 1990s paint a general picture of the occupational mortality rate from conventional sources (see Table 17-4. Summary of Occupational Mortality in Power Generation). Note that these mortality rates don't include the effects on the public from air pollution emitted by coal-fired power plants or latent deaths from large-scale nuclear accidents, but they do include latent occupational deaths from disease and cancer.

The mortality rate for wind energy is comparable to or an order of magnitude less than that from nuclear or large hydro. Wind's occupational death rate is comparable to that from natural gas and an order of magnitude less than that from oil-fired generation. As one might intuitively expect, the mortality rate of wind energy during the past decade is two orders of magnitude less than that from coal-fired generation—and this doesn't include any of the widely acknowledged mortality due to coal-fired air pollution.

Hazards

Falling from the tower is the single most apparent occupational hazard of working with wind energy. Industry practice and what some would argue to be common sense suggest that McCartney and Mehrkam all made the same fatal mistake: they did not use any form of fall protection.

Falls

While fall-protection terminology can be arcane, the principles are not. Fall protection comprises both tools to prevent falls as well as devices to arrest falls that do occur. Where falls can occur, for example, in the absence of a work platform and railing, a fall-protection system must include three elements: body support, lanyard, and anchorage. The lanyard—short sections of rope or webbing—connects the body support to a sturdy attachment on the nacelle or tower. Body or work belts and lanyards in conjunction with a suitable anchorage are tools used to restrain a windsmith (technician) from falling while allowing him to position himself to perform a given task. If a windsmith does fall, a fall-arresting

Table 17-4. Summary of Occupational Mortality in Power Generation

	Deaths/TWh	
	Low	High
Coal	0.100	1.318
Oil	0.247	
Gas	0.032	
Nuclear	0.020	0.227
Hydro	0.023	0.409
Wind	0.020	0.027

Coal & nuclear includes latent mortality.
Wind: Average for 5 and tens years to 2013.
Data compiled from multiple sources. See the appendix for more detail.

system is designed to prevent serious injury. Like fall restraint, a fall-arresting system includes a lanyard and a suitable anchorage. Unlike fall restraint, a fall-arresting system requires a full-body harness.

Little is known about how McCartney died. He obviously was not using a fall-arresting system. But Mehrkam's death was investigated by California's Department of Occupational Safety and Health. They concluded that Mehrkam climbed to the top of the tower without any form of fall protection and either fell or was thrown off the tower to his death.

In another case, Richard Zawlocki fell to his death in 1992 while descending a tower near Palm Springs. Unlike Mehrkam, Zawlocki was wearing his work belt when he fell while descending the tower. However, he was not using his positioning lanyard as protection against a fall from the tubular tower's interior ladder. His lanyard was later found atop the tower—holding the nacelle cover open.

Zawlocki's death is troubling for what it says about the human factor in all accidents. It is evident that Zawlocki was aware of the risk of working on the tower because he was wearing his work belt. But he failed to use his positioning lanyard. We will never know why. We do know that when safety equipment is inconvenient or uncomfortable, there is a tendency to avoid using it.

All large wind turbines and most medium-size turbines as well include a fall-arresting system designed specifically to prevent the kind of accident that killed Zawlocki. This system uses a metal sleeve that slides along a steel cable or rail that runs the length of the tower. When ascending or descending the tower, windsmiths attach their body harness to this sleeve. Should they slip, the sleeve grips the cable, arresting their fall.

The tower Zawlocki was descending had recently been installed in a repowering project where turbines had been moved from Northern California to the San Gorgonio Pass in Southern California. The tower lacked a fall-arresting cable or "ladder-safety device" in the jargon of the trade. As an alternative, technicians were instructed to attach their lanyards to the ladder when climbing the tower. This entails climbing a few rungs and then removing and reattaching their lanyards. Anyone who has used this system knows that it is awkward and time-consuming or, in other words, inconvenient. As a consequence, this technique is more honored in the breech than observed in practice, which led to Zawlocki's death.

Spinning Rotors

Another lesson learned from previous accidents is that no one should ever work atop the tower when the wind turbine rotor is turning during high winds. This simple proscription would have prevented the deaths of Terry Mehrkam in California, Dirk Hozeman in the Netherlands, and Bernhard Saxen in Germany.

Mehrkam died when he tried to stop the rotor on one of his wind turbines. The brakes had

TERRY MEHRKAM THROWN TO HIS DEATH

BOULEVARD, CA (UPI)–Terrence Mehrkam, 34, owner of a Hamburg, Penn., windmill manufacturing company, was struck and killed by the blade of one of his own windmills at a "wind farm" in this San Diego County community.

The coroner's office said Mehrkam was struck by one of the blades after falling from a platform.

Sadly, Terry Mehrkam's death in 1981 wasn't the first, nor was he the last person killed working with wind energy. His accident and those of others should serve as constant reminders of the danger inherent in working on a power plant high above the ground. California's Department of Occupational Safety and Health concluded that Mehrkam climbed to the top of the tower without using any form of fall protection and either fell or was thrown off the tower to his death.

Terry Mehrkam. Mehrkam stands atop the nacelle on one of his 40-foot (12.2-meter) diameter wind turbines near Hamburg, Pennsylvania, circa 1980. In this photo, Mehrkam has a rope tied around his waist. He was thrown from a similar wind turbine near San Diego, California, in 1981 and killed. At the time of his death, Mehrkam, who was not wearing a work belt, climbed to the nacelle and tried to manually stop a runaway rotor.

failed in high winds. Subsequently, the rotor "ran away" and went into uncontrolled overspeed. This is the worst scenario imaginable for going anywhere near the wind turbine. Under these conditions, the rotor on Mehrkam's machine would have been a blur. As insane as it seems now, it was Mehrkam's practice with runaways such as this to climb the tower and manually brake the rotor to a halt by wedging a pry bar into the brake calipers!

Mehrkam made two mistakes: he climbed the tower while the rotor was spinning uncontrollably, and he did so without any fall protection, not even a work belt and lanyard. He must have been frantic to save his machine because he had used a rope restraint in the past. The rope restraint, crude even by the lax standards of the day, could have prevented his fall and saved his life. It was utter foolhardiness to mount the turbine and straddle the nacelle like Slim Pickins riding a nuclear bomb to its target in the movie *Dr. Strangelove*. Mehrkam should have simply walked away, cleared the site, and waited for the wind to subside.

Similarly, Dutch homeowner Dirk Hozeman was killed when he vainly attempted to stop the runaway rotor on his wind turbine from destroying itself in a violent winter storm. Tragically, the Polenko turbine had been inoperative for two years and had just recently been returned to service when—against professional advice—Hozeman climbed to the nacelle to bring the rotor under control. After squeezing into the small nacelle, Hozeman became snagged on a turning shaft. Rescue crews retrieved his body the next day—after the wind subsided.

Hozeman, like Mehrkam, should have walked away from the wind turbine and waited for the storm to pass. There is little one can do when a wind turbine rotor becomes unloaded in high winds, as when the grid goes down and the overspeed control devices fail.

In the industry's "wild west" days there are accounts of windsmiths throwing ropes through the rotors on runaway wind turbines in Denmark as well as the San Gorgonio Pass. Such attempts were often unsuccessful, and wind companies eventually abandoned efforts to rescue turbines in overspeed as being too risky. As longtime Tehachapi windsmith Jon Powers explains, "Wind turbines are replaceable, people are not."

Another lesson from these accidents is the danger inherent in working around poorly designed wind turbines. The Mehrkam turbine and its clones were notorious for self-destructing in California's windy passes. The Dutch Polenko was not much better.

On occasion, early Danish turbines also failed catastrophically. Yet Danish turbines failed far less often than the flimsy US-designed machines of the day. The pitchable blade tips on Danish turbines could typically be depended upon to protect the rotor from destroying itself. "We rely on the centrifugal tips on our stall-regulated turbines to account for brake, generator, and drivetrain failures," says Powers. When the wind subsides and the turbine is safe to approach, the nacelle is yawed out of the wind and the rotor eventually stops.

In contrast to Danish designs, Mehrkam's turbines had no aerodynamic means of overspeed control. If the drivetrain or brakes failed and the winds were high, his machines often spectacularly self-destructed.

In 1997, Bernhard Saxen was crushed to death inside the nacelle of an experimental wind turbine when it flew off the top of its tower at the Kaiser-Wilhelm-Koog test center in Germany during a storm—with him in it. Though the turbine that killed Saxen was built by a reputable manufacturer and was a far cry from the crude contraptions built by Mehrkam, it remained experimental. By its nature, an experimental turbine has not proven that it is safe to operate under all conditions and as such poses a potential hazard.

On large turbines that have a proven record of reliability, there may be occasions where it is acceptable for a technician to remain inside the nacelle when the turbine is operating, say site managers. The caveats are that no one enter or exit the nacelle while the turbine is operating, that there is sufficient space inside the nacelle to work safely, that all rotating shafts are covered, and that this never be attempted during high winds or storms when the control systems are under stress. Bernhard should not have been in the nacelle of an operating experimental turbine during a storm. Whatever his reason for doing so, he took an unnecessary risk that cost him his life.

No one should approach a small wind turbine rotor when it is operating—for whatever reason.

Common in other industries, all contemporary commercial wind turbines now have shaft guards to prevent the kind of accident that killed Hozeman in the Netherlands and John Donnelly a year earlier in the San Gorgonio Pass. (Such safety measures always seem obvious after an accident has occurred.)

Large wind turbines and many—but not all—medium-size turbines now have sufficient space inside the nacelle for someone to safely stand and observe the turbine's operation. Regardless of the turbine's size, site managers warn that no one should enter or leave the nacelle unless the turbine has been brought to a full stop and the nacelle is prevented from yawing in response to changes in wind direction. Access to even large wind turbine nacelles is tight and often requires some scrambling.

Some turbines have automatic emergency stop switches on the upper access hatch leading into the nacelle from the top of the tower. Should anyone reach the upper platform and raise the hatch while the turbine is operating, the switch activates the control system and brings the turbine to a stop. Such switches are intended to prevent potentially hazardous entry into the nacelle while the wind turbine is operating.

When anyone is working on the rotor or drivetrain, the rotor must be locked in place. How this is done varies from turbine to turbine but typically involves placing a pin, suitably rated for the task, through a rotating component of the drivetrain. Such a locking pin would have prevented an accident in Denmark, where the brake was inadvertently released, allowing the rotor to begin turning. The blades clipped a suspended basket sending Leif Thomsen and Kai Vadstrup to their deaths while a colleague hung on for dear life.

A locking pin would have also prevented John Donnelly's death near Palm Springs. During a seemingly calm day, Donnelly climbed the turbine to repair a damaged brake. The turbine was off-line, but without the brake the rotor was able to freewheel. A slight breeze started the rotor turning, catching Donnelly off guard. By the time he realized what was happening, his lanyard was snagged on the main shaft and it was too late to react.

Locking pins themselves are not foolproof. There are limits to their effectiveness. Manufacturers may specify wind speeds above which such devices should not be used. In one case, two windsmiths were servicing the hub on a large wind turbine. They placed the locking pin in position as instructed, but they disregarded the onset of increasing wind speeds. While they were inside the hub, the locking pin failed, and the rotor began to spin, trapping them inside the hub. With their tools flying about their heads they must have thought they were on a deadly *Mr. Toad's Wild Ride*. Eventually their coworkers saw that something was amiss and braked the rotor to a halt. If the blades had not been pitched to feather, and the rotor only capable of spinning slowly, this humorous tale would have taken a deadly turn.

Unfortunately, a locking pin, or "rotor fixing bolt" in news accounts, failed in a large wind turbine in Germany, killing one worker and injuring another as recently as 2005.

Electrical

Next to falls and spinning shafts, the most serious hazard with wind energy is working around electricity. The most common form of serious, nonfatal accidents in California wind plants is injuries from electrical burns. In one case, a Tehachapi woman was maimed when she touched energized equipment inside a transformer cabinet with a vacuum cleaner. Again, the hazards are similar to those in the electric utility industry, and the precautions developed during the past century for safely generating and transmitting electricity are applicable. If they had been followed, they would have prevented the Tehachapi accident and others like it.

Construction

As noted, more than half of those killed through 2014 died in construction-related accidents. McCartney was dismantling a turbine, and Mehrkam was trying to rescue a turbine he had just recently installed. Construction is not an ongoing activity, and the risk associated with construction normally occurs only twice in the life of the wind turbine: during installation and then during removal.

Analysis

The deaths in the wind energy industry, while alarming, may not accurately reflect what can be expected from the technology as it matures. Many of the hazards encountered in building and operating a wind turbine are not unique to wind energy, and a considerable body of knowledge has accrued over the decades for how to work safely around rotating machinery, with electricity, and at heights. If the early wind industry had followed practices common in other industries, several of the recorded deaths would have been avoidable.

However, no passerby has ever been injured or killed by a wind turbine. The German parachutist, though a member of the public, was not a passerby: someone who walks or drives by a wind turbine and is inadvertently injured. The wind turbine that killed the parachutist could easily have been a building, tree, or cell-phone tower. While terribly tragic, the child killed in Ontario was playing on a component of a wind turbine. The lesson here is that children shouldn't play around machinery of any kind. Similarly, the teenager killed in Ohio was climbing the wind turbine as a prank, reinforcing the need to prevent unauthorized people from climbing the tower.

There remains some, albeit minor, risk for neighbors and passersby. For example, some wind turbines have thrown their blades. Few have, it is true, but as turbines have grown larger, the consequences of any accident have grown as well.

Despite their hazards, wind machines are no more dangerous than many other aspects of modern life. We have all grown to accept the hazards of the electricity and natural gas that flow through our homes. Yet accidents with these common energy sources, though not frequent, are certainly not rare. Common do-it-yourself projects, such as painting the eaves or repairing the family car, are just as dangerous as working on a wind turbine.

Treat wind systems with the same respect you would give any machine, and work as though your life were at stake—because it is. Safety equipment must be used, and used properly, before it's of any value. A safety harness, lanyard, or hard hat is no good when left on the ground.

Tower Safety and Fall Protection

Work on the tower poses the most risk to those who install or service wind turbines: the possibility of accidents is greatest, and the severity of possible injuries is highest. Anyone who has hung from the top of a slender guyed tower in a strong wind readily appreciates this. So, let's turn to work on the tower.

The most reliable way to avoid accidents is to avoid the hazard. Manufacturers of small wind turbines should strive to eliminate working on the tower altogether. Wherever possible, for example, assembly should be completed on the ground and the wind turbine and tower erected as an entire unit. Similarly, performing maintenance on small wind turbines while they are atop the tower should be avoided as much as possible by either using a hinged tower that can be lowered to the ground or by using an aerial lift.

Raising and lowering hinged towers create their own not-insignificant hazards. But tilt-up towers and integrated micro and mini wind turbines with few moving parts are beginning to make the goal of minimal maintenance on top of a tower a reality. Still, there are some small wind turbines installed on fixed guyed towers or on freestanding lattice towers, requiring inspection—if not maintenance—at the top of the tower.

Positioning Belts and Full-Body Harnesses

The term *safety belt* is often used incorrectly to describe any belt or harness used where a fall hazard exists. The so-called safety belts once used on construction sites and lineman's belts are more correctly labeled *positioning* or *work belts* (see Figure 17-2. What's wrong with this picture?). These belts free the hands, and, when used with care, can restrain falls.

Work belts were once made from wide leather straps; today they use synthetic webbing buckled around the waist. On each side of the belt are large metal D rings. One or two lanyards link the wearer of the belt to the tower via attachments at the D rings and to anchorages on the tower. When you're using a positioning belt, your legs carry most of your weight. However, these belts were never designed to safely stop or arrest a fall. They are simply a tool to help perform the task at hand. Should a fall

Figure 17-2 What's wrong with this picture? When the Alternative Energy Institute's Ken Starcher showed the author how to replace the brushes on a rebuilt Jacobs in 1979, we gave little thought to workplace safety. There was no fall-arrest system, no work platform, and no anchors for the positioning belts; nor were we were trained in how to work on the tower safely. The positioning belts used here were intended to restrain the wearer from falling. These belts were not designed to safely arrest a fall, if one occurred, but to aid working at heights. Today full-body harnesses and energy absorbing lanyards would be used in addition to the positioning belts. A work platform would have made this job much easier–and safer. To climb this tower now, both of us would have had to take and pass a training program in tower safety.

occur, you'll need more protection than a work belt can provide.

Two decades ago, many of us didn't know the difference between a belt and a harness. We thought a work belt was sufficient for working safely at heights. Since then we've learned that during a fall a tremendous shock is transmitted though the lanyard or lifeline to the work belt and, hence, to the pelvis. An unconscious wearer may flip upside down after a fall and slip out of the belt completely.

Today, anyone who works around wind turbines should use a full-body harness (see Figure 17-3. Full-body harness). Harnesses have been required as part of fall-arresting systems in the United States since 1995 and are also required in Europe.

Full-body harnesses are designed to safely arrest a free fall, using leg and chest straps to distribute the shock of the fall over the entire torso instead of only the pelvis. These harnesses incorporate D rings on the back and on the chest. In combination with a lanyard and appropriate anchorage, the D ring on the back is designed to keep the wearer upright after a fall. The chest D ring is used when climbing a ladder.

Figure 17-3. Full-body harness. Combination full-body harness for fall protection (back and chest D ring) and positioning belt (side D rings). The chest D ring is used when ascending and descending a ladder. (MSA)

Combination harnesses include both fall-arresting and positioning components. Work on the towers of small wind turbines often requires standing for uncomfortably long periods on a narrow ladder rung or cross-girt. Some harnesses include a seat sling or strap across the buttocks that allows hanging partially suspended from the positioning lanyards. These positioning belts take some of the load off the legs and make working at height more tolerable. (A better solution than a sit harness is a full work platform or an aerial boom.)

In the United States, the Occupational Safety and Health Administration (OSHA) sets standards for work belts and personal fall-arresting systems, including lanyards, carabiners, and anchors. Sit harnesses used for sport climbing have been used by some homeowners and do-it-yourselfers while servicing their wind turbines. However, anyone working on a tower or a wind turbine in the United States should use only OSHA-rated equipment.

Lanyards, Lifelines, and Anchorages

For a full-body harness to safely arrest a fall or for work belts to free the hands, they must of course be connected to a suitable anchorage on the wind turbine or tower. To do so often requires a lanyard (see Figure 17-4. Lanyards). To comply with OSHA in the United States, a fall-arresting system must be capable of limiting the forces on the body to less than 1,800 pounds (~ 820 kilograms) when stopping a free fall. Such a system must be used when the hazard of falling can't be controlled by railings or floors. To meet this standard, lanyards should use an energy-absorbing element.

Unlike lanyards, which can be used for either positioning or to arrest a fall, lifelines are used only for arresting a fall. On a wind turbine, a lifeline, or fall-arresting cable, runs the length of the tower and is anchored at the top and the bottom of the tower.

For maximum fall protection, make sure that lanyards and lifelines are always at their peak strength. Rope and cable serve many functions during the installation and service of a wind system. For this reason, lifelines and lanyards used for fall protection should be dedicated to that application and not used for any other purpose.

Repeated use, dirt, chemicals, sunlight, oil, and grease all weaken rope and synthetic webbing. Keep lanyards and harnesses clean and replace them regularly.

Lanyards or lifelines must be attached to a fixed structural element rated for fall protection. On modern medium-size and large wind turbines, anchorages for fall-protection lanyards are easily identifiable and appropriately labeled (see Figure 17-5. Lanyard anchorage).

Note that on some large wind turbines, a low rail surrounds the top of the nacelle, and similar rails can be seen on the hubs of some machines as well. The rails on some turbines, although possibly useful for positioning, are not anchorages rated for fall protection. Never assume that a rail is rated for fall protection unless the manufacturer has assured that it will arrest a fall. Even then inspect the anchorage first before using it. German technicians have found severely corroded bolts in some anchorages on wind turbines at coastal locations. The German windsmiths concluded that some of these bolts would not arrest a free fall.

Wind plant managers instruct their windsmiths that they must be connected to an anchorage whenever there's a fall hazard. On top of the nacelle of a modern large turbine, for example, the windsmith must use two lanyards, says Mike Kelly, a longtime operations manager

Figure 17-4. Lanyards. Top: Positioning lanyard is used to restrain a fall and allow use of the hands. Twin snap hooks attach to side D rings of a positioning belt. Single snap hook attaches to anchorage. Bottom: Energy-absorbing lanyard is used as part of a fall-arresting system. One snap hook is attached to rear D ring of a full-body harness, the second to the anchorage. (MSA)

Figure 17-5. Lanyard anchorage. Clearly labeled lanyard anchorage inside the tower of a Vestas V82 in the Manawatu Gorge near Palmerston North, New Zealand.

in the wind industry. When moving from one position to the next on top of the nacelle, for example, one lanyard always remains attached to an anchorage. An aide-mémoire, says Mick Sagrillo, is "One [lanyard] for me, one for the wife and kids."

Small wind turbine towers, except for those built in Denmark and Germany, seldom offer ready lanyard anchorages. Ian Woofenden, a senior editor at *Home Power* magazine, argues that small wind turbines in North America, when not installed on tilt-up towers, often use guyed lattice masts or freestanding lattice towers, which offer ample anchorages. "The whole top of the tower is an anchorage," says Woofenden. Possibly, but give careful thought to how you use those anchorages. Woofenden is a skilled arborist; he's learned how to work safely at heights.

Avoid the temptation to simply wrap your lanyard around the top of the tower like a lineman on a utility pole. Should a fall occur, you'll slide down the mast until you reach the first set of guys. This is a common hazard to linemen on wooden utility poles in North America when their climbing gaffs "cut-out" and they drop straight to the ground with wooden splinters flying. Make sure you pass your lanyard through the tower—not around it.

Snap Hooks, Carabiners, and Slings

As the name implies, *snap hooks* are shaped like hooks and can be snapped onto an anchorage, a tower leg, the cross-girt of a lattice tower, or the D ring of a body harness. Riggers, windsmiths, and others in the construction industry use snap hooks for positioning and fall protection. *Carabiners* or snap links may also be used if they're rated for fall protection. Carabiners are useful for clipping gear onto a work belt. Any time a rope must be fastened and unfastened a number of times, a carabiner can make the task easier.

All snap hooks and carabiners used for positioning and fall protection should have both a spring-loaded gate (keeper) and a latch or fail-safe lock, says CalWind's Jon Powers (see Figure 17-6. Locking snap hooks). The fail-safe or double-acting lock prevents the gate from opening inadvertently and releasing from the anchorage during a fall.

Sewn loops of synthetic webbing or slings were developed after climbers learned that a great deal of valuable rope was being lost when constructing rope harnesses. Slings can be used to carry equipment up the tower, tie down the rotor on small wind turbines, and perform any number of other tasks. Wider and stronger polyester slings are used extensively in rigging.

Figure 17-6. Locking snap hooks. Snap links (carabiners) and snap hooks with locking gates can be applied and released with one hand. The lock prevents the gate from unexpectedly opening and releasing. Left: Double-acting snap hook for attaching to anchorage ring. Right: Rebar snap hook. (MSA)

An aide-mémoire, says Mick Sagrillo, is
"ONE [LANYARD] FOR ME, ONE FOR THE WIFE AND KIDS."

Figure 17-7. Fall-arresting cable and sleeve. Windsmiths attach the chest D ring on their safety harness to the carabiner before climbing the ladder inside this Vestas V44 tower in Traverse City, Michigan. When windsmiths ascend, the sleeve is below their harness (lever points up), and the sleeve slides freely up the cable. Should a windsmith fall, the sleeve locks onto the cable (lever points down as shown here), arresting the fall. Many fall-arresting cables and rails are positioned in the center of the ladder, and the chest D ring is attached directly to the sleeve with a carabiner. All towers that must be climbed to service the wind turbine should have a fall-arresting system. Note that the carabiner uses a locking gate.

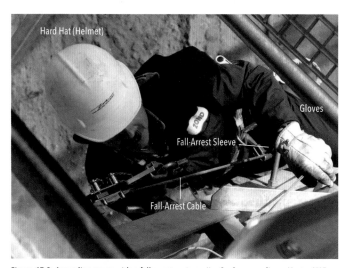

Figure 17-8. Ascending tower with a fall-arrest system. Jim Spohn ascending a Vestas V15 on the Zond site in 1984 using a fall-arrest system. Note the fall-arrest sliding sleeve and the fall-arrest cable. There is an empty sleeve at the top of the cable that the author had used to reach the nacelle. Spohn is fully outfitted with gloves and a helmet. Today, he would wear a full-body harness and not just the belt seen here.

Where erection jigs or fixtures are not handy, slings (appropriately rated) are a good choice for lifting loads such as small wind turbines or a completely assembled turbine and tower.

Fall-Arresting Systems

The most dangerous activities involved in working on a wind system are ascending and descending the tower. Amateurs often compound the risk by scaling the tower, then securing their lanyards after they have reached their workstation. Such a practice is even more dangerous when descending. After several hours on the tower, you're tired and your timing can be off, particularly in winter when biting winds quickly sap your strength.

Several manufacturers offer devices designed to mitigate this climbing hazard. These fall-arresting systems employ a sleeve that slides along a taut steel cable that spans the height of the tower (see Figure 17-7. Fall-arresting cable and sleeve). You attach your harness directly to the sleeve with a snap hook or carabiner. The sleeve rides up the cable as you climb the tower. Should you slip, it locks onto the cable, arresting your fall.

Fall-arresting systems on wind turbine towers typically use steel cable (see Figure 17-8. Ascending tower with a fall-arrest system). However, some fall-arresting systems use a steel rail. The latter has the advantage that it can be installed on tower sections along with the access ladder before the tower sections are raised into place. As the tower sections are bolted together, the rails are also joined, providing continuous protection for those installing the tower and turbine. A fall-arresting rail permanently a part of tubular tower sections could have prevented Richard Zawlocki's death near Palm Springs. No fall-arresting system was in place when he descended the tower, and he failed to use his lanyard to protect himself. However, the joint between sections of fall-protection rail are the weak link in this system, and there has been a least one fatal fall from a fall-protection rail when the sliding sleeve slipped out at a joint between sections of rail.

In practice, these fall-arresting systems protect windsmiths as they climb the ladder to the enclosed work platform found on most medium-size wind turbines and the enclosed nacelle on large turbines.

The use of two lanyards can be an alternative

"YOU MUST NEVER RELY ON YOUR HANDS ALONE,"
says Mike Kelly, who has climbed his share of towers.

to fall-arresting cable systems. When ascending, one lanyard is attached above as far as possible. When it is reached, the second lanyard is attached above and then the first removed, and so on. This ensures that you are always tied to the ladder rung or tower cross-girt—even when reattaching a lanyard. However, a fall-arresting cable and sleeve system is always preferable because it is so much easier to use and, therefore, more likely to be used.

As noted above, the two-lanyard technique illustrates a good overall safety practice—always keep one lanyard attached. "You must never rely on your hands alone," says Mike Kelly, who has climbed his share of towers. Like other fall-arresting lanyards today, these lanyards must be capable of absorbing the energy of a fall.

When servicing most small wind turbines, because they often lack fall-arresting systems, you will need to use the 100% tie-off technique.

Fall-arresting systems are used to prevent serious injury should you fall. They should not be used as a work tool for positioning, say fall safety experts. Where a lanyard is used for positioning, then a separate fall-arresting system should be employed. Work tools and fall-arresting systems should be independent. F. Nigel Ellis explains in his book *Introduction to Fall Protection* that "if you put your weight on it, it's for positioning."

To recap, for tower work, positioning lanyards and a work belt allow you to tie off, freeing your hands. You'll still need a full-body harness and fall-arresting system, such as a self-retracting lifeline anchored above you, to work safely. Although these principals are acknowledged among professional windsmiths, few homeowners and farmers are aware of their relevance to working safely with wind energy.

Work Platforms

Most medium-size wind turbines incorporate a work platform at the top of the tower to aid in

Figure 17-9. Work platform. Even on this early Vestas V17, the Danish manufacturer provided a sturdy work platform with railing. This wind turbine has operated for nearly three decades at Ringkøbing's sewage treatment plant and was operating when this photo was taken in 2012. Note the tubular tower used for its aesthetic appeal in this urban location. Most Vestas turbines of the day were installed on lattice towers.

TOWER WORK AND DO-IT-YOURSELFERS

Any wind turbine and tower that cannot be safely lowered to the ground for servicing should have a fall-arresting system for ascending, descending, and working atop the tower, a sturdy work platform, and safe, clearly identifiable anchorage points for attaching your lanyard. Homeowners, farmers, and do-it-yourselfers should stay off towers of any type unless they've received training in tower safety.

servicing the machine (see Figure 17-9. Work platform). Contemporary large turbines are now big enough that most work is performed from safely inside the nacelle. On these turbines, the body harness and energy-absorbing lanyard are used principally to protect against a fall while ascending and descending the tower. They are also used when working on the top of the nacelle or wherever there's the risk of a fall.

Small wind turbines, especially those manufactured in North America, seldom provide a work platform at the top of the tower. As a result, working on the tower of a small wind turbine demands a combination harness that can safely arrest a fall,

Figure 17-10. Wooden work platform. Even the most rugged wind turbines in the most out-of-the-way places need occasional service, and a work platform makes the job easier. Here, an old Aerowatt stands on a makeshift tower at a ranger station guarding rare Magellanic penguins in Chile's Patagonia region in 1998. (Nancy Nies)

Figure 17-11. Minimalist work platform. The simple work platform on Endurance's E3120 includes lanyard anchors but no guardrail. This 50-kW turbine is operating in a sheep paddock on Draffanmuir Farm near Glasgow, Scotland, in 2012.

while also freeing the hands for assorted tasks. Working at the top of the tower without a platform also demands stamina, as anyone who has done it can attest.

Ty McNeal knows. He's installed a host of wind turbines in Iowa, from household-size machines, like the Bergey Excel and Jacobs 20 kW, to medium-size Danish wind turbines. But McNeal has tired of installing and servicing the smaller turbines, in part because they lack a work platform. "I just don't want to hang on to a toothpick anymore," says McNeal. "At least with the 65-kW [Danish] turbines, there's a solid platform to stand on."

However, household-size turbines manufactured in Europe often do include work platforms, modest as they may be. For example, there's a work platform with handrail on Gaia's 11-kW, 13-meter (43-foot) diameter turbine (see Figure

Figure 17-12. Boom or aerial lift. Technicians service an experimental small wind turbine at the Folkecenter for Renewable Energy in Denmark using an aerial lift. The platform on the boom lift includes a railing, lanyard anchorages, and toe guard. (One technician is on the work platform.) 2012.

2-6. Small commercial-scale wind turbine), reflecting its European origin.

When towers don't have work platforms, enterprising owners make their own (see Figure 17-10. Wooden work platform).

In consideration of the technicians who have to service the machines, and in the interest of safety, any wind turbine that requires on-tower service—regardless of the size—should provide a work platform of some kind—even if minimalist. Two examples of the minimalist approach are North American manufacturers Northern Power and Endurance. Northern Power's NPS 100 employs a simple work platform on either side of the nacelle for reaching the rotor. They also provide handholds and lanyard anchorages (see Figure 5-33. Direct-drive, medium-size turbine). Similarly, Endurance provides a simple work platform on its medium-size turbine (see Figure 17-11. Minimalist work platform).

In the absence of a work platform, a boom or aerial lift provides a comfortable and secure working environment (see Figure 17-12. Boom or aerial lift). Lift platforms include a guardrail, toe guard, and lanyard anchorages. These lifts have become increasingly common for servicing wind turbines of all types and sizes. They also eliminate hazards associated with climbing the towers of small wind turbines.

Ladders

As with work platforms and fall-arresting systems, small wind turbines installed in North America on guyed lattice masts lack a ladder. You're expected to use the mast's cross-girts to scale the tower. Freestanding lattice towers, on the other hand, often provide climbing pegs.

Tubular towers on medium-size and larger wind turbines include an interior ladder. Some manufacturers place the access ladder near the tower wall. Windsmiths climb the ladder with their back to the wall. This enables them to rest by leaning back and placing their shoulders against the inside wall of the tower.

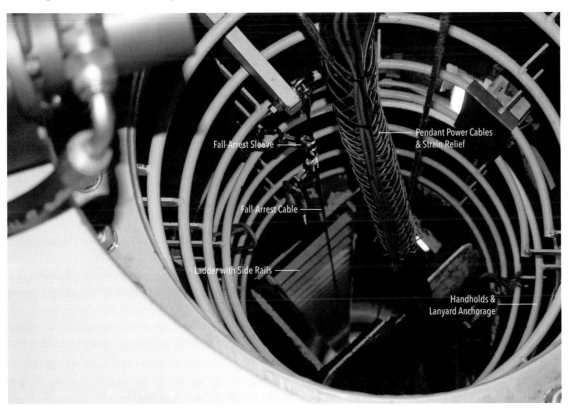

Figure 17-13. Tower top interior. This Windflow 500 turbine near Christchurch, New Zealand, illustrates care for those who have to service the machine: two fall-arrest cable and sliding sleeves, handholds and lanyard anchorage, ladder with side rails, and pendant cables supported with a strain relief at the top of the tower. Two people are in the nacelle, so there are two empty sleeves on the fall-arrest cable.

Ladders should include side rails—so your foot doesn't slip off the rung and into midair. Foot pegs should be avoided as should ladders without full side rails. Climbing a tall tower is challenging enough without worrying about your foot slipping off the end of the peg. In the safety trade, foot pegs and ladders without side rails are called "gut rippers" for a reason. They are particularly dangerous when wet or if your boots are muddy. It's not that much more expensive to install an approved ladder with a fall-arrest system than it is to use foot pegs, whether access to the nacelle is on the outside or the inside of the tower (see Figure 17-13. Tower top interior).

Most manufacturers of commercial-scale turbines now provide rest platforms at regular intervals. This detail is much appreciated when you're climbing a 50-meter (160-foot) tower. Many large turbines now also include elevators. In this—as in so many other aspects of wind energy—the Tvind turbine was way ahead of its time. They've used an elevator since 1978.

More Tower Tips

Besides combination harnesses and lanyards, there is other gear that can make tower work both safer and more comfortable. Boots are one, gloves another. Always wear boots with firm, nonslip soles. Your feet tire less and are less likely to slip from a girt or ladder rung than with street shoes. Gloves do more than protect the hands: they help get a better grip, and a good grip is paramount. Leather is best. The galvanizing used on towers forms droplets on the steel before it cools. These droplets can be sharp as a knife and cut through cloth gloves with ease.

Hard hats or helmets are also essential attire. Admittedly, they are uncomfortable—particularly in winter—and they're difficult to wear in a high wind unless fitted tightly or used with a chin strap. However, their value becomes apparent when you're working around small wind turbines that lack parking brakes.

Most micro and mini wind machines do not have parking brakes. Even when furled or dynamically braked, the rotor may still spin slowly. Those blades may not look like much, but they can easily knock you off the tower. The rotor drivetrain contains a lot of inertia when it is turning, and this inertia can drive a lightweight blade with damaging force.

A similar problem is the unexpected yawing of the turbine in gusty winds. Just when you think you're clear of all that machinery, the wind will change direction and bring everything swinging your way. It's then that a hard hat and a fall-protection system are truly important. Larger, commercial turbines feature parking brakes, rotor-locking pins, and yaw brakes to prevent the nacelle from yawing unexpectedly. For small wind turbines that must be serviced on the tower, rotor locking pins and yaw locks should be required.

DYNAMIC BRAKING OR STOP SWITCHES FOR SMALL WIND TURBINES

One fundamental rule of working with wind turbines is to never go near a spinning rotor. Period. Unfortunately, many small wind turbines, even some household-size turbines, lack a mechanical brake that can stop and hold the rotor. If you have to work near a small wind turbine that doesn't have a mechanical brake, furl the rotor or apply a dynamic brake, and then only approach the rotor in calm weather. Occasionally, neither furling nor dynamic braking will entirely stop the rotor from spinning, but these measures can bring the rotor under a semblance of control.

Manufacturers of mini and micro turbines seldom provide the ability to manually furl the rotor. Dynamic braking, effected by shorting the phases in the stator of permanent-magnet alternators, is the only means available for controlling the rotor on many micro and mini turbines. However, the effectiveness of dynamic braking or stop switches in bringing the rotor to a halt depends upon the magnets used. In the past, the strength of ferrite magnets in permanent-magnet generators was insufficient to stop the rotor under all conditions. The introduction of powerful neodymium magnets makes dynamic braking more effective than before.

However, Mick Sagrillo warns that dynamic braking may not work when you need it most: during a storm's high winds. Strong wind may overpower the generator, says Sagrillo, spinning the rotor and causing potentially damaging current in the generator's windings. Like Sagrillo, Mike Bergey of Bergey Windpower cautions that dynamic brakes must be used with care. If the rotor doesn't reduce speed quickly, the switch must be reopened immediately to avoid damage to the alternator.

Scoraig Wind Electric's Hugh Piggott suggests putting an ample dump or resistive load directly on a battery-charging wind turbine instead of shorting the phases together. Piggott says this will provide more braking torque than a dead short. Once the rotor begins to slow down, the brake switch can then be applied to short the windings.

The National Electrical Code (2014 NEC) now requires a stop switch for wind turbines greater than 50 m². This is equivalent to a household-size wind turbine with a rotor 8 meters (26 feet) in diameter.

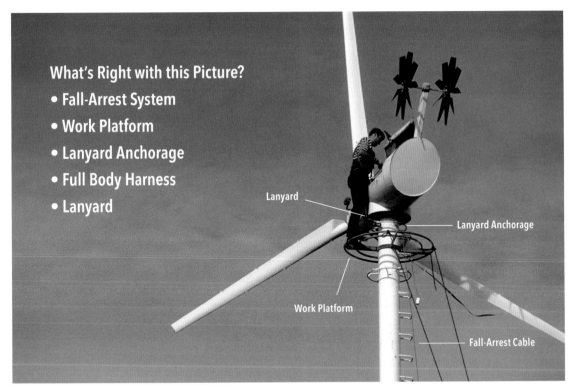

Figure 17-14. What's right with this picture? Niels Ansø servicing the Folkecenter for Renewable Energy's 7-kW turbine used for domestic heating in 1998. Nearly two decades after the scene pictured in Figure 17-2, working conditions for technicians servicing small wind turbines had greatly improved–at least in Denmark. Ansø used a fall-arrest system to climb the tower. He's standing on a work platform. His full-body harness is attached to a lanyard and lanyard anchorage, and he's wearing leather gloves. Ansø has also parked and tied off the rotor so it won't rotate, and he has two hand lines for bringing parts up to the work platform. He should also wear a helmet, the tower should have an approved ladder with side rails, and the turbine's rotor should have a rotor locking pin (he then wouldn't need to tie off the rotor) and a yaw locking pin.

Never work alone. Always have someone nearby who can go for help if you need it. Hand-held radios are a useful tool for talking with your ground crew. Even a slight breeze makes it hard to hear commands from the top of a tower.

If the tower doesn't have a cable and sleeve device, climb exterior ladders on the windward side whenever possible. The wind will force you into the tower, not off it. Never climb the tower in high winds. At California wind plants, for example, no work on exposed lattice towers is performed in winds above 25 to 30 mph (~ 12 m/s).

Keep the base of the tower clear, in case a tool or some lost parts come hurtling to earth. No one should work at the base of the tower while someone is working above. One California windsmith—who wasn't wearing a hard hat—received a serious head injury when his colleague dropped a bolt from the top of the tower!

As for tools, always carry them in a tool belt; keep them on a tool loop, or in a bucket. All other items should be hoisted up with a rope once you are safely in place. Take a hand line up with you on your first trip. Then use a nylon or canvas bucket to ferry small parts up and down the tower or to hold parts while you're working (see Figure 17-14. What's right with this picture?).

When around rotating machinery, whether it's a wind machine at the top of a tower or a bicycle in the garage, don't wear rings, watches, necklaces, loose clothing, or long hair (tuck it under your hard hat if need be). Hugh Piggott in *Windpower Workshop* recounts a "hair-raising" tale told by Mick Sagrillo of the encounter between Sagrillo's former ponytail and the slowly turning shaft of a Jake brake in an Alaskan shop. Sagrillo lived to tell the tale, minus his ponytail.

Never climb a guyed tower that is not properly secured to its anchors. This may seem patently obvious, but Steen Aagaard's accident shows that it's not. Stay clear of the tower during ice storms or freezing rain. If operating, the rotor will shed ice by throwing it to the ground. Ice projectiles typically strike directly below the wind turbine and can be a hazard in the immediate vicinity of the tower. Never climb a wet or ice-covered

STEEN AAGAARD'S CRIPPLING FALL

"Two Hurt in Turbine Accident" blared the headline in the *Bakersfield Californian* on October 23, 1999. "The tower toppled while it was being erected, horrifying onlookers," continued the article by a correspondent at the scene. The article was accompanied by photos of the crumpled tower, a crushed pickup truck, and a close-up of Steen Aagaard, 38, clinging to the slender tower as it fell over.

The details of what happened remain sketchy, but the general outline of events is clear from eyewitness accounts and from an accident report prepared by California's Division of Occupational Safety and Health (CalOSHA). Aagaard was installing a Bergey Excel on an 80-foot (24-meter) guyed lattice tower. The Excel uses a 7-meter (23-foot) rotor and weighs 1,000 pounds (460 kilograms). Two sets of three guy cables and anchors are designed to hold the lattice mast of solid steel rod upright.

The turbine, tail vane, and rotor had been mounted while the tower was on the ground. Aagaard, an experienced Tehachapi windsmith, was directing a crane and his crew while raising the assembled turbine and tower when the accident occurred.

Aagaard made sure the mast had been placed on its pier and the guy cables attached, as he had done in the past. He then climbed the tower to release the lifting sling from the crane boom. Prior to climbing the tower, he personally inspected the guy cables, Aagaard told CalOSHA. He climbed the tower wearing a full-body harness and clipped his positioning lanyard to the tower. Aagaard then released the crane without incident. While remaining on the tower, he directed his ground crew in tensioning the guy cables.

At this point one of the cables "came loose from the come-along device" being used, says CalOSHA. The remaining guy cables "caught and held the tower precariously for a moment," says the newspaper account, before "the tower pivoted through the air and crashed to the ground."

Erik Slocum, part of Aagaard's ground crew, was tensioning one of the guy cables with a come-along tool, when it released, says CalOSHA's report. Slocum instinctively grabbed the guy cable and was pitched 15 feet (~ 5 meters) into the air. He suffered minor injuries and was taken to Kern Medical Center, where he was treated and released.

According to CalOSHA, the guy cables were never directly attached to the guy anchors or the guy anchor turnbuckles. Instead, the guy cables were attached to a cam-actuated cable grip. This grip was then attached to a tensioning device, what CalOSHA calls a come-along device. This tensioning tool was then attached to the guy anchor turnbuckle.

Cam-actuated cable grips grasp the cable under tension and release the cable under compression. They are designed for ease of use in rigging to allow quick and frequent take-up of slack in a cable or wire rope. They are not designed or intended for use as anchors or where someone would be at risk should the grip unexpectedly release. In other words, they are not rated for human loads.

The CalOSHA report notes only that the guy cable "came loose" from the tensioning tool. It's not clear whether the cable slipped through the come-along device or the cable grip used to hold the cable. In either case, the guy cable was never securely attached to the anchor's turnbuckle before Aagaard climbed the tower. The turnbuckles are normally used to tension the guy cables after the tower has been set on its pier, but before anyone ascends the tower.

Bergey Windpower's installation manual explains the sequence to be used. Item #16 states "Attach each of the guy wires to its turnbuckle." Subsequently, item #17 advises using "the turnbuckles to move the tower towards vertical and set tension in the guy wires." Finally, item #18 says: "After the guy wires are secure and adjusted, the crane rigging can be released."

Other factors may have played a part. Aagaard was experienced in servicing commercial wind turbines. Installing household-size turbines was a sideline. He and his crew may have been overconfident; the Bergey turbine was a fraction of the size of the turbines they normally serviced. The news media was present, and there was a host of onlookers. Even the most experienced crew can be distracted by curious passersby. And when the media is present, it takes willpower not to "perform."

The onlookers were also too close to the tower and the installation crew. "Many of those present had to rush clear of the tower as it fell," the *Californian* reported. No one except the installation crew should ever be within the tower's fall zone. Item #1 under tower safety in Bergey's installation manual advises that "persons not involved in the installation should stay clear of the work area."

Aagaard survived his fall but suffered crippling injuries. The fall broke his back in two places, paralyzing him below the waist. Eight months after the accident, CalOSHA fined Aagaard's company $450 for violations of its regulations: one of these was for not following the manufacturer's installation instructions.

On July 4, 2006, Aagard challenged Cal-OSHA's description of the accident. "For the record we did use a quick grip and come-along but also had the guy cable looped through the ground anchor and secured with the supplied cable clamps. I would never have let go of the crane trusting only a come-along, they are notorious for failing. Also we checked and rechecked everything because we were making an instructional video for future use."

On September 20, 2011, Steen Aagaard took his own life with a gunshot to the head. He was 50 years old.

---

ladder or tower. Remember, the wind turbine's expendable, you're not.

Dave Blittersdorf, one of North America's more successful wind entrepreneurs, warns that raising and lowering hinged towers demand your full attention. His rules are no bystanders, no news media—and no distractions. He should know. Blittersdorf was one of the pioneers in lightweight tilt-up towers. With hinged towers, the lifting loads are greatest when the tower is near the ground. This is when any component that can fail, will. Though advantageous when you're raising the tower, as it provides the opportunity to test all components under full load, this can be disastrous when you're lowering the tower.

An experienced field crew at the Alternative Energy Institute at West Texas A&M was lowering a 25-kW Carter turbine at their test field outside Canyon, Texas. They had done so many times before. They were good at it, but this time was different. There was a miscommunication. Worker number one thought the turbine was ready to be lowered after checking that the hoisting cable was secured to the gin pole and the tow vehicle, and proceeded to release the turnbuckle connecting the gin pole to the anchor. Meanwhile, worker number two decided to replace the pin connecting the hoisting cable to the tow vehicle without telling worker number one. As a consequence, when the turnbuckle was released, the guy cable was not attached to its anchor, and the hoisting cable was not attached to the tow vehicle. And to make matters worse, there was a photographer in the path of the tower. The tower whizzed by inches from the photographer's head and crashed to the ground. No one was hurt, but there were a lot of deep breaths and red faces. The photographer quickly left Texas and never returned.

Blade Root Doors

One of the characteristics of the Hütter flange for attaching fiberglass blades to the hub of a wind turbine is its large diameter relative to the blade root and the hub. This large diameter opening at the root end of the blade poses its own unique hazard to windsmiths servicing the rotor hub. Wind turbine blades are not solid but hollow, and on large wind turbines they are very long. When working inside the hub of a large wind turbine, for example to service the pitch mechanism, one or two of the blades will be pointing down toward the ground. The root end of the blades are now large enough to swallow a technician who stumbles or falls into the blade root. To eliminate this hazard, the root end of the blade should have a sturdy door or panel that closes off access to the interior of the blade (see Figure 17-15. Blade root door).

Manufacturers or operators of large wind turbines must also ensure that all large-diameter openings in the nacelle or hub are covered. The

Figure 17-15. Blade root door. Panel closes off the blade-root end of a Vestas V47 blade to prevent windsmiths from inadvertently falling into the blade when servicing the pitch mechanisms in the hub. This is not a hypothetical hazard! These doors or access panels are found on the blades of all large wind turbines today.

cavernous blade roots on large turbines are only one such hazard. If a door covers a hatch in the floor of the nacelle, it must be capable of supporting the weight of an adult without failure. There are harrowing tales of floor hatches—used to raise small components to the nacelle—giving way. It's not good enough to just have a door or access panel. It must be designed for the purpose.

Small Turbine Electrical Safety

Micro, mini, and many household-size wind turbines use permanent-magnet alternators.

These alternators produce a voltage whenever the rotor turns—even when disconnected from the load or control panel! Open-circuit voltage can be up to five times nominal voltage. This can be both a hazard to safety and to sensitive electronic equipment. Reduce the risk when servicing the control panel by disconnecting the power supply from the turbine. Install a fused disconnect switch for this purpose (see Figure 17-16. Disconnect switch.) It will come in handy and is often required to meet building codes.

In a battery-charging system, disconnect the batteries as well. If the turbine is interconnected with the utility, also disconnect utility power.

Before you disconnect the turbine from the control panel or the load, check with the turbine

Figure 17-16. Disconnect switch. Throwing the switch to off and putting a lock through the handle ensures that no current is flowing from the switch to the load. This disconnect switch was installed by Dave Blittersdorf on the guyed, tilt-up tower supporting the Bergey Excel at his home near Burlington, Vermont. Because the Bergey Excel uses a permanent-magnet alternator, there is a potential electrical hazard whenever the rotor is turning. Note the junction box beneath the switch box, flexible conduit from the tower to the disconnect switch, bare copper grounding wire connecting tower to a buried ground rod, and hitch pins on tower hinge. Instead of the hitch pins shown here, use a through bolt with locknut.

manufacturer. Unlike Bergey Windpower's designs, few small turbines were ever intended to operate unloaded. Scoraig Wind Electric's Hugh Piggott warns that waiting for a calm to do repairs may not alone be sufficient to protect the wind turbine. Restrain the rotor physically, or by shorting the alternator's phases together to prevent the rotor from generating damaging voltages.

Off-the-grid power systems can experience high current draws and high charging rates. Both conditions require that for safe operation all cabling be amply sized and the connections terminated correctly.

To protect against overcurrent damaging components or creating a fire hazard, fuse all power sources in any installation—off the grid or interconnected. In a hybrid wind and solar system, fuse both sources and fuse AC and DC loads as well as connections to the batteries. For electricians and those who've worked with small wind turbines for many years, this admonition may seem obvious, but it's not.

Several manufacturers offer preengineered, preassembled power panels that include all necessary fusing or circuit breakers for both the DC and AC side of off-the-grid power systems. These panels are part of an encouraging trend toward more standardized and professional DC to AC power systems (see Figure 12-7. Cabin-size power center). Use them.

If you have any doubts about how to properly fuse a part of your power system, or how to make sound terminations, consult the manufacturer or supplier of the component or hire a licensed electrician.

Wind generators produce high voltages. Use extreme caution anytime you open the control box or the nacelle cowling or work around the slip rings. Always turn off the power from all sources before working around electrical components. In any wind system, there is power from both the wind turbine side and from either the utility side or from the batteries.

Use insulated tools whenever possible. Remember, electricity can kill. But if you're working on the tower, the shock itself may not be the greatest danger. Electric shock can cause you to lose your grip. Even if you fall only a short distance, you could be seriously injured.

Before poking your insulated screwdriver into a control panel, check the circuits with a multimeter. You could have thrown the wrong switch by mistake, or you could wrongly assume that someone else has deenergized the circuit. This is a particular hazard during the installation of multiple turbines when there is a lot of activity on the construction site and it may not be clear which turbines have been energized. This is also a problem in wind-PV hybrids with multiple power sources. Always test first.

Take your time, and think about what you're doing. Never wear metal jewelry when working around electricity. Don't wear rings—even inside gloves—watches, or necklaces.

Avoid constructing a tower near utility lines. If you have any doubts about clearance between the tower or the boom of a crane and a power line, call the utility company before you start to erect the tower. This precaution could have prevented Pat Acker's electrocution in Bushland, Texas, when the rebar cage for an Enertech E44 he was moving came in contact with a high-voltage distribution system.

Stay clear of the tower if a storm is threatening, especially an electrical storm. A lightning strike anywhere near the tower will energize all metal components.

Loss Prevention

Like swimming pools, wind turbine towers are considered an "attractive nuisance" in many communities. There's always the possibility that the tower will be scaled by thrill seekers, children, vandals, or those with suicide on their mind. Prevent a tragedy by providing some provision that prevents unauthorized access.

Large wind turbines on tubular towers always include a massive steel door that is normally locked (see Figure 9-19. Tower exterior access).

On small wind turbine towers, anticlimb guards can be purchased from the tower supplier. Or you can improvise. On freestanding, truss towers, and tubular towers with exterior access, you can remove the lower rungs of the ladder or climbing pegs (see Figure 17-17. Anticlimb guard). On guyed lattice masts, you can also wrap the lower section with sheet metal. This is nearly always sufficient. There's no need to fence wind turbine towers. Period.

Protect manual controls at the base of the tower by removing winch handles or chaining them down.

The massive doors on the tubular towers of medium-size wind turbines include locks for preventing unauthorized entry. Some doors, for example some of those in vandal-prone California, incorporate a blind metal cover protecting the lock, discouraging even the most ardent troublemaker.

Because the attachments of the guy cables on guyed towers are so tempting to vandals, treat the threads of bolts to prevent the nuts from being removed. (One way to do this is to peen the exposed threads.) Also install a safety cable through the turnbuckles. This prevents both vandals and normal vibrations from loosening the turnbuckles and releasing the cable.

Figure 17-17. Anticlimb guard. On exposed ladders and climbing rungs, there must be a provision for preventing unauthorized access. The lower section of the ladder has been raised and locked on this Endurance 3120, a 50-kW downwind turbine. There is no need for a fenced exclosure and barbed wire that would mar this pleasing installation in a parking lot on the campus of Taylor University in Upland, Indiana. Note also the fall-arrest cable, the work platform, and the ladder with side rails. 2015.

Similarly, ensure that the hinges on tilt-up towers can't work free. Use hinge shafts or bolts with a bored hole in both ends. Retain the hinge shaft with washers and a through bolt with locknut. Avoid using the retaining pins found on trailer hitches. They were designed to be quickly and easily removed, which is the last thing you want for a pin holding up your valuable wind turbine.

Avoid placing guy anchors in pathways, but where you must do so, consider planting low shrubs or bushes around the anchors. People tend to detour around hedgerows rather than go charging through them. Shrubs can also soften the line between the tower and the anchor. Slip fluorescent plastic guards over the guy cables to make them more visible.

Tubular towers that are to be installed during the winter should be shipped with covers on both ends to prevent ice and snow from accumulating inside. This simple precaution would have prevented Eddie Ketterling's death on an icy winter day in southwestern Minnesota when a chunk of ice broke free from the inside of a tower section killing him.

All fall-arresting components must be inspected prior to use by the end user and should be periodically inspected by someone trained to detect defects. Harnesses and lanyards should always be replaced after they have arrested a fall. Harnesses and lanyards soiled with grease or oil should also be replaced. And periodically inspect bolts supporting fall-arresting anchorages for corrosion or missing fasteners.

This chapter describes why it's important to take safety seriously, where the industry stands relative to other sources of electricity generation, and mentions some of the measures needed to work with wind energy safely. However, this chapter provides only a few general guidelines for working safely around wind turbines, large and small. This material is by no means exhaustive: entire books have been written on fall protection alone (see F. Nigel Ellis' *Introduction to Fall Protection*). If you're prudent and cautious, you should be able to install, operate, and maintain a wind turbine in relative safety. If you have any doubts about your ability to perform tasks safely, don't hesitate to seek professional help or attend a workshop where safe working practices are part of the class syllabus. Our goal should always be a zero fatality rate from wind energy.

18

Operation and Maintenance

Der Wind ist für alle da. (The wind is there for everyone.)

—Gerald Hüttner, German master miller and operator
of a 1.5-MW wind turbine for the past two decades in
Saschsen-Anhalt of the former East Germany

"WIND TURBINES ARE NOT TOASTERS. YOU CAN'T JUST PLUG them in and walk away," says Wisconsin's Mick Sagrillo. And therein lies the problem. Even seemingly simple devices, such as micro wind turbines, require periodic attention—some would say care and nurturing—in order to perform reliably over long periods of time.

Once in service, wind turbines operate automatically, whether charging batteries or delivering megawatts of electricity to the utility network. Small wind turbine rotors typically freewheel, and when there's sufficient wind, spin up to speed and begin generating electricity or pumping water with seeming effortlessness. In high winds, the rotor furls or regulates without any user intervention.

Similarly, the far more complex large wind turbines think for themselves. Each turbine monitors environmental conditions (such as wind speed and direction) and internal parameters (such as the presence of utility power, voltage, frequency, generator temperature, and oil pressure) and responds accordingly. The controller—the turbine's brain—compares its measurements with its programming and, when conditions are acceptable, signals the turbine to begin or continue operation. When any condition exceeds the manufacturer's programmed limits, the controller orders the turbine to begin the sequence necessary to take the turbine out of service: to feather the blades and then apply the parking brake. If this is an "emergency stop," the controller calls the owner or responsible windsmith and transmits a message of what went wrong. The operator can electronically tell the controller to restart the turbine, or the operator can send someone to investigate.

This ability to operate for days, weeks, and months on end without human intervention can give a false impression: first, that the turbines are indestructible—they are not—and second, that they need little or no maintenance. Small wind turbines are particularly misleading in this regard.

For micro and mini wind turbines, there are very few "owner-serviceable" components. However, careful periodic inspection can catch problems before they become serious. Household-size turbines may require more actual maintenance; less for designs using direct drive, and more for those using transmissions, blade-pitching mechanisms, or blade tip brakes.

Large turbines require the same degree of maintenance as any piece of heavy machinery, like a farm tractor or large diesel engine. Again, careful inspection pays dividends in detecting problems before they become costly.

The ruggedness of modern wind turbines also suggests that initial start-up, or first rotation, is less critical than it really it is. Most defects, in both large and small wind turbines, occur during the first few months of operation. This is a critical period in the life of any rotating machine but especially true for wind turbines.

The start-up procedure for a large wind turbine resembles the preflight checklist for an aircraft. A lot's riding on getting it right the first time. Fortunately, the steps are well defined, and the professionals who install these turbines have the experience to do it right.

Small Wind Turbine First Rotation

Installation of small wind turbines is sometimes more haphazard and the installers less experienced. For this reason, try not to let the thrill of starting your new wind turbine get the better of you. As Jack Park advised more than three decades ago, avoid "fire-'em-up-itis," a disease sometimes fatal to wind turbines. True, small wind turbines are ruggedly built to withstand the elements. However, they're still electrical machines. As such, they're sensitive to proper installation. As with most electrical components, you don't get a second chance. An improperly wired inverter or control panel can lead to costly repairs before the wind system has generated its first kilowatt-hour.

Take your time and look over the entire installation. Go over the manufacturer's checklist: all bolts correctly torqued, cables secured, wires connected to the proper terminals, and so on. If everything meets your satisfaction, go through the suggested start-up procedure. Avoid starting the machine for the first time in a high wind. If a problem develops, light wind minimizes the potential for damage and makes it easier to bring the turbine under control. It's also a good idea to keep a watchful eye on the entire wind system the first few days, to make certain that it operates as expected.

Interconnected Wind Systems

You should notify the utility—in writing—several months before installation that you plan to interconnect a wind turbine with its lines. This gives your letter ample time to move through the utility's bureaucracy and get the proper clearances. If the utility requires any special switches or metering, it's better to find out as early as possible in order to minimize costly modifications once the turbine is in place. Expedite the process by calling first and finding the person responsible for "small power producers" or "independent generators." Include in your notification the brand name and model number of the wind turbine, the maximum power output in kilowatts, the operating voltage, the number of phases, and a line drawing of the proposed installation, including any disconnect switches.

The line drawing should show the location of all disconnect switches and protective relays (contactors) relative to the service panel, the kilowatt-hour meter, and the point where the utility's service enters the building or property. Describe how the wind system functions under both normal and emergency conditions (such as a power outage on the utility's lines). The utility wants to know how the design of the wind turbine guards against energizing a downed line and endangering its personnel.

Your letter should address the utility's concerns. It should explain how the wind turbine's controls will automatically isolate the wind turbine from the utility's lines whenever a fault on the wind turbine side of the interconnection is detected, a fault on the utility's line is detected, or when abnormal operating voltage, frequency, or phase relationship is detected.

Additional information on the wind system's power factor or VAR (volt-ampere reactive), current harmonics, and maximum inrush currents may be helpful to the utility, but is unnecessary to guarantee a safe interconnection. The turbine's manufacturer will provide this information to you if the utility insists on it. Most utilities in North America and Europe should have enough experience with wind

For interconnected wind turbines, both large and small,
MONITORING PERFORMANCE CAN BE AS SIMPLE AS
reading the kilowatt-hour meter on a regular basis.

turbines—of all sizes—so that these questions have been answered many times over.

The utility is responsible for determining whether the interconnection poses a safety hazard to its personnel and to the customer, and whether the interconnection will interfere with the utility's service to other nearby customers. To determine this, the utility will want to inspect the installation before the wind system begins operation. Don't panic; this is a reasonable action. In some cases, the utility will accept the inspection report of the local fire underwriter or code enforcement officer and issue you an approval to begin operation until inspectors can get out to your site themselves. Sometimes the utility engineer can offer valuable advice on how the wind turbine can best meet both your needs and those of other utility customers.

Battery-Charging Wind Systems

Stand-alone power systems are much more complex than interconnected wind machines. There are more components (batteries, inverters, and backup generators), more cables, and more connections. There's a lot more room for error than in interconnected systems. Each component demands special attention and should be carefully checked before initial start-up. Consult the manufacturers' service or operation manuals for a start-up checklist on each component. Again, follow the checklists carefully.

Monitoring Performance

One key to overseeing any wind turbine is to regularly monitor its performance. In short, does it do what it is supposed to do? Is it producing the amount of electricity expected? Has its performance degraded over time? Operators of commercial wind turbines in wind power plants use automatic data-acquisition systems for this purpose. They pore over their digital records looking for any sign of abnormality, like a doctor examining X-rays. And as for a deadly disease, early detection and treatment can prevent more serious consequences later.

For interconnected wind turbines, both large and small, monitoring performance can be as simple as reading the kilowatt-hour meter on a regular basis. Records of monthly and annual generation are the bare minimum needed to determine trends. When you're examining the production of a specific wind turbine, it's useful to compare it to like machines in similar wind regimes. Although there's little published data on the performance of small wind turbines, there are reams of data on large wind turbines. For example, private wind turbine owners in Denmark regularly record monthly production from their wind turbines and mail the results to a central clearinghouse. Many owners of wind turbines in Germany do as well.

Data from such a source can't be used to say with certainty how well a specific wind turbine will perform at a specific site during a specific period; however, such historical information remains invaluable. Analysts use this data to estimate a range of production that a given model might reasonably be expected to generate. Such comparisons are particularly helpful in cutting through the marketing hype surrounding some wind turbines. At the least, such data can be helpful in determining the upper limits of what can be expected.

For example, historical performance data from thousands of existing wind turbines can illustrate that a manufacture's claims about a new wind turbine are outlandish in comparison to all the wind turbines that have gone before it. The data doesn't prove that the new wind turbine can't deliver as much electricity as promised but that it is unlikely. Similarly, if the manufacturer asserts that the new wind turbine will produce orders of magnitude more electricity than other wind turbines of comparable swept area, the data suggests that the claims are wildly exaggerated.

Small Wind Turbines

The simplicity of small wind turbines has one drawback. When they're running properly—with no unusual sounds or vibrations—it's difficult to know how well they're performing. Unlike their larger brethren, small wind turbines have little or no metering to indicate their performance. You can make up for this deficiency by installing your own monitoring system and periodically checking the wind system's performance. At a minimum, every interconnected wind system should include a dedicated kilowatt-hour meter. Surprisingly, many don't.

You will also want to install an anemometer on the tower near the rotor (but not so close that the rotor interferes with the anemometer). And you'll also need some means to log the data from the anemometer. The expense of household-size and larger wind turbines justifies the purchase of commercial data loggers, such as Second Wind's Nomad used at the Wulf Test Field, APRS World's data logger, or the many other similar products now on the market. At a minimum, what you want is the ability to collect wind speed data and average it over a period of time. For more sophisticated analysis, you will want to install a watt transducer to measure instantaneous power.

One simple test is to observe the wind speed when the turbine first starts generating. You can observe when it reaches peak power and, subsequently, when it begins to furl or otherwise curtail generation. You're only seeking a gross approximation. First, the wind measured by the anemometer will never be the same as that striking the rotor. Second, wind speeds are always fluctuating. There's a lag between the time when a gust hits the rotor and when the wind turbine responds. Likewise, there's a lag when the wind ebbs and the rotor begins to slow down. Don't be alarmed if your initial observations of wind speed and power don't match the manufacturer's specifications. You need to analyze a great number of observations to make any sense of what's happening.

The data logger stores, processes, and summarizes these electronic observations into something meaningful. What you want the data logger to do is tally numerous measurements of the average of both power and wind speed over either 1-minute or 10-minute periods. The logger then sorts these average measurements by wind speed, which can then be used to compare with the manufacturer's advertised power curve.

Maintenance

"Take good care of your tools, and they will take care of you" is a good practice to live by. The adage can certainly be applied to maintaining your wind turbine, which is unlike most other machines we're familiar with. Consider that most wind turbines will operate more hours per year than an automobile will during its entire lifetime. The 100,000-mile life span of a typical American car is equivalent to 2,000 hours of operation at 50 mph. Wind turbines exceed that during their first year of operation. Most wind turbines operate from 3,000 to 6,000 hours per year. As Mick Sagrillo tells his workshop students; "Life expectancy [of a wind turbine] is directly related to your involvement with the machine."

Small Wind Turbines

There's no concise answer to the question of how much maintenance is required on small wind turbines. Some wind machines are marvels of simplicity and appear nearly maintenance-free. Others are more complex and the level of service required is more obvious. There are some small wind turbines that never seem to work right, or for very long, without a repair. Others operate day in, day out with no problem. The amount of maintenance required depends on the type of wind turbine, its size, and the approach of its designers.

Here are a few guidelines. Rotors with fixed-pitch blades require less maintenance than those using variable-pitch governors. Machines using direct drive require less maintenance than those

> "TAKE GOOD CARE OF YOUR TOOLS, AND THEY WILL TAKE CARE OF YOU" **is a good practice to live by. The adage can certainly be applied to maintaining your wind turbine.**

using transmissions. Freewheeling drivetrains require less maintenance than those where the rotor must be motored up to speed. And those turbines using passive yaw to orient the rotor require less maintenance than those using active yaw drives. If minimizing maintenance is one of the designer's top priorities, it's reflected in the final product. Wind machines destined for remote, battery-charging applications, where maintenance is not only infrequent but also costly, are designed to be as maintenance-free as possible. On the whole, it can be said that small, integrated, direct-drive wind turbines require far less maintenance than large wind turbines. However, the cost of maintenance on a small wind turbine can represent a much greater portion of the small turbine's revenue than that for large wind turbines because small turbines produce so many fewer kilowatt-hours.

Many small wind turbines are nearing a state of hands-off operation. But we're not there yet. In an age where we're accustomed to automatic everything, we should hardly be expected to run up and down a tower in foul weather carrying a grease gun. Yet after developing autos for 100 years, we still service them regularly. We shouldn't expect a machine operating in an environment as punishing as the wind to be as maintenance-free as a refrigerator in your kitchen.

Sagrillo warns his students against being lulled into the false notion that small wind turbines are trouble-free. "Sure, you can go out only once a year and look at your turbine," he says tongue in cheek. "If it's not on the top of the tower, it's on the ground, and you definitely have a problem." Danish wind turbine designer Claus Nybroe agrees. "Of course it has to be maintained, and even repaired a bit from time to time. Even the best [wind]mill will experience some trouble."

You should inspect the turbine and tower at least twice each year, says Sagrillo: once in the spring, after the turbine has withstood winter storms, and once in the fall, in preparation for winter. This gives you the opportunity to detect any problems that need to be corrected. "This is an essential consideration, when it comes to small turbines," says Nybroe, since owners often have to do everything themselves. For example, Robert Gutowski has operated his Bergey 1000 nearly 20 years at his home in Northampton, Massachusetts. Gutowski installed the turbine and tower himself, and his experience has come in handy. He's had to remove the turbine from his 125-foot (38-meter) tower seven times during the past two decades to replace rotor bearings and re-epoxy the turbine's magnets.

If you like, you can do this inspection in a cursory manner from the ground, on a calm day, using binoculars. Check the rotor for symmetry. See if all blades look alike. If they don't, obviously, you have a serious problem. Watch how the turbine changes direction as the wind shifts. Note if the turbine yaws smoothly or abruptly. Erratic yawing can be due to turbulence, in which case the only solution is to install a taller tower. Erratic yawing can also occur if the tower is not vertical. (You can check the plumb of the tower with a level, plumb bob, or transit.) While on terra firma, check that the tower is still properly grounded and that the guy cables (where used) are tensioned correctly. If the turbine was installed on a hinged tower, lower the tower to ground level for a more detailed inspection. Look for cracks, worn fittings, and excessive grease or oil.

Check that all bolts on the turbine and tower are snug. If any bolts or nuts are missing, replace them immediately. (Usually, if there are any missing, you will already know about it.) Check the yaw assembly and whether the turbine can yaw freely or whether it binds in one position. If it does bind up, you may need to grease or replace the yaw bearings.

Maintenance may include little more than tightening loose bolts. Occasionally it may also entail cleaning slip rings, where used, or replacing worn components. If the wind machine has grease fittings, they will have to be greased semiannually or quarterly. Micro and mini wind turbines, however, use sealed bearings and bushings that are designed to last the life of the machine.

If the turbine uses a gearbox, as many household-size and larger turbines do, the oil will have to be changed periodically. This can be messy but need not be. If it's required, make sure it gets done. Commercial wind farm operators have developed techniques that enable them to change transmission oil in a safe and environmentally sound manner. Manufacturers can specify which system works best with their particular product.

Visually check for corrosion and secure connections at all wiring terminations on the turbine, in junction boxes, at disconnect switches, and in the control panel. Use extreme caution anytime you open the control panel, synchronous inverter, or disconnect switch. If the connections look good, leave well enough alone. When closing the door on any electrical enclosure, make sure you don't pinch any conductors between the door and the box.

Balance of Remote Systems

In stand-alone systems, batteries are the single-most maintenance-intensive component, followed by the backup generator. The batteries' state of charge should be routinely monitored. If the state of charge indicates that the batteries are low, remember that lead-acid batteries should never be discharged to more than 20% of their capacity; when they are, permanent damage can result.

Make sure that the backup generator operates properly in order to recharge the batteries when needed. For this reason, it may be necessary to start the backup generator occasionally to keep it well lubricated and to ensure that it will work when you need it.

Periodically it will be necessary to produce an equalizing charge on lead-acid batteries to balance out any voltage discrepancies between cells. Most battery chargers are capable of providing an equalizing charge.

Occasionally top off battery electrolyte fluid with distilled water. Protect battery terminals from corrosion, and keep battery tops clean. Also keep the battery storage area tidy, and don't store anything above the batteries. This will prevent metal objects from falling across battery terminals.

For more information on proper handling of batteries and their maintenance in a remote power system, see *The New Solar Home Book* by Joel Davidson. *Home Power* magazine also

Figure 18-1. Blade service by abseiling. Installation of vortex generators on a GE 1.5 wind turbine blade by technicians abseiling (rappelling) down the blade. This is one way to inspect and repair minor damage to the blade–but it requires a strong constitution. 2013. (Robert Bergqvist, Wikimedia Commons)

carries articles with helpful tips on how to get the most use out of batteries and inverters (see the Appendix for details.)

Large Wind Turbines

Large wind turbines require more, and more highly skilled, maintenance than small wind turbines but less than they did in the early 1980s (see Figure 18-1. Blade service by abseiling). Mike Kelly explains that early commercial turbines required quarterly maintenance while turbines now need only biannual inspections and service. Kelly, who has managed wind turbines in California, Italy, Iowa, and elsewhere, says that scheduled maintenance ranges from simple visual inspection and general housekeeping to more sophisticated measures, such as analysis of the vibration of critical bearings and examination of pitting on the teeth of transmission gears. Kelly also says good housekeeping, while occasionally overlooked, is critical. "If there's oil everywhere, you can't find the leaks" that need to be addressed. "On a multimillion-dollar machine, housekeeping more than pays for itself," he says.

With the advent of sophisticated monitoring equipment, notes Kelly, managers are now able to predict the life remaining on critical components. These components can then be replaced before they fail, on a schedule that minimizes lost production. Such advanced techniques transform costly, unscheduled maintenance into routine maintenance.

Unscheduled maintenance results when a fault takes the turbine out of service, or when the operator detects a problem with the potential to seriously damage the turbine. During a period of strong winds, lost production from unscheduled maintenance costs the operator lost revenue. Wind plant managers, such as Kelly, schedule maintenance to minimize lost production, ideally when the turbines are idle in light winds.

Though manufacturers provide a maintenance schedule, says Kelly, each wind farm operator modifies it based upon his or her actual experience with the wind turbines. He pays special attention to high wear components in the braking and yaw system and to visual inspection of the blades both from the ground and from the nacelle. More rigorous annual

Figure 18-2. Torquing tower bolts. Windsmiths tightening bolts on a truss tower supporting a Zond Z-750 kW wind turbine near Palm Springs, California. Though this turbine is only a few months old, there is grease on the blades from leaky pitch bearings. If the problem worsens, the blades will need to be cleaned.

blade inspections require specialized equipment, cranes, lifts, or rigging that allows both examination and minor repairs with the blade on the rotor.

Kelly says his crews test the torque on 10% of all bolts in a tower annually. They test the flange bolts between sections of tubular towers for example, as well as the foundation bolts. If all meet specifications, he takes no further action. He argues that truss towers, properly designed and assembled, should not require any more frequent bolt tightening than a tubular tower. With more than three decades working in the wind industry, Kelly is the first to admit that many truss towers have not met this standard (see Figure 18-2. Torquing tower bolts).

Tvindkraft

There are now many wind turbines that have been operated and maintained for more than two decades, some for more three decades. Yet few have quite the amazing background of the turbine as the Tvind school in the northwest corner of Denmark's Jutland Peninsula. Tvind's turbine has been in regular service since 1978 (see Chapter 4, "The Great Wind Revival").

Allan Jensen began servicing the Tvind turbine in 1982 (see Figure 18-3. Monitoring Tvind's yaw control). He is probably the longest-serving windsmith on any single wind turbine in the world. Allan, along with his team of coworkers, including students from the Tvind school, perform daily, biannual, and annual inspections.

Though they have detected some pitting on the gear teeth, the gearbox has never needed repair or replacement, nor has the generator. This is striking. Many much smaller wind turbines in commercial service in California, Denmark, and other countries have needed their drivetrain components repaired or replaced more than once.

Each day one of Tvind's technicians dons bright blue Danish coveralls, rides the elevator to the nacelle, and makes a simple visual inspection to ensure that all's well. They check that all lights and meters are functioning. They also inspect the gearbox filter and clean when necessary.

Biannually they test all safety systems and perform emergency stops. They also grease bearings and fittings, and from inside the hub, they look for cracks in the blades and make visual inspection of bolts in the nacelle.

Each year they check bolt torque as part of a scheduled 10-year rotation. That is, every bolt is inspected and torqued once every 10 years. Critical bolts are inspected for corrosion and fatigue. They also inspect and clean the blades and paint metal components and the tower as needed. Every 3 years, they inspect hidden areas for corrosion and check the lightning rods on the nacelle. "We have modified our maintenance regime based on our practical experience with the components at certain intervals. Everything has been very reliable," says Britta Jensen (no relation).

Tvind has taken care of their history-making turbine, and it's taken care of them. In more than 36 years, they've only had to replace the blades and blade-pitch bearings.

Blade and Tower Cleaning

One surprising lesson learned in California during the mid-1980s was the necessity of periodically giving the rotor a good bath. Many early wind turbines installed in California performed poorly, even when operating as the manufacturer expected. The search for lost production reads like a good detective novel. Eventually, engineers identified the culprit: dirty blades.

Nearly all wind turbines had been designed in rainy climates: Denmark, Germany, or the northeastern United States. But it seldom rains in sunny Southern California. In the explosion of new life that takes place every spring, millions of insects hatch and fly off to find a mate. Some find the leading edge of a wind turbine blade instead. And like the windshield of your car as you barrel down the highway at high speed, splattered insects begin to accumulate, leaving a sticky goo. On your windshield, this goo obscures your vision. On a wind turbine blade's leading edge, it disturbs the airflow. A good hard rain washes the accumulated dust and dead bugs off the blade as the rotor turns.

In arid environments, such as Southern California, there's not enough rain to clean the

Figure 18-3. Monitoring Tvind's yaw control. Since 1982, Allan Jensen has been servicing the Tvind turbine near Ulfborg, Denmark. Here he's seen in 2005 monitoring the turbine's yaw control in the nacelle of the 1-MW turbine built by volunteers in 1978.

blades. Once this was realized, regular maintenance began to include washing the blades. On some wind turbines, dirty blades cut production nearly in half. Thus, washing the blades became a necessity.

Some companies developed spray bars that were permanently attached to the tower. Water trucks would then be used to pump large volumes of water through the spray bars while the turbine was in operation, simulating rain. Some firms specialized in washing wind turbine blades. They adapted the high-pressure nozzles used in the utility industry to clean insulators on electric lines so that they sprayed a powerful stream of water at the blade's leading edge. Other firms specialize in rappelling down the blade from the hub with a high-pressure wand, cleaning the blade as they descend (clearly not a job for the faint of heart). Today's airfoils are less sensitive to soiling than they were 20 years ago. Nevertheless, blades need periodic cleaning if hydraulic fluid, gearbox oil, or grease leaks onto the blade.

Towers may need cleaning as well. Some wind turbines' designs—there's no delicate way to put this—are incontinent. Mitsubishi's 250 kW model (of which there were more than 600 at one time near Mojave, California) were notorious for their leaky drivetrain and the oil-stained towers that were the result. Leaking fluids pose a potential environmental hazard if they soak into the soil. Leaks also pose a risk to technicians if they make access to the turbine hazardous. While fluid leaks are a design defect in this Mitsubishi model, other turbines occasionally suffer from a spill and need cleaning as well (see Figure 18-4. Regular bathing). It's always cheaper to prevent leaks in the first place or provide drip pans (for collecting unavoidable leaks) than it is to clean a wind turbine tower.

Painting

At coastal sites where corrosive salt-laden air envelopes the turbine and tower every day, metal surfaces may require periodic painting. Many tubular towers for commercial wind turbines

Figure 18-4. Regular bathing. Cleaning a tubular tower on a large wind turbine. Leaks of oil or grease from the nacelle are difficult and costly to clean. Good design tries to avoid the problem. Note use of gloves, helmet, and eye protection by Mike Martin in 2006, near Tehachapi, California. (Rope Partner)

Figure 18-5. Painting. Treating galvanized surfaces on a freestanding lattice tower near Montefalcone, Italy. Note use of boom or aerial lift. Contrast this with Figure 18-2 where the windsmiths are suspended from their harnesses inside the lattice tower, tightening the bolts.

are shipped with special coatings to ward off corrosion. Some older towers and all freestanding lattice towers use galvanizing to protect the underlying steel. Unfortunately, the coatings on all towers are not of equal quality, and the effectiveness of the best coatings is compromised when nicked or scraped to bare metal. One need only see the lattice and tubular towers in India's Gujarat state on the Gulf of Kutch to see how quickly metal coatings deteriorate in corrosive environments. Even in California's relatively benign Altamont Pass, the marine air has disfigured the ocean-facing surface of some cheaply painted towers.

Under normal conditions, galvanized metal should rarely need treatment (see Figure 18-5. Painting). Painted or coated surfaces may periodically need a touch up, especially if they have been washed repeatedly to remove grease or oil. At some sites, or with lower-quality coatings, towers may need to be repainted during their lifetime.

Cost of Operations and Maintenance

Always budget for maintenance. Commercial wind developers understand this. It's the job of managers like Mike Kelly to devise complex budgets to keep their turbines in service at minimum expense. However, small wind turbines and even medium-size turbines in distributed applications have repeatedly fallen prey to the misperception that the wind is free. Indeed, the wind is free, but the turbine and the cost of maintaining it are not.

Many a wind turbine has been installed under a foreign-aid or publicly supported demonstration program, only to operate intermittently. Grants pay for the turbine and its installation but seldom for its maintenance. Often it is not clear who bears responsibility for the turbine and who will operate it and pay for its upkeep. Consider a cautionary tale from New England.

The high school in Hull, Massachusetts, won a state government grant for an Enertech E44 to help offset the school's electricity bill. The school installed the household-size turbine at the appropriately named Windmill Point near the entrance to Boston Harbor in early 1985. During this period, the turbine only twice produced as much electricity as expected. The turbine's poor performance was attributed in part to the harsh marine environment and in part to poor maintenance by the school. Maintenance and repair of the E44 cost Hull High School $17,000 during its decade of fitful operation. Prorated, this is equivalent to about 2% of the turbine's installed cost, annually, but was insufficient to maintain the turbine properly. The Enertech operated sporadically until mid-1996, when it was removed. Fortunately, Hull's experience with the E44 didn't sour the community on wind energy, and the saga has a happy ending as explained in Chapter 20 on community wind.

Small Wind Turbines

As Claus Nybroe explained, most owners repair their own micro and mini wind turbines. Consumers internalize the cost in the form of their time and effort. Hiring a professional for simple tasks can cost a significant proportion of the benefits these wind turbines generate. On the other hand, the greater generation from household-size turbines, as well as their greater complexity, often justifies professional help.

The cost of operation and maintenance is often given in units of currency relative to kilowatt-hours produced. Some costs are fixed: they remain constant, regardless of how hard the wind turbine works or how much electricity it generates. Other costs vary with how much time the wind turbine is in operation and how hard it works when it is. Costs per kilowatt-hour are often higher in areas of moderate winds and lower in areas with more wind (and thus, more total generated kilowatt-hours).

In a 2010 presentation, Paul Kühn, Fraunhofer Institut für Windenergie und Energiesystemtechnik (IWES), summarized the operations, performance, and cost of 51 small wind turbines in Germany. He found that the average annual reoccurring cost for small and household-size turbines was €0.16/kWh ($0.22/kWh) and for small commercial-scale turbines €0.035/kWh ($0.05/kWh). Of course, this was the average cost during 14 years of operation; the cost was higher for some turbines, less for others. Annual reoccurring costs also include insurance and taxes, costs not always included under the rubric of operations and maintenance. For comparison, Kühn noted that the annual reoccurring costs for large wind turbines in IWES's extensive monitoring program were only €0.015/kWh ($0.021/kWh).

Pacific Northwest National Laboratory, in a study for USDOE in 2013, estimated that the cost to keep a small wind turbine in reliable operation ranged from $0.01 to $0.07/kWh. They found that the costs for small commercial-scale turbines of 100 kW in size averaged $30/kW to $35/kW per year, or roughly $0.02/kWh to $0.025/kWh.

Large Wind Turbines

In the early 2000s, Danish authority BTM Consult estimated that operations and maintenance costs for wind turbines in the 55-kW class—the first truly commercial wind turbines—averaged about €0.03/kWh ($0.042/kWh) during their first two decades of operation through the 1990s. BTM says turbines in the 200- to 500-kW class cost about €0.01 to €0.015/kWh ($0.014-0.021/kWh) in the decade they were in service in the 1990s.

The BTM data is comparable to that reported annually by the Lawrence Berkeley National Laboratory (LBNL) on large wind projects in the United States. In their 2013 report, LBNL found that the cost of operating wind turbines installed in the 1980s averages $0.034/kWh during their third decade of operation (see Table 18-1. LBNL Operations & Maintenance Cost for Large Wind Turbines in Commercial Wind Plants). The cost to operate wind turbines installed in the 1990s averages $0.023/kWh, while the cost of those installed during the 2000s is $0.01/kWh. It's safe to say that the cost to operate and maintain large wind turbines has been at least cut in half, if not reduced by two-thirds, since the early 1980s.

The next chapter on investing in wind energy explores the annual reoccurring costs further. Annual reoccurring costs include operations, maintenance, repairs, insurance, and more. We'll also look at the longevity of wind turbines—how long they will last—as this also affects their long-term economic performance.

Table 18-1. LBNL Operations & Maintenance Cost for Large Wind Turbines in Commercial Wind Plants

When Installed	$/kWh	$/kW-yr
1980s	0.034	66
1990s	0.023	55
2000s	0.010	25

2012 Wind Technologies Market Report, Ryan Wiser and Mark Bollinger, Lawrence Berkeley National Laboratory, August 2013.

http://emp.lbl.gov/sites/all/files/lbnl-6356e.pdf

19

Investing in Wind Energy

Ill blows the wind that profits nobody.
—William Shakespeare, *Henry VI, Part 3,* Act II, scene 5

WIND TURBINES, LARGE OR SMALL, ARE NOT CHEAP. In the language of economics, wind energy is capital intensive. The fuel — the wind—may be free, but buying the machinery to harvest it is not. You have to pay nearly all of the cost of wind energy up front. Yet wind energy can be worthwhile, in both economic and environmental terms. The challenge is finding the right combination of factors that makes a venture into wind energy profitable and beneficial for both you and your community.

When *Wind Energy* was first published three decades ago, it focused on how a homeowner could buy and install a household-size wind turbine. This edition, while continuing to include small wind turbines, has broadened the scope to include large wind turbines that may be owned by farmers or groups of people in a venture of shared ownership. Similarly, the subject of this chapter has been expanded from simply *buying* a wind turbine to how you can *invest* in wind energy.

Because wind turbines must be operated over such long periods, cost so much up front, and demand our attention and care for such long periods, it's much better to view them as an investment rather than simply as a purchase. This is more than a philosophical difference. It governs not only how we design wind turbines and how we care for them, but also how we relate to them. As with marriage, it's important to get this relationship right because, if all goes well, you're going to be together for a long, long time.

In this chapter we'll examine the factors that determine the profitability of an investment in wind energy, whether it's an investment in a household-size wind turbine or a cluster of large wind turbines owned by a group of people. We also cover what you need to know if you're buying a small wind turbine. But before we do, let's revisit the question of how the generating capacity of wind turbines has been rated in the past and how this inexact term can be deceptive to investors in wind energy.

Power Ratings and Cost Effectiveness

The practice in the past has been to describe a wind turbine's size by referring to its generator capacity in watts or kilowatts at some rated wind speed. This power rating was then used extensively in product promotion.

For a moment, though, let's take an excursion into yesteryear. Let's go to Denmark—Roskilde, to be specific—to the Danish Test Station for Small Wind Turbines in 1980 during the first European Wind Energy Association conference. The occasion was a visit to the test station by the conference attendees, including a high-level delegation of Americans.

"Helge, that's a beaut. What size is it?"

"Ten meters."

"What? No, I meant how big is it. My Danish isn't too good."

"It's a 10-meter, but there's a bigger 12-meter down the road. Would you like to see it?"

"We're not going another step until you tell me how big that is!"

"You Americans are so demanding."

"All right, one more time, Helge, how big a generator does it have?"

"Which one?" Helge asked quizzically.

Boy, what a case of jet lag, the American said to himself.

"I want to know how big the generator is. You know, the thing that generates the electricity, the guts of the machine."

Helge's patience was beginning to wear thin. "It has two generators, the largest of which is 30 kilowatts, but it has a 10-meter rotor driving it."

"Phew, I thought I'd never get it out of you," said the exasperated Yankee.

"*Tak* [Thanks]," replied Helge.

This hypothetical exchange probably took place often in the early 1980s as the Danish test center hosted frequent visitors from North America eager to learn about Danish wind turbines.

We find ourselves in this predicament—identifying wind machines by their generator size—because many of the early pioneers in wind technology came from electric utilities or were designing wind turbines for power companies. In common parlance, we refer to power plants by the combined size of their generators. For example, an Inuit village runs a 50-kW diesel generator; it is the plume from a 500-MW coal-fired power plant that clouds the valley; and it was the 900-MW Unit 2 reactor that was damaged at Three Mile Island. (Yes, there was a major reactor accident before Chernobyl and Fukishima.)

Utilities try to run their power plants as close to full load as possible—when they are in operation—so the generators perform as efficiently as possible. As a result, a 500-MW generator nominally produces 500 MW. Engineers understandably use this power rating when they talk to one another.

Wind turbines are different because the wind is a variable resource. Fluctuations in wind speed cause generator output to vary as well. Seldom does a wind generator produce its rated output for any extended period of time. Moreover, optimum rotor and generator combinations depend on the wind regime. A wind turbine with a 10-meter rotor may, for instance, perform most efficiently (deliver the most energy) matched with a 10-kilowatt generator in regions with low average wind speeds, but the same rotor may work better with a 25-kilowatt generator where it's windier (see Chapter 8, "Silent Wind Revolution," for more on this).

Rotor diameter and swept area are better measures of a wind turbine's capability than its generator rating because it's the rotor and not the generator that captures the wind and converts it to a useful form. The generator comes later in the conversion process. Because nearly all wind turbines today use conventional rotors that sweep a circle, rotor diameter becomes ready shorthand for the area swept by the rotor and that's why this measure is used so frequently in *Wind Energy*.

Most Danes, as Helge in our fictional anecdote, refer to the size of their wind turbines by rotor diameter. For example, Danish manufacturer Vestas has consistently designated its various models by rotor diameter in meters. Vestas shipped hundreds of its V15 to California during the state's great wind rush. In the late 1990s, hundreds more of its V27 were installed. By the mid-2010s, the seaside community of Hvide Sande (White Sands) was celebrating the installation of its three V112s—wind turbines that intercept nearly 10,000 m² of the wind sweeping in from the North Sea (see Figure 2-8. Very large wind turbine).

Similarly in Germany, Enercon sold thousands of its pioneering E40 direct-drive wind turbines. It's the E40 because it uses a 40-meter (130-foot) diameter rotor. Enercon continues to offer a full spectrum of wind turbines from the E30 up to its top-of-the-line E126, one of the largest wind turbines in the world.

The awareness of a rotor diameter's importance in designating a wind turbine's size, coupled with a reluctance to part completely with traditional generator ratings, at one time led to hybrid designations that used both rotor diameter and generator capacity. In the early 1980s, for example, Enertech introduced its model 44-40, which used a rotor 44 feet (13 meters) in diameter to drive a 40-kW generator. Lagerwey sold hundreds of its 18/80 model to Dutch farmers in the late 1980s, its designation for the 18-meter-diameter, 80-kW turbine.

The rated power system is not only confusing; it can be dangerously misleading. Until recently there was no wind-speed reference to compare one wind turbine to another. Rated speeds range from 10 m/s (22 mph) to over 15 m/s (30 mph). Further, some manufacturers rated their machines at peak power output; others did not. As noted earlier, wind turbines designed to use aerodynamic stall to regulate peak power often exceeded their rated capacity, sometimes by up to 30% (see Chapter 4, "The Great Wind Revival").

In a few notorious cases, manufacturers took advantage of the emphasis on generator size by adding large generators to relatively small rotors. By using this rating system, it's possible to slap a 6-foot plank on the shaft of a 25-kW generator and call it a 25-kilowatt wind turbine. In one particularly egregious example, Fayette Manufacturing built a 10-meter diameter turbine and saddled it with a 95-kW generator. Most other manufacturers at the time would have rated a turbine of this size 25–35 kW. Years of results from some 1,000 of these machines in California proved that they performed no better and often much worse than other turbines with rotors of similar swept area driving 25-kW generators. After years of poor performance, the Fayette machines were eventually scrapped.

In a few notorious cases, manufacturers took advantage of the
ILL-INFORMED EMPHASIS ON GENERATOR SIZE
by adding large generators to relatively small rotors.

FANTASY WIND TURBINES: IF IT'S TOO GOOD TO BE TRUE . . . OR HOW TO SPOT SCAMS, FRAUDS, AND FLAKES

German engineering professor Robert Gasch calls them fantasy wind turbines. These are the inventions—or contraptions—that bedevil serious wind turbine advocates. They are the "revolutionary" inventions that periodically rise up from the dead whenever the price of oil goes up or there's a power crisis somewhere in the world.

Very few of these ideas are new, and certainly none of them are "revolutionary." While some of these devices may spring from well-intentioned inventors, others are the conceptions of fast-buck artists. David Sharman, a serious wind turbine designer, calls the proponents of such machines "bozos and shysters."

It's often difficult for the uninitiated to tell the difference between the real and the imaginary, the fraudulent from the worthwhile, and that's the problem.

Here are some tips for spotting questionable products. The most important tip to keep in mind comes from Professor Gasch: if there is a new wind turbine, no one should pay the slightest attention to it—and that includes the media—until they "build it, measure it, and publish" the results. Until then, it's just hot air—marketing hype—and nothing more.

How can you identify a questionable wind turbine design? Here are some tips.

- Hype high, low or nonexistent experience
- Aggressive marketing (Look for multilevel or "pyramid" schemes of the "get in on the ground floor now" variety.)
- Celebrity endorsements (What do celebrities know about wind turbines?)
- New design, "not like the others"
- "New" patents
- Drag devices (squirrels in a cage)
- Ducted turbines (funnels)
- No hardware, but a fancy website (Websites are always cheaper than hardware.)
- "Works in low wind"
- Is silent (No wind turbine is silent.)
- Does not kill birds (All wind turbines can kill birds or bats.)

What lessons have we learned from more than 30 years of modern wind turbine development?

- There are no panaceas.
- There are no cheap solutions.
- There are no breakthroughs—no miracles.
- Numbers matter. (Wind energy is always about numbers.)
- Experience matters. (If they haven't been building these things for years, then how do you know that it works like they say it does?)
- Size matters. (You can't get a lot of electricity from a small rotor.)

What we learned, for example, from the Vortec ducted turbine disaster in New Zealand:

- Always check the numbers. (The numbers didn't add up.)
- Always check the references. (The references were discredited in the United States.)
- Always google. (There are ducted turbine critics on the web.)
- Always go to the library—or to your neighborhood bookstore!

In sum, always be wary of "new" designs. There's rarely anything truly new under the sun—or in the wind—and if it seems too good to be true, it probably is.

Giving a wind turbine a high power rating—regardless of its swept area—is a favorite tactic of hustlers and charlatans because it makes their products "appear" more cost effective. They manipulate a simple metric that many people erroneously use to gauge cost effectiveness: installed cost relative to the turbine's generator size—$/kW in North America. By artificially boosting the generator size, they can charge more for their wind turbine while making their product appear more cost effective than wind turbines from legitimate manufacturers.

One of the lessons we've learned from three decades of modern wind energy is you can't get a lot of electricity from a small rotor. It's a tip-off that the product is a *fantasy wind turbine* if it has a high power rating and a small rotor.

Because of questionable power ratings by some manufacturers and general confusion among consumers as to what the numbers mean, the American Wind Energy Association (AWEA) attempted to clarify performance ratings in the 1980s by calling for a standard list of parameters. These included maximum power (not rated) and, most importantly, the annual energy production at various average speeds. AWEA hoped to eliminate the rated power at rated speed nomenclature with values that made more sense. They've had a modest degree of success by at least establishing a standard wind speed, 11 m/s (24.6 mph), at which small wind turbines can now be compared.

Efficiency or Cost Effectiveness

Along with the relative cost per kW as a measure of cost effectiveness, many people also have a disturbing fondness for the word *efficiency*. Invariably, they will say "Yeah, all those calculations are fine. But what's the most efficient turbine built today?" Some promoters cater to this obsession by hyping their wind turbines as the "most efficient" on the market. One that did so was Mag-Wind.

What many overlook when they focus on efficiency is that energy generation is more sensitive to wind speed (because of the cube law) and swept area (because of the square of the rotor's radius) than efficiency. Wind turbines must first be reliable; second, they must be cost effective. Efficiency is important, but it's not the sole criteria for judging the performance of a wind machine.

Wind energy's raison d'être is the generation of clean electricity cost effectively. You could install the most inefficient wind machine ever built, the kind that would bring a tear to an engineer's eye, but if it worked well and was cheap enough, it could be more cost effective than a modern engineering marvel that was unreliable or cost too much. If you deliver lower cost electricity with an inefficient wind machine than with an efficient one, so be it.

As emphasized throughout *Wind Energy*, to gauge the potential of a wind turbine, ignore the size of the generator or its purported efficiency and get right to what matters most: the annual energy production for your site. But if the AEP of a turbine is not certified, for example the wind turbine is intended for a battery-charging system, nothing outside the wind itself, no other single parameter, is more important in determining a wind turbine's capability of capturing the energy in the wind than the area swept by the rotor.

Measures of Cost Effectiveness

There are several measures of cost effectiveness in use: cost per kilowatt of installed capacity, cost per rotor swept area, and cost per kilowatt-hour of energy generated. The most frequently used measure of cost effectiveness is cost per kilowatt. However, it's about as meaningful as the power rating in describing the size of a wind machine. This measure came into use the same way—utility engineers were accustomed to using it. Like power ratings, the cost per kilowatt works well for power plants that run at constant output. But for wind machines, the cost per kilowatt just confuses matters. Worse, it can be easily manipulated.

The cost per kW remains helpful in public policy debates where data on the swept area of wind turbines isn't always available. It's commonly used to track price trends over the years because

Table 19-1. Installed Wind Project Relative Cost in the US: LBNL 2012			
	Average	Low	High
	$/kW	$/kW	$/kW
Large	$ 1,940	$ 1,500	$ 4,250
2012 Wind Technologies Market Report, Ryan Wiser and Mark Bollinger, Lawrence Berkeley National Laboratory, August 2013. http://emp.lbl.gov/sites/all/files/lbnl-6356e.pdf			

Table 19-2. Relative Costs for GE 1.6 MW Platform

		Standard	Low Wind
Platform	MW	1.62	1.62
Diameter	m	82.5	100
Operating Costs	$/kW/yr	$ 60	$ 60
Relative Installed Cost	$/kW	$ 1,600	$ 1,850
Derived			
Swept Area	m²	5,346	7,854
Operating Costs	of Installed Cost	3.8%	3.2%
Relative Cost per Swept Area	$/m²	500	400

Adapted from Ryan Wiser LBNL 2012
http://emp.lbl.gov/sites/all/files/wind-energy-costs-2-2012_0.pdf

Table 19-3. Representative Relative Installed Wind Turbine Costs

	Relative Installed Cost*	
Size	Low ($/m²)	High ($/m²)
Micro	1,500	2,500
Mini	1,250	2,500
Household	1,250	2,250
Small Commercial	800	1,250
Large**	400	600

*Gross approximation only.
**Several turbines in small clusters.

data on installed cost and capacity are more readily reported than swept area. For example, Ryan Wiser at Lawrence Berkeley National Laboratory annually reports on the average relative installed cost of large wind turbines in the United States. In 2012, he found the average cost was about $2,000 per kW (see Table 19-1. Installed Wind Project Relative Cost in the US: LBNL 2012).

With the advent of the silent wind power revolution's large-diameter rotors, the utility of cost per kW of capacity as a measure of cost effectiveness is waning—even for the purpose of tracking price trends. Wiser's 2012 report on the US market reflects this by noting the difference in the relative installed cost of two different GE wind turbines (see Table 19-2. Relative Costs for GE 1.6-MW Platform). In this comparison, it appears that the large-diameter, low-wind turbine is less cost effective in $/kW than the standard turbine, but this is misleading.

Though the turbines appear to be the same—because they have the same rated power—they are not. The low-wind turbine has a much larger rotor (see Figure 8-5. GE 1.6 MW wind turbine). It intercepts nearly 50% more of the wind stream than the standard turbine. The higher relative cost in $/kW incorporates the cost for the larger rotor.

As noted repeatedly throughout *Wind Energy*, a truer measure of a wind machine's size is the area swept by its rotor. Thus, a more useful measure of cost effectiveness is the cost per swept area. As seen in Table 19-2, GE's low-wind turbine is 20% less costly per unit of swept area than the standard turbine. For sites where either turbine might be used, the low-wind turbine would be more cost effective despite its cost per kW being greater.

As a rule, cost effectiveness increases with swept area. Small wind turbines, though they cost less, are relatively more expensive than large turbines. The installed cost of small wind turbines relative to their swept area is three to as much as six times greater than that for large wind turbines (see Table 19-3. Representative Relative Installed Wind Turbine Costs).

The limitation on using cost per swept area is the assumption that all wind machines are equally efficient at converting the energy in the wind to electricity. They are not, but the differences among large commercial wind turbines are not significant despite manufacturer claims to the contrary.

The cost per swept area is just a shortcut for what counts most: the relative cost per kilowatt-hour generated at your specific site. This cost per kilowatt-hour isn't the same as the cost per kilowatt-hour you pay for electricity from the utility or the cost of energy used in public policy debates comparing wind energy to other forms of generation. The cost per kilowatt-hour measure should only be used for comparing one wind machine to another at a specific site. It's not appropriate for comparing a wind turbine to other forms of energy because it doesn't account for all the costs and benefits from the wind turbine over its entire life cycle. It's merely a measure or "figure

AS A RULE, COST EFFECTIVENESS INCREASES WITH SWEPT AREA.

Table 19-4. Hypothetical Relative Costs for GE 1.6 MW Platform

		Standard	Low Wind
Platform	MW	1.62	1.62
Diameter	m	82.5	100
Swept Area	m^2	5,346	7,854
Relative Installed Cost	$/kW	$ 1,600	$ 1,850
Installed Cost	$	2,600,000	3,000,000
Specific Area	m^2/kW	3.30	4.85
~Yield at 7.5 m/s	kWh/m^2/yr	1,000	850
~AEP	kWh/yr	5,300,000	6,700,000
Relative Cost	$/kWh	0.49	0.45

of merit" for comparison shopping—nothing more.

Let's use the example of GE's 1.6-MW platform again to illustrate this. We'll use a typical yield for wind turbines with comparable specific area as the two GE turbines (3.3 m²/kW and 4.9 m²/kW respectively) at an average hub-height wind speed of 7.5 m/s where the two IEC class service ranges overlap (see Figure 11-4. Annual specific yield by IEC class). The smaller turbine yields about 1,000 kWh/m²/yr and the larger turbine about 850 kWh/m²/yr (see Table 19-4. Hypothetical Relative Costs for GE 1.6 MW Platform). The relative cost of the larger diameter turbine is $0.45 per kWh. It is more cost effective than the smaller turbine even though—and this is the point of this exercise—the smaller turbine has a lower cost per kW of installed generator capacity.

It may seem arcane at first, but this is a technique used by professionals to compare the cost effectiveness of various wind turbine models in different wind regimes. Cost-effective turbines at windy sites produce lower costs per kilowatt-hour.

To illustrate how this measure of cost effectiveness is not the same as the cost of energy, LBNL's Wiser and his team, when accounting for all costs over the two-decade expected life of the GE wind turbines used in his example, estimate that the cost of electricity from the new large-diameter wind turbines can be as low as $0.08 to $0.10 per kWh without tax subsidies.

Investors have lost staggering amounts of money, not just homeowners but savvy executives who should have known better, because they didn't grasp these simple concepts of what constitutes cost effectiveness. Even after years of experience in operating wind turbines, and reams of technical documents, there's always an inventor or promoter who has made a startling discovery of how to make a wind turbine more efficient or with a lower cost per kW of installed capacity than any other wind turbine that has gone before. As sure as the rain will fall, there'll be someone willing to part with his or her savings before punching a few numbers into a calculator or scribbling on the back of an envelope to cut through the hype of shysters or bozos.

Small Wind Turbine: Testing and Standards

The example illustrating cost effectiveness used a large wind turbine, specifically an example of the new wind turbines in the silent wind power revolution. As explained in Chapter 11, "Estimating Performance," small wind turbines are less productive than large wind turbines for the same wind conditions. They are also often installed on much shorter towers than large wind turbines, reducing their actual yield from that expected for a given wind regime. Investing in small wind turbines, then, presents a special challenge.

Small turbines make economic sense most notably in battery-charging systems, where they are essential. They also continue to offer promise in interconnected applications. The good news is that the reliability and performance of small wind turbines has made great progress during the past decade. They had a long way to go.

SHYSTERS AND BOZOS

Mick Sagrillo blames "bozos and shysters" for the continual reappearance of VAWTs, DAWTs, and other revolutionary "new" windmills that will produce electricity "too cheap to meter." Bozos, says Sagrillo, are simply clueless when it comes to the engineering and physical limitations of their inventions. "They don't know what they don't know," he adds disdainfully. Shysters, on the other hand, know that their claims are false and are out to milk up-front subsidy programs as quickly as they can or lure unsuspecting and often well-intentioned investors. Shysters always make money, regardless whether their "invention" performs as promised or not. They are adept at moving on—with the money—before the house of cards comes tumbling down.

In the year 2000, researchers at Spain's national energy center, CIEMAT (Centro de Investigaciones Energeticas, Mediombientales y Tecnologicas) took the pulse of the small turbine industry. They were mostly interested in battery-charging turbines, but their findings are relevant to those connected to the grid as well.

CIEMAT's Ignacio Cruz examined the status of small turbine technology, relying in part on an extensive survey of trade literature and in part on measurements at Spain's Soria test field in the highlands northeast of Madrid. He reached several conclusions mirrored elsewhere at the time.

- Small turbines cost more relative to large turbines.
- Small turbines are less productive relative to large turbines.
- Lack of standardized testing has hindered progress.
- There is a need for consumer labeling and certification.

Conventional small turbines cost more than twice as much as large wind turbines relative to the area swept by their rotors. Vertical-axis wind turbines and ducted turbines often cost considerably more than conventional small wind turbines.

Fortunately, all of the other conditions have greatly improved in the intervening years. As discussed in Chapter 11, the performance of small wind turbines has risen dramatically with the introduction of rare-earth magnets. Small turbines using neodymium magnets can deliver 50% more yield in kWh/m²/yr than the older generation of small turbines.

At the time, Cruz had decried the absence of test fields where the "performance and feasibility" of small wind turbines could be measured in accordance with internationally accepted procedures. Such test fields are essential, Cruz said, for optimizing and improving performance. Standardized testing—like that done on large turbines—was critical for providing consumers with the confidence that the turbines will perform as advertised. It's in testing and certification that significant progress has also been made.

Standardized Tests

For years there has been an international consensus that the only way to properly test the performance of small wind turbines is in the field, that is, on a tower at a windy site. Some promoters have hyped questionable products as "wind tunnel tested" to gullible journalists and consumers.

Truck or wind-tunnel testing is useful to designers for observing and fine-tuning the furling or regulating behavior of their small turbines but not for measuring performance. Truck testing, where a small wind turbine is mounted on top of a pickup truck and the truck then driven down an abandoned airport runway, has sometimes also been used in North America. Truck testing is "just one tool in the [designer's] toolbox, but only one," says Mick Sagrillo.

Standards for small wind turbines specifically exclude performance measurements of wind turbines in wind tunnels or through truck tests. Neither replicates how a wind turbine will perform under real-world conditions (see Figure 19-1. Testing small wind turbines).

Wind-tunnel tests usually overstate performance. This phenomenon was seen several times in the 1970s, most notably with McDonnell Aircraft's giromill (see Chapter 6, "Vertical-Axis and Darrieus Wind Turbines"). Consumers never see the performance measured in a wind tunnel.

To limit the damage to the industry's reputation by hucksters and their ever-revolutionary new wind turbines, a committee of the American Wind Energy Association (AWEA) issued the first draft of a proposed performance standard in 1979. After more than two decades of bickering, AWEA finally approved the standard.

AWEA's performance standard was intended "to provide consumers . . . with an equitable basis for comparing the energy production performance and operating characteristics" of different wind turbines. It avoided any discussion of reliability or durability, except as it affected testing of the power curve. (You can't measure a power curve if the turbine's not operating.) While the standard was specifically compiled for small wind turbines, the method is similar to that used to measure the performance of wind turbines of all sizes.

Figure 19-1. Testing small wind turbines. There are several sites worldwide capable of testing small wind turbines to international standards. One such site is the Folkecenter for Renewable Energy on Denmark's northwest Jutland Peninsula. Here is a small wind turbine at the test site in 1998.

At the heart of the standard was a detailed description of how power curves should be measured and how the results were reported. Further, the standard required all manufacturers to prepare a "test report" describing the techniques used to measure their power curves and the results of their measurements. This report would then be available to the public upon request.

American manufacturers' reluctance to embrace any standards whatsoever goes back to the revival of wind energy in the United States in the mid-1970s. This recalcitrance results from a uniquely American fear that standards stifle creativity. Standards never hurt manufacturers of large wind turbines. Very few, if any, large wind turbines sold on the international market can obtain financing without a performance test conducted by an independent testing laboratory. Still, it was decades before standards became the norm for small wind turbines.

Certification and Labeling

The purpose of both standards and certification is straightforward: to verify manufacturer's claims by reporting the results of standardized testing in a consistent manner. Certification is the step manufacturers have to go through to confirm or certify that the tests conducted on their wind turbines conform to national or international standards.

Certification is also necessary in certain jurisdictions in order for a turbine owner to qualify for subsidies and incentive programs. In Britain, for example, certification of small wind turbines is also a requirement for exemption from certain planning requirements that apply to almost everything else in that crowded isle.

There are three classes of standards: design standards, performance testing standards, and noise measurement standards. Complying with all three and certifying the results is an expensive but necessary process for reputable small-turbine manufacturers to reduce the influence of over-hyped, unreliable, and sometimes fraudulent wind turbines.

The design standard is the most confusing to consumers. Compliance with the design standard doesn't mean that the wind turbine won't fail in use, require maintenance, or otherwise not experience problems. The design standard simply says that according to existing engineering knowledge, the wind turbine and its components seem right for the application. Compliance with the design standard is not a guarantee. It certainly is not a guarantee that the wind turbine will operate flawlessly for decades, nor is it a guarantee that the wind turbine will deliver the performance promised.

Electrical standards are a subset of design standards. Most wind turbines that are designed for connection to the grid must have the relevant electrical components, such as inverters, tested and certified to a separate electrical standard. Nearly all building codes and most insurers require conformance to an electrical standard. The primary US standard is the National Electrical Code.

SMALL WIND TURBINE CERTIFICATION

In the United States, the Small Wind Certification Council (SWCC) certifies performance for small and medium-size wind turbines to the American Wind Energy Association performance standard. By late 2015, there were nine wind turbines certified by SWCC in the United States. SWCC publishes both a certificate summarizing the data most useful to consumers as well as provide detailed technical reports on performance and noise measurements (see SWCC Certification for Bergey Excel 6).

Great Britain's Microgeneration Certification Scheme (MCS) certifies performance for wind turbines up to 50 kW in the United Kingdom. In 2014, 31 wind turbines had qualified for certification. MCS also certifies installation companies to ensure the microgeneration products have been installed correctly.

British manufacturers, such as Ampair, publish the MCS certificate as well as detailed technical reports on performance and noise testing.

There are equivalent certification programs in other countries, notably Japan.

NEVER BUY A SMALL WIND TURBINE WHOSE PERFORMANCE HAS NOT UNDERGONE TESTING AND CERTIFICATION.

It has taken three decades of struggle for standardized testing and certification of small wind turbines to arrive. Don't subvert it. Never buy a small wind turbine whose performance has not undergone testing and certification. This is the bare minimum necessary to assure a reasonable probability that the wind turbine will perform as advertised. Standardized testing and certification are the only way we can eliminate the shysters and bozos who plague the small wind turbine industry.

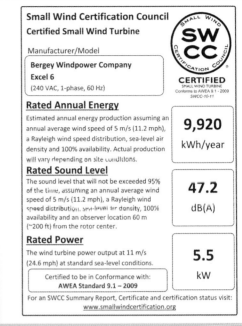

SWCC certification for Bergey Excel 6. Small Wind Turbine Certification Council certificate for Bergey Windpower's 6-meter diameter turbine. The certificate lists power form, rated annual energy production (9,920 kWh/yr) at an average annual wind speed at hub height of 5 m/s, rated sound level (47.2 dBA), and rated power (5.5 kW) at 11 m/s. The current version of all certificates and labels can be found on the SWCC website, www.smallwindcertification.org. (Small Wind Certification Council)

Wind turbines using components that are in compliance with an electrical standard, such as those with a UL label, are not necessarily in compliance with any of the other standards. Some fast-talking promoters have tried to falsely link compliance with electrical standards to a broader approval or endorsement of the wind turbine's basic design. Compliance with electrical standards only suggests that when used in the manner prescribed the equipment won't cause a fire.

Standards and certification for small wind turbines have been a long time coming, but they're finally here. In early 2008, the British Wind Energy Association (BWEA) adopted a certification requirement for small wind turbines. AWEA followed in 2009.

The absolute minimum a consumer should demand of any manufacturer are the results of performance tests conducted in accord with standard international practice. This information should be available either in product literature or on the manufacturer's website. The manufacturer should clearly state whether the tests were done to the international standard. The resulting data can be presented in tabular form, as a power curve, or as a curve of estimated annual energy production (AEP).

Most importantly, data collected on the performance of a wind turbine must be "averaged" over a period of time—10 minutes for large wind turbines, and often 10 minutes for small wind turbines, though 1-minute averaging periods may be acceptable.

It is the averaging of test data that winnows the chaff from the grain in wind turbine testing.

For marketing purposes, some manufacturers of noncertified turbines like to report only the maximum instantaneous power measured at a given wind speed. The manufacturer of the Air Breeze was notorious for emphasizing the instantaneous power, not the average. In reality, the average measured power is often only a fraction of that reached instantaneously.

And the average measured power must include a minimum number of 1-minute averages because the 1-minute averages themselves vary dramatically. For example, measurements at the Wulf Test Field in the Tehachapi Pass found that for the Air Breeze, the maximum 1-minute average power was wildly different from the minimum. Within the wind speed bin of 22 mph (10 m/s), the Air Breeze would produce 1-minute averages from a low of 40 W to a high of more than 200 W. The average of all the 1-minute samples in this wind speed bin was about 170 W.

It was questionable practices, such as using instantaneous measurements, that ultimately led to standardized testing, certification, and labeling. Remember, it's the average power generated over a period of time that determines the amount of energy a wind turbine will produce (energy being the product of power and time)—not the instantaneous power.

Certification does not ensure that the wind turbine will work well in general or work well at a particular site. Performance certification only provides assurance that the wind turbine can produce the stated amount of electricity under the conditions given. Nevertheless, certification is a major step toward raising the quality of small wind turbines and providing consumers the confidence they need to buy the machines.

Buying a Small Wind Turbine for the Home

Many homeowners have no idea how much electricity they consume. Sure, they may know how much money they spend. But that's not the same as knowing how much electricity they consume. Many mistakenly think a micro or mini wind turbine will power their entire home. "Don't expect to spend $500 and become energy independent," cautions Jason Edworthy, one of Canada's foremost experts on wind energy. "You have to spend enough money to do it right," he says, and that's usually a lot more money than many people expect.

The best place to begin evaluating the cost of a small wind system is to realistically determine what it is you want. Do you want a wind turbine to meet the limited needs of a vacation cabin used only on weekends? Or do you want the wind turbine as a complement to the photovoltaic panels and the stand-alone, battery-charging power system you already have? For a battery-charging system, sum the anticipated electrical consumption you want the wind turbine to meet, estimate the wind available, and then determine the size of wind turbine you need.

Next, compare the various wind turbines in the size class you need. This is quite subjective because you must weigh not only price but less tangible factors such as quality and reliability. Avoid buying any product on the basis of price alone; instead, look at relative cost to determine what may be more cost effective.

This task has been made much easier since the introduction of standardized testing and certification of small wind turbines designed for interconnection to the grid. All certified products report their performance in the same manner, so you can compare apples to apples (see Table 11-5. Table of Estimated AEP for Bergey Windpower's Excel 6).

And always weigh total installed costs, not just the cost of the turbine alone. For micro turbines, a quality tower will cost as much as the turbine—sometimes more. The least costly and most user-friendly tower option for micro and mini wind turbines is a tilt-up, guyed tubular mast.

You'll also need all the electronics, if the turbine is not packaged with them. And it's essential to know how the wind turbine is performing, so that when it's not, you can get it fixed. Consequently, all small turbines should have some kind of recording meter so you can track performance over a period of time. Simple meters measuring instantaneous power or current, while helpful, are insufficient for anything more than determining whether the turbine is generating or not. They're better than nothing—but not by much.

For smaller, less sophisticated wind turbines, it's also essential to have some means of stopping the turbine when needed. In micro and mini wind turbines, this is often a brake switch that places a large electrical load on the generator, stalling the rotor and dramatically reducing its speed. Unbelievably, many micro turbines don't have this feature.

Controls

There are times when you need to stop the rotor of a small wind turbine or otherwise bring it under control. Imagine that something goes wrong at 3:00 a.m. in the middle of a gale, and you're standing at the base of the tower in your underwear wondering what to do. It's a time like that when you wish all small turbines included a reliable means of stopping or slowing the rotor to a safe speed.

Many small turbines use dynamic braking of the generator to slow the rotor when desired—with varying degrees of success. Often dynamic braking is most effective in moderate winds and least reliable when you really need it.

"If a wind turbine's more than 1 kW," says North Dakota wind enthusiast Mike Klemen, it should have "a brake that will work under any wind conditions." For Klemen, if the generator is not sufficiently robust to stop or significantly slow the rotor through dynamic braking, then there needs to be some other means, mechanical or aerodynamic, of controlling the rotor under emergency conditions. For a small wind turbine, this could be as simple as a furling winch.

Operational History

Check the operational history of the turbine model. Determine what kinds of tests have been conducted, for how long, and the highest winds experienced. Not all small wind turbines are built to the same standard, though this is improving with certification. There was one case during the early 1980s where a wind turbine designer sized his machine to withstand a maximum wind speed of no higher than 90 mph in a parked condition. He asserted that no winds above that speed had ever been measured near his site in western Pennsylvania. This claim was false, and consequently the design of the wind machine was suspect, more so when considering that all other US manufacturers at the time were designing their products to withstand a maximum wind speed of 120 mph (54 m/s).

A SMALL WIND SYSTEM IS MUCH MORE THAN A WIND TURBINE

Buying a wind system is much more than just buying a wind turbine. For micro and mini wind turbines, the "balance of system" costs for the tower, cables, switches, and connectors can amount to a substantial part of the total cost. Take the installation of the popular Air series of micro turbines, for example. Properly installing the turbine on a relatively short tower close to the load it will power can easily cost more than twice the cost of the wind turbine alone (see Equipment Cost for Adding Micro Wind Turbine to Existing Off-Grid System). You can certainly do it for less, but why should you? If you want to do it right, you'll need these components to minimize problems in the long term.

Equipment Cost for Adding Micro Wind Turbine to Existing Off-Grid System

SWP Air model on 45-foot (14 m) tower with 100 foot (33 m) cable run

AirX	$ 700
45-ft tiltup tower	$ 700
3 NSI Connector blocks	$ 30
Wire mesh strain relief	$ 30
10 ft Flexible, liquidtight conduit	$ 50
Post for disconnect switch	$ 15
Fused, raintight disconnect/stop switch	$ 100
50 ft Sch 40 PVC conduit	$ 30
Fittings	$ 30
2x100 ft # 8 Insulated Cu conductor	$ 70
100 ft #8 Bare Cu ground	$ 30
Ground rod	$ 15
30 A DC Circuit breaker (for load center)	$ 15
	$ 1,815
Turbine/Total	39%

Assumes tower is tower length from battery location.
Assumes existing approved load center and batteries.

Knowing how long the tests were run or how long a particular model has been in service is especially important. Unscrupulous manufacturers have frequently resorted to touting their products as "extensively tested" when they haven't been. In one case, the new product had been in operation during only a mild summer in Ohio, an area of moderate winds. The first time this "extensively tested" product was installed at Alternative Energy Institute's windy West Texas test field, it suffered severe blade flutter, experienced a brake failure, and the tower fell over—all within the first hour.

In another notorious case, the "extensive tests" were conducted on a bench-scale model! The manufacturer hadn't even bothered to test the turbine outdoors before it began selling the turbine to wind farm developers in California. Eventually this company was prosecuted for fraud, but not before walking off with their investors' money.

Reputable, well-tested products, in contrast, have been in unattended operation for months, if not years, at numerous sites in widely different wind regimes. Well-tested products have endured hurricane-force winds without damage. Manufacturers with well-tested products can provide documentation on the performance of their machines over time, under harsh conditions.

When evaluating operational history, don't be alarmed by occasional reports of defects. You're looking for trends. If every wind turbine of a particular model has thrown a blade and is still throwing them, then there's a good chance the one you're looking at will too.

Wind turbines shouldn't be held to any higher standards than we hold other machines. After more than 100 years of development, automobiles are still being recalled by the thousands for manufacturing and design defects. Yet we continue to buy and use them. We try to minimize the risk of buying a lemon by trying to select a model with the least potential for problems. Reputable manufacturers of wind machines make mistakes like everyone else. Your challenge is to find one that makes fewer mistakes than the rest.

Design defects usually appear within the first year of operation. Like automobiles, new products must undergo a period of debugging. Unexpected problems will undoubtedly arise and must be corrected. These problems are greatest when the product is first introduced and decline thereafter. You want a wind turbine that has been on the market for several years, and one that operates successfully in a range of environments. Let someone else pay for the field testing that all wind turbines require.

Product Specifications

Examine the promotional literature describing the wind turbine. Are the estimates of energy production reasonable? Do they stress generator size, while ignoring energy output altogether? Most manufacturers will present a list of parameters that succinctly describe how the wind machine performs, how it functions, and what it can be used for (see Table 19-5. Product Specifications for a Small Wind Turbine). This will include the annual energy production, power curve, and power form.

Estimates of annual energy production could ultimately replace the rated power at rated wind speed currently used to describe the size of a wind turbine. But, it hasn't happened yet. More often than not, the AEP is presented as a graph of estimated generation at various wind speeds at hub height but will also be presented as a table of annual generation at various average wind speeds (see Figure 11-7. Chart of annual energy production).

Power form indicates how the power will be used. For small wind turbines interconnected with the utility, this will be given as the nominal voltage and frequency. For battery-charging wind systems, power form should indicate the DC voltage. For interconnected wind turbines, power form should also include the number of phases. In North America, for example, most homes and many small farms use single-phase service. Large farms and businesses use three-phase service.

Most product specifications should also include the cut-in and cut-out wind speeds, maximum power, noise level in dBA, maximum design wind speeds, rotor speed, and overspeed control.

Table 19-5. Product Specifications for a Small Wind Turbine	
Rated power	5.5 kW @ 11 m/s
Peak power	6.7 kW
Annual Energy Production	9,920 kWh
Cut-in wind speed	2.5 m/s
Cut-out wind speed	None
Maximum design wind speed	60 m/s
Nominal rotor speed	0–400 rpm
Overspeed control	Furling
Power form	240-VAC, 1-phase, 60 Hz
Rated sound pressure level	47.2 dBA
http://bergey.com/documents/2013/10/excel-6-spec-sheet_2013.pdf	

Product literature should always specify the type of noise data presented. Preferably, noise data will be shown as source emission strength or sound power levels. Specifications for most large wind turbines include sound power levels. However, small wind turbines intended for general consumers will typically use sound pressure levels because that's what most regulators expect to see.

The *maximum design wind speed* is the speed the turbine was designed to endure unattended without suffering damage. Notations may also indicate the maximum measured or tested wind speed the turbine has survived without damage.

Rotor speed is the number of revolutions per minute of the wind turbine's rotor. For a wind turbine driving a two-speed induction generator, there are two rotor speeds: the rpm when operating on the low-power generator or windings, and the rpm when operating on the primary generator or full-power windings. In a variable-speed wind turbine, rotor speed is given as a range of values.

Overspeed control is a concise description of the method used to protect the wind turbine in high winds or during a loss of load.

Evaluating Vendors

After you have dissected the technology, you must evaluate less tangible factors such as the longevity of the manufacturer and the reputation of the dealer.

Manufacturers

How long has the company been in the wind business? What's its track record? Does it have sufficient financial resources to honor its warranty commitments? There's no easy way to find answers to these questions. In most cases, you'll be dependent on the dealer for information. Even when you do get the answers, it's hard to determine what's important and what isn't. For example, a well-established company that has been in business for several years is a better risk than one just starting out. Likewise, if a company is partially or wholly owned by a major national corporation, that usually indicates it has ample financial reserves to survive a major warranty recall. Nevertheless, a corporate executive who has no personal stake in the company can much more quickly decide to cut his losses and cease production during hard times than can the owner-entrepreneur who has put his own sweat and blood into the business. Only you can decide on which business you want to place your bets.

Dealers

The dealer you choose, if you use one, is determined primarily by the wind machine you want and where you live. Most dealers, to round out their product line, represent more than one company. Even so, within a certain locale there will be only one dealer for each brand. (Manufacturers want to ensure a healthy dealer network so they limit the number of dealers selling their product.) Proximity is important. If repairs or service are needed, particularly during an emergency, you don't want a dealer who lives on the other side of the continent.

Determine if dealers are reputable by checking with their previous clients. Have they been prompt in making repairs, or have they taken their time while hustling new sales? Dealers should have

VENTILATORS AND SQUIRRELS IN A CAGE

Like ducted turbines (see Chapter 7, "Novel Wind Systems"), a perennial favorite of hucksters and charlatans is, for lack of a better word, squirrel-cage rotors (see Figure 6-3. Squirrels in a cage). Many are nothing more than rooftop ventilators repackaged as "wind turbines." As ventilators, they work fine. It's when someone tries to couple them to a generator that they quickly learn why wind turbines use two or three slender, airfoil-shaped blades. Most hucksters, however, never progress that far. They never build actual wind turbines, and if perchance they do, they never measure the "wind turbine's" performance. Of course, they wildly exaggerate the potential of these breathtaking new inventions.

Doug Selsam, himself an inventor, has tried to understand why consumers—and the news media—are so gullible. His explanation: ventilators and squirrel-cage rotors are easy to understand, modern wind turbines much less so. After all, a rooftop ventilator with its entire swept area covered with blades looks like it will capture more wind than a modern wind turbine with only a few blades, some with—unbelievably—only one.

In a 2002 Internet scam, a company peddling ventilators as "wind turbines" claimed its product would produce nearly five times more electricity than a conventional wind turbine of the same size. Naturally, for this "superior" performance they would charge two to three times more than for a real wind turbine. The company asserted that it was "thinking outside the box," a catchphrase of 1990s management gurus. It certainly was. This company was not even close to the box. It was on another planet where the laws of physics don't apply.

references available for such an inquiry. If not, are they willing to provide them?

Also call the manufacturer and check whether the "dealer" is authorized to sell its product. In one instance, a so-called dealer was selling a popular brand without the manufacturer's authority to do so. This dealer had declared bankruptcy previously, leaving a number of clients high and dry without spare parts or service for their ailing wind machines. In this case, he was selling a used wind machine—that is, until the authorized dealer blew the whistle. The whole sad affair could have been avoided by a single phone call to the manufacturer.

Don't be misled by membership in various organizations as a claim of legitimacy. Some dealers and manufacturers use membership in a trade association as a promotional tool. It's one of the oldest marketing tools in the book, and the one most often used when no other credentials exist. Anyone can join an association, and most associations don't police their ranks. Suspect anyone trying to cash in on checkbook credibility.

Contracts and Warranties

To ensure that you get what you pay for, put it in writing. Demand a written contract and warranty, and consider having an attorney look them over. Installing a wind system is a major investment. It's worth the added cost of getting good legal advice. You may need an accountant's advice as well.

You need to know specifically what is included in the price you have been quoted. If you plan to do any of the work yourself, the contract must spell out exactly where the dealer's responsibilities end and yours begin. The contract should also describe exactly what you must do to meet the terms of the warranty. Who has the final say, for example, as to how the work should be done? How will disputes be resolved, should they arise? What is covered by the warranty? What isn't? How long does it last?

Most small wind machines come with a limited warranty. Extended warranties are sometimes available—for a price. Because of the difference in warranties between manufacturers, it's wise to read the fine print. Check whether the warranty is transferable or assumable by the manufacturer, if the dealer goes bankrupt. Ask who pays for shipping or for the fieldwork on warranty repairs, and whether damaged or defective parts must be returned before replacement parts are shipped.

When Southwest Windpower introduced its Air series, the early turbines were unreliable and owners had to frequently send the machines back for repairs. Though the company paid for the repairs, it didn't pay for shipping or the cost of the turbine's removal and reinstallation.

Another aspect of the contract is the terms of sale. The contract should state the amount of payments, how and when they should be made. In general, you will pay most of the cost for the wind machine and its installation in advance. You don't drive off the lot with a new car until you have handed over your check. Similarly, you shouldn't expect the dealer to install the wind system without you first paying a hefty deposit.

Terms vary from one dealer to the next. Usually a down payment is made to secure your order. Then full payment is required for the turbine and tower when they are ready to be shipped. Some dealers require payment for the turbine, tower, and installation in advance. In most cases 5% to 10% of the total contract is held by the buyer until the wind machine has been installed and operates properly.

In multiple machine purchases such as for a commercial wind farm, the buyer has more leverage with the manufacturer and can obtain written assurance that the wind machine will generate power as advertised and that it will be available to generate power a minimum percentage of the time. In Denmark and Germany, where there's a strong association of wind turbine owners, manufacturers have been held responsible for their performance claims. Such assurances are often not offered to purchasers of small and medium-size wind turbines in North America. And there's no association of wind turbine owners in North America like those in Germany and Denmark to protect the interests of users.

What to Expect

If you buy a small wind turbine through a dealer or use a contractor to install the turbine, you have a right to expect that all work will be performed according to standard practices and local building and electrical codes. The work should also be performed in a timely manner, and all construc-

Study after study has concluded that large wind turbines, working reliably at good sites,
ARE ONE OF THE LEAST-COST SOURCES OF NEW GENERATION.

tion debris removed from the site before the job is considered finished.

Don't expect overnight miracles. If delivery of a component has been delayed due to circumstances beyond the dealer's control, you shouldn't hold the dealer accountable. The dealer should make a reasonable effort to expedite the installation of your wind system or its repair, but don't expect the dealer to jump at your every request. Keep in mind that dealers operate a business and that they may have other commitments. At the same time, the dealer should fulfill those obligations stated in the contract or implied during negotiations.

Financial and Economic Models

There are several methods for evaluating whether wind energy makes an economic investment. Each serves a different purpose. Some, such as *payback*, are well suited to a quick, back-of-the-envelope assessment. Others, such as *cash flow*, enable investors and financiers to weigh their return on investment. Public policy debates often revolve around the *cost of energy*.

Cost of Energy

Determining the cost of energy (COE) produced by a wind system is one approach popular with engineers and policy makers. The COE accounts for initial cost, maintenance, interest rates, and performance over the life of the wind system. The COE formula can be found in any engineering textbook. It produces an estimate of cost over the entire life cycle of the power plant in cents per kilowatt-hour. The results can be compared with the cost of energy from other new sources of energy.

Yet COE has limited utility to individuals, businesses, and even policy makers because it reveals only whether the wind system's generation will cost more or less than that from other sources using comparable assumptions. It doesn't tell you how much of a bargain—or cost—the wind system might be, and it doesn't tell you the real cost. It's just one metric. The COE, consequently, cannot be used to judge whether or not your money would be more productive in an interest-bearing account at the local bank or some other investment or whether society would be better off investing in wind energy or some other technology.

Tables listing the cost of energy for different technologies imply that the technologies are otherwise equal. Of course they are not. A coal-fired power plant is not the same as a wind power plant in terms of its long term costs to society and the environment. Decisions about the future of our power system are typically not made—and should never be made—solely on the basis of the cost of energy. Moreover, the COE calculation makes some very sweeping assumptions about the future, which may or may not reflect reality. The cost of each technology responds differently to how the future unfolds. The cost of energy could be higher from wind energy than other technologies, and it would still make sense to invest in wind energy because investments in renewable energy are a hedge against possible nasty surprises waiting in decades to come. Investments in wind energy are a form of insurance against the rising cost of fossil fuels or the astronomical costs of a major nuclear accident.

Regardless of these limitations, study after study has concluded that large wind turbines, working reliably at good sites, are one of the least-cost sources of new generation. In all cases today, commercial wind generation is less expensive than generation from new nuclear power plants and often less expensive than generation from new coal-fired power plants as well. This is why wind energy is expanding so rapidly around the world.

Similarly, small wind turbines are nearly always an economically attractive choice for off-the-grid systems. The economics of small wind are more problematic when interconnected to the grid in a net-metering application because they have to compete with electricity from power plants amortized long ago.

Payback

Finding the payback, or the time it will take for an investment to pay for itself, is an easy way to gauge an investment's worth. For that reason, it is the first approach most people turn to. They simply divide the wind system's cost by its projected revenue. If the time to payback is less than the life of the wind turbine, the turbine has paid for itself.

Calculating payback is simple, but it can be misleading. Payback doesn't account for effects that take place after payback occurs. It gives no indication of the earnings a wind turbine will produce after it has paid for itself. Since wind generators are designed to last 20 years or more, much of their return takes place in later years after the initial cost has been recouped.

Payback is well suited for low-cost items like storm doors, weather stripping, and added insulation but not for costly long-term investments such as wind turbines.

Payback is related to return on investment. A short payback offers higher returns than a long payback. To maximize the return—to get the most for your money—you want as short a payback as possible. Most people are overly concerned about payback. They fail to realize that some wind machines will pay for themselves many times over. Wind systems are long-term investments and should be treated as such.

Like the cost of energy, payback is just one metric used to gauge the relative merits of an investment. Community-owned projects using large wind turbines in distributed applications yield simple paybacks of less than 10 years under conditions in Germany. The payback for small wind turbines is often greater.

Cash-Flow Models

Financial analysts use complex *cash-flow* models to analyze the income and expenses of a project from one year to the next. These are used to determine how the project will perform financially over time. Spreadsheets for these models can be a dozen or more pages with hundreds of rows on each page. Analysts argue over who has the better spreadsheet to capture all the variables.

There's no better way to gain a sense of how a wind system will perform financially than to estimate the cash flow from one year to the next.

Constructing a cash-flow table is the only way to examine the economics over the long term with any degree of realism. It's what businesses use when they're considering investments in new equipment.

However useful to bankers and project developers, cash-flow models are overkill for policy making by burying the most important parameters of a wind project in a blizzard of detail.

Profitability Index Method

Developed by French engineer Bernard Chabot, the profitability index method is a relatively simple economic model that uses the capital recovery factor to calculate what price—or tariff—a wind turbine or wind project should be paid to achieve a targeted profitability expressed as the net present value per dollar invested. This ratio, the profitability index, should be in the range of 0.1 to 0.3—sufficient to attract private capital but not so lucrative as to create a stampede like the great California wind rush of the early 1980s. From this ratio, it's easy to calculate all the other economic parameters of a project, including simple and discounted payback, internal rate of return, and the required margin between the manufacturing cost of electricity and its selling price. For this reason, Chabot's approach is most useful to policy makers and regulators trying to determine a fair price for wind energy.

Chabot's system was used for the preliminary design and calculation of the French wind tariffs first published in mid-2001. The tariffs launched wind energy development in France. More important, the French approach demonstrated how Chabot's profitability method could be used to design *differentiated* wind tariffs. To prevent overpayment for wind energy in windy areas of France, the payment per kWh for wind energy is reduced on a sliding scale as a function of the wind resource. To prevent underpayment—and the resulting lack of investment—the Chabot system calculates the higher rate necessary to make less windy regions financially attractive. You could characterize this as Goldilocks pricing: not too much, not too little, just right.

We'll look at examples of both a cash-flow model and the Chabot method in a moment. Now let's turn to some of the factors affecting a wind investment.

Economic Factors

There are two sides to the economic equation: cost and revenue. We'll look at both in turn. The cost of generating electricity from wind energy is primarily a function of the installed cost and the all-inclusive term, annual reoccurring costs. The revenue from a wind turbine is determined by the amount of electricity generated and the tariff or rate paid for the electricity.

Installed Cost

The *installed cost* of a wind system is simply the cost of the wind turbine, tower, wiring, and installation, less any state or federal investment tax credits, or capital grants. Where they exist, tax credits reduce an individual's tax liability, instead of merely reducing taxable income as with tax deductions. Thus, investment tax credits effectively reduce the initial cost. More on tax credits in a moment. Grants or subsidies based on the capital cost of the wind turbine also reduce the total installed cost.

In the mid-2000s, the relative cost of large turbines, which had been steadily declining for two decades, rose dramatically. German operators were reporting price increases of 20% to 30%. For large turbines in small numbers, relative costs ranged from $2,000/kW to $2,500/kW and about $700/m² to $1,000/m² for a 2-MW turbine. Though relative installed costs have dropped, there remains a wide spread between the least costly large projects and the more expensive projects (see Table 19-1. Installed Wind Project Relative Cost in the US: LBNL 2012). The relative cost for the two GE turbines in the example used previously may not be representative of the industry as a whole (see Table 19-2. Relative Costs for GE 1.6 MW Platform).

As with a small wind turbine, the total installed cost of a large wind turbine includes much more than the wind turbine itself. In some cases, the electrical connection of the wind turbine to the grid may add significantly to the cost. This was the case in Ontario in the early 2000s where utilities larded on unnecessary requirements to discourage private ownership of wind turbines (see Table 19-6. Sample Cost Estimate for a Large Wind Turbine Relative to Total Installed Cost).

Table 19-6. Sample Cost Estimate for a Large Wind Turbine Relative to Total Installed Cost (As a Percentage of Total Cost)

	Project Size	
	1 MW	100 MW
Pre Development Engineering, Site Assessment and Land Acquisition	9%	5%
Site Engineering and Equipment Selection	5%	6%
Blade Procurement	16%	17%
Turbine/Generator/Nacelle Procurement	21%	22%
Tower Fabrication and Delivery	18%	19%
Electrical Control Panels	3%	3%
Unit Transformer	3%	3%
Site Works Including Switchyard	14%	15%
Foundation for Tower Including Reinforcing	5%	5%
Crane Service and Turbine Erection Cost	1%	1%
Commissioning, Warranty and Interest During Construction	6%	4%

Source: APPrO, Ontario Wind Power Task Force, February 2002, pages 93 & 94.

Subsidies and Incentives

Subsidies, or the more politically correct incentives, are payments or grants given to buyers of wind turbines to encourage them to invest. Typically, subsides represent some portion of the equipment cost. Often these grants come from taxpayers and are subject to the give-and-take of periodic budgeting by the legislature.

Sometimes the grants are funded from a pool of money collected from a surcharge on sales of electricity. These public goods funds or system benefit charges were created in the aftermath of the deregulation craze that swept North America in the 1990s. Proponents reasoned that when electric utilities were "restructured," there were public goods, like research and development, that would no longer be funded as a part of the new, leaner, and meaner utilities that resulted.

Electric utility regulation—and as a result, public goods funds—are administered at the state and provincial level in Canada and the United States. Consequently, subsides for wind energy are a hodgepodge of programs. It's enough to make your head swim.

Legions of lawyers in both countries have carved out profitable niches parsing the complex regulations governing subsidies and the even more arcane tax implications of various forms of wind investment. At least in the United States, there's a database that tracks all these programs

US FEDERAL TAX CREDITS

Small Wind Turbines

Small wind turbines up to 100 kW installed at a home, whether the principal residence or not, qualify for a federal Residential Renewable Energy Tax Credit of 30% of the total installed cost, not including specific state subsidies. The credits are available for installations through the end of 2016. Small wind turbines must meet performance and quality standards to qualify for the credit.

Large Wind Turbines

The Production Tax Credit (PTC) for commercial sale of wind-generated electricity was $0.023 per kilowatt-hour in 2016 for a period of 10 years from the date when the wind turbine was placed in service. The PTC has been extended to the end of 2019. Thus, if a wind turbine of any size had been installed in 2016 and delivered electricity to the grid, it would qualify for $0.023 for each kilowatt-hour produced through 2026. The PTC is reduced 20% for projects beginning construction in 2017, 40% for those beginning construction in 2018, and 60% for those beginning in 2019. Wind projects beginning construction before 2020 can also elect to take a 30% Investment Tax Credit (ITC) instead of the PTC. The ITC phases out in the same increments as the PTC.

For up-to-date information on federal and state subsidies, consult the Database of State Incentives for Renewables and Efficiency (DSIRE) at www.dsireusa.org. For details on the federal tax credits, click on the Federal Incentives icon.

PAYING FOR PERFORMANCE

Tom Starrs, an attorney with an illustrious career in renewable energy that began during California's famed wind rush, has observed that during the early 1980s most renewable incentives in the United States were based on capital grants or tax credits as a percentage of installed costs. "The consensus among analysts," says Starrs, "is that the use of capacity-based incentives in the US during the 1980s contributed significantly to performance-related problems and in some cases to outright fraud." He goes on to add that "these problems contributed to the federal government's abandonment of wind energy incentives in 1986."

Starrs's remedy? Paying for performance. Good public policy pays only for performance, he says. It is in the best interests of ratepayers–electricity consumers–as well as taxpayers that incentives paid for renewable energy result in the actual generation of electricity. This can only be achieved when payments are coupled directly with the number of kilowatt-hours produced. This places the responsibility on the manufacturer of the wind turbine and on the investor, where it should be, not on ratepayers or taxpayers.

and their requirements. There's nothing equivalent in Canada.

Federal Policy

The US Production Tax Credit (PTC) is a much ballyhooed, on-again, off-again subsidy program funded by Congress from the federal treasury. The PTC grew out of an older tax subsidy program of the early 1980s. In the earlier program, the subsidy was based on the installed cost or the capital cost of the wind project. This led to widespread abuses that gave renewables a poor reputation for years afterward. Unscrupulous developers would inflate the cost of the wind turbine so they could collect more subsidies. (This was easy to do with wind turbines by simply increasing their "rated power" as explained earlier.)

Subsequently, when subsidies were eventually restored in the late 1990s, they were offered on only actual generation, that is, paid only on the amount of electricity produced. While this eliminated fraud and the tendency to inflate costs simply to increase the subsidy payment, Congress remained timid and placed onerous restrictions on who could use the PTC. As a result, only the most profitable companies in the nation—those with the most taxes to pay—could participate in the program. Homeowners, farmers, and small businesses need not apply.

Further, Congress was never certain it really wanted the program, or wind energy, at all. The PTC has been extended, and then allowed to expire several times in the past decade. This uncertainty has led to a boom-and-bust pattern that has characterized the US wind industry for decades. As one wag put it, "This is no way to run a business."

There is no federal subsidy in Canada.

State and Provincial Programs

Some states and provinces in North America operate their own incentive programs. These are typically built around capital or up-front subsidy payments. Sometimes called "buy-downs," these programs seek to lower the initial cost of small wind turbines to make them more affordable.

Because these are grants (rather than payment for actual generation), the money is paid out whether the wind turbine works or not. Thus, there's opportunity for abuse. Consequently, legislators and regulators place numerous requirements on participants to prove that they qualify for the grant. They also typically limit the size of the wind turbines that qualify and the total amount given out. In many cases, the wind turbine model chosen must be "certified" in some manner in order for the owner to participate in these programs. While not foolproof, certification weeds out the "Internet wonders."

Cost of Capital

The cost of financing the purchase of a wind system can add significantly to overall costs. You immediately become aware of financing costs if you have considered using a loan to buy your wind system. It's much like building an addition onto your house with a loan from the bank. You pay for the use of the bank's money. You can't avoid financing costs simply by paying with cash. The cash invested in the wind system could have been earning interest at the bank. For homeowners and farmers, the installation of a wind turbine can be financed by increasing the mortgage on the property instead of by taking out a short-term loan. The financing or interest cost of the loan will then reflect the current mortgage rate.

For commercial projects, or a community-owned project using large wind turbines, there are two forms of capital: equity and debt. Equity is the portion raised from the project's investors. German farmers and homeowners are willing to put up risk capital to earn a 5% to 7% return for a project that they will own. Commercial projects in North America may demand as much as 12% to 14% return for their capital.

Debt is provided by a bank or other lender. For small household-size wind turbines, homeowners in North America could finance debt at 6% to 8% interest for a 15-year term using their home as collateral in 2014. Commercial projects may have to pay higher interest rates over a shorter term because the project itself is the collateral. This "project financing" can appear as a higher risk to the bank.

Banks see projects with a predictable revenue stream, for example those who are paid for their electricity through a power purchase agreement or a feed-in tariff, as a better risk than projects that have volatile payments for their electricity. Commercial projects with feed-in tariffs in Germany can obtain as much as 80% or more of the project's financing through debt. In Nord Friesland, the local development bank has never had a wind project default in the two decades it has been financing locally owned wind projects. Due to Germany's system of feed-in tariffs, the bank can reliably predict how the projects will perform under wind conditions near the border with Denmark.

The proportion of equity to debt and the returns required for each determine the weighted average cost of capital (WACC).

Annual Reoccurring Costs

Annual reoccurring costs are a catchall category that includes the cost of operations, maintenance, parts replacement, and insurance lumped together.

Operations and Maintenance Cost

Maintenance costs are expenses for servicing or repairing the wind system. Operating costs may include maintenance and repair. These costs can be expressed in cents per kilowatt-hour as discussed in the previous chapter or relative to the initial cost (see Table 19-2. Relative Costs for GE 1.6 MW Platform). This can be expressed in terms of cost per year relative to installed cost or as a percentage of installed cost.

Insurance

Insurance is an often overlooked cost. There should be insurance not only for the wind system itself but also for any accidents due to the wind machine's operation. Since most wind systems require an extended period to pay for themselves, an owner would be foolhardy to operate an uninsured machine. An unexpected event could wipe out an expensive wind turbine before the investment had been recouped. Wind systems are also a potential hazard. Personal injuries and property damage can occur in a multitude of ways: the tower can fall over, or (the most likely form of accident) someone can fall off the tower.

Royalties and Leases

For community-owned or commercial projects using large wind turbines, there is also an expense for renting the land on which the turbine stands. Payment for the *lease* is usually based on a percentage of the revenue generated by the turbine. This *royalty rate* for use of the land may also include a minimum payment per

Table 19-7. Royalties & Land Rents for Wind Turbines as Percent of Gross Revenue (Sorted by Royalty Rate)						
				Years		
Country	Location	Year	Project Size (MW)	1-10	11-20	21-30
Germany	Coastal, Average	2004		5-8%		
Germany	Interior, Average	2004		3-5%		
Germany	Bürgerbeteiligung Lüdersdorf	2004	23	7.0%		
USA	Cielo Wind Power	2004	80	6.0%		
Germany	Bürgerwindpark Cappeln	2000	4	4.0%	5.85%	5.74%
USA	Indian Mesa, Texas Land Commission	2002	34	4.0%	6.00%	8.00%
USA	Montana State Trust Lands			4.0%	6.00%	8.00%
USA	Woodward Mesa, Texas Land Commission	2002	32	4.0%	6.00%	
Germany	Bürgerwindpark Freiburg	2003	11	3.8%	5.40%	
Portugal	Portugal	2004		2.5%		
Canada	Ontario #2	2004		2.5%		
Canada	Ontario #1	2004		2.0%		
Canada	Ontario #3	2004		1.8%		
Note: Bürgerbeteiligung is a citizen- or community-owned project.						

year to assure the landowner that he or she will be paid even if the wind turbine is not operating properly.

The leases cover the entire period the project will be in operation, typically 20 years or more. Often the royalty rate is constant through the entire period. However, in some cases the royalty rate increases after a period of time sufficient for the developer to recover project costs, usually after year 10. For example, the royalty rate may increase in years 11 through 20, and again in years 21 through 30.

Royalties or land rents vary widely from one country to the next. They range from a low of less than 2% in Canada to a high of 8% in Germany (see Table 19-7. Royalties & Land Rents for Wind Turbines as Percent of Gross Revenue).

The rates also vary from one type of landowner to the other. In North America, where data on royalty rates is often considered confidential, large landowners with sophisticated staff often negotiate much better leases with wind projects than individual farmers. Because of nondisclosure agreements in North America, farmers and individual landowners frequently have no idea what their neighbors are being paid.

In contrast, community-owned projects in Germany list the lease terms (*pacht*) in their offering documents. Royalty rates are also published in German trade magazines, such as *Neue Energie* (New Energy magazine). German community wind project developers may also be willing to pay more for leases than in North America because the landowners are their neighbors and often part owners in the projects.

Though German projects pay relatively high royalty rates, not all of the royalty may go to the landowner. Some of the lease may be set aside for others in the community. Heinrich Bartelt, a German community-wind project developer and one of the country's wind pioneers, reported in an interview that in his projects in Sachsen Anhalt, he sets aside 1% of the 5% royalty for the nearest villages. Of this, a portion is set aside for community groups, sports halls, and so on. For example, some of the royalty may go for the village *Mannechor* (choir) or brass band. Of the 4% remaining,

Table 19-8. Annual Reoccurring Costs as a Percentage of Installed Cost		
	Low	High
O&M	0.5%	1.0%
Repairs	1.0%	2.0%
Insurance	0.5%	1.0%
Lease	0.5%	1.0%
Management	0.5%	1.0%
Total	3.0%	6.0%
Data from the early 2000s.		

WIND TURBINE ENVY AND LAND LEASE POOLING

Often in North America wind project developers lease land from individual landowners. Developers lease the land to install their wind turbines or ancillary structures such as roads, transformers, substations, and power lines. Unfortunately, landowners often only receive rents or royalties when these structures are placed on their land. Neighboring landowners without wind turbines or the ancillary structures sometimes receive little or nothing.

Though some leases include clauses for sharing of revenue on land without wind turbines or other facilities, most do not. This breeds resentment among neighboring landowners who receive no financial benefit, yet must live with the sight and sound of the wind turbines in their midst and leads to "turbine envy."

Turbine envy, not surprisingly, has led to opposition to some wind projects by neighboring landowners who feel left out–or cheated out–of potential revenues.

European projects, where each parcel is much smaller than in North America, first encountered this dilemma as a result of the large number of neighbors at each site. Down at the local pub in Wale's Cemaes Bay, for example, neighbors of EcoGen's Ryd y Groes project grumble: "Why couldn't they have put them on my land?"

Henning Holst faced the same problem in Germany. He envisioned compensation based on a series of concentric zones around the turbines. Those with turbines on their land would receive compensation as they do today. But turbine owners could also pay peripheral landowners more modest sums based on the parcel's distance from the project. Those nearest the turbines would receive more than those farther away.

German developers did devise a means for compensating landowners who didn't have a turbine on their site. Typically, they create pooling arrangements among neighboring landowners. In one example in the region of the German city of Paderborn, the developers combined all neighboring landowners in a land association or pool. The land association then negotiated the terms of the land lease collectively with the developers (see Royalty sharing among farmers). Subsequently, the land association leased their combined land to several investor-owned cooperatives (Bürgerbeteiligung).

Some wind developers in Canada have proposed similar arrangements where neighboring landowners receive a portion of the royalties as a percentage of their land in the pool. Landowners with wind turbines or other structures receive payments from the pool in addition to the land lease payment.

In 2005, Prince Edward Island's then Minister of Energy Jamie Ballem devised a lease model adapted from Germany. Ballem offered the landowners 2.5% of the gross revenue from the project. The royalties are then divided into three zones. Landowners with turbines on their property receive 70% of the royalties. Those landowners within 100 meters (330 feet) receive 20%. Those within 300 meters (1,000 feet) receive 10%.

French project developer Erélia followed a similar approach. It paid landowners with wind turbines on their land 70% of the royalties paid for land leases in their Le Haut des Ailes project in northeastern France. The remaining 30% was paid to adjoining landowners who were not fortunate enough to have a turbine or other structure on their own land.

To spread out the royalty benefits to as many landowners as possible, Erélia moved some of the turbines slightly from their preferred locations to neighboring properties so that more landowners had turbines on their land. They made preferential payments to those farmers with adjoining property or those within 80 meters (250 feet) of each turbine. They also made payments to landowners for the passage of roads and cables.

Although in no way assuring that all landowners will be amenable to such an agreement or won't be resentful about some landowners receiving more payments than others, land pooling does increase the overall acceptance of renewable projects. Overall, pooling arrangements are more equitable to all landowners in an area than leases with only some of the landowners.

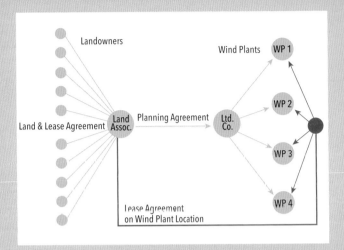

Royalty sharing among farmers. The Sintfield project near Paderborn was one of the largest of its kind in Germany if not Europe. It encompassed 765 hectares (~ 2,000 acres) and four different wind power plants–all owned locally or by residents of the region. The 111-MW project also pioneered the use of royalty sharing among the many different landowners involved. All landowners were organized into a land association that negotiated the land lease and royalty terms collectively. The land association then leased the combined land to the different projects so that all landowners, including those without wind turbines on their land, would share in royalty revenues from the project greatly reducing "turbine envy."

half is paid to the landowner (2%) and half is distributed among surrounding landowners who don't have a turbine on their land.

The latter payment may appear surprising but reflects one of the reasons for German success with wind energy. Neighboring landowners see and hear the wind turbines even if one is not sited on their land. Because parcels are much smaller than is typical for North America, this is a common condition. To compensate neighboring landowners, they are included in the royalty. This addresses *turbine envy* where a neighboring landowner may resent not being selected to host one of the project's turbines.

Altogether, the annual reoccurring costs can range from a low of 3% to a high of 6% of the installed cost (see Table 19-8. Annual Reoccurring Costs as a Percentage of Installed Cost).

Taxes

Taxes have a profound effect on the economics of any investment, largely on the cost side of the ledger. For homeowners in the United States, the costs of financing a wind system through a home mortgage are tax deductible. Because the interest costs are high in the early years of a loan, this deduction dramatically improves the attractiveness of small wind turbines to homeowners over what would be apparent from a simple payback calculation. The tax bonus is even more important for businesses because they can deduct the cost of maintenance and other expenses.

The *tax bracket* of the investor determines the value of any deductions. Those in higher tax brackets save more by reducing their tax burdens than those who pay a lower tax rate. If you're in the 30% tax bracket, you save 30 cents for every dollar in tax deductions.

Similarly, homeowners receive more than 1 cent in value when they offset 1 cent worth of electricity with a wind turbine because such savings are not taxed. (No one taxes you for using less electricity —at least not yet.) If a homeowner offsets 1 cent's worth of electricity purchases, they create 1.43 cents in value after taxes for someone in the 30% tax bracket (1/[100%–30%]). For those in lesser tax brackets, the benefits are not as great, but they're still considerable. Conversely, where taxes are higher, the benefits are greater.

There may be some hidden costs as well. A wind turbine could potentially increase property taxes, though in some areas they are tax exempt. For this reason financiers and accountants look at the attractiveness of a wind investment both before—and after—taxes.

Revenue

The income derived from a wind project is a product of the annual energy production (AEP) and the value of the electricity generated. Like costs, the value of the electricity generated depends upon a number of factors, the most important of which is the *rate* or *tariff* per kWh.

Net Metering

In a *net-metering* application, the wind turbine displaces electricity that otherwise would be bought from the utility. Sometimes this is clearly stated on your electricity bill—sometimes not. For a homeowner, the cost of electricity per kilowatt-hour is easy enough to estimate. To get an average cost in cents per kilowatt-hour, simply pull out the last 12 months of your electric bills, sum them, and divide by the total amount of electricity consumed. For farms and businesses, it may not be so clear.

Commercial customers pay not only for the energy they use but also for the demand they put on the utility system. This charge is based on the maximum power drawn during the billing period in relation to their total energy consumption. This compensates the utility for maintaining generators online to provide power at the demand of the customer. By installing a wind turbine, a business lowers its total consumption while possibly not reducing its peak demand. Thus the wind system could actually increase the demand charge while lowering costs for the energy consumed. This isn't a problem with all utilities, but farmers and businesses need to watch for it.

Once the rate or tariff for the electricity offset has been found, the value of annual revenues is simply the product of the rate or tariff times the amount of electricity generated.

Net metering is not limited solely to small turbines at homes and business. Large turbines are being increasingly installed at commercial and industrial sites in Europe and North America. In the United States, many states have policies that specify how large a wind turbine can be installed under net metering and offset consumption at the prevailing tariff—regardless of whether that is the retail or commercial tariff. There are several such installations in California where regulatory policy limits the wind turbine to a maximum of 1 MW.

Germany doesn't have net metering as it's known in North America, but it allows wind generation for on-site consumption where it offsets the commercial tariff. Notably, there are no size limits. In the summer of 2013, Hamburg's container terminal, Eurogate, installed a Nordex 2.4-MW, IEC Class III turbine. The 117-meter (380-foot) diameter turbine is visible to thousands of commuters crossing the architecturally stunning suspension bridge over the Elbe above the port. At this industrial site in the heart of the Hamburg conurbation, the turbine is generating nearly 10 million kWh annually. The turbine has proven so successful that the port authority plans to add seven more by 2015.

Feed-in Tariffs and PPAs
Commercial wind projects using large wind turbines more typically feed all their electricity to the grid, selling their electricity to the utility. In North America, most of these sales are made through a power purchase agreement (PPA) between the project owner and the utility. There are a few jurisdictions in North America, notably Ontario and Nova Scotia, where the electricity is sold under a feed-in tariff (FIT) contract. In much of continental Europe, wind generation is sold through FIT policies. Either the PPA or the FIT establishes how much the project is paid for every kWh delivered to the grid.

For example, a wind turbine installed under Nova Scotia's Community Feed-in Tariff or COMFIT is paid $0.131 Canadian dollar (CAD) per kWh for 20 years. Similarly, a wind turbine installed in Ontario in 2014 would be paid $0.115 CAD per kWh for a duration of 20 years.

The tariffs paid for wind-generated electricity in France and Germany under their FIT programs are more complicated because the rate is partly based on the resource intensity where the turbines are located. Projects are not paid the same rate across the entire 20 years at windy sites. They are paid one price per kWh during an initial period, then a second much lower price for the remainder of the contract.

PPAs in North America may also reduce the tariff per kWh in later years, but most of these contracts are considered secret and not open to the public.

Neither FITs or PPAs are subsidies or incentives. They are simply contracts for the sale of electricity that specify the price that will be paid for that electricity.

Electricity Price Escalation
One of wind energy's chief advantages over generating electricity by conventional means is that the fuel (the wind) is free. The bulk of the cost for a wind system occurs all at once. Once paid for, the energy produced costs little over the remaining life of the wind system. Conventional generation, on the other hand, consumes nonrenewable fuels whose costs continue to escalate. Thus, our analysis would be incomplete if we didn't account for the rising price of electricity. Like other aspects of energy production, the rate at which utility prices will rise over time is hotly debated.

Utility rates escalated sharply during the 1970s and early 1980s. These price hikes were due in part to the rapid rise in the cost of oil caused by two oil embargoes and in part due to completion of expensive new nuclear power plants. Oil-dependent utilities were hit the hardest, and their rates jumped dramatically. Energy costs at other utilities eventually also rose because coal and natural gas prices often track price increases in oil. Further, many utilities in the 1970s committed themselves to massive construction programs to meet expected growth in electricity consumption. These plants, both coal fired and nuclear powered, were completed during the early 1980s and were enormously expensive, particularly the nuclear reactors.

After these price shocks, electricity prices stabilized. Rate increases were much less severe during the late 1980s. In real terms, after accounting for inflation, the rates at some utilities declined. In the 1990s, electric utility rates began to rise modestly. However, rates have begun a steady climb once again as the power plants built in the 1980s near retirement. Electricity from new power plants always costs more than that from plants that have already been paid off.

The most important aspect of utility rate escalation is its relationship to inflation. Critics of utilities, or those hawking energy-saving devices, tend to overemphasize the effect that rising utility rates have over time. They fail to mention, or conveniently ignore, the effect of inflation on rising rates in real terms. If utility rates were rising 10% per year, an *inflation rate* of 5% erodes the potential cost—or value—of each kilowatt-hour.

Currently, the utility industry is undergoing radical change worldwide. No one can predict what effect these changes will have on utility rates in real terms during the years to come. During the 2000s, the price of electricity in parts of the United States rocketed to heights once considered unimaginable, illustrating that electricity prices, like prices in the broader energy

How long do wind turbines last? Some turbines operate for only a few years.

OTHERS ARE STILL OPERATING AFTER THREE DECADES.

market, will remain highly volatile. This volatility is all the more reason why the fixed-cost of wind energy and other forms of renewable energy is a hedge against inflation in electricity prices.

Longevity

On the revenue side of the equation is the term of the contract and the implication that the turbine will continue to generate electricity during the entire contracted period. As Ken Starcher of the Alternative Energy Institute at West Texas A&M University notes, wind energy can be both affordable—and a good investment—but only over the long term. To recover those high up-front costs and earn a profit, the wind turbine has to operate for a long period of time—often two decades or more. Thus, the longevity of the turbine and its reliability are critical to keeping the cost of wind energy down and the profits up. This is one of the reasons why *Wind Energy* has repeatedly emphasized the length of time some wind turbines have remained in service. After its installed cost and its productivity, a wind turbine's longevity is probably the next most critical factor in whether a wind turbine investment is profitable or not.

How long do wind turbines last? Some turbines operate for only a few years. Others are still operating after three decades.

Some small wind turbines, notably lightweight machines, may survive for as few as five years before requiring replacement. Others, with regular maintenance and the occasional repair, can operate for decades.

Longevity of Micro and Mini Wind Turbines

Beech Mountain, near Boone, North Carolina, is famous as the host of one of the United States' early and ill-fated experiments with large wind turbines during the 1970s. It's in a Class 5 wind regime in the American system with an average wind speed at its 1,560-meter (5,100-foot) summit of about 7 m/s (16 mph). Beech Mountain is a good place to put small wind turbines through their paces.

Appalachian State University (ASU) operates a test site atop Beech Mountain, and ASU has tested several well-known small wind turbines. For example, the university had no problems with an Air X, a micro turbine from now-defunct manufacturer Southwest Windpower. It ran reliably for four years before failing.

Results similar to those on Beech Mountain were also found at the Wulf Test Field in California's Tehachapi Pass. There, a mini wind turbine, a Whisper H40, operated for four years, reliably charging batteries, until it failed. Four years is not a long time as wind turbines go.

Mick Sagrillo has long had a reputation among small wind aficionados for his preference for "heavy metal"—that is rugged, more massive turbines than lightweight, flimsy machines. Heavier in this sense is the weight or mass of the turbine relative to the area swept by the rotor. Brent Summerville's experience with small wind turbines on Beech Mountain suggests he's a heavy metal fan too. "We like quiet and reliable because they're the best to live with," says Summervile. And reliability is another word for robustness, the ability to withstand gusty winds for years on end.

One of the problems limiting the effectiveness of small wind turbines in particular is that they are, well, small and consequently don't produce a lot of electricity. All wind turbines eventually need repair or service, some more than others. As Scoraig Wind Electric's Hugh Piggott points out, what's the point of repairing a small wind turbine if it doesn't generate enough electricity to justify the cost of repair? "Where is the motivation to get it fixed?" asks Piggott.

When there's no good reason to fix a small wind turbine, it invariably stands idle. While Europeans might be inclined to remove the turbine out of public embarrassment, North Americans are more likely to leave the turbine standing derelict than to invest in the effort to remove the machine. There are some small wind turbines in California that have stood inoperative, that is, derelict, for more than two decades.

Longevity of Household-Size Wind Turbines

The Bergey Excel is a rugged household-size wind turbine that has been on the market since the early 1980s. Some of these units have been in near continuous operation since then. West Texas A&M's Alternative Energy Institute has operated a Bergey Excel in the Texas Panhandle for more than 20 years, with only one major repair to the turbine itself and three repairs to its inverter. AEI's Bergey has consistently generated about 9,000 kWh per year. With a little bit of luck, careful monitoring, and maintenance when needed, the more rugged of the household-size turbines—like the Bergey Excel—should last 15 to 20 years.

Longevity of Small Commercial-Size Turbines

For commercial wind turbines, the rule of thumb has been a 20-year life span, and many machines have demonstrated that this is a realistic expectation. BTM Consult reports that many early turbines in Denmark are being replaced before they reach the end of their useful lives. As the technology rapidly advances, a turbine's economic life becomes shorter than its technical life. Consequently, many used turbines from California, Denmark, and Germany are finding new homes, showing that these early machines still have many more years to spin wind into electricity.

Even some of the old turbines from the 1970s and early 1980s continue to spin out the kilowatt-hours. For example, the US Department of Agriculture's experiment station near Bushland, Texas, operated an early version of the Enertech E44 for more than two decades (see Figure 19-2. Enertech E44 at Bushland, Texas).

Thankfully, the USDA kept careful records chronicling the turbine's performance. Overall, the turbine logged more than 100,000 hours of operation, and—on average—was available for

Figure 19-2. Enertech E44 at Bushland, Texas. The USDA operated an Enertech E44, an early American-designed turbine, for more than two decades at its Bushland experiment station.

operation nearly 90% of the time in this distributed application. The USDA's E44—a moniker designating the rotor diameter in feet—is a 13.4-meter diameter, small commercial-size turbine operated in three different configurations: first 20 kW, then 40 kW, and finally 60 kW. About every five years, the USDA took the turbine down and overhauled it.

This particular design was notorious for being sensitive to blade soiling, which would substantially reduce its performance. Nevertheless, the USDA's long-term record keeping shows that this early turbine delivered about 500 kWh/m² per

Table 19-9. Bushland, Texas Enertech E44				
	Average Annual Wind Speed	AEP	Annual Specific Yield	Efficiency
	m/s	kWh/yr	kWh/m²/yr	η
Maximum	5.6	91,732	649	36%
Average	5.7	71,232	504	27%
Minimum	5.7	43,750	310	23%
13.4 m diameter; 141 m², 20 years of operation				

year at the Texas Panhandle site where the average wind speed for the period was 5.7 m/s (~ 13 mph) (see Table 19-9. Bushland, Texas E44).

Similar size turbines in California's windy passes and on the coast of Denmark have delivered equivalent performance over a comparable period of time. Some of these early turbines have been in near continuous operation for more than three decades and they're still operating. Corky Brittan, a philosopher who ranches near Livingston, Montana, has operated his Windmatic 15S on the family homestead for 30 years.

Longevity of Large Turbines
Large wind turbines are designed to last 20 to 25 years, though few multimegawatt turbines have been around long enough to prove that with the exception of Tvind in Denmark. The oldest multimegawatt turbines are still in operation after nearly two decades of service in Germany, and if the longevity of early Danish turbines is any guide, large turbines will last as long as well.

Of course, large wind turbines require regular professional care and service. Major components, such as gearboxes, generators, and blades, will have to be replaced or repaired periodically. But this is true of all large machinery—and these machines are indeed large.

Manufacturers of gearless or direct-drive turbines emphasize that their designs eliminate one of the more troublesome components of large turbines, the transmission. It is noteworthy that Bear Mountain Wind chose Enercon's direct-drive turbines for its large project near Dawson Creek in British Columbia. Enercon is Germany's largest wind turbine manufacturer. The direct-drive machine, considered the "Mercedes Benz" of wind turbines, is frequently the choice of German farmers and *Bürgerbeteiligungen* (citizen-owned projects).

More than half of Enercon's sales in Germany are to small, locally owned projects. Observers have suggested that local or "community" owners want wind turbines that last for the long term without the need for expensive gearbox replacements and are thus willing to pay a premium price for longevity. Corporate owners, they argue, have a shorter time horizon—especially in the United States where tax subsidies are limited to 10 years—and opt for turbines with lower up-front costs. These turbines may be more problematic over the long term.

For commercial wind turbines, the rule of thumb has been a 20-year life span, and many machines have demonstrated that this remains a realistic expectation.

Putting It All Together

To recap on the cost side of the ledger, if you lower the installed cost, or lower the long-term running costs, profitability will increase. On the revenue side, if you increase the amount of electricity produced or if you increase the payment for the wind-generated electricity—the tariff—profitability will increase.

Let's put all this together by considering two hypothetical turbines: one large commercial wind turbine and one small commercial wind turbine. We'll first examine a simplified cash flow for the large wind turbine.

Simplified Cash Flow: Large Turbine
Note that this is an extremely simplified example to illustrate how a cash-flow model is constructed. We'll use a large-diameter wind turbine of the silent wind revolution with some of the parameters identified in the LBNL survey from 2012.

The 100-meter (328-foot) diameter turbine is rated at 1.6 MW and costs about $3.5 million as part of a large project. We'll assume that its annual reoccurring costs are 4% of its installed cost. To simplify the spreadsheet further, we'll unrealistically assume that the entire cost is financed with debt at 7.5% interest that must be repaid within 10 years. At a 7 m/s site, the yield for this type of turbine is around 800 kWh/m^2 per year after accounting for losses of 10% due to array effects and losses in the electrical conductors. So the turbine will generate nearly 6.3 million kWh annually.

The turbine operates under a contract that pays $0.10 per kWh through either a PPA or a feed-in tariff. This is a comparable to payment for wind-generated electricity at an interior site in Germany and France and in Ontario, Canada.

Some PPAs and feed-in tariffs inflate the

Table 19-10. Sample Cash Flow for Single Large Wind Turbine in Commercial Application Before Tax							
Assumptions: Rotor Diameter, 100; Swept Area, 7,854; Avg. wind speed, 7; Yield, 800; Net Generation, 6,283,185; Installed Cost, 3,500,000; Annual Reoccurring Costs, $140,000; Tariff or PPA Rate, $ 0.10; Inflation Rate, 2%; Loan Term, 10; Loan Interest, 7.50%.							
Year	Gross Revenue	Annual Reocurring Costs	Loan Interest	Loan Principal	Annual Revenue (Loss)	Cumulative Revenue (Loss)
0						
1	$628,319	($140,000)	($262,500)	($247,401)	($21,582)	($21,582)
2	$640,885	($142,800)	($243,945)	($265,956)	($11,816)	($33,398)
3	$653,703	($145,656)	($223,998)	($285,902)	($1,854)	($35,252)
4	$666,777	($148,569)	($202,556)	($307,345)	$8,307	($26,945)
5	$680,112	($151,541)	($179,505)	($330,396)	$18,671	($8,274)
6	$693,714	($154,571)	($154,725)	($355,176)	$29,242	$20,968
7	$707,589	($157,663)	($128,087)	($381,814)	$40,025	$60,993
8	$721,740	($160,816)	($99,451)	($410,450)	$51,024	$112,017
9	$736,175	($164,032)	($68,667)	($441,234)	$62,242	$174,259
10	$750,899	($167,313)	($35,574)	($474,326)	$73,685	$247,944
11	$765,917	($170,659)			$595,258	$843,202
12	$781,235	($174,072)			$607,163	$1,450,364
13	$796,860	($177,554)			$619,306	$2,069,670
14	$812,797	($181,105)			$631,692	$2,701,363
15	$829,053	($184,727)			$644,326	$3,345,688
16	$845,634	($188,422)			$657,212	$4,002,901
17	$862,547	($192,190)			$670,357	$4,673,258
18	$879,798	($196,034)			$683,764	$5,357,021
19	$897,394	($199,954)			$697,439	$6,054,461
20	$915,341	($203,954)			$711,388	$6,765,848
For illustrative purposes only.							

payment per kWh with inflation, though Germany does not. In this example, we'll assume that the tariff increases with the rate of inflation.

We'll consider cash flow before tax only. There are no subsidies—no ITC, no PTC—nor deductions for expenses, nor depreciation deductions included in this example (see Table 19-10. Sample Cash Flow for Single Large Wind Turbine in Commercial Application Before Tax).

Under these conditions, the project will see a positive cash flow in year 5. Over the 19-year life of the project, the turbine will earn nearly $7 million, or somewhat less than twice its initial cost.

Tariff Calculation: Large Turbine

Now let's take a look at how we would apply the Chabot profitability index method for calculating what tariff or rate a project should be paid using many of the same assumptions as in the sample cash flow. We'll also perform a sensitivity analysis on two of the key parameters that most directly affect the tariff: annual specific yield and relative installed cost. The annual yield reflects the amount of electricity that will be generated, while the installed cost is the amount that must be paid off before the project can become profitable. Other factors are important, of course, but none have so direct an influence on the price of electricity that a project must be paid to be profitable as annual yield and installed cost.

We'll begin with the same hypothetical large turbine used in the simplified cash flow (see Table 19-11. Feed-In Tariff Calculation Using Chabot PI Method Hypothetical Large Wind Turbine without Tax Credits or Subsidies). For the Chabot model, we'll provide a little more detail. The project will use 30% equity that must earn a 10% return to attract capital from investors. The remainder of the project's cost will be financed with debt that pays 7.25% interest. To simplify the calculations, we'll assume that the term of the debt equals the life of the turbine. Unrealistic yes, but some projects have found debt for surprisingly long periods. This gives the project a weighted average cost of capital of 8.1% and a real WACC of 6.0% after accounting for inflation.

The profitability index target is a factor to account for the relative attractiveness of the investment. It is the net present value divided by the installed cost. Large, capital-intensive projects in high-risk industries, such as oil and gas, have a profitability index considerably greater than 0.3. For our example, we'll use a ratio of 0.2.

Under these conditions, our hypothetical large turbine should be paid slightly more than $0.07 per kWh with the rate annually adjusted for inflation.

The assumptions in this example are reasonable but somewhat optimistic. If the installed cost was higher, say $500/m², and the yield lower, for example, 700 kWh/m² per year, the project would need to be paid $0.10 per kWh to meet the other conditions in our example (see Table 19-12. Sensitivity Analysis Feed-In Tariff Calculation Using Chabot PI Method Hypothetical Large Wind Turbine without Tax Credits or Subsidies in $/kWh). This is about what wind turbines were paid on average at interior sites in Germany and France in 2014.

Tariff Calculation: Small Commercial Turbine

Next, let's examine a small commercial wind turbine being installed on a large farm. Like the large turbine example, this turbine is representative of the silent wind revolution, but in the small wind turbine class. It uses a relatively large rotor to drive a 50-kW generator, giving it a specific area of 5.8 m²/kW.

Like other small turbines, though, its relative cost is considerably greater than that of the large turbine at $1,380/m².

We'll assume that the farmer has access to low-cost debt at 5.5% interest but who wants to earn a 7.5% return on his down payment—the equity in the turbine.

With the same yield as in the previous example, this wind turbine would need a tariff of nearly $0.22 per kWh to make the project profitable (see Table 19-13. Feed-In Tariff Calculation Using Chabot PI Method Hypothetical Small Commercial Wind Turbine without ITC or Subsidies).

At a less windy site, where the turbine would produce a quarter less generation (600 kWh/m²/yr), the turbine would need to be paid about $0.30 per kWh (see Table 19-14. Sensitivity Analysis Feed-In Tariff Calculation Using Chabot PI Method Hypothetical Small Commercial Wind Turbine without ITC or Subsidies in $/kWh). This is about the price that turbines in this size class are paid under Great Britain's precedent-

Table 19-11. Feed-in Tariff Calculation Using Chabot PI Method – Hypothetical Large Commercial Wind Turbone without ITC or Subsidies		
Asssumptions		
Rated Power	kW	1,600
Rotor Diameter	m	100
Specific Installed Cost	$/m²	$400
Federal ITC Rate		0%
Annual Expenses	on installed cost	4.0%
Equity		30%
Return on Equity		10%
Interest on Debt		7.3%
Loan Term	years	20
Inflation		2.0%
Annual Specific Yield	kWh/m²/yr	800
Profitability Index Target	NPV/I	0.2
Results		
Swept Area	m²	7854.0
Installed Cost		$3,141,593
Federal ITC		$0
Total Less ITC		$3,141,593
Specific Installed Cost after ITC	$/m²	$400
Debt		70%
Nominal WACC		8.1%
WACC real	Discount rate real	6.0%
Generation	kWh/yr	6,283,185
Capital Recovery Factor		0.0869
Tariff Required	$/kWh	$0.072
Simple Payback	years	9.6
See the Appendix for further details.		

Table 19-12. Sensitivity Analysis Feed-in Tariff Calculation Using Chabot PI Method Hypothetical Large Wind Turbine without Tax Credits or Subsidies in $/kWh

		Sensitivity Analysis						
		$/m²						
		$ 400	$ 450	$ 500	$ 550	$ 600	$ 650	$ 700
Yield	600	$ 0.10	$ 0.11	$ 0.12	$ 0.13	$ 0.14	$ 0.16	$ 0.17
kWh/m²/yr	650	$ 0.09	$ 0.10	$ 0.11	$ 0.12	$ 0.13	$ 0.14	$ 0.16
	700	$ 0.08	$ 0.09	$ 0.10	$ 0.11	$ 0.12	$ 0.13	$ 0.14
	750	$ 0.08	$ 0.09	$ 0.10	$ 0.11	$ 0.12	$ 0.13	$ 0.13
	800	$ 0.07	$ 0.08	$ 0.09	$ 0.10	$ 0.11	$ 0.12	$ 0.13
	850	$ 0.07	$ 0.08	$ 0.08	$ 0.09	$ 0.10	$ 0.11	$ 0.12
	900	$ 0.06	$ 0.07	$ 0.08	$ 0.09	$ 0.10	$ 0.10	$ 0.11

Table 19-13. Feed-in Tariff Calculation Using Chabot PI Method Hypothetical Small Commercial Wind Turbine without ITC or Subsidies

Assumptions		
Rated Power	kW	50.0
Rotor Diameter	m	19.2
Specific Installed Cost	$/m²	$ 1,382
Federal ITC Rate		0%
Annual Expenses	on installed cost	4.0%
Equity		40%
Return on Equity		7.5%
Interest on Debt		5.5%
Loan Term	years	20
Inflation		2.0%
Annual Specific Yield	kWh/m²/yr	800
Profitability Index Target	NPV/I	0.3
Results		
Swept Area	m²	289.5
Installed Cost		$ 400,000
Federal ITC		$ 0
Total Less ITC		$ 400,000
Specific Installed Cost after ITC	$/m²	$ 1,382
Debt		60%
Nominal WACC		6.3%
WACC real	Discount rate real	4.2%
Generation	kWh/yr	231,623
Capital Recovery Factor (n,t)		0.0750
Tariff Required	$/kWh	$ 0.224
Simple Payback	years	11

*See the Appendix for further details
Adapted by Paul Gipe from Bernard Chabot*

setting program of feed-in tariffs for small wind turbines.

In the United States, where federal tax subsidies for small wind turbines (the 30% ITC) have been the rule, the tariff necessary to make this small wind turbine profitable would need to be $0.17 per kWh (see Table 19-15. Sensitivity Analysis Feed-In Tariff Calculation Using Chabot PI Method Hypothetical Small Commercial Wind Turbine with ITC in $/kWh).

We've learned in this chapter a number of the factors to consider when buying a wind turbine or investing in wind energy. We've also seen that contemporary large wind turbines in commercial applications can be operated profitably when paid a fair rate for their electricity. And the tariffs necessary to do so are modest by today's standards. Small wind turbines remain more expensive relative to their larger brethren, but provide valuable services in the distributed applications where they are best suited.

Table 19-14. Sensitivity Analysis Feed-in Tariff Calculation Using Chabot PI Method Hypothetical Small Commercial Wind Turbine without ITC or Subsidies in $/kWh

		Sensitivity Analysis $/m²						
		$1,000	$1,100	$1,200	$1,300	$1,400	$1,500	$1,600
Yield	600	$0.22	$0.24	$0.26	$0.28	$0.30	$0.32	$0.35
kWh/m²/yr	650	$0.20	$0.22	$0.24	$0.26	$0.28	$0.30	$0.32
	700	$0.19	$0.20	$0.22	$0.24	$0.26	$0.28	$0.30
	750	$0.17	$0.19	$0.21	$0.23	$0.24	$0.26	$0.28
	800	$0.16	$0.18	$0.19	$0.21	$0.23	$0.24	$0.26
	850	$0.15	$0.17	$0.18	$0.20	$0.21	$0.23	$0.24
	900	$0.14	$0.16	$0.17	$0.19	$0.20	$0.22	$0.23

Table 19-15. Sensitivity Analysis Feed-in Tariff Calculation Using Chabot PI Method Hypothetical Small Commercial Wind Turbine with ITC in $/kWh

		Sensitivity Analysis $/m²						
		$1,000	$1,100	$1,200	$1,300	$1,400	$1,500	$1,600
Yield	600	0.16	0.18	0.19	0.21	0.22	0.24	0.26
kWh/m²/yr	650	0.15	0.16	0.18	0.19	0.21	0.22	0.24
	700	0.14	0.15	0.16	0.18	0.19	0.21	0.22
	750	0.13	0.14	0.15	0.17	0.18	0.19	0.21
	800	0.12	0.13	0.14	0.16	0.17	0.18	0.19
	850	0.11	0.12	0.14	0.15	0.16	0.17	0.18
	900	0.11	0.12	0.13	0.14	0.15	0.16	0.17

20

Community Wind

> Wind is a local resource. It is our resource. We want renewable
> energy, and we want to make money out of it. So we do it ourselves.
>
> —Wolfgang Paulsen, *stromrebel*

WIND CAN BE—IS TODAY—A PRODUCER OF BULK ELECTRICITY by the massing of tens, sometimes hundreds, and occasionally thousands of wind turbines in large arrays. These wind power plants are the visible manifestation of wind energy's remarkable progress during the past three decades. But wind energy can be much more. Wind energy can become an integral part of the communities where the wind turbines are located by offering farmers, as well as both rural and urban residents, the ability to build, own, operate—and profit from—their own renewable source of electricity. And by doing so, wind energy can help revitalize local economies by putting more money in local pockets.

This chapter examines wind energy's third way. What it is and where it has been most successful. We'll touch on some examples from three continents to illustrate that this concept isn't unique to any one people or nation. And we'll meet some of the rebels that disregarded the naysayers and made the third way possible.

The Third Way

Unlike conventional power plants, wind energy is modular; each wind turbine is a power plant unto itself. Wind power plants can be of any size and located anywhere. A wind power plant can contain from one turbine to thousands. They can be concentrated in large arrays or dispersed across the landscape. They can be installed at the end of long transmission lines or employed where the electricity is needed. Wind turbines, for example, can be placed in one's backyard, near a factory, on a farm, at a harbor, within cities, or even offshore from cities, as in Copenhagen.

North Americans have been exposed to only a few of wind energy's many possibilities. For most North Americans, wind energy is either giant wind farms at the end of long transmission lines great distances from the cities they serve, or small wind turbines found at rural homesteads.

There is, however, a third way. In Denmark and Germany—world leaders in wind energy development—many commercial-scale wind turbines are installed as single units or in small clusters distributed across the countryside, or scattered around and sometimes within urban agglomerations.

Figure 20-1. People power–revolution from below. German symbolic take on a famous photo of raising a wind turbine in the 1970s, itself a take on the photo of US Marines raising the American flag on Iwo Jima. Of course, it's how all towers and utility poles were raised in the days before mobile cranes. During the 1930s, rural-electric cooperatives raised tens of thousands of poles this way across the breadth of North America. The message is that everyone metaphorically can raise his or her own wind turbine–and should. (Agentur für Erneuerbare Energien)

And many of these turbines are either owned by the farmers on whose land the turbines stand, or by groups of local residents. They are, in effect, owned by the community, by its citizens.

It's the modularity of wind energy that makes possible the generation of electricity where the electricity is most valuable, minimizing the need for long high-voltage transmission lines. Each wind turbine, though physically large, is small relative to fossil-fuel-fired and nuclear-fueled power plants. This modularity enables placing wind turbines nearer the load than possible with large, centralized power plants, providing the resiliency necessary for stable electricity networks that was identified decades ago by Amory and Hunter Lovins in their book *Brittle Power*.

Wind energy's modularity also permits local participation in the ownership and operation of the wind turbines. With ownership comes a sense of control and a stake in the future. To farmers and villagers buffeted by globalization and the industrialization of agriculture, community or local ownership of wind energy can offer a ray of hope for rural communities around the globe.

Community Wind

Community wind, as Wisconsin activist Mike Mangan calls it, is wind development on a human scale. Mangan promotes small clusters of wind turbines, like those found in Europe. To Mangan, community wind signifies local turbines, locally owned.

They may be owned individually, cooperatively, or mutually through numerous mechanisms. The key, says Mangan, is for the community to identify the turbines as its own. Doing so may avoid the all-too-common conflicts encountered when developers, viewed as outsiders, propose projects that primarily benefit absentee owners—sometimes located on the other side of the continent or on another continent entirely.

Wind energy's general acceptance in Denmark and Germany has often been attributed to the dispersal of the wind turbines across the landscape and the distribution of ownership across hundreds of thousands of individual participants. Through the late 2000s, some 90% of all the wind-generating capacity installed on land in Denmark has been developed by windmill

Table 20-1. Co-Op & Farmer-Owned Wind Turbines in Europe			
	Farmer	Co-Op	Corporate
The Netherlands	60%	5%	35%
Germany	10%	40%	50%
Denmark*	64%	24%	12%
Spain	0%	0%	100%
Great Britain	1%	1%	98%

Source: Dave Toke, University of Birmingham, 2005, updated to Toke 2008.
*Onshore.

Table 20-2. Ownership of Wind in Germany in 2010			
	%	MW	TWh
Individuals	51.5%	14,015	20.6
Farmers	1.8%	490	0.7
Developers	21.3%	5,797	8.5
Utilities	7.4%	2,014	3.0
Investment Funds	15.5%	4,218	6.2
Industrial	2.3%	626	0.9
Others	0.3%	82	0.1
Total		27,241	40.0

Source: Marktakteure Erneuerbare: Energien Anlagen In der Stromerzeugung, Trend Research, August 2011.
http://www.kni.de/media/pdf/Marktakteure_Erneuerbare_Energie_Anlagen_in_der_Stromerzeugung_2011.pdf.pdf

guilds (*vindmølleaug*) or, what we call in the English-speaking world, cooperatives (see Table 20-1. Co-Op & Farmer-Owned Wind Turbines in Europe).

As much as one-half of all wind capacity in Germany—one of the world's largest producers of wind-generated electricity—has been built by farmers, and associations of local landowners and nearby residents called *Bürgerbeteiligung*. Individual German investors and landowners installed nearly 15,000 MW of wind-generating capacity in Germany through 2010 (see Table 20-2. Ownership of Wind in Germany in 2010). These German citizens, the *Bürger* of *Bürgerbeteiligung*, have invested nearly $30 billion of their own money in wind energy, in the process creating an energy rebellion in Germany (Figure 20-1. People power—revolution from below). While for many it was the right thing to do, it was also a good investment, earning them collectively more than $2 billion per year in revenues. They can use the profits from their wind energy investments to pay for their retirement, their children's education, or however they wish.

Though no reliable data currently exists, half a million Europeans may own a share in local wind turbines. Thus, community wind is wind energy for the rest of us. It's wind energy that everyone can participate in, regardless of where someone lives—whether in a city high-rise or in a small town. You don't have to own a windy plot of land in a rural area to invest in wind energy. You can develop and own a share of wind energy wherever you live.

Much like investing in a mutual fund, participating in community wind can spread the cost—and risk—of developing wind energy over a large number of shareholders. Instead of installing a small wind turbine in your backyard, the cost and maintenance of which are entirely your responsibility, you can join with your neighbors and together invest in a commercial-scale wind turbine in your community or nearby.

What Is Community Wind?

There's no hard and fast definition of "community wind." This in part results from our attempt to describe in English a European phenomenon found notably in Denmark, Germany, and the Netherlands. The term typically brings to mind shared ownership structures that allow local investment in renewable energy projects. But farmers and small businesses individually also own a large portion of the wind and biogas projects in these countries, and in some cases local ownership is mixed with corporate or municipal ownership (see Figure 20-2. Joint ownership).

Similarly, community wind often implies small groups of large wind turbines, but this isn't always the case. While many community power projects fall into the category of distributed generation, some community wind projects are quite large and fall into the category of wind power plants connected at transmission voltages, albeit owned by those nearby or within the neighboring region.

Wind turbines are significantly more cost effective on a commercial scale, and the community

Figure 20-2. Joint ownership. Two of five 600-kW NEG-Micon turbines along a drainage canal in the Wieringemeer polder of Noord Holland. At the time, farmers on either side of the canal owned two of the turbines jointly, one was owned by the local utility, and one was owned by the manufacturer. Late 1990s.

Figure 20-3. Community wind for a school. Dedication of a Northern Power Systems' 100-kW wind turbine at Camden Hills Regional High School in Rockport, Maine. The wind turbine is an example of community wind because the project was initiated by the school's students: they raised the money for the project, and they chose the turbine to use. The turbine benefits the community indirectly by directly benefiting the school. (Bill Hopwood)

wind projects using large wind turbines in their communities.

It's often easier to identify what isn't a community wind project than what is. For example, the Los Angeles Department of Water and Power (LADWP) built a large wind farm in the Tehachapi Mountains far outside its service area. For those nearby, the utility—even though municipally owned—is an absentee landowner with no connection to the area where the turbines are located. In this sense, LADWP is no different from a traditional corporation whether it's Florida Power & Light Company or Electricité de France, both of which also operate commercial wind farms in the Tehachapi Pass.

In short, you'll know community wind when you see it. Often it is locally initiated, locally developed, and locally owned. Sometimes ownership is broadened to surrounding regions to raise enough funds to build the project, but the focus always remains local.

Why Community Wind?

Community wind enables everyone to participate in the renewable energy revolution sweeping the globe, not just corporate CEOs or those fortunate enough to live on a windy plot of land. For wind energy to reach its full potential, it must be accepted by the communities of which it is a part, and that will only happen when more people have an ownership stake in wind energy—in *their* wind energy.

Large conventional power plants, because there are so few of them, can be—and often are—built over the objections of local residents. Not so with renewables. Distributed renewables are different because there must be thousands, possibly millions, of wind turbines, solar panels, and biogas plants scattered across the vast breadth of North America to meet our energy needs. To succeed at the scale necessary, distributed renewables in some form or another must be welcomed into nearly every community. If members of the community have a stake in a group of wind turbines, they are less prone to suggest putting them somewhere else—or worse—arguing that wind turbines are not needed at all.

wind movement in Europe has focused exclusively on commercial-scale turbines. However, community wind does include medium-size turbines installed at schools or community centers where the objective is to provide direct benefits to the community (see Figure 20-3. Community wind for a school).

Local projects that are closely identified with the community but developed by municipal utilities also fall under the rubric of community wind. Some municipal utilities have been leaders in developing local projects for local benefits not only in Germany and Denmark but also in North America. Hull, Massachusetts, and Toronto, Ontario, are but two cities where municipal utilities have developed or aided small

THE RESIDENTS OF GIGHA AFFECTIONATELY ADDRESS THEIR TURBINES AS THE "THREE DANCING LADIES."

Greater Acceptance

Community wind development can never guarantee total community acceptance, but it does offer a means to take full advantage of wind's modular attributes while maximizing wind's public acceptance. Academic studies and public opinion polls have consistently confirmed what has been known anecdotally for some time: give someone a financial stake in a wind turbine, and he or she will more likely accept its place in the community.

In one study, Dutch researchers examined the attitudes toward wind energy in two otherwise similar villages in the East German state of Saxony. In the village of Zschadraß, the wind turbines were locally owned; in the other, Nossen, they were not. Not surprisingly, the researchers found that the residents of Zschadraß were "consistently more positive towards local renewable energy as well as renewable energy in general than the residents of Nossen" where the wind turbines were owned by an outside company. The main elements for the success of wind energy in Zschadraß, they found, were that the project was initiated locally by people trusted in the community and that the profits of the project were reinvested in the community to the benefit of the local population.

It's not just Germans or Danes who feel this way. It's universal. Give people a measure of control over their community and their energy future, and they will respond positively.

Researchers found in a 2008 study there was general support for wind energy regardless of ownership in southwest Scotland. However, their data showed that the residents of the Island of Gigha, where there was a community-owned wind farm, were consistently more positive about wind energy than those in a community where local people didn't own the turbines. Interestingly, residents in both communities found that the wind turbines did have a visual or aesthetic impact on the landscape—an unexpectedly positive one. The residents of Gigha affectionately address their turbines as the "three dancing ladies." The researchers concluded that local ownership created a "positive psychological effect" and speculated—like others before them—that increasing local ownership will improve public attitudes toward wind energy in Scotland. (No doubt to the dismay of the likes of Donald Trump who vociferously opposed wind turbines near one of his Scottish golf courses.)

Greater Economic Benefits

In community wind, less of the revenue leaves the community for distant financial centers, such as Toronto, New York, or Chicago, than in traditional corporate development. Community wind puts the revenue—the profit—from renewable generation into the pockets of the communities where the wind turbines will be installed. More of the earnings stay in the community, circulating among residents and shopkeepers, creating more local jobs than otherwise.

This is not speculative theory. Studies have consistently shown that the economic benefits of wind energy are greater if the turbines are owned locally than when they are not. In 2009, NREL not only surveyed the existing academic literature on the economic impact of wind development, but it undertook its own analysis of existing and proposed wind projects in the heartland of the United States. Its findings, while more conservative than others, confirmed that community-owned or locally owned wind projects delivered

Table 20-3. Economic Impact of Community Ownership Relative to Conventional Absentee Ownership NREL 2009		
	Multiplier	
	Construction	Operation
Existing Projects		
Local ownership	1.1-1.3	1.1-2.8
New Projects		
Local ownership	3.1	1.8
Absentee ownership	1	1

Source: Economic Development Impacts of Community Wind Projects, NREL/CP-500-45555, April 2009.
http://www.nrel.gov/docs/fy09osti/45555.pdf

two to three times more economic benefits than traditional corporate ownership (see Table 20-3. Economic Impact of Community Ownership Relative to Conventional Absentee Ownership NREL 2009).

Here's a concrete way to look at it. Consider the example of one 80-meter (260-foot) diameter wind turbine at a modest site in the heartland of North America with an average wind speed of 6 m/s (13.4 mph). Let's say the 2-MW turbine generates 4 million kilowatt-hours per year, and let's assume we can sell that electricity for $0.10 per kWh. This one lone wind turbine will earn gross revenues of $400,000 per year. If a farmer leased his land to an outside wind company, he might, at best, receive a royalty of 5%, or $20,000 per year. This is a new cash crop that the farmer didn't have before. It's far better to earn 5% than to just curse the wind.

Now, let's assume that the farmer is willing to take the risk of managing his wind investment. And let's assume that the total cost of the project, after accounting for all operating expenses, can be recovered within 10 years. During the second 10 years of the 20-year life of the turbine, the farmer potentially could earn $4 million, less annual operating expenses.

Over the same 20 years, the farmer who leased his land would earn royalties of about $400,000. Certainly leasing land to a commercial wind developer is easier and less risky than making an investment in a 20-year, capital-intensive project like a wind turbine. And farming is certainly a risky business to begin with. On the other hand, most wind leases in North America pay only a small fraction of the 5% royalty common in Europe. Many pay only half that—some pay even less.

In this example, one commercial-scale, locally owned wind turbine could potentially pump millions of dollars into the local economy—an order of magnitude more than a traditional land lease. In practice, direct ownership of a large wind turbine is simply too great a financial risk for most farmers. However, they can reduce their individual risk by owning the turbine together with their neighbors in a community wind project.

The community-owned project will then also lease the land where the wind turbine stands from the farmer, but the farmer or landowner can also invest in the project. In this way, the landowner shares the project's risk with others in the community or region but also shares in the profit. Meanwhile, the landowner receives lease payments from the community wind project—often on better terms than from an absentee owner because the landowner is negotiating with his neighbors.

In both cases, the profit from the wind turbines flows directly to local owners and then on to the community where they live, shop, and pay taxes.

None of this is new, especially in Denmark and Germany. It harkens back to the early days of the cooperative movements in both countries. Friedrich Wilhelm Raiffeisen (1818–1888), the small-town financier whose name is on cooperative banks all across Germany, put it succinctly when he said "*Das geld des Dorfes dem Dorfe!*" (The money of the village for the village!). Martin Hoppe-Kilpper, the head of the Institute for Decentralized Energy Technologies (Institut dezentrale Energietechnologien gemeinnützige) in Kassel explains Raiffeisen's motto more prosaically by noting that the key to regional value creation is investment and finance from the region itself.

Cooperative and Mutual Investment

Despite the vastness of North America, not everyone can install a wind turbine—of any size—in his or her own backyard. The majority of North Americans live in urban agglomerations or tract housing where a personal wind turbine is out of the question. Even in rural areas, farmers and other small businesses may find installing their own wind turbine to be too great a personal risk.

Because of their higher population density, continental Europeans long ago confronted a similar dilemma. Their ingenious solution was to join together and buy one or more large wind turbines as a group and place them to best advantage in or near their community. This simple but

---

Friedrich Wilhelm Raiffeisen put it succinctly when he said
"DAS GELD DES DORFES DEM DORFE!"
(The money of the village for the village!).

revolutionary idea powered first Denmark, then Germany, to world dominance in wind energy development.

The advantages of shared investment over individual ownership were manifold, but most importantly it spread the risk to a larger group and minimized the investment any one family needed to make. Many Danish cooperatives, like the Middelgrunden co-op outside Copenhagen, deliberately offer shares for less than the equivalent of $1,000 to attract as many participants as possible. By doing so, they not only spread the risk but also the opportunity to profit from wind energy to the entire community.

An investment group, whether a limited partnership, a mutual society, or a cooperative, could also raise a larger pool of risk capital than any single homeowner or farmer. The group or partnership could afford to hire the professional engineers and meteorologists necessary to ensure a sound investment in one or many large wind turbines. These are the professional skills that are prohibitively expensive for a homeowner or small landowner installing his or her own wind turbine. Thus, shared investment enables the owners to gain the economies of scale that accrue from using a large wind turbine as opposed to a small wind turbine.

An unsung benefit of taking on such a daunting task as installing a wind turbine is the mutual support and encouragement a group of like-minded investors provides. There's a certain amount of institutional inertia or stubbornness necessary to install a wind turbine—of any size. Large corporations are successful at developing wind energy in part because they can financially support a staff who can keep plugging away month after month to see a project to fruition. And it takes nearly as much effort to push a project with 10 turbines through to completion as it does a project with 50 or 100.

Shared ownership can take many forms. Some are cooperatives; others are limited partnerships. In a cooperative, the members may individually provide all the capital, both equity and debt. The members may borrow the money from their bank to make the investment, but their purchase of shares represents their total investment in the project. For example, co-op investors in Middelgrunden's 20-MW share of the 40-MW offshore project raised $38 million. Or the cooperative may raise the equity portion of the investment from its members and finance the rest of the project with a loan. Similarly, most shared-ownership projects in Germany use limited liability partnerships. In these projects, the owners raise the equity portion of the project and finance the remainder of the project with debt. The equity portion of these investments is typically 20% to 30% of the project's total cost.

Characteristics of Community Wind

There are several characteristics that separate community wind projects from conventional corporate investment. Most noticeable is a difference in attitude or philosophy on the part of developers. Rather than a desire to maximize their own profit by restricting who can own the project in a traditional investment, a community wind project is more egalitarian, opening its doors to as many investors as possible to share in the benefits.

A common allegory in business is a difference in negotiating styles. Aggressive, hard-nosed corporate investors look at a pot of money left on the table at the end of the year or at the end of negotiations. Their instinctive response is to grab all the money left over for themselves and let the others at the table shout all they want. They have possession. They decide whether or not to share with the others at the table and to what degree. For community wind organizers, the pot of money left on the table is everyone's to share.

Another characteristic of community wind is a low minimum investment to enable as many people as possible to participate. Similarly, there may be limits on the maximum number of shares owned by any one individual; again, the intent is to spread the opportunity to profit from the wind to as many people as possible.

Because projects are organized locally for local benefits, community wind projects typically give priority or first right of refusal to those living within the vicinity of the projects. If more capital is needed to fund the project than can be raised in the community, the circle of investors is gradually expanded until sufficient funds are invested. Thus, larger projects will entail region-wide investments, but those in the neighboring communities are always given the first opportunities to invest.

There is also a willingness to share information among community wind projects that's typically not found in traditional absentee ownership. There is both open communication with members as well as with the public. In the past, most locally owned projects, such as Kennemerwind in the Netherlands, published regular newsletters. Now most information is posted online for everyone to see—and this includes production data.

Cooperatives in particular are much more open to sharing production data than traditional corporate projects. Most cooperatives post their annual production; some go so far as to post monthly production or production by individual wind turbines. Danish co-ops, or *vindmølleaug*, take their responsibility to share information further than most and often post information in English for non-Danish speakers.

Another difference appears in how the projects are administered. While many community wind projects are not organized as true cooperatives, those that are grant each person one vote in the running of the cooperative regardless of the number of shares owned. Co-op members vote for who is elected to the board of directors and thus have a say in how the business is run (see Figure 20-4. Local democracy in action).

Figure 20-4. Local democracy in action. Basia Pioro collects votes for new officers from WindShare members at the 2006 annual general membership meeting in Toronto. Each member of the cooperative has one vote regardless of the number of shares he or she owns.

Typically community wind investors are also willing to earn a lower rate of return on their invested capital than traditional equity investors. The hurdle rate for traditional equity investors, said a 2014 study by the Deutsches Institut für Wirtschaftsforschung (DIW), a German economic think tank, was a 7% to 9% annual return. Whereas, the citizen owners of community wind projects were willing to accept only 4% to 6% annual returns in part for the opportunity to invest in their own community. This in turn translates into a lower weighted average cost of capital for German community wind projects than a traditional corporate investment, enabling wind turbines to be profitably installed in areas with lower winds—and closer to the loads needing the electricity—than more conventional projects.

Denmark's *Fællesmølle* and *Vindmøllelaug*

Denmark has long been known for the high concentration of jointly owned wind turbines (*fællesmølle*) or wind cooperatives (*vindmøllelaug*), as well as commercial-scale wind turbines owned individually by farmers (see Figure 20-5. Modern Danish wind cooperative). Cooperative and shared ownership of wind development is a natural outgrowth of Danish cultural and agricultural interests says Asbjørn Bjerre of Danmarks Vindmølleforening (the Danish Wind Turbine Owners Association). It's also an outgrowth of the movement to stop nuclear power through citizen action in building and owning their own renewable generation.

Following the example set by Tvind, the movement pushed Danish lawmakers to make it easier for Danes to invest in and own their own wind turbines through mutual action. The Folketing, or Danish parliament, responded, says Bjerre, by exempting owners of wind turbines from taxes on the portion of the wind generation that offsets a household's domestic electricity consumption. The objective, he explains, was to encourage individual participation in community action toward meeting Danish energy and environmental policy.

Through this program, nearly any Danish household could effectively generate all its own electricity with wind energy without necessarily

Figure 20-5. Modern Danish wind cooperative. Four of the eight turbines in this project in Denmark's Limfjord are owned by the Thyborøn Harboøre Vindmøllelaug, or wind turbine cooperative. The Vestas V80s were installed in 2003 and "have their feet wet" as the Danes say. That is, the turbines are offshore at high tide but connected to shore at low tide. The turbines generate from 29 to 35 million kWh per year at the windy site on Denmark's northwest coast. For an aerial photo of the site, a gallery of installation photos, and current and past production, visit www.roenland.dk.

capacity the member has installed, the association is managed like a cooperative. Unlike traditional trade organizations where money talks—those with the most money call the shots—each member in Danmarks Vindmølleforening has one vote and one vote only. Utility companies that own giant offshore wind farms, despite their financial and political clout, are entitled to the same one vote as a pig farmer from Jutland.

While there are many cooperatives and locally owned wind projects in Denmark, here are a few examples.

Lynetten Vindmøllelaug

The term *cooperative* is used loosely. In the Danish context, many cooperative ventures were assembled as limited liability companies—investment co-ops—in response to the vagaries of Danish law and tax policy. These associations funded installation of single turbines, clusters of turbines, and sometimes small wind farms, such as the Lynetten cooperative, which owns four out of the seven 600-kilowatt wind turbines on a breakwater in Copenhagen's port. (The other three turbines are owned by Copenhagen's municipal utility.) The turbines are visible from the Christianborg Palace, the seat of parliament, the Folketing. The turbines are also visible from many of the most visited tourist attractions in the city (see Figure 20-6. Lynetten Vindmøllelaug). Information on the production from the turbines is publicly available on the cooperative's website.

putting the wind turbine in the family's backyard. Typically, a wind cooperative would buy a wind turbine, site it to greatest advantage, sell all the electricity to the utility, and share the revenue among its members. This enabled a group to buy the most cost-effective turbine available, even though it would generate more electricity than any one individual member needed.

The concept was a resounding success. In 2014, Denmark generated 40% of its electricity by wind energy alone. For many Danes, a share in a wind turbine is most rewarding when it's cold and windy outside—and that's often the case in Denmark. According to Bjerre, they could be heard saying "Oh, it's a good day today—it's cold and windy."

The association has been successful too. Today, 80% of all wind-generating capacity in Denmark is represented by members in the association. Though membership dues are based on the amount of

Figure 20-6. Lynetten Vindmøllelaug. The cooperative's four wind turbines are owned by residents of Copenhagen. The other three turbines are owned by the city's municipal utility. The turbines are visible from many points in Copenhagen, including from Christiansborg Palace, the seat of Denmark's parliament. Here, picnickers enjoy the historical park of Trekroner in 1998 with the Lynetten wind turbines in the background.

Middelgrunden Vindmølleaug

Despite changes in Danish policy toward renewable energy during the past decade, cooperative action remains an important avenue toward local ownership. Half of the Middelgrunden wind farms of twenty 2-MW turbines installed just offshore from Copenhagen were developed cooperatively (see Figure 20-7. Offshore Middelgrunden Vindmølleaug). The remaining ten turbines are owned by the municipal utility, as in the Lynetten example.

Despite numerous obstacles, including vehement opposition by wealthy beachfront property owners, the project was eventually completed and has since become world famous as an example of how cooperatives can take on even intimidating projects such as an offshore wind farm within view of a major city. Organizers believe the project would not have been built without the public support engendered by local ownership.

To enable as broad a participation in the project as possible, the cooperative set the share price at only €570 ($750) each. Altogether 8,500 investors in Denmark's capital city bought shares. Though half of the project is owned by the city utility, the project was initiated and developed by the cooperative.

Middelgrunden is also a shining example of how to aesthetically integrate a wind project into a cityscape with its graceful alignment. Both the project's physical beauty and the people's sense of ownership of the turbines are two reasons that the city and its inhabitants have proudly adopted the turbines as symbols of the city, which hosted the climate summit in 2009. Copenhagen, says leading Danish landscape architect Frode Birk Nielsen, now uses the Middelgrunden project to market itself to the world.

Hvidovre Vindmøllelaug

The Hvidovre Vindmøllelaug or cooperative installed two 3.6-MW Siemens turbines in a near-shore environment just outside a harbor breakwater in 2009. The 120-meter (390-foot) diameter turbines are in an industrial area near a large combined heat and power plant in the community of Hvidovre, population 50,000.

As with other cooperative wind projects within Copenhagen's densely populated urban area, the co-op owns one turbine, the local utility owns the other. The project was organized by many of the same people behind the famous 40-MW Middelgrunden offshore project.

The €7.8 million (~$10 million USD) project generates about 12 million kWh annually. As is common in Denmark, shares are based on 1,000-kWh units, and each share was sold for €670 (~$900). The co-op estimates that the project will earn its shareholders an average of 6.7% annual return during its 25-year life. Each co-op investor has one vote, and the average investor bought three to five shares.

The co-op has 2,300 investors of which 440, or nearly one-fifth, are from the local area. Local ownership keeps the profit from the turbines local as well, say organizers. The co-op has engaged local schools to use their wind turbine for science education and hosts an open house for the public at the turbine every two years.

Dutch Cooperatives

Denmark has no monopoly on cooperatives. Many of the first wind turbines installed in

Figure 20-7 Offshore Middelgrunden Vindmølleaug. One-half of this famous and oft-photographed offshore wind farm, 20 MW of the 40 MW total, is owned cooperatively by people living in Denmark's capital city. The other half is owned by the municipal utility. The turbines stand in a gentle arc in the Øresund to make the turbines aesthetically more pleasing when seen from shore. The turbines are only 3.5 km (2 miles) from the center of Copenhagen and clearly visible in satellite imagery. 2009. (Kim Hansen, Wikimedia Commons)

the Netherlands were installed by co-ops, though conditions were far less favorable than in Denmark. There are now wind cooperatives throughout the country. One of the earliest is Coöperatieve Windenergie Vereniging Kennemerwind.

The co-op's original Lagerwey 15/75 installed in 1989 is still producing after 25 years of operation. They added 9 Lagerweys a few years later as part of a group of 15 turbines along a canal in Noord Holland (see Figure 20-8. Kennemerwind cooperative). These too are also still operating after more than two decades in service.

Like cooperatives elsewhere in Europe, the Dutch cooperative prominently—and proudly—posts its production data online. You can track the performance of each individual wind turbine the co-op owns. Kennemerwind generates 13 to 14 million kWh annually, earning its owners annual returns of from 4% to 7%.

The co-op has now grown to 800 members and paid off all its original investors. Kennemerwind also continues to expand, adding two Vestas V52s to its fleet in 2009. For the new turbines, the co-op raised one-fourth of the project's cost from co-op members and residents near the turbines through a bond yielding 8% and the remainder financed with a loan from social investment lender Triodos Bank. The co-op plans to repower its older site as soon as it can get planning approval. Kennermerwind remains a pioneer in wind energy in the Netherlands, and a pioneer in local ownership in northern Europe.

Germany's Electricity Rebels

As in Denmark and the Netherlands, wind energy in Germany grew out of a grassroots rebellion by those that wanted to break with conventional ways of generating electricity. In German they're called *stromrebellen* (electricity rebels), and one such rebel is Josef Pesch. He's the man behind Fesa, a small firm that develops locally owned renewable energy projects near Freiburg in southern Germany's state of Baden-Württemberg (see Figure 20-9. *Stromrebel* Josef Pesch).

Pesch organizes *Bürgerbeteiligungen* around solar-, wind-, and wood-pellet powered district heating plants in the foothills of Germany's famed Schwarzwald or Black Forest. "Farms here are not very rich," says Pesch in explaining why southern Germany's farmers were quick to consider renewables as a new source of income. "Farmers were the pioneers of wind energy throughout Germany," he adds. Despite their low incomes, farmers were more easily able to find the money to invest than others because farmers have land they can use as collateral.

By engaging the public in developing renewable energy, "We're creating a new class of entrepreneurs," Pesch notes. "Wind wouldn't exist

Figure 20-8. Kennemerwind cooperative. The four Lagerwey turbines (right) along the Noord Holland Kanaal are part of the Kennermerwind coop. Each turbine generates 150,000 to 200,000 kWh per year, earning their owners from 4% to 7% return annually. 1990s.

Figure 20-9. *Stromrebel* Josef Pesch. Two Enercon E66 turbines on a hilltop, Holzschlägermatte, in Germany's Schwarzwald or Black Forest. The view is from Schauinsland, a popular tourist destination with a scenic overlook. The turbines are in a *Bürgerbeteiligung*, a project owned by local citizens. Inset. *Stromrebel* Pesch is responsible for organizing more than 60 MW of locally owned wind, solar, and biomass projects in the vicinity of nearby Freiburg. 2005.

> "We're creating a new class of entrepreneurs," Pesch adds. "Wind wouldn't exist today if it relied only on the investments of the big utilities."

today if it relied only on the investments of the big utilities. The earnings [from wind energy] are too low [here in Germany] because the winds are too low [to interest utilities]. Wind would never have been developed," in Germany, he explains, if they had waited for the utilities to do it.

Yet private citizens are willing to invest for such modest earnings. Half of the investors in Fesa's Freiamt project, for example, probably never invested in anything before, Pesch suspects. Now they are co-owners of a small wind farm.

In another Fesa project, Regiowind Freiburg, the 521 owners invested $6 million in a $20 million project comprising six turbines. Four of the turbines stand atop Roßkopf and are clearly visible from Freiburg, the other two rise above a hill, Holzschlägermatte, on the route to a popular tourist destination, Schauinsland, so named for its beautiful view. Half to two-thirds of the investors were local, another 10% were from the state of Baden-Württemberg, and the rest came from elsewhere in Germany.

In this manner, small local projects have been developed all across Germany.

In the *Bürgerwind* movement, says Henning Holst, the citizens provide the equity in a limited liability company. Typically, the organizers raise as much equity from the local community as

FULL SPEED AHEAD SAYS FRIENDS OF THE EARTH GERMANY: WIND ENERGY IS THE WORKHORSE OF THE ENERGY TRANSITION

I was strolling down the street in Kassel, a medium-size town in Germany's heartland, looking for a bookstore when I stumbled upon a storefront for BUND (Bund für Umwelt und Naturschutz Deutschland), the Friends of the Earth affiliate in Germany. That's not so remarkable, Friends of the Earth is much more visible in Europe than in North America.

What was remarkable was the storefront display on the *energiewende*, Germany's transition from nuclear and fossil fuels to renewable energy. I was attending a conference on communities that have gone to 100% renewable energy so I was aware what has already been accomplished.

The display was simply beautiful–bright, cheery, positive. Gone were those scary images of terrible environmental calamities that are befalling us. Friends of the Earth chose to emphasize what could be done to make a better world not frighten us with the terrible things happening to it now.

The theme was explained in a colorful brochure "*Volle Kraft voraus! Für die ökologische Energiewende von unten*," or full speed ahead for an ecological energy transition from below, that is, from the people. BUND wasn't saying, "Let's force the utilities to install renewable energy–a little bit at a time." No. BUND was saying, "Let's start an energy revolution now–ourselves. we're the only ones who can do it, so let's get started and go full speed ahead!"

The brochure depicted cheerful demonstrators holding placards calling for the "*Energiewende* now" and "no brakes on the *Energiewende*." In other words, they were not urging a campaign for 10%, 20%, or 30% renewable energy. Instead they were saying emphatically let's make the transition to 100% renewable energy–now, not tomorrow.

BUND celebrates Germany's citizen-led energy revolution by noting that more than 50% of all the renewable energy in Germany is owned by farmers and individuals like BUND's members. And a big chunk of the remainder of German renewable energy is owned by small to midsize companies–not utilities.

For BUND, wind energy is the "workhorse" of Germany's *Energiewende*.

They make clear that there are places where wind energy shouldn't be developed, national parks for example, and that careful planning is required near other protected areas. Nevertheless, they don't wring their hands in despair. BUND goes on to note that only 2% of Germany's land area is required to generate 400 TWh of electricity per year with wind energy–or 60% of current consumption. The amount of land actually used by the wind turbines themselves is even less, and the turbines, towers, and foundations can then be removed after 20 years without any threat of radiation or toxic wastes. Thus, wind energy is an efficient and environmentally desirable use of land. Certainly wind energy is not something to be feared.

Moreover, says BUND, wind energy is the cheapest form of renewable energy, and it pays back the energy used in its construction within three to six months. Better yet, wind energy can be developed, owned, and operated by cooperatives, municipal utilities, or partnerships of local people. Thus, wind energy can be developed from *unten*, that is from below. Multinational companies are not needed. The people can do it–themselves.

In fact, the wheels turn slowly in Berlin, Brussels, and New York, says BUND. Often those in these centers of power put the brakes on what needs to be done. Climate change demands immediate action, and that can only happen when the people take the *Energiewende* into their own hands and see that it gets done–now.

Though it is known for its opposition to fossil fuels and nuclear power, BUND proudly notes that its members have always been at the forefront, if not the pioneers, of renewable energy in Germany–not unlike the rebels at Tvind in Denmark in the 1970s. They close their brochure by issuing a revolutionary call to arms: "The *Energiewende*: we can do it [ourselves]. Join [the revolution] now."

For a North American renewable energy activist simply searching for a book store, BUND's small storefront, its cheery display, and its little brochure filled with hope and optimism grounded in the German people's real world accomplishments were an inspiring lift to the spirits.

possible, and then expand their net to the region. If there's still not enough capital to build the project, they open investment up to the entire country.

Holst develops citizen-owned wind projects in Germany's northernmost state, Schleswig-Holstein. His engineering office is located in Husum, the one-time center of the German wind universe where for a quarter century its biennial wind extravaganza, Husum Wind, was the world's largest wind energy trade fair.

One of Holst's most significant projects was repowering a portion of the wind plants on the island of Fehmarn off Germany's North Sea coast. His task was to upgrade 46 MW of older wind turbines with fewer larger turbines. In doing so, Holst substantially raised capacity to 136 MW.

Unlike smaller projects connected at the distribution voltages elsewhere in Germany, the Fehmarn turbines are grouped in clusters and connected at transmission voltages. Holst had to maintain the existing connection to the grid while adding the new generation via a 31-km, 110-kilovolt subsea cable to the German mainland. He estimates the project will cost nearly $200 million when completed, including the new subsea cable.

Holst's Fehmarn repowering project is a good example of how small entrepreneurial companies combined with the *Bürgerwind* movement have built major projects. Size is not a limiting factor. It also illustrates the can-do attitude common among German engineers like Holst and community wind organizers like Pesch.

Another *stromrebel* in Germany's far north is Wolfgang Paulsen, one of the owners of a *Bürgerwind* project near the Danish border. Paulsen lives in Bohmstedt, a village with a grand total of 650 people, 20 farms, and, as Paulsen likes to emphasize, one *Kneipe* or pub—an essential element of German life.

In 1998 Bohmstedt's residents installed nine 600-kW turbines. The turbines produce 11 million kilowatt-hours per year, enough electricity for 3,000 north German households—far more than that consumed in tiny Bohmstedt itself.

"Our objective was to involve as many citizens as possible," says Paulsen. Each participating family invested risk capital of about $2,500. Together they raised $75,000 for the first phase: planning and development. If the project didn't move forward, their initial investment would be lost. Even in the favorable renewable energy climate of Germany, investment in community wind requires a willingness to take risks.

They then needed to raise 15% of the $7.5 million cost of the project. The remainder was financed with bank loans. Ultimately, each family invested a minimum of $35,000 per share.

In explaining the philosophy behind the *Bürgerwind* movement in northern Germany, Paulsen is frank. "Wind is a local resource. It is our resource. We want renewable energy, and we want to make money out of it. So we do it ourselves."

It's a corollary of the can-do attitude so common among the *stromrebels*. "Of course we can do this. Why wouldn't we."

Fesa's Pesch has similar views. "Those who consume electricity should also be part of the solution," say Pesch. They should develop renewable energy themselves and not leave it to the utilities or multinational companies.

Friedrich-Wilhelm-Lübke-Koog

The Friedrich-Wilhelm-Lübke-Koog is an unlikely place to foment revolution. It's a municipality located on a polder (*Koog*) in the far northwest corner of Germany buffeted by gales from the North Sea. Yet the unassuming citizens of this artificial landscape have unknowingly become an inspiration to others worldwide who want to develop wind and solar energy for their own benefit.

It all started in the *Koog* where its 176 residents, nearly all farmers and their families, live within the rectilinear confines of the polder's 1,350 hectares (~ 5 square miles). The *Koog* has had a long history of innovative practices since the polder was settled in 1958. Many of the descendants of those pioneers still live in the polder, including Hans-Detlef Feddersen (see Figure 20-10. Electricity rebellion).

It was Feddersen and his neighbors in the *Koog* who had the chutzpah to ask why they couldn't install their own wind turbines when they saw them being installed in the polder for the first time.

The *Koog* saw the construction of its first wind farm in 1991. The project consisted of fifty 250-kW wind turbines, manufactured at the old

Figure 20-10. Electricity rebellion. Germany's electricity rebellion began in the far northwestern corner of Germany near the Danish border in Friedrich-Wilhelm-Lübke-Koog. Farmers in the *koog* or polder believed in the then radical concept that they could develop wind energy themselves for their own benefit. The concept has now spread throughout Germany. The monument marks the event in 1992 when the citizen-owned wind turbines went into operation. Financing was provided by the local *Raiffeisen* (cooperative) bank. Inset: Hans-Detlef Feddersen, a charismatic leader in the *Bürgerwind* movement in northern Germany. 2012.

shipworks in Husum farther south. At the time, it was one of Europe's largest wind power plants. Shares in the project were sold across Germany and—in an unusual departure for the time—shares were also made available to the *Koog*'s residents.

The year was historic for other reasons. At the end of 1990, the Bundestag, Germany's parliament, passed a law on feeding electricity into the grid—the now famous Stromeinspeisungsgesetz, the first electricity feed law.

Bürger Windpark Lübke-Koog

As the privately financed project was going into the ground, farmers in the *Koog* began meeting to discuss installing their own wind turbines, which they could now do under the new law. At first only 10 farmers were interested, and after just two meetings they concluded it was only going to be possible if they banded together with others. They formed an association, and then set out to inform their neighbors what they were considering. They wanted as many of the residents to participate as possible, both financially and in spirit, so they could build a *Bürger* or "citizen-owned" project. Their model was the many community-owned wind turbines just a few kilometers across the border in Denmark.

Two decades ago, when this was all new, there was some opposition to the wind turbines, says Feddersen. Some were afraid of what the turbines might do to tourism, or the noise they would make. Feddersen made it a point to involve as many people as possible from the very beginning so their concerns could be addressed early on.

In only three months, the small group had grown to 44 shareholders and had won the support of the community in the *Koog*. They invested 10% of the cost of the project directly and raised the remainder through a loan from the region's cooperative bank. Courageously, they used their land and homes as collateral to secure the loan.

After two years of planning, the *Bürgerbeteiligung*, or citizen-owned limited partnership, installed its first 14 turbines. That project was so successful that it was expanded to 32 turbines in 1999, more than doubling the revenue from the first project.

The initial fear that the wind turbines would discourage tourism was turned on its head. The *Koog*'s residents have used the existence of the turbines to promote tourism. Tourist brochures encourage visitors to see the wind turbines as the start of a local adventure, including windmill climbing—a form of recreation not recommended for the faint, or weak, of heart.

Community Economic Development

Feddersen is an active participant in his community. He's been vice mayor of the *Koog* for the past two decades. He has the self-assured manner of someone accustomed to speaking in public about an idea close to his heart. In fact, Feddersen is a stirring speaker about the need for and the role of community-initiated renewable energy development. He emphasizes how revenue from the turbines flows back into the pockets of local owners, increasing their incomes and purchasing power, which they then spend in the regional economy.

The *Koog*'s residents benefit in several ways, explains Feddersen. Landowners with turbines on their property receive payments for land leases in the form of royalties on the revenue generated from the turbines. The *Bürgerbeteiligung* also pays as much as €13,000 ($17,000) per MW of capacity in commercial property taxes, which flows directly into the community's coffers. For a community of less than 200 people, that's a substantial source of revenue.

The wind turbines also provide climate-protection benefits—an important factor for everyone in the polder because they live below sea level. Climate change is not an esoteric threat to them.

Feddersen continues to grow wheat, rapeseed, and sugar beets on his 90-hectare (225-acre) family farm. Today, he spends less time farming than in the past. He spends more and more of his time managing the existing renewable projects and planning new projects for other communities.

Renewable Success

Once the *Koog*'s farmers succeeded with their first project, they realized they could do even more. The experience was empowering. One has only to drive through the polder to see the result of the residents' efforts. While the wind turbines are the most visible form of the *Koog*'s renewable development, there are 1.6 MW of biogas plants and 1.8 MW of rooftop solar in the polder as well.

Sixty years after his family's house was built, Feddersen added 32 kW of solar photovoltaic panels to the roof—as did many of his neighbors. In this part of Germany, the house and the barn are integrated into one building. Thus, there's room for such large arrays on most of the structures in the *Koog*.

The *Koog*'s pioneering success is recognized throughout Germany. The residents won the top prize in a national competition for the highest concentration of solar energy per capita in 2009, 2010, and again in 2011. And in a region with the highest concentration of wind energy worldwide, the *Koog* remains a leader in wind too.

From High Risk to Project Finance

In the early 1990s when Feddersen and his neighbors sat around their kitchen tables considering how they could use wind energy just like the absentee owners of those early turbines, locally owned wind was a high-risk venture. Up until then, wind development in the region was limited, and what had been done by a few regional utilities hadn't been successful. Their heart wasn't in it, explains Feddersen about the utilities' failures.

Yet, the *Koog*'s farmers persevered and "learned by doing." Their expectations were also lower than those of the utilities. They were farmers, after all—experienced at eking out an existence from the land. So they were willing to accept lower returns than a conventional corporate enterprise that had to report to shareholders in Frankfurt.

They put more at stake too. Their farms and their livelihoods were on the line. Thus, they had every reason to make as few mistakes as possible, and when their turbines were out of service, they made sure they were put back in service as soon as possible.

Klaus Rave, a former director of Investitionsbank Schleswig-Holstein, the regional development bank, says that it's this local ownership—where the residents are themselves responsible for their investments—that makes them such good risks for a bank. In the two decades he has financed locally owned wind projects, the bank has never had a default even during the 2007–2008 financial collapse.

Wind is successful in the state of Schleswig-Holstein, says Rave, because it is accessible and adaptable and has local acceptance. Wind is accepted because it is owned locally by the people who have to live with it and because they profit from it both directly and indirectly.

NORDFRIESLAND: GERMANY'S COMMUNITY WIND CAPITAL AND AN ELECTRICITY REBEL STRONGHOLD

What started in Friedrich-Wilhelm-Lübke-Koog didn't stay in the *Koog*. What seemed to the residents of the polder a reasonable and uncontroversial idea—that the wind was theirs and they should use it themselves—soon spread over the dikes to the mainland and struck a chord with Germans from the Danish border in the north to conservative Bavarians in the south.

Friedrich-Wilhelm-Lübke-Koog is located within the *kreis* or county of Nordfriesland in the north German state of Schleswig-Holstein. And nowhere in Germany has the development of wind energy, and especially community-owned wind, been more intensive than in Schleswig-Holstein.

There were 3,300 MW of wind-generating capacity in Schleswig-Holstein in 2011, almost as much as in Denmark (4,000 MW) at the time. But Schleswig-Holstein is much smaller than the nation of Denmark with a fraction of the land area and about half the population. Consequently, Schleswig-Holstein has more than twice the density of wind-generating capacity of Denmark and more than seven times the density of wind generation in Iowa, the state with the highest concentration of wind turbines in North America (see Table: Density of Wind Generating Capacity in Schleswig-Holstein).

Yet few places anywhere compare to Nordfriesland, the wind energy capital of Germany. With its county seat of Husum–*die graue Stadt am grauen Meer* (the gray city on the gray sea)–and the surrounding countryside in northwest Schleswig-Holstein, Nordfriesland has one of the highest concentrations of wind-generating capacity on record (see Table: Density of Wind Generating Capacity in Northwest Schleswig-Holstein). Nordfriesland is only surpassed by the smaller district bordering it on the south, Dithmarschen.

What is most startling of all, more than 80% of all the wind-generating capacity in Nordfriesland is citizen owned. The wind turbines of Nordfriesland generated 1.3 TWh of electicity in 2007, spinning out €7.6 million (~ $10 million) in business taxes, €3.7 million (~ $5 million) in landlease revenues, and sales revenue of €2.4 million (~ $3 million) per year.

For comparison, the locally owned wind turbines of Nordfriesland–many of which were developed by farmers like Hans-Detlef Feddersen–generated more electricity in 2007 than the wind farms of California's famed San Gorgonio Pass, more than the Altamont Pass, and nearly as much as the Tehachapi Pass where none of the wind turbines are locally owned.

What began as a kaffeeklatsch among farmers in Friedrich-Wilhelm-Lübke-Koog in the early 1990s has grown to one of the highest densities of wind-generating capacity in the world–100 times that in Iowa and 30 times that in Denmark, demonstrating to anyone willing to look that residents can live with the wind turbines they own.

With such a successful build out, one could be excused for thinking that the citizens of Schleswig-Holstein were finished with wind energy. No, not at all. In 2012, the state parliament approved a doubling of the land area (1.8%) devoted to wind energy and set a target of doubling the amount of wind generation on land. The state's minister-president (or governor), Torsten Albig, announced that Schleswig-Holstein plans to generate 300% of their electricity consumption with renewable sources by 2020. Wind generation on land will provide more than one-third of that. Most of the wind turbines will likely be owned by its own citizens–the electricity rebels that started an energy revolution.

Density of Wind Generating Capacity in Schleswig-Holstein

	Capacity	Land Area	Density	Population	
	MW	km^2	kW/km^2	Millions	kW/capita
Schleswig-Holstein	3,271	15,763	208	2.8	1.2
Denmark	3,951	43,098	92	5.6	0.7
Iowa	4,400	145,743	30	3.0	1.4

Data for 2011.

Density of Wind Generating Capacity in Northwest Schleswig-Holstein

	Capacity	Land Area	Density	Population	
	MW	km^2	kW/km^2		kW/capita
Friedrich-Wilhelm-Lübke-Koog	47.5	13.5	3,519	176	270
Dithmarschen Kreis	718	1,404	511	135,136	5.3
Nordfriesland Kreis	847	2,047	414	165,707	5.1

Data for 2010.

Wind is successful in the state of Schleswig-Holstein, says Rave, because it is

ACCESSIBLE AND ADAPTABLE AND HAS LOCAL ACCEPTANCE.

Once a project has proven successful, the banks as well as the farmers know the resource, and with the transparent payments from Germany's feed-in tariffs, they can predict the revenue stream from the turbines with a high degree of accuracy. This makes possible the repowering of older projects with newer, more powerful turbines—without collateral. The new projects use "project finance" where the project itself is sufficient collateral for a loan.

Wind, says Rave, has become a safe investment in northern Germany. Investitionsbank prefers investing in community-owned wind projects rather than more conventional investments because *Bürgerbeteiligung* pump more into the local economy than traditional absentee ownership, and the purpose of Investitionsbank is to spur local economic activity.

Repowering

By 2004 when the polder's residents began to replace their earlier installations, there were 32 wind turbines in the *Koog* with a combined capacity of 18.5 MW. Because they live among the turbines they own, the *Koog*'s residents may take more care than others elsewhere in how they site the turbines and how the turbines fit into the landscape.

All the wind turbines are at least 400 meters from the farmers' homes, as required in the state of Schleswig-Holstein, says Feddersen. If the turbines create more than 45 dBA of noise at night, they must be sited farther away than the minimum.

The residents deliberately chose shorter towers than those that would have normally been used. With total heights above 100 meters, wind turbines in Germany must be marked to warn aircraft of their presence. During the day, the nacelles must be illuminated with white strobe lights or the tips of the blades must be painted red. At night the nacelles are illuminated with flashing red lights. The *Koog*'s residents chose turbines and towers that didn't require flashing lights at night, says Feddersen, because "The night should stay dark."

Each time the residents have repowered an existing project with new turbines, they have substantially raised the installed capacity. In 2009, the old wind turbines were removed and new ones installed. The repowerings reduced the number of turbines to 30 units, 25 of which are owned by *Bürgers* of the community. (Five turbines are owned by outside investors from that first project installed in the early 1990s.)

Today, there are four community-owned wind plants in the *Koog*, representing a total of 48 MW. They are currently repowering a site that has Enercon E-66 turbines installed as recently as 2004. The 1.5 MW turbines will be replaced with fewer, but substantially larger, turbines. This will allow residents to buy even more shares in the turbines.

Recipe for Success

None of what the *Koog*'s residents accomplished was easy, says Feddersen, but "it shows what citizens from a small community are capable of when everyone pulls together." Now, more than 95% of all households in the polder have invested in renewable energy in some form.

There are several key themes of the *Bürgerwind* movement, Feddersen says.

- We want renewable energy.
- We can do it ourselves.
- We bring our own risk capital and we invest in our region.
- We take the risk together.
- We accept the change to the landscape that results.

The recipe to accomplish this, says Feddersen is well known. Most importantly, there must be transparency. (This explains why so many community wind projects not only make their production data readily available to everyone; they also reveal the royalties paid to landowners.) People within the region must also want it, he adds, and have the desire to make it happen. They must have the willingness to cooperate and compromise with their neighbors and creatively solve the many problems that they will encounter. But most critically, they must believe that "together we're strong." Because of these characteristics, Friedrich-Wilhelm-Lübke-Koog's success has become a model for community renewable power development worldwide.

Saterland Bürgerbeteiligung

Community-owned projects need not be small as shown by the 72-MW Saterland Bürgerbeteiligung in Lower Saxony (Niedersachsen). Enercon's in-house newsletter boasted of the project that installed 24 of their E-101 direct-drive turbines in what is the largest community-owned wind farm in Germany. It may also be the largest community-owned wind plant in the world. More than 700 regional property owners and residents invested in the project.

"This is a totally community-owned project," said Hubert Frye, *Bürgermeister* or mayor of Saterland. "Everybody was on board during the planning [of the project] as well as for operation [of the turbines]. This was the winning formula," he was quoted as saying. Stephan Kettler, the project's manager, explained that everyone had to be in the same boat to ensure smooth sailing through the complex planning process.

As an indication of how strongly the idea of community-owned wind has put down roots in northern Germany, Kettler said that the local council would only grant planning permission if the project were owned by local residents.

Because of the size of the project and the communities affected, a compromise was reached to establish two operating companies. Out of the 24 3-MW machines, 13 are operated by one wind farm partnership with 666 local owners. The remaining 11 turbines are operated by a second partnership with 50 landowners who provided the land on which the project rests.

The community will earn €350,000 ($450,000) per wind turbine in business taxes over the 20-year period of the project. The organizers also awarded several local firms contracts during the construction phase. The project was financed by DZ Bank, the central bank for the Raiffeisenbank in Scharrel, itself locally owned by farmers and residents. The local cooperative bank managed the project and arranged the financing.

German Genossenschaft or Cooperatives

While most locally owned wind projects in Germany are in the form of limited partnerships, or *Bürgerbeteiligung*, the number of true cooperatives, or *Genossenschaft*, began growing rapidly in the early 2010s. They formed for the same reasons and with the same goals, principally to take control of the *energiewende* "into their own hands."

Their philosophy is simple: "Those who see a wind turbine should also use it." And like Feddersen, they emphasize that the key to successfully organizing a cooperative is many meetings among possible participants, open public information, and full transparency. The cooperatives campaign across Germany under the rallying cry: "By citizens, for citizens."

While the typical community wind project in Germany was a cluster of five wind turbines or fewer, organizers are not intimidated by the size or cost of modern wind turbines. Communities work together across local boundaries to build projects much bigger than any one community could undertake alone as in the Saterland example. And they're willing to work with local utilities as needed.

There are hundreds of local distribution utilities in Germany. Rather than fight the introduction of a community wind project that they don't own directly, many are working together with citizen-owned cooperatives to make projects a reality. They gain projects without direct investment, help their communities, which, after all, is their mission, and they maintain some control or influence on the direction of the project. That's the case outside Kassel in the village of Wolfhagen.

There a cooperative works with Wolfhagen's *Stadtwerke*, or municipal utility. They began with a district heating plant and then expanded to solar. Now they're developing a wind plant. The co-op has 600 members and owns 25% of the projects with the utility that serves the city's 18,000 residents. The cooperative sold shares for €500 ($650), each with a limit on the maximum number of shares any one person can own. The cooperative's investors expect a return of 6% per year. The municipal utility gained new generation —and the cooperation of the community— without spending any of its own capital.

Genossenschaft Odenwald's Christian Breunig explains that in the past small towns and rural areas looked to the cities for support and economic development. With Germany's feed-in tariffs and the advent of community renewables, they can develop their communities themselves. "We look to our own communities and to our own regions" ourselves says Breunig. The Odenwald cooperative's 2,100 members raised enough capital for a 7.2-million investment. "*Wir tragen die energiewende*" (We're making the energy transition happen) is the cooperatives catchphrase.

Community Wind in Britain

Cooperatives or the idea of communities banding together to help themselves is not a Nordic or Germanic innovation. The first successful cooperatives were created in the coalfields of central England in the mid-19th century. It was from there that the idea rapidly spread across Europe. It's fitting then that the modern concept of community wind would return to Britain via Scandinavia.

Baywind

The first community-owned wind project in the British Isles was built by the Baywind Energy Co-operative in early 1997 patterned after similar ventures in Sweden. Situated on Harlock Hill in Cumbria's scenic Lake District, the project is so well known in the community that the local visitors' bureau in Ulverston can direct tourists to the site. There they will find the five 500-kW Wind World turbines with their characteristic upturned boat-shaped nacelles standing amid green swards dotted with grazing sheep.

As repeated in subsequent community wind projects throughout the English-speaking world, the dedication of the cooperative's turbines was a cause for celebration, attracting 400 guests to the windy hilltop site overlooking the Irish Sea. The turbines generate about 5 million kWh per year, earning the cooperative's 1,300 members a 6% annual return (see Figure 20-11. Baywind Energy Co-operative).

Like most cooperatives and shared-ownership projects in Europe, Baywind returns a portion of its profit to the community of which it is a part.

Figure 20-11. Baywind Energy Co-operative. Baywind became the first wind cooperative in Great Britain when it began operating five 500-kW WindWorld turbines in early 1997. The cooperative is seeking planning approval to repower the site in Britain's scenic Lake District with fewer, larger turbines. 2012.

For example, Baywind donates 0.5% of its profit to a local energy conservation trust fund.

Baywind is currently planning to repower the site with much larger turbines. The new turbines will increase production to nearly 20 million kWh per year.

Westmill Wind Farm

As the signboard welcoming visitors to Westmill Wind Farm Co-operative proudly proclaims, its nearly 2,400 members worked hard for four years to raise the £4.3 million (~ $7 million) necessary to complete the project in the spring of 2008. Not only was Westmill the first project of its kind in southern England, it had to overcome the opposition of wealthy landowners in what community power expert David Toke calls "stockbroker country" west of London. Like the American aristocracy in Hyannis Port, Massachusetts, who have fought Cape Wind for so many years, those from the City (London's financial heart) didn't want wind turbines in their midst.

Nevertheless, the locals prevailed in part due to the dogged determination of Adam Twine, one of Britain's electricity rebels (see Figure 20-12. Westmill's Gusty Gizmo). Twine organized the cooperatively owned project for a site his family owned at an old RAF air base overlooking the

Figure 20-12. Westmill's Gusty Gizmo. Adam Twine, Liz Rothschild, and their faithful sidekick Max at one of the five Siemen's 1.3-MW turbines in the Westmill Wind Farm Co-operative in 2012. Developed on an old RAF air base near Swindon, South Oxfordshire, in 2008, the cooperative was the first of its kind in southern England. Since then, the site now also boasts a 5-MW, cooperatively owned solar power plant. The whimsical name, Gusty Gizmo, was given by local schoolchildren from Southfield Junior School. Note that access to the turbines is at the equivalent of the second story because the transformers are placed inside each tower, and that the tower foundation bolts are capped. These details give the Westmill site a clean and tidy appearance, typical of wind turbine installations in Great Britain.

historic Vale of Whitehorse. The five 1.3-MW Siemens turbines stand prominently on the road from Swindon to Oxford near the village of Watchfield. Today, a site once swirling with the roar of warplanes during the Battle of Britain now peacefully generates 10 million kWh annually from the wind.

As in cooperatives on the continent, Westmill gave preference to investment from those living nearby. Westmill also made participation as accessible as possible by setting the minimum investment at only £250 (~ $400) and limited the maximum amount anyone could invest to £20,000 (~ $30,000).

In 2011, Twine was instrumental in developing an adjoining site for a 5-MW solar power plant. The Westmill Solar Co-operative believes that it operates the largest single community-owned photovoltaic power station in the world.

For historical, political, and regulatory reasons, community-owned wind represents only a small fraction of the installed wind-generating capacity across the British Isles. As in other English-speaking countries, it is more difficult to install community-owned wind turbines than to build any other kind of wind project. Despite this, community groups continue to forge ahead. Westmill suggests their success proves that "ordinary people can co-operate to achieve what politicians merely talk about."

Community Power Down Under

Hepburn Wind pioneered community wind Down Under with Australia's first community-owned wind farm. The six-year-long project followed the pattern set in Europe for maximizing local participation by giving local residents priority for investing in the turbines, becoming a model for other Australian community wind projects. The cooperative's nearly 2,000 members invested $9.0 million AUS ($8 million USD) in the turbines and secured more than $13.1 million AUS ($12 million USD) in additional funding through government grants and bank financing.

The cooperative brought the two 2-MW RePower turbines online in mid-2011 with great fanfare, proving that it was possible to raise the financing from local and regional investors in the state of Victoria—investors who want a direct stake in renewable energy development (see Figure 20-13. Hepburn Wind).

Like the organizers in Friedrich-Wilhelm-Lübke-Koog, Hepburn Wind made a special effort to integrate the turbines into the community. They specifically prohibited logos on the nacelles as well as aviation warning lights. The cooperative sponsored a turbine-naming contest and provides public tours of the turbines, and their annual general membership meetings are a celebration of what the cooperative's members were able to accomplish together.

Hepburn's success has now been followed with a second cooperatively owned wind plant in Australia. Appropriately enough, the project is named after the adjacent local village. The Denmark Community Windfarm is located on the opposite side of the continent in Western Australia.

The project began in 2003 as a series of workshops in the coastal town of Denmark to build a wind farm as a local response to global climate change. The decadelong effort finally succeeded in winning approval for up to four turbines on a seaside cliff. Two Enercon turbines finally came online in 2013.

The project, the largest in the town's history, was celebrated with 150 guests at its dedication and in its first year produced 5.4 million kWh, slightly more than projected. Denmark Community Windfarm's progress and stunning photographs of the site can be monitored online.

Community Wind in North America

Community wind energy as found in Europe represents a small niche of the wind industry in North America—today. It need not always be so. As the preceding examples illustrate, groups in many different countries, working in several different languages, have overcome what at the time seemed insurmountable obstacles. They were undeterred, demanding changes to the laws standing in their way and campaigning for the policies that would make it easier for others to follow the path they blazed.

Figure 20-13. Hepburn Wind. Visitors and brass band leaving the dedication ceremony for Australia's first community-owned wind farm in the spring of 2011. Hepburn Wind's two 2-MW turbines became the model for other cooperative wind projects in Australia. (© Karl von Muller, Hepburn Wind)

COMMUNITY WIND NORTH AMERICAN SOURCES OF INFORMATION

Windustry: www.windustry.org
Ontario Sustainable Energy Association: www.ontario-sea.org
Wind-Works.org: www.wind-works.org
The Citizen-Powered Energy Handbook by Greg Pahl: www.chelseagreen.com
Toronto Renewable Energy Cooperative: www.trec.on.ca

The seminal research on community wind development in the United States was conducted by Mark Bolanger for Lawrence Berkeley National Laboratories (LBNL) in 2001. The report focused on how community wind was being developed in Europe and described various approaches to its implementation in the United States. Since then, Bolanger has tracked community wind development, and his data appears in LBNL's annual assessments of wind energy's growth in the United States.

Bolanger defines community wind as projects using turbines over 100 kW in size and completely or partly owned by towns, schools, commercial customers, or farmers, but excluding publicly owned or municipal utilities. According to these criteria, he calculates that only 2% of the wind-generating capacity operating in the United States was community owned. This contrasts with more than 50% in Germany, the world's fourth-largest industrial nation.

Community wind represents an even smaller percentage of wind capacity in Canada. But groups in Ontario and Nova Scotia have put the idea on the political map.

Ontario

Nowhere in North America has any jurisdiction undertaken a more aggressive effort to develop community ownership than in the Canadian province of Ontario—the nation's most populous province and its industrial heart. It all began with one lone wind turbine in the province's capital city, Toronto.

WindShare

The story begins in 1998 when a group of hardy—and necessarily stubborn—organizers banded together as the Toronto Renewable Energy Cooperative (TREC). They intended to replicate the European cooperative wind model. They had no idea how difficult it would be to plant that seed in the soil of one of North America's largest cities.

The learning curve was steep. Initially TREC planned to build a turbine in North Toronto, but it didn't take too long for them to learn that a small turbine located far from the lakefront surrounded by turbulence-inducing buildings was not the best choice. Sites along the waterfront quickly rose to the top of the list due to a cleaner, stronger wind resource. Finally, TREC's organizers settled for up to three sites on the lakeshore; however, one by one they were eliminated. The first two sites were vetoed by the island airport authority. The third site ran into opposition from an environmental organization in partnership with an adjacent yacht club who filed for an injunction. At the bleakest point, direct talks with the opponents provided a way forward. Opposition ceased with an agreement that TREC would explore, in earnest, siting the turbines at Exhibition Place, the site of the Canadian National Exhibition.

ExPlace, as it's called, is no small player in Toronto. It hosts more than five million visitors per year on its expansive grounds. Early on in TREC's planning, ExPlace had been passed over due to the perceived levels of bureaucracy that would stand in the way. It took one meeting with the General Manager of ExPlace to realize that she was a strong supporter of the project and wanted the first turbine to be at ExPlace. Finally, a site was secure.

TREC negotiated a contract with the city's municipal utility, Toronto Hydro, to jointly develop and own the wind project like those in Copenhagen. Immediately, they had to confront the problem that bedevils all such projects in North America: there were no policies that enabled the profitable sale of electricity from such a venture. Their solution was a form of virtual net metering that would allow members of the cooperative to offset their own consumption. Then began a chain of events that nearly killed the newborn in its cradle. Toronto Hydro informed TREC that it couldn't offer virtual net metering after all. Reluctantly, TREC switched strategies and decided to sell its power into Ontario's newly deregulated electricity market.

COMMUNITY WIND

Table 20-4. Representative Shared Ownership Wind Projects

	Region	Country	MW	million kWh/yr	Investors	Equity million	Total Investment million
WindShare[1]	Ontario	Canada	0.75	1	425	$ 0.8	$ 0.8
Regiowind Freiamt	Baden-Württemberg	Germany	3.6	5.7	142	$ 2.1	$ 6.5
Minwind I & II	Minnesota	USA	3.8	11.1	66	$ 1.1	$ 3.5
Regiowind Freiburg	Baden-Württemberg	Germany	10.8	16.8	521	$ 6.7	$ 20.6
Paderborn	Nordrhein-Westfalia	Germany	18.2	31.4	91	$ 6.0	$ 28.5
Middelgrunden[2]	Zealand	Denmark	20	44	8,500	$ 38.2	$ 38.2
Hvidovre	Zealand	Denmark	3.6	12	2,300	$ 10.0	$ 10.0

1. Only half of the turbine is owned by the cooperative.
2. 20 MW of 40 MW project developed by the cooperative, 1,000 kWh/share.

With Ex-Place's vocal support, TREC won planning approval. However, the approach to a nearby island airport limited the project to only one turbine rather than the two once envisioned. Determined to proceed, TREC organized a cooperative dubbed WindShare to own and operate the turbine in partnership with Toronto Hydro.

The cooperative's 425 members raised the equity for their share of the 750-kW Lagerwey wind turbine (see Table 20-4. Representative Shared Ownership Wind Projects). WindShare installed the distinctive direct-drive turbine in late 2002 in a park-like setting overlooking the Gardiner Expressway—the route into the city from Toronto's international airport. It began operation in early 2003 (see Figure 20-14. Toronto's WindShare cooperative).

Figure 20-14. Toronto's WindShare cooperative. WindShare's more than 400 members own one-half of the wind turbine. Toronto Hydro, the municipal utility, owns the other half. The Lagerwey 52-meter (170-foot) diameter wind turbine stands at a prominent location overlooking the Gardiner Expressway in Canada's largest city. The first urban wind turbine in North America, it has became a beacon for renewable energy advocates across the continent. 2009. Inset: Ed Hale, one of Ontario's electricity rebels who made the WindShare turbine possible. Hale is in a climbing harness for the annual open house and summer picnic beneath the wind turbine in 2004. Visitors were welcome to climb the tower.

Meanwhile, Ontario's experiment with deregulating its electricity market—not unlike that in California at the time—proved disastrous. The provincial government panicked and reverted to previous policies that effectively subsidized electricity consumption. WindShare's business case was built on the deregulated market. The situation looked bleak.

There was no alternative. For WindShare to succeed and for community power to grow beyond one single wind turbine, Ontario needed a policy like that in Germany where there was a right to sell electricity to the grid and a fixed-price for the electricity that was delivered. WindShare needed a feed-in tariff.

TREC spun off the Ontario Sustainable Energy Association (OSEA) to promote community power across the province. OSEA then turned its attention to bringing feed-in tariffs back to North America. Though feed-in tariffs had been used in California during the mid-1980s to power the great wind boom, they had been abandoned. Europeans, especially Germans, took up the baton and ran with the idea.

OSEA led the campaign for community power in Ontario and the feed-in tariffs that would make it possible. They were remarkably successful. In 2006 the province launched its first crude attempt at a feed-in tariff with the awkward title of the Renewable Energy Standard Offer Program or RESOP.

Finally in 2009—six years after WindShare's wind turbine had gone into operation—the province launched the Green Energy and Green Economy Act, the central element of which was North America's first comprehensive system of feed-in tariffs. The ambitious policy was designed to put Ontario on the renewable energy map by spurring the growth of not only wind energy, but solar, biomass, and hydro as well.

Ontario's policy singled out community power for favorable treatment, putting WindShare and the projects to follow on a sound financial footing. And for the first time in North America, Ontario's policy had specific provisions for renewable projects owned by Ontario's First Nations and Ontarians of aboriginal and Métis descent.

M'Chigeeng First Nation

By the spring of 2012, Ontario had received applications for nearly 3,600 MW of renewable capacity from more than 400 community and aboriginal projects. Despite the promise, however, few community wind projects had entered operation two years later. Feed-in tariffs are only one part of the community wind puzzle. Access to the grid remains problematic for many community projects.

One example where Ontario's feed-in tariffs have fulfilled their intent is a pioneering 4-MW wind project by the M'Chigeeng First Nation (see Figure 20-15. M'Chigeeng First Nation). In mid-2012, a who's who of Ontario's first nation leaders dedicated two Enercon E-82 wind turbines installed on Manitoulin Island in Lake Huron in a traditional ceremony. The Mother Earth Renewable Energy Project was the first of its kind in Canada to be entirely owned by a first nation. The band invested in wind energy, said

Figure 20-15. M'Chigeeng First Nation. The Mother Earth Renewable Energy Project was developed by and is owned in its entirety by the M'Chigeeng First Nation of Manitoulin Island in Lake Huron. Here, Alma Jean Migwans, a first nation elder, dedicates the two Enercon turbines in a traditional ceremony attended by dignitaries from across the Canadian province of Ontario in 2012. The project was the first of its kind and was made possible by Ontario's precedent-setting system of feed-in tariffs. (Kris Stevens, Ontario Sustainable Energy Association)

economic development officer Grant Taibossigai, because "renewable energy meets our goals for economic development and it does so in an environmentally and socially sustainable way."

Generation from the turbines will be paid under a 20-year contract with the Ontario Power Authority (OPA) of $0.135 CAD per kWh. The project qualifies for an additional $0.015 CAD per kWh designed specifically to aid aboriginal groups. Altogether, OPA will pay the band $0.15 CAD (~ $0.14 USD) per kWh for its wind generation.

Ontario's program has been described as the most progressive renewable energy policy in North America in three decades. By 2016, there may be as much as 13,000 MW of new renewable generating capacity installed in the province. Its success has in part enabled the province to close all its coal-fired power plants. That act alone is the single-most significant policy to reduce carbon emissions in North America. Despite the good intentions, only a small portion of Ontario's new generation will be community owned.

Nova Scotia

Nova Scotia, one of Canada's Maritime Provinces, responded to the clamor generated by Ontario's bold move with its own program. The province took a more conservative and, some say, timid approach to renewable energy policy than Ontario by implementing a feed-in tariff program limited solely to community-based renewables. And to make sure the program wouldn't threaten the incumbent utility, it limited the total program to only 100 MW. Yet in meeting its restricted ambitions, Nova Scotia inadvertently installed more community-owned wind with its Community Feed-in Tariff, or COMFIT, than Ontario.

Considering the program's strict requirements, COMFIT has awarded a surprising number of contracts since its launch in 2011. Though some may argue over the province's definition of "community" renewables to include a biomass combined heat and power plant at a pulp mill, more than 100 MW of wind power projects have been awarded contracts and many have already started operation. One example is the turbines installed atop Spiddle Hill.

Colchester-Cumberland Wind Field

At a session of the Canadian Wind Energy Association's 2012 conference devoted to community-owned projects, David Stevenson engagingly described how he and 300 local investors overcame numerous obstacles to install their very own Enercon E-53 in Nova Scotia (see Figure 20-16. Spiddle Hill). The turbine joined two 50-kW wind turbines they also installed under the province's COMFIT program.

Completed in 2011, the 800-kW Enercon wind turbine on Spiddle Hill overlooks the small town of Tatamagouche and the Northumberland Strait separating Nova Scotia from Prince Edward Island. Like sister projects in Germany and Denmark, the Colchester-Cumberland Wind Field was initiated and developed by the community itself.

Two provincial programs made the project possible: Nova Scotia's COMFIT and a special tax credit for investment in community development. Unique among Canadian provinces, Nova Scotia offers Community Economic-Development Investment Funds (CEDIF), an initial tax credit of 35%, and subsequent tax credits of 20% and 10%. Organizers of the group qualified to issue shares under the program in 2007.

The projects won 20-year contracts under Nova Scotia's COMFIT program, which pays $0.13 CAD (~ $0.12 USD) per kWh for all the electricity that the 800-kW turbine generates

Figure 20-16. Spiddle Hill. A technician stands inside the nacelle of Colchester-Cumberland Wind Field's Enercon E-53 turbine, awaiting the nose cone. The 800-kW turbine overlooking the Northumberland Strait near Tatamagouche, Nova Scotia, is locally owned by 300 investors. 2011. (Colchester-Cumberland Wind Field)

and $0.50 CAD (~ $0.46 USD) per kWh for generation from the 50-kW turbines.

By the end of 2013, the province had awarded 200 MW of COMFIT contracts for community-owned projects—or the equivalent of 200 wind turbines the size of Colchester-Cumberland's E-53.

As in polls of those living near wind turbines in Germany and Scotland, there was strong support for community-owned projects in Nova Scotia. In a public opinion survey of those within 10 kilometers (6 miles) of three projects, Dalhousie University graduate student Tiffany Voss found that 80% of those polled in the vicinity of Colchester-Cumberland's Spiddle Hill turbines supported the project. She attributed the high support to the group's efforts at inclusion. Significantly, more respondents near the Spiddle Hill project believed they had the opportunity to invest in the turbine than those near other wind turbines. One-quarter of those polled had already invested in the turbines. Of the three projects surveyed, those near Spiddle Hill felt the greatest sense of "community ownership."

Not satisfied to rest on their laurels, Colchester-Cumberland's organizers have added another Enercon E-53 to their site. The project also inspired the installation of five EV-charging stations in their community, moving the group beyond simply generating electricity to "using the wind to drive our own wheels."

Massachusetts

Policies in the United States have been even less conducive to local ownership than those in Ontario and Nova Scotia. Renewable Portfolio Standards (RPS) in the United States have effectively excluded community-owned projects. There are some noteworthy examples, however, where farmers, community groups, and small towns have fought the tide and installed wind turbines that they own, including several in Massachusetts.

Hull

Hull is a small coastal town 13 kilometers (8 miles) across the bay from Boston and is only 8 kilometers (5 miles) south from Logan International Airport. The village has a long history with wind since settlement. But its acquaintance with modern wind energy begins in 1985 when it installed an Enertech E44 near Hull High School at the appropriately named Windmill Point.

Despite a good wind resource, the E44 turned in a checkered performance, and the turbine was eventually removed. Surprisingly, that experience didn't turn the community off from the idea of using the wind. The climate for wind remained favorable, and local wind advocates championed a new project.

On December 27, 2001, John MacLeod, the manager of the Hull Light Plant, and a group of community activists dedicated a new Vestas V47 at the same site (see Figure 20-17. Hull 1). They intended for the 660-kW turbine to provide municipal lighting for the town of 10,000. The dedication took place during the darkest week of the year, symbolically saying to the community at large that the lights of Hull would shine brightly without the burning of oil, like the lamps in the temple at Jerusalem celebrated during Hanukkah, the festival of lights.

Hull's V47, one of a growing number of urban wind turbines in North America, has become the pride of the community. Hull Light's MacLeod reports that tourists to Windmill Point have increased threefold since the commercial wind turbine was put into service. The Vestas turbine not only powers all of Hull's streetlights, traffic lights, and other municipal lighting needs—more than 400,000 kWh annually—but also generates more than sufficient revenue, says MacLeod, to pay for its maintenance. Hull was so satisfied with its V47 that the city considered expanding further.

The city surveyed residents about their attitudes toward the turbine and the possibility of adding another wind turbine. More than 90% approved of the proposal.

The city installed Hull 2, a Vestas V80, in 2006 on a reclaimed landfill designated the Wier River Estuary Park, making it the first wind turbine installed on a landfill in North America. The 1.8-MW wind turbine stands on pilings driven 80 feet into the landfill.

The two turbines proved so popular that the community weighed installing more wind turbines offshore, but as of 2012 there was no program in Massachusetts that paid a sufficient amount for the electricity to justify the venture.

Hull's success soon came to the attention of city fathers across the bay.

Figure 20-17. Hull 1. Vestas V47 at Windmill Point in Hull, Massachusetts. The 660-kW wind turbine, known now as Hull 1, is within 100 meters (330 feet) of Hull High School and alongside a service road that circles the peninsula where the school is located. The site is 8 km (8 miles) south of Boston's Logan International Airport. The turbine provides power to the municipality's street and traffic lights, 2009. (Doc Searls, Wikimedia Commons)

Boston

While there are no large wind turbines in so-called green cities such as San Francisco or Los Angeles, several large wind turbines have been installed in Boston. Massachusetts's Water Resource Authority installed two 600-kW turbines at its sewage treatment plant on Deer Island in Boston Harbor and followed those with a 1.5-MW turbine at its headquarters in central Boston. Altogether the three turbines are expected to save the authority 6 million kWh per year.

In fact there are more medium-size and large wind turbines in distributed applications, such as those at Hull, scattered across Massachusetts than in California, a state 16 times larger. Some of these installations are at noteworthy sites, such as the 100-kW turbine at the International Brotherhood of Electrical Workers just south of Boston on I-93, or the commercial-scale wind turbines at the Maritime Academy and the three units at the Cape Cod Air Force Station.

Minnesota

In the United States, probably no state has done more to make local ownership possible than Minnesota. Perhaps it was the numerous farm cooperatives started by the descendants of Scandinavian settlers that provided ready-made examples of how it could be done. Perhaps it was campaigns by local ownership advocates such as Windustry or the Institute for Local Self Reliance that made it possible. Maybe it was the bumbling of the regional electric utility as an environmental group with deep roots in the Midwest, the Izaak Walton League, nipped at its heels. For whatever reason, the state of Minnesota had the only policy specifically designed to aid community ownership of wind energy in the United States. Its C-BED, or Community-Based Energy Development, program has resulted in the most community wind in the country, more than 100 MW, and most of that is farmer owned. Altogether, Minnesotans have installed nearly 300 MW of locally owned wind under various programs—the most for any region in North America. Not only farmers have taken advantage of local ownership, several of the state's colleges and universities have as well.

Northfield: Twin Turbines for Twin Colleges

Outside the since discontinued C-BED program, the only way for communities to use wind is to offset their own consumption. In Hull and Boston, Massachusetts, the turbines were used to generate electricity for municipal services. Similarly, a number of small towns and villages in the Midwest looked at wind to offset the consumption of utility-supplied electricity.

One such community was the college town of

Figure 20-18. St. Olaf's own. Vestas V82 on the campus of St. Olaf College, Northfield, Minnesota, in 2009. The turbine is a prominent feature from anywhere on the campus. It offsets consumption from the regional electric utility. Northfield's other private college, Carleton, has installed two similar wind turbines for the same purpose.

Northfield, Minnesota. Northfield is unique in that it has twin private colleges: Carleton and St. Olaf. In the fall of 2004, Carleton College dedicated a 1.65-MW Vestas V82, at the time the first installation of a large wind turbine for a college or university in the United States. The turbine offsets about 40% of Carleton's electrical consumption. Carleton installed a second wind turbine that provides power directly to the campus in late 2011, providing for an additional 30% to 40% of the college's electrical needs.

Nearby St. Olaf College installed its own Vestas V82 (see Figure 20-18. St. Olaf's own). The turbine is prominently visible from throughout the campus. The turbines have become landmarks for both colleges and for the town.

Nevada, Iowa

Community-owned wind turbines have been an integral part of Nevada, Iowa's townscape since late 1993 when prominent public-spirited citizens Harold and Marjorie Fawcett, along with Harold's sister Josephine Tope, donated three turbines to the community. Two Wind World turbines were installed at the local high school and one Vestas V27 was installed for the benefit of the Story County Medical Center (see Figure 20-19. Nevada's twin Wind World turbines). Though the turbines have not been without problems, the trust fund set up by the Fawcett's has covered the cost of repairs for the past two decades.

Figure 20-19. Nevada's twin Wind World turbines. Local philanthropists donated these two turbines, one 200 kW, the other 250 kW, to the community of Nevada, Iowa, where they were installed at the local high school–near the football field. There are few things more important in the life of a small midwestern town during the autumn than Friday night at the football game. That the turbines were installed where they are indicates the importance that the community placed on them.

Who Owns the Wind?

Although community wind is not overtly prohibited in North America, there are very few opportunities for it to flourish. Organizations, such as the Institute for Local Self-Reliance in Minnesota and OSEA in Ontario, hope to change that, but it's an uphill fight.

There are few policies specifically targeted at community wind. And some renewable energy incentives, such as the production tax credits in the United States, are an obstacle to local ownership, not an aid. The tax credits were intended for projects developed by large, multinational corporations and consequently are of little use to community wind projects, farmers, and small businesses, which have limited tax liability.

Like Denmark's early programs, some jurisdictions provide incentives specifically for community wind and restrict participation to those that meet specific criteria.

It's noteworthy that in Germany there are no ownership requirements to qualify for the feed-in tariffs that make German-style community wind possible. When OSEA chose to replicate the German model of community wind development in North America, it chose a similarly open policy. OSEA realized early on that existing policies in Ontario favored typical absentee owners, and in order to move beyond one community-owned wind turbine, new policies needed to be put in place.

In 2004, OSEA launched a campaign to bring electricity feed laws, like those used so successfully in Germany, to Ontario. At the time, OSEA deliberately chose not to place any ownership requirement on the program. It was OSEA's intent that the program be open to all participants: farmers, homeowners, cooperatives, indigenous communities, and traditional corporations. OSEA chose this route because it could not envision all the possible combinations of corporate ownership, individual ownership, and mixed-ownership models that would serve the needs of both community economic development and the rapid growth of renewable energy.

Others have chosen different models. To qualify for C-BED tariffs in Minnesota, 51% or more of the equity in a project must be owned by Minnesotans. Moreover, no single owner may be allowed to own more than 15% of a project. Minnesota wanted to spread the benefits of the C-BED program to as many participants as possible, rather than see development concentrated in the hands of a small number of wealthy individuals.

Nebraska followed a path similar to Minnesota's. Nebraska's Rural Community-Based Energy Development Act requires that 33% of the payments from a wind project over the 20-year contract period must go to Nebraskans, for the project to qualify.

Elsewhere in North America, there are no specific criteria on what is not community wind. At one end of the spectrum is the WindShare cooperative that owns half of the wind turbine at Toronto's Ex-Place. Like projects in Denmark, the cooperative initiated and led development, and like the *vindmøllelaug* in Denmark, WindShare is 100% owned and controlled by investors living in the region. Each shareholder has as much say as another in the management of their wind turbine. WindShare has become a North American model for engaging the public's imagination and active participation in developing renewable energy. Their wind turbine is a fixture on Toronto's urban skyline.

At the other end of the spectrum is Peace Energy Cooperative's Bear Mountain Wind project. When commissioned in 2009, the project represented a total investment of nearly one-quarter billion dollars and was among the largest wind projects in Canada. The 250-member Peace Energy Cooperative initiated the project by obtaining leases on public land and measuring the wind resource themselves. However, the project of 60 Enercon E-82 turbines was built by a private developer. Unlike WindShare, Peace Energy Cooperative doesn't have a controlling interest in nor have managerial control of the project.

Farmers on the south shore of Quebec's Lac St. Jean have tried to walk a path between these two approaches. The farmers formed a cooperative, Val-Éo, and a limited partnership to develop their wind resource. The farmers provided all the up-front risk capital for developing a 25-MW project. Like Peace Energy Cooperative in British Columbia, Val-Éo chose to work with an experienced private partner to eventually build

> Paulsen is channeling Abraham Lincoln at Gettysburg by implying that community wind can be the
POWER OF THE PEOPLE, BY THE PEOPLE, FOR THE PEOPLE.

and operate the wind plant. However, the cooperative plans to act as the general partner of the limited partnership.

In the Val-Éo model, the cooperative can provide up to 50% of its own equity in the project. However, in an unusual twist, Val-Éo will maintain 50% of the voting shares regardless of the percentage of equity actually owned by community members once the project is built.

Unfortunately, Quebec, like many other jurisdictions in North America, has not been supportive of community wind. Quebec prefers to keep wind energy in the hands of the provincially owned monopoly Hydro-Québec. While not overtly hostile to local ownership, these jurisdictions have chosen a different development model that effectively bars community wind by transferring the right to develop wind energy to a few chosen corporate partners.

What's Required to Make Community Wind Happen

Nowhere is it easy to develop community wind, but it's particularly difficult in North America.

Due to the arcane tax laws in the United States, and Congress's restrictions on use of the federal tax subsidies, communities in the upper Midwest developed complex financial transactions that allowed farmers and others to build projects while using the tax credits. Minnesota farmers, for example, created the "Minnesota flip." In these transactions, farmers partner with a corporate entity that can use the tax credits. The partner then "owns" the project for the first 10 years while the tax credits remain in effect. After they exhaust the tax subsidies, they then "flip" ownership back to the group of farmers who originated the project, often for a fee.

The flip—where it works—is only useful if the project has a contract to sell its electricity. And in the United States, where most contracts are awarded through byzantine bidding programs, multinational corporations gobble up the contracts before cooperatives and local groups can organize their first kaffeeklatsch. It's as though the answer to that rhetorical question of "Who owns the wind?" is an arrogantly dismissive "It's not you."

Few ask the question, Why is this so? Why is this the way it's done? Even fewer boldly state, as Wolfgang Paulsen does at the beginning of this chapter, that the wind is ours. It is our resource and we want to use it ourselves. It's as though Paulsen is channeling Abraham Lincoln at Gettysburg by implying that community wind can be the power of the people, by the people, for the people.

The demand for renewable energy in North America is so great that we'll need a mix of corporate development of absentee-owned wind farms, community wind, and individually owned wind turbines. We need it all. In the meantime, we need more community wind to redress the imbalance of ownership of today.

Europeans have been successful in the rapid development of wind energy for two reasons. They have broad public support, in large part due to community participation. Communities and individuals (farmers as well as homeowners) have been able to participate because of electricity feed-in laws that open up the market to players of all sizes, regardless of financial clout. Feed laws not only allow a connection to the grid, they also offer a contract to sell the electricity that the wind turbines will generate. Further, they set a fair price for the electricity that will be produced.

These payments for renewable energy generation, or feed-in tariffs, are transparent. They are set in the open in a public process, and they are posted for all to see. Prices are not set behind closed doors and sealed in confidential contracts that only a privileged few regulators are permitted to see. With feed-in tariffs, everyone knows what is paid for each kilowatt-hour and for how long he or she will be paid.

And feed-in tariffs are equitable. Everyone can receive the payments, regardless of his or her position in society, rich and poor alike. With information on the tariff that will be paid, any homeowner, farmer, or community group can calculate whether an investment in a wind turbine will be profitable—or not.

And just as important, every banker can make the same calculation as well. Bankers can quickly determine whether an investment is a good risk or not. Where the payments are high enough, or the wind is strong enough, the banks know they can lend money to community wind projects and that they will be paid back. With bank financing, a community wind project begins to look much like a traditional corporate wind project, except that more of the benefits stay in the local community.

To redress the imbalance in North American wind development, to give local organizers of community wind projects the opportunity to develop "their" resource, a comprehensive feed-in tariff policy with the right to connect to the grid will be necessary. A "feed-in tariff is absolutely essential to make this happen" on a significant scale, says the Institute for Decentralized Energy's Martin Hoppe-Kilpper. It can be done without feed-in tariffs as the Australians have shown, but it can't be done at scale.

We'll need a lot of investment for the energy transition; money that need not—and cannot—come from the government. It has to come from the people, says Hoppe-Kilpper. "Everyone can be an entrepreneur with feed-in tariffs."

Though only Ontario has implemented a comprehensive program of feed-in tariffs to date, some targeted specifically at community ownership, there is a growing movement across North America demanding the right to develop the community's wind resource—our wind resource—for our own profit and the profit of our communities.

North Americans are no different than Europeans. Both Canadians and Americans have proved equal to formidable challenges in the past. We brought electricity to our rural neighbors across the breadth of the vast North American continent at a time when many said it couldn't be done—or that it shouldn't be done.

And just as the revolutionaries at Tvind demanded that Danish politicians allow their people to develop wind energy and Germany's *stromrebellen* told their government they needed feed-in tariffs to pay for the wind generation they planned to produce for their community, North America will forge its own rebels from the fields of Minnesota to the great urban centers of the continent. Fortunately, we have many inspiring examples to emulate.

21

The Challenge

> Make no little plans. They have no magic to stir men's blood.
> —Daniel Burnham, American architect and urban planner (1846–1912)

IN THE PRECEDING CHAPTERS WE'VE LEARNED THAT WIND energy works, it works reliably, it can be integrated into the grid, it's safe, and it's a good investment for individuals, communities, and society at large. Four decades after the modern wind revival began, we can say with certainty that wind energy has come of age as a commercial generating technology. Wind energy is here to stay.

No miracles or media-grabbing breakthroughs are necessary before we put wind energy to use pumping water, powering remote homesteads, or generating commercial quantities of electricity to make the transition to a renewable future a reality. As Green Energy Ohio's Bill Spratley argues, we don't need more studies, we need to seize the day or, as he puts it, *carpe ventum*, seize the wind and put more of this technology to use. The sooner the better.

As wind energy continues to grow, new applications are appearing that at one time seemed far-fetched, the province of dreamers like naval architect Bill Heronemus who envisioned wind turbines riding offshore platforms, or others who saw a future with wind-powered electric cars, or communities powered by their own wind turbines.

Still, there many pitfalls wind energy must avoid—some due to its remarkable success—for it to continue its rapid expansion. And there remains plenty of room for improvements in the technology and how we use it.

Pitfalls to Avoid

Wind energy's success, its steady technological progress, and its rapidly growing deployment worldwide can lure the industry into a false sense of security. There are pitfalls to avoid, or wind's march to the future could be derailed. Most dangerous of all is the lure of panaceas.

The Lure of Panaceas

Planners have long looked offshore as a means of reducing siting conflicts between wind turbines and their neighbors in densely populated Europe. In the midst of sometimes protracted planning battles, wind companies and

OFFSHORE AND NEAR SHORE WIND

In the early 1970s, an obscure professor at the University of Massachusetts proposed an outlandish scheme, or so it was thought at the time. Bill Heronemus, who as a young engineer helped construct nuclear submarines, broke ranks with his former colleagues, concluding that plans to lace Boston Harbor with submerged nuclear power plants were not only too costly but also too dangerous. Heronemus instead proposed fleets of offshore wind turbines.

At a time when nuclear power was seen as the wave of the future, Heronemus was decried as an unrealistic, wild-eyed dreamer. Nuclear faltered, and then all the king's men couldn't put it back together again. Meanwhile, wind energy proved itself, and Heronemus's vision has taken form, not in Boston Harbor as yet but in the shallow waters off Denmark and Great Britain (sees Offshore wind). Unlike those envisioned by Heronemus, these offshore wind plants were not built on floating platforms but firmly rooted to the ocean floor–and to reality.

Offshore wind has the potential to produce large quantities of electricity near urban markets. An early example of this was the 40-MW Middelgrunden project installed in 2000 that serves Copenhagen, the Danish capital. While the offshore market has been slow to develop–small near-shore projects were completed in waters off Denmark and Sweden in the 1990s–it has now gathered steam. However, Middelgrunden remains the only offshore project that's community owned.

By mid-2014 there were 7,000 MW of capacity from more than 70 different wind projects throughout Europe. Another 5,000 MW of offshore projects are on the drawing boards. Nevertheless, for all the attention offshore wind garners, it represents only 2% of wind generation worldwide.

Offshore wind is certainly no panacea. Because turbines offshore require more expensive infrastructure, such as wave-proof towers and foundations, projects have longer lead times and are much more costly than wind turbines on land. Typically, offshore wind costs twice as much as wind energy on land and in some cases three times as much.

The cost and complexity of offshore wind has precluded community ownership since Middelgrunden. The electricity rebels of Nordfriesland, like Wolfgang Paulsen, tried their hand at offshore. The Butendiek *Bürgerbeteiligung* raised risk capital from 8,400 participants in the region. They overcame numerous planning obstacles, but the massive 240-MW project was beyond their reach. They eventually walked away, selling their project to a commercial developer.

Since the attempt at Butendiek, offshore wind has become the private playground of the world's biggest utilities and integrated oil companies. Yet this could change.

Danish researchers at the University of Aalborg uncovered surprising results in a 2012 study of offshore wind. Their study, "Evaluation of offshore wind resources by scale of development," suggests that small wind projects developed near shore may generate lower-cost electricity than massive wind projects far offshore.

The study examined the costs of wind-generated electricity from existing and proposed offshore projects and found that, contrary to expectations, there were no apparent economies of scale as the projects grew bigger. Instead, offshore wind "has shown ever-increasing costs of installation, and an inverse economy-of-scale." In other words, the bigger offshore projects became, the more costly their electricity.

The Danish researchers concluded that their findings reopened the debate about what kind of offshore wind is best. For example, smaller projects closer to shore may be more cost effective even though it's less windy near the shore than far out to sea. They warned, however, that Denmark's near-shore wind resource may "only" be capable of generating 40% of Danish electricity supply in the year 2020.

Another disadvantage, they noted, was that near-shore projects would be more visible to people onshore than those far offshore. This, the Danish researchers suggested, obviates the need for an alternative form of ownership to that currently practiced. "The high visibility of these installations may be offset by local ownership . . . and higher public involvement," they concluded.

Despite these "limitations," near-shore wind plants could reopen the question of community ownership for offshore wind. Smaller projects with lower infrastructure costs, such as easier access to the grid on land, could bring wind projects near shore within reach of cooperatives as demonstrated in Copenhagen. Locally owned turbines near shore, like community-owned projects on land, would pump money into harbors and coastal communities hard hit by the loss of shipbuilding jobs to Asia. This in turn would generate a "warm shower" of income flowing into the rural areas surrounding these ports, benefiting entire regions.

Offshore wind. Large offshore wind plant composed of two separate projects: 207 MW Rødsand II (foreground), containing 90 Siemens 93-meter (300-foot) diameter turbines, and 166 MW Rødsand I (background), containing 72 Siemens 82-meter (270-foot) diameter turbines. The massive project was developed by Danish electric utility Dong Energy, 10 km (6 miles) west from Gedser on the island of Falster, the site of Johannes Juul's famous wind turbine of the 1950s. 2011. (Wikimedia Commons)

developers eye offshore wistfully as an answer to all their ills. "Out of sight, out of mind" they muse. Not so. No place is out of sight or out of mind in today's interconnected world. If wind energy doesn't solve problems with public acceptance on land, the problems will only follow them out to sea. One need only examine the vicious fight over Cape Wind's project in Nantucket Sound, Massachusetts, to know that offshore is no panacea for the problems wind energy faces on land.

Public Relations Puffery

Wind's explosive expansion has assured us that the technology is here to stay, but this very success has also shone a light on one of wind energy's most striking features, the rapidity with which wind turbines can be added to the landscape. One day there are none, but within days there can be tens of machines, and within weeks hundreds. Wind energy can change the landscape literally overnight.

Regardless of our zeal to see wind energy succeed, we should never overlook wind's warts, such as the rapaciousness of an irresponsible developer or the fraudulent hype of an Internet swindler. Honest advocacy will move us further in the direction we want to go than public relations puffery.

Wind energy is no environmental cure-all. It's just one of many technologies we must use to build a sustainable future. Wind energy is a relatively benign technology, when used with care and developed with sensitivity for the community of which it will be a part. Wind energy, more than other technologies, depends upon public acceptance. We ignore this at our peril.

Too Cheap to Meter

One unfortunate aspect of North America's on-again, off-again love affair with renewable energy has been the sometimes desperate attempt by wind energy's proponents to promise more than they can deliver. To curry favor with politicians, researchers, manufacturers, and their trade groups, such as the AWEA, project that the cost of wind energy will continue to decrease far into the future.

This is a dangerous game. Politicians typically respond by saying in effect, "Well, if wind energy will be that cheap in the future, why should we bother with it now? We'll use wind energy later when it's cheaper still."

To those with a historical perspective, the industry's projections of ever lower costs sound eerily like those of the nuclear industry in the 1950s. When the chairman of the US Atomic Energy Commission, Lewis Strauss, promised that nuclear power would become "too cheap to meter," he created an expectation that could never be met. The failure to reach that unrealistic target called into question the basic tenets of the nuclear program. For wind to avoid this trap, proponents must never oversell the technology. Wind will never be "too cheap to meter," in part because it produces a higher value product than that of fossil and nuclear fuels.

Wind is clean and renewable. It offers a hedge against the volatility of fossil fuels, doesn't require cooling water, and poses no risk of catastrophic accidents requiring the evacuation of entire regions. Those are sufficient grounds to demand that wind be paid a fair price that reflects its cost of generation. We shouldn't settle for anything less—or promise anything more.

The North American Challenge

Unfortunately, North Americans have been dabbling around the edges of energy policy for more than three decades. Few have acknowledged the seriousness of the challenges facing the continent. We're vulnerable. We're vulnerable to our overdependence on oil for so much of our transport. We're vulnerable to volatility in the supply and cost of natural gas because so many of us use it to heat our homes. We're vulnerable to

WIND IS CLEAN AND RENEWABLE.
It offers a hedge against the volatility of fossil fuels, doesn't require cooling water, and poses no risk of catastrophic accidents requiring the evacuation of entire regions.

Wind energy, more than other technologies, depends upon public acceptance.
WE IGNORE THIS AT OUR PERIL.

climate change, as our summers become hotter and dryer with drought stalking our arid western states and provinces.

We've seen what can happen elsewhere. Though no one can say with certainty that the killer heat wave that hit Europe in 2003 was due to climate change, it did fit the pattern that's expected in a hotter world. Whatever the cause, the result was disastrous. More than 50,000 excess deaths were attributed to the heat, half of those in France alone. Nuclear power plants were forced to shut down across the continent, as cooling water exceeded their safe operating temperatures.

Houston, we have a problem. The public in both Canada and the United States understands this. They're living it daily now. While politicians in both countries have been hamstrung by futile ideological struggles, their constituents have been trying to figure out what to do. They want action.

North American Consumption

First, the bad news. North Americans consume more electricity per capita, per unit of gross domestic product, per almost any unit of measure, than any other people on the planet. We're the world's energy hogs. Of course, this isn't new. Environmentalists and energy security analysts have been saying this for years. Environmentalists have been concerned because of the pollutants that result from squandering electricity. Energy security analysts have become increasingly concerned because our profligate consumption makes us vulnerable to supply disruptions and the volatility in the global price of fossil fuels as they become increasingly subject to political instability.

Now, the good news. We have lots of room for cutting our consumption of electricity dramatically. Energy-efficiency guru Amory Lovins has been saying for decades that we can cut our consumption in half. Indeed, we can. We have only to look to Europe to see that it has already been done elsewhere.

Lots of books have been written comparing electricity consumption in North America to that of the rest of the world, so there's no need to repeat that work here. Instead, let's look at how many wind turbines it takes to meet the electrical needs of a North American household relative to those of a European household. We'll use rotor swept area as our unit of measurement. This is useful for gaining a sense of scale. Then we'll turn our attention to meeting the challenge facing us.

Swept Area Needed to Meet Consumption

From a wind energy perspective, it takes much more of the wind stream—for a given wind speed—to meet the energy needs of a North American home than it does for a home in Germany. Let's use our example from the preceding chapter and assume we'll use a large wind turbine installed at a modest site in the interior of the continent with an average wind speed at a hub height of 6 m/s (13.4 mph). An 80-meter diameter wind turbine will generate about 4 million kilowatt-hours per year at such a site. We could calculate how many homes such a wind turbine would serve, and this is exactly what most reporters want to know when they're writing a story about wind energy. It should be clear that the answer to this question is it depends on where the wind turbine is located. The same wind turbine in Germany will provide electricity for nearly four times more homes than it will in Texas, for a given wind resource.

But let's turn the question around and ask how much swept area of the wind stream, effectively how much of a wind turbine, do we need to meet the demand. This is useful for visualizing how reducing our consumption reduces the amount of wind turbines we must build to meet our needs. Fewer wind turbines needed means fewer environmental impacts, less steel and fiberglass required to build the turbines—all in all simply less to do the job. And if the job is smaller—that is fewer wind turbines—we can do it much more quickly than otherwise (see Figure 21-1. Swept area needed per household).

At our hypothetical site, it would take 4.5 m^2 of the wind stream to meet the needs of the typical German household. To meet the needs of a household in Texas, we would need more than 15 m^2 of the wind stream—3.5 times more area than that in Germany—to do the same job.

This should be a sobering thought to those who worry that massive development of wind energy will ravage the continent. If we are going to use wind energy in North America—and we are—and if we're concerned about the impact of wind

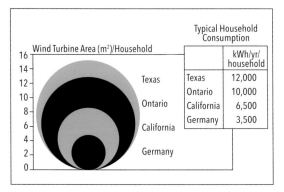

Figure 21-1. Swept area needed per household. The amount of rotor swept area needed to meet average residential electrical consumption for an average annual wind speed of 6 m/s (13.4 mph) at hub height. The typical Texas household consumes nearly four times more electricity than that in Germany and nearly twice that of a household in California.

energy on the landscape, there's no more powerful way to reduce that impact than by reducing our consumption of electricity. Less consumption means fewer wind turbines. It's that simple.

Why this is important becomes clearer, when we consider what it will take—how much wind-generating capacity—to meet the challenge facing North America.

The Challenge

In the summer of 2008, former vice president Al Gore laid before Americans a bold challenge. He said that the United States could meet 100% of its electricity from renewable energy and truly clean carbon-free sources within 10 years. He likened it to President John F. Kennedy's challenge to go to the moon, which galvanized the nation at the height of the space race with the Soviet Union.

Despite being immediately dismissed by his critics as a dreamer, Gore's bold vision stirred Canadians and Americans alike. Could it be done? people asked. And if it can be done, could such a national—even continental—endeavor address the dual problems of climate change and our dependence on fossil fuels? Just as importantly, could such a vast undertaking revitalize the industrial heartland of the continent, bringing jobs and economic prosperity back to once proud but now decimated communities?

Clearly, the scale of the task is enormous. But the first question remains, can it be done? Let's examine what it will take and whether it's even possible.

Offsetting Fossil-Fuel-Fired Generation

The United States consumes ~ 4,000 TWh per year of electricity; Canada, proportionally less. (Canada has about one-tenth the population of the United States.) About three-quarters of US electricity is produced by burning fossil fuels. Canada, because of its abundant hydroelectric power, uses "only" 150 TWh of fossil-fuel-fired electricity generation per year. To eliminate the air pollutants and global-warming gases from the electricity sector, we would need to generate about 3,200 TWh per year with renewable sources to offset fossil-fuel-fired generation in North America (see Table 21-1. North American Wind Energy Challenge: Eliminating Fossil-Fired Generation).

Most wind turbines operating in North America today have been installed on the windiest sites possible. These turbines are highly productive. However, as the industry expands across the continent, it will use increasingly less windy sites like those in the Midwest. Conservatively, one 2-MW wind turbine will generate about 4 million kWh per year at a modest interior site. This is the historical performance we see in large fleets of wind turbines on a regional or national scale, like those in Germany, Denmark, or California. These large fleets generate approximately ~ 2 TWh per year for every 1,000 MW of wind capacity installed.

Thus, to offset fossil-fuel-fired generation in North America we would need a fleet representing ~ 1.6 million MW. At the beginning of 2014, there were about 60,000 MW of wind-generating capacity operating in the United States and 5,000 MW operating in Canada.

Daunting? Yes. But before we move on to determining whether it can be done or not, let's up the ante.

Table 21-1. North American Wind Energy Challenge: Eliminating Fossil-Fired Generation		
	~North American Fossil-Fired Thermal Generation	~Equivalent MW of Wind Turbines at 6 m/s*
	TWh/year	MW
Canada	150	75,000
USA	3,000	1,500,000
Total	3,200	1,600,000
*Assumes 2 million kWh/MW/year.		

Offsetting Oil in Passenger Vehicle Transport with EVs

If we're serious about reducing North America's contribution to climate change, and if we're serious about North America's dependence upon oil for transport—and the vulnerability that entails—we need to look at transportation, specifically passenger vehicles and light trucks.

Using wind and solar energy to reduce North America's seemingly insatiable appetite for liquid transportation fuels has long been a dream of environmentalists. While refueling electric vehicles with wind energy at one time seemed fanciful, that's not the case today. Sales of electric vehicles (EVs) are booming. Though they still represent a very small segment of the market, their rate of adoption exceeds that of the Prius, Toyota's wildly successful gasoline-powered hybrid, when it was introduced.

EVs produce fewer nitrogen oxides, carbon monoxides, reactive organic gases, and carbon dioxide than gasoline-powered vehicles. The Union of Concerned Scientists (UCS) found that EVs generate fewer emissions per mile than the average fossil-fueled compact car in all regions of the United States, even when taking into account emissions from the conventional power plants that would normally generate the electricity to charge them. In regions with a high penetration of renewables, EVs are cleaner still. For example, in California with its high concentration of renewables, the environmental performance of an EV is equivalent to a conventional vehicle delivering nearly 100 miles per gallon (mpg). Upstate New York with its abundant hydroelectricity does even better at more than 110 mpg.

Electric cars could absorb large amounts of wind generation that utilities might otherwise have difficulty using. When wind energy is abundant and demand weak, excess generation could be directed into electric vehicles batteries: a kind of mobile storage system.

The performance of EVs is steadily improving too. UCS, a nationwide environmental group, found in 2014 that Nissan's battery-electric model, the Leaf, improved its performance from 0.34 kWh/mile (0.21 kWh/km) to 0.30 kWh/mile (0.19 kWh/km) since its introduction in 2011. (This is equivalent to an improvement from 2.9 miles/kWh—units that EV drivers are more familiar with—to 3.3 miles/kWh.) They calculated that the fleet average of new EVs sold in the United States was 0.325 kWh/mi (0.20 kWh/km) in 2013 or 3.1 miles/kWh (see Table 21-2. Energy Efficiency of Electric Vehicles).

What would happen if we were able to convert our light vehicles to electric cars? How much more electricity would we need? North Americans drive their cars and trucks nearly 5 trillion kilometers (~ 3 trillion miles) per year. We drive more than anyone else on the planet.

For our case study, let's assume that the fleet average is only 0.25 kWh/km (0.40 kWh/mi) to account for both passenger vehicles and light trucks. Converting North America's light vehicles to EVs would require an additional 1,200 TWh per year of wind-generated electricity. We'd need another 600,000 MW of wind capacity to meet that need (see Table 21-3. North American Wind Energy Challenge: Passenger Vehicle Transport).

It looks like we'll need somewhat more than 2 million MW of wind capacity to replace fossil-fuel-fired electricity generation and convert passenger vehicle transport to electricity (see Table 21-4. Wind Capacity Required in North America to Offset Thermal Generation and Light Vehicles).

Table 21-2. Energy Efficiency of Electric Vehicles

kWh/mi	mi/kWh	km/kWh	kWh/km	
0.34	2.9	4.9	0.20	Early Nissan Leaf
0.30	3.3	5.6	0.18	2013 Nissan Leaf
0.35	2.9	4.8	0.21	Tesla Model S
0.33	3.1	5.1	0.20	2013 US EV fleet average

Source: How do EVs Compare with Gas-Powered Vehicles? Better Every Year by Don Anair, Union of Concerned Scientists, September 2014. http://blog.ucsusa.org/how-do-electric-cars-compare-with-gas-cars-656?_ga=1.62037544.529347279.1412002686

Table 21-3. North American Wind Energy Challenge: Passenger Vehicle Transport

	Light Vehicles trillion km/year	Electric Vehicle Consumption TWh/year*	Equivalent MW of Wind Turbines @ 6 m/s**
Canada	0.30	75	37,500
USA	4.50	1,125	562,500
Total	4.80	1,200	600,000

*Assumes 0.25 kWh/km travelled.
**Assumes 2 million kWh/MW/year.

Table 21-4. Wind Capacity Required in North America to Offset Thermal Generation and Light Vehicles

	Thermal Generation	Light Vehicles	Total
	MW	MW	MW
Canada	75,000	37,500	112,500
USA	1,500,000	562,500	2,062,500
Total	1,600,000	600,000	2,200,000

Other forms of electrified transportation could also use wind energy. Wind-generated electricity powers Calgary's light rail system in its Ride the Wind! program. Similar projects could be used to offset the consumption of subway systems in major cities, or of high-speed electrified rail systems like those in France and Germany. For example, promotional materials for California's high-speed rail authority show the sleek train passing fields of wind turbines, which would be used to power the system (see Figure 21-2. Electrified transport). Electrified high-speed trains running from Los Angeles to San Francisco would represent as much as 1% of the state's electricity consumption.

Manufacturing Capacity

Does North America have the manufacturing capability to build this much wind capacity in any reasonable amount of time? Is it hopeless? No, not at all. Skilled workers in Canada and the United States churn out 365,000 heavy trucks every year. The steel and other components used in a heavy truck and the skill necessary to design, manufacture, and assemble it are roughly comparable to the components and skill necessary to produce a wind turbine. Each heavy truck represents about 500 kW of capacity, or half a MW each. Thus, every year the existing heavy truck industry produces 180,000 MW of equivalent power machinery.

What earlier seemed a daunting task now looks far more achievable. Gore was right. We have the manufacturing capacity. If we were to convert our heavy truck industry to building wind turbines in a crash program to offset our fossil-fueled power plants, it would only take a decade. To reduce our dependence upon oil for passenger transport would take only another four years (see Table 21-5. North American Heavy Truck Manufacturing as Proxy for Large Wind Turbine Manufacturing).

No one is proposing that we stop manufacturing heavy trucks. This was merely an exercise to determine if we had the manufacturing capacity to do it—and we do.

Figure 21-2. Electrified transport. Artist's rendering of a high-speed train passing through California's Tehachapi Pass. The image's message is that the train will be powered by wind energy. When fully built, the rail system will consume 3 TWh of electricity per year, or 1% of California's electricity consumption. (California High-Speed Rail Authority)

Table 21-5. North American Heavy Truck Manufacturing as Proxy for Large Wind Turbine Manufacturing

	Trucks/year	Equivalent MW of Wind Turbines/year*	Years to Meet Fleet Required to Offset Fossil-Fired Generation	Years to Meet Fleet Required to Offset Passenger Vehicle Transport	
Canada	65,000	30,000	2.5	1.3	3.8
USA	300,000	150,000	10	3.8	13.8
Total	365,000	180,000	8.9	3.3	12.2

*Assumes 0.5 MW/unit.

Stanford University professor Mark Jacobson likes to use another example of our manufacturing might: aircraft production during World War II. Airplanes are far more complex and sophisticated than wind turbines, and airplanes use large quantities of costly materials, such as aluminum. In 1940 when the United States entered the war, we built less than 4,000 military aircraft. Within four years we were building 20 times that amount. Canada had similarly stepped up production for the war effort.

In 2012, wind companies in the United States installed more than 10,000 MW of new wind-generating capacity. If we were to scale up installations as rapidly as aircraft manufacturers did during the early 1940s, within a few years we could be installing as much 200,000 MW per year. So, yes, we could do it.

Land Area Required
There's also more than ample land area in North America for such a large number of wind turbines. If we use NREL's conservative estimate of the land area occupied by wind turbines of 35 ha/MW of installed capacity, the United States will need more than 700,000 square kilometers (280,000 square miles), or about 9% of the lower 48 states. Canada would need 39,000 square kilometers (15,000 square miles) or 0.4% of its vast land area (see Table 21-6. Land Area Required in North America to Offset Thermal Generation and Light Vehicles).

While an open spacing of the large diameter wind turbines in the silent wind revolution currently underway appears to require even more land area than the example here, this will not be the case. The large diameter turbines will generate far more electricity—possibly as much as 50% more—per MW of installed capacity than the wind turbines used in the NREL study, requiring far fewer turbines than in the example. Thus, the NREL estimate of the land area occupied by wind turbines remains conservative.

NREL calculates that the wind turbines, their roads, and ancillary structures use only 1% of the land that they occupy. Thus, the actual land area used in the United States would represent only 0.09% of the lower 48 states. The remainder of the land would continue in its existing use, whether in row crops or for grazing.

Wind turbines would likely be visible in every state and province across the continent, but they would take very little land area away from other productive purposes.

100% from Renewables
This example illustrates that we have the land area and the manufacturing capacity to install the equivalent of 100% of the US electricity supply with wind energy. (It would be even easier and quicker in Canada where the abundant hydro is a perfect match for wind energy.) In this example, wind-generation provides 75% of current electricity consumption to offset fossil fuels. On top of that, we've added the equivalent of 25% of

Table 21-6. Land Area Required in North America to Offset Thermal Generation and Light Vehicles

	Total				
	MW	km²	mi²	% Land Area Occupied	% Land Area Used
Canada	112,500	39,000	15,000	0.4%	0.004%
USA Lower 48	2,062,500	722,000	279,000	8.9%	0.09%

Assumes 0.35 ha/MW of land area occupied, 1% land area used for the turbine, roads, and ancillary structures.

100% RENEWABLE VISION BUILDING: TREND TOWARD NEW TARGETS OF 100% RENEWABLE ELECTRICITY—AND HIGHER

Increasingly countries and regions are leapfrogging timid renewable targets and moving aggressively toward full 100% integration of renewables into the electricity supply. Some advocates are moving even further, suggesting 150%, even 300%, renewable electricity generation to meet not only electricity supply but also heat and transport.

How times have changed.

When I began my career three decades ago, our demands were modest if not meek. We could hardly imagine wind supplying more than 10% of electricity consumption. Then the California wind rush arrived in the early 1980s, and we realized that wind energy had indeed come of age as a commercial generating technology.

Our expectations increased accordingly. Wind penetration of 20% then began to seem a reasonable objective. But we stumbled badly here in the United States post California. Meanwhile, the Danes continued to erect ever more wind turbines throughout the 1990s. Soon Denmark was closing on 20% of supply from wind energy alone and it became apparent—again—that our targets were too modest.

Even then I remember writing that we were not suggesting that renewables would completely replace fossil fuels. No, I said, we'd always need fossil fuels for some portion of supply. Wind—and solar too—would just be parts of the resource mix, maybe a big part, but still just a part.

"Facts on the ground," as they say, were changing faster than my thinking of what was possible. Renewables were capable of growing much faster than any of us had ever anticipated. Reality was overtaking our imaginations.

In 2014 wind turbines generated 40% of Danish electricity. But of course that's not all. The Danes didn't stop with just wind. They've also been building hundreds of biogas digesters and waste-to-energy plants as well. Together, wind and biomass provide more than 50% of the electricity consumed by Denmark's nearly six million inhabitants.

All of this was accomplished with policies implemented before the climate crisis was fully felt—and well before Fukushima.

In retrospect, none of this should be surprising. After all, Norway has generated 97% of its electricity with hydro for decades, and tiny Iceland already generates 100% of its electricity with a mix of hydro and geothermal sources.

What is different is the realization that the rapid growth of new renewables can cut through the Gordian knot of what to do about fossil fuels in transportation and heating.

This was brought home to me when I listened to presentations by two longtime renewable energy pioneers, Preben Maegaard from Denmark and Johannes Lackmann from Germany. Independent of each other, both had come to the same conclusion. To address both climate change and energy security, we must move well beyond 100% renewable energy in electricity supply and build an integrated network capable of using more than 150% renewable energy, up to as much as 300% renewable energy to offset fossil fuels in transportation, and heating.

This is the kind of bold, visionary thinking that is now being debated. As more countries and regions adopt what was once unthinkable—100% renewable energy targets in electricity supply—others are asking questions about what it will take to go even further.

The most famous example of an ambitious target is Denmark. In the spring of 2012, the Danish energy minister and then holder of the EU presidency, Martin Lindegaard, issued the country's 100% Renewable Energy Declaration. Denmark proposes to meet 50% of its electricity supply with wind energy and 35% of its total energy consumption with all renewables by 2020. That's not all. The Danes plan to produce 100% of their electricity and heat with renewables by 2035 and 100% of the energy used in transport by 2050. "I think it's doable, I think it's necessary, and I think it's good for the economy," said Lidegaard in the declaration.

Closer to home, dissatisfaction with the typically timid targets found in state renewable portfolio standards has led new players in the renewable arena to challenge the traditional incremental approach. They argue that the times demand more aggressive action—and targets that are ambitious enough to elicit the dreams and hopes of North Americans with the policies to match them.

Some communities, such as Greensburg, Kansas, are taking matters into their own hands. After a tornado leveled the city in 2007, the community decided to do things differently when they rebuilt. One of their objectives was to rebuild with 100% renewable energy and most of that will be from wind.

Consequently, imagining what it would take to generate 75% of the US electricity supply with wind energy and another 25% for offsetting fossil fuels in transportation, in this chapter's example, is not as avant-garde as it seems. It's now respectably mainstream.

The budding movement toward 100% renewable energy could be just what's needed to reawaken North America's lagging can-do spirit.

supply to account for replacing fossil-fueled light vehicles with EVs.

Of course, we won't just use wind energy. We'll use other renewables as well. As energy analyst Toby Couture says, "It's all hands on deck." We'll need them all. Wind will be the workhorse, picking up about 60% of the load; solar and the other renewables will supply the remainder. This will cut the land area required for wind energy by about half. And we can cut the land area required by half again if we cut our consumption by 50%.

Sharing the load with other renewables and reducing the amount of generation required makes moving the United States toward 100% renewable electricity far more manageable. It also reduces the amount of land needed for wind energy to slightly more than 2% of the lower 48 states. This is in line with studies in Germany that meeting 80% of their electricity supply there will require about 2% of their land area.

As mentioned in Chapter 15 on siting, Germany's Environment Agency (Umweltbundesamt) reported on a 2013 study of the country's wind energy potential on land. They found that nearly

Can we afford to develop such a staggering amount of wind energy?
MANY WOULD ARGUE THAT WE CAN'T AFFORD NOT TO.

14% of Germany was suitable for wind energy. With then contemporary wind turbines, they concluded that this land area could support an astonishing 1,200,000 MW capable of generating 2,900 TWh—equivalent to nearly five times Germany's current consumption of electricity.

The results from Germany are quite similar to those in the example for North America used here. Interestingly, the German study didn't use modern wind turbines in the silent wind revolution. Using these more productive machines, Germany would need even fewer units and less land area than that in the Environment Agency's report. This is equally true in North America. We wouldn't need as many wind turbines or as much land area as in our hypothetical case study if we deployed more modern wind turbines.

Affordable

Can we afford to develop such a staggering amount of wind energy? Many would argue that we can't afford not to. We are spending enormous sums to import fossil fuels—money that is

ONSHORE WIND RETURNS THREE TIMES MORE USABLE ENERGY IN TRANSPORTATION THAN INVESTMENT IN OIL

French investment bank Kepler Cheuvreux rocked the energy and financial community in late 2014 when it issued a report saying that $100 billion invested in wind energy would deliver more usable energy when used to power passenger vehicles and light trucks than the same amount of investment in oil.

The report by Mark Lewis, *Toil for Oil Spells Danger for Majors*, issued a warning to oil companies and their investors that oil could become a stranded asset as renewables–especially onshore wind energy–become more cost effective. Lewis, a seasoned analyst and former head of energy research at Deutsche Bank, argues that if the price of oil stays low, getting it out of the ground is increasingly unprofitable, but if prices increase, oil becomes ever more uncompetitive with other resources such as wind and solar energy.

Lewis devises a new take on the common energy return on energy invested (EROEI): EROCI for energy return on capital invested. He compares what you get by investing $100 billion in the oil industry or making the same investment in various renewable technologies.

In the short term, the gross return on investment in extracting oil is more beneficial. However, over the typical 20-year life of the projects Lewis weighed, onshore wind was competitive with oil at $75 per barre, and 40% more cost effective than oil at $100 per barrel.

Yet it was when Lewis turned to what the oil is used for, transportation, that Kepler Cheuvreux reached its startling conclusions. Lewis calculated the net energy yield from various investments relative to driving light vehicles (passenger cars and light trucks). He took into account the low conversion efficiency of oil to mobility from internal combustion engines as well as the losses in powering electric vehicles (EVs). The losses in EVs occur both in the transport of electricity and in the conversion of electricity in the vehicle. Nevertheless, Lewis found that onshore wind energy is nearly three times as cost effective as oil at $75 per barrel in powering transportation.

The cost effectiveness of renewables and EVs will only continue to increase, Kepler Cheuvreux's report emphasizes, making further investment in extracting oil for transport a losing proposition.

Energy Return on Capital Investment (EROCI)						
	Gross EROCI			Net EROCI		
	Annual		Cumulative	Annual		Cumulative
	10 yrs	20 yrs	20-year lifetime	10 yrs	20 yrs	20-year lifetime
	TWh	TWh	TWh	TWh	TWh	TWh
Oil $75/bbl	225	113	2,250	56	28	563
Oil $100/bbl	169	85	1,694	42	21	424
Wind onshore	117	117	2,336	76	76	1,518
Wind offshore	62	62	1,246	39	39	779

Source: Toil for oil spells danger for majors by Mark Lewis, Kepler Cheuvreux, 2014, page 8.
http://www.longfinance.net/images/reports/pdf/kc_toilforoil_2014.pdf

lost to our economy, money that enriches others, not North Americans. If we spent these same amounts on developing our own indigenous renewable resources, the money would circulate through our economy, enriching our own industries and our own communities.

Again, for a sense of scale, let's look at another big recent expenditure and see how it stacks up to reindustrializing our manufacturing industries with massive wind development. The United States has spent $700 billion in stimulus funds to bail out the banking industry after the 2008 financial collapse. That's more than $2,000 for every man, women, and child in the country. Consider what that means. Every person in the United States would now—not tomorrow—have 1 kW of their very own wind-generating capacity. The average American household would have the equivalent of a 2.5-kW wind turbine in their backyard—or the equivalent of a 2.5-kW investment in a large wind turbine that they owned together with their neighbors.

Imagine if they were paid a modest $0.10 per kilowatt-hour for the generation from such a wind turbine at a modestly windy site. Every family would earn $500 per year for the next 20 years for a total of $10,000 over the life of the wind turbine. Instead of pouring money into Wall Street, we could have invested $5,000 in our own people and earned a profit of $5,000 in the bargain.

Here's another way to look at it. If we had invested that $700 billion in direct costs, we could have installed 350,000 MW of wind capacity, enough to meet nearly 18% of the US current electricity consumption. Or we would have been one-fifth of the way to our goal of 2 million MW of wind capacity in North America.

Doable

Not only is such an ambitious undertaking possible, it's eminently doable. No, it certainly won't be easy, or without controversy, but it can be done. The question becomes, do we want to rise to the challenge and in doing so transform our society?

North Americans have risen to great challenges in the past, and we can do so again. We've built great public work projects to pull ourselves out of the Great Depression. The vast hydroelectric projects on the Columbia, the Colorado, the Niagara, and the St. Lawrence rivers are witness to what we can accomplish when we put our minds to it. We rose to the challenge of fighting fascism in World War II despite those who said the Axis couldn't be beat. We belatedly granted civil rights to all our citizens in the 1960s, and in the modern era, we have pushed cigarette smoking to the fringes of society when critics said it was a waste of time because it couldn't be done.

We did it. We proved the naysayers wrong. And we can do it again. But we certainly can't do it at the plodding pace of the present, with the timid programs currently on the books. We will have to ramp up development dramatically, and we won't be able to do that without the support of the people, the citizens of both the United States and Canada.

Everyone has a part to play. We will have to pay for it, surely—we always do in the end. We'll have to grow accustomed to seeing solar panels and wind turbines in new places, maybe in our backyards, maybe down the street, certainly nearby. We will only gain that acceptance—no doubt grudging at times—if we grant everyone the opportunity to participate, to profit from the renewable energy revolution. And there's only one renewable energy policy mechanism that offers opportunity to all: electricity feed laws.

Electricity Feed Laws

The concept of electricity feed laws or the feed-in tariffs that make them work is simple: they permit the interconnection of renewables with the grid, and they specify the price paid for the electricity delivered. Via a public policy debate, society (a parliament, congress, or state assembly) determines a rate to be paid for every kilowatt-hour generated with renewables. That's all there is to it. No cumbersome bureaucracy. No secret bidding. No sweetheart deals. A level playing field for all: farmers, homeowners, small businesses, municipal governments—everyone.

Feed laws pay only for actual generation; they don't pay for crackpot inventions that don't work or, equally as damaging, the tax scams once so common in the United States. Feed laws don't guarantee a profit. They only guarantee that if the project is built, and it generates electricity, the owners will be paid for their electricity at the price—the tariff—advertised. The burden remains on the owner to make sure the wind turbine operates as expected. If it doesn't, or if it doesn't produce as much electricity as planned, the owner suffers the loss, not consumers.

No other program has delivered more wind energy than electricity feed laws. None. Neither net metering, nor renewable portfolio standards, nor tax credits, nor even PURPA (Public Utility Regulatory Policies Act) has produced more wind-generated electricity than feed laws. To put it simply, feed laws work. More than half of all the wind-generating capacity in the world today is the result of feed laws; more than 85% of the wind capacity installed in Europe since 1997 has been built using various systems of feed-in tariffs.

Feed laws are also the most egalitarian method for determining where, when, and how much renewable generating capacity will be installed, enabling people from all walks of life to become renewable energy entrepreneurs—today's electricity rebels. Everyone can sell electricity to his or her utility company for a profit, whether it is homeowners installing solar photovoltaic systems on their rooftops, or farmers installing large wind turbines on their land, or cooperatives building small wind farms in their communities. John Geesman, a former commissioner on the California Energy Commission, suggests that it is this feature of feed laws that makes them so powerful; they can democratize the generation of electricity by distributing the opportunity for ownership to all. You don't have to be a multinational electric utility to own your own wind turbines or solar panels.

FEED LAWS ARE ALSO THE MOST EGALITARIAN METHOD

for determining where, when, and how much renewable generating capacity will be installed, enabling people from all walks of life to become renewable energy entrepreneurs— today's electricity rebels.

Feed laws, like all policy mechanisms for spurring the renewable generation of electricity, must, at a minimum, include measures for priority access to the grid and payment of a fair price for the electricity produced. These elements are the essential parts of the development equation. One without the other will not lead to the kind of rapid growth required to meet any but the most timid target.

Germany's groundbreaking feed law, the Stromeinspeisungsgesetz (or law on feeding electricity into the grid), provided both elements: access and price. But the idea isn't foreign to North America. In 1978, Congress passed PURPA, which permitted interconnection of renewable energy generators with the grid. It was revolutionary in its day. Many wondered at the time how it passed through the gauntlet of congressional special interests. Yet it did.

Unfortunately, PURPA didn't specify the price, only the means for calculating it. Feed laws, like PURPA, grant priority to the interconnection of renewable sources of electricity with the electric-utility network. Unlike PURPA, however, feed laws specify the *tariffs*, or rates, that renewable generators are paid for their electricity.

Germany's system of advanced renewable tariffs introduced in 2001, the Erneuerbare Energien Gesetz (or Renewable Energy Sources Act) clearly provides for access to the grid in its preamble. The law is formally known as the "act on granting priority to renewable energy sources" and goes on to specify in detail the prices that will be paid for different sources of renewable generation.

Successful feed laws—those that produce the rapid growth of a diverse mix of renewable technologies at a modest cost—provide a payment for feeding electricity into the grid that is both fair and reasonable. Balancing these demands is less difficult than it first appears. Geesman, a former regulator himself, points out that utility commissions in both Canada and the United States have been making just such judgment calls for decades.

The objective is to calculate a tariff based on the cost of the renewable generation plus a reasonable profit. This is in fact how we regulated the price of electricity for decades. Regulators were charged with ensuring that utility companies made a fair profit while at the same time protecting ratepayers from price gouging by utilities through their monopoly power.

The difference with past practice is that we determine the appropriate price up front, and then make it available to all comers. We say, in effect, "Here's the tariff. If you can build a wind project and make a profit, go right ahead. The sooner you can build it, the better. We want clean, renewable generation—and we want it now."

Advanced renewable tariffs, like those developed in Germany and elsewhere, differ from simpler feed laws by differentiating the tariffs according to several factors. Tariffs within each technology can be differentiated by project size and application or, in the case of wind energy, by the productivity of the resource. There can be several different prices for wind energy, several different tariffs for solar, and so on. What makes the tariffs "advanced" is the increased sophistication required to fine-tune the tariffs to achieve the desired objectives. Where we want rapid growth, for example, we increase the profitability by raising the tariff. Investors take it from there.

Such tariffs are not a subsidy, explains Jérôme Guillet, a French banker who specializes in financing offshore wind power plants. Feed laws offer a fair transaction between wind turbine owners, for example, and consumers. In effect, consumers purchase a hedge against not only the volatility of fossil-fuel prices—and the cost to society this volatility entails—but also a hedge against the inevitable increase in the cost of fossil fuels. It's also a hedge against the risk of large-scale accidents from nuclear power. That's the grand bargain, says Guillet. Owners of wind turbines get a fair price for their generation, and consumers get a secure, safe, and stable supply of electricity at a reasonable cost.

Because renewable sources of generation, such as wind turbines, are capital intensive, they require long periods of time to return their investments and earn a profit. Contract terms under feed laws are typically 20 years, sometimes more for long-lived hydro projects, sometimes less for other technologies.

Consequently, the prerequisite for a successful renewable energy program, above all else, is the political desire—the political will—for the program to succeed. And for it to succeed there must be the willingness to pay what it costs. Where the will exists, there is the stability of public policy that ensures investors, and the banks who loan to them, that there will be a fair opportunity to earn a return on their investment. Homeowners, farmers, and community groups must have certainty that there will be no abrogation of the contracts once they've invested in wind energy in their communities.

Deutsche Bank summarized the elements of good public policy for developing renewable energy through feed-in tariffs by playing on a familiar acronym: TLC. In the investment world it stands for transparency, longevity, and certainty. Successful feed-in tariffs must be simple, comprehensible, and transparent. Feed laws must provide simplified interconnection requirements with priority access to the grid. They must provide a price high enough to drive rapid development, and they must provide a contract length sufficiently long to reward investment.

Aggressive targets require aggressive programs. Feed laws and their feed-in tariffs are not for the politically faint of heart. We won't offset fossil-fuel-fired generation in North America or substitute renewably generated electricity for oil in light vehicle transport with incremental change in response to timid targets. As one political observer wryly noted, "Incremental change will only get you incremental results."

Importantly, renewable tariffs must be sufficiently differentiated to deliver the kind of renewable development from the technology desired in the location desired. There must be tariffs for each technology under a variety of conditions, such as a suite of tariffs for small wind turbines as well as tariffs differentiated by resource intensity for large turbines if everyone is to participate in the renewable energy revolution.

Small Wind Tariff

As an example of how aggressive programs must become, consider the special case of small wind. As we've seen, small wind turbines must be paid a relatively high price for their generation in order for their owners to earn a profit.

Britain has shown that it's politically viable to do so. In 2010, they launched the world's most sophisticated program supporting small wind turbines with a system of feed-in tariffs based on size. Though they targeted their policy toward microgenerators, they included projects up to 5 MW—hardly what most would call small wind turbines. Nevertheless, the program has been a resounding success. By late 2014, Britain had installed more than 260 MW of capacity, rivaling that installed in the United States after more than three decades of on-off programs for small wind development.

Small wind turbines—those less than 100 kW in size—accounted for about half of total capacity in the British program. Medium-size wind turbines contributed the other half. All the capacity was used in distributed applications, such as on farms or at hospitals and other institutions.

Fortunately, distributed wind energy is not limited only to small wind turbines. Nearly all the wind turbines installed on land in Denmark and Germany can be found in distributed applications scattered across the landscape.

Differentiated Tariffs for Distributed Wind

The good news is that the feed-in tariffs needed for large wind turbines in distributed applications developed and owned by those living nearby are a fraction of that necessary for small and medium-size wind turbines. However, feed laws for wind energy work best when they take into account the unique characteristics of the wind resource. One size (or tariff) doesn't fit all.

The intent of feed-in tariffs is to pay for the cost of generation plus a fair profit, nothing more, nothing less. In this way we get the wind development we need without paying more than we need to. It's Goldilocks pricing: not too high, not too low, just right.

As we've seen, the cost of wind generation is largely a function of the cost of the wind turbine and its productivity. Performance varies dramatically with slight changes in average wind speed. Thus, the cost of wind generation changes with the resource available. If there is only one price—or tariff—for wind generation, and it's designed for the windiest sites, wind can't be developed at less windy sites near where most people live. If the tariff is high enough for less windy sites, those with access to the windier sites will earn "undue" profits at the expense of consumers. Neither is fair.

To limit undue profits while paying what's needed elsewhere, the tariffs for wind energy must vary with the wind resource. Without tariffs for wind energy differentiated by the wind resource, we'll never spur the broad geographic development of wind that we demand.

Commercial developers can move their projects to wherever they like. They will concentrate only on the windiest sites to maximize their profits. Farmers, other landowners, and community groups, in contrast, are landlocked. Unlike a commercial developer, they can't move to a windy location to install their own turbines. A one-price-for-all policy favors some, but denies opportunity to everyone else. It can also allow some developers to earn excessive profits—revenues above and beyond that needed to encourage profitable development.

In California, wind turbines are concentrated in its windy passes. Germany and France wanted to avoid massing wind development only at its windiest locations—along scenic coastlines or atop mountain ridges. Instead, they wanted to distribute wind development across the landscape, to gain more of the benefits of wind energy by moving the turbines closer to the load—the people. They also wanted to spread opportunity geographically, so that those living in the interior of the two countries could develop their own wind resource. As a result, they pioneered renewable tariffs differentiated by wind resource intensity.

Feed-in tariffs for large wind turbines in Germany and France vary by the productivity of the wind turbine. The turbine's productivity, or yield, is a surrogate for the strength of the wind at the site—its resource intensity.

The objective was twofold: to lessen development pressure on the windiest sites by enabling development in other, less windy sites, and to provide siting flexibility. The programs in Germany and France have been successful in spreading development across the landscape of each country. While development still favors the windier regions, development is not solely concentrated on the windiest areas. Today, nearly 60% of German wind development is in the interior of the country and has moved away from the coastline. Wind turbines can now be seen in almost every part of Germany.

Germany and France each use a different mechanism for determining site productivity. However, both use a trial period after which the productivity is calculated and a subsequent tariff is determined. Thus, the maximum tariff is fixed, in order to provide a targeted profitability at the targeted sites, but the final tariff paid for more productive, that is, windier sites, declines on a sliding scale as a function of the site's productivity. Critics of wind energy often erroneously—or deliberately—focus only on the initial price, missing the significant savings that occur in later years.

For example, the base tariff in France was €0.082 ($0.11) per kWh during the trial period. After the trial period, the tariff fell to €0.28 ($0.038) per kWh at the windiest sites. That's less than large wind power plants were being paid in the United States through renewable portfolio standards, a tenth of that for small wind turbines in Britain, and one-sixth that for offshore wind.

Similarly, Germany was paying €0.0893 ($0.12) per kWh during the trial period. At its windiest sites, German tariffs fall to €0.071 ($0.066) per kWh.

In both countries, the tariffs were high enough and the contracts for long enough (15 years in France, 20 years in Germany) that investors could profitably install wind turbines at moderately windy sites. Yet at the same time, neither country overpaid for wind generation at the windiest sites. Both nations got what they wanted where they wanted without paying more for it than necessary.

The specific prices paid are less important than the principle. The price per kilowatt-hour for wind generation will vary from country to country and region to region based on actual costs. However, differentiating the tariffs is necessary if we want to open the market for wind energy to everyone, including farmers, homeowners, and cooperatives, not just a few favored commercial developers. This principle must be part of any aggressive renewable energy policy. If we are to see wind development from the windy Great Plains to modestly windy sites around the Great Lakes, the Midwest, and Ontario and the Eastern Seaboard, we'll need differentiated wind tariffs to make it possible.

Tariffs differentiated by site productivity are a powerful tool for encouraging wind development where it's needed most, near the people. Wind tariffs differentiated by productivity can increase distributed generation by distributing wind development across a broad geographic area. These tariffs can reduce development pressure on the windiest sites, though they won't eliminate it. Such a system will dampen social friction by spreading wind development across many sites and opening up the opportunity for more people to profit from the wind. It will also increase program flexibility by lessening pressure to get the price for wind energy exactly right the first time.

Banks prefer differentiated tariffs over a single fixed price because it lowers the development risk. Estimating the performance of a wind project—for all its sophistication—remains an inexact science. After all, we're dealing with a natural resource that varies from one year to the next. With differentiated tariffs, banks can be assured a project will receive a reasonable tariff if the site is less productive than first thought, earning enough to pay off its loan. And for that assurance, they can offer lower interest rates on their loans as seen in Germany. Consequently, the overall cost of wind generation can be lower when projects have access to lower-cost financing.

Most importantly, tariffs for wind energy differentiated by resource intensity spread opportunity to all, not just to those fortunate enough to live in the windiest locales. They enable fair profits at medium-wind sites while limiting excessive profits at windy sites.

No jurisdiction in North America has ever offered tariffs for wind energy differentiated by resource intensity. For us to do so, we must overcome policy inertia, our penchant for incrementalism, and our reluctance to share the opportunities that a more aggressive renewable energy policy will bring. We need to rethink our approach to energy policy.

Energy for Life: The Pursuit of an Ethical Energy Policy

Let's get this out of the way first. There are no panaceas, no quick fixes for North America's energy challenge. It's taken us decades to get into our current predicament, and it will take us a long time to get out of it. Electricity feed laws are one very important solution, but only one.

Energy policy should be built upon the framework laid down by Grundtvig more than a century ago:
ALL PUBLIC POLICY, IN FACT ALL ENDEAVORS, SHOULD BE LIFE AFFIRMING.

The title of this section, "Energy for Life," is a play on the words and the work of the 19th-century Danish theologian, N. F. S. Grundtvig. A contemporary of Kirkegaard, Grundtvig's powerful influence helped shaped the democratic, pluralistic, and egalitarian Denmark that exists today. For many Danes, their country's success with wind energy grew from seeds planted a century earlier by Grundtvig.

What does Grundtvig have to do with energy? The answer is at the same time simple and complex. Energy policy should be built upon the framework laid down by Grundtvig more than a century ago: all public policy, in fact all endeavors, should be life affirming. It's a simple idea that's difficult to put in practice.

By 2035 Denmark expects to produce 100% of its electricity and heat from renewable resources. Similarly, Germany plans to generate 80% of its electricity from renewables by 2050. These are not idle "aspirational" goals. Danes and Germans appear to have the will to meet these targets.

Why? In part, because these nations believe they will create long-lasting jobs for their citizens. When the rest of the world decides that it wants wind turbines and solar panels, Denmark and Germany will be only too happy to sell them their products. But they also believe it's the right thing to do. It is not just a question of abstract economics. They feel a moral imperative to meet their aggressive renewable targets.

Thirty years ago, President Jimmy Carter asked Americans to embark on a venture that was "the moral equivalent of war." He may have picked the wrong metaphor, but he was right about the moral implications of how we use energy, who has access to it, and where we get it.

The environmental community's interest in wind energy grew out of the ethical dilemma posed by, for example, the strip mining of coal. If we were to condemn strip mining for what it does to the land and people of Appalachia, then what would we substitute? Wind and solar energy are a relatively more benign choice. *Relative* is the operative word. There's no environmental free lunch.

Fortunately, there are a number of priests, rabbis, ministers, and imams who have called upon us as North Americans to examine the ethics of our consumptive lifestyle, just as clerics in the past have called upon us to examine our behavior in the more traditional realms of religious teaching. For example, should we raise a moral red flag when a suburbanite drives his Lincoln Navigator from his "modest" 3,000-square-foot, cathedral-ceilinged house to pick up a loaf of bread? Does paying for the gasoline used so inefficiently absolve consumers of responsibility for the choices they've made?

Ethics, relative to the consumption of natural resources, is in part, but not solely, the difference between responsible and irresponsible use. We need a lamp to light the darkness so we can curl up with a good book. The lamp in a dark, occupied room serves a purpose; it meets a need. An office building with all the lights ablaze long after everyone has gone home doesn't meet a need. It wastes or squanders a valuable resource, one costly to extract, refine, produce, and deliver.

Implicitly then, there are values at work here. Should we gouge out great furrows in the earth of Montana, or chop the top off a mountain in Tennessee, so that a careless or irresponsible person in Chicago can leave a light on in an unoccupied room? That's certainly not the way many North Americans were raised. Many were taught "waste not, want not" by parents who lived through the Great Depression and World War II. They learned frugality long before the environmental movement proffered new reasons for being miserly.

Within the environmental movement itself, the debate rages over the amount of emphasis to place on the word *efficiency* on the one hand, and on the word *conservation* on the other. Efficiency means, as Buckminster Fuller would say, doing more with less. Conservation implies doing without or using less. In the American context, conservation has been a hard sell. Those on the front lines of energy policy are often chided for mentioning the word *conservation*. "Americans don't want to hear it," we're warned.

Yet efficiency alone, while important, won't take us where we want to go. Conservation or—if that word is unpalatable—stewardship of a bountiful but finite earth is critical to our collective future. For example, which of the following solutions is superior? Is it better to modestly improve the efficiency of the Lincoln Navigator from 14 mpg to say, 18 mpg? That 30% improvement is easily and immediately obtainable. Or is it better to conserve gasoline by not building, buying, or using the behemoth in the first place and driving a Toyota Prius instead? The Prius gets nearly 260% better mileage than the Navigator and yet often does the same job: it carries the driver and the loaf of bread. Or is it better still to build real cities where you can walk down the street, pick up the loaf of bread yourself, and talk to your neighbors along the way? Now, how do you calculate the improvement in fuel efficiency of walking?

Paying the monetary price for a Lincoln Navigator—the largest, nonmilitary, personal transport vehicle in the world—doesn't absolve the driver of responsibility for the effluent it spews out, the carbon dioxide it emits, or the military establishment necessary to protect its fuel supply. These costs are borne by us all, including those of us who drive small cars and those of us who walk to pick up our loaf of bread.

In electricity, this conflict between efficiency and conservation has come back to bite efficiency advocates in the derrière. California building standards require all new homes to meet certain standards: the amount of insulation, type of windows, and so on. Good, we say, we've cut the household energy consumption significantly. But what has happened? Our industrial housebuilders began building bigger and bigger houses. You've seen them. They're monster homes, McMansions. Their ads gloat that with all the energy-saving features, "You too can now afford twice the house you could before."

The result is that we're not just running in place, we're losing ground. This is the social price for the disconnect between efficiency, conservation, and values.

Let's revisit our suburbanite. And let's put her in affluent Palo Alto and propose a new power plant there to meet rising demand from ever more, ever bigger McMansions. "No way," our suburbanite will say. "Put those power plants in Kern County. They like power plants down there [snicker, snicker]."

What's unsaid, but understood, is that Kern County is plebian to Palo Alto's patrician world. This cultural gap isn't unique to California. It can be found among Ontario's lakeside "cottagers" and their disdainful view of farmers and rural communities who want wind farms and biogas plants erected in their midst.

Is there something wrong with those who want or at least accept renewable energy in their backyard? Not at all. There should be a renewable power plant in everyone's backyard—or on his or her roof, or down the street, or nearby.

NIMBY, or not in my back yard, is just another manifestation of trying to pass the social costs of energy choices—as in the Lincoln Navigator or the McMansion—on to other, and often less politically powerful, groups.

North Americans are victims of a myth, a belief so all pervasive that few question it, and even fewer still realize it is a belief—a veritable secular religion. Deregulation, restructuring, market liberalization—whatever you call it—was a craze, a long-running one that lasted three decades. But unlike the harmless hula-hoop craze of the 1950s, this craze hurt real people and will continue hurting their descendants for years to come. This fad resembled the tulip mania that engulfed the Dutch in the 17th century, bankrupting the country.

Neoliberalism's chief tenet is that all things good come from the market, that there's a price for everything, and that everything can be monetized. This was Enron's mantra. Unfortunately, an unregulated free market is a fool's paradise where the values most important to us—love, family, clean air and water, a secure future—are either undervalued or not valued at all, and where price—but only the apparent price—is raised onto an altar and we are asked to worship it.

No North American politician has been immune to this siren's song, and we have crashed the ship of state on her shoals. Deregulation's failure offers us an unparalleled, if unfortunate, opportunity to rethink how we produce, consume, and—most importantly—value electricity in our society.

Electricity is a means to an end—a tool for

Figure 21-3. Energy for life. Wind energy can and should be a celebration of life–for the people of today and for those of tomorrow. Adam Twine epitomizes the new breed of renewable energy advocate–positive, upbeat, a doer, and, at heart, an electricity rebel.

meeting the needs of people—and not an end in itself.

Therefore, we need to envision an electricity system that is sustainable, meets the needs of people today as well as those of tomorrow, and is built upon sufficiency for all, equitably distributed. We need to envision a system that enhances the quality of life for all, rich and poor alike. Such a system is built upon services rendered and needs met, not upon a constant and never-ending growth of supply. We must envision an energy system for people or, as Grundtvig would say, an energy system for life.

If we choose to do so, we can regain control of our energy destiny and construct an electricity supply system that emphasizes using energy responsibly and with respect for our neighbors and for the environment.

One achievable near-term objective fitting this vision is lowering our residential per capita consumption of electricity to that of Europeans, who share our standard of living.

Per capita, Europeans consume a fraction of the electricity of North Americans, while enjoying the same standard of living. By several measures, Europeans enjoy a higher standard of living than many in Canada and the United States yet use less electricity, less oil, less energy in general. They often can walk to pick up a loaf of bread, for example.

A young California woman said, "What we waste today, we steal from our children." She meant it literally as a future mother, but she could easily have meant it metaphorically: what we squander today by our excessive consumption, we steal from future generations.

We can do as well as Europeans. Many families in Canada and the United States have shown that it can be done without hardship. What would happen if everyone did this? Well, for starters, we could cut our greenhouse emissions in half. That alone makes it a worthy goal.

Of course, we still need new sources of supply, especially to replace the old, dirty power plants that are still operating. And one way to do that is to offer the people of North America the same deal that Germany, France, and now even Switzerland —yes, even the conservative Swiss—offer their citizens, by instituting an electricity feed law. Such a law would simply state that you can connect your solar or wind power system to the grid and, more importantly, spell out the price you would be paid for your electricity, a price high enough to make it profitable for you, your neighbors, and your entire community.

That's expensive some say. Maybe. Maybe not. It depends upon your values. Germans, Danes, the Swiss—even capitalist China—thought it was the right thing to do. To them, it was worth it (see Figure 21-3. Energy for life).

We can bring back the luster of North America's industrial heartland, if we have the will to do so. We can build a bright renewable future for North America and for North America's children and grandchildren. We can move from a culture of consumption to a culture of conservation, from a nation of consumers to—once again—a nation of producers. This is the challenge—a challenge worthy of great nations.

There's no time to lose, no time for half measures.

Let's put wind energy in everyone's backyard.

Let's create an energy system for life.

Appendix

Constants and Conversions

Constants

ICAO Standard Atmosphere
- Air pressure at sea level (p) = 29.92 in. Hg
  - = 760 mm Hg
  - = 1013.25 mb
  - = 1.01325×10^5 N/m² or Pa
- Temperature at sea level = 15°C
  - = 288.15 K
  - = 59°F
- Acceleration due to gravity (g) = 9.807 m/s²
- Gas constant for dry air (R) = 287.04 J/kgK
- Absolute temperature (T) = 273.15 K
- Normal atmospheric lapse rate (Γ) = 6.5°C/1000 m
- Density of dry air, standard atmosphere (ρ) = 1.225 kg/m³

Conversions

Speed
- 1 mph = 0.447 m/s
- 1 mph = 0.870 knots
- 1 mph = 1.61 km/h
- 1 knot = 1.15 mph
- 1 knot = 0.514 m/s
- 1 knot = 1.85 km/h
- 1 m/s = 2.24 mph
- 1 m/s = 1.94 knots
- 1 m/s = 3.60 km/h
- 1 km/h = 0.621 mph
- 1 km/h = 0.54 knots
- 1 km/h = 0.278 m/s

Length
- 1 inch = 2.54 centimeters
- 1 centimeter = 0.394 inches
- 1 meter = 3.28 feet
- 1 foot = 0.305 meters
- 1 kilometer = 0.620 miles
- 1 mile = 1.61 kilometers

Area
- 1 square kilometer = 0.386 square miles
- 1 square kilometer = 100 hectares
- 1 square mile = 2.59 square kilometers
- 1 hectare = 10,000 square meters
- 1 hectare = 2.47 acres
- 1 acre = 0.405 hectares
- 1 acre = 4,047 square meters

Volume
- 1 cubic meter = 35.3 cubic feet
- 1 cubic meter = 264 US gallons
- 1 cubic foot = 0.028 cubic meters
- 1 liter = 0.264 US gallons
- 1 US gallon = 3.78 liters
- 1 US gallon = 0.004 cubic meters

Flow Rate
- 1 liter/second = 0.264 US gallons/second
- 1 cubic meter/minute = 264 US gallons/minute
- 1 gallon/minute = 3.79 liters/minute
- 1 gallon/minute = 0.063 liters/second
- 1 gallon/day = 0.158 liters/hour

Weight
- 1 metric ton = 1.10 short tons
- 1 metric ton = 1,000 kilograms
- 1 long ton (Imperial) = 2,240 pounds
- 1 short ton (US) = 2,000 pounds
- 1 kilogram = 2.21 pounds
- 1 pound = 0.454 kilograms

Power
- 1 kilowatt = 1.34 horsepower
- 1 horsepower = 0.746 kilowatts

Torque
- 1 Newton-meter = 0.738 foot-pounds
- 1 foot-pound = 1.356 Newton-meters

Temperature
- degrees Fahrenheit = 9/5 °C + 32
- degrees Celsius = 5/9 (°F - 32)
- degrees Kelvin = 273.15 + °C

Units of Energy

1 kWh	=	3,412 Btu
1 kWh	=	0.1 therm
1 kWh	=	860 kilocalories
1 kWh	=	3,600 kJ
1 kWh	=	3.6 MJ
1 Therm (EU)	=	100,000 Btu
1 Therm (US)	=	99,976 Btu
1 MMBtu (thousand thousand)	=	1,000,000 Btu
1 MMBtu	=	1 MBtu
1 Therm	=	29.3 kWh
	=	25,194 kilocalories
1 GigaJoule	=	277.8 kWh
		947,818 Btu
		9.48 Therm
1 PJ	=	0.278 TWh
1 TWh	=	1,000,000,000 kWh
1 TWh	=	3.6 Peta Joule (PJ)
1 Quad		Quadrillion Btu

Energy Equivalence of Common Fuels

1 kWh	~	3.41 cu. ft. of natural gas
	~	0.034 gallon of oil
	~	0.00017 cord of wood
975 cu. ft of nat. gas	~	1,000,000 Btu
1 Mcf of natural gas	=	1,000 cu. ft. of natural gas
1 MMBtu of natural gas	~	975 cu. ft. of natural gas
1 Therm		100 cu. ft. of natural gas
	~	1 Ccf of natural gas
	~	1 gallon of oil
1 Therm		0.005 cord of wood
1 Therm		1 Ccf of natural gas
1 Quad		Tcf of natural gas
		Trillion cubic feet
1 barrel of oil		5,487 cu. ft. of natural gas
1 gallon of oil		100,000 Btu
1 cord of wood		20,000,000 Btu
1 kWh	~	10,000 Btu (Conventional Plant)
1 kWh	~	7,500 Btu (Combined Cycle Plant)
1 kWh	~	2,520 kilocalories (Conventional Plant)
1 kWh	~	10,551 kJ (Conventional Plant)
600 kWh	~	1 barrel of oil equivalent
60 W	~	1 Human sitting
1 kW	~	1 Human running
28 kW	~	1 Auto @ 55 mph
88 Quads	~	USA Consumption
1000 MWe	~	9,000 tons/day or 100 90-ton gondola cars/day
1000 MWe	~	40,000 bbl/day or 1 tanker per week
1000 MWe	~	2.4 x 10^8 SCF/day natural gas

Scale of Energy Equivalents

Energy	Approximate size wind turbine to provide the same annual energy output by Rotor Diameter*		
	Rotor Diameter		
kWh	m	ft	
1			auto battery, 100 watt light for 10 hours
10			electric space heater for 10 hours
50	0.25	1	average per capita consumption in India
1,000	1.5	5	
3,500	2.5	8	average residential consumption in Northern Europe
4,000	3	10	
6,500	4	13	average residential consumption in California
12,000	5	16	average residential consumption in Texas
15,000	6	20	
20,000	7	23	typical electric home consumption in Oklahoma
75,000	10	33	
250,000	18	60	
500,000	25	80	
1,000,000	35	110	
2,000,000	45	150	
3,000,000	55	180	
4,000,000	65	210	
6,000,000	80	260	
8,000,000	90	300	
10,000,000	100	330	

* Class 4 wind resource, 50 meters (164 feet) hub height, 7.5 m/s (16.8 mph).

Scale of Equivalent Power

Rotor Diameter		Typical Wind Turbine Peak Power Rating	Equivalent
m	ft	kW	
1.5	5	0.25	⅓ horsepower electric motor, electric drill
2	7	0.50	⅔ horsepower electric motor
3	10	1	hair dryer, electric space heater
7	23	10	garden tractor
10	33	25	
18	59	100	passenger car engine
25	82	250	
40	131	500	heavy truck engine
50	164	1,000	race car engine, small diesel locomotive
100	328	3,000	diesel locomotive
		500,000	coal-fired generator, small nuclear reactor
		1,000,000	large nuclear reactor

Battelle Classes of Wind Power Density									
	Wind Speed and Power at 10 m (33 ft)			**Wind Speed and Power at 30 m (100 ft)**			**Wind Speed and Power at 50 m (164 ft)**		
	Power Density	Speed		Power Density	Speed		Power Density	Speed	
	W/m²	m/s	mph	W/m²	m/s	mph	W/m²	m/s	mph
	50	3.5	7.8	80	4.1	9.2	100	4.4	9.9
1									
	100	4.4	9.9	160	5.1	11.4	200	5.5	12.3
2									
	150	5.0	11.5	240	5.9	13.2	300	6.3	14.1
3									
	200	5.5	12.3	320	6.5	14.6	400	7.0	15.7
4									
	250	6.0	13.4	400	7.0	15.7	500	7.5	16.8
5									
	300	6.3	14.1	480	7.4	16.6	600	8.0	17.9
6									
	400	7.0	15.7	640	8.2	18.4	800	8.8	19.7
7									
	1000	9.5	21.3	1600	11.1	24.7	2000	11.9	26.7

Note: Increase in speed and power with height assumes 1/7 power law. Wnd speeds are based on the equivalent of a Rayleigh distribution.

American Wire Gauge to Metric Conversion

In the United States, manufacturers of small wind turbines specify conductor size in American wire gauge (AWG). The rest of the world uses cross-sectional area in square millimeters (mm^2). Converting from AWG to mm^2 is, unfortunately, not straightforward.

US manufacturers sometimes list the metric equivalent. For example, the Electric Vehicle Service Equipment installed at our home includes a charging cable with three #10 conductors. Because the manufacturer sells the cable internationally, it included the metric cross-sectional area of 5.26 mm^2 on the identifying label.

Note that the resistance (ohms per measure of distance) in the following table is rated at two different temperatures. The rated resistance for AWG sizes is from the National Electrical Code in the United States. Increasing the temperature increases resistance.

American Wire Gauge to Metric Cable Conversion & DC Resistance in Ohms							
		Area		Resistance Copper@ 75° C	Area	Nearest Metric Size*	Resistance Copper@ 20° C
AWG	Strands	cmils	in²	Ohms/1000 ft	mm²	mm²	Ohms/km
10	7	10,380	0.0082	1.2900	5.26	6	3.080
8	7	16,510	0.0130	0.8090	8.37	10	1.830
6	7	26,240	0.0206	0.5100	13.3	16	1.150
4	7	41,740	0.0328	0.3210	21.1	25	0.727
3	7	52,620	0.0413	0.2540	26.7	35	0.524
2	7	66,360	0.0521	0.2010	33.6	35	0.524
1	19	83,690	0.0657	0.1600	42.4	50	0.387
0	19	105,600	0.0829	0.1270	53.5	70	0.268
2/0	19	133,100	0.1045	0.1010	67.4	70	0.268
3/0	19	167,800	0.1318	0.0797	85.0	95	0.193
4/0	19	211,600	0.1662	0.0626	107.2	120	0.153
250	19	250,000	0.1963	0.0535	126.7	150	0.124

1 inch = 25.4 mm
*Note: direction of conversion is from AWG to metric sizes, reversing the direction could result in undersize conductors.
Source: AWG, NEC 1999 Edition; Metric, Folkecenter for Renewable Energy, Denmark

Nongovernmental Organizations (NGOs)

The following organizations can be contacted for further information. Most of the national wind energy associations hold annual conferences on the latest developments in wind turbine technology within their respective countries. These conferences deal mostly with the commercial use of large wind turbines, though some have sessions on small wind turbines as well.

Outside English-speaking countries, there's usually a distinction between trade associations that promote the interests of businesses working with wind energy, such as wind turbine manufacturers, and groups that represent the interests of consumers or the environment. In Germany, Bundesverband Windenergie represents wind turbine owners and Fördergesellschaft Windenergie represents the interests of manufacturers. This distinction is important to know when searching for information. Trade associations exist to promote the interests of their members and not necessarily the public interest. Don't expect the American Wind Energy Association (AWEA) or the Danish Wind Turbine Manufacturers Association (Vindmølleindustrien) to help you locate information on German wind turbines.

American Wind Energy Association
www.awea.org

Asociación Empresarial Eólica (Spanish Wind Energy Association)
www.aeeolica.org

Austrian Wind Energy Association (IG Windkraft)
www.igwindkraft.at

British Wind Energy Association (see RenewableUK)

Bundesverband Windenergie (German Wind Turbine Owners Association)
www.wind-energie.de

Canadian Wind Energy Association
www.canwea.ca

Centre for Alternative Technology (Great Britain)
www.cat.org.uk

Danmarks Vindmølleforening (Danish Wind Turbine Owners Association)
www.dkvind.dk

Distributed Wind Energy Association (USA)
distributedwind.org

Chinese Wind Energy Association
www.cwea.org.cn

European Wind Energy Association
www.ewea.org

Fördergesellschaft Windenergie (German Wind Turbine Manufacturers Association)
www.wind-fgw.de

France Énergie Éolienne (French Wind Energy Association)
www.fee.asso.fr

Global Wind Energy Council (Belgium)
www.gwec.net

Irish Wind Energy Association
www.iwea.com

Japan Wind Power Association
http://jwpa.jp/index_e.html

New Zealand Wind Energy Association
www.windenergy.org.nz

Nordisk Folkecenter for Vedvarende Energi (Nordic Folkecenter for Renewable Energy Denmark)
www.folkecenter.dk

University of Massachusetts Wind Energy Center
www.umass.edu/windenergy

RenewableUK (formerly the British Wind Energy Association)
www.renewableuk.com

Small Wind Certification Council (USA)
http://smallwindcertification.org/

South African Wind Energy Association
www.sawea.org.za

Vindkraftföreningen r.f. (Finnish Wind Energy Association)
www.vindkraftforeningen.fi

Vindmølleindustrien (Danish Wind Turbine Manufacturers Assoc.)
www.windpower.dk

Suisse Eole (Swiss Wind Energy Association)
www.suisse-eole.ch

Svensk Vindkraftförening (Swedish Wind Energy Association)
svensk-vindkraft.org/

World Wind Energy Association (Germany)
www.wwindea.org

Government-Sponsored or Affiliated Laboratories

These research centers concentrate on highly technical aspects of large wind turbine technology. However, several have tested small and medium-size wind turbines.

Wind Energy Institute of Canada
www.weican.ca

Deutsches Windenergie Institut (German Wind Energy Institute)
www.dewi.de

CIEMAT, División de Energias Renovables (Division of Renewable Energy, Spain)
www.ciemat.es

Energy Research Centre of the Netherlands
www.ecn.nl/expertises/wind-energy/

Fraunhofer-Institut für Windenergie und Energiesystemtechnik (Fraunhofer IWES) (Fraunhofer Institute for Wind Energy and Energy System Technology, Germany)
www.iwes.fraunhofer.de

Institut dezentrale Energietechnologien gemeinnützige (IdE) (Institute of Decentralized Energy System Technology, Germany)
www.ide-kassel.de

National Engineering Laboratory (Great Britain)
www.tuvnel.com

National Renewable Energy Laboratory, National Wind Technology Center (USA)
www.nrel.gov/wind

DTU Institute for Vindenergi (formerly Risø National Laboratory) (Denmark)
www.vindenergi.dtu.dk

Sandia National Laboratories (USA)
energy.sandia.gov/energy/renewable-energy/wind-power/

US Department of Agriculture, Agricultural Research Service, Conservation and Production Research Laboratory
www.cprl.ars.usda.gov

Websites

There are literally hundreds of sites on the World Wide Web discussing the topic of wind energy. For tips on the design of small wind turbines visit the sites of Scoraig Wind Electric's Hugh Piggott, scoraigwind.co.uk, and Windmission's Claus Nybroe, www.windmission.dk.

Most national wind energy associations sponsor sites with information on programs in their countries. Many of the European websites are offered in several languages, including English. Remember that trade associations are in business to promote their member's products.

As with any source of information, be wary. There are many websites offering bogus, misleading, and sometimes fraudulent material about wind energy and proffering "products" of doubtful provenance. Some of these sites are "here today, gone tomorrow."

Electronic Forums on Small Wind Turbines

There is an English-language newsgroup that deals with small wind turbines on Yahoo: https://groups.yahoo.com/neo/groups/small-wind-home/info. There is also a discussion group on small wind turbines, *petit éolien*, in French: https://fr.groups.yahoo.com/neo/groups/petit-eolien/info. These electronic forums are accessible worldwide.

Before posting any questions or comments, it's always good "netiquette" to monitor the lists for some time to become familiar with the topics and the participants. It's also wise to read the FAQ or frequently-asked-questions files first. Participants on these lists have answered many of the most common questions so many times, they may bristle at seeing them again. Be forewarned that discussions on these lists can be quite lively.

Workshops

Both Mick Sagrillo and Hugh Piggott offer workshops on small wind turbines. Sagrillo targets sizing, proper siting, and installation. Piggott teaches how to build a sturdy, functioning small turbine from scrap parts.

Mick Sagrillo
Sagrillo Power & Light
msagrillo@wizunwired.net

Hugh Piggott
Scoraig Wind Electric
scoraigwind.co.uk

Derivative workshops based on Piggott's designs are also found in France, Italy, Canada, and the United States.

Several organizations offer short courses on wind energy for professionals, including ECN in the Netherlands and DEWI in Germany.

Deutsches Windenergie Institut (DEWI)
www.dewi.de

Energy Research Centre of the Netherlands (ECN)
www.ecn.nl/expertises/wind-energy/

Community Wind Organizations

There are several organizations in North America actively encouraging community-owned or farmer-owned wind development. the Institute for Local Self-Reliance and Windustry in Minnesota and the Ontario Sustainable Energy Association in Canada. There are also groups throughout Europe.

Community Energy Scotland
www.communityenergyscotland.org.uk

Community Power EU (European Union)
www.communitypower.eu

Community Power Network (Washington, DC)
communitypowernetwork.com

Deutscher Genossenschafts und Raiffeisenverband, Energiegenossenschaften (German organization promoting renewable energy cooperatives)
www.genossenschaften.de/bundesgesch-ftsstelle-energiegenossenschaften

Energy4All (Great Britain)
energy4all.co.uk

Institute for Local Self-Reliance (USA)
www.ilsr.org

Ontario Sustainable Energy Association
www.ontario-sea.org

Organisatie voor Duurzame Energie (Netherlands)
www.duurzameenergie.org

REScoop.EU (European Union)
www.rescoop.eu

Toronto Renewable Energy Cooperative (Ontario)
www.trec.on.ca

Windustry (Minnesota)
www.windustry.org

Community-Owned Projects Mentioned

Australia

Denmark Community Windfarm
www.dcw.org.au

Hepburn Wind
hepburnwind.com.au

Canada

Colchester-Cumberland Wind Field (Nova Scotia)
www.ccwf.ca

M'Chigeeng First Nation (Ontario)
www.mchigeeng.ca

WindShare Co-operative (Ontario)
www.windshare.ca

Europe

Baywind Energy Co-operative (Great Britain)
www.baywind.coop/baywind_home.asp

Hvidovre Vindmøllelaug (Denmark)
www.hvidovrevindmollelaug.dk

Kennemerwind (Netherlands)
www.kennemerwind.nl

Lynettens Vindkraft (Denmark)
lynettenvind.dk

Middelgrundens Vindmøllelaug (Denmark)
www.middelgrunden.dk

Paderborn (Germany)
http://de.wikipedia.org/wiki/Windpark_Sintfeld

Regiowind Freiburg (Germany)
www.oekostrom-freiburg.de/index.php?id=46

Thyborøn Harboøre Vindmøllelaug (Denmark)
www.roenland.dk

Westmill Wind Farm Co-operative (Great Britain)
www.westmill.coop

USA

Camden Hills Regional High School (Maine)
www.fivetowns.net/subsites/windplanners/

Carleton College (Minnesota)
http://renewnorthfield.org/?page_id=5

Forest City Schools Wind Turbine (Iowa)
https://sites.google.com/a/forestcity.k12.ia.us/forest-city-schools-wind-turbine/

Hull Wind (Massachusetts)
www.hullwind.org

St. Olaf College (Minnesota)
www.stolaf.edu/map/WindTurbine.html

Prowind Groups

There are a number of organizations and environmental groups that support wind energy globally. Here are a few in the English-speaking world. One website that tries to debunk many of the myths surrounding wind power is Greenpeace's Yes2Wind. Another is Pro Wind Alliance—a voice of reason in the British Isles not connected to any trade association. There is a similar group in Australia as well.

Australian Wind Alliance
www.windalliance.org.au

Pro Wind Alliance
www.prowa.org.uk

Yes2Wind
www.yes2wind.com

Historical Sites and Museums

Museums with Wind Exhibits

Deutsches Technikmuseum Berlin. A superb exhibit on wind, meteorology, and early wind energy technology, *WindStärken*, ran through October 2013 at the museum. Other exhibits on wind energy may be produced in the future.
www.sdtb.de

Museo Nazionale della Scienza e della Tecnologia "Leonardo da Vinci." The Milan technical museum had an exhibit of a 5-kW Riva Calzoni one-blade wind turbine in the basement in the 1990s.
www.museoscienza.org/english/

Panhandle-Plains Historical Museum, Texas A&M University, Canyon, Texas. See the *Capture the Wind* exhibit. The museum also has an extensive literature archive collected by T. Lindsay Baker.
http://panhandleplains.org/pages/windmills_49.asp

York County Heritage Trust, York, Pennsylvania. The trust contains a collection of shop models, photos, logbooks, and other records of the Smith-Putnam turbine.
www.yorkheritage.org

Open-Air Museums: North America

American Wind Power Center, Lubbock, Texas. An open-air museum for the American-style water pumping windmill and related exhibits on wind electricity.
www.windmill.com

Batavia Historical Museum Windmills, Illinois. Several original windmills manufactured in Batavia have been located, purchased, renovated, and erected along the city's river walk.
www.bataviahistoricalsociety.org/wmills.htm

Canadian National Historic Windmill Centre, Alberta. Fifteen miles east of Foremost, the center features outdoor restored examples of windmills encompassing over 200 years of Canadian wind power.
www.foremostalberta.com

Kregel Windmill Factory Museum, Nebraska City, Nebraska. This factory museum demonstrates how Americans have used and can continue to use the renewable power of the wind to enhance the quality of human life.
kregelwindmillfactorymuseum.org

Mid-America Windmill Museum, Kendallville, Indiana. A major purpose for this museum was to display windmills manufactured by Flint and Walling Company. The museum has expanded its display to 52 windmills.
www.midamericawindmillmuseum.org

Shattuck Windmill Museum. Fifty-one vintage water-pumping windmills and one wind generator are on display at the junction of Oklahoma Highway 283 and Highway 15.
http://www.shattuckwindmillmuseum.org/

Windmill State Recreation Area, Gibbon, Nebraska. Eclipse windmills are on display.
outdoornebraska.ne.gov/parks/places/campmaps/maps/windmill.pdf

Open-Air Museums: Europe

De Vereniging Zaanse Molen, the Netherlands. The Society of Zaan Mills was founded in 1925. The society features 12 industrial windmills and a museum. The mills are an important part of the

cultural heritage of Noord Holland and continue to this day to delineate the Zaan skyline. Some historians argue that the industrial revolution began with the windmills at Zaan.
zaanschemolen.nl

Energimuseet (The Energy Museum), Bjerringbro, Denmark. The museum's exhibition on wind power has the title "As the wind blows." It tells the story of how Denmark became the world's leader in wind energy technology.
energimuseet.dk

Folkecenter for Renewable Energy, Jutland, Denmark. At the center is a showroom for historic Danish wind turbines with an extensive collection of nacelles from the early days of Danish wind power. The Windmill Blade Expo is nearby.
www.folkecenter.net/gb/tour/blade_expo
www.folkecenter.net/gb/tour/expo

Frilandsmuseet (National Museum of Denmark). This open-air museum north of Copenhagen is one of the largest and oldest in the world. The museum houses more than 50 farms, windmills, and houses from the period 1650–1950.
natmus.dk/en/the-open-air-museum/

Germania (molen), Groningen, the Netherlands. This platform grain-grinding mill is one of the nearly 1,000 windmills in the Netherlands that is periodically open to the public.
nl.wikipedia.org/wiki/Germania_(molen)

Internationales Muhlenmuseum, Gifhorn, Niedersachsen, Germany. An open-air windmill museum.
www.muehlenmuseum.de

Museummolen Schermerhorn (Museum Mill), the Netherlands. An open-air museum in Noord Holland, north of Amsterdam.
www.museummolen.nl

Museum Park, Deutsches Technikmuseum, Berlin, Germany. The museum displays a historic stage mill, a mechanical farm windmill, and a micro wind turbine.
www.sdtb.de/Museum-park 1293.0.html

Poul la Cour Museum, Vejen, Denmark. The place where modern wind energy began—the laboratory of Poul la Cour, the Danish Edison.
www.poullacour.dk/engelsk/museet.htm

Schloss Sanssouci, Berlin, Germany. A reconstructed stage or gallery windmill that has served the palace (Schloss) since 1787 is on display. It was a mill on this site that served in the famous legend of the Miller of Sanssouci who challenged kingly power.
en.wikipedia.org/wiki/Historic_Mill_of_Sanssouci

World Heritage Site of Kinderdijk, the Netherlands. The 19 windmills of Kinderdijk symbolize the way in which the Dutch have managed water drainage with one of the world's oldest wind farms.
www.kinderdijk.nl

Open-Air Museums: Elsewhere

Fred Turner Museum, Loeriesfontein, South Africa. The museum displays 27 water-pumping windmills.
en.wikipedia.org/wiki/Loeriesfontein#Windmill_Museum

Morawa District Historical Society and Museum, Australia. The small rural town of Morawa with its museum is approximately 400 kilometers north of the state capital, Perth, in the northern wheat belt of Western Australia. The museum's collection includes Australian-made as well as imported water-pumping windmills.
http://members.iinet.net.au/~caladenia@westnet.com.au/

History of Wind Power Additional Sources

North American Focus
American Wind Charger Industry History. Website maintained by author Craig Toepfer on battery-charging wind turbines from the 1930s through the 1950s.
www.windcharger.org

Wincharger.com. Website celebrating the battery-charging wind turbine that was manufactured in Sioux City, Iowa.
www.wincharger.com

Wind: An Energy Alternative. A video produced in 1979 by the Solar Energy Research Institute. Almost corny by today's standards. Includes Terry Mehrkham's Dorney Park turbine before switching to DOE's Mod 0A turbine at Clayton, New Mexico. Vintage footage of the Smith-Puthnam turbine atop Grandpa's Knob near Rutland, Vermont, and several other well-known wind experiments in the United States.
http://images.nrel.gov/viewphoto.php?albumId=614873&imageId=7680104

International Focus

Winds of Change. Website maintained by Erik Grove-Nielsen, one of Denmark's wind pioneers. Twenty-five years of wind power development in photos from the years 1975 to 2000.
www.windsofchange.dk

The International Molinological Society (TIMS). Website for the organization that fosters interest and understanding of wind, water, and animal-driven mills worldwide.
www.molinology.org

Periodicals

Windpower Monthly is the commercial wind industry's principal trade publication and is an unaffiliated observer of wind energy worldwide. *Home Power* is the principal magazine covering small wind turbine topics. *Home Power*, as the name implies, caters to homeowners using small off-the-grid power systems or small wind turbines interconnected with the utility. *Naturlig Energi*, a publication of the Danish Wind Turbine Owners Association, is a good source for statistical data on wind turbine performance. *Neue Energie* (in German) and its sister publication, *New Energy* (in English), are published by the German Wind Turbine Owners Association and are good sources for developments in Europe. *DEWI Magazin* is a quarterly technical publication of the German Wind Energy Institute. *DEWI Magazin* is written in German but includes summaries in English. *Systèmes Solaires* (in French) devotes at least one issue per year to wind energy developments in Francophone countries.

DEWI Magazin
www.dewi.de
Home Power magazine
www.homepower.org
New Energy magazine (Neue Energie)
www.wind-energie.de/en/association/new-energy
Renewable Energy World Magazine
www.renewableenergyworld.com
Sun & Wind Energy magazine (Sonne Wind & Wärme)
www.sunwindenergy.com
Systèmes Solaires, le Journal des Énergies Renouvelables
www.systemes-solaires.com/accueil-systemes-solaires.asp
Wind Engineering
www.multi-science.co.uk/windeng.htm
Windpower Monthly
www.windpower-monthly.com
WindTech International
www.windtech-international.com

Selected Sources

Wind Energy was never intended as an academic work and for the most part has relied on my knowledge of the industry and that of my colleagues who are quoted in the text. The book has continued to grow over the years without recourse to citations. However, this version includes far more technical material and background information than before. For the student of wind energy or those curious about the sources I've consulted, I am listing some of the references I've used.

I borrowed heavily from my own 1995 book *Wind Energy Comes of Age* for Chapters 3, 4, and 16. For the chapters on the history of wind energy and its modern development, I've relied on historians Etienne Rogier and Robert Righter; pioneers in the Danish wind power movement Preben Maegaard, Benny Christensen, and Erik Grove-Nielsen; and academic specialists in the development of technology, Peter Karnøe, Matthias Heymann, and Rinie van Est.

Chapter 3: Where It All Began and Chapter 4: The Great Wind Revival

Allgaier-Group, "History," Accessed November 5. 2014. www.allgaier.de/en/content/allgaier-werke/history.

Bonnefille, B. "Les Réalizations d'Electricité de France Concernant l'Energie Eolienne." *La Houille Blanche* 1 (1975).

Borish, Steven. *Land of the Living: The Danish Folk High Schools and Denmark's Denmark's Non-Violent Path to Modernization*. Nevada City, CA: Blue Dolphin Publishing, 1991. ISBN: 1-57733-108-7.

Christiansen, Benny, and Jytte Thorndahl. *Fra Husmøller til Havmøller: Vindkraft i Danmark i 150 år*. Bjerringbro, Denmark: Energimuseet, 2012. ISBN: 978-87-995167-0-4.

Dörner, Heiner. *Drei Welten—Ein Leben: Prof. Dr. Ulrich Hütter*. Heilbronn, Germany: H. Dörner, 1995. ISBN: 3-00-000067-4.

Eggleston, David, and Forrest Stoddard. *Wind Turbine Engineering Design*. New York: Van Nostrand Reinhold, 1987.

Gipe, Paul. *Wind Energy Comes of Age*. New York: John Wiley & Sons, 1995. ISBN 0-471-10924-X.

Golding, E. W. *The Generation of Electricity by Wind Power*. London: E. & F.N. Spon, 1955. Reprinted by John Wiley & Sons, 1976.

Gougeon, Meade. Personal communication, June 8, 1993.

Grove-Nielsen, Erik. "25 Years of Wind Power Development on Planet Earth." Winds of Change. Accessed November 3, 2014. www.windsofchange.dk.

Haka, Andreas. "Flügel aus 'Schwarzem Gold': Zur Geschichte der Faserverbundwerkstoffe." *NTM Zeitschrift für Geschichte der Naturwissenschaft, Technik und Medizin* 19, no. 1 (2011): 69–105. Accessed November 3, 2014. http://download.springer.com/static/pdf/482/art%253A10.1007%252Fs00048-011-0047-4.pdf?auth66=1415044741_27f f94a3a857774ab8acf69d2c09ecf0&ext=.pdf.

Handschuh, Karl. *Windkraft Gestern und Heute: Geschichte der Windenergienutzung in Baden- Württemberg*. Staufen, Germany: Ökobuch Verlag, 1997. ISBN: 9783922964339.

Heymann, Matthias. *Die Geschichte der Windenergienutzung 1890–1990*. Frankfurt, Germany: Campus Verlag, 1995. ISBN 3 539 35278 8.

———. "Why Were the Danes Best? Wind Turbines in Denmark, West Germany and the USA 1945–1985." Paper presented at the SHOT annual meeting, Cleveland, Ohio, October 18–21, 1990.

Jameison, Peter. *Innovation in Wind Turbine Design*. Chichester, UK: John Wiley & Sons, 2011. ISBN: 9780470699812.

———. Personal interview, February 5, 1993.

Karnøe, Peter. "Approaches to Innovation in Modern Wind Energy Technology: Technology Policies, Science, Engineers, and Craft Traditions in the United States and Denmark 1974–1990." Unpublished paper, Institute for Organization and Industrial Sociology, Copenhagen Business School, December 1992.

———. "Entrepreneurial Organization and the Accumulation of Knowledge in Modern Wind Technology." Paper presented at Røsnæs Workshop on the Process of Knowledge Accumulation and the Formation of Technology Strategy, Institute of Political Studies, University of Copenhagen, May 1990.

Koeppl, Gerald W. *Putnam's Power from the Wind*. 2nd ed. New York: Van Nostrand Reinhold, 1982.

Maegaard, Preben, Anna Krenz, and Wolfgang Palz, eds. *Wind Power for the World: The Rise of Modern Wind Energy*. Singapore: Stanford Publishing, 2013. ISBN-10: 9814364932, ebook: ISBN-13: 978-9814364935.

Nissen, Povl-Otto, Therese Quistgaard, Jutte Thorndahl, Benny Christensen, Pregen Maegaard, Birger T. Madsen, and Kristian Hvidfelt Nielsen. *Wind Power the Danish Way: From Poul la Cour to Modern Wind Turbines*. Askov, Denmark: Poul la Cour Foundation, 2009. ISBN: 978-87-993188-0-3.

Orkney Sustainable Energy. "Costa Head Experimental Wind Turbine." Video of John Brown installation on Costa Head Orkney. Accessed October 30, 2014. www.orkneywind.co.uk/costa.html.

Petroski, Henry. "From Slide Rule to Computer: Forgetting How It Used to Be Done." Chap. 15 in *To Engineer Is Human*. New York: Vintage Books, 1992.

Price, Trevor J. "James Blyth." In *Oxford Dictionary of National Biography*. Oxford, UK: Oxford University Press, 2014. Accessed November 3, 2014. www.oxforddnb.com/view/printable/100957.

Putnam, Palmer Cosslett. *Power from the Wind*. New York: Van Nostrand Reinhold, 1948. Reprinted 1974. ISBN: 0-442-26650-2.

Rapin, Marc, and Jean-Marc Noël. *Énergie Éolienne: Principes—Études de cas*. Paris: Dunod, 2010. ISBN: 9782100508013.

Rasmussen, Bent, and Flemming Øster. "Power from the Wind." In *Wind Energy in Denmark: Research and Technological Development*, edited by F. Øster, and H. M. Andersen. Copenhagen: Danish Energy Agency, 1990.

Righter, Robert. *Wind Energy in America: A History*. Norman: University of Oklahoma Press, 1996. ISBN 0-8061-2812-7.

Rogier, Etienne. "Constantin, precursor des èolieene modernes." *Systèmes Solaires* 136 (2000).

———. "Georges Darrieus, père des éolienne á axe vertical." *Systèmes Solaires* 139 (2000).

———. "La Belle Epoque des moteurs á vent." *Systèmes Solaires* 131 (1999).

———. "L'électricité éolienne de la Belle Epoque à EDF." *Cahiers D'Eole* no. 2 (November 2000): 8–20. Accessed October 31, 2014. eolienne.cavey.org/documents/cahiereole2.pdf.

———. "Les pionniers de l'électricite éolienne." *Systèmes Solaires* 129 (1999.)

———. "1944: L'Eolien vu par Pierre Ailleret." *Systèmes Solaires* 145 (2001).

Shaltens, R. K., and A. G. Birchenough. "Operational Results for the Experimental DOE/NASA MOD-0A Wind Turbine Project." Paper presented at the Sixth Biennial Wind Energy Conference and Workshop, Minneapolis/St. Paul, Minnesota, June 1–3, 1983.

Shepherd, Dennis. "Historical Development of the Windmill." In *Wind Turbine Technology: Fundamentals Concepts of Wind Turbine Engineering*, edited by David Spera. New York: American Society of Mechanical Engineers, 1994. ISBN 0-7918-1205-7.

Stoddard, Forrest (Woody). "Wind Turbine Blade Technology: A Decade of Lessons Learned." In *Energy and the Environment into the 1990s: 1st World Renewable Energy Congress*. Plenary papers, vol. 3. Edited by A. A. M. Sayigh. Tarrytown, NY: Pergamon Press, 1990.

Thorndal, Jytte. *Gedsermøllen: den første modene vindmølle*. Bjerringbro, Denmark: Elmuseet, 2005. ISBN: 87-89292-45-6.

Toepfer, Craig. "The American Wind Charger Industry History: 1916 to 1960." Wind Charger. Accessed October 31, 2014. www.windcharger.org/Wind_Charger.

Tvindkraft. Accessed October 30, 2014. www.tvindkraft.dk/.

Twidell, John. Personal communication (biographical notes on James Blyth), April 3, 2014.

van Est, Rinie. *Winds of Change: A Comparative Study of the Politics of Wind Energy Innovation in California and Denmark*. Utrecht, the Netherlands: International Books, 1999. ISBN 90-5727-027-7.

Vestergaard, Jørgen. "Wind Power—A Danish Story." Denmark: JV Film & TV, 2003. Film, 75 mins.

Wikipedia. "Tvindkraft windmill." Accessed October 30, 2104. en.wikipedia.org/wiki/Tvind#Tvindkraft_windmill.

Chapter 6: Vertical-Axis and Darrieus Wind Turbines

Darrieus, G. J. M. "Turbine Having Its Rotating Shaft Transverse to the Flow of the Current." US Patent 1,835,018, page 4. Accessed December 9, 2013. http://tinyurl.com/zjz3k8f.

FloWind Corporation. "Final Project Report: High-Energy Rotor Development, Test and Evaluation." Sandia report no. SAND 96-2205. Sandia National Laboratory, Albuquerque, New Mexico, September 1996. Accessed October 31, 2014. prod.sandia.gov/techlib/access-control.cgi/1996/962205.pdf.

Freris, L. L., ed. *Wind Energy Conversion Systems*. London: Prentice Hall, 1990. ISBN-10: 0139605274/ISBN-13: 978-0139605277.

Gasch, Robert, and Jochen Twele, eds. *Wind Power Plants: Fundamentals, Design, Construction and Operation*. Berlin, Germany: Springer Verlag, 2012. ISBN 978-3-642-22937-4.

Johnston, Sidney F. "Proceedings of the Vertical Axis Wind Turbine (VAWT) Design Technology Seminar for Industry." Sandia report no. SAND80-0984. Sandia National Laboratory, Albuquerque, New Mexico, July 1982. Accessed October 31, 2014. energy.sandia.gov/wp/wp-content/gallery/uploads/Sand80-0984.pdf.

Khammas, Achmed. "Vertikalachsen-Rotoren." In *Buch der Synergie*. Accessed December 9, 2013. www.buch-der-synergie.de/c_neu_html/c_08_08_windenergie_senkrechtachser.htm.

Lacroix, G. "Les Moteurs a Vent: Les Eoliennes Electriques Darrieus." In *La Nature*. Paris : Masson et Cie, 1929. References 550–553. Description of Darrieus's wind turbines, 8 meters, 10 meters, and 20s meter in diameter. Accessed December 9, 2013. http://cnum.cnam.fr/ILL/4KY28.117.html.

Legendre, M. Robert. "Notice nécrologique sur Georges Darrieus." Darrieus's obituary, March 1980, page 7. January 7, 2016. http://www.academie-sciences.fr/pdf/eloges/darrieus_cr1980.pdf

Malcolm, David. "Market, cost, and technical analysis of vertical and horizontal axis wind turbines Task #2: VAWT vs. HAWT technology." Global Energy Concepts, Lawrence Berkeley National Laboratory, 2003. pp 23. Accessed January 7, 2016. www-eng.lbl.gov/~rasson/windsail/gec/Task2report_17Ssept03.doc.

Martin, Paul. "Welsh government's £48k wind turbine creates £5 of power a month." BBC News website; November 7, 2013. Accessed November 6, 2014. www.bbc.com/news/uk-wales-24844182.

Moran, W. A. *Giromill Wind Tunnel Test and Analysis: Volume 1—Executive Summary*. Final report, June 1976–October 1977. McDonnell Aircraft Company. Accessed December 9, 2013. wind.nrel.gov/public/library/gwt.pdf.

Musgrove, Peter J. "Vertical Axis WECS Design." In *Wind Energy Conversion Systems*, edited by L. L. Freris. London: Prentice Hall, 1990. ISBN-10: 0139605274/ISBN-13: 978-0139605277.

NREL. "Mariah Power's Windspire Wind Turbine Testing and Results." National Renewable Energy Laboratory. Accessed

October 31, 2014. www.nrel.gov/wind/smallwind/mariah_power.html.

Paraschivoiu, Ion. *Wind Turbine Design: with Emphasis on Darrieus Concept*. Montréal: École Polytechnique de Montréal, 2002. ISBN: 2-553-00931-3590.

Price, Trevor J. "UK Large-Scale Wind Power Programme from 1970 to 1990: The Carmarthen Bay Experiments and the Musgrove Vertical-Axis Turbines." *Wind Engineering* 30, no. 6 (2006): 19. Accessed December 9, 2013. http://dspace1.isd.glam.ac.uk/dspace/bitstream/10265/236/1/Musgrove%20Wind%20Engineering%2030-3.pdf.

Sagrillo, Mick. "The Myths & Mysticism of Vertical Axis Wind Turbines." *Windletter* 24, no. 2 (February 2005). Accessed December 9, 2013. www.renewwisconsin.org/wind/Toolbox-Homeowners/VAWT%20myths.pdf.

Spera, David, ed. *Wind Turbine Technology: Fundamental Concepts of Wind Turbine Engineering*. New York: American Society of Mechanical Engineers, 1994. ISBN-10: 0791812057/ISBN-13: 978-0791812051.

Sutherland, Herbert J., Dale E. Berg, and Thomas D. Ashwill. *A Retrospective of VAWT Technology*. Sandia report no. SAND2012-0304. Sandia National Laboratories, Albuquerque, New Mexico, January 2012. Accessed October 31, 2014. energy.sandia.gov/wp/wp-content/gallery/uploads/SAND2012-0304.pdf.

Vadot, Louis. "The Design and Testing of Wind Power Plants." In *Proceedings of the United Nations Conference on New Sources of Energy*. Vol. 7, *Wind Power*. Rome, August 21–31, 1961. New York: United Nations, 1964.

Vosburgh, Paul. *Commercial Applications of Wind Power*. New York: Van Nostrand Reinhold, 1983. ISBN 10: 0442290365 / ISBN 13: 9780442290368.

Windward Engineering. "Power Performance Test Report for the Windspire." Windward Engineering, February 28, 2013. Accessed December 9, 2013. windwardengineering.com/wp-content/uploads/2012/11/2013-02-28-PP-Test-Report-Windspire_vfs5.8-DJL-DAD-pg80.pdf.

Chapter 7: Novel Wind Systems

ARPA. "FloDesign Wind Turbine: Breakthrough High-Efficiency Shrouded Wind Turbine." Advanced Research Projects Agency, US Department of Energy. Accessed November 4, 2014. arpa-e.energy.gov/?q=slick-sheet-project/mixer-ejector-wind-turbine.

Barnard, Mike. "Airborne Wind Energy: It's All Platypuses Instead of Cheetahs." CleanTechnica, March 3, 2014. Accessed November 5, 2014. http://cleantechnica.com/2014/03/03/airborne-wind-energy-platypuses-instead-cheetahs/.

Bullis, Kevin. "A Design for Cheaper Wind Power: A design that draws on jet engine technology could halve the cost of generating electricity from wind." *Technology Review*, December 1, 2008. Accessed November 4, 2014. www.technologyreview.com/news/411274/a-design-for-cheaper-wind-power.

Bureau of the Census, US Department of Commerce. "Energy." Chap. S in *Historical Statistics of the United States: Colonial Times to 1970; Part 2,* pages 811, 818. Washington, DC: US Dept. of Congress, 1975. Accessed November 4, 2014. fraser.stlouisfed.org/docs/publications/histstatus/hstat1970_cen_1975_v2.pdf.

Dörner, Heiner. "Concentrating Windsystems—Sense or Nonsense?" Accessed November 4, 2014. www.heiner-doerner-windenergie.de/diffuser.html.

Enercon. "Rotor sail ship 'E-Ship 1' saves up to 25% fuel." Enercon: Energy for the World. Accessed November 5, 2014. www.enercon.de/p/downloads/PM_E-Ship1_Ergebnisse_DBU_en.pdf.

Glenn Research Center, NASA. "Lift of Rotating Cylinder." NASA. Accessed November 5, 2014. www.grc.nasa.gov/WWW/K-12/airplane/cyl.html.

Greenbird. "Record set at Ivanpah in time for the 2009 America's cup landsailing regatta." Accessed November 4, 2014. www.greenbird.co.uk/land-record.

Hansen, Martin. *Aerodynamcis of Wind Turbines*. 2nd ed. London: Earthscan, 2008. ISBN: 978-1-88407-438-9.

Ogin. Ogin: The New Shape of Energy. Accessed November 4, 2014. www.oginenergy.com.

Phillips, Derek Grant. "An Investigation on Diffuser Augmented Wind Turbine Design." PhD diss., University of Auckland, 2003. Accessed November 4, 2014. https://researchspace.auckland.ac.nz/handle/2292/1940.

Phillips, Derek Grant, R. G. J. Flay, and Trevor Nash. "Aerodynamic analysis and monitoring of the Vortec 7 diffuser-augmented wind turbine." *IPENZ Transactions* 26, no. 1/EMCH (1999): 13–19. Accessed November 4, 2014. www.ipenz.org.nz/ipenz/publications/transactions/Transactions99/EMCh/Phillips.PDF.

Righter, Robert. *Wind Energy in America: A History*. Norman: University of Oklahoma Press, 1996. ISBN 0-8061-2812-7.

Rotorflugzeug. Interessengemeinschaft. "Literatur über den Magnus Effekt." Rotorflugzeug. Accessed November 5, 2014. www.rotorflugzeug.de/literatur-download.html.

Sailrocket. Website on the Vestas Sailrocket 2. Accessed November 4, 2014. www.sailrocket.com/.

Satchwell, C. J., ed. *Windship Technology*. Amsterdam: Elsevier, 1985. Proceedings of the International Symposium on Windship Technology (Windtech '85), Southampton, England, April 24–25, 1985.

Schenzle, Peter. "Wind Propulsion for Solar Ship Operation." Paper presented at 25th International Conference on Offshore Mechanics and Arctic Engineering, Hamburg, Germany, June 4–9, 2006. Accessed November 5, 2014. www.omae2006.sea2ice.com/pdfs/OMAE200692701Schenzle-condensed.pdf.

SkySails. "Powerful—Unlimited—Free." SkySails Propulsion for Cargo Ships. Accessed November 4, 2014. http://www.skysails.info/english/skysails-marine/skysails-propulsion-for-cargo-ships/.

Vadot, Louis. "The Design and Testing of Wind Power Plants." In *Proceedings of the United Nations Conference on New Sources of Energy*. Vol. 7, *Wind Power*. Rome, August 21–31, 1961. New York: United Nations, 1964.

Whitford, Dale, and John Minardi. "Utility-Sized Madaras Wind Plants." *International Journal of Ambient Energy* 2, no. 1 (January 1981). Accessed November 5, 2014. http://www.tandfonline.com/doi/abs/10.1080/01430750.1981.9675748#preview.

Wikipedia. "E-Ship 1." Accessed November 5, 2014. en.wikipedia.org/wiki/E-Ship_1.

———. "Flettner-Rotor." Accessed November 5, 2014. de.wikipedia.org/wiki/Flettner-Rotor.

———. "Rotor Ship." Accessed November 5, 2014. http://en.wikipedia.org/wiki/Rotor_ship.

Chapter 8: Silent Wind Revolution

Armbrust, Sepp. "Regulating and Control System of an Experimental 100 kW Wind Electric Plant Operating Parallel with an AC Network." In *Proceedings of the United Nations Conference on New Sources of Energy*. Vol. 7, *Wind Power*. Rome, August 21–31, 1961. New York: United Nations, 1964.

Chabot, Bernard. "New IEC3A Wind Turbines: Bright Strategic and Economic Perspectives for Onshore Wind in Medium to Low Wind Speed Areas." Windtech-International, August 29, 2013. Accessed October 31, 2014. www.windtech-international.com/editorial-features/features/articles/new-iec3a-wind-turbines.

———. "Wind Power Silent Revolution: New Wind Turbines for Light Wind Sites." Renewables International, May 6, 2013. Accessed October 31, 2014. www.renewablesinternational.net/turbines-in-low-wind-areas/150/505/62498/.

De Veries, Eize. "Finding the optimum low-wind design combination." Windpower Monthly, July 1, 2013. Accessed October 31, 2014. www.windpowermonthly.com/article/1187461/finding-optimum-low-wind-design-combination?HAYILC=TOPIC.

Early, Catherine. "Low wind focus opens up new markets." Windpower Monthly, July 1, 2013. Accessed October 31, 2014. www.windpowermonthly.com/article/1187460/low-wind-focus-opens-new-markets?HAYILC=RELATED&HAYILC=INLINE.

Henderson, Geoff. "Technical Comparators: Swept Area Preferred over Rated Power." Windflow Technology Newsletter, June 2011. Accessed October 31, 2014. www.windflow.co.nz/news/newsletters/Newsletter%2031.pdf.

Molly, Jens-Peter. "Design of Wind Turbines and Storage: A Question of System Optimisation." DEWI Magazin, February 2012. Accessed October 31, 2014. www.dewi.de/dewi/fileadmin/pdf/publications/Magazin_40/04.pdf.

———. "Rated Power of Wind Turbines: What Is Best?" DEWI Magazin, February 2011. Accessed October 31, 2014. www.dewi.de/dewi/fileadmin/pdf/publications/Magazin_38/07.pdf.

Chapter 9: Towers

Energy Saving Trust. Location, Location, Location: Domestic Small-Scale Wind Field Trial Report. London: Energy Saving Trust, July 2009. Accessed November 7, 2014. www.wind-works.org/cms/fileadmin/user_upload/Files/Reports/Location_Location_Energy_Saving_Trust.pdf.

Chapter 10: Measuring the Wind

Elliott, Dennis, C. G. Holladay, W. R. Barchet, H. P. Foote, and W. F. Sandusky. Wind Energy Resource Atlas of the United States. Richland, WA: Pacific Northwest Laboratory, 1986. DOE/CH 11093-4.

Frost, Walter, and Carl Aspliden. "Characteristics of the Wind." In Wind Turbine Technology: Fundamentals Concepts of Wind Turbine Engineering edited by David Spera. New York: American Society of Mechanical Engineers, 1994. ISBN 0-7918-1205-7.

Golding, E. W. The Generation of Electricity by Wind Power. London: E. & F.N. Spon, 1955. Reprinted by John Wiley & Sons in 1976.

Huler, Scott. Defining the Wind: The Beaufort Scale, and How a 19th-Century Admiral Turned Science into Poetry. New York: Three Rivers Press, 2004. ISBN: 1-4000-4885-0.

Molly, Jens-Peter. Windenergie: Theorie, Anwendung, Messung. Karslruhe, Germany: Verlag C.F. Muller, 1990. ISBN 3-78807269-5.

Park, Jack. The Wind Power Book. Palo Alto, CA: Cheshire Books, 1981.

Park, Jack, and Dick Schwind. Wind Power for Farms, Homes, and Small Industry. Nielsen Engineering and Research, Mountain View, California. Washington, DC: US Department of Energy, 1978. RFP-2841/1270/78/4.

Chapter 11: Estimating Performance

Idaho National Laboratory. January 7, 20167, 2014. http://www.wind-works.org/cms/fileadmin/user_upload/Files/Reports/Idaho_National_Laboratory_Power_Curves_2007.xls

Naturlig Energi: Månedsmagasin. September 2012.
    Online Danish magazine on wind power. Accessed December 31, 2015. www.naturlig-energi.dk/cms/gode-gamle-nyheder/195-september-2012.

Chapter 12: Off-the-Grid Power Systems

Davidson, Joel. The New Solar Electric Home: The Complete Guide to Photovoltaics for Your Home. Ann Arbor, MI: Aatec Publications, 1987. ISBN: 0-937948-09-8.

Holz, Richard G., E. Ian Baring-Gould, David Corbus, Larry Flowers, and Andrew McAllister. "Initial Results from the Operation of Vilalge Hybrid Power Systems in Chile." Paper presented at Windpower '97, Austin, Texas, June 15–18, 1997. NRELICP-440-23436. Accessed November 7, 2014. www.nrel.gov/docs/legosti/old/23436.pdf.

Pratt, Doug. The Real Goods Solar Living Sourcebook: The Complete Guide to Renewable Energy Technologies & Sustainable Living. 10th ed. White River Junction, VT: Chelsea Green, 1999.

Chapter 13: Interconnection and Grid Integration

Budischak, Cory, DeAnna Sewell, Heather Thomson, Leon Mach, Dana E. Veron, and Willett Kempton. "Cost-minimized combinations of wind power, solar power and electrochemical storage, powering the grid up to 99.9% of the time." Journal of Power Sources 225 (March 2013): 60–74. Accessed November 1, 2014. www.sciencedirect.com/science/article/pii/S0378775312014759.

Chabot, Bernard. "May 11, 2014: A Record Production from [Wind + PV] in Germany." SolarServer, May 12, 2014. Accessed November 13, 2014. www.solarserver.com/fileadmin/user_upload/downloads/bc2RecordG11-5-14.pdf.

———. "Overview of German renewable electricity from 1990–2013." Renewables International, May 30, 2014. Accessed November 13, 2014. www.renewablesinternational.net/overview-of-german-renewable-electricity-from-1990-2013/150/537/79182/.

DeCarolis, Joseph F., and David W. Keith. "The Costs of Wind's Variability: Is There a Threshold?" The Electricity Journal 18, no. 1 (January/February 2005): 69–77. Accessed November 1, 2014. http://keith.seas.harvard.edu/papers/72.Decarolis.2005.Threshold.e.pdf.

Dena Project Steering Group. "Summary of the Essential Results of the Study, Planning of the Grid Integration of Wind Energy in Germany Onshore and Offshore up to the Year 2020 (Dena Grid study)." Deutsche Energie-Agentur, Berlin, March 15, 2005. Accessed November 1, 2014. www.dena.de/fileadmin/user_upload/Publikationen/Energiedienstleistungen/Dokumente/dena-grid_study_summary.pdf.

Gross, Robert, Philip Heptonstall, Dennis Anderson, Tim Green, Matthew Leach, and Jim Skea. The Costs and Impacts of Intermittency: An assessment of the evidence on the costs and impacts of intermittent generation on the British electricity network. Report of the Technology and Policy Assessment, UK Energy Research Centre, London: UKERC, March 2006. ISBN 1 90314 404 3. Accessed November 1, 2014. www.ukerc.ac.uk/Downloads/PDF/06/0604Intermittency/0604IntermittencyReport.pdf.

Guillet, Jérôme, and John Evans. "Part III: The value of wind power—balancing costs." Renewables International, August 20, 2014. Accessed November 13, 2014. www.renewablesinternational.net/part-iii-the-value-of-wind-power-balancing-costs/150/435/81055/.

Kirby, Brendan, Michael Milligan, Yuri Makarov, Howard Hawkins, Kevin Jackson, and Henry Shiu. California RPS Integration Cost Analysis Phase I: One Year Analysis of Existing Resources; Results and Recommendations. Sacramento: California Energy Commission, December 10, 2003. CEC 500-03-108C. Accessed November 1, 2014. www.energy.ca.gov/reports/2004-02-05_500-03-108C.PDF.

Lang, Matthias "Energy Transition may proceed without electricity storage for 20 years— transport, heat and chemicals markets will drive growth." Energy Transition: The German Energiewende, September 23, 2014. Accessed November 1, 2014. energytransition.de/2014/09/energy-transition-may-proceed-without-electricity-storage-for-20-years/.

Pape, Carsten. "Scenarios with High Shares of Renewable Energies." Fraunhofer Institut für Windenenergie und Energiesystemtechnik. Presentation to the Ontario Sustainable Energy Association, Kassel, Germany July 2012. Accessed May 11, 2016. http://www.wind-works.org/cms/index.php?1d=91.

Sterner, Michael, Norman Gerhardt, Kurt Rohrig. "Energie speichem, Zukunft meistern Welche Speicher brauchen wir auf dem Weg zur regenerativen Vollversorgung?" Fraunhofer Institut für Windenergie und Energiesystemtechnik. Ostwind Forum, Husum 2010. Accessed May 11, 2016. http://tinurl.com/zmhkhgl.

Wikipedia. "Fukushima Daiichi Nuclear Power Plant: Operating History." Accessed November 13, 2014. en.wikipedia.org/wiki/Fukushima_Daiichi_Nuclear_Power_Plant#Operating_history.

Chapter 14: Pumping Water

Baker, T. Lindsay. *A Field Guide to American Windmills.* Norman: University of Oklahoma Press, 1985 ISBN: 0-8061-1901-2.

Clark, Nolan R., Vaughn Nelson, Robert E. Barrieu, and Earl Gilmore. "Wind Turbines for Irrigation Pumping." *Journal of Energy* 5, no. 2 (March–April 1981): 104–105. Accessed November 1, 2014. www.cprl.ars.usda.gov/pdfs/1981.%20Clark,%20R.N.,%20and%20V.%20Nelson,%20and%20R.E.%20Barieau,%20and%20E.%20G.pdf.

Fraenkel, Peter, Roy Barlow, Francis Crick, Anthony Derrick, and Varis Bokalders. *Windpumps: A Guide for Development Workers.* London: Intermediate Technology Publications, 1993. ISBN: 1-85339-1263.

Webb, Walter Prescott. *The Great Plains.* Boston: Ginn and Company, 1931. Reprinted in 1972. ISBN: 0-448-00029-6.

Chapter 15: Siting and Environmental Concerns

Atkinson-Palombo, Carol, and Ben Hoen. *Relationship between Wind Turbines and Residential Property Values in Massachusetts.* Joint report of University of Connecticut and Lawrence Berkeley National Laboratory. Boston, January 9, 2014. Accessed November 1, 2014. http://images.masscec.com/uploads/attachments/2014/06/Relationship%20between%20Wind%20Turbines%20and%20Residential%20Property%20Values%20in%20Massachusetts.pdf.

Australian Medical Association. Wind Farms and Health—2014. AMA, March 18, 2014. Accessed November 2, 2014. ama.com.au/position-statement/wind-farms-and-health-2014.

Barnard, Mike. "Property Values Not Hurt by Wind Energy." Energy and Policy Institute, April 2014. Accessed November 1, 2014. www.energyandpolicy.org/tags/property_values.

———. "Wind Health Impacts Dismissed in Court." Energy and Policy Institute, August 14, 2014. Accessed November 1, 2014. www.energyandpolicy.org/wind-health-impacts-dismissed-in-court.

Blanca, Palomo, Claire Michaud, and Bastien Gaillardon. "Life Cycle Assessment of a French Wind Plant." *JEC Composites Magazine* 90 (June/July 2014). Accessed November 1, 2014. rescoll.fr/blog/wp-content/uploads/2014/03/LCA-French-plant-Full-Paper.pdf.

Boulanger, Vincent, ed. *Le bruit de l'éolien: Rumeurs, cancans, mensonges et petites histoires.* Paris: Systèmes Solaires, 2004.

Chapman, Simon, Alexis St. George, Karen Waller, and Vince Cakic. "Spatio-temporal differences in the history of health and noise complaints about Australian wind farms: evidence for the psychogenic, 'communicated disease' hypothesis." *PLoS ONE* 8, no. 10 (October 16, 2013). Accessed December 31, 2015. DOI:10.1371/journal.pone.0076584. journals.plos.org/plosone/article?id=10/371/journal.pone.0076584.

Chatham-Kent Public Health Unit. *The Health Impact of Wind Turbines: A Review of the Current White, Grey, and Published Literature.* Chatham-Kent Municipal Council, Chatham, Ontario, Canada, June 1, 2008. Accessed November 2, 2014. www.wind-works.org/cms/fileadmin/user_upload/Files/Health_and_Wind_by_C-K_Health_Unit.pdf.

Clarke, Alexi. "Windfarm Location and Environmental Impact." NATTA report. Open University, Milton Keynes, England, June 1988.

Colijn, Helen. *The Backroads of Holland: Scenic Excursions by Bicycle, Car, Train, or Boat.* San Francisco: Bicycle Books, 1992, 71–72. See also the cycling itineraries in *A Closer Look at Wind Energy* by Chris Westra. Amsterdam: Chris Westra Produktie, 1993.

Connor, Steve. "Exclusive: Billionaires secretly fund attacks on climate science." *The Independent,* January 24, 2013. Accessed November 11, 2014. www.independent.co.uk/environment/climate-change/exclusive-billionaires-secretly-fund-attacks-on-climate-science-8466312.html?origin=internalSearch.

———. "Top climate scientist denounces billionaires over funding for climate-sceptic organizations." *The Independent,* January 25, 2013. Accessed November 11, 2014. www.independent.co.uk/news/science/top-climate-scientist-denounces-billionaires-over-funding-for-climatesceptic-organisations-8467665.html?origin=internalSearch.

Denholm, Paul, Maureen Hand, Maddalena Jackson, and Sean Ong. *Land-Use Requirements of Modern Wind Power Plants in the United States.* National Renewable Energy Laboratory, Technical Report NREL/TP-6A2-45834. Golden, CO: NREL, August 2009. Accessed November 1, 2014. www.nrel.gov/docs/fy09osti/45834.pdf.

"Development standards and conditions." Kern County, California: Code of Ordinances. Ordinance No. G-8601, passed October 13, 2015. Accessed November 1, 2014. https://library.municode.com/index.aspx?clientId=16251&stateId=5&stateName=California&customBanner=16251.jpg&imageclass=D&cl=16251.txt.

Ellenbogen, Jeffrey M., Sheryl Grace, Wendy J. Heiger-Bernays, James F. Manwell, Dora Anne Mills, Kimberly A. Sullivan, and Marc G. Weisskopf. *Wind Turbine Health Impact Study: Report of Independent Expert Panel.* Massachusetts Department of Environmental Protection and the Massachusetts Department of Public Health, January 2012. Accessed November 2. 2014. www.mass.gov/eea/docs/dep/energy/wind/turbine-impact-study.pdf.

Elsam. "Life Cycle Assessment of offshore and onshore sited wind farms." Elsam Engineering A/S, translated by Vestas Wind Systems, October 20, 2004, page 40. Accessed November 2, 2014. www.vestas.com/files%2Ffiler%2Fen%2Fsustainability%2Flca%2Flca_v80_2004_uk.pdf.

Fritzsche, Andrew F. "The Health Risks of Energy Production." *Risk Analysis* 9, no. 4 (1989): 565–577.

Fulton, R., K. Koch, and C. Moffat. "Wind Energy Study, Angeles National Forest." Graduate studies in landscape architecture, California State Polytechnic University, Pomona, California, June, 1984.

Ginsberg, Steve. "The Wind Power Panacea: Is There Snake Oil in Paradise?" Audubon Imprint, Santa Monica (Calif.) Bay Audubon. 17:1, 1–5.

Gipe, Paul. "Public Acceptance of the Potato and What It Tells Us about the Acceptance of Wind Energy." Wind-Works, March 7, 2013. Accessed November 1, 2014. www.wind-works.org/cms/index.php?id=43&tx_ttnews[tt_news]=2228&cHash=5b7218c411b6f7b571d6575df8b8532f.

Goldenberg, Suzanne. "Media campaign against wind-farms funded by anonymous conservatives." *The Guardian*, February 15, 2013. Accessed November 11, 2014. www.theguardian.com/environment/2013/feb/15/media-campaign-windfarms-conservatives?CMP=twt_gu.

———. "Secret funding helped build vast network of climate denial thinktanks." *The Guardian*, February 14, 2013. Accessed November 11, 2014. www.theguardian.com/environment/2013/feb/14/funding-climate-change-denial-think tanks-network.

Greenpeace. "Koch Industries: Secretly Funding the Climate Denial Machine." Greenpeace. Accessed November 11, 2014. www.greenpeace.org/usa/en/campaigns/global-warming-and-energy/polluterwatch/koch-industries/.

Grum-Schwensen, Erik. "The Real Cost of Wind Turbine Construction." *Wind Stats* 3, no. 2 (Spring 1990): 1–2.

Guezuraga, Begoña, Rudolf Zauner, and Werner Pölz. "Life cycle assessment of two different 2 MW class wind turbines." *Renewable Energy* 37, no. 1 (January 2012): 37–44. Accessed November 1, 2014. www.ewp.rpi.edu/hartford/~ernesto/S2013/MMEES/Papers/ENERGY/6AlternativeEnergy/Guezuraga2012-LCAWindTurbines.pdf.

Gupta, Ajay K., and Charles A. S. Hall. "A Review of the Past and Current State of EROI Data." *Sustainability* 3, no. 10 (2011): 1796–1809. Accessed December 31, 2015. doi:10.3390/su3101796

Gydesen, A., D. Maimann, and P. B. Pedersen. "Renere Teknologi pa Energiomradet." Energigruppen, Fysisk Laboratorium III, Danmarks Tekniske Højskole, Miljøministeriet, Miljøprojekt Nr. 138, Denmark, 1990, 123–127.

Haapala, Karl R., and Preedanood Prempreeda. "Comparative life cycle assessment of 2.0 MW wind turbines." *International Journal of Sustainable Manufacturing* 3, no. 2 (2014). Accessed November 1, 2014. www.ourenergypolicy.org/wp-content/uploads/2014/06/turbines.pdf.

Hagedorn, Gerd. "Kumulierter Energieaufwand von Photovoltaik und Windkraftanlagen." Lehrstuhl fur Energiewirtschaft und Kraftwerkstechnik, Technische Universitat, Munich, 1992.

Hagedorn, G., and F. Ilmberger. *Kumulierter Energieverbrauch fur die Herstellung von Windkraftanlagen.* Forschungsstelle fur Energiewirtschaft, Im Auftrage des Bundesministeriums fur Forschung und Technologie, Munich, August 1991. In this study, the primary energy used to construct the wind turbine was given in units of kWh. However, only 35% of the energy burned in a power plant is converted to useful work. To present the data from this study in a format consistent with that from the other studies, the kWh consumed has been multiplied by 0.35.

Health Canada. "Wind Turbine Noise and Health Study: Summary of Results." Health Canada. October 30, 2014. Accessed November 6, 2014. www.hc-sc.gc.ca/ewh-semt/noise-bruit/turbine-eoliennes/summary-resume-eng.php.

Hoen, Ben, Ryan Wiser, Peter Cappers, Mark Thayer, and Gautam Sethi. "The Impact of Wind Power Projects on Residential Property Values in the United States: A Multi-Site Hedonic Analysis." Lawrence Berkeley National Laboratory, December 2009. Accessed November 1 2014. emp.lbl.gov/sites/all/files/REPORT%20lbnl-2829e.pdf.

Howe, Brian. "Low Frequency Noise and Infrasound Associated with Wind Turbine Generator Systems: A Literature Review." Ontario Ministry of the Environment, December 10, 2010. Accessed November 2, 2014. http://tinyurl.com/zl4xym3.

Hubbard, Harvey H., and Kevin P. Shepherd. "Wind Turbine Acoustics." In *Wind Turbine Technology: Fundamentals Concepts of Wind Turbine Engineering,* edited by David Spera. New York: American Society of Mechanical Engineers, 1994. ISBN 0-7918-1205-7.

James, Ross D., and Glenn Coady. *Exhibition Place: Wind Turbine Bird Monitoring Program in 2003.* Report to Toronto Hydro and WindShare, December 2003. Accessed December 31, 2015. www.windshare.ca/wp-content/uploads/2014/01/Bird-Monitoring-Study.pdf.

Knopper, Loren D., Christopher A. Ollson, Lindsay C. McCallum, Melissa L. Whitfield Aslund, Robert G. Berger, Kathleen Souweine, and Mary McDaniel. "Wind turbines and human health." *Frontiers in Public Health* 2, no. 63 (June 19, 2014). Accessed November 2, 2014. doi: 10.3389/fpubh.2014.00063. journal.frontiersin.org/Journal/10.3389/fpubh.2014.00063/full.

Kubiszewski, Ida, Cutler J. Cleveland, and Peter K. Endres. "Meta-analysis of net energy return for wind power systems." *Renewable Energy* 35, no. 1 (January 2010): 218–225.

Lambert, Jessica, Charles Hall, Steve Balogh, Alex Poisson, and Ajay Gupta. "EROI of Global Energy Resources Preliminary Status and Trends." State University of New York, College of Environmental Science and Forestry, November 2, 2012. Accessed November 2, 2014. www.roboticscaucus.org/ENERGYPOLICYCMTEMTGS/Nov2012AGENDA/documents/DFID_Report1_2012_11_04-2.pdf.

Larwood, Scott, and C. P. van Dam. "Permitting Setback Requirements for Wind Turbines in California." California Wind Energy Collaborative. California Energy Commission, November 2006. CEC-500-2005-184. Accessed November 1, 2014. www.energy.ca.gov/2005publications/CEC-500-2005-184/CEC-500-2005-184.PDF.

Leventhall, Geoff. "Basic Acoustics." In *Wind Turbine Noise,* edited by Dick Bowdler and Geoff Leventhall. Brentwood, UK: Multi-Science Publishing, 2011, page 215. ISBN: 978-1-907132-30-8.

Mandel, Jenny. "Gas flare draws thousands of birds to their deaths, and ignites questions." *EnergyWire*, October 11, 2013. Accessed November 1, 2014. www.eenews.net/stories/1059988683/print.

Moomaw, William, Peter Burgherr, Garvin Heath, Manfred Lenzen, John Nyboer, and Aviel Verbruggen. "Annex II: Methodology." In *IPCC Special Report on Renewable Energy Sources and Climate Change Mitigation.* Cambridge, UK, and New York: Cambridge University Press, 2011. Accessed November 1, 2014. srren.ipcc-wg3.de/report/IPCC_SRREN_Annex_II.pdf.

National Health and Medical Research Council, *NHMRC Draft Information Paper: Evidence on Wind Farms and Human Health.* NHMRC, Australia, February 2014. Accessed November 2, 2014. consultations.nhmrc.gov.au/files/consultations/drafts/nhmrcdraftinformationpaperpublicconsultationfebruary2014.pdf.

Neddermann, B., J. Raabe, and P. Paysen. "Plenty of Potential Still Available for Wind Energy Use in Northern Saxony." *DEWI Magazin* 45 (August 2014): 31. Accessed November 1, 2014. www.dewi.de/dewi_res/fileadmin/pdf/publications/Magazin_45/06.pdf.

Orloff, S., and A. Flannery. 1992. *Wind turbine effects on avian activity, habitat use, and mortality in Altamont Pass and Solano County WRAs.* Prepared by BioSystems Analysis, Inc., Tiburon, California, for the California Energy Commission, Sacramento. Accessed November 11, 2014. www.energy.ca.gov/windguidelines/documents/2006-12-06_1992_FINAL_REPORT_1989-1991.PDF.

Public Health Division. *The Strategic Health Impact Assessment on Wind Energy Development in Oregon.* PHD, Oregon Health Authority, March 2013. Accessed November 2, 2014. public.health.oregon.gov/HealthyEnvironments/TrackingAssessment/HealthImpactAssessment/Documents/Wnd%20Energy%20HIA/Wind%20HIA_Final.pdf.

RenewableUK. *The effect of wind farms on house prices.* RenewableUK & Cebr study, March 2014. Accessed November 1, 2014. www.renewableuk.com/en/publications/index.cfm/RenewableUK-Cebr-Study-The-effect-of-wind-farms-on-house-prices.

Sathaye, J., O. Lucon, A. Rahman, J. Christensen, F. Denton, J. Fujino, G. Heath, S. Kadner, M. Mirza, H. Rudnick, A. Schlaepfer, and A. Shmakin. "Renewable Energy in the Context of Sustainable Development." Chap. 9 in *IPCC Special Report on Renewable Energy Sources and Climate Change Mitigation.* Cambridge, UK, and New York: Cambridge University Press, 2011. Accessed November 1, 2014. srren.ipcc-wg3.de/report/IPCC_SRREN_Ch09.pdf.

Sterzinger, George, Fredric Beck, and Damian Kostiuk. "The Effect of Wind Development on Local Property Values." Renewable Energy Policy Project, May 2003. Accessed November 1, 2014. www.wind-works.org/cms/fileadmin/user_upload/Files/Reports/REPP-The_Effect_of_Wind_Development_on_Local_Property_Values.pdf.

Tayor, Derik, and Marcus Rand. "How to Plan the Nuisance Out of Wind Energy." *Town and Country Planning,* May 1991, 152–155.

Thibault, Benjamin, Tim Weis, and Eli Angen. "Survey of Complaints Received by Relevant Authorities Regarding Operating Wind Energy in Alberta." Edmonton, Alberta. Pembina Institute. July 26, 2013. Accessed November 1, 2014. www.pembina.org/reports/documenting-wind-energy-complaints-alberta.pdf.

Renewable Energy World. "Toronto Wind Turbine Proves Bird Friendly." Renewable Energy World, March 16, 2004. Accessed November 1, 2014. www.renewableenergyworld.com/rea/news/article/2004/03/toronto-wind-turbine-proves-bird-friendly-10727.

van den Berg, Frits. "Effects of Sound on People." In *Wind Turbine Noise* edited by Dick Bowdler and Geoff Leventhall. Brentwood, UK: Multi-Science Publishing, 2011. ISBN: 978-1-907132-30-8.

van der Borg, N. C. J. M., and J. W. Stam. "Acoustic Noise Measurements on Wind Turbines." European Wind Energy Conference, Glasgow, 1989.

Vyn, Richard J., and Ryan M. McCullough. "The Effects of Wind Turbines on Property Values in Ontario: Does Public Perception Match Empirical Evidence?" *Canadian Journal of Agricultural Economics* 62, no. 3 (January 23, 2014): 365–392. DOI: 10.1111/cjag.12030. Accessed November 1, 2014. onlinelibrary.wiley.com/doi/10.1111/cjag.12030/abstract?campaign=woletoc.

Wagner, Siegfried, Rainer Bareiß, and Gianfranco Guidati. *Wind Turbine Noise.* Berlin: Springer-Verlag, 1996. ISBN: 3-540-60592-4.

Williams, Wendy, and Robert Whitcomb. *Cape Wind: Money, Celebrity, Energy, Class, Politics, and the Battle for Our Energy Future.* New York: Pubic Affairs, 2007.

Zimmerling, J. Ryan, Andrea C. Pomeroy, Marc V. d'Entremont, and Charles M. Francis. "Canadian Estimate of Bird Mortality Due to Collisions and Direct Habitat Loss Associated with Wind Turbine Developments." *Avian Conservation and Ecology* 8, no. 2 (2013): 10. Accessed November 1, 2014. http://dx.doi.org/10.5751/ACE-00609-080210.

Chapter 17: Safety

Aubrecht, Gordon J. III. "Comparing nuclear and fossil-fuel energy risks: Risks to workers." Chap. 20 in *Energy.* New York: Prentice-Hall, 1995. 3rd ed. published 2003. Accessed November 2, 2014. wps.prenhall.com/wps/media/objects/2513/2574258/pdfs/E20.12.pdf.

Ellis, J. Nigel. *Introduction to Fall Protection.* 2nd ed. Des Plaines, IL: American Society of Safety Engineers, 1993. ISBN 0-939874-97-0.

Fritzsche, Andrew F. "The Health Risks of Energy Production." *Risk Analysis* 9, no. 4 (1989): 565–577. Accessed November 2, 2014. onlinelibrary.wiley.com/doi/10.1111/j.1539-6924.1989.tb01267.x/abstract.

Hamilton, L. D. "Health and Environmental Risks of Energy Systems." Proceedings of an International Symposium on the Risks and Benefits of Energy Systems, International Atomic Energy Agency, Vienna, 1984.

Harrison, H. C. *Story of Sprowston Mill.* London: Phoenix House, 1949. ASIN: B00105SCY0.

International Energy Agency. *Environmental Health Impacts of Electricity Generation: A Comparison of the Environmental Impacts of Hydropower with those of Other Generation Technologies.* IEA Hydropower, June 2002. Accessed November 2, 2014. www.ieahydro.org/reports/ST3-020613b.pdf.

Morris, Samuel C. "Health Risks of Coal Energy Technology." In *Health Risks of Energy Technologies,* edited by Curtis C. Travis and Elizabeth L. Etnier, Boulder, CO : Westview Press 1983.

Rowe, Michael. "Health Risks in Perspective: Judging Health Risks of Energy Technologies." Brookhaven National Laboratory, Upton, New York, September 18, 1992.

Chapter 18: Operation and Maintenance

Kühn, Paul. "Kleine Windenergieanlagen Betriebserfahrungen & Ertragsabschätsung." Fraunhofer Institut für Windenergie und Energiesystemtechnik, Kleinwindanlagen Symposium, Husum, Germany, March 20, 2010. Accessed November 2, 2014. www.wind-works.org/cms/fileadmin/user_upload/Files/presentations/2010-03-20_Kuehn_Husum.pdf.

Millborrow, David. "Breaking down the cost of wind turbine maintenance." *Windpower Monthly*, June 15, 2010. Accessed November 2, 2014. www.windpowermonthly.com/article/1010136/breaking-down-cost-wind-turbine-maintenance.

Orell, A. C. "2012 Market Report on Wind Technologies in Distributed Applications." Pacific Northwest National Laboratory. USDOE, August 2013. Accessed November 2, 2014. www1.eere.energy.gov/wind/pdfs/2012_distributed_wind_technologies_market_report.pdf.

Wiser, Ryan, and Mark Bollinger. "2012 Wind Technologies Market Report." Lawrence Berkeley National Laboratory, August 2013. Accessed November 2, 2014. http://emp.lbl.gov/sites/all/files/lbnl-6356e.pdf.

Chapter 19: Investing in Wind Energy

Association of Power Producers of Ontario. "Ontario Wind Power Task Force Industry Report and Recommendations." APPrO, February 2002. Accessed November 12, 2014. www.appro.org/Wind_Power_Task_Force_Report,_February_2002.pdf.

Wiser, Ryan, and Mark Bollinger. "2012 Wind Technologies Market Report." Lawrence Berkeley National Laboratory, August 2013. Accessed November 2, 2014. http://emp.lbl.gov/sites/all/files/lbnl-6356e.pdf.

Chapter 20: Community Wind

Bolanger, Mark. "Community Wind Power Ownership Schemes in Europe and their Relevance to the United States." Lawrence Berkeley National Laboratory, May 2001. LBNL-48357. emp.lbl.gov/sites/all/files/REPORT%2048357.pdf.

Bund für Umwelt und Naturschutz Deutschland (BUND). "Volle Kraft voraus! Für die ökologische Energiewende von unten." Brochure, modified March 2014. Accessed November 3, 2014. www.bund.net/fileadmin/bundnet/publikationen/

energie/130429_bund_klima_energie_volle_kraft_voraus_energiewende_von_unten.pdf.

Coöperative Windenergie Vereniging Kennemerwind website. Accessed November 2, 2014. www.kennemerwind.nl/.

Denmark Community Windfarm website. Accessed November 2, 2014. www.dcw.org.au/index.html#2.

Farrell, John. "Advantage Local: Why Local Energy Ownership Matters." Institute for Local Self Reliance, September 2014. Accessed November 3, 2014.www.ilsr.org/wp-content/uploads/downloads/2014/09/Advantage_Local-FINAL.pdf.

Gute Nachbarn: Starke Kommunen mit Erneurbaren Energien website. Accessed November 2, 2104. www.kommunal-erneuerbar.de.

Hepburn Community Wind Park Co-operative website. Accessed November 2, 2014. www.embark.com.au/display/public/content/Hepburn+Community+Wind+Park+Co-operative.

Hvidovre Offshore Wind Farm website. Accessed November 2, 2014. www.hvidovrevindmollelaug.dk/wp-content/uploads/2013/05/original_hvidore_wind_farm_originaleng.pdf.

Lantz E., and S. Tegen. "Economic Development Impacts of Community Wind Projects." National Renewable Energy Laboratory. NREL/CP-500-45555, April 2009. Accessed November 2, 2014. www.nrel.gov/docs/fy09osti/45555.pdf.

Lovins, Amory R., and L. Hunter Lovins. *Brittle Power: Energy Strategy for National Security.* Andover, MA: Brick House Publishing, 1982. Accessed November 2, 2014. www.rmi.org/Knowledge-Center/Library/S82-03_BrittlePowerEnergyStrategy.

Musall, Fabian David, and Onno Kuik. "Local Acceptance of Renewable Energy: A Case Study from Southeast Germany." Institute for Environmental Studies, University Amsterdam, May 2011. Accessed November 2, 2014. www.sciencedirect.com/science/article/pii/S0301421511001972.

Nestle, Uwe. "Marktrealität von Bürgerenergie und mögliche Auswirkungen von regulatorischen Eingriffen." Bündnis Bürgerenergie e.V. (BBEn) und dem Bund für Umwelt und Naturschutz Deutschland e.V. (BUND), Leuphana Universität Lüneburg, Institut für Bank, Finanz, und Rechnungswesen, April 2014. Accessed November 3, 2014. www.bund.net/fileadmin/bundnet/pdfs/klima_und_energie/140407_bund_klima_energie_buergerenergie_studie.pdf.

Tisdale, Mathew, Thilo Grau, and Karsten Neuhoff. "Impact of Renewable Energy Act Reform on Wind Project Finance." Deutsches Institut für Wirtschaftsforschung, Berlin, 2014. Accessed November 2, 2014. www.diw.de/documents/publikationen/73/diw_01.c.466289.de/dp1387.pdf.

Toke, Dave. "Are green electricity certificates the way forward for renewable energy? An evaluation of the United Kingdom's Renewables Obligation in the context of international comparisons." *Environment and Planning* C: Government and Policy 23, no. 3 (2005): 361–374.

Trend:Research. *Marktakteure Erneuerbare-Energien-Anlagen In der Stromerzeugung.* Klaus Novy Institut, August 2011. Accessed November 12, 2014. www.kni.de/media/pdf/Marktakteure_Erneuerbare_Energie_Anlagen_in_der_Stromerzeugung_2011.pdf.pdf.

Vass, Tiffany. "The influence of local project initiation, participation, and investment on local perceptions of small-scale wind energy projects in Nova Scotia." Dalhousie University, Halifax, Nova Scotia, March 2013. Accessed November 2, 2014. //sites.google.com/site/nswindenergystudy/.

Warren, Charles R., and Malcolm McFadyen. "Does community ownership affect public attitudes to wind energy? A case study from south-west Scotland." *Land Use Policy* 27 (2010): 204–213. Accessed November 2, 2014. www.embark.com.au/download/attachments/2889510/Warren+-+Does+Community+Ownership+Affect+Public+++++Attitudes.pdf.

Westmill Wind Farm Co-operative website. Accessed November 2, 2014. www.westmill.coop/westmill_home.asp.

Wikipedia. "Hepburn Wind Project." Accessed November 2, 2014. en.wikipedia.org/wiki/Hepburn_Wind_Project.

Wiser, Ryan, and Mark Bollinger. "2010 Wind Technologies Market Report." Lawrence Berkeley National Laboratory, June 2011. Accessed November 3, 2014. emp.lbl.gov/sites/all/files/lbnl-4820e-ppt.pdf.

Chapter 21: The Challenge

Anair, Don. "How do EVs Compare with Gas-Powered Vehicles? Better Every Year…" Union of Concerned Scientists, September 16, 2014. Accessed November 3, 2014. //blog.ucsusa.org/how-do-electric-cars-compare-with-gas-cars-656?_ga=1.62037544.529347279.1412002686.

Anair, Don, and Amine Mahmassani. "State of Charge: Electric Vehicles' Global Warming Emissions and Fuel-Cost Savings across the United States." Union of Concerned Scientists, June 2012. Accessed November 3, 2014. www.ucsusa.org/sites/default/files/legacy/assets/documents/clean_vehicles/electric-car-global-warming-emissions-report.pdf.

"Global Climate Change Policy Tracker: An Investor's Assessment." Deutsche Bank (DB Climate Change Advisors), October 2009. Accessed November 13, 2014. www.wind-works.org/cms/fileadmin/user_upload/Files/Reports/Global_Climate_Change_Policy_Tracker_Exec_Summary.pdf.

Lewis, Mark C. "Toil for oil spells danger for majors: Unsustainable dynamics mean oil majors need to become energy majors." Kepler Cheuvreux, Paris, September 15, 2014. Accessed November 3, 2014. www.longfinance.net/images/reports/pdf/kc_toilforoil_2014.pdf.

Möller, Bernd, Lixuan Hong, Reinhard Lonsing, and Frede Hvelplund. "Evaluation of offshore wind resources by scale of development." *Energy* 48, no. 1 (2012): 314–322.

Annotated Bibliography

Aesthetics and Noise

Bowdler, Dick, and Geoff Leventhall, eds. *Wind Turbine Noise*. Brentwood, UK: Multi-Science Publishing, 2011. 215 pages. ISBN: 978-1-907132-30-8. Addresses claims about health impacts and annoyance of wind turbine noise by experts who work in the field of noise and public health.

Nielsen, Frode Birk. *The Nature of Wind Power*. Denmark: Landart, 2009. 184 pages. ISBN: 978-87-993240-0-2. This book is a joy to behold, beautifully illustrated, with passages that are almost poetic in their description of the importance of wind energy and its role in the landscape. This is exemplified when Nielsen opens the book by paraphrasing H. C. Andersen, known to us as the storyteller Hans Christian Andersen, by describing wind energy as the "ugly duckling that became a beautiful swan."

———. *Wind Turbines & the Landscape: Architecture & Aesthetics*. Arhus, Denmark: Birk Nielsens Tegnestue, 1996. 63 pages. ISBN 87-985801-1-6. A beautifully illustrated book showing how wind turbines can exist harmoniously on the landscape.

Pasqualletti, Martin, ed. *Wind Power in View: Energy Landscapes in a Crowded World*. San Diego: Academic Press, 2002. 234 pages. ISBN 0-12-546334-0. An entire book devoted to how wind turbines appear on the landscape.

Wagner, Siegfried, Rainer Bareiß, and Gianfranco Guidati. *Wind Turbine Noise*. Berlin: Springer-Verlag, 1996. 205 pages. ISBN 3-540-60592-4. Details on the sources of wind turbine noise and how to treat them.

Modern Wind Energy History

Asmus, Peter. *Reaping the Wind*. Washington, DC: Island Press, 2001. 277 pages. ISBN 1-55963-707-2. The book sings the praises of California's hitherto unsung energy heroes, the men and women who made the sometimes faulty technology work in the state's windy passes and the eccentrics that made the pioneering days so colorful. A novelist would be hard-pressed to create the cast of characters that Asmus assembles: Zen Buddhists, Jungian dreamers, and a host of crooks, charlatans, and hucksters.

Baker, T. Lindsay. *A Field Guide to American Windmills*. Norman: University of Oklahoma Press, 1985. 528 pages. The definitive history of the American farm windmill.

Borish, Steven. *The Land of the Living: The Danish Folk High Schools and Denmark's Non-Violent Path to Modernization*. Grass Valley, CA: Blue Dolphin Publishing, 2005. 516 pages. ISBN-10: 0931892627, ISBN-13: 978-0931892622. No one can understand Denmark's outsize influence on the development of modern wind energy without understanding the role of the *folkehøjskol* movement and the influence of theologian N. F. S. Grundtvig on Danish culture.

Dörner, Heiner. *Drei Welten—Ein Leben: Prof. Dr. Ulrich Hütter*. Heilbronn, Germany: H. Dörner, 1995. 336 pages. ISBN: 3-00-000067-4. Detailed account of the life and technical contributions of Ulrich Hütter by someone who worked with him.

Handschuh, Karl. *Windkraft Gestern und Heute: Geschichte der Windenergienutzung in Baden-Württemberg*. Germany: Ökobuch Verlag, 1997. 115 pages. ISBN: 9783922964339. Description of the early work by Ulrich Hütter and the development of Allgaier's commercial product in the 1950s.

Heymann, Matthias. *Die Geschichte der Windenergienutzing 1890–1990*. Frankfurt am Main, Germany: Campus Verlag, 1995. 518 pages. ISBN 3-539-35278-8. It's worth learning German to read this controversial book on the history of wind energy in Germany from the late 19th century to the early 1990s. It contains the most extensive discourse to date on the Third Reich's interest in wind energy and an unflattering, matter-of-fact description of attempts by some of the grand names in German wind energy to curry favor with the Nazis. This book is a powerful reminder that technology is not divorced from politics.

Maegaard, Preben, Anna Krenz, and Wolfgang Palz, eds. *Wind Power for the World: The Rise of Modern Wind Energy*. Singapore: Stanford Publishing, 2013. 676 pages. ISBN-10: 9814364932, ebook: ISBN-13: 978-9814364935. *Wind Power for the World* tells an exciting story of hope and promise: how a small band of activists, dreamers, and entrepreneurs built one of the world's fastest growing and dynamic industries. It's a must-read for anyone who wants to understand how we got to where are today.

Nissen, Povl-Otto, Therese Quistgaard, Jutte Thorndahl, Benny Christensen, Pregen Maegaard, Birger T. Madsen, and Kristian Hvidfelt Nielsen. *Wind Power the Danish Way: From Poul la Cour to Modern Wind Turbines*. Askov, Denmark: Poul la Cour Foundation, 2009. 88 pages. ISBN: 978-87-993188-0-3. *Wind Power the Danish Way* describes the struggle to liberate rural Denmark from poverty by bringing electricity to even the poorest of villages using wind power. This was Poul la Cour's mission.

Oelker, Jan. *Windgesichter: Aufbruch der Windenergie in Deutschland*. Dresden, Germany: Sonnenbuch, 2005. 400 pages. ISBN 3-9809956-2-3. Translated as "The Face of Wind: Dawn of Wind Energy in Germany," Oelker's text is one of those rare cases where you can indeed tell a book by its cover. The cover photo is a portrait of a bearded Karl-Heinz Hansen in his blue work jacket astride an early Vestas wind turbine.

Hansen is looking to the right and off into the distance. He could be gazing into the future that he and the others in the book helped make a reality. He beckons the reader to turn the page.

Rapin, Marc, and Jean-Marc Noël. *Énergie Éolienne: Principes—Études de cas*. Paris: Dunod, 2010. 304 pages. ISBN: 9782100508013. The book is also available electronically and includes an excellent and comprehensive history of wind energy in France, including many photographs.

Righter, Robert. *Wind Energy in America: A History.* Norman: University of Oklahoma Press, 1996. 366 pages. ISBN 0-8061-2812-7. A thought-provoking account of the people and ideas behind the use of wind energy in the United States. The book emphasizes the conflict between centralization and distributed use of wind-generated electricity.

Thorndal, Jytte. *Gedsermøllen: den første modene vindmølle*. Bjerringbro, Denmark: Elmuseet, 2005. 99 pages. ISBN: 87-89292-45-6. Early Danish wind turbine design up to and including Juul's Gedser mill.

Torrey, Volta. *Wind-Catchers: American Windmills of Yesterday and Tomorrow*. Brattleboro, VT: Stephen Greene Press, 1976. 226 pages. ISBN 0-8289-0292-5. This is an engagingly written history of wind energy in the United States with chapters on the Smith-Putnam turbine, the American water-pumping windmill, and 1970s pioneers of modern wind turbines. Righter's *Wind Energy in America: A History* adds his own original contributions and brings the story up to the 1990s.

van Est, Rinie. *Winds of Change: A Comparative Study of the Politics of Wind Energy Innovation in California and Denmark*. Utrecht, Netherlands: International Books, 1999. 368 pages. ISBN 90-5727-027-7. This monumental work is destined to become a classic in its field. It paints a detailed, carefully researched picture of why the development of wind energy technology failed in the United States during the 1970s and 1980s but succeeded in Denmark.

Large Wind Turbines

The following is only a sample of the books that have been published on wind energy in the past 50 years. There are hundreds of books on wind energy in various languages. Many are out of print and available only in large libraries or private collections. Each author makes a unique contribution to the subject. Reviews of some of these books, including their table of contents, can be found on the author's website at www.wind-works.org.

Ackermann, Thomas, ed. *Wind Power in Power Systems*. Chichester, UK: John Wiley & Sons, 2005. 691 pages. ISBN: 0470855088. As wind power expands dramatically, the question of integration with the existing utility system becomes more critical—and, more important, how to design the grid of the future to take best advantage of renewable resources.

Burton, Tony, David Sharpe, David Jenkins, and Ervin Bossanyi. *Wind Energy Handbook*. Chichester, UK: John Wiley & Sons, 2002. 617 pages. ISBN 0 471 48997 2. One of the reference engineering texts on wind turbine design in English. First such work to include chapters on design of wind farms, including discussion of environmental impacts as well as cash flow.

Earnest, Joshua, and Tore Wizelius. *Wind Power Plants and Project Development*. New Delhi: PHI Learning Private, 2011. 496 pages, ISBN: 978-81-203-3986-6. Naturally directed at the Indian market, it nonetheless has taken a comprehensive approach to explaining wind turbine theory, design, and project development to Indian engineers.

Eggleston, David, and Stoddard, Forrest. *Wind Turbine Engineering Design*. New York: Van Nostrand Reinhold, 1987. 352 pages. For professional wind turbine designers. Not for the faint of heart. The authors warn readers that a working knowledge of differential equations is essential to understanding the text. Includes an extensive discussion of aerodynamics and structural dynamics.

Freris, L. L., ed. *Wind Energy Conversion Systems*. Hemel Hempstead, UK: Prentice-Hall, 1990. 388 pages. ISBN 0-13-960527-4. Engineering text by the leaders in British wind energy. A good academic reference book on wind technology.

Gasch, Robert, and Jochen Twele, eds. *Wind Power Plants: Fundamentals, Design, Construction and Operation*. Heidelberg, Germany: Springer Verlag, 2012. 2nd ed. 548 pages. ISBN 978-3-642-22937-4. Although it's difficult to pick one of the new engineering texts (there are two by John Wiley & Sons alone) as "best in the class," Gasch's book is clearly one of the best. In a sign of the times, this English-language version drops the chapter on vertical-axis wind turbines, adding instead a chapter on off-shore wind.

Gipe, Paul. *Wind Energy Basics*. White River Junction, VT: Chelsea Green Publishing, 1999. 122 pages. ISBN 1-890132-07-1. Cursory introduction to wind energy and small wind turbines.

——— *Wind Energy Comes of Age*. New York: John Wiley & Sons, 1995. 613 pages. ISBN 0-471-10924-X. A chronicle of wind energy's progress from its rebirth during the oil crises of the 1970s through a troubling adolescence in California's mountain passes in the 1980s to its maturation on the plains of northern Europe in the 1990s. Selected as one of the outstanding academic books published in 1995.

Golding, E. W. *The Generation of Electricity by Wind Power*. London: E. & F.N. Spon, 1955. 332 pages. Reprinted by John Wiley & Sons in 1976. Still a classic of English language books on wind technology. Recounts British research on wind energy during the early 1950s.

Hau, Erich. *Windkraftanlagen: Grundlagen, Technik, Einsats, Wirschaftlichkeit*. 2nd ed. Berlin: Springer Verlag, 1996. 460 pages. ISBN 3-540-57430-1. A thorough perspective on Germany's development of multimegawatt wind turbines during the early and mid-1980s. The second edition includes expanded coverage of medium-size wind turbines that were beginning to appear in large numbers on the German market in the mid-1990s. Also available in English.

Jamieson, Peter. *Innovation in Wind Turbine Design*. Chichester, UK: John Wiley & Sons, 2011. 298 pages. ISBN: 9780470699812. The title says it all. This is a book by one of the world's leading wind turbine designers.

Johnson, Gary L. *Wind Energy Systems*. Englewood Cliffs, NJ: Prentice-Hall, 1985. 360 pages. An engineering textbook strong on the wind resources of Kansas and on electrical engineering.

Koeppl, Gerald W. *Putnam's Power from the Wind*. 2nd ed. New York: Van Nostrand Reinhold, 1982. 470 pages. The first half of the second edition is a reprint of Putnam's original book. The second half examines large wind turbine development programs in the United States and Europe during the 1970s.

Le Gouriérés, Désiré. *Les Éoliennes: Théorie, conception et calcul pratique*. 2nd ed. Goudelin, France: Éditions du Moulin Cadiou, 2008. 306 pages. ISBN: 978-2-9530041-0-6.

Manwell, J.F., McGowan, J.G. and Rogers, A.L. *Wind Energy Explained: Theory, Design, and Application*. Chichester, United Kingdom,: John Wiley & Sons, 2002. 590 pp. ISBN: 0-471-49972-2. Manwell and McGowan are old wind hands. They are well-known in the American wind industry for their work keeping UMass' Renewable Energy Research Laboratory alive when similar programs were closing across the country.

Molly, Jens-Peter. *Windenergie: Theorie, Anwendung, Messung*. Karlsruhe, Germany: Verlag C.F. Muller, 1990. 316 pages. ISBN 3-78807269-5. German development of medium

medium-size wind turbine technology through the late 1980s. One of the most thorough and well-illustrated books on modern wind technology in any language. Includes useful German-English and English-German translations of common technical terms.

Nelson, Vaughn. *Wind Energy: Renewable Energy and the Environment*. Boca Raton, FL: CRC Press, 2014. 2nd ed. 328 pages. ISBN-13: ISBN-13: 978-1420075687, ISBN-10: 146658159X. This is a book by one of the industry's pros, targeted toward students of engineering and physics. He has been teaching the importance of swept area on wind turbine performance for more than three decades. Nelson's no-nonsense style and his emphasis on the fundamentals, such as how to understand orders of magnitude, are refreshing.

Paraschivoiu, Ion. *Wind Turbine Design: With Emphasis on Darrieus Concept*. Montréal, Canada: École Polytechnique, 2002. 590 pages. ISBN: 2-553-00931-3. Paraschivoiu seeks to document for posterity the development of Darrieus technology in Canada.

Putnam, Palmer Cosslett. *Power from the Wind*. New York: Van Nostrand Reinhold, 1948. 224 pages. Reprinted in 1974. ISBN 0 442 26650-2. The classic account of constructing the 1.25-megawatt Smith-Putnam turbine during the early 1940s. Like Golding, many of Putnam's observations still apply.

Saulnier, Bernard, and Réal Reid. *L'éolien au cœur de l'incontournable révolution énergétique*. Quebec, Canada: Éditions MultiMondes, 2009. 432 pages. ISBN: 978-2-89544-145-8. Saulnier and Reid take on the question of wind's intermittency. Tellingly, they title the section "variability." As they explain, wind generation is not really intermittent but variable and isn't radically different from other power plants. They illustrate how wind works, and how wind can and does work well in electric utility systems—not only in theory but in practice.

Spera, David, ed. *Wind Turbine Technology: Fundamentals Concepts of Wind Turbine Engineering*. New York: American Society of Mechanical Engineers, 1994. 650 pages. ISBN 0-7918-1205-7. The book is noteworthy for a chapter on NASA/DOE's large turbine development program by the program's principal proponent, Lou Divone.

Walker, John F., and Nicholas Jenkins. *Wind Energy Technology*. Chichester, UK: John Wiley & Sons, 1997. 161 pages. ISBN 0-471-96044-6. A handy introduction to wind energy for engineering students.

Wizelius, Tore. *Developing Wind Power Projects: Theory and Practice*. London: Earthscan Publications, 2006. 296 pages. ISBN 1844072622. Longtime wind energy advocate Tore Wizelius is one of the Swedish pioneers of the technology. Finally, his thoughtful book on wind energy and its integration into communities is now available in English.

Wizelius, Tore. *Windpower Ownership in Sweden: Business models and motives*. Abingdon, UK: Routledge, 2014. 224 pages. ISBN: 978-1-13-802111-2. Tore Wizelius helps Anglophones understand how Swedes have taken a sizable ownership of wind energy in spite of their government. In this, his book can serve as an inspiration to community wind advocates worldwide who face many of the same challenges as those in Sweden.

Small Wind Turbines

Bartmann, Dan, and Dan Fink. *Homebrew Wind Power: Hands-on Guide to Harnessing the Wind*. Masonville, CO: Buckville Publications, 2009. 320 pages. ISBN: 978-0-9819201-0-8. In the mold of Michael Hackleman's series *Wind and Windspinners*, Bartmann and Fink take a romp through wind theory, technology, and, yes, how to build your own axial-flux wind turbine.

Chiras, Dan, with Mick Sagrillo and Ian Woofenden. *Power from the Wind: Achieving Energy Independence*. Gabriola Island, Canada: New Society Publishers, 2009. 288 pages. ISBN: 9780865716209. Subtitled *A Practical Guide to Small-Scale Energy Production*, the book follows ground well trod before and seeks to capitalize on the odd American desire to run kilowatt-hour meters backward.

Clark, Nolan. *Small Wind: Planning and Building Successful Installations*. Waltham, MA: Academic Press, 2014. 224 pages. ISBN-10: 0123859999, ISBN-13: 978-0123859990. Clark summarizes a lifetime of developing and testing wind turbines for agricultural use. As one of the early proponents of standards and certification to weed out the hustlers and charlatans in the small wind business, Clark offers a nearly complete certification report for one small wind turbine as an example of what consumers should expect.

Hackleman, Michael. *The Home-Built, Wind-Generated Electricity Handbook*. Mariposa, CA: Earthmind, 1975. 194 pages. Explains how to find, lower, and rebuild pre-REA wind generators.

Park, Jack. *The Wind Power Book*. Palo Alto, CA: Cheshire Books, 1981. 253 pages. ISBN 0-917352-06-8. A useful reference for experimenters with an aversion to metric units.

Piggott, Hugh. *It's a Breeze*. Machynlleth, Powys, Wales, UK: Centre for Alternative Technology, 1995. 36 pages. Straightforward advice on how to use small wind turbines.

———. *Windpower Workshop*. Machynlleth, Powys, Wales, UK: Centre for Alternative Technology, 1997. 160 pages. ISBN 1-898049-13-0. Chock-full of tips on building your own windcharger.

Woofenden, Ian. *Wind Power for Dummies*. New York: John Wiley & Sons, 2009. 384 pages. ISBN: 978-0-470-49637-4. As in other Dummies books and certainly as found in the pages of Home Power Magazine, Woofenden uses homespun aphorisms to drive home his points. One such example is his advice about "thinking before you act." "Wearing a hard hat doesn't mean a lot if you don't have much to protect in the first place. Your number one piece of safety gear is on your shoulders. You need brains, determination, knowledge, and experience to be safe."

Shea, Kevin, and Brian Clark Howard. *Build Your Own Small Wind Power System*. New York: McGraw-Hill/TAB Electronics, 2011. 512 pages. ISBN-10: 0071761578. There have been a slew of new books in the past few years on small wind. Some are by pros, such as Ian Woofenden, and some are by professional writers, such as Dan Chiras. This is one of the first by both.

Rigging

Carter, Paul. *Backstage Handbook: An Illustrated Almanac of Technical Information*. New York: Broadway Press, 1994. 310 pages. ISBN:0-911747-39-7. Packed with useful information on tools and rigging and has an especially good section on wire rope.

Ellis, J. Nigel. *Introduction to Fall Protection*. 2nd ed. Des Plaines, IL: American Society of Safety Engineers, 1993. 228 pages. ISBN 0-939874-97-0. Informative discussion of the development of fall safety standards for engineers and managers. Should be required reading for anyone working at heights or asking others to do so.

Kurtz, Edwin, and Thomas Shoemaker. *The Lineman's and Cableman's Handbook*. Ninth edition. New York: McGraw-Hill, 1997. 1,056 pages. ISBN: 0070360111. An extremely useful handbook on setting poles, stringing conductors, and installing earth anchors. This is the source for the raising of utility poles without cranes mentioned in Figure 20-1.

Index

Aagaard, Steve, 431–432
AC. *See* alternating current (AC)
acceptance, community ownership and, 483
access to towers, 222–223, *223;* unauthorized, 435–436
accidents, fatal, 415–417, *416, 417*
accumulators, 247, 311
Acker, Pat, 435
active yaw, *86,* 87
Adecon turbines, 140, *140*
Adolfsen turbine, 64
advertising. *See* marketing, aggressive and misleading
Aermotor windmill, *93,* 93–94, 303, *303,* 306
aerodynamics: blades and, 89–101; brakes and, 131–134; history of wind energy and, 33, 35, 36–37; noise and, 338–340, *340;* stall and, 129–130
Aeropower turbines, 108
aerospace industry, 75–76
Aero-Star, 68
AeroVironment, 101, 217
Aerowatt, *127, 428*
aesthetics, 221, 346, 360–365
affordability, 420–521
Agora Energiewende, 300
Agricco (40-kW) turbine, **35–36**
ailerons (flaps), 133, *133*
Ailleret, Pierre, 47
Airborne Wind Energy Systems (AWES), 183–185
air brakes, 131, *131,* 144–145
Airbreeze turbine, *99,* 458
air-conditioning, 268–269
air density, 231–233, 238
airfoils, 28, 35, 87, 89–90, *90*
air-gap generators, 117
airlift (bubble) pumps, 308, *308*
Air series turbines, 115, 117, 122, 205, 215, 472; micro, 273, *273;* noise and, 331, 337–338, *338,* 340, 343; repairs and, 462
Alberta, 330, *330*
Alcoa turbines: aluminum blades and, 103; blade number and, 139–140, *140;* overspeed control and, 169; silos and, 218; VAWTs and, 142, 146–147
Allgaier Company, 42–45, *43*

Altamont Pass (CA): bird fatalities and, 348–350; grazing and, 365; turbines in, *60, 106, 128,* 148, 446; wind in, 227, 239, 243, 296
alternating current (AC), 35–36, 44; generators and, 114, 116–118; safety and, 434; systems of, 269–270, 272–273
Alternative Energy Institute (West Texas A&M), 308–309, *312,* 312–313, *313,* 433, 459
alternators, *116,* 116–118, 433
aluminum blades, *103,* 103–104, 142, 146, 170
Amarillo (TX), 267
ambient noise, 335–336
American Wind Energy Association (AWEA), 341, 452, 455–457
American Wire Gauge (AWG) to metric conversion, 532, *532*
Ampair turbines, 108, *109, 234,* 238, 272, *337,* 338
amplitude modulation, 342
amps, 19
anchorages, 424–425, *425, 431*
anchor knuckle, *382*
anchors, 375–381, *379,* 432, 435–436
ancillary structures, 363, *364*
Anderson, Dick, 348–349, 352
anemometer, *224,* 246–247, 440
anemometer masts, 247, *248*
anemometer towers, 247–248
angle of attack, *90,* 90–91
annual energy production (AEP): certification and, 457; estimating performance and, 252–256, *254, 255, 256;* manufacturers' estimates and, 261–262, 460; power curve and, *260;* web-based calculators of, *262*
annual reoccurring costs, 467–469, *468*
annual specific yield: estimating performance and, 254, 263, *263;* by IEC classes, 257–258; productivity and, 193–194, 200; of VAWTs, *169*
annual wind energy density, 238, *238*
Ansa, Niels, *431*
anticlimb guards, 435, *435*
anti-wind groups, 316, 320–321
Appalachia State University (ADU), 472
apparent wind, *90,* 90–91

area of wind stream, 83, *84,* 192–194; estimating performance and, 252–258; measures of relative swept, 194–195; power and, 234, *234,* 450, 452–453
Areva Wind, 113
Armbrust, Sepp, 199
Armstrong, John, 95, 100
arrays, 3–7, *5–6,* 254, 269, 361
articulating blades, 152
Artig, Rory, 242
asynchronous generators, 119–120, 285
Atlantic Orient turbines, 74
Atlantic Wind Test Site (PEI), 145–146, *146,* 158
attractive nuisance, turbine as, 326–327
Audubon Society, 350
augmented (ducted) turbines (DAWTs), 176–183
Australia, *85,* 279, 344, 498–499, *499*
Austria, *370*
automatic disconnect, 287, *287*
aviation, wind research and, 36–37
aviation obstruction markings, 327–328, *328*
AWEA. *See* American Wind Energy Association (AWEA)
AWG. *See* American Wire Gauge (AWG) to metric conversion
Axial-flux generators, 117, *117*

back gearing, 305, *305*
Baden Baden (ship), 188
Ballem, Jamie, 469
Barbara (ship), 188
Barnard, Mike, 184–185, 317, 344, 352
Bartelt, Heinrich, 468
Bat Conservation International, 352
bats. *See* birds and bats
Battelle Pacific Northwest Laboratories, 236, 243–244, 248
batteries: charging of, *8,* 22, 24, 30, 38, 118, 134; estimating performance and, 251, 262; interconnection and, 288, *288;* maintenance and, 442–443; off-the-grid systems and, 273–275, *274;* sailboats and, 10, *10,* 122
battery charge controller, 275
battery-charging wind systems, 439
Baupin, Denis, 179

INDEX

Baywind Energy Co-operative (UK), 497, *497*
Bear Mountain Wind (BC), 474, 507
Beaufort scale of wind force, 230–231
Beech Mountain (NC), 472
belts, positioning or work, 418, 422–424
Bendix-Schacle turbine, 113, 220, *220*
Bergey, Mike, 155, 259, 292, 324, 343, 430
Bergey Windpower turbines, 75, 104, 108, *117*, 117–118, 134–135; certification of, *457*; compare to VisionAIR, 163; conductor size and, 397, 399; estimating performance and, *261*, 261–262; installation of, *406–407*; interconnection and, 285; longevity of, 473; noise and, *338*, *339*, 343, 345–346; off-the-grid systems and, 277; overspeed control and, *122*, 124, *124*; safety and, 432, 434, *434*; swept area and, *234*; towers and, *204*; VAWTs and, *152*; water pumping and, 308–310, *309*, *309*, *310*
Bérubé, Jean-Yves, 151
BEST-Romani turbine, 47–48
Betz limit, 94, 154, 182–183, 252
billboards and logos, 361, *361*
BioSystems, 350
birds and bats, 167, 348–352, *351*, *351*
Bjerre, Asbjørn, 486
Bjervig, Leon, 143
Black, Bryce, 314
Black Island (McMurdo Sound, Antarctica), 276
blade-activated governor, *127*
blade number, 34, 83, 95–100, 139–140
blade root doors, 433, *433*
blades: attachment of, *105*, 105–106; Danish concept and, 68; falling, 325–326; feathering of, 126–127; flutter of, 122; Gougeon, 74, 102; Grove-Nielsen (Økær) and, 64–66, *66*, 78; hazard of spinning, 419–421; installation of, *411*; maintenance of, 440, *442*, 444–445; materials used in, 44, 60, 63, 101–105; pitchable tips of, 45, 47, 66, 133–134; pitch of, 91, 126–129; tapered, 89, *89*, 91; Tvind, 61, 61–64, 63; twist and, 91, *92*; Voland, 58, *59 See also* rotors
Blitterdorf, Dave, 213, 343, 432, *434*
Blyth, James, 29–30, 34
Boeing turbines, 54, *55*, *59*, 102, 112, 134, 239
Bogø prototype turbine, *45*, 45–46
Bohmstedt project (Germany), 491
Bolanger, Mark, 500
bolts: capped foundation, 223, *223*; maintenance and, 436, 441, 443, *443*; split, 395, 397
Bonus turbines, *67*, 67–68, 70, 77, 112, 412
boom or aerial lift, *428*, 429
boots and gloves, 430
Bornay, Juan, 125
Boston (MA), 505
bottom-up design strategy, 77–80
brakes, 67–68, 130–134, 169, 430 *See also* dynamic braking
brake switch, 459

Breunig, Christian, 497
Britain: certification in, 456–457; community wind in, 497–498; compatible land uses in, 365–366, *368*; evolution of wind energy and, 36; FITs and, 523–524; fixed-pitch straight-blade VAWT and, 155–157; grid integration in, 297, *298*; rooftop mounting and, 217; towers and, 327
British Wind Energy Association (BWEA), 457
Brittan, Gordon (Corky), 101, 474
Brothers, Carl, 279, 332
Brown, John, 47
Brush, Charles, 31–32, 34
Brush Dynamo, *31*
BTM Consult, 447, 473
Buffalo Ridge (MN), 242
building and electrical codes, 322
building permits, 322–325
bull pins, 386
BUND, 351, 490
Bürgerbeteiligung, 492–493, 495, 496–497, 512
Bürgerwind movement, 490–496
Bürger Windpark Lübke-Koog, 492
Bushland (TX) USDA test station, *145*, *337*, 435; turbines at, *473*, 473–474; water pumping and, 306, 308–310, *312*, 312–313
Butendiek Bürgerbeteiligung, 512
"buy-down" subsidies, 466

cable grips, *381*, 381–382, 432
cables: aboveground and buried, 396; guy, 140, 145, 165, 170, 381–382; sleeves and, 426–427, *431*
Cade, Tom, 348, 352
Calgary (AB), 517
California: bird fatalities in, 348–350; capacity credits in, *296*; Fayette Manufacturing in, 451; high-speed electrified rail systems and, 517, *517*; land area required, 354; regulation cost in, *297*; salvage value in, 359; setback requirements in, 327–330, *329*; turbines in, 4, *6*, 110, 147–150, 163, 170; two-blade turbines in, 97, *97*; wind rush in, 68–69, 72–74, 77, 243–244 *See also* Altamont Pass (CA); Kern County (CA); San Gorgonio Pass (CA); Tehachapi Pass (CA)
California Public Utilities Commission (CPUC), 296–297
Calley, Dave, 215, 343
Camden Hills Regional High School (ME), 482
Cameron Ridge. *See* Tehachapi Pass (CA)
Campbell, Patrick, 365
Canada: aviation obstruction markings in, 327; bird fatalities in, 349, 351; community wind in, 500–504, 507; compatible land uses in, *368*, *370*; FITs in, 471; lease pooling in, 469; noise research and, 342–344; setback requirements in, *330*, 330–331, 345 *See also* by province or city
Canadian National Exposition (CNE), 349, *368*, 500–501
capacitors, 288–289

capacity credits, *296*, 296–297
capacity factor (CF), 193–194, *199*, 199–201, 201–202
Cap Chat, Gaspé Peninsula (PQ), 151, *151*
capital, cost of, 467
capped foundation bolts, 223, *223*
carabiners, 425, *426*
Caribbean, winds in, 236–237, *237*, 279
Carstensen, Uwe, 75
Carter, Jimmy, 526
Carter turbines, 70, *70*, 86, 125, 129
cash-flow models, 463, 464, 474–475, *475*
C-BED. *See* Community-Based Energy Development (C-BED) (MN)
Centre for Economics and Business Research (CEBR), 353
certification and labeling, 456–458
Chabot, Bernard, 191–192, 195, *202*, 257, *299*; profitability index method of, 464, 475–477
challenges, 511–528; electricity feed laws and, 521–525; ethical energy policy and, 525–528; in North America, 513–521; pitfalls and, 511–513
Chapman, Simon, 344
Château de Lastours test site (France), *389*
Cheuvreux, Kepler, 520
chinooks, 228
Chiu, Graham, 336
Christopher Point, Vancouver, 145
CIEMAT, 455
Clark, Nolan, 100, 145, 306, 308–309, 313
Clarke, Alexi, 329
Cleanfield turbines, 155, 157–159, *158*, 166
Cleveland (OH), *331*
Clipper Windpower, 113
cloth blades, *101*
cluster arrays, 4, *5*
Coal Clough (UK), 366
cogging, 118
cohesive and noncohesive soils, 376, 379
Colby, David, 344
Colchester-Cumberland Wind Field (NS), *503*, 503–504
Colijn, Helen, 365–366
Colorado, *13*, *327*
Columbia River Gorge (OR), 227
come-along tool, 432
Commoner, Barry, 115
Community-Based Energy Development (C-BED) (MN), 505, 507
community concerns, 315–317, 323
Community Economic-Development Investment Funds (CEDIF) (NS), 503
community ownership, 342–343, 483, *483*, 483–484, 484–485, *501*
community wind, 479–509; in Australia, 498–499; in Britain, 497–498; characteristics of, 485–486; cooperative and mutual investment in, 484–485; Danish Fællesmølle and Vindmøllelaug and, 486–488; description of, 480–482; Dutch cooperatives and, 488–489; Germany's electricity rebels and, 489–497,

507; in North America, 499–506, 507–509; reasons for, 482–484; requirements for, 508–509; third way and, 479–480
compression connectors, 395, 396–397
concrete foundations: dismantling of, 413–414; installation of, 376–378, 377, 378, 402, 408
concrete towers, 210–212, 211, 212, 221
conductors, 394–399; sizing of, 397, 397–399, 399; strain relief for, 396–397
conduits: electrical metallic tubing (EMT) as, 395–396; PVC (polyvinyl chloride), 396; size of, 400
conical wind catcher, 33, 33
coning, 125–126
conservation, ethics and, 526–527
Constantin, Louis, 36, 54
constants and conversions, 529–530
construction accidents, 421
consumption: cutting, 268–269; in Europe, 528; in North America, 514–515; residential energy, 268
continuous output rating, 273
contracts, 462
control panel, 18
convective circulation, 226–227, 229
conventional turbines, 83–135; aerodynamics and, 89–101; blade materials and, 101–105; drivetrains and, 106, 108–114; generators and, 114–121; hubs and, 105–107; large, 111–113, 135; lift and drag and, 87–89; medium-sized, 110–111; orientation and, 84–87; overspeed control and, 121–134; small, 108–110, 134–135
conversions: American Wire Gauge to metric, 532; constants and, 529–530
cooperatives: British, 497–498; Canadian, 500–504, 507–508; Danish, 485, 487–489; Dutch, 486, 488–489; German, 496–497; U.S., 504–506
Coöperative Windenergie Vereniging Kennemerwind (Denmark), 486, 489, 489
Copenhagen (Denmark), 487, 487, 488
cost: installed, 452, 453, 465, 465; insurance, 467; interconnection and, 297–298; investing and, 452–454, 459; of maintenance, 446–447; per kW, 452–454; of regulations, 297, 297–298, 298 See also investing in wind energy
cost of energy (COE) formula, 463
Cousineau, Kevin, 133
Cousteau, Jacques, 188
Couture, Toby, 519
covenants, legal, 323–324
CPUC. See California Public Utilities Commission (CPUC)
cranes, installation and, 384, 388, 406–407
creep, in soils, 376
crimp connectors, 395
Crotched Mountain (NH), 72
Cruz, Ignacio, 455
cube factor, 236

curing, of concrete, 376–377
current: in-rush, 289–290; as term, 19 See also alternating current (AC); direct current (DC)
cut-in speed, 258
cut-out speed, 258
CWATs, 181
cyclers in wind turbine sites, 365–366, 368
Cycloturbines, 152, 154–155

DAF-Indal, 103, 141–146, 144, 145, 169, 312, 313
Danish Wind Power Society, 34
Danish Wind Turbine Owners' Association, 79–80, 121–122, 143, 263, 345, 486–487
Danmarks Vindmølleforening, 486–487
Dansk Vindkraft griomill, 153
Darrieus, Georges, 36, 137–139, 166
Darrieus turbines (VAWTs), 84, 84, 137–173; aluminum rotor/blades of, 103–104; blade number and, 139–140; cantilevered, 140–141, 141; claims and counterclaims about, 164–167; configurations of, 138, 140; conspiracy against, 171, 173; design characteristics of, 167–170; fixed-pitch H-Rotor, 155–162; future of, 137, 162–164, 170–171; guy cables and, 140, 145, 165, 170; H-configuration (straight-blade), 151–155; lift and drag and, 138–139; Φ configuration (eggbeater) development and, 142, 142–151, 165, 170; rooftop, 145, 157–159; specific mass of, 167, 167; test standards and, 171; towers and, 139–140, 140–141
Database of State Incentives for Renewables and Efficiency (DSIRE), 466
data logger, 247, 247, 440
Davis, Jim, 324
DAWTs. See ducted (augmented) turbines (DAWTs)
DC. See direct current (DC)
dealers, 461–462
debt, 467
DeCarolis, Joseph, 297–298
decibels, 332–333
decommissioning wind plants, 359, 402
De Feltre, Duc. See De Goyon, Charles
De Goyon, Charles (Duc de Feltre), 30, 32, 34
demand, reducing, 267–269
Dempster windmills, 302, 305, 309, 309–310, 310
Denmark: cluster arrays of, 4, 5; community wind in, 480–481, 485–488, 487; compatible land uses in, 365–366, 371; "Danish concept" and, 68, 77–81; design philosophy compared to US, 69–72, 70, 71, 75–77, 80–81; drivetrains of, 110–112; early wind-electric, 40; evolution of wind energy in, 27–28, 32–35, 37, 40–42, 44–47; generators in, 119–121; grid integration and, 294–295, 298; interconnection and, 35–36, 292–293; large-diameter turbines in,

192–193, 195, 450; noise and, 345; offshore and near shore wind and, 512; overspeed control and, 121–122, 133–134; performance claims in, 462; rebels in, 59–68; renewables and, 298, 519, 526; siting in, 318; top-down R&D program in, 58–59, 80
Denmark Community Windfarm (Australia), 499
deregulation, 527
design standards, 456
Deutsche Bank, 523
Deutsches Windenerie Institut (DEWI) (Germany), 354–355
diameter: of turbine rotors, 22, 43, 234, 252, 254; of windmill pipes, 304; of windmill rotors, 302, 305–306
diesel, twinning wind power with, 278–279, 279
differentiated wind tariffs, 464, 524–525
Difrancisco, Lisa, 171, 173
direct acting windmills, 305
direct current (DC), 35–36, 44; generators and, 114, 115, 116; safety and, 434; systems of, 269–270, 270
direct drive, 108, 109, 113, 474
disco effect, 347
disconnect switches, 396, 433–434, 434, 438
dismantling wind turbines, 402, 412–414
distributed wind generation, 2, 7–10; interconnection and, 282, 282, 292–293; tariffs and, 524–525
district-heating systems, 300
diurnal hourly wind speed, 229
DOE (US Department of Energy): DAWTs and, 176, 181; VAWTs and, 153–154; wind R&D and, 52–54, 74, 77
Donnelly, John, 421
Dörner, Heiner, 177
Dornier turbines, 142
double-fed induction generators, 120, 120
downwind rotors/turbines, 69–70, 85–86, 86, 126, 130
drag: conventional turbines and, 83, 87, 87–90; DAWTs and, 183; defined, 206; ratio of lift to, 90, 90; VAWTs and, 138–139
Drees, Herman, 154–155, 170
dress, proper, 262–263, 263
drift pins, 386
D rings, 422–423, 423, 425, 426
drivetrains, 106, 108–114, 441
dual generators, 120–121
ducted (augmented) turbines (DAWTs), 176–183; rebranding of, 181; summary of, 180
dynamic braking, 430, 459
Dynergy turbines, 110
DZ Bank, 496

eagles, golden, 348–350
easements, legal, 324
EasyWind turbines, 108, 109, 121
Eclipse windmill, 302
École turbines, 151, 151
economics: community ownership and, 483, 483–484; energy return on

investment (EROI) and, 356–357; interconnection cost analysis and, 297–298; investing and, 465–474; maintenance costs and, 446–447; off-the-grid power systems and, 276 *See also* investing in wind energy
Edwards, Peter, 121
Edworthy, Jason, 458
effective ground level, 240
efficiency: estimating performance and, 252–256, *253, 255, 256;* ethics and, 526–527; fondness for, 452
eggbeater (Φ configuration) Darrieus turbines, 142–151
Eggleston, David, 76, 89
Eggleston, Eric, 308, 310
Eiffel, Gustave, *208*
electrical accidents/safety, 421, 433–435
electrical metallic tubing (EMT), 395–396
electrical standards, 456
electrical systems. *See* alternating current (AC); direct current (DC)
Electricité de France (EDF), 47–48, 279
electricity feed laws, 521–525, 528
electricity from wind energy, 2–3
electricity price escalation, 471–472
electric vehicles (EVs): charging, *10,* 10–11; offsetting oil for transport and, 516, *516,* 520
electric winch, 392, *393*
electrification, 34–35
electromagnets, 112–113
electronic inverters, 285–286, *286*
Eléna Energie, *178,* 178–179
elevation and air density, 232
Ellis, F. Nigel, 427
Elsam, 59
emissions of **CO2** equivalent gases, 357–358, *358*
Endurance turbines, 86, 100, *111,* 121, 199, *428,* 429
Enercon: community wind and, *489,* 496, 499, 502, *503,* 503–504; drivetrains in, 112–113, *113;* E-Ship 1 of, 188–190, *189;* interconnection and, 286; large-diameter turbines and, 450; longevity and, 473, *473;* siting and, *342, 355, 370, 372;* towers and, 211–212, *212;* turbines of, 78, 101, 412, 474
Energiesystemtechnik, 300
energy crisis of 1970s, 51–52
energy density, 238, *238*
energy equivalents, scale of, *531*
Energy Research Centre (ERC) (UK), 297
energy return on capital invested (EROCI), 520, *520*
energy return on energy investment (EROEI), 520
energy return on investment (EROI), 356–357, *357*
Energy Savings Trust (UK), 217
energy *vs.* power, 18–19
EnerKite, 184
Enertech turbines, 74, 86, 100; community wind and, 504; drivetrains and, 108–110, *109;* interconnection and, 285, 290; maintenance and, 446;

overspeed control and, 130, *132,* 133; water pumping and, 313
Enflo Windtec turbines, *177,* 177–178
England, southern, 497–498 *See also* Britain
environmental concerns, 315–317; birds and bats, 348–352; emissions of **CO2** equivalent gases and, 357–358, *358;* energy return on investment (EROI) and, 356–357; water consumption, 358, *358*
equations, 19
equity, 467
equivalent power, scale of, *531*
erection wrenches, 386, *386*
Erélia, 469
EROCI. *See* energy return on capital invested (EROCI)
EROEI. *See* **energy return on energy investment (EROEI)**
EROI. *See* **energy return on investment (EROI)**
erosion control, 363, *363*
ESI turbines, 70–71, *72,* 100, 110, 130
Esperance (Australia), 279
estimating performance, 251–264; fleet performance and, 262–263; manufacturers' estimates and, 261–262; power curve method of, 258–261; power plants losses and, 262; swept area method of, 252–258
ethical energy policy, 525–528
Eurogate, turbine in, 470–471
Europe: aviation obstruction markings in, 327–328; community wind in, *480,* 480–482, 486–498, 508; compatible land uses in, 365, *371;* energy consumption in, 528; FITs in, 471, 522; land area required, 354–356; lease pooling in, 469; noise and, 341; offshore and near shore wind and, 511–513; siting in, *317,* 318, 330–331; wind in, 226, 236, 243 *See also by country or city*
Evans, John, 297
EVs. *See* electric vehicles (EVs)
exceedance level of noise, *333,* 333–334
excitation. *See* self-excitation
ExPlace. *See* Canadian National Exposition (CNE)
exposure, siting and, 318, *318*

fall-arresting systems, 418–419, 423–424, *426,* 426–427, *429,* 430, 436
Fællesmølle (Denmark), 486–488
fall-protection, 422–427
false precision, avoiding, 19–21
fantails, 84, *85*
fantasy (fraud) turbines, 451–452
farm windmills. *See* windmills
"fast runners," 28, 166
Fatal Light Awareness Program (FLAP) (ON), 349
Fawcett, Harold and Marjorie, 506
Fayette Manufacturing, 195–196, 214, 451
Fedderson, Hans-Detlef, 491–495, *492*
feed-in tariffs (FITs), 36, 471, 475–477; challenge of, 521–525, 528; community wind and, 502–504, 508–509

Fehmarn (Germany) repowering project, 491
fencing, 326–327
Fesa (Germany), 489–490
Feuchtwang, Julian, 94
fiberglass blades, 44, 63, *104,* 104–105
fire protection, storage and, 310–311
first rotation of wind turbines, 438–439
FITs. *See* feed-in tariffs (FITs)
fixed-pitch rotors/turbines, 91, *106;* H-rotor VAWTs and, 155–162
FLAIR prototype, 96
flange: Hütter, 63–64, 66, 433; installation and, *409*
Flettner rotors, 186–190, *189*
Flettner Ventilator, 188
FloDesign (Ogin), 180–182, *181*
FloWind turbines, 142, 169; failure of, 165, 170; success of, *147,* 147–150, *148, 150,* 163
F. L. Smidth turbines, *40,* 40–42, *41*
flyball or Watt governor, 126, *126*
Folkecenter for Renewable Energy (Denmark): history collection of, 63, 67, *133;* test site at, *456;* turbine development by, 64; turbines at, 67, *67, 428*
folkehøjskole movement, 32, 34–35
Forecast turbines, 147
Forest City (IA), *7,* 8
fossil-fuel-fired generation, offsetting, 515, *515*
foundations: dismantling of, 402, *413*–414; installation and, 375–381, *408;* nomenclature and, *18*
Fox Chapel Township (PA), 316–317
Fraenkel, Peter, 313
France: compatible land uses in, *368, 369, 370;* evolution of wind energy and, 30–31, 36, 47–49; FITs in, 524–525; lease pooling in, 469
fraud. *See* marketing, aggressive and misleading
Fraunhofer Institute for Wind Energy (Germany), 198, 300
Freere, Peter, 298
free standing towers, 207–208; installation of, 375, 380, 385–388, 401; safety and, 429; space and, 221–222
friction, 238; windmills and, 304–305, *305*
Friedrich-Wilhelm-Lübke-Koog (Germany), 491–496
Friends of the Earth, 351, 490
Fritzche, Andrew, 328
frost heaves, 376
Frye, Hubert, 496
Fukushima nuclear power plant, *295,* 295–296
Full-load hours, 199–201
full-span pitch control, *128,* 129
furling, *122,* 122–125, *123,* 430
furling speed, 258

Gaia turbines, 86, 106, 110, 121, 134, 199
Gasch, Robert, 88, 171, 451
gearboxes, 108, 111–112; maintenance of, 441; noise and, 332, 338, *341,* 342; pumping water and, 305

Gedser turbines, *46, 46–47*, 60–61, 63, 76
Geesman, John, 522
General Electric (GE) turbines, 193, *193, 194,* 195, 258, 264; cost and, 453–454; noise and, 331–332, 340–341; towers and, 221, *221*
generating hours for Raleigh distribution, *121*
generation of wind energy, 1–3, *2*
generator ratings, 192–193, 450–451
generators, 114–121; backup, 275–276, 442; interconnection and, 285; wiring and, 394 *See also* power ratings
Genossenschaft, 496–497
Gerath, Ruth, 347
German Environment Agency (UBA), 331, 519–520
Germania Molen turbine, *92*
German Wind Turbine Owners' Association, 79
Germany: aviation obstruction markings in, 328; community acceptance and, 483; community wind in, *480,* 480–481, 485, 489–497, 507; compatible land uses in, 365–366, *368;* electricity rebels in, 489–497; evolution of wind energy and, 36, 38–40, 42–45, *43;* FITs in, 522–523, 524–525; grid integration and, 294–297, *299,* 299–300; interconnection and, 293; land required for, 354–356, *355;* large-diameter turbines in, 450; lease pooling in, 468–469; performance claims in, 462; post wind rush market in, 77–78; removal bonds in, 359; renewables and, 298–299, 526; shadow flicker and, 346–347; silent wind revolution and, 198; siting in, 318, 319, 331, *331,* 519–520; top-down R&D program in, *57,* 57–58, 75, 77, 80; winds in, 237, 243
ghosting, 346
gin poles, installation and, 384–385, 389–390, *390,* 404
Ginsberg, Steve, 366
Gipe, Paul, 60, 179, *423*
giromills, *152,* 152–155, *153, 156*
Gjerding, Jens, 64
Global Energy Concepts, 173
gloves and boots, 430
golden eagles, 348–350
Golding, E. W., 229
Google, 183–184
Gore, Al, 515, 517
Göttingen, 412
Gougeon blades, 74, 102
grazing and turbines, 365, *367*
Great Britain. *See* Britain
Great Plains: irrigation in, 312; small wind turbines in, 9, 31, 60; towers and, 203, 239; water pumping windmills and, 301–303; wind shear and, 242
Green, Jim, 327
Greenbird (Sailrocket), 186
Green Energy and Green Economy Act (ON), 502
Greensburg (KS), 519
grid, off-the. *See* off-the-grid power systems

grid integration, 293–300
Griggs-Putnam index of Deformity, *245*
griphoists, *391,* 391–392, *392,* 394, 403–405
grounding, 401
Grove-Nielsen, Erik, 64–66, **79**
Growian turbines, *57,* 57–58, 77, 96
Grumman Aerospace, 180, 183, 211
Grundtvig, N. F. S., 32, 526
Guillet, Jérôme, 297, 523
Gusty Gizmo, *498*
Gutowski, Robert, 441
guy cables, 381–382; safety and, 431–432, 435–436; VAWTs and, 140, 145, 165, 170
guyed towers, 208, 212–215, *213, 214;* anchors and, 375, *375,* 378–380; assembly and erection of, 381–385; guy radius and, 214, *214;* maintenance and, 222; safety and, 431; space and, 221–222, 319; tilt-up, 390–391; VAWTs and, 140, 145, 165, 170

Halladay rosette or umbrella mill, *123, 123,* 302
Haller, Mark, 220
Hamilton Standard WTS-4 turbines, *56,* 56–57
Hammett, Bruce, 288–289
Hansen, Martin, 182
hard hats, 430, 431
harmonics, 290
harnesses, full-body, *423,* 423–424, 436
Harrison, H. C., 415
HAWTs, compared to VAWTs, 141, 163, *168, 169*
hazards and safety, 418–422
H-configuration (straight-blade) Darrieus turbines, 151–162
head, pumping water and, 304–305
headless horsemen, 361, *361*
health and noise, 344–345
Health Canada, 342–343, 345
heating, 11, *11,* 13
height, wind measurement and, 238–242, *240, 241,* 246
Helical Wind Turbines, 160–162, *161*
Henderson, Geoff, 180
Hepburn Wind (Australia), 498–499, *499*
Herborg Vindkraft (HVK), 66, 143
Heronemus, Bill, 511–512
Herr, Jonathan, 324–325
Heymann, Matthias, 39–40, 76, 77, 79
high-speed electrified rail systems, 517
hikers in wind turbine sites, 366–367, *369*
Hinesburg (VT), *9*
hinged (tilt-up) towers, *210,* 213, *213;* installation of, 381, 383, *388,* 388–394, 389, *389, 393;* maintenance and, 222; safety and, 422, 432–433, 435
history of wind energy, 27–81; aerospace industry and, 75–76; American early 1980s designs and, 69–75; beginning of, 29–35; boom and bust survivors in, 77–78; bottom-up strategy in, 77–80; Denmark's rebels and, 59–68, 81; design standards and, 80–81; energy

crisis of 1970s and, 51–52, 68–69; importance of, 52; increase in power and, 28–29; in interwar years, 35–37; modern era and, 81; in postwar years, 42–49; small scale technology and, 37, 42, 54, 77; top-down strategy in, 52–59, *55,* 75; in war years, 37–42; Wind Turbine Owners' Associations and, 78–79
Holst, Henning, 469, 490–491
home light plants, 9, 265
Hoppe-Kilpper, Martin, 509
Hopwood, Bill, 317
horizontal-axis turbines, 84, *84, 86 See also* conventional turbines
horizontal furling, 123–124, *124*
household-size turbines, 23, *23;* installation of, 373–374, 380, 384, 396; longevity of, 473; maintenance of, 437–438, 441; monitoring performance and, 440; safety and, 428–429
Hozeman, Dirk, 420
hub-height, 247, 252
hubs, *18,* 105–106, *106, 107, 110,* 433
Hull high school (MA), 446, 482, 504, *505*
hunting in wind turbine sites, 366
Husum wind energy exhibition, 177–178, 491
Hütter, Ulrich: design principles of, 54, 57, 69, 75, 86, 96; flange of, 63–64, 66, 433; turbine development by, 39–40, 42–44, 199
Hvidovre Vindmøllelaug, 488, *501*
HVK turbines, 66
hybrids (wind and solar), *266,* 266–275; cabin-size, 274, *274, 275;* household-size, 275; micro, 272, *272*
hydraulics, 113
hydro-electricity, 294, *295*
Hydro-Québec, 144–145, 508
hydro systems, integrated with winds, 47
hype. *See* marketing, aggressive and misleading

ice throw, 226, *226*
IEA propagation model, 334, 337, 340
IEC classes, *196,* 196–199, *197, 198,* 257–258
Îles de la Madeleine (PQ), 144–145
incentives, 465–466
Indiana wind resource map, *264*
induction generators, 116, *116,* 119–120, *120,* 121; interconnection and, 285–287, 289; overspeed control and, 129–130
inflation and rising electrical costs, 471
information kiosks, 364, *364*
infrasound, 343–344
in-rush current, 289–290
installation, 373–411; foundations and anchors and, 375–381; of free standing towers, 385–388; of guyed towers, 381–385; parts control and, 375; of tilt-up (hinged) towers, 381, 383, 388–394, 389, *389, 393;* wiring in, 394–402
installed cost, 452, *453,* 465, *465*
insurance cost, 467

integrated drivetrains, 108–110, *109*, 112
interconnection, 281–293; community wind and, 502; dealing with utilities and, 291–292, 322, 438–439; distributed generation and, 282, *282*, 292–293; models of, 282–283; net metering and, 290–291; power quality and safety of, 286–290; self-use and, 291, *291*; technology of, 283–286, *284*
interference, radio and television, 346
Intergovernmental Panel on Climate Change (IPCC), 357, *357*
intermittent generation, 294
International Energy Agency (IEA). *See* IEA propagation model
international standard atmosphere, 232
inverters, 272–273, *274*, 285–288, *286*, *288*, 290
investing in wind energy, 26, 449–478; certification and labeling and, 456–458; cost effectiveness and, *452*, 452–454, *453*, *454*; economic factors and, 465–474; equipment costs and, 459; financial and economic models for, 463–464; power ratings and, 449–452; in small home turbines, 458–463; small turbine testing and standards and, 454–456, *456*
Iowa, 294, 363, *364*, 506, *506*
IPCC. *See* Intergovernmental Panel on Climate Change (IPCC)
irrigation and drainage, *312*, 312–313
islanding, 287
Italy, 8, 96, *319*, 328, 363, *364*, 372
Ivestitionsbank, 493, 495
IWES, 446

Jackson, Kevin, 100
Jacobs, Marcellus and Paul, 127
Jacobson, Mark, 518
Jacobs Wind Electric Company, 37, 108, 115–116, 126–127, 207, 343
Jameison, Peter, 76
Jarvis, James, 176
Jensen, Allan, 444, *444*
Jeumont, 117
Johannink, Robin, 179
Jørgensen, Karl Erik, 66
junction boxes, 400, *400*
Juul, Johannes, 44–47, *45*, *46*, 65–66, 76

Kahuka Point (HI), 227
Kansas, 519
Karnøe, Peter, 76, 77, 78, 80
Keith, David, 298
Kelley, Neil, 338–340, 342
Kelly, Mike, 224–225, 227, 443, 446
Kelvin, Lord. *See* Thomson, William
Kennedy clan (MA), 317
Kennemerwind. *See* Coöperative Windenergie Vereniging Kennemerwind
Kern County (CA), 327, 329–330, 336, 360
Ketterling, Eddie, 436
Kettler, Stephan, 496
kilowatt-hours (kWh), 19
kinetic energy of wind, 229, 231–232

King, Arlene, 344
kites, 183–187
Klemen, Mike, 384, 391, 459
Koch, Bill, 317
Komanoff, Charles, 173
Kreithen, Melvin, 350
Krogsgaard, Per, 296
Kruse, Andy, 205
Kühn, Paul, 446–447
Kuriant turbine, 64

labeling, *457*
Lackmann, Johannes, 519
La Cour, Poul, 32–35
ladders and safety, *429*, 429–430, 431, *435*
Lagerway turbines, 113, *368*, 489, *489*, *501*
Lake District (UK), 497
land: compatible uses of, 365–372; required for siting, 354–356; required in North America, 518, *518*
lanyards, 418, 422–425, *424*, 426–427, *431*, 436
large-diameter turbines, 191–202, 271, 518
large turbines, 24–25, *25*, 111–113, 135; cash-flow model and, 474–475, *475*; estimating performance of, 251, 256, *256*, 259–261; installation of, 373–374, 394, *408–411*; interconnection and, *284*, 293; longevity of, 474; maintenance of, 337–338, *442*, 443, *443*, 447, *447*; noise and, 341; safety and, 427, 433; silent wind revolution and, 191–198; siting and, 319, 325, 328; tariff calculation and, 475–476, *476*, *477*
Larson, Paula, 12
lattice towers, 212–213, *213*, 221, 375, 381, 429
Lawrence Berkeley National Laboratory (LBNL), 352–353, 447, *447*, 500
lead-acid batteries, 274
leases: cost of, 467–469, *468*; pooling and, 468–469
Le Désirade (Guadeloupe), 279
LED lighting, 266, 268
Lee, Harley, 392
Leventhall, Geoff, 335
Lewis, Mark, 520
LiDAR devices, 247–248
lifelines, 424
lift, 83, 87–90; to drag ratio, 90, *90*; VAWTs and, 138–139
lighting and reducing demand, 268
lightning protection, 400–401
Lilienthal, Peter, 271–272
Lindegaard, Martin, 519
linear arrays, 3–4, *5*
line synchronization, 287
litter and boneyards, 364, *364*
Littler, Phil, *287*
LM Glasfiber, 68
loads: blade material and, 194; circuits and, 18–19; DAWTs and, 180, 182, 183; generators and, *118*, 119–120; novel wind systems and, *177*; off-grid systems and, 266–267, 269–270, 273, 278; performance and, 251; silent

revolution and, 194, 196, 198–200; towers and, 140–141, 205–206, 207, 214, 216, 219
locking pins, 421, 430
locking snap hooks, 425, *425*
Lodge, Malcolm, 145
logos, 361, *362*
longevity and revenue, 472–474
Los Angeles Department of Water and Power, 292, 482
Lovins, Amory, 480, 514
Lovins, Hunter, 480
Lower Saxony (Germany), 496
low-frequency noise, 343–344
low wind *vs.* high wind sites, 201–202, 254
Lundsager, Per, 80
Lykkegaard Machine Works (Denmark), 34, 36
Lynetten cooperative (Denmark), *371*, 487, *487*

MacLeod, John, 504
Madaras, Julius, 188
Maegaard, Preben, 59, *63*, 79, 519
Magenn blimp, 184
magnets: neodymium, 255, 455; permanent, 114–115; safety and, 430
Magnus effect, 187–188, *189*
Mag-Wind turbine, 172, *172*, 452
Maine, *482*
maintenance of wind turbines, 337–338, 440–447, 467
Maison de l'Air (France), 178–179
Makani's flying wind turbine, 183–184
Maltese Falcon (ship), 186
Mangan, Mike, 480
MAN turbines, *57*, 58
manufacturers, evaluating, 461
manufacturing capacity in North America, 517–518, *518*
Manwell, Jim, 182
Mariah Windspire turbines, 155, *159*, 159–160, 165, 171
marketing, aggressive and misleading: DAWTs and, 179, 181–182; investing in wind energy and, 451–452; public relations puffery and, 513; rooftop mounting and, 215; VAWTs and, 148, 158–159, 160, 163–164, 171
Marlec turbines, 122, 272, *272*, 343
Massachusetts, *11*, *482*, 504–505
Massachusetts Clean Energy Center, 353–354
Massachusetts's Water Resource Authority, 505
mass of wind, 229
maximum design wind speed, 461
Mayhew, Mark, 173
Mays, Ian, 156–157
McAlpine, Sir Robert, 156–157, *157*
McCartney, Tim, 415, 418–419
McDonnell Aircraft's giromill, 153–154, *154*, 166, 170, 455
M'Chigeeng First Nation, *502*, 502–503
McMansions, 527
McNeal, Ty, 428
measurement, units of, 21
measurement of wind, 225–250; defining wind and, 226–228, *227*; height and,

238–242, *240, 241;* instruments for, 246–249; power in the wind and, 229–238; published data and, 242–244, 273; surveying and, 244–250; wind distributions and, 235–238; wind speed and power and, 234–235; wind speed and time and, 228–229, *229,* 231
mechanical brakes, *130,* 130–131
mechanical governors, 129
mechanical noise, 338–340, *340,* 341–342
medium-size turbines, 24, *25,* 110–111; installation of, 373–374, 394; interconnection and, *287;* irrigation and, 312; safety and, 435; silent wind revolution and, 197–199, *198;* work platforms and, 427
Mehrkam, Terry, 74–75, 131, 415, 418–420, *419*
Messerschmidt-Bölkow-Blohm turbines, 95–96
metal blades, 102–104, *103*
method of bins, 259
Metropolitan Edison (PA), 291
micro and mini turbines, 22, *23;* direct drive and, 108, *109;* estimating performance and, *253,* 253–254, *254;* installation of, 373–411; longevity of, 472; maintenance of, 337, 441; towers and, 204–205, *205*
Microgeneraton Certification Scheme (MCS) (UK), 457
Middelgrunden Vindmøllelaug (Denmark), 488, *488, 501,* 512
Miller, Tad, 12
mini turbines. *See* micro and mini turbines
Minnesota, 37, 505–506, 507
Minwind l & ll, *501*
Mitsubishi, 129, 445
Mod-0 and Mod-0A turbines, 52–53, *53,* 77, 103
Mod-1 and Mod-2 turbines, 52–54, 77, *103,* 239
Mod-5b turbines, 54, *55,* 112
Møller, Torgny, 60, 79
Molly, Jens-Peter, 202
monitoring performance, 439–440, 458
Monopteros turbines, 95–96, *96*
mortality rate of wind, 417–418, *418*
Mother Earth Renewable Energy Project (ON), 502–503
Mott, Lawrence, 278
mountain-valley winds, 226–227, *227*
Mount Washington (NH), 227
MP5 turbines, 96, *96*
MS Beluga SkySails (ship), 186
Multibrid, 113
multimeter, 434
Musgrove, Peter, 89, 142, *156,* 156–157, *157*
MV Michael A. (ship), *188*

nacelle, 18, *18;* aesthetics of, 221; dismantling of, *412;* Enercon, 78, *414;* hazards of, 420–421; installation of, *410*
NASA, 52–57, 142
National Electrical Code (NEC), 394, 430, 556

National Renewable Energy Laboratory (NREL): data of, 273, 337, 518; land area required and, *354,* 354–355; testing by, 171; water consumption and, 358; wind measurement by, 249, 264
near shore wind, 512
Nebraska, 507
NEC-Micon turbines, 102
Nedwind turbines, 97, 155
Nelson, Greg, *288*
Nelson, Vaughn, 94, 192, 303, *313*
Neoliberalism, 527
Netherlands: community acceptance and, 483; community wind in, 481, *481,* 488–489; compatible land uses in, 365–366, *369;* cooperatves in, 488–489; linear arrays in, 3, *5;* noise research and, 342; siting and, *317,* 330; water pumping and, 307, 313, *313*
net metering: breaking free from, 283; as income, 470–471; interconnection and, *282,* 282–283, 290–291
Nevada (IA), 506, *506*
Newfoundland, 279
New Zealand, 179–181, *324, 367, 430,* 451
Neyrpic Company, 48–49
Nibe turbines, *58,* 58–59
Nielsen, Frode Birk, 488
Nierenberg, Ron, 242
Nies, Nancy, 212
NIMBY, 527
Nissan Leaf, 516
NIVE (Nordvestjysk Institut for Vedvarende Energi) (Denmark), 64, 65, *65*
Nixon, David, *272*
nocturnal jet, 242
Noël, Jean-Marc, 47
noise, 331–346; ambient, 335–336; annoyance of, 53, 86, 342–344; community standards and, 336, *336;* components of, 332–334; consequences of, 345–346; estimating levels of, *340,* 340–341, *341,* 461; health and, 344–345; lowering of, 341–342; overspeed control and, 124, *133;* propagation of, 334–335, *335;* sound power levels and, 337–338, *339;* sources of wind turbine, 338–340, 343; tip-speed ratio and, 95, 338, 340, *340,* 341, 343; urban legends about, 320–321, 343; VAWTs and, 167
noise measurement standards, 456
nomenclature: of power curve method, 258; of pumping water, *304;* of turbines, 17–19, *18*
Nordex turbines, 193, *193,* 412, 470–471
Nordfreisland (Germany), 494, 512
Nordic Wind Power, 97
Nordtank, 67–68
North America: challenge of, 513–521; community wind in, 499–506, 507–509; energy consumption of, 514–515; FITs and, 525 *See also* by state, province or city; Canada; United States

North Dakota, 12
Northern Power Systems turbines: drivetrains, 111, *111;* generators and, 116; off-the-grid systems and, 276; overspeed control and, 124–125, *125, 133, 133;* safety and, 429; silent wind revolution and, 197, 199; towers and, 219
Northfield (MN), 505–506, *506*
Nova Scotia, 503–504
novel wind systems, 175–190; Airborne Wind Energy Systems (AWES) as, 183–185; ducted or augmented turbines (DAWTs) as, 176–183; wind ships as, 185–190
Noyes, LaVerne, 302–303
NRC (National Research Council) (Canada), 140, 142
NREL. *See* National Renewable Energy Laboratory (NREL)
nuclear energy: Denmark and, 59; energy crisis of 1970s and, 51; Germany and, 490; heat wave in Europe and, 514; Sweden and, 59; United States and, 512, 513; variability and, 295–296, 298
Nybroe, Claus, 441, 446

Obermeier, John, 77
O'Brock, Ken, 218, 312
obstructions, towers and, 319, *319*
Occupational Safety and Health Administration (OSHA), 424
Odenwald cooperative, 497
odometer, 247
offshore and near shore wind, 511–513, *512*
off-the-grid power systems, *265,* 265–279; distributed wind and, 8–10, *9;* economics of, 276; hybrids and, *266,* 266–275; return on investment of, *268;* start up of, 439; telecommunications and, 276–277, *277;* village electrification and, *277,* 277–278; wind-diesel twinning and, 278–279, *279*
Ogin, 180–182, *185*
oil embargoes, 51–52, 68–69, 471
oil for transport, offsetting, *516,* 516–517, *517*
oil generation, wind *vs.,* 520, *520*
Økær Vind Energi, 64–66, 68, 78
Oliver, Dew, 176
omnidirectional VAWTs, 164
onboard battery charging, 10, *10*
one-seventh power law, 239–240, 242
Ontario, 500–503, 507
Ontario Power Authority (OPA), 503
Ontario Sustainability Energy Association (OSEA), 502, 507
operation of wind turbines: cost of, 467; first rotation and, 438–439; history of, 459–460; monitoring performance and, 439–440
Oregon, *328,* 344–345
orientation of turbines, 84–87, *86*
Orkney Islands (UK), 47
OSEA. *See* Ontario Sustainability Energy Association (OSEA)
overspeed control, 121–134;

aerodynamic controls and, 131, *131*; aerodynamic stall and, 129–130; brakes and, 130–134; changing blade pitch and, 126–129; coning and, 125–126; furling, 122–125; product specification and, 461; self-destruction and, 420; VAWTs and, 169–170
Oxford University (UK), 36–37

Pacific Gas and Electric (PG&E), 8, 54, 239, 268, 296, 348
packer head, 305
PacWind turbines, 155
Paderborn (Germany), 469, *501*
painting, *62,* 68, *211,* 222, 444, *445,* 445–446
Palm Springs (CA), *6, 97,* 220, 332
panaceas, lure of, 511–513
panemones, *87, 142,* 176
Park, Jack, 206, 235–236, 438
parking brakes, 430
Parris-Dunn wind charger, 124–125, *125,* 203
passive yaw, 85–86, *86*
Paulsen, Wolfgang, 479, 491, 508, 512
payback, 464
Peace Energy Cooperative (Canada), 507
peak power, 258
pendant conductors, 394
penetration of wind, 298
performance: estimating (see estimating performance); monitoring (see monitoring performance); paying for, 466
performance testing standards, 456
Perry, Thomas, *93,* 302–303, 305
Pesch, Josef, *489,* 489–491
phase angle, 288
Phillips, Derek, 183
photovoltaic (PVs) arrays: community wind and, 493, 498; off-the-grid systems and, 266, 269–271, *271;* resources on, 271
piers, installation and, 375–378, *378,* 380–381
Piggott, Hugh: aerodynamics and, 94, 100; generators and, 118; installation and, 382–383, 392, 395, 397, 398; investing and, 472; off-the-grid systems and, 272; overspeed control and, 122, 124; safety and, 430, 431, 434; tail vanes and, 85; towers and, 205; workshops of, 24, *24*
pilot vane, *123,* 123–124
Pinson Cycloturbine, 154–155, *155,* 170
pipe towers, *205,* 213, *213*
pitch, 333
pitchable blade tips, 45, 47, 66, *133,* 133–134
pitch control systems, 47, 63, 91, 126–129
pitch weights, 127, *127*
pitfalls, 511–513
Pittsburgh (PA), 267
plant factor, 194
pneumatics, 113–114
pole towers, 210, *210*
Porsche, 39
Portugal, 294
power: generators and, 114; quality of, 286–290; true and apparent, 288–289; *vs.* energy, 18–19; in watts, kilowatts, or megawatts, 19; in wind, 229–238; wind speed and, 234–235
power cable routing, 319, 322
power curve method, 258–261, *259, 260*
power density, 231, 235, 236, *236,* 238, 252, *532*
power electronics, 118–119
power factor, 288–289
power form, 460
power law, 239
power lines, 362, *362*
power plants, wind, 3–7, 4, *5–6,* 34, *282,* 283
power purchase agreements (PPAs), 471
power ratings, 28–29; conventional turbines *vs.* DAWTs and, 182; cost effectiveness and, 449–454; manipulation of, 148–149, 195–196; stall regulation and, 45 *See also* generator ratings
Powers, Jon, 162, 420, 425
Prandtl, Ludwig, 36
pressure tanks, 311, *311*
Preus, Robert, 165, 169, 170
Prius, 527
Production Tax Credit (PTC), 466
productivity, metrics of, 193–194
product specifications, *460,* 460–461
profitability index method, 464, 475–476, *476, 477, 478*
Prölss, Wilhelm, 186
promotions. *See* marketing, aggressive and misleading
property values, 352–354, *354*
Proven Engineering, 126
Public Utility Regulatory Policies Act (PURPA), 522
pumping water, *13,* 13–14, 301–314; capacity and, 305–307, *306, 307;* counterbalancing and, 307–308; electrical wind pumps and, 308–310; on the Great Plains, 301–303; irrigation and drainage and, 312–313; mechanical wind pumps and, 304–305, 307; nomenclature of, *304;* storage and, 310–312; worldwide data on, *304*
PURPA. *See* Public Utility Regulatory Policies Act (PURPA)
Putnam, Palmer, 38, 86
PVC (polyvinyl chloride) conduit, 396
PV systems. *See* photovoltaic (PVs) arrays
pyramidal power, 172

Quebec, 508
Quiet Revolution turbines, 160–162, *161,* 165, 171

radios, hand-held, 431
Raiffeisen, Friedrich Wilhelm, 484
rails, safety and, 424, 426, 430
rail systems, high-speed electrified, 517, *517*
Ramea Island (Newfoundland), 279, *279*
Ranji, Raj, 142
Rapin, Marc, 47
rated speed and power, 258
Rave, Klaus, 493, 495
Rayleigh distribution, 235–238, *236,* 261

reactive power, 289
rebar cage, 377–378, *408, 413–414*
rectilinear arrays, 4, *6*
red flags, 184
regional winds, 228, 236
Regionwind Freiamt project (Germany), 490, *501*
Regionwind Freiburg project (Germany), 490, *501*
regulations: costs of, *297,* 297–298, *298;* siting and, 328–329; subsidies and, 465–466
reliability of wind, 294–296
remote sites, 8–10, *9*
removal bonds, 358–360
renewable energy: demand for, 2; Denmark and, 298, 519, 526; Germany and, 298–299, 526; in Kansas, 519; vision of 100%, 518–520
Renewable Energy Policy Project (REPP), 352
Renewable Energy Sources Act (Germany), 78
Renewable Portfolio Standards (RPS), 504
RenewableUK, 353
repowering, 359, 402, 491, 495
Residential Renewable Energy Tax Credit, 466
resource maps: Indiana wind, *264;* North American wind, 249
revegetation of site, 363, *363*
revenue, wind project, 470–474
Ride the Wind! program (AB), 517
ridges and wind speed, 227, *227*
ridgetop arrays, 4, *6*
Riisager, Christian, 60–61
Riisager turbines, *60,* 60–61, 65
Risø National Laboratory (Denmark), 59, 63, 76, 80, 129, 143
Riva Calzoni, 96, *96, 97*
Robertson, Robert, 415
Rocky Flats, 228
Rogier, Etienne, 32, 47, 166
Romani, Lucien, 47–48
Romanowitz, Hal, 70
rooftop turbines: DAWTs and, 178, *178;* mounting and, 215–218, *216, 217;* VAWTs and, 145, 157–159, *158*
Rothschild, Liz, *498*
rotors: bearingless, 129; controlling, 459; diameter of, 22, 43, 234, 252, 254, 450–453; dismantling of, *412;* hazard of spinning, 419–421; installation of, *410–411;* locking pins for, 430; maintenance and, 444–445; noise and, 335, *335, 338,* 338–341; orientation of, *84,* 84–87; self-starting, 100–101; as term, 17–18, *18 See also* blades; overspeed control
rotor speed, 28, *118,* 118–119, 121, 461
Rowan, James Alan, 172
Royal Society for the Protection of Birds (UK), 350–351
royalties, cost of, 467–469, *468*
rpm (revolution per minute), 28
Rural Community-Based Development Act (NE), 507
Russia, 37, 220

Rutland 913F, 122
RWE, 160–161

Sacred Winds (ND), 12, *12*
safety, 415–436; blade root doors and, 433; electricity and, 421, 433–435; fall protection and, 422–427; fatal accidents and, 415–417; gears and, 430–431; hazards and, 418–422; interconnection and, 286–290; ladders and, 429–430, 431; mortality rate and, 417–418; siting and, 325–331; unauthorized access and, 435–436; work platforms and, 427–429
Sagan, Carl, 175
Sagrillo, Mick: aerodynamics and, 94; blades and, 101–102; height estimation and, 246; installation and, 382, 391, 397; investing and, 455, 472; noise and, 332, 343; off-the-grid systems and, 271; operation and maintenance and, 437, 440, 441; overspeed control and, 129; safety and, 425, 430, 431; towers and, 204, 215, 219; VAWTs and, 162, 164, 171, 173; water pumping and, 310; wind data and, 246, 247; workshops of, 24, *24*, 374
sail power, 185–186, *186*
Salmon, Jim, 145, 232, 392
salvage value, 359
San Bernardino Mountains pass (CA), 227
Sandia National Laboratories, 139, 142, 146, 168–169, 273
San Gorgonio Pass (CA), 227, 236, 331, 349, 366
San Joaquin Valley (CA), 227, 349
Santa Ana winds, 228
Saterland Bürgerbeteiligung, 496
Saulnier, Bernard, 201
Savonius rotor, *88*, 88–89, 138–139, *139*, 164, *164*
Saxen, Bernhard, 420
scale of energy equivalents, *531*
scale of equivalent power, *531*
Schacle, Charles, 220
Schenzle, Peter, 189
Schleswig-Holstein (Germany), 493–495, *494*
Scotland, 29–30, 47
screw anchor installation, 378–380, *379*, *403–404*
Seaforth Energy, 74
SEAS, 44–46
seas breezes, 226
self-excitation, 287–289
self-generation. *See* net metering
self-starting rotors, 100–101, 144, 170
Selsam, Doug, 461
Sencenbaugh, Jim, 108, 401
series production, 48–49
setback requirements, 328–331, *329*, *330*, 331, 345
shadow flicker, 346–347
Sharman, David, 218, 451
shear strength of soils, 376, 379–380
Sherwin, Bob, 130, 133
shipping, 185–190, *186*, *187*, *190*
Siemans turbines, 111, *112*, 113, *341*, 342, 498, *498*, 512

Sierra Club, *325*, 348, 350, 366
silent wind revolution, 191–202; large-diameter turbines and, 191–198, *192*; small and medium-size turbines and, 197–199, *198*
silos as towers, 218
Silver Eagle turbines, 12
Sinclair, Karin, 350
siting, 315–372; aesthetics of, 221, 346, 360–365; building permits for, 322–325; community concerns and, 315–317; compatible land uses and, 365–372; disco effect and, 347; erosion control and revegetation of, 363; land area required for, 354–356; noise and, 331–346; planning permission and, 322; property values and, 352–354, *354*; radio interference and, 345; removal bonds and, 358–360; safety and, 325–331; shadow flicker and, 345–347; tower placement and, 317–319, 322; urban legends about, 320–321
size, turbine: classes of, *22*; off-the-grid systems and, 270–272; relativity of, 21–25, *22*
SkySails, 186–187, *188*
Skystream turbines, 86, 216, *216*
sleeves, cable and, *426*, 426–427
slings, 425–426
Slocum, Erik, 432
small turbines, 24, *24*, 134–135; aesthetics and, 365; buying of, 458–463; drivetrains and, 108–110; estimating performance of, 251, *253*, 253–256, 259–260; first rotation of, 438–439; installation of, 373–411; interconnection and, 285, 293; longevity of, *473*, 473–474; maintenance of, 437–438, 440–443, 446–447; monitoring performance of, 440; noise and, *339*, 341, 343; safety and, 427–428, 433–435; silent wind revolution and, 197–199, *198*; siting and, 319, 328; tariff calculation and, *476*, 476–477, *478*; testing and standards for, 454–456; towers and, 203, 327; wiring and, 394
Small Wind Certification Council (SWCC), 337, 338, 341, 457
small wind tariff, 523–524
Smeaton, John, 88
Smedemestermølle turbine, 64, *65*
Smith, Don, 11, 296
Smith-Putnam turbine, *37*, 37–38, *38*
snap hooks, 425, *425*
snap links, 425
SODAR devices, 247–248, *248*
soil, 376, 379–380
solar energy, 266, 273, 299
solidity, 93–94
Soria test field (Spain), 455
sound power levels, *337*, 337–338
sound pressure levels, *332*, 332–333, *339*, 341
South, Peter, 142
Southern California Edison, 113, 146, 220
Southwest Windpower (SWP), 115; blades and, *104*; generators and, 117;

interconnection and, 285–286; noise and, 331; overspeed control, 122; repairs and, 462; rooftop mounting and, 215, *216*; towers and, 205
Spain, 294–295, 455
specific area, 194–195, *195*, 201–202
specific power (capacity), 194–195, *195*, *199*, 199–202
speed of wind: average, 228, *229*, 245, 249; avoiding confusion of average, 259; biological indicators of, *245*, 245–246; calculators for measuring, 244; distribution of, 235–238, *236*, 252–253; estimating performance and, 252; height and, 238–242, *240*; power and, 234–235; surveying, 244–250; time and, 228–229, *229*, 231; turbulence and, *318*; units of, 228
Spiddle Hill (NS), *503*, 503–504
split bolts, 395, 397
Spohn, Jim, *426*
spoilers, 131, 133
Spratley, Bill, 511
spud wrenches, 386
squirrels in a cage turbine, 138, *139*, 461
stall, 90, 126, 129–130, 169–170
stall regulation, 44–45, 47
Starcher, Ken, 374, *423*, 472
Starrs, Tom, 466
start-up speed, 258
steel pipe towers, 218
stepped tubular tower, *18*
Stern, Walter, 188
Stevenson, David, 503
StGW-34 turbine, 43–44
Stiesdal, Henrik, 64, 66, 67–68, 79, 101; noise and, 341–342
Stoddard, Forrest (Woody), 76–77, 89
stop switches, 430
storage: grid integration and, 299–300; water pumping and, 310–312, *311*
Storm Master turbines, 70–71, *71*, 86
St. Paul Island (AK), 278–279
straight-blade Darrieus turbines, 151–155
strain relief, 396–397, *405*
Strauss, Lewis, 513
stromrebellen, 489–497
struts and stays, 105–106
Struzik, Cynthia, 348
stub tower, 222
subsidies and incentives, 465–467
sucker rod, 305, 307
Summerville, Brent, 472
surge protection, 400–401
surveying wind speed, 244–250
Swan, Tony, *369*
SWCC. *See* Small Wind Certification Council (SWCC)
swept area, 83, *84*, 192–195; calculating, *255*; estimating performance and, 252–258; North American consumption and, 514–515, *515*; power and, 234, *234*, 450, 452–453; rules of thumb of, 257
synthetic natural gas, 300

tables, usefulness of, 19
Tacke turbines, *130*
Taibossigai, Grant, 503

tail vanes, 84–85, *85,* 123, 125
Tangler, Jim, 168
tariffs, 471; calculation of, 475–477; differentiated, 464, 524–525; small wind, 523–524 *See also* feed-in tariffs (FITs)
tax brackets, 470
tax credits, 466, 503, 507, 508
taxes, investing and, 470
Taylor, Derek, 329
technological development: aerospace industry and, 75–76; American early 1980s designs and, 69–75; boom and bust survivors in, 77–78; bottom-up strategy in, 77–80; California wind rush and, 68–69; centralized *vs.* incremental, 51–52, 54, 77, 80; "Danish concept" and, 68; Denmark's rebels and, 59–68; design philosophy of US and Danes, 69–72, *70, 71,* 75–77, 80–81; modern era and, 81; small scale, 37, 42, 54, 77; standardized parts and, 68; top-down strategy in, 52–59, *53, 54, 55,* 75; Wind Turbine Owners' Associations and, 78–79
teetering hub, 106, *107*
Tehachapi Pass (CA): bird fatalities in, 349; erosion and, *363;* grazing and, 365; hikers and, 366–367, *369;* interference and, 346; safety and, 327–329, *329;* turbines in, 3, *10,* 99, 122, 147, *147,* 221, *325;* wind in, 102, 227–229, 232–233, 242, 243, 273
telecommunications, wind-powered, 276–277, *277,* 346
temperature and air density, 232
temperature inversions, 228
Templin, Jack, 142
Ten Mile Lagoon (Australia), 279
Tera Kora (Curaçao) wind plant, 237
terminations, good, 394–395
test fields: absence of, 455; ASU, 472; Atlantic, *146,* 158; Bushland, *145,* 145–146, 459, *473;* Château de Lastours, *389;* Danish, 59, *59,* 64, 220, *341,* 456; Oxford University, 36–37; Risø, 80; Wulf, 253–254, 273, *340,* 458, 472
tests, standardized, 455–456
Test Station for Small Windmills (Risø) (Denmark), 80
Texas Panhandle, *145,* 308–310, *309,* 312, *313*
thimbles, *382,* 382–383
Thomsen, Lief, 421
Thomson, William (Lord Kelvin), 27, 29–30
thrust, 214
thrust, defined, 206
Thy Møllen turbines, 110, 134
tilt-up towers. *See* hinged (tilt-up) towers
tip brace, *18*
tip brakes, 131, *132,* 133
tip-speed ratio, 28, 37, 91, 94–95, *95;* Hütter on, 40, 44; noise and, 95, 338, *340, 341,* 343
tip vanes, 101, *101*
Tjæreborg (Denmark) test site, 59, *59,* 64

Toke, David, 497
tool belts, 431
top-down design strategy, 52–57, *53, 54, 55*
Tope, Josephine, 506
Toronto (ON), 7, 349, *368,* 500–501
Toronto Hydro, 500, *501*
Toronto Renewable Energy Cooperative (TREC), 157–158, 349, 500–501
torque: blade area and, 93; on bolts, 443, *443*
total dynamic head, 304–305, *305*
tower adapter, 222
towers, 18, 203–223; access to, 222–223, *223,* 435–436; aesthetics of, 221; buckling strength of, 205–207; height of, 203–205, 239; installation of, 222, 375; maintenance and, 222, 445, *445;* Nordtank and, 68; rooftop mounting and, 215–218; safety and, 422–427; siting of, 317–319, 322; space and, 221–222; types of, *207,* 207–215; unconventional, 218–220; VAWTs and, 140–141
traction kites, 186–187, *188*
Tranæs, Flemming, 79
tranformers, interconnection and, 289–290
transmissions, 111–114, 441
TREC. *See* Toronto Renewable Energy Cooperative (TREC)
trees as towers, 219
tripod tower and platform, 220, *220*
truck testing, 455
truss towers, 206, 207–208, *208,* 219, 221; installation of, 380, 385–386, 389; maintenance and, 222, 443
tubular towers, 206, 207–208, *209,* 210, 221; dismantling of, *412–413;* guyed, *214,* 215; installation of, 380–381, *387,* 387–388, *388,* 390; maintenance and, 222; safety and, 429, 436
turbine envy, 469
turbines: Brush Dynamo as, *31;* conical wind catcher as, 33, *33;* definition of, 192; downwind dominant, 69–70; first interconnected, 35–36; horizontal-axis, 84, *84, 85;* increase in power of, *28,* 28–29, *29;* in interwar years, 35–37; large-diameter, 191–202; orientation of, 84–87; productivity metrics and, 193–194; size of rotor and, 114; as term, 17; at turn of 19th century, *30;* in war years, 37, 37–42, *38, 40, 41;* wind stream area of, 83, *84;* worldwide operation of, 2–3 *See also* by individual names; conventional turbines; large turbines; medium-size turbines; small turbines; vertical-axis turbines (VAWTs)
turbulence, siting and, *318,* 318–319
Turby turbines, 160, 166
turnbuckles, *383,* 383–384, 432, *433,* 435
tvindkraft, *61,* 61–64, *62,* 81, 430, 444, *444*
Twidell, John, 30
Twine, Adam, 497–498, *498,* 528
twin tails, *85*

UBA. *See* German Environment Agency (UBA)
UMass Wind Furnace, *11*
Union of Concerned Scientists (UCS), 516
United Kingdom. *See* Britain
United States: annual average wind speed in, 249; aviation obstruction markings in, 327; bird fatalities in, 348–350; certification in, 457; community wind and, 504–506; compatible land uses in, 365–366; design philosophy compared to Danes, 69–72, *70, 71,* 75–77, 80–81; designs of in the early 1980s, 69–75; evolution of wind energy and, *31,* 31–32, *37,* 37–38, *38;* FITs in, 522; geographic size and, 78–79; grassroots movement in, 60–61; interconnection in, 293; land area required, 354–356; post wind rush market in, 77–78; royalties and leases in, 468; setback requirements in, 328–230, *329;* tax credits in, 466, 477; top-down R&D program in, 52–57, 75–76 *See also* by state or city
United Technologies, 71, 125
University of Connecticut research, *353,* 353–354
University of Delaware research, 299
University of Rhode Island research, 352–353
up-front subsidies, 466
up-tower block connectors, 395, *395*
upwind rotors/turbines, 86, *86;* overspeed control and, 126, 130
Urban Green Energy (UGE), 163
urban wind, *7,* 7–8, *331, 368*
US Energy Information Administration (EIA), 294
uses of wind, current, 1–14; distributed wind and, 7–10; electric vehicle charging and, 10–11; heating and, 11, *11,* 13; pumping water and, 13–14; specialty applications and, 10; wind power plants and, 3–7
USWindpower (USW) turbines, 72–74, *73,* 77, 79
utility poles as towers, 219
UTRC (United Technologies Research Center), 71, 125, *125,* 129

Vadot, Louis, 48–49, 54; DAWTs and, 183; VAWTs and, 169–170
Vadstrup, Kai, 421
Val-Éo, 507–508
Van de Borg, Nico, 340, *340*
Van den Berg, Frits, 342–343
variability of wind, 294–300
variable-pitch hub, 106, *107*
variable pitch rotors/turbines, 91; overspeed controls and, 126, *128*
variances, legal, 323
Vawtpower turbines, 147, 169
VAWTs. *See* vertical-axis turbines (VAWTs)
Venco Twister helical VAWT, *161*
vendors, 461
ventilators, 461
Ventimotor Company, 38–40
Vergnet turbines, 116, *127,* 279, 389, *389*

Vermont, 9, 35, 37–38
vertical-axis turbines (VAWTs), 29, 84, *84*, 137–173; annual specific yield of, *169*; blade number and, 139–140; claims and counterclaims about, 164–167; compared to HAWTs, 163, 168, *168*; configurations of, *138*, 140; conspiracy against, 171, 173; design characteristics of, 167–170; fixed-pitch H-Rotor, 155–162; future of, 137, 162–164, 170–171; H-configuration (straight-blade), 151–155; lift and drag and, 138–139; Φ configuration (eggbeater) development and, *141*, 142–151, 165, 170; rooftop, 145, 158, *158*; rotor performance of, *169*; specific mass of, 167, *167*; test standards and, 171; towers and, 139–140, 140–141; web sources on, 141
vertical furling, 124–125, *125*
Vertica "salad spinner," 138, *138*
Vestas Sailrocket, 186, 190
Vestas turbines: aesthetics and, 346; community wind and, *487*, 504, *504*, 506, *506*; drivetrains and, 111, 113; estimating performance and, 260–261; generators and, 119, 121; interconnection and, *287*; large-diameter, 192–193, *193*, 195, 450; noise and, 340; off-the-grid systems and, 278–279; overspeed control and, 129, 130; safety and, *427*; siting and, *324*, 331; VAWTs and, 142–143, *143*; wind energy development and, *66*, 66–67, 68, 70
Vester Egsborg turbines, 44–45
Vick, Brian, 308, 310
village electrification, wind-powered, *277*, 277–278
village wind, 7–8
Vindmøllelaug (Denmark), 486–488
Vindtræf (wind) meetings (Denmark), 64, 79
vineyards and wind turbines, 369
VisionAIR turbine, 163
visual uniformity, *360*, 361
Voland blades, 58, *59*
voltage: conductors and, 397–398; generators and, 114–115; off-the-grid systems and, 270; safety of small turbines and, 433–435
voltage flicker, 289–290
Vortec turbines, 179–180, 451
Vosburgh, Paul, 146–147, 168
Voss, Tiffany, 504

warranties, 462
water consumption, 358, *358*
water pumping windmills. *See* pumping water
Watt, Alan, 306
Watt governor, *126*
WE-10 turbines, 42–43
Webb, Walter Prescott, 301
Wehr, Gottfried, 359, *413*
Weibull distribution, 235–238
weighing scales, noise and, 333, 343
Weimar (Germany), 39
Wenco turbine, *325*
Western Pennsylvania Conservancy, 317
Westinghouse 600 turbines, 56, *56*
Westmill Wind Farm (UK), 497–498, *498*
West Texas A&M. *See* Alternative Energy Institute (West Texas A&M)
Wheeler, Leonard, 123–124, 302
Whisper turbines, *337*, 340, 472
Williams, Wendy, 317
Wilshire, Howard, 356
Wilson, Robert, 88
Winchargers, 104, 108, 115, 203
wind: defined, 226–228, *227*; distribution of speed of, 235–238, 252–253; height and, 238–242, *240*, *241*; power in, 229–238; published data about, 242–244; speed of and power, 234–235; speed of and time, 228–229, *229*, 231; surveying speed of, 244–250; variability of, 294–300
wind calculators, 244
windchargers, 37, 60, 215
Wind Energy Collaborative, 329
Wind Energy Group (WEG), 97, *128*, 129
wind farms, 4, 72, *371*
Windflow turbines, 97, 100, 102, 106, 180, *429*
Windkraft Diemarden (Germany), 359, 412
wind machine, as term, 17
Windmatic turbines, 106, *106*, 133, 279, *279*
windmills: conversions and, 308; high-solidity, *93*, 93–94, 310; overspeed control and, 123–124; power of traditional, 28, *28*; rotor orientation and, 84; as term, 17; as towers, 218, 220; water pumping, 301–314
Windmission's Windflower rotor, *100*
wind parks, as term, 4
wind propulsion: average force of, *190*; modern, 186–190, *187*; traditional sail power and, 185–186, *186*
wind regimes and design standards, 196–197
wind revival, 51–81; aerospace industry and, 75–76; American early 1980s designs and, 69–75; boom and bust survivors in, 77–78; bottom-up strategy in, 77–80; California wind rush and, 68–69; Denmark's rebels and, 59–68, *60*, *61*, *62*; design standards and, 80–81; energy crisis of the 1970s and, 51–52, 68–69; modern era and, 81; small scale technology and, 37, 42, 54; top-down strategy in, 52–59, *53*, *54*, *55*, 75; Wind Turbine Owners' Associations and, 78–79
WindShare, *7*, *368*, *486*, 500–502, *501*, 507
wind shear, 239–242; exponent, *241*, 243
wind ships, 185–190
wind speed bin, 258
wind system, as term, 17
Windtech turbines, 70, 71, *71*, 110
wind tunnels, 33, 455
Wind Turbine Company, 97, *214*
Wind Turbine Industries, 116
wind turbines. *See* turbines
Wind World turbines, 497, *497*, 506, *506*
winglets, 101, *101*
wire nuts, 395
wire rope clips and thimbles, *382*, 382–385, *383*
wiring in installation, 394–402
Wiser, Ryan, 201, 453–454
Wobben, **Alois/Aloys,** 78, 187, 188–190
Wolfhagen (Germany), 496
wood blades, 60, 101–102, *102*
wood towers, 218–219
Woofenden, Ian, 162, 219, 425
work platforms, *18*, *427*, 427–429, *428*, *431*
workshops, 24
World Health Organization, 343, 345
Wortman, Franz, 96
Wright, Eric, 145
Wulf Test Field, 253–254, 273, 458, 472; installation and, 397, 399, *403–405*; siting and, *337*, 337–338, *340*
www.wind-works.org, 16

yaw (orientation), 85–87, *86*; brakes and locks for, 430; maintenance and, 441, 442

Zawlocki, Richard, 419
ZGF, *216*, 216–217
Zilkha Renewable Energy, 363, *364*
zone of disturbed flow, *318*
zoning appeal board, 323
zoning laws, 322

PARISIAN INTERIORS

Copyright © 2008 Filipacchi Publishing, a division
of Hachette Filipacchi Media U.S., Inc., for the present edition.
This is an updated edition of *Insider's Paris* published in 2003.

Copyright © 2003, EDITIONS FILIPACCHI, ELLE DÉCORATION, for the original French edition

ISBN: 1-933231-51-3
Library of Congress Control Number: 2008931133

Translation: Fern Malkine-Falvey
Editing: Jennifer Ditsler-Ladonne

All rights reserved. No part of this book may be reproduced or transmitted in any form or by any means, electronic or mechanical, including photocopying, recording, or by any information storage and retrieval system, without permission in writing from the publisher.

Printed in China

The stores listed in the "Strolling" sections are meant to represent the spirit of an area and are by no means comprehensive.
Addresses, telephone numbers and Web sites listed in this book are accurate at the time of publication, but they are subject to frequent change.

PARISIAN INTERIORS

Jean Demachy - François Baudot

CONTENTS

7	Preface
9	Introduction
16	From Saint-Germain-des-Prés to Montparnasse
124	From the Marais to the Madeleine
192	From the Beaux Quartiers to Montmartre
250	The Marché aux Puces and the Suburbs
276	Photographic credits
279	Acknowledgements
I-VIII	Addresses

"What is essential to French art is that it absorbs all others."
André Chastel

For centuries, Paris seems to have inhaled, absorbed and assimilated all styles…. Anyone who loves to stroll around Paris, crisscrossing the streets of the capitol (particularly those with an interest in decorating), would be fascinated to know what is happening behind the city's elegant facades. Only a team from an eminent publication like *Elle Deco* could gain access to these inner sanctums and be permitted to photograph them in stunning detail. In covering everything from family dwellings to lofts, *Elle Deco* has established a rare documentation on the manner in which Parisians actually live—from the Left Bank to the Right, from neighborhood to neighborhood.

We have mixed private residences with restaurants, hotels, as well as the shops and boutiques that actually provide the beautiful wares on view in these pages. Although we do not pretend to offer exhaustive coverage, we have collected here a wonderful selection of our favorite areas of Paris. After years of research and discovery, we are happy to include the secret, never-before-published places that capture the imagination and awaken the senses.

We were privileged to be given an insider's view of these places. Remaining true to the *Elle Deco* style, this compilation demonstrates that there is more to the art of living than just the exterior trappings of wealth and luxury—a certain style, found here, under the Parisian skies.

J. D. - F. B.

INTRODUCTION

Ah, to live in Paris! Only a Parisian could down play her mythological status in the eyes of the world, yet an amazing variety of people share the honor of residing behind the city's elegant walls. In Insider's Paris by Elle Decor, we have divided this diverse city into four major sections. The first section, From Saint-Germain-des-Prés to Montparnasse, covers the entire Left Bank, with its ancient buildings and artist's studios. The second section, From the Marais to the Madeleine, reveals the secret places from the Madeleine to what was once the ancient prison of the Bastille, destroyed so long ago. The third section, From Montmartre to the Beaux Quartiers, includes some of Paris's most exclusive neighborhoods, where the entrenched bourgeoisie lives alongside the artists who resurrected the legendary neighborhood of Montmartre. In our last section we find ourselves on the fringe of the city where the Marché-au-Puces and the suburbs draw city dwellers to their outer limits. Subdividing the city into four sections has simplified matters; and, as they say, to divide is to conquer! Parisians love to classify and analyze, and each individual district, a village unto itself, carries its own subtle references and social connotations. But rather than each of Paris's arrondissements—which start at the center at Ile de la Cité and spiral outward toward the suburbs—having its own unique style, we have found the opposite to be true; many things have changed in Paris over the last twenty years, and within each arrondissement we find some element of the diversity that characterizes the Paris of today.

It is interesting to note, however, that with all the social changes that have swept over Paris, people living side-by-side in neighboring districts continue to identify strongly with their own neighborhood. Each arrondissement generates an ambiance all its own: All good Parisians know, for example, that those living in the Palais-Royal area do not live by the same morays as those living near the Quai Voltaire, though they face each other across the Seine every day. Similarly, those living in Montparnasse will admit to knowing little if anything about the people in the Plaine Monceau right next door. This is also true for the faubourg Saint-Antoine neighborhood too, where they seem to deny that their neighbors to the west in the faubourg Saint-Honoré practically live on the same street. Everyone in Paris is a fan of her own parish. But in the end, the people who live within the city "walls" are, if somewhat reluctantly, supportive of each other, connected as they are by a deep, abiding—and mutual—love for their city. Though built on just a few acres, the entire world fixes its gaze on Paris's splendors. Neither war, nor short-sighted architects, city planners with political ambitions, modernity or even the global economy has been able to alter Paris's essential spirit. Its private homes, if not national treasures, often display distinctive, and distinguished features worthy of attention. Although this book gives only a small taste of the feast that is Paris, it nevertheless offers views rarely seen by those "outsiders" that make up most of the world's population. We offer here our modest contribution to revealing some of her hidden charms—a labor of love that surely lends support to the old French adage, "Paris is worth any sacrifice."

Sometimes it seems like everything has already been said about Paris. We have already been shown all of its beauty and have been apprised of its faults—one as numerous as the other: from the Eiffel Tower to the Jussieu skyscraper, the Butte aux Cailles to the hillside of Montmartre. Despite its infinite attractions, Paris is still, in reality, only a small city easy to cross on foot. From the immigrant section of la Goutte-d'Or to the faubourg Saint-Germain; from the Pont-du-Jour to Bercy; from the arch at Place du Carrousel to the Arc du Triomphe at l'Etoile—even if you happen to get lost in Paris, you can never really be "lost," because it is always easy to speak with Parisians. (It is especially easy for foreigners, as Parisians prefer to speak with them than to their own fellow citizens.) During the 20th century, only Paris has been as adept as New York City in welcoming, housing and in integrating such a variety of races and cultures. Perhaps the preamble to the French Constitution, "The Rights of Man...," does stand for something after all. Parisians speak every language but English, which seems to be relegated to coursework in school. Home to a wide variety of ethnic groups, all religions are practiced and every kind of lifestyle is available in Paris. And with its historical and cultural sites, its architecture and gorgeous settings, sightseers on both sides of the Seine are treated to a visual feast. Whether you're reading a long or a condensed version of the last 2,000 years of Paris history, so many people have written, sung about or critiqued Paris that many of them have, over time, been lost or omitted from successive texts. To give you an idea of how many there have been, the marvelous writer Jean Favier, of l'Institut de France, in his brilliant *Paris* (Fayard, 2001) has compiled in 1,008 pages an "incomplete" collection of these often passionate, almost forgotten narratives. No one who is interested in our capital will be left indifferent by it. About those who are eventually forgotten by history, Apollinaire wrote in 1918, the year of his premature death: "Men never leave things behind without some regret, even the places, things and people that made them the most unhappy; they do not let them go without feeling pain."

Indeed, the very fabric of Paris is made up of regrets that refuse to die, of dear ones who have departed, of masterpieces in constant danger, and of a collection of silent grievances. And yet the city as a whole, with all its terror and beauty, has learned how to transform itself over the years without renouncing what it essentially is—Paris has understood how to be middle class without being boring. We Parisians live rather well, and if we appear to protect ourselves a little too much from those distant "barbarians" from beyond the beltway, it's because our personal secrets are our surest guarantee for maintaining our individuality. How many invaluable masterpieces (a little Corot here, a portrait of Queen Marie-Antoinette there), still hang in some old, now-decrepit salon, where a doddering old coachman still guards the gates where carriages used to enter? How many potential museums exist at the end of some dank alleyway? What kinds of eccentricities must still exist behind those massive facades designed by Eugene Haussmann for the upper crust? Paris, initially built on a limited piece of terrain, has not only expanded physically, but has evolved socially as well. What caused Paris's social stratification, where each floor of a building, as Zola so aptly criticized in *Pot-Bouille*, was reserved for a particular social class; where the rich only crossed the poorer classes in the stairwell on their way to their attic rooms? Over the last 50 years, things have greatly changed. Those with more moderate incomes have been pushed out to the suburbs; the number of singles continues to rise, as does the average

age of the population in general. The perpetual bustle of the city, fashion and peoples' passions…all these are fuel for an engine that never stops pulsing with the rhythm of the times. Today, Paris is still the most desirable city in the world to live in, and is the best suited to represent the "heart" of Europe.

Another legendary aspect of Paris is the Parisian woman—a species that still stands out from the rest. Their particular kind of seduction is so well known that it needs no further comment. As for the Parisian male, he embodies all of the characteristics that people commonly associate with him without mentioning them aloud—the suavity and style that make him a Parisian male. Then there are those Parisians who are actually from Paris—there aren't that many, so they tend to brag about it. And then there's everyone else—all those who have come from the four corners of the planet to live here. These new Parisians consistently reinforce our distinctive identity, our wealth and our differences. Out of this melting pot, they have created wonders. But you are either from Paris or you're not, and you will never be. Just accept it; those who were there before you have decreed it so.

And then of course all these fine people need lodgings. Finding a place to live in Paris is no picnic. Many Parisians live and die in the same area, perhaps even in the same house. They have spent their entire lives looking out at their city without ever noticing themselves aging. Those who have found a place they like rarely move. In the really nice houses there are always "more behinds than chairs" as the saying goes. In Paris, we give each other the once-over. We criticize one another. The rites we observe are so subtle and arcane that even the most perceptive ethnologists become discouraged. Who among them, what Maeterlinck in their midst, will write about the Parisian ants, and describe their nests, their riches and their secret retreats? *Elle Decor* is happy for the opportunity just to photograph them, as though photographing the chimney smoke curling upward from the city that pilgrims coming from afar used to love to observe. Curiosity, it is said, killed the cat. But that has been our photographers' objective—to satisfy our insatiable curiosity about a legendary place. And these Thebans are not only full of energy; they're also full of ideas.

From African art to art deco, from Madame de Pompadour to the designer Madeleine Castaing, from bohemian chic to cosmopolitan sophistication…by simply strolling along the banks of the Seine, one can find curiosities of all kinds, collections of all kinds and a greater variety of treasures than one could ever hope to find in shops or bazaars anywhere else in the world: Kilim rugs, coral, crystals, silver, tapestries, religious icons…. African art, modern art, Gothic sculptures of Christ, Japanese stamps, antique treasures from every era and the recent resurgence of "modern" items from the 1960s…all this bric-à-brac, luxuries and objects one only dreams of, can be found in the windows and galleries of these labyrinthine streets that our reporters have tirelessly covered. Taking special care when exploring sites that have no equal, they have summoned the ancient god Asmodée to lift the roofs off these dwellings so we can observe, acquaint ourselves and finally understand what transpires inside these Parisian homes. For this book we set out in search of Parisians who live well—in true Parisian style—no matter what their circumstances. *Insider's Paris* is an homage to diversity, to people's fantasies and to their creative free will that, surprisingly, often manifests itself in the form of simple common sense. Even with all the diverse inclinations and never-ending alterations, a certain unity emerges. Perhaps it is what people refer to as "the spirit of Paris."

Under the Mirabeau Bridge there flows the Seine
Must I recall
Our loves recall how then
After each sorrow joy came back again
Let night come on bells end the day
The days go by me still I stay.

("*The Mirabeau Bridge*" Guillaume Apollinaire, trans. Richard Wilbur, Random House)

    Paris wasn't born yesterday; it has 2,000 years of history behind it. It has experienced wave after wave of immigrants and has evolved slowly from the time of the Gallo-Romans, culminating with the authoritarian urbanization of the city under Napoleon III by the civil engineer Eugene Haussmann. Despite the ruthless destruction that has taken place in modern-day Paris, the basic layout of the streets and the initial shape of the city has not changed a bit. This is perhaps the secret to its uniqueness and to its enduring charm. Its urbanization is based on its history, beginning with the Hôtel de Ville (Paris's City Hall) to Place de la Concorde. From there, it continues in a straight line from the Champs-Elysées to the modern complex of buildings known as la Défense; on out to the terraces of Saint-Germain-en-Laye. The Seine, in contrast, gradually meanders away from this straight line, with the city's greatest monuments embellishing both sides of its banks: the Université de Paris, the Palais des Beaux-Arts, the Grand Louvre, the Hôtel des Invalides, the Ecole Militaire and the Palais de Chaillot. Symmetrically aligned are the Assemblée Nationale and the church of the Madeleine.

    The walls that once demarcated the periphery of this magnificent city (which was then only a fortified village) were built and rebuilt over the ages, each time farther away from its center. Today there is almost nothing left of that original wall built around the city in 1190 by King Philippe Auguste. By the 19th century, the working-class areas expanded out to the exterior boulevards, which had been built over the top of the city's ancient fortifications. From that point outward, there is "la zone" where a mixture of people live. The outlying beltway, constructed under de Gaulle's presidency in the 1960s, came to represent by its circle of concrete just where the city limits were. From that point on, the city could grow only via its suburbs. The metro system, built in 1900, did not go out that far. (Originally the fortifications around Paris had doors that could be closed when necessary to keep intruders out. Today, these doors exist in name only.) Paris institutions have all developed within parameters that have not changed since the 19th century. Beyond the beltway, we find a bustling world—the working-class residential areas—in other words, the suburbs. Paris either controls things or it resists them, for in Paris all business activities are centrally operated.

    Within such a geographically restricted area, greenery is a luxury, a lawn is an event and private gardens are a rarity. Paris is populated by people who dream of wide-open spaces. The luckier ones will move out toward the western part of the city, where the land, being less populated, gives them the illusion of living in the country. At the beginning of the 20th century, far from the foul-smelling smoke that belched forth from industries developing in the eastern part of the city, the wealthier denizens got to breathe the clean air of the Bois de Boulogne. Another kind of segregation, that began at the beginning of the 20th century, came about by the invention of the elevator. Whether a building had an elevator or not came to affect its real-estate value. After the easily accessible first floor, comes all the other floors which are less so. As time passed, buildings were built higher and with more floors. During the 19th century, the city got running water and gas heat…but the maids still had to climb to their rooms under the eaves on the very top floors. In 1954, garbage chutes suddenly appeared, but by then it didn't really matter what floor you lived on; what mattered was whether you lived in the city or not. Transportation problems still existed, though, and if anything, with the snarl of traffic and metro strikes, got worse. In the past, when there was no public transportation, people in the suburbs had to leave their homes early, sometimes in the middle of the night to get to their jobs in the city on foot. This shadowy mass that we call the working class often found themselves mingling with crowds of late-night revelers.

Today, this metropolis called Paris is almost a whole new city. Around 1900, at the time of the famous doorman at the exclusive nightclub Chez Maxim's, the 200,000 or so Parisians were on familiar terms with one another, frequented the same clubs, flirted with the same beautiful women, died side-by-side in the heroic battles of world wars and ended up in the Père-Lachaise cemetery. That Paris no longer exists, although the "upper crust" (which is as ancient as the Jurassic age) which has always existed, still lives on today, followed by the "nouveaux riches," who came from all over the world. Today, they have evolved into that category of people who are old and venerable and though often poor are generally well housed. During the 1920s, 1950s and 1980s, an infusion of the young flocked to Paris and upset the established order of things. In the '20s, Cubism started influencing the world of decorating; in the '50s, there was a craving for all the modern products that the Americans were touting. In the '80s, a "politically correct" culture appeared whose byword, paradoxically, was "caviar." And although Louis XVI himself was guillotined, his style proved impossible to dethrone. The French, it appears, always seem to return to the neoclassic.

Now, silk-screen prints by Andy Warhol and works by avant-garde artists available only to the rich lend a little craziness to the most conservative homes. "I'm not just another pretty face," Marilyn seems to say from where she hangs on an 18th-century paneled wall. Sculptures by César, Niki de Saint-Phalle and Fontana blend well in ancient townhouses. Here we find a chest of drawers by Boulle; idols from the Cyclade islands and easy chairs by Mies Van der Rohe. Since WW II, eclecticism has ruled—a mixture of decorating styles, ancient and modern. To add to the mix, the flea market and the Surrealists joined forces to, in their words, "cross umbrellas with sewing machines on multiple dissecting tables." First came the invasion of Majorelle's art nouveau furniture and Gallé's creations in glass, then came those art deco divans set at right angles. With the 1960s came the advent of "design" with all its new expressions and ways of marketing. All followed by the arrival of housing for people of moderate incomes: "something nice for everyone," as they said. Unfortunately, this resulted in no one having anything unusual. The '70s brought "retro"—from laquerware by Ruhlmann to calendars of Vargas pin-ups, the film *Victor-Victoria*, Borsalino and Company, boas worn by Régine and fake Chapirus. The reissuing of Mallet and Stevens's 1934 book on modern architecture (along with silk-screens designed and conceived of by young artists) resulted in a new style for a new age—a style that harmonized well with American loft-style living. The kitchen became a bar. The bed was king-size. The bathroom opened out onto the living quarters, like you might find in an emergency ward. While Mozart's music was still a mainstay, art deco works by Reynaud were considered contemporary. As for the last part of the century, it seemed to detach itself from its earlier emotions to go way back to things nearly forgotten altogether: chandeliers, ornamental screens and canopied beds all made a comeback, as did the word "charm." But contemporary art was still without question paramount. This was especially true in photography, an art easily accessible to everyone. The new styles were influenced by and mixed in with the styles of the 18th century. The 20th century began and ended under the influence of the neo-Louis XVI style of decoration, first made popular by the writers Helleu, Proust and the brothers Goncourt. But by the end of the century, it was stripped of its lavishness. With the new millennium came the second death of Marie Antoinette. Once again, lavish apartments were emptied of their treasures, but this time by American millionaires. Today, we no longer clamor for their return. Instead, we borrow the American style—like their raw-brick walls, for example, that goes so well with sketches by Cy Twombly.

And while we fuss around trying to decide whether to hire an upholsterer or an interior decorator, whether we prefer practical women or women of the world, men of taste or—but what taste are we talking about? The Seine river continues to flow beneath the bridges of Paris. It crosses the capital in the shape of a crescent moon. The length of its arc extends seven and a half miles from Boulogne in the west to Bercy in the east. Thirty-nine bridges and footbridges link the opposite banks (and their often-antagonistic residents). And even though one doesn't have to be a bohemian agitator to live on the Left Bank, the myth, the legend no doubt contributes to ones sense of being there. And at great expense, mind you, as the price one now pays per square foot of living space in the Saint-Germain-des-Prés area is greater than just about anywhere else in Paris. As for living on the Right Bank…no one brags about it, but it is not completely void of originality. The very exclusive apartments of the 16th arrondissement that were detested only yesterday, have been rediscovered by the trendy bourgeoisie who now live there, unless they've decided instead to move to the fashionable lofts of the Saint-Antoine neighborhood. Open space has become synonymous with reclaiming a kind of freedom.

Although Paris's "melting pot" is limited to some degree by its external boundaries and by the forests that encircle it, it has nevertheless been able, despite its small size, to redefine its social topography through its geography. People with diverse lifestyles observe each other and oppose each other in order to, in the end, understand one another. Paris remains a city of contrasts, with its luxuries, its office buildings; its well-fed citizens and its homeless; its shopping malls and health food stores, traffic jams and public works. It's a city of deep hurts and fleeting moments of happiness. Parisians scurry about on this square patch of land with all their contradictions, their conformity and their quirkiness. They love to be at home, but at the drop of a hat they can be found marching in solidarity with strikers in the streets. They have read everything, seen everything and have done just about everything. And at the end of the day, they know that Paris is the worst place to live—except for anywhere else.

François Baudot

from SAINT-GERMAIN-DES-PRÉS *to* MONTPARNASSE

The Left Bank, as we say in Paris, is above all a state of mind. In the Paris of old there were those who wouldn't dream of crossing the Seine to visit the Right Bank. Parisians, of course, often choose to live on the Right Bank, yet consider the cafés, bookstores and the houses of Saint-Germain-des-Prés their territory as well. In the 1970s, when the famous designer Yves Saint Laurent decided to open his first boutique carrying his ready-to-wear collection in rue de Tournon, he named it "Rive Gauche" (Left Bank). The connotation was clear—there was no better way to express the panache and style that came to define the legendary Left Bank.

The Right Bank, in contrast, is a vast business sector, refuge to the upper-middle classes and the lumpen proletariat. The Left Bank, home to the aristocracy, was also claimed by those other patricians revered and canonized in the great Parisian tradition: philosophers, anarchists, artists, loafers and foreigners in love with the Paris mystique. But the intellectuals of the old Left Bank have increasingly given way to a new bourgeoisie and the rooms and cafés that harbored agitators of the avant-garde, now sport boutiques with all manner of luxury items.

Today, a strong sense of an elegant kind of "neglected comfort" has replaced the rooms of old where people sought to change the world. The isolationist stance of the Left Bank has not stopped an invasion from the Right Bank. The 1980s saw the arrival of jewelry and clothing stores and those high-end accessories usually associated with the most exclusive neighborhoods and luxury hotels. Some retailers, longing for a certain cultural (to say nothing of artistic) legitimacy migrated to the Left Bank for its instant caché. This trend signaled the evolution of a fashion where authenticity became a byword for success. When a well-known trade name installed itself in Saint-Germain-des-Prés, it signalled a distancing from the world of flashy consumerism to align itself with things that endure. The historical buildings, bookstores for intellectuals and bibliophiles, the legendary cafés all represent the spirit of the Left Bank. Now, as a sign of the times, one

Opposite: The Eglise Saint-Germain.

Next spread:
1. The cloister at the Ecole des Beaux-Arts.
2. The Palais du Sénat.
3. The café Les Deux Magots.
4. Quai Voltaire, the building's courtyard is hidden behind a carriage entranceway.
5. The four seasons fountain.
6. L'Institut de France seen from the garden.
7. The Panthéon.
8. The famous brasserie La Closerie des Lilas, behind the statue of Maréchal Ney.

from SAINT-GERMAIN-DES-PRÉS

frequently sees books being used as props in luxury stores; but not just any books—books by such authors as Lacan, Cioran and Duras, who haunted the Left Bank in days gone by and now serve as indicators of the standards of numerous establishments all madly in search of times past.

Another sign of the times is the renewed interest in the 19th-century passion for natural history and taxidermy; for shells, stuffed creatures, curios and scientific objects. This mixture of things that evince a curiosity about the natural world and the life of the mind, like a jumble of items displayed in a mirrored curio cabinet, seem to reflect back to Parisians their own changing image.

What still remains, and will always endure are the ancient houses and garrets that despite their being increasingly taken over by the boutiques of the boulevard Saint-Germain, still maintain their integrity and charm—some dating back to the 17th and 18th centuries. Their influence on newer buildings is evident; they have inspired both the design and the layout of newer apartments in their emphasis on light, and more importantly, in their preoccupation with plants and gardens in their design—an element found only in this section of Paris. These are some of the factors that have contributed to the 6th arrondissement's growing status over the last few years as the most expensive place in the city to live. Many fashionable foreigners want to have a pied-à-terre here—add to this already cosmopolitan atmosphere French nobles and elderly dowagers dressed in black, all living side-by-side with moneyed sophisticates and you get a good idea of the eclectic and interesting mix that is Saint-Germain-des-Prés.

While business lunches are all the rage across Paris, in the Saint-Germain-des-Prés to Montparnasse area they have reached a pinnacle; literary feasts, celebrations of the spirit, the "commerce" of ideas. In the famous cafés like Les Deux Magots, Café de Flore and Brasserie Lipp, this genre of lunches have flourished more than in the larger cafés of Montparnasse, despite the growth of tourism. In the 1920s, Montparnasse cafés were the preferred haunts of the artists and writers who frequented them. Although café society disappeared long ago, the legends live on—legends like the one of Saint-Michel. On the strength of its reputation alone, it has been embraced by young suburbanites looking for a remedy for the relative quiet (and relative boredom) of suburban life. Unfortunately, the boutiques that followed suburban

to MONTPARNASSE

shoppers to the Left Bank displaced those that had until then carried wares that exhibited an original mixture of cultures and tastes. Up until 1968, eclectic shops existed side-by-side like nowhere else in Paris. The famous student revolts in that year left their mark on Paris's Left Bank. Whether it was an elitist or a populist revolution back then, the repercussions of that period are still being felt today.

Those who live in Montparnasse have a bit of the artist in them and their lodgings reflect that element. We see more sky and more trees there, and the Place Denfert-Rochereau has more of a feeling of the countryside about it than anywhere in Paris. Hidden among more modern buildings are villas, villages and marketplaces hidden to outsiders who are unaware of the life behind. So the inhabitants, if they prefer, can remain insulated from the outside world.

Another bastion of the Left Bank is the sacred (and arduously steep) Montagne Sainte-Geneviève. First, you cross the boulevard Saint-Michel taking care to avoid some very ugly thoroughfares lined with what appear to be modest homes with sandstone facades. Upon reaching the summit, in the shadow of the Panthéon, one discovers ancient back streets with houses and porches so old that they might well have sheltered the Three Musketeers, heard the poems of Verlaine or the convent school pupils reciting their Latin declensions.

Montmartre's more famous Place du Tertre is the Right Bank equivalent to Montagne Sainte-Geneviève with its Place de la Contrescarpe. Over the years, it too has been flooded with sightseeing buses, but this should not deter a stroll along the hillside—as one does in Montmartre—with the same feeling of being someplace far away. And indeed, when these two places are mentioned people say, "It's so far away," even though they have no real cause to say so. For nothing in the capital, no matter how far from the center of Paris (about where the Pyramid stands in the courtyard of the Louvre), is farther than three miles away. Even today Paris, and the Left Bank in particular, remains a city best experienced on foot.

A COLLECTOR'S ELEGANT ABODE

Left: François-Joseph Graf designed the library, dining room and den in this townhouse in the 7th arrondissement. In one room, a rich celadon green enobles the walls; in the other, red flannel serves as the backdrop for a series of early 19th-century botanical engravings. The armchairs are made of ebony and gilded bronze (Westenholz, England, 19th century).

Above: An alcove in the library with a series of 18th-century English engravings by Thomas Frye.

SAINT-GERMAIN

An 18th-century English mahogany bureau stands between two French windows in a living room on Faubourg Saint-Germain which overlooks the garden. The comfortable sofa and armchairs are by Decourt. Sage green walls bring out the tones of a striking late 17th-century painting by St. John Baptiste Medina. The curtain fabric is from Fortuny in Venice. Room designed by François-Joseph Graf.

SAINT-GERMAIN

Right: In this living room-library, the woodwork is in the Louis XVI style and the parquet is in the "Versailles" style. The library's English mahogany stepladder is from the 18th century (Westenholz). A 17th-century Turkish rug from Oushak lends the room vibrancy.

Below: A painting by Brazilian artist Reynaldo Fonseca.

STROLLING THROUGH SAINT-GERMAIN-DES-PRÉS

1. LES MARRONNIERS A small, shaded courtyard adjacent to what was once Delacroix's studio; one of the area's most lovely and best kept secrets.
21, rue Jacob, 75006.
Phone: +33 1 43 25 30 60.
www.hotel-marronniers.com

2. LA MAISON DE BRUNE Lamps in bronze patina, heavy oak furniture, a drawing, woven cloth with a color scheme of dark brown, green and plum.
23, rue du Cherche-Midi, 75006.
Phone: +33 1 42 22 49 05.

3. GALERIE DENIS DORIA
This merchant's austerity and preference for rationalist architecture has made him one of the foremost specialists on modern art.
16, rue de Seine, 75006.
Phone: +33 1 43 54 73 49.

4. HUILERIE LEBLANC This small boutique carries an extensive variety of pure oils, from olive oil to pine cone oil.
6, rue Jacob, 75006.
Phone: +33 1 46 34 61 55.

5. ROYAL ARROW
Benches, lounge chairs, plant tubs... all made of teak, the outdoor material par excellence.
206, boulevard Saint-Germain, 75006.
Phone: +33 1 45 49 49 89.
www.royal-arrow.com

6. DEBAUVE ET GALLAIS The setting here is magnificent—a palace for discerning palates. A chocolate store that has satisfied sweet-toothes from the time of the First Empire.
30, rue des Saints-Pères, 75007.
Phone: +33 1 45 48 54 67

7. MUSÉE MAILLOL A beautiful museum founded by Dina Vierny, the master's model and heir, with her own funds. Along with the permanent collection, the museum hosts changing exhibitions.
61, rue de Grenelle, 75007.
Phone: +33 1 42 22 59 58.
www.museemaillol.com

8. GALERIE CATHERINE MEMMI
The lines of the house, the furniture and accessories, all adhere to the same rigorous simplicity.
11, rue Saint-Sulpice, 75006.
Phone: +33 1 44 07 02 02.
www.catherinememmi.com

9. LA CASA DEL HABANO
A Cuban enclave just steps away from the Brasserie Lipp. Here one can buy any hard-to-find cigars.
169, boulevard Saint-Germain, 75006.
Phone: +33 1 45 49 24 30

10. MAISON DE FAMILLE Assembled here are all those practical things we'd like to have in our own homes; linens, kitchen accessories, furniture, both sensible and charming.
29, rue Saint-Sulpice, 75006.
Phone: +33 1 40 46 97 47.

11. FRAGONARD A gift shop for every occasion, from fragrant things to bric-à-brac...tasteful temptations for the epicure.
196, boulevard Saint-Germain, 75007.
Phone: +33 1 42 84 12 12
www.fragonard.com

12. ARCADE OF RUE DE VAUGIRARD
A sheltered arcade where one can stroll alongside the Luxembourg, the Left Bank's largest garden, between the rue de Rennes and de l'Odéon.

1

2

3

4

Opposite and below: Beneath the blue skies of Saint-Germain, on the corner of rue Jacob and rue Bonaparte, lies the famous Ladurée. Who would ever guess, with its old-fashioned facade and traditional delicacies, that it was actually built in this century?

RUE BONAPARTE

DELICACIES FOR EVERY SENSE

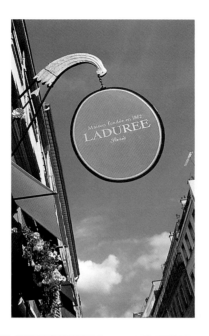

Right: This was once the office and boutique of Madeleine Castaing—one of the greatest interior decorators and antiques dealers of the postwar era. With such elegant delicacies the pastry chef has sought to recapture the Balzacian atmosphere that Castaing so loved. Interior decorator Roxane Rodriguez created this delightful ambiance.

Between the first and second floors of the old houses that lead to the Seine, mezzanines, with their low ceilings, make cozy intimate cocoons where the romanticism affiliated with richly upholstered furniture from the 19th century sets the mood—as seen here on the second floor of the Ladurée tea salon. This exquisite room, with sumptuous fabrics and late 19th-century photographs, transports guests to another era.

SAINT-GERMAIN

STRENGTH AND SIMPLICITY

Designer and interior decorator Catherine Memmi lives here amid her own creations, and practices the same austere sense of luxury at home that is found in her boutiques. The long sofa is covered in pristine white linen, on the low Wenge-wood table a ceramic dish tastefully displays unusually colored candles. Between the two windows is a painting by Hilton McConnico.

SAINT-GERMAIN

This office also serves as a dining room. Catherine Memmi designed the wood table and the bookcase in bleached ash, where multicolored ceramics are exhibited along with files and boxes created by Marie Papier. The chairs are covered in Memmi's signature natural linen and the cotton curtains are backed in percale. Unique adjustable floor lamps are by Christophe Delcourt.

SAINT-GERMAIN

THE REIGN OF THE UNUSUAL

An Italian client who loves the unusual entrusted the interior design of her pied-à-terre on a second floor of this townhouse on rue de Verneuil to Stéphanie Cauchoix. The woodwork is painted in a pale gray wash. The life-size horse is made of papier mâché—a 19th-century saddler's prop to advertise harnesses (Actéon collection). A startling mannequin with adjustable limbs sits on the couch (Cauchoix). In the foreground, a Napoleon-era camp bed is covered in linen. A 17th-century Venetian mirror hangs unexpectedly over a wall mirror (Galerie Meyer).

SAINT-GERMAIN

In Saint-Germain, a certain bourgeois bohemia allows for an array of styles—classic and modern, minimal and ornate all harmonize in the mix. Here, ancient furniture goes well with the more industrial-style railway racks from the 1950s used here as bookcases. An 18th-century day bed is upholstered with an Indian quilt. The eclecticism exhibited here produces a kind of fantasy world, awash in the light streaming in through the high windows.

STROLLING THROUGH SAINT-GERMAIN-DES-PRÉS

1. LE SAINT-GRÉGOIRE Cozy, elegant and not too big, this refuge has more than earned its reputation as a "hotel of charm."
43, rue de l'Abbé Grégoire, 75006.
Phone: +33 1 45 48 23 23.
www.paris-hotel-saintgregoire.com

2. DAUM A prestigious name since the early 20th century, the Left Bank branch of this store, which specializes in crystal and sintered glass, also carries work by such luminaries as Philippe Starck and Hilton McConnico.
167, boulevard Saint-Germain, 75006.
Phone: +33 1 42 22 16 12.
www.daum.fr

3. LE JOUR ET L'HEURE A little piece of England; stationery, pens, famous Smythson notebooks from Bond Street, soap, plaid throws and small but exquisite leather goods.
6, rue du Dragon, 75006.
Phone: +33 1 42 22 96 11.

4. À SAINT-BENOÎT-DES-PRÉS
A mini-bookstore with a remarkable, eclectic selection of books and posters.
2, rue Saint-Benoît, 75006.
Phone: +33 1 40 20 43 42.

5. CABINET DE CURIOSITÉS
A paradise of the unexpected; here rare and puzzling objects converge. A little of everything, but not just anything.
23, rue de Beaune, 75007.
Phone: +33 1 42 61 09 57.

6. LA PALETTE An institution suspended in time. A rough but friendly reception, inexpensive red wines and cured meats delight tourists and regulars alike.
43, rue de Seine, 75006.
Phone: +33 1 43 26 68 15.

7. LE CACHEMIRIEN The great woven fabrics of India, from Pashmina to Benares brocades—a boutique that specializes in items that keep the cold at bay.
13, rue de Tournon, 75006.
Phone: +33 1 43 29 93 82.
www.lecachemirien.com

8. MOISSONNIER Builders of fine furniture since 1885. A repertoire of objects with finely detailed features exquisitely rendered inspired by the 18th-century style.
28, rue du Bac, 75007.
Phone: +33 1 42 61 84 88.
www.moissonnier.com

9. FARROW AND BALL Enough colors to satisfy any and all tastes and wallpaper available in the most refined tones: the caviar of housepaints.
50, rue de l'Université, 75007.
Phone: +33 1 45 44 47 94.
www.farrow-ball.com

10. RUE MONSIEUR-LE-PRINCE
An exquisite 18th-century door.

11. HERVÉ LORGERÉ This relatively small space—in relation to the clutter of objects—only serves to accentuate the air of discovery in this paradise of the unexpected.
25, rue des Saints-Pères, 75006.
Phone: +33 1 42 86 02 02.

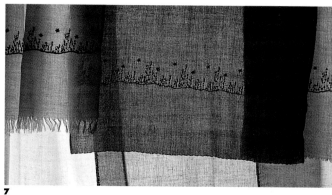

SAINT-GERMAIN

NEOCLASSICISM IN CLOSE QUARTERS

In the antiques dealer Alexandre Biaggi's library, situated over his store, we find his books, catalogues and drawings from every era arranged snugly; the sculptures are from the late 19th century. The Charles X desk of blond wood goes perfectly with a 1930s armchair. The side table is by Jacques Adnet.

SAINT-GERMAIN

All the charm of this cozy dining room overlooking the rue de Seine emanates from the rapport between its small size, the delicate lines of the furniture (in the Directoire style), and the colossal bust of Minerva (Biaggi collection). The antique silverware is by Christofle, the cut-crystal candleholders are from the 19th century and the plates are by Christian Lacroix for Christofle.

SAINT-GERMAIN

AN EXERCISE IN STYLE

Yves Gastou took a gamble when he introduced 1940s and '50s furniture in this elegant 18th-century townhouse owned by a Parisian collector. In the foreground a casual pair of oak stools are offset by a bronze armchair with a leather seat (André Arbus, 1952) and a coffee table made by the iron craftsman Gilbert Poillerat, all resting on a rug by Arbus. The vases are Venetian glass.

SAINT-GERMAIN

THE COUNTRYSIDE IN PARIS

This duplex has a double theme: flowers and music (jasmine, lavender, mimosa, traditional roses…and the cello). The arched bay window in the bedroom opens onto a garden terrace encircled by a zinc flower box. The teak lounge chair is resistant to Parisian rain.

Opposite: The bathroom is both simple and refined, with its traditional linens and reassembled doors dating back to the 19th century.

Above: In the large studio-music room, light filters through a skylight of frosted glass. The wide oak-plank floors were specially made in the Aveyron region of France. A pair of Renaissance-style lead flower basins grace on either side of the steps. The elegant Bechstein piano and the cello belong to the master of the house.

RUE DES BEAUX-ARTS
L'HÔTEL

This beautiful hotel, located on rue des Beaux-Arts, needs no other name than L'Hôtel. It became famous when Oscar Wilde, then exiled in Paris, died there in abject poverty—very incongruous indeed to the sumptuous Empire-style dining room that characterizes the hotel today. L'Hôtel was recently restored by Jacques Garcia.

SAINT-GERMAIN

A LARGE COLLECTION IN A SMALL SPACE

Interior decorator Jacques Leguennec's library reflects his wide variety of literary interests—each section is arranged by subject matter. The inviting sofas (Schwartz) are covered in raw linen (Lauer). In front of the window a lecture stand made of white-painted oak holds the interesting book of the moment (studio Leguennec). A Picasso drawing lends sophistication; the unique sculptural lamp behind the sofa is by Alain Ozanne.

55

Ozanne and Leguennec both have a fondness for collecting and have devised ways to show their collections to their best advantage in a small space. Paintings, curios in wood, autographs and a variety of delightful objects are cleverly arranged on buttress-shelves, thereby avoiding the sometimes tedious symmetry of hanging things. The mantle becomes a display shelf where a 19th-century articulated mannequin holds court above a trompe-l'oeil drawing by Pascale Laurent.

SAINT-GERMAIN

RUE DE L'ÉCHAUDÉ

A FLAIR FOR FASHION AND PHOTOGRAPHY

When Karl Lagerfeld and this old notions shop *(above)* met up, it was immediate synergy. Lagerfeld installed his personal office here, facing a gallery converted by Andrée Putman that houses a variety of perfumes, clothing and photographs signed by the artist. In the front of the space a staircase is cleverly concealed; a custom display table *(right)*, and wash basin *(left)* were both designed by Andrée Putman.

STROLLING THROUGH SAINT-GERMAIN-DES-PRÉS

1

1. BRASSERIE LIPP
The smart set and followers all show up here, not so much for the quality of the comestibles as for the interesting clientele. The 1900 decor is a historical landmark.
151, boulevard Saint-Germain, 75006.
Phone: +33 1 45 48 53 91.
www.brasserie-lipp.fr

2. ATELIER MARC DEKEISTER He loves Dürer, Vermeer and…comics. Strongly influenced by all, his portrait paintings of houses are of the highest quality.
32, rue Saint-Guillaume, 75007.
Phone: +33 1 45 44 99 75

3. MADELEINE GÉLY This highly respected vendor won't sell you an umbrella, but she'll honor you by allowing you to buy one!
218, boulevard Saint-Germain, 75007.
Phone: +33 1 42 22 63 35.

4. GALERIE 54 Works by Prouvé and other designers from the 1950s occupy this austere but perfect space.
54, rue Mazarine, 75006.
Phone: +33 1 43 26 89 96
www.galerie54.com

5. UPLA The shoulder bags that once made this name famous now rest alongside a new line of luggage and tableware in a large welcoming loft.
5, rue Saint-Benoît, 75006.
Phone: +33 1 40 15 10 75
www.upla.fr

6. THE FOUNTAIN DE MÉDICIS
In the spring lovers rendezvous here in the shade of the Luxembourg Gardens.

7. LA GALERIE MODERNE
Once "modern," these works are now considered antiques. All those things that only yesterday were old-fashioned are now sought after collector's pieces.
52, rue Mazarine, 75006.
Phone: +33 1 46 33 13 59.

8. CLAUDE NATURE There isn't anyone else quite like this naturalist when it comes to locating just the right little critter. Taxidermy madness!
32, boulevard Saint-Germain, 75005.
Phone: +33 1 44 07 30 79.
www.claudenature.com

9. TECTONA A teak paradise—from traditional designs for English gardens, to new creations by young designers.
36, rue du Bac, 75007.
Phone: +33 1 47 03 05 05
www.tectona.net

10. JEAN-CLAUDE GUÉRIN & PHILIPPE RAPIN We find quality antiques and the sound judgement of two dealers who are masters in the art of mixing styles.
25, quai Voltaire, 75007.
Phone: +33 1 42 61 24 21

11. LOUIS XV FACADE on rue de Seine. Easy to sculpt, limestone is one of the materials that characterized Paris of old.

12. L'HÔTEL DE BRANCAS The Institut Français d'Architecture organizes thematic exhibitions in the annex of this grand old house just next door to the Palais du Sénat.

13. L'ÉCOLE NATIONALE SUPÉRIEURE DES BEAUX-ARTS
Architecturally, each pavilion has its own style. All together, they constitute a museum of styles.
14, rue Bonaparte, 75006.
Phone: +33 1 47 03 50 00.
www.ensba.fr

4 5

7

9 10

61

SAINT-GERMAIN

RUE DE L'ABBAYE PLACE FURSTENBERG

A NAME SYNONYMOUS WITH STYLE

Right: One might think they'd walked right into someone's home, but everything here is for sale. Flamant is an unusual boutique with a certain provincial style, a touch of England, a breath of Scandinavia; the ensemble of this potpourri opens onto one of Paris's prettiest squares *(above)*. Not far away is the Delacroix museum-studio with its charming little gardens.

SAINT-GERMAIN

HARMONY OF HUES

Yves and Michèle Halard—a designer and an interior decorator—frequently change addresses, yet each of their residences gives the feeling that they've always lived there. Their passion for fabrics inspires them to mix different materials in unexpected ways. They are a study in independence, smack in the middle of the ultra-traditional neighborhood of Saint-Sulpice.

SAINT-GERMAIN

The dining room, with its purple walls and windows dressed with antique curtains, brings to mind the old-fashioned atmosphere of the ancient streets nearby. A wall light in zinc, tablecloths, plates and silverware all by Yves Halard.

SAINT-GERMAIN

RUE JACOB
LIKE SNOW IN THE SUN

Monic Fisher, Blanc d'Ivoire's designer, mixes quilts with decorative objects. She loves to juxtapose high quality with refined simplicity. Combining the classic with the exotic, the traditional with the modern, her house in Saint-Germain-des-Prés is an extension of her popular boutique *(above)*.

Next spread: In the living room an ornamental screen made of shutters allows light to filter through from the bay window. On the low table, a taffeta quilt changes color with the light (Blanc d'Ivoire) and contrasts with a sofa covered in canvas (Frey) and a comfortable leather armchair (Flea market, 1930s).

SAINT-GERMAIN

THE CHARM OF TIME SUSPENDED

Near the Quai d'Orsay, designer Michelle Joubert includes taxidermy, for which she has a soft spot, in her eclectic decor (see next spread). At the far end of the living room an 18th-century Italian banquette complements a 1950s Baguès chandelier. In the foreground on the left, a library table accommodates a candelabra by Christian Tortu. To the right, a whimsical wooden elephant (Vivement Jeudi).

Left: In a large neo-Gothic viewing case, a collection of stuffed birds contemplates an armchair in the Directoire style. Above: The library staircase—hold on to the banister!

SAINT-GERMAIN

RUE DU BAC

DEYROLLE'S BESTIARY

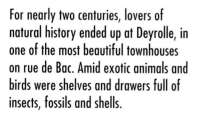

For nearly two centuries, lovers of natural history ended up at Deyrolle, in one of the most beautiful townhouses on rue de Bac. Amid exotic animals and birds were shelves and drawers full of insects, fossils and shells.

STROLLING THROUGH QUAI VOLTAIRE

1, 2, 4 & 5. *Plaques, like the ones pictured here on the Quai Voltaire, adorn the dwellings of old Saint-Germain, helping us to imagine the district's illustrious past.*

3. *The Restaurant Voltaire, in a converted tabac, is now one of the most popular spots in this district of antiques dealers.*

6. *The Pont des Arts footbridge was one of the first examples of metal architecture in Paris. On beautiful days, many people traverse the bridge for its magnificent views. François Le Vau's masterpiece extends the axis of the chapel of the Académie Française.*

7. *The Seine, and many a love affair, flow beneath the bridges of Paris....*

8. *Created by the iron craftsman Raymond Sube at the end of the 1930s, these telescopic obelisks, situated at the four corners of the Pont du Carrousel, dramatically illuminate the night.*

9 & 10. *Two views of the prestigious Camoin gallery. Ring the bell and famous decorator Alain Demachy will invite you into his space. He juxtaposes furniture and objects from all eras, chosen with two criteria in mind: originality and quality.*
9, quai Voltaire, 75007.
Phone: +33 1 42 61 82 06.

1

2

3

4

5

6

7

THE MAJESTY OF THE 18TH CENTURY REVISITED

Conjuring up the past without nostalgia; displaying beautiful old books to their greatest advantage; mixing a touch of the spirit of Provence with the elegant orderliness of the Place du Palais-Bourbon—that is what Michelle Joubert does to impart that feeling of a family manor to her interiors.

Top, left: A large bookcase with dove-gray shelves, a Dagobert armchair and a studio table of bleached walnut create a distinctive ambiance (Vivement Jeudi).

Bottom, left: An eclectic collection of globes.

Opposite: In the dining room a mirror rests on an easel, and a carafe from the 18th century in the Spanish "Infante" style graces the table.

HIGH DRAMA AND ECLECTICISM UNITE

Theatricality is an understatement at this basement-cum-gallery on boulevard Saint-Germain.

SAINT-GERMAIN

Here, each item is unique: a velvet-covered ladder for the delicate-footed. On the shelves, unusual drawings by Laurent de Commines mingle with items of unprecedented flair and imagination.

The great designer Yves Saint Laurent transformed this garden-level room into a salon-library that juxtaposes and harmonizes a variety of masterpieces from his personal collection.

Far left: A tapestry by the Pre-Raphaelite painter Edward Burne-Jones (c. 1860) hangs majestically over a table conceived by the sculptor Marcial Berro. Antique crater and torso, candlestick holders made of rock crystal, bronzes from the 17th-century, Régence seats and a rug by Da Silva Bruhns all blend magnificently.

ART DECO OPIUM

Top, left: In front of the fireplace and its overmantel mirror (c. 1930s), rests a unique boot rack made of python by Dunand (c. 1920s). The painting is by Mondrian; the bronzes from the Fontainebleau school and the bulls are by Gianbologna.

Bottom, left: Beneath a large *kouros* sit portraits of Yves Saint Laurent's dog Moujik by Andy Warhol.

Top: A collection assembled by Yves Saint Laurent was given a stunning arrangement by Jacques Grange in compartments with security glass backed in velvet.

Above: On the terrace, the bird chairs are by François-Xavier Lalanne.

Right: The living room's oak bookcases painted in lead white were designed by Grange, and the chairs in apricot leather were upholstered by Decourt. To the left, a male nude by Géricault.

A few steps from rue Mouffetard is the house of two antique dealers who open their doors every Thursday to sell to a clientele of initiates. Hence the name of their small company, "Vivement Jeudi," (I can't wait for Thursday!)

Above: In the garden, a prolific vine, a few acacias, verbena and gooseberries. The terrace is done in brick.

Opposite: In the kitchen, an artistic mix of still lifes with sweets.

A TASTE OF THE COUNTRY BEHIND THE PANTHEON

Opposite, clockwise from top left: In Dominique and Pierre Bénard-Dépalle's living room the mantel serves as a landing for their various discoveries. A Louis XVI-style table whose surface is made up of sections of stone fit together; some light furniture with openwork makes for a kind of organized disorder; a graceful Italian chest of drawers (c. 1760); a still life of things from the sea.

Above: A late 18th century oval table upon which garden implements and a notably large wooden vase with Chinese decorations, among other curiosities, provide interesting clutter.

SAINT-GERMAIN

IN THE SPIRIT OF THE 1930S

Joëlle Mortier Vallat lives in a quiet apartment that opens onto the large gardens of the Ecole des Beaux-Arts. In the inter-connecting rooms, painted in pale tones, African art and art deco live side by side with a touch of the Baroque. The large mirror is a cast from a gilded 18th-century Italian-style wooden frame. Between the two windows, a "witch's mirror" by Line Vautrin.

Here is a glimpse of antiques dealer Joëlle Mortier Vallat's collection on rue des Saints-Pères.

Clockwise, from top left: A console in painted metal with a removable top made of travertine, a vase by René Butheau and painting by Marquet; a painting by Levy-Dhurmer above a card table by Ruhlmann; in the kitchen, a Russian armoire from the early 20th century, a lamp by Serrurier-Bovy, and silver by Jansen and Christofle (c. 1900); a neoclassic drawing and a 19th-century lecture lamp with a shade of pleated silk.

Opposite: In a small living room above a mantel painted in trompe-l'oeil leopard, a stuffed bison head dominates the room. The photos of Indian chiefs are by Curtis (mid 19th century).

STROLLING THROUGH SAINT-GERMAIN-DES-PRÉS

1. LA MÉDITERRANÉE The restaurant specializes in fish dishes and dates back to the 1930-40s era. The decor from that period by Cocteau, Bérard and Vertès has been restored.
2, place de l'Odéon, 75006.
Phone: +33 1 43 26 02 30.
www.la-mediterranee.com

2. HUMEURS Despite its austere look, this is one of the most prestigious stores in the area.
4, rue de l'Université, 75007.
Phone: +33 1 42 86 89 11.

3. DAVID HICKS FRANCE
A particular modern English style from the 1960s adapted here to that of the Left Bank bourgeoisie.
12, rue de Tournon, 75006.
Phone: +33 1 55 42 82 82.

4. CAP SHORE In this temple of tableware, the dishes and accessories balance one's taste with one's purse.
26, rue des Plantes, 75014.
Phone: +33 1 49 27 90 00.

5. MURIEL GRATEAU The designer of this space has pushed the art of detail to the point of perfection; pure lines, muted colors—with just a few things she says a lot.
37, rue de Beaune, 75007.
Phone: +33 1 40 20 42 82.
www.murielgrateau.com

6 & 7. PLACE DE L'ODÉON AND THÉÂTRE DE L'EUROPE
All around the Théâtre de l'Europe, built by Peyre de Wally, a typical arched circle from the end of the neoclassic 18th century. Here one can escape from the crowds of the Latin Quarter.

8. PIERRE BURG Passed down from father to son, this business makes superb garden ornaments in terracotta and ceramic, but their specialty is trompe-l'oeil.
Espace Buffon
27, rue Buffon, 75005.
Phone: +33 1 47 07 06 79.
www.espacebuffon.com

9. FABIENNE VILLACRÉCÈS From pillows embroidered with sequins to necklaces made of pearl-gray Venetian glass, this designer ignores fashion in order to make the most of the styles from the turn of the last century.
18, rue du Pré-aux-Clercs, 75007.
Phone: +33 1 45 49 24 84.

10. SAINT-SULPICE CHURCH Symbol of the "Saint-Sulpician style" of art because of the vendors of religious souvenirs who in the past swarmed the area, some of whom remain today. It is also one of the largest Baroque churches in the capital.

11. LIBRAIRIE LE MONITEUR
Everything you ever wanted to know about world architecture from any era can be found in this bookstore.
7, place de l'Odéon, 75006.
Phone: +33 1 44 41 15 75.

2

3

4

5

8

9

10 11

SAINT-GERMAIN

PLACE SAINT-GERMAIN
THE LOUIS VUITTON BOUTIQUE

Interior decorator Anouska Hempel created the decor for the Vuitton store in Saint-Germain-des-Prés. Old trunks, the smell of wax and exotic perfumes, all compete to create the incomparable atmosphere.

Left: This stained mahogany writing desk has unusual vertical drawers.

Opposite: The space is arranged in such a way that the luggage appears to be furniture. The suspended lamp is made of black rubber. The table dates from the 1940s.

SAINT-GERMAIN

Left: An imposing bookcase with wire-net doors punctuated by stucco half-columns (Oberlé-Laurent). On the desk a quilted tablecloth from Provence, a crystal girandole in the style of Louis XIV and a bronze vase converted into a lamp.

Below: On the mantel, 19th-century photos and a zebu tusk serve to augment the exotic flavor around a mirror that reflects a maharajah painting (Rajastan).

CLASSIC OR EXOTIC

SAINT-GERMAIN

A GLOBE-TROTTER'S BACHELOR PAD

Near the Seine at the house of decorator Tino Zervudachi, a mixture of objects found while bargain hunting at flea markets and on his many travels.

Top: The living room is in the Directoire style.

Center: In the office, functional furniture from the 1930s. The two mirrors and the rug are based on a design by Zervudachi.

Bottom: The bedroom overlooks a sunlit courtyard.

Opposite: The dining room, located between the foyer and the bedroom, is functional for any occasion.

SAINT-GERMAIN

REAR WINDOWS

Above: Agnès Comar has taken over these buildings with glass facades overlooking an old cobbled courtyard in the faubourg Saint-Germain. Once you've passed through the double door that leads to the courtyard, two ecru-colored raw linen curtains with black borders help give the feeling of being in a living room whose "ceiling" is the open sky.

Opposite: Above the table in the kitchen-dining room are three identical lanterns made of a striped fabric with trimmings (Thomas Boog). The table runner and plates were designed by Agnès Comar. Painted-wood mirrors reflect the building's glass facade.

Next spread: A winter garden living room, designed by architect Anne-Cécile Comar allows the courtyard to become part of the house itself. On the right, a gold-leaf cat and two Zulu bridal headdresses rest on a lacquered Japanese table.

STROLLING THROUGH MONTPARNASSE

1. LA MAISON-ATELIER DE ROBERT COUTURIER One of the last artists to witness the quarter's heyday, the sculptor worked here until the end of his life. His studio is just steps away from the Parc Montsouris.

2. LES ATELIERS DE JEAN PERZEL In this atelier, founded in 1923, the craftmanship still lives up to its original slogan: tradition and perfection. It is the Rolls Royce of art deco light fixtures.
3, rue de la Cité-Universitaire, 75014.
Phone: +33 1 45 88 77 24.
www.jean-perzel.com

3. FRANÇOIS HUBERT, HORLOGER-PENDULIER
What is more challenging than repairing or restoring an antique clock? Here, there is a solution for everything; some rare pieces are even for sale.
43, rue Madame, 75006.
Phone: +33 1 45 44 22 00.

4. HÔTEL SAINTE-BEUVE
A small, charming place to meet only steps away from the Vavin intersection.
9, rue Sainte-Beuve, 75006.
Phone: +33 1 45 48 20 07.
www.hotel-sainte-beuve.fr

5. LA COUPOLE All the great celebrities from the 1930s to the 1970s have shown up here at one time or another. The historic decor, now restored to its original glory, is well worth seeing.
102, boulevard du Montparnasse, 75014. Phone: +33 1 43 20 14 20.

6. LA CITÉ INTERNATIONALE UNIVERSITAIRE The university's international building, inaugurated in 1936, houses a remarkable library, as well as foreign students, with its many rooms available for rent.
19, boulevard Jourdan, 75014.
Phone: +33 1 43 13 65 00.

7. THE OBSERVATOIRE DE PARIS, opened by Louis XIV in 1667, is still in use and is fascinating to visit.
61, avenue de l'Observatoire, 75014.
Phone: +33 1 40 51 22 21.
www.obspm.fr

5

6

7

MONTPARNASSE

BOURGEOIS-BOHEMIAN CHARM

This artist's studio is compact, but still reflects all the bourgeois-bohemian charm of Montparnasse.

Top: In the office, that occupies a corner of the living room, is a pastel by Leon Lhermitte and a wash drawing by J.-B. Huet sits on an easel. The art nouveau-style armchair is English.

Center: In the foyer, a Regency umbrella stand and a stepladder discovered at the flea market. The portrait is of the owner's grandfather.

Bottom: A reverse shot of the corner office.

Far right: In this bedroom, under a mansard roof, the drawings and landscapes are English. The bed linens are embroidered and come from the flea market at Saint-Ouen. A favorite resting place for Phillips the cat.

ROBERT COUTURIER'S HOME-STUDIO

Just steps away from the Parc Montsouris, this student of Maillol pursued his art with sensitivity and humor.

Above: In the living room, 17th century furniture—referred to as the "sheepbone" style—with natural slipcovers. Seen here from the back, is a Couturier sculpture from 1950 titled *Fillette sautant à la corde*.

Right: On the desk in the bedroom, the sculpture *Amandine* (1984), and on a nearby shelf rests *La Fillette au cerceau* (1952).

Opposite: A view of the garden with (from left to right) *La Pensée* (1948), *Janus* (1950), *La Jeune Fille lamelliformée* (1950) and *Août 1996*. On the table, *17 ans ou l'adolescent* and *Moitié de couple* (1981). The collected works of a prolific life.

MONTPARNASSE

A GARRET UNDER THE MANSARD ROOF

A beautiful light illuminates these cozy rooms, which architect Laurent Bourgois transformed into a discretely luxurious nest. In the corner of the living room the couches are covered in flannel by Pierre Frey with kilim-covered throw pillows. On the 1960s coffee table is displayed a collection of Bakelite, wood and ivory balls. In the foreground, a duck by Lalanne.

INDUSTRIAL MEETS MODERN ELEGANCE

Nathalie Decoster, who sculpts people in motion, loves the contrast of black and gray steel against white walls.

Above: In the living room, furniture in leather and metal and a reissued Godin stove. In the foreground, a piece by Nathalie titled *L'Air du temps.*

Right: Light colored wood lends an airiness to the bedroom of the owner who enjoys gazing at the fire from her bed.

Opposite: For this artist's kitchen only professional equipment. The country table is made of old oak planks, and the armchairs are a 1940s design reissued by Habitat.

Sixty years ago, Japanese painter Foujita lived and worked here. Now restored, this retreat still carries with it memories of the 1920s when Montparnasse was in its heyday.

Below: Ordered chaos dominates this book-lover's bedroom; the armchair is by Le Corbusier, the lithograph is by Hartung.

Right: The old studio, as seen from above. The mezzanine is now a kitchen-dining room with a table by Knoll; the copper fish is by Lalanne.

IN FOUJITA'S FOOTSTEPS

HOME AMERICAN-STYLE

A studio atmosphere, with a love for all things American. This historic abode has seen George Sand and other renowned artists and craftsmen pass through.

Above: For a group of Parisian artists who love to relax, an old linden tree in the courtyard fills their summers with fragrance and provides plenty of shade.

Left: A jazz trio and singer (American, 1940s).

Opposite, top: In the foreground, the opposite of a torso painted in the style of Magritte by Charles Matton. On the wall, a sign for an old English pub.

Opposite, below: In a corner of this white-paneled retreat, the staircase leads to a mezzanine.

A GARDEN RETREAT

Here, in the Parc Montsouris, a grand view of what must be one of the most beautiful open spaces in Paris. This country-like corner provides the perfect retreat for work and for dreaming. The park itself is a paradise for children and for poets.

from the MARAIS to the MADELEINE

Long ago, a famous omnibus linked the Bastille—that ancient fortress which was later so unceremoniously destroyed—to the Madeleine, an ornate temple dedicated to the glory of Napoleon's army. Just as the Bastille later became a prison, the Napoleonic temple eventually became the church it is today. King Louis Philippe dreamed of transforming the Madeleine into Paris's first train station. But in 1837, it was finally determined that the first Paris-Saint-Germain train, bellowing smoke as it went, would instead originate at Saint-Lazare. The train station's location, the train line itself and the rather haphazard expansion of this section of Paris were all part of the big push westward. It was as if a movement in this direction would allow the fresh winds from Deauville or Dieppe to dispel the smog that cloaked the capital.

Today, despite successive upheavals, those living on the Right Bank—undoubtedly shrewder than those inhabiting the other side of the Seine—remain very attached to the old historic buildings in their arrondissement. This is especially true of the 19th century dwellings—now 200 years old and by far the most romantic feature of this area. But what distinguishes the Right Bank most of all is the demographic detachment of its people. In the 1970s, it was *de rigour* among the Right Bank bourgeoisie to go slumming in Saint-Germain, with all its plotting and insurrections. Then there was the exodus from the center of the city to the suburbs, where urbanites sought open spaces and an escape from the endless expansion of office buildings into the area. Finally, there was a general push toward the west—Trocadéro, the Bois de Boulogne, Neuilly, even as far as the heights of Saint-Cloud—by well-to-do families seeking manor houses.

Of the million inhabitants who lived between the 1st and the 11th arrondissements in 1954, only 600,000 still remained at the end of the 20th century; a decrease of 42%. During the same time period, the number of people living in the 11th arrondissement was three times greater than that of the 8th arrondissement, where in 1954, the ratio had been 1 to 2*. This evolution, which appears even now to continue unabated, empties the heart of a city through its lack of concern for its citizens—whether they

Opposite: The Place des Vosges.

Next spread:
1. The hôtel de Sully.
2. The Centre Georges Pompidou.
3. The Conservatoire National des Arts et Métiers.
4. The Eglise Saint-Eustache.
5. The Place des Victoires.
6. The Marché Saint-Honoré.
7. The Galerie Vivienne.

from the MARAIS

are ecological or didactic concerns. Singles, who are often trendsetters, are the people who mostly inhabit the center of Paris. They have reinvested in a district besieged all day and all week by the service industry. They have renovated apartments that once belonged to the upper middle class and have invested in former warehouses and converted them into lofts.

Although central to Parisian business activity, the Madeleine still falls within the sphere of influence of the Beaux Quartiers, where it sits imposingly among sandstone buildings whose facades tantalize the real estate brokers. In contrast, the Place de la Bastille opens out onto a neighborhood of factory workers and artisans. It is a place where many a revolt evolved into a revolution. The Bastille-Madeleine route links two worlds where the rich and poor, factory workers and financiers, blue and white collar workers each look upon the other with suspicion. These two important neighborhoods, where cohabitation is lived out at arm's length, are both an integral part of the complex fabric that is Paris.

Over the years, the Grands Boulevards that connect these two historic bastions have remained basically unchanged, at least in their liveliness and traffic jams. They form an arc that follows the ruins of the encircling wall built by Charles V at the end of the 14th century—a wall later fortified by Louis XIII in the 16th century, which passes through the archways of Saint-Martin and Saint-Denis built by Louis XIV. Beyond the arc lie the working-class areas; villages and small clusters of houses that were not incorporated into the city until the mid-19th century. There is no better place to observe this route than from the crescent-moon-shaped beltway that surrounds these ancient districts of Paris. From the Marais to Saint-Honoré, we can almost see the different layers of "skin" that make up the city. La Villette, Barbès, Pigalle, La Ville-l'Evêque. Like the opposition of north and south, the Left Bank and the Right Bank—this area forms the capital's east-west axis. Over time, the Boulevard evolved into the Grands Boulevards, which Yves

* These figures come from *Paris Mosaïque*, by Michel Pinçon and Monique Pinçon-Charlot, Ed. Calmann-Lévy, Paris 2001.

to the MADELEINE

Montand made famous in his song, "I love to stroll along the grand boulevards. There are so many things, so many things to see!" Later on, via Place de la République, the boulevards became the axis for union rallies and demonstrations.

Perpendicular to the Grands Boulevards are the boulevards Sébastopol and Strasbourg that, attached end to end, serve to lengthen the boulevard Saint-Michel. Here, students mix with immigrants from the Goutte-d'Or region of Paris. Although Paris is neatly divided into "arrondissements" or districts—like a pie cut into equal parts—not all of the districts are valued equally. What are considered fashionable districts today may not be tomorrow. And, according to Parisians, there are also districts to avoid. Those living in the 8th arrondissement are well aware of what distinguishes them from their fellow Parisians living in the 19th. Some who inhabit the 18th arrondissement may go so far as to differentiate themselves from neighbors living only yards away. There are even streets and avenues where one side of the street is favored over the other—usually based upon which side gets direct sunlight. And then there are those streets, like rue Saint-Florentin on the western edge of the 1st arrondissement, which has the east side of the street included in its jurisdiction, but not the west side.

In Paris, landmarks that help define the spirit of each district consistently punctuate the east-west axis from Place de la Bastille, to Place de la Madeleine. Place des Vosges, Place de l'Opéra, Place de la République, Place de la Concorde and Place Vendôme are the most famous. Each represents something different: the area surrounding Place des Victoires, known long ago for the production of upholstery fabric, and was later pretty much abandoned. The area saw new life when the famous fashion designer Kenzo moved there in the 1970s, ten years later he was joined by other fashion designers. Les Halles, which had become an urban blight, was completely transformed. In the Marais, beautiful hôtels particuliers built in the 16th and 17th centuries provide a contrast to the townhouses of the Left Bank. As of late, the prestige of the 3rd arrondissement has risen greatly due in part to the presence of the Picasso Museum which opened there in 1985, and the numerous contemporary art galleries. Living in the Marais district today is almost like joining a religious sect where a certain image is strictly observed. Beyond the Bastille, the faubourg Saint-Antoine is also being rediscovered and reinvented. Everywhere one looks—above, beyond and all around Paris's 20 districts, in areas still disdained by many—one can feel a new sense of vibrancy and life. More likely than not, it is from these new frontiers that the newest ideas in decorating will come.

THE MARAIS

Thanks to an eclectic auctioneer, a museum-like amalgam of periods and styles enlivens this 17th-century townhouse in the heart of the Marais.

Top: In the foyer, a pair of stenciled columns surrounds a fresco mural of Hercules by Farnèse.

A CROSS-CULTURAL MIX OF PERIODS AND STYLES

Bottom: In a corner of the living room a Kota reliquary made of wood and copper contrasts with an art deco Syrian vase. The portrait is by Morvan (1960).

Opposite: Interconnecting rooms open onto the garden. The double doors are painted in faux mahogany and the moldings are accented with gold leaf.

STROLLING THROUGH THE MARAIS

1. SQUARE LÉOPOLD-ACHILLE
A peaceful location situated between the rue Payenne and rue de Sévigné.

2. LE PAVILLON DE LA REINE
Hidden behind an imposing building on the Place des Vosges, this graceful townhouse hotel is a haven of peace and tranquility, only steps away from the bustling activity of the district.
28, place des Vosges, 75003.
Phone: +33 1 40 29 19 19.
www.pavillon-de-la-reine.com

3. LES INDIENNES Not a place for wild Indians, this boutique offers examples of a sophisticated art— 19th-century Indian cashmere shawls.
10, rue Saint-Paul, 75004.
Phone: +33 1 42 72 35 34.

4. LES CARRELAGES DU MARAIS
Glazed terracotta tiles in any shape, color or size, from floor to ceiling. These tiles are the very best.
46, rue Vieille-du-Temple, 75004.
Phone: +33 1 42 78 17 43.
www.carrelagesdumarais.com

5. SENTOU exhibits contemporary works that are both beautiful and stylish. This place practically defines creativity.
24, rue du Pont-Louis-Philippe, 75004.
Phone: +33 1 42 71 00 01.
www.sentou.fr

6 & 9. MUSÉE CARNAVALET The setting of this museum devoted to the history of Paris is magnificent. Housed in three period townhouses, it is undoubtedly the most charming museum in the capital.
23, rue de Sévigné, 75003.
Phone: +33 1 44 59 58 58.
www.carnavalet.paris.fr

7. PLAISAIT An illustrious Paris goldsmith since 1820, here one finds replicas of rare old pieces along with contemporary designs —all crafted with the greatest precision and artistry.
9, place des Vosges, 75004.
Phone: +33 1 48 87 34 80.

8. MUSÉE PICASSO During the time of Louis XIV, this townhouse was called "Salé," the home of a rich salt tax collector. Both the building and its contents are jewels of an area already glittering with architectural treasures.
5, rue de Thorigny, 75003.
Phone: +33 1 42 71 25 21.
www.musee-picasso.fr

10. ARREDAMENTO
Devoted to contemporary design, this shop carries works by such illustrious names as Ettore Sottsass and Gilles Derain. The items date from the period between yesterday and tomorrow.
18, quai des Célestins, 75004.
Phone: +33 1 42 78 71 77.
www.arredamento.fr

11. LE GÉNIE DE LA BASTILLE
The French Revolution's ideals have never been portrayed so lightly.

12. IZRAËL It's more than a grocery store; it's a museum. We would list this boutique for its scent alone. The variety of products in this shop makes it a genuine Promised Land.
30, rue François-Miron, 75004.
Phone: +33 1 42 72 66 23.

1

5

6

9

2 3

4

7 8

10 11 12

THE MARAIS

WHERE ELEVATION MEETS SECLUSION

This space was created out of an old industrial building and has all of the dramatics of a loft, while still providing its occupants with their own private space. The master designer of this house, made entirely of wood, is the young architect Hervé Vermesch. Here in this ancient tannery, a perfect balance is achieved between the original structure and the new spaces.

Left: By removing a section of the roof, Vermesch made room for a private indoor garden.

THE MARAIS

Right: For the bathroom walls, plain white tiles and a sink made of a block of damascened stone (Robinetterie Chavonnet).

Below: In the kitchen, a soapstone worktable; the range hood is made of cathedral glass. All cabinetry is by Ikéa. On the walls, sections of damascened stone.

Opposite: The living room beams were exposed by the bachelor himself. In the background a partition separates the living room from the TV room. The central partition, with its built-in photos, was once a wall used for jai alai. It was found in a salvage shop.

STROLLING THROUGH THE MARAIS

1. KIMONOYA K as in kimono, I as in Ikebana, M as in of the moment.... It has all the mystery of the Empire of the Rising Sun, falling somewhere between the traditional and the innovative.
11, rue du Pont-Louis-Philippe, 75004.
Phone: + 33 1 48 87 30 24.

2. LES MILLE FEUILLES Flowers and other ephemeral objects live side-by-side in perfect harmony. A garden of the senses, fleeting yet always renewed by Pierre Brinon and Philippe Landri.
2, rue Rambuteau, 75003.
Phone: + 33 1 42 78 32 93.
www.les-mille-feuilles.com

3. MADE IN JAPAN MINIATURE Here, everything comes from far off. An endless supply of exotic gifts.
11, rue de Béarn, 75003.
Phone: + 33 1 42 77 03 63.

4. GALERIE TAVOLA Their specialty is table art. Connoisseurs come to find a rare and lovely assortment of wares from the great potters and ceramists of the 19th and 20th centuries.
19, rue du Pont-Louis-Philippe, 75004.
Phone: + 33 1 42 74 20 24.

5. MAISON LAURENÇAT The ultimate in glass restoration and repair since 1894. New objects in glass can also be found in this studio-boutique devoted to fragile, beautiful works.
19, rue des Gravilliers, 75003.
Phone: + 33 1 42 72 96 45.

6. LE COMPTOIR DES ÉCRITURES An exceptionally wide variety of paper, brushes and inks; a temple to calligraphy in the age of computers.
35, rue Quincampoix, 75004.
Phone: + 33 1 42 78 95 10.

7. GALERIE FIFTEASE This gallery pays homage to the 1950s, dedicating itself exclusively to the French decorative arts.
7, rue du Perche, 75003.
Phone: + 33 1 40 27 04 40.
www.fiftease.com

8. ENTRÉE DES FOURNISSEURS Two thousand different varieties of buttons, an abundance of anything else one finds in a notions shop.
8, rue des Francs-Bourgeois, 75003.
Phone: + 33 1 48 87 58 98.

9. L'ESCARGOT MONTORGUEIL In this, one of the oldest restaurants in les Halles, the decor vies with the food. In the entranceway a ceiling originally painted for Sarah Bernhardt's kitchen. A must-see.
38, rue Montorgueil, 75001.
Phone: + 33 1 42 36 83 51.

10. LE GEORGES Named after the French President Georges Pompidou, founder of the Centre Pompidou, this restaurant atop the museum offers the perfect place to rest a while, with one of the most beautiful vistas of Paris.
19, rue Beaubourg, 75004.
Phone: + 33 1 44 78 47 99.

11. TEISSO ANTIQUITÉS Airport terminal lamps by Jieldé, chairs by Cadestin de Bertoïa, a bed by Prouvé, armchairs by Eames....need we say more?
81, rue Vieille-du-Temple, 75003.
Phone: + 33 1 48 04 59 07.

12. MICHEL GERMOND Whether the job takes 15 minutes or 1,500 hours, for Michel Germond, restorer and maker of fine furniture, each time it's a labor of love. To him, what matters most is the artistry and timelessness of each piece he makes.
78, quai de l'Hôtel-de-Ville, 75004.
Phone: + 33 1 42 78 04 78.

13. THE STAIRCASE OF THE HÔTEL LE PELETIER-SAINT-FARGEAU All the nobility of the Age of Enlightenment can be found in this hotel, recently linked to the Carnavalet townhouse.

2

3

4

7

8

13 10

THE MARAIS

PAST AND PRESENT

In this old apartment in the Marais, the owners mix a bit of everything—bric-à-brac, mirrors and works in wood. Despite the extreme contrasts they blend beautifully alongside important examples of modern art. An eclectic yet harmonious decor designed by Frédéric Méchiche. The living room is filled with woodwork from a stagecoach station dating back to the 18th century.

Who said the Directoire style wasn't compatible with a taste for the sporty?

Above: This black-and-white room is embellished by a work in black ink by painter Charles Blais; the delicate string of lights is from the Galerie Sentou. In the foreground, a plaster vase rests on a pedestal made of bleached pine.

Below: Frédéric Méchiche's workout room looks more like a salon; the wooden sculpture is by Sofu Teshigahara, the original plaster work is by Laszlo Szabo. Beneath an engraving by Pincermin, a compression sculpture by César.

Opposite: An unusual Directoire-style bathtub made of deep-gray painted sheet metal. The bathtub pedestal is made of stone; the decorative marble floor is from the 18th century.

THE BEAUTY OF TRUE SIMPLICITY

Decorator and designer Didier Gomez is very much at home in the Marais. He prefers simple forms, and likes the contrast between shadow and light. In his apartment, he has eliminated hallways to accentuate the open space. With an emphasis on symmetry, he has placed all of the doorways in the same line of vision. The panels are made of Wenge-wood, the hinges are mounted on linchpins. French windows provide a mixture of rusticity and refinement.

STROLLING THROUGH THE BASTILLE

1. L'ART DU TEMPS Bird cages, curiosities in cast iron, an armadillo... this surreal inventory is as captivating as the Bastille itself.
63, rue de Charonne, 75011.
Phone: + 33 1 47 00 29 30.

2. TROPICAL TRADING
A good place to rest a while while exploring the Bastille.
This ancestral stronghold represents the finest in Parisian craftsmanship.
23, rue Basfroi, 75011.
Phone: + 33 1 44 93 93 44.
www.tropical-trading.fr

3. FLORENCE DUFIEUX
She is wild about batiks and paints large motifs on silk, each a one-of-a-kind design to adorn screens, seats, trellises, slip covers, you name it.
9, rue de la Fontaine-au-Roi, 75011.
Phone: + 33 1 42 72 87 79.

4. LA QUINCAILLERIE AU PROGRÈS
Just about anything imaginable can be found in this paradise of a store, where time stands still. Established in 1873, it specializes in home furnishings for craftspeople or anyone who loves both the beautiful and the practical.
11 bis, rue Faidherbe, 75011.
Phone: + 33 1 43 71 70 61.
www.auprogres.net

5. PASSAGE One hundred different doors, from traditional to modern. As its name suggests, it is housed in an open space. An equally huge variety of door handles, both practical and decorative.
9, rue Véga, 75012.
Phone: + 33 1 55 78 20 30.
www.passage-porte.com

6. LE MASSON BRUNOT In this studio, painted furniture is not only restored, but returned to its original glory by owner Gwenola Masson.
171, rue du Fbg-Saint-Antoine, 75011.
Phone: + 33 1 49 28 00 38.

7. GALERIE SÉGUIN
This man just loves the 1950s, and is an expert on furniture by Jean Prouvé: he's held several exhibitions of Prouvé's work, and has published a remarkable catalogue on his oeuvre.
5, rue des Taillandiers, 75011.
Phone: + 33 1 47 00 32 35.
www.patrickseguin.com

8. MAISON DELISLE Here, all things luminous—from lamps to sconces to chandeliers—come to life. A distinguished member of the entrepreneurial Colbert Committee, he bases the designs for these beautiful lamps on antique models.
4, rue du Parc-Royal, 75003.
Phone: + 33 1 42 72 21 34.
www.delisle.fr

9. ARZINC, FRANCIS ARSÈNE
This one-time roofer has a passion for zinc. He makes amazing things out of the material that once graced Paris's rooftops, creating a wide variety of objects for the home.
84, rue de la Chapelle, 75018.
Phone: + 33 1 40 09 74 46.

10. FÉES D'HERBE In this very civilized jungle, Sylvie Aubry and Dominique Bernard display their knowledge of interior decorating and their imaginations with panache.
23, rue Faidherbe, 75011.
Phone: + 33 1 43 70 14 76.

11. SOUBRIER This family den extends to several floors. The company, which rents items out to filmmakers, has accumulated over the last 150 years objects that they sometimes agree to part with but only with the greatest reluctance.
14, rue de Reuilly, 75012.
Phone: + 33 1 43 72 93 71.
www.soubrier.com

1

4 5

7

9 10

2

3

6

11

8

PALAIS-ROYAL

OFFICE ELEGANCE IN THE HEART OF PARIS

The internationally known decorator Alberto Pinto wanted to house his business in a typically Parisian house. He found the perfect atelier in the rear courtyard of this 17th-century townhouse, in a space measuring 3,600 square feet and covering four floors. Here Pinto exhibits his genius for the art of gracious living.

Below: Plenty of glass overlooking a spacious courtyard.

Opposite: The original stairwell was painted to look like stone. The chandelier is a 17th-century original.

In the atelier's central workroom, documentation of a huge variety of colors, fabrics and materials provide inspiration for future designs. Here assistants research the essential elements that will eventually result in a creation of distinction and flair.

Alberto Pintos's relaxed conference room: slipcovers on the armchairs and sofa are made of raw linen (Pierre Frey), a rug designed by Pinto for Sam Laïk, a large painting by Manolo Valdès and chairs and a round table designed by Ed Tuttle.

The Ritz hotel houses hundreds of objects and antiques and is one of the most prestigious showplaces, for the most famous people in the world.

Clockwise, from top left: The Windsor suite. A detail of the ceiling in the Chopin suite. In the attic, candelabras and lamp bases sit side by side with busts of Hemingway. The swimming pool in the hotel's health club.

Opposite: The Tea Room in the Vendôme salon.

PLACE VENDÔME
IN THE WINGS AT THE RITZ

154

PALAIS-ROYAL

Left: These fanciful Indian cotton fabrics lend a certain whimsy and exoticism to this otherwise classic apartment. The drapes are by Braquenié and the objects are from the Paul Bert market.

Below: On the mantel a pair of lamps by Nicole Mugler and a Chinese cup. Above, a classic Louis XVI gilded mirror. To the left, a series of four drawings in red chalk by Hubert Robert.

A PASSION FOR PRINTS

PALAIS-ROYAL

Right: This salon, tucked beneath a mansard roof, overlooks the Tuileries Gardens. The sofas and padded chairs are original Napoleon III (from the Drouot auction house); a romantic floral screen highlights a marble fireplace with a painting by Anne Vincent hanging above.

Below: A collection of gouaches depicting Chinese vases hang side-by-side on a large paisley-motif wallpaper (Rubelli). On the Louis XV armchair, fabric by Le Manach.

This small apartment with its stunning view, nestles under the eaves of a building on the corner of rue de Rivoli overlooking the Tuileries Gardens. Each element was chosen to show the view to its greatest advantage. In the distance, the Jeu de Paume, the Invalides, the Eiffel Tower.

Top: The padded leather armchair is by Yves Halard; the side table, lamp, fruit bowl, tray and the French quilted bedspread are by Blanc d'Ivoire.

A PIED-À-TERRE IN THE SKY

Bottom: In the entranceway, the flooring is made of cement tiles; the clock and the sofa are from the era of Napoleon III, the mirror and the chest of drawers are in the Dutch style.

Opposite: This apartment was originally an immense living room-dining room. Today, it's been transformed into a vast studio loft. Brick-colored walls lend a theatrical air to this garret space.

PALAIS-ROYAL

RUE SCRIBE

A NIGHT AT THE OPERA

The Opera Garnier was first unveiled to the public at the 1867 World's Fair, and was officially inaugurated in 1875. Throughout the Second Empire, Charles Garnier hired only the very best artisans from each different guild to create this exquisite interior, with its gilded masks and garlands, bronzes, colored marble and red-velvet upholstery. At the turn of the 21st century—after two years of restoration—this sumptuous edifice has reclaimed all of its former splendor. A feast for the eyes, under a ceiling painted by Chagall (not shown).

Below: The Opera's interior, as seen from the boxes.

Opposite: A phalanx of the Opera's famous armchairs as far as the eye can see.

CLASH OF STYLES, HARMONY OF TASTE

Interior decorators Maurice Savinel and Roland Le Bevillon have ingeniously mixed together objects which could be considered bourgeois with other more humble pieces. Put together with imagination and taste, this is a quintessentially Parisian apartment.

Above: Kitchen furniture designed by the decorators themselves.

Right: Through the doorway, we can just make out the dining room-kitchen.

Opposite: In the living room, the couch, also designed by the owners, is covered in a prune-colored velvet (Pierre Frey). On the art deco-inspired rug sit taborets made of ironwood from Burma. In the background, two plaster columns flank a piece of Chinese furniture, above which hangs a 1940 oil painting by Soungourof.

163

STROLLING THROUGH THE PALAIS-ROYAL

1. SOFITEL DEMEURE HOTEL CASTILLE
At this private hotel near Place Vendôme—where Cocteau once lived—we find an elegant intimacy. Jacques Grange's redecoration exudes a sense of warmth and comfort.
33, rue Cambon, 75001.
Phone: + 33 1 44 58 44 58.
www.castille.com

2. SALONS DU PALAIS-ROYAL SHISEIDO
Designed by perfumer Serge Lutens, these mauve and plum colored salons give off a hint of the fragrances found within.
25, rue de Valois, 75001.
Phone: + 33 1 49 27 09 09.
www.salons-shisheido.com

3. DIDIER LUDOT He knows the art of transforming secondhand clothes into works of art, or at least very wearable fashions. An antiques dealer who specializes in dresses, he investigates grandmothers' closets and resells the contents to their granddaughters. Clever!
24, galerie de Montpensier, 75001.
Phone: + 33 1 42 96 06 56.
www.didierludot.com

4. CAFÉ DE L'ÉPOQUE Just a simple bistro except for its unusual charm and its upper-crust clientele. Situated at the entrance of the marvelous Véro-Dodat passageway.
2, rue du Bouloi, 75001.
Phone: + 33 1 42 33 40 70.

5. LE GRAND VÉFOUR This temple to gourmet cooking maintained the marvelous decor that brought it from the early 19th century—when Balzac was writing A Harlot High and Low.
17, rue du Beaujolais, 75001.
Phone: + 33 1 42 96 56 27.
www.grand-vefour.com

6 & 11. THE GARDENS OF THE PALAIS-ROYAL As soon as the weather breaks, businessmen from the surrounding area enjoy a sandwich here at noon, and the birds eat the crumbs. In modern-day Paris, there is no place quite as convivial.

7. MARCHÉ SAINT-HONORÉ
The old market's iron arcades have been replaced by an impressive structure of glass and steel designed by Ricardo Bofill. This, one of his most beautiful constructions, is Bofill at his best.
6, rue du Marché-Saint-Honoré, 75001.

8. LE PRINCE JARDINIER A descendent of the illustrious Broglie family, the owner sells a variety of garden implements that carry his own label.
37, rue de Valois, 75001.
Phone: + 33 1 42 60 37 13.
www.princejardinier.fr

9. GALERIE COLBERT
Built in 1826, it was once a high-society gathering place. Today, it is a modest, charming passageway between the Palais-Royal and the Bibliothèque Nationale.

10. LA GALERIE DE VALOIS
Built by Victor Louis for Philippe d'Orléans—the first real estate deal—this prominent Parisian building remains one of the most charming places in the heart of the capital to stroll past.

1

3

4

6

8

9

5

7

10 11

PALAIS-ROYAL

Left: There's a romantic feeling to this living room created out of a hallway. The long tablecloths help define the space, while identical table treatments create a mirroring effect. The mahogany chairs are in the 19th-century Russian style. The hanging oil lamps are made of sheet metal. On each table a small pagoda made by master craftsmen. In the background, a 17th-century portrait of an English gentleman.

Below: A hallway is cleverly transformed into a photo gallery. On the fabric-covered divider, "Retour d'Egypte" (Braquenié), and a four-leaf clover by Louise de Vilmorin.

LIFE BY LENGTH

Right Bank style: Above an English table, an oil painting by Garel surrounded by lesser masters of the 19th century.

Interconnecting living rooms with a prominently displayed portrait by Christian Bérard.

PALAIS-ROYAL

IN A WRITER'S HOME

In Paris each arrondissement can be viewed as a village with a distinct personality all its own, where people of discernment can choose their location according to taste and temperament. Writer Marie-France Pochna chose this distinguished residence near the Place des Victoires. Here she cultivates a mixture of styles, while at the same time clearly defining her own. In the living room, the walls are floor-to-ceiling bookshelves. The two facing sofas are by Dominique Kieffer. The screen is by David Webster, the table is from the 1940s, and the Egyptian stool is a contemporary piece.

PALAIS-ROYAL

Below: An unlikely encounter between a bathtub and a leather Chesterfield couch. The stool is Egyptian and the cabochon goblets are in the baroque style.

Opposite: In this kitchen-dining room the table is set with porcelain dishes designed by Raymond Loewy for Rosenthal.

A BOOKSTORE THAT'S NOT BY THE BOOK

Frédéric Castaing collects and sells notable letters, manuscripts, autographs, dedications and photos from such diverse characters as Francis I to the French actress Arletty.

Above: Claude Monet at Giverny, shown here alongside one of his letters.

Right: This handsome shop, that overlooks the Palais-Royal Gardens, has a subdued and distinguished atmosphere befitting its merchandise.

Opposite: The guardian of this establishment, with its wall-to-wall framed pictures, likes to evoke a sense of nostalgia by pairing letters with a photo of their author—putting a name with a face, so to speak.

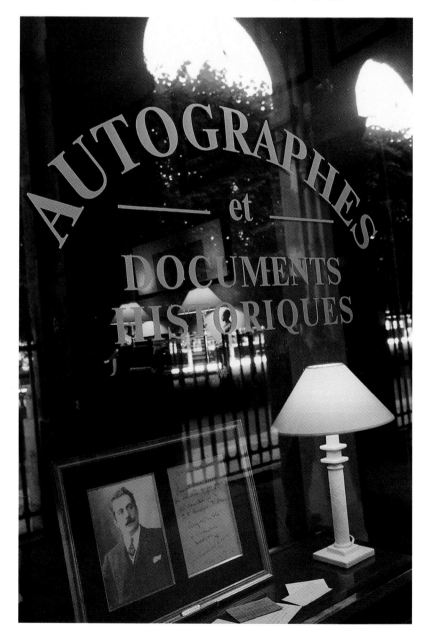

FAUBOURG SAINT-HONORÉ

TROMPE-L'OEIL ELEGANCE

In an unusual touch Anne Gayet commissioned Nathalie Mahiu to paint the walls of this cozy living room to resemble woodwork. The sofa's white-cotton slipcover is by Rebecca Campeau; cashmere kilim rugs and a romantic day bed, all serve to recreate the eclecticism of 19th-century Paris.

These days eat-in kitchens are no longer the exception but the rule. Here, a dining room with a kitchen alcove. Both the original molding and the fireplace, dating back to the early 19th century, were restored. A quilt from Provence and rattan chairs, a chandelier and a wicker tea cart harmonize in this small room.

The Hôtel Costes, redecorated by Jacques Garcia, has become one of the most fashionable hotels in Paris. In spite of its diminutive size, the ambiance is warm and welcoming—a comfortable haven at the corner of the Place Vendôme. The decor is completely original with comfortable upholstered chairs and flea-market finds, creating a bohemian-style elegance. It's this combination of imagination and chic that attracts clients weary of the usual decor one finds in luxury hotels.

Top: In the courtyard, two large stone porticos support antique statues in the neoclassic style.

Right: In the entrance, velvet consoles are sheathed in studded fabric.

Far right, and opposite, bottom: Jacques Garcia used antique furniture both for the bathrooms and the private rooms (here, neo-Indo-Portuguese from the time of Napoleon III), but not in the capacity for which they were originally intended.

Opposite, top: In one of the living rooms, a framed herbarium. A Napoleon III-style sofa was redesigned according to the patterns from that time.

PLACE VENDÔME

A RAVISHING BOUTIQUE HOTEL

Pastel tones, a collection of whimsical drawings from the turn of the century, souvenirs and mementoes create a soft, feminine ambiance, reminiscent of a young Parisian girl's room at the turn of the century.

Left: On the walls of this salon wide pine paneling painted in pale tones, a collection of pastels and "three-pencil" drawings by Helleu blend right in.

Below: Here, a collection of decorative vases where works by Daum, Galée and Lalique mingle with less valuable works.

TRUE TO THE BELLE ÉPOQUE

FAUBOURG SAINT-HONORÉ

Right: In this living room in the 8th arrondissement, the exposed beams have been painted glossy white. A handsome Louis XVI chest of drawers and a sleek roll-top desk from the 19th century blend well with the chairs' clean white slipcovers.

Below: A collection of small clocks and old watches positioned around an alabaster vase converted into a lamp. In the foreground, a glass inkwell.

STROLLING DOWN THE FAUBOURG AND THE RUE SAINT-HONORÉ

1. HERMÈS The most distinguished leather-works store in the Faubourg, now offers an extensive collection of household items. Watch who comes and goes during the holidays and you'll see a very wide range of the privileged class. 24, rue du Faubourg-Saint-Honoré, 75008.
Phone: + 33 1 40 17 47 17.
www.hermes.com

2. GRANGE Comfort, tradition and style define this establishment that gives the flavor of an old family home.
116, boulevard Haussmann, 75008.
Phone: + 33 1 45 22 07 72.
www.grange.fr

3. GALERIE ARIANE DANDOIS This merchant recently opened her store of beautiful, decorative furniture at Place Beauvau.
92, rue du Faubourg-Saint-Honoré, 75008. Phone: + 33 1 43 12 39 39.

4. GALERIE AVELINE If the objects and furniture here weren't so rare and so expensive, we'd encourage everyone to visit this gallery where everything is both enchanting and exceptional, and often amusing. The "curator," Jean-Marie Rossi, is himself an international star. We visit him as we would a frequently discussed, rare historical monument.
94, rue du Faubourg-Saint-Honoré, 75008. Phone: + 33 1 42 66 60 29.

5 & 6. FRETTE Not simply sheets and bedding here, but the most opulent and voluptuous bedroom accessories imaginable.
49, rue du Faubourg-Saint-Honoré, 75008. Phone: + 33 1 42 66 47 70.
www.frette.com

7. HÔTEL ASTOR Removed from the hustle and bustle of the city, yet centrally located. Frédéric Méchiche redecorated this luxury hotel in the style of the late 1930s.
11, rue d'Astorg, 75008.
Phone: + 33 1 53 05 05 05.
www.hotel-astor.com

8. COLETTE Simply the place to shop or just visit. No further explanation needed. Colette is a phenomenon; that is its greatness and its limitation.
213, rue Saint-Honoré, 75001.
Phone: + 33 1 55 35 33 90.

9. THE RITZ The reception and service one receives here is unsurpassed in all of Paris. Designed to please the rich and the powerful, one can still have lunch or tea here for a reasonable price.
15, place Vendôme, 75001.
Phone: + 33 1 43 16 30 30.
www.ritzparis.com

10. PLACE VENDÔME is a masterwork by Hardouin-Mansart, architect to Louis XIV. A statue of the King once stood center square. The column that replaced the monarch's statue is made from cannons used at the battle of Austerlitz.

11. LA MADELEINE Built by Napoleon as a temple to the glory of his armies, the chic now come to this church to get married. It matches the neoclassic style of the Assemblée Nationale.

2

3

4

11

7

9

NATURALISM IN THE ART DECO STYLE

Thomas Boog, the grand master of shell collecting, is also an amateur collector of Oriental art and hunting trophies. A prolific traveler and explorer, his shop and his apartment are filled with the spoils of his travels. The settees are by Le Corbusier, the chairs by Eames, the zebra skin is from South Africa, a Chinese-style clock was designed by Boog himself, and the antlers come from an Egyptian buffalo.

FAUBOURG SAINT-HONORÉ

Right: A tableau of objects against a background of enlarged botanical reproductions.

Below: In this room, a 19th-century silk-embroidered wall hanging; the lantern is by Thomas Boog.

Opposite: The ambiance of this room evokes the Far East, with its mixture of Chinese furniture and lacquered objects. The hanging lamp is made of white silk; the candlestick holder of wrought iron in a coral design was created by the master of the house.

BEATAE · MARIAE · VIRGINI · LAVRETANA

from MONTMARTRE to the BEAUX QUARTIERS

n 1860, Montmartre—the hill that dominates Paris, with its vineyards, windmills and garrets—was annexed to Paris, along with a myriad of other outlying villages that stretched from Passy to La Villette. Among all of them, however, it was Montmartre that stood out. The statesman Georges Clemenceau was once the mayor of this community—a place where the inhabitants leave their enclave with great reluctance, and only when obliged to do something in the city.

No one passes *through* Montmartre; one goes right *to* it. Montmartre is reached by way of a network of small streets that Edith Piaf immortalized in song. Today, the steep staircases are just as difficult to climb as they always were, but maneuvering cars on this hillside—transformed long ago by a flood into a veritable mount—poses a whole different set of problems. The Sacré-Coeur basilica with its imposing white dome tops Montmartre and dwarfs the jumble of historic, slate-gray houses nearby that people will snap up at any price. Montmartre is like a fortified city, an entrenched camp with even its own cable car that lords over the bustling boulevards below. Isolated as it may be, few districts are as linked to the legend of Paris as Montmartre. All true Parisians carry a piece of Montmartre in their soul. The feeling of solidarity with Montmartre radiates far beyond the district itself—this "island" has over time represented independence, insurrection, bohemia and the simple pleasures of life. At the turn of the 20th century, it also became a cradle of modern art. One can imagine la Goulue, the cabaret dancer made famous by Toulouse-Lautrec, crossing paths with Picasso's *Demoiselles d'Avignon*, Manet's top hats and Poulbot's blue-collar workers.

The slopes that begin at the Sacré-Coeur de Jésus and run down to Bonne Nouvelle and Montmartre's huge fabric store, the Marché Saint-Pierre, down to the Galeries Lafayette are very long. It's only when we near Pigalle, Trinité and Nouvelle Athènes on the side of Notre-Dame-de-Lorette and over to the Opera, that we once again come upon those kinds of old buildings with modest, melancholy facades. Many

Opposite: Notre-Dame-de-Lorette and the basilica of the Sacré-Coeur.

Next spread:
1. Place Saint-Georges. *2.* Musée de la Vie Romantique. *3.* Passage Jouffroy and bargain bookstores. *4.* Cité Trévise. *5.* Hôtel Vernet's dining room, on the street with the same name. *6.* Musée d'Art Moderne now renamed the Palais de Tokyo. *7.* Cité Léandre. *8.* Place de l'Alma, the flame here is the same size as on the French Statue of Liberty, by Bartholdi.

from MONTMARTRE

of them recall Paris's romantic period when tragedians, orientalist painters, bohemians in love with young shop girls, powerful financiers or those on the brink of success...an entire cast of characters out of some Balzacian play resided behind the neoclassic walls of Montmartre.

The first houses in Montmartre were built cheaply and quickly by factory workers who came, for the most part, from Limousin. Originally built to house the more modest elements of society, they were nevertheless dependable and worthy of the prestigious sandstone of which they were built. In general, they are attractive, modestly proportioned 18th-century-style buildings, reminiscent of a time when life was so much simpler. Their interiors too are modest, with none of those frenzied patterns found on stucco walls. Often the only decoration on these houses is latticework window sills, a few serrated cornices, an ancient porch here a stone post there, and quaint lanterns made of sheet metal. But what history! Behind the old cast-iron fountain, there is a tree so old that it must remember having seen Rubempré and Coralie, Balzac's star-crossed lovers*, passing by. Farther west is the Quartier l'Europe where trains from the station at Saint-Lazare (named after an ancient prison for women) carried the first Impressionist painters, eager to paint in the *plein air* genre, out towards Saint-Germain-en-Laye. The trains also carried those first tourists, also desirous of taking in the fresh air.

From the Quartier l'Europe, we move on to the Beaux Quartiers, the most exclusive sections of Paris, situated in the westernmost part of the city which includes the 16th, 17th, 7th and 8th arrondissements, as well as Neuilly-sur-Seine. The people who live here are not exactly what you would call defenders of a particular aesthetic, but compared to all the other denizens of Paris, who derive their identity from its history a legend or from their geographical placement, the people here (who came to refer to the buildings in this area, almost all constructed in one fell swoop, as Paris's "Haussmanian" district) base their traditions on a particular state of mind, rather than on anything they've inherited. The idea here, basically, is to live among

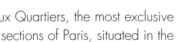

* Honoré de Balzac, *A Harlot High and Low*, 1838-1847.

to the BEAUX QUARTIERS

one's own.

In 1826, a finance company mapped out a new group of streets which would fan out in the pattern of a star across what was still then the countryside. All the streets intersected at Place de l'Europe and each street was named after a European capital in what was to be a congenial collaboration of cosmopolitanism established on an architectural plan. The Boulevard Malesherbes plows through a network of interwoven streets. Its temperament is slightly different than bohemian Montmartre, the light-heartedness of Pigalle, or the very respectable 17th arrondissement—where, many years ago, loose women discreetly entertained the wealthy tenants living on the ground floor. (These women now go by other names, but the Beaux Quartiers are still full of them.) On the second and third floors of these buildings lived famous doctors who received their guests in those bourgeois apartments with all their vast connecting rooms; like Doctor Proust, for example, who lived on rue de Courcelles, and was father to one particular Marcel. Above them, members of the middle middle class resided, and one floor above that lived members of the lower classes. Finally, in tiny rooms just under the eaves of the building, lived the maids, who were relegated to using the back stairs to empty the chamber pots.

Today, these houses have been transformed into comfortable apartments. Finally freed from the influence of their elders, the younger generation of Parisians has invested a great deal here. If one finds a place that isn't exactly one's style, all that's required is imagination. Dare to diverge from the beaten path and you can successfully turn your place into

"Home, Sweet Home." Paris was once very structured—different sections of the city, it was understood, housed different kinds of people. Today all of that has changed. Places once considered "out of the loop," too "historical" or "bourgeois" are now considered fair game. Despite its propensity for classification, Paris still remains a city where anything is possible, as long as imagination reigns.

MONTMARTRE

NEW LIFE FOR AN OLD STUDIO

Two young architects took a whole new look at this artist's house located at the foot of the hill Montmartre. They tore down the walls and ceiling to maximize both light and space.

Left: In the garden patio, the main pathway is made of Iroko-wood slats, and sandstone cobblestones define the borders of the carpet-like squares of stone laid in grass. The trees are Japanese maples.

MONTMARTRE

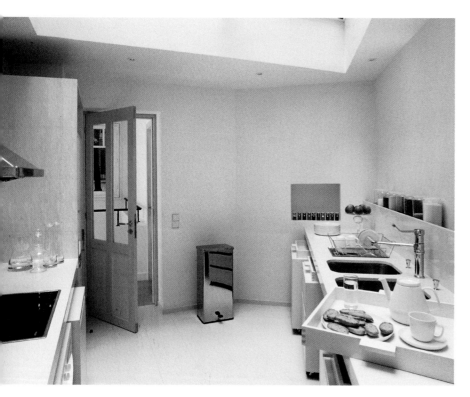

A large oak door separates the kitchen and dining area and a "withdrawing" room in which to rest.

Above: In the kitchen, lots of light, stone and sycamore wood.

Right: In the living room, the serenity of Asiatic wickerwork, a paper lantern by Noguchi and a low, 19th-century Chinese lacquered table.

STROLLING THROUGH MONTMARTRE

1. À LA MÈRE DE FAMILLE
Few households today can boast cabinets as well stocked with such delicacies as we find here. Dedicated to the gourmet and the gourmand alike, this is a boutique devoted to the good life.
35, rue du Faubourg Montmartre, 75009.
Phone: +33 1 47 70 83 69.
www.lameredefamille.com

2. CLAUDE HUOT This builder of fine furniture—whose den is frequented by the top antiques dealers—also restores paintings on wood.
13, rue de Montyon, 75009.
Phone: +33 1 47 70 58 19.

3. LIBRAIRIE CHAMONAL
In this 100-year-old bookstore, exclusively devoted to books on travel and science, one can also find rare editions. For the expert and amateur alike.
5, rue Drouot, 75009.
Phone: +33 1 47 70 84 87.

4. SERGE PLANTUREUX This eccentric with a mania for photography has his own museum; the annex is in the Galerie Vivienne. He's a collector and a dealer of a wide variety of curios.
4, galerie Vivienne, 75002.
Phone: +33 1 53 29 92 00.
www.sergeplantureux.fr

5. HÔTEL DROUOT Visit here on a regular basis and waste plenty of time, as time can always be made up later. A truly exceptional establishment in France, this auction house is also unique in the world. The best items sit side by side with the worst, just like in the real world.
9, rue Drouot, 75009.
Phone: +33 1 48 00 20 20.
www.drouot.fr

6. THOMAS BOOG
Previously located in Passage Jouffroy (photo opposite, Thomas Boog, an expert on shells, he recently opened a boutique on rue de Bourgogne, although more than just shells reside here.
52, rue de Bourgogne, 75007.
Phone: +33 1 43 17 30 03.
www.thomasboog.fr

7. L'ACADÉMIE DE BILLARD
If for no other reason, one should visit this repository of green felt tables for the decor alone. It's essential.
84, rue de Clichy, 75009.
Phone: +33 1 48 78 32 85.

8. HUBERT DUCHEMIN A tiny gallery where the owner has amassed 18th-century drawings and studies from different periods. Interesting to all knowledgeable collectors who are guaranteed to find something unusual.
Passage Verdeau, 75009.

9. ATELIER GUSTAVE MOREAU
For a long time it was Paris's best-kept secret. Rarely does one find a place like this where time stands still; where one can penetrate the universe of a great artist. The master of symbolism, Moreau left all his works to the city of Paris: paintings, drawings…everything exactly as it was at the moment of his death. An extraordinary place.
14, rue La Rochefoucauld, 75009.
Phone: +33 1 48 74 38 50.

10. ÉCOLE DE BRODERIE D'ART LESAGE Having outlived the great silk embroiderers of the last century, François Lesage is looking toward the future and is nourishing new talent. This conservatory, whose items are both beautiful and finely crafted, is open to the public.
13, rue de la Grange-Batelière, 75009.
Phone: +33 1 48 24 14 20.

11. SOPHA At Sopha, one finds the most beautiful and contemporary faucets and sinks. Here, hygiene has risen to an art form.
44, rue Blanche, 75009.
Phone: +33 1 42 81 25 85.

1

5

7

9

2

3

4

10

6

8

11

MONTMARTRE

THE RETURN OF THE ROMANTIC

The 9th arrondissement shelters mysteries of its own. Here, a young lawyer mixes neoclassic symmetry with a touch of modernism, along with the spirit of objects from Africa and Asia. When the weather is fine, he gives sophisticated dinners in his tiny jungle garden.

Left: Wide stripes painted on the wall contrast with the finer stripes of an 18th-century Gainsborough chair covered in velvet (Lelièvre). On the wall a pair of terracotta neo-Egyptian vases in the neoclassic style and a pair of late 18th-century engravings depicting Cupid with his swans. In the center, a work in ink by Jean-Marc Louis; the adjustable lamp on the right is by Philippe Starck.

Opposite: In the entranceway-dining room, an early 19th-century plaster Bacchus finds its place in a niche formerly reserved for a stove. The whitewashed-gray double doors are trimmed in orange; a metal chandelier from the time of Charles X.

MONTMARTRE

In the living room, the Empire style fraternizes with a bit of everything else: a tiger skin (from the Saint-Ouen flea market), a low, armless fireside chair by Jacob covered in ocelot skin, Louis Philippe-era armchairs inspired by the 17th century, a silk Asian-inspired lantern by Thomas Boog. On the Empire pedestal table, a chocolate-colored porcelain tea set by Utzschneider of Austria. The couch is in the Empire style; the 19th-century portraits depict Chinese dignitaries. The majority of the books were found bargain hunting at Béatrice Bablon.

A MAN OF MANY TALENTS

Above: In Christian Astuguevieille's kitchen, a table and a chair of braided hemp rope ingeniously designed by this decorator and innovator. The wall's ocher tones, the monochromatic earthenware plates (by Creil and by Wedgewood), and the cupboard made of chestnut planks all harmonize with the natural color of hemp.

Left: Both functional and decorative, the cupboard provides the perfect display for a collection of plates. In the foreground, two slightly different chairs made of raw hemp rope.

Opposite: Against the back wall, a large sideboard made of cord was specifically designed to hold this collection of antique dishware.

MONTMARTRE

A minimum of color and a sparsity of furniture serve to enhance the beautiful lines and high ceilings of this old artisan's studio.

Left: The counter is hewn out of a block of white stone, "Crème des Dunes." The faucets and sink are from Robinetterie Dornbracht. The table and two benches are made of heavy teak, the bookshelf is of oak —all stained a rich chocolate brown (design by Antonio Virga).

Below: The stained oak kitchen furniture was made to order and stained the same color as the parquet floors.

WHEN LESS IS MORE

MONTMARTRE

Right: Diaphanous white curtains allow light to flood the room without obscuring the beautiful old frame windows. The salvaged radiator is from Le Radiateur en Fonte. The teak armchair was designed by Antonio Virga.

Below: In the bathroom the double-pedestal sink comes from the Grand Hôtel de Cabourg (La Baignoire Délirante). The tinted made-to-order oak cabinets swivel around to become mirrors and drawers.

MONTMARTRE

This apartment, obviously inhabited by a collector, mixes 1940s design with neo-baroque furniture and interesting pieces of plaster.

Left: The 1950 dining room table is by Jean Royère. The paintings are by Pierre Buraglio and Jan Voss. The ceramic anthropomorphic vase is by Guéden. The floor lamp is by Poillerat. In the foreground, the coffee table made of mirror and bronze is by Arbus, all from the 1940s.

Below: On the mirrored bureau, a 19th-century porcelain bust by Thomas-Victor Sergent. The taboret is by Olivier Gagnère.

A LIGHT TOUCH

MONTMARTRE

Right: The late-1930s straw screen on the left by Jean-Michel Frank is a centerpiece of the room. Two taborets from around 1990, by Gagnère, lend a touch of the contemporary. Hanging dramatically above the bed is a plaster-framed mirror attributed to Serge Roche; the gilded iron chair is by René Drouet; the pair of bedside tables are by André Arbus.

Below: In the bathroom, a wrought-iron easy chair from the 1980s by André Dubreuil. Two stucco mirrors with scallop shells are from Galerie Kartica.

BEAUX QUARTIERS

Here we see the home environment of a collector who revels in the classic—souvenirs from travels abroad, classical statuettes, beautiful French and Swedish furniture from the 1930s and 1940s.

Left: The 1930s armchairs and taborets are by Jean-Michel Frank; the vases are by Ernst Boiceau (from Galerie Eric Philippe), and the 1940s portrait is by Christian Bérard.

Below: Atop the pale oak bookcase, *La Flûte de Pan* is by Lurçat; the reclining nude is by Eric Grate; the bronze lamp is by Arbus. Decor by Francis Roos.

THE COMFORT OF WOOD ALL AROUND

The study of a seasoned traveler and avid reader.

Left: The 1920s mahogany desk inlaid with holly is from Galerie Eric Philippe; porcelain Sèvres vase from the 1940s; the lamp with the astrolabe base is by Gilbert Poillerat, as is the standing lamp in the corner of the library.

Opposite: This room's centerpiece is a large, still life by Francis Jourdain. The Italian two-tiered library table from the 1940s is by Gio Ponti. A caned birch chair (Carl Malmstein); the rug is by Sue and Mare. Everything comes from Galerie Eric Philippe.

RUE MONCEAU
BEHIND THE SCENES AT THE HÔTEL DE CAMONDO

This recently restored home was named after its owner, Moïse de Camondo, who donated this property for a museum dedicated to the decorative arts of the 18th century. Although inspired by the architecture of the Trianon at Versailles, this beautiful house in the Plaine Monceau district had what was in 1900 considered an ultra-modern kitchen.

BEAUX QUARTIERS

THE PARC MONCEAU REVEALED

A well-known publisher lives here, at the other end of the book district. "This Haussmanian apartment, so often described in novels from the end of the 19th century, remains for me—once freed from its superfluous stuffiness—an ideal place to live," confesses the owner.

Right: In the living room, Roberto Bergero decorated the bookshelves in alternating patinas and Venetian stucco.

Below: In the dining room, a sienna patina gives the walls their buffed-gold look. The panels and resin wall lights were designed and painted by Bergero.

BEAUX QUARTIERS

Bill Pallot is a prominent specialist of 18th-century furniture and objects, an art historian and an associate of the great antiques dealer Didier Aaron. He's not exactly trying to create a "tasteful" decor in his home. Rather, he seeks shock value through an audacious mixture of styles. In his living room, the centerpiece Is an oil painting by Jean-Michel Basquiat,

WHEN CONTEMPORARY ART MEETS GREAT OPULENCE

flanked by two magnificent 18th-century wall lights in gilded bronze designed for the duchess of Parma during the time of Louis XV. To the right of the chimney, a portrait by Antoyan. On the left, a dance mask from the Bobo tribe of Burkina-Faso. The couches, which date from the time of the Second Empire, are covered in embossed, silk velvet (Lelièvre).

Clockwise, from top left: In front of a 1990s painting by Combas titled *Adam et Eve*, a 19th century clock, a Charles X candlestick holder with a reflecting screen; a 19th-century Austrian horn and the head of a cheetah made of papier-mâché. The 1940s table, decorated with lacquered panels, is from China. In the entrance, a large rain drum from the Vanuatu tribe (Oceania). A Second Empire-patinated bronze griffin. Beneath a contemporary sculpture titled *Momie*, by Iommim, a remarkable 1750s armchair by the fine furniture maker Nicolas Heurtaut.

Opposite: A green neo-Gothic chair by George Alphonse Jacob, and an 1830 armchair with the original upholstery.

IN A BIBLIOPHILE'S HOME

For his personal library, José Alvarez, founder of French publishing company Editions du Regard, designed three levels of shelves for a 22-foot high room. Four French windows and wall lights attached to oak supports all help to brighten this room devoted to culture of all kinds. On the concrete floor, a lead sculpture by Anselm Kiefer. The blond-wood library table is from the 1930s. On the walls photographs by such varied artists as Gisèle Freund, Rogi-André, Rodchenko, and Helmut Newton.

BEAUX QUARTIERS

THE PINNACLE OF ART DECO

Designed by architect Boileau in 1925, the Prunier restaurant was rescued and restored by Pierre Bergé and transformed into a fish restaurant whose specialty is caviar from the Aquitaine region of France. The romantic, dusky atmosphere will no doubt play a part in its success.

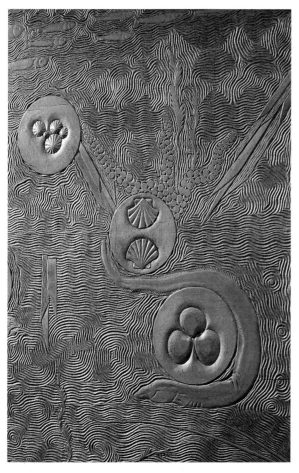

Clockwise from top left: The gilded art deco wooden panel evokes a mysterious undersea world; the marble staircase is inlayed with onyx, sintered glass, and gold leaf; at the end of the bar, a bronze Breton fisherman who fished in Iceland, reminds one of the writer Pierre Loti and his work, a contemporary of Emile Prunier and the author of *Fishermen of Iceland;* a view of the bar, left intact from the days when Prunier was considered one of the most elegant restaurants of the 16th arrondissement—a time when few restaurants existed in the area.

Opposite: On the double doors, the sea horse motif etched in glass is the restaurant's trademark. The restaurant is registered with the Bureau of Historic Monuments.

BEAUX QUARTIERS

Sculptor Hubert Le Gall designed this space to accommodate items from the 1930s and 1940s that the owners had assembled in their 16th-arrondissement townhouse.

Left: Between two windows, a work by Tamara de Lempicka. In the foreground, furniture from the 1940s (Christian Sapet) on an art deco-style rug.

Below: Closely hung works with a Picasso drawing in the center suspended from a bronze leaf.

A COLLECTOR'S HOME

BEAUX QUARTIERS

In this duplex near the Eiffel Tower, black and white starck predominate. The stunning interior, by Frédéric Méchiche, looks out onto the Parisian sky.

Left: In this opulent 1950s building, the terrace runs the length of the apartment—the terrace furniture was designed by Frédéric Méchiche for Hugonet.

Below: In the austere living room, a cauldron-shaped sculpture by Eric Schmidt; the striped sofas and the day bed are covered in raw linen. In the background, to the right, is the dining area.

A DUPLEX'S WHITE MAGIC

BEAUX QUARTIERS

Below: In this bedroom, the slipcovers are made of white linen (First Time).

Opposite: The kitchen's black-and-white motif matches the living room. The mosaic checkerboard flooring is made of antiqued marble. The ceiling is painted in black lacquer; the kitchen range and hood and the work areas are of stainless steel (Gaggenau). The wall lights are from the 1950s, the chair is by Bertoïa (Knoll), and the black dishes are by Calvin Klein. On the counter, a ceramic work by the American artist Sol Lewitt.

BEAUX QUARTIERS

This triplex, which overlooks the Bois de Boulogne, makes use of transparency, natural tones and sparsity. Julia Errera, a real estate agent, entrusted Yves Taralon with the redecoration of her triplex.

Left: "City" couches (B et B), drapes by Rubelli, a wool rug by Lauer, all beautiful shades of white.

Below: In this passageway through which has to walk when going from the kitchen to the living room, an oak bar designed by Yves Taralon.

Right: The wrought iron handrail along the staircase is by Olivier Gagnère.

BEAUX QUARTIERS

Right: In Julia's room, the bed is set at a diagonal. The linens are from Postel-Vinay; Egée lamps (Artoff), teacup by Bernardaud; the painting is by Claude Lepoitevin.

Left: The etched-glass doors are dotted with the same gilded bronze as these specially designed door handles.

Below: A view of the bathroom, as seen through a door of corded ash. The sink is by Philippe Starck.

BEAUX QUARTIERS

LET THERE BE LIGHT

Bright light and raw materials are the theme for this loft-like duplex apartment. The living room's pine plank floors are painted lead white. The 1925 cupboard is made of burled walnut; the pillar sculpture to the far left is from the Dogon tribe of Mali (galerie de Monbrison). The closet doors and table, of the African wood fraké, were made to order. The armchair is by Pescatore; the kitchen chairs are made of zinc (Fenêtre sur Cour). The walls are covered in flat white paint (Emery et Cie).

Left: In the entrance, a large bookcase made of Fraké-wood covers the wall. The polished concrete staircase leads to an open bedroom.

Opposite: A glass wall allows for magnificent unobstructed views of the rooftops.

the MARCHÉ AUX PUCES *and the* SUBURBS

Speaking of the most famous of all the flea markets…it was previously called the Saint-Ouen flea market; today it is known as the flea market at Clignancourt. In its first iteration, it referred to a village on the outskirts of Paris. In the second, it goes by the name of one of the numerous "gates" to the city that once allowed outsiders through the city's encircling wall. Clignancourt is at the city limits, where the metro line ends for a long time, between this gate and the village of Saint-Ouen, it was a wasteland. As late as the 1960s, gypsies, rag pickers and junk peddlers camped there. They played the guitar, burned unidentifiable debris and sold the pitiful belongings of people who had reached the end of their wretched lives to others who were as destitute as they (but for the fact they still had their lives!). After some haggling, these buyers would purchase such necessities as a tabletop gas stove or a chamber pot.

"Cleanliness is the luxury of the poor—the rich, go ahead and be dirty," wrote André Breton in his first Surrealist Manifesto. The original members of the Surrealist movement, displaying through their art the tenets Breton laid down in his manifesto, went bargain hunting in the area between Clignancourt and Saint-Ouen for wicker mannequins, nightingale cages, umbrellas and sewing machines. Soon it had a certain bohemian caché and photographs from the late 1930s show celebrities such as Bérard, Kochno, Henri Sauguet and Christian Dior strolling through the cast-off remnants of this world, drawing inspiration from objects they found there. It was called the "flea" market because there was always the chance that you'd get some.

Fifty years later, it's still called the flea market. Flea markets are a metaphor for our own lives—they represent what is lost and gained; as characterized by the famous phrase "one man's trash is another man's riches." Browsers always leave something that they hadn't been looking for. People speak with familiarity to merchants who, everyone supposes, are cheating you. And perhaps one day, if they do well enough here, these merchants will be able to open up some fancy establish-

the MARCHĒ AUX PUCES

ment in the Beaux Quartiers area. For the real miracle of the flea market is seeing such an enormous number of people pass through at such a furious pace, exchanging money (and considerable sums, at that) in the absence of any commercial controls.

The flea market could have become just another chic, trendy place—like the Village Suisse, or the original second-hand stores around the Place Furstenberg. The secret to its success is that it remained a place where it was possible to find just about anything—and something for everyone. Not unlike days long past when at the openings of Jean Genet's plays, socialites mingled with pimps. The contrasts can be stark: from antiques once passed down through generations to cell phones; here, we find mirrors; over there, second-hand smoking jackets. It's for this very reason that we love our flea markets. They are veritable showplaces that represent the evolution of a species that faithfully follows the formula of a famous Nobel Prize winner: "Chance, and necessity."

As the name indicates, the suburbs begin where the city ends—not a very enticing description. Although long a source of numerous problems, they are also the source of many services. Up until the 1960s, suburban produce farmers and florists still drove to Paris in their horse-drawn carts to deliver their wares. After that, developers came in and took over most of the farms and orchards—it had become necessary to house the masses of people who flocked to Paris for jobs or whatever draws people to the city.

As mentioned earlier, Paris, over the centuries, grew outward from its center at the Ile de la Cité (then known as Lutèce) in a spiraling fashion, expanding as its fortifications expanded. Just beyond the final outer walls, built by Minister Thiers between 1841 and 1844, lay what was referred to as "la zone." That term, which became part of the popular vernacular, came to describe something very specific: neither the city, the country, nor the suburbs, but a place of upheaval and deterioration.

In the past, this area outside the city walls supposedly allowed our courageous defenders to shoot their enemies out in this open area like rabbits. Although the countryside has never been very far away from Paris, "la zone"—although topographically closer—is still considered today, some 150 years later, far removed from the center of Parisian life. Situated for the most part to the north and to the east of Paris, those who lived there, bohemians and the destitute, constructed their shantytowns after the war with whatever fate provided them. Little by little, these shantytowns were replaced by moderately priced housing projects and stadiums, or were left as green open spaces. Beyond

"la zone" lie the suburbs. Without a doubt, "la zone" represents the future of greater Paris.

Beyond the beltway, which was built around the city during de Gaulle's administration, a variety of communities have emerged, each one different in character and each with its own individual charm—unless ravaged by city planners. But Parisians who are familiar only with the Neuilly region or the Marché

aux Puces should be reminded that our suburbs too are worth knowing and should be recognized as such; like Levallois, Asnières, Issy-les-Moulineaux, Garches, Saint-Cloud, Saint-Mandé and Charenton. The people who live there are often former Parisians who decided to leave because of the exorbitant price of housing, because the air was cleaner outside of the capital or because their families were growing and they needed more space. Today Paris is populated by the rich, the elderly and by singles (who now make up more than 50% of the population). The rest of the city is mostly offices and boutiques.

These words from a famous song, "It isn't love, it's only its suburbs," illustrate the Parisian contempt for their suburbs. But in a truly French paradox, these bucolic regions outside of Paris's "walls" seem to nevertheless attract city-dwellers. The tradition of fleeing the miasma and crush of the city on beautiful Sunday afternoons dates back to the 18th century and still continues today—Parisians went there to row boats and fill the country bistros.

Rich in cellars, attics, arbors and borders of flowers, these homes have become the dream of the new bourgeois-bohemian set whose ancestors came from the nearby countryside themselves. As of late, young couples have taken to dreaming of owning their own English cottage or a small farmhouse, or joining a retreat for artists and unrecognized composers…not to mention the seaside resorts reachable by Parisian train lines.

Things from the 1970s that were once considered kitsch, though still considered tacky, are now viewed as charming. We don't throw away grandmother's armoire, or her cuckoo clock, or even her old metal box for bouillon cubes. Everything has its purpose, and its price. With time, all things change and evolve and what could not have been imagined ten short years ago is now all the rage. Flea markets attest to that fact as much as the evolution of the suburbs: Tomorrow is always another day.

AN ANTIQUES DEALER'S SHOWCASE HOUSE

Having left Paris to live close to the flea market, Annick Clavier, who deals in antiques and flea-market finds, operates her business every weekend from her picturesque suburban home swathed in greenery on rue Paul-Bert in Saint-Ouen. The Virginia creeper is her only inviolate secret.

Right: The garden where everything is for sale.

Annick Clavier's basement serves as a typically French kitchen, dining room and living room.

Above: In a corner of the dining area, exposed brick has been painted white.

Right: The corner kitchen is separated from the living room by a wooden counter. In front of the couch (First Time), a pair of English leather armchairs.

SAINT-OUEN

In this living room, a portrait of a 1930s pugilist and neoclassic leather chairs. The pine paneling was found in a salvage shop.

SAINT-OUEN

A TRIPLEX WITH CONFIDENCE

Christian Sapet, one of the most popular antiques dealers of the Paul Bert flea market, maintains that he was only being practical when he elected to buy for his weekend getaway this big old three-story house.

Left: Below the second floor staircase, a small bedroom-sitting room. Above the Directoire-style bed on the left, a mirror gives the illusion of being in the crosshairs.

SAINT-OUEN

Above, right: In this bachelor's bedroom, a wrought iron bed covered by a 19th-century quilt from Provence. The bookshelves are archival shelving. The walls are paneled in raw wood planks whitewashed in Blanc d'Espagne.

Below, right: The bathroom sink comes from a luxury hotel and dates back to the turn of the century. An ultra-flat radiator serves to camouflage the bathtub.

Opposite: Louvered shutters filter the light from above falling on this loggia. The bronze sculpture is by Hiquili; the 1910 oil painting is by Paul Vera; the 1910 table is by Serrurier-Bovy and the 1950 seats are by Subes.

ISSY-LES-MOULINEAUX

AN OASIS UNDER THE VIADUCT

In the suburbs, amidst a tangle of vegetation, some strange islets can be found, like this unusual sheltered villa near a railroad track in Issy-les-Moulineaux.

Opposite: A lovely view overlooking the willow trees in a garden configured by Alain-Frédéric Bisson.

Left: A log table by David Hicks; the "log" pottery is by Lopez. Everything here enhances the sense of calm, even the dog relaxing under an open window.

Above: In this corner kitchen, a painted butcher block was turned into a console; the candlestick lamps have been electrified. The screen is covered in damask (Pierre Frey).

Opposite: In this house in Issy, the kitchen-dining room has a neo-Gothic mantelpiece. Between two windows, a rustic sideboard from the 17th century. The chairs are from the Arts and Crafts period, and the carpet is in needlepoint.

Above: A romantic bedroom setting with a faux leopard rug (Casa Lopez), and a pair of mirrored mahogany armoires in the Louis Philippe style. A French quilted bedspread (Blanc d'Ivoire).

Opposite: In the living room a three-sided ottoman couch in the style of Louis XV is flanked by two unusual painted metal shelves resembling Christmas trees, and a rug with floral designs (all from Casa Lopez). A rustic Italian table (Vivement Jeudi).

MEUDON

Just minutes away from the bustling French capital, a calm, verdant enclave dominates the heights of Meudon with its very British style of charm. Nad Laroche and her architect husband moved there some 20 years ago in order to raise their three daughters and they've never regretted it.

Below: In the entrance, the marble flooring is from Carrara, Italy. The painted console table is in the Directoire style.

ENGLISH STYLE AT THE GATES OF PARIS

MEUDON

Left: In this half-paneled dining room, the ebony table is circled by Arne Jacobsen chairs (1950s). The toile de Jouy wallpaper in manganese is by Braquenié; the chandelier is from the 19th century.

Below: The kitchen opens onto a winter garden. Living in the suburbs, one can take advantage of the bounty of both city and country.

MEUDON

Right: In the bedroom-cum-library, skillfully ordered disorder. The 19th-century furniture is mahogany; the Nad Laroche tapestry is after a design by Paul Klee; the chairs are English, and the jumble of objects, many bestowed as gifts, are testimony to numerous friendships.

Below: In the bathroom, fruitwood and marble add distinction.

Photographs by:
Alexandre Bailhache: pp. 164-165 *(ph. 4, 6, 7, 10, 11, 12)*, 192, 194, 200-201 *(ph. 1–6 and 8–12)*.
Joseph Benita: p. 186 *(ph. 11)*.
Guy Bouchet: p. 187 *(ph. 12)*, back cover *(4th row ph. 3)*.
Gilles de Chabaneix: pp. 48–51, 88–91, 162–163, 242–245, 270–275, back cover *(4th row ph. 4)*.
Stephen Clément: pp. 210–213, 246–249, back cover *(1st row ph. 3, 3rd row ph. 2)*.
Xavier Coton: pp. 16, 18-19 *(ph. 1, 3–8)*, 26 *(ph. 1)*, 28 *(center)*, 60-61 *(ph. 1, 3, 13)*, 78-79, 109 *(ph. 5)*, 126-127, 131 *(ph. 12)*, 136-137 *(ph. 9, 10)*, 145 *(ph. 8)*, 187 *(ph. 10, 13)*, 195 *(ph. 8)*, back cover *(3rd row ph. 3)*.
Vera Cruz: pp. 254–259, back cover *(1st row ph. 1)*.
Pierre-Olivier Deschamps: pp. 174-175, 200 *(ph. 7)*, 252 *(bottom left)*, 253 *(top left)*.
Pierre-Olivier Deschamps/Vu: pp. 61 *(ph. 8)*, 82–83, 97 *(ph. 2)*, 108 *(ph. 2)*, 136 *(ph. 6)*, 145 *(ph. 11)*, 187 *(ph. 3, 4, 5)*, 195 *(ph. 6)*, 230-231, back cover *(4th row ph. 1)*.
Deyrolle: pp. 76 *(top left and bottom right)*, 77.
Jacques Dirand: pp. 46-47, 64–67, 92–95, 109 *(ph. 9)*, 118-119, 138–141, 196–199, 208-209, 214–217, 238–241, back cover *(1st row ph. 2)*.
Alain Gelberger: p. 97 *(ph. 8)*.
G. Guérin: p. 253 *(bottom right)*.
Patrice de Grandry: pp. 116–117.
Marianne Haas: endpapers *(ph. 2)*, pp. 26-27 *(ph. 4–11)*, 32–39, 40–41 *(ph. 3, 5, 6, 9, 10)*, 42–45, 68–71, 76 *(top right and bottom left)*, 84–87, 96 *(ph. 4, 5)*, 98-99, 114-115, 120-121, 128-129, 130-131 *(ph. 1, 2, 4, 6–8, 10, 13)*, 136-137 *(ph. 1, 8, 12)*, 154–157, 164–165 *(ph. 1, 3, 5, 8, 9)*, 226–229, 236-237, back cover *(2nd row ph. 2, 4th row ph. 2)*.
Pierre-Laurent Hahn: pp. 160-161.
Olivier Hallot: pp. 144 *(ph. 5)*, 186-187 *(ph. 2, 6, 7)*.
Éric d'Hérouville: endpapers (except ph. 2), 58-59, 61 *(ph. 2)*, 136 *(ph. 5)*, 144-145 *(ph. 1, 2, 4, 6, 7 10)*, 186 *(ph. 9)*.
Noëlle Hoeppe: pp. 137 *(ph. 2)*, 224-225.
Seline Keller: pp. 132-135.
Vincent Knapp: pp. 104-107, back cover *(2nd row ph. 1)*.
Joël Laiter: pp. 52-53, 80-81, 202–207.
Guillaume de Laubier: cover, pp. 18 *(ph. 2)*, 20–25, 27 *(ph. 13)*, 40 *(ph. 1)*, 41 *(ph. 11)*, 54–57, 61 *(ph. 6, 11, 12)*, 72–75, 96-97 *(ph. 1, 3, 6, 7, 13)*, 102-103, 108-109 *(ph. 1, 4, 6, 7, 8)*, 110-111, 112-113, 124, 131 *(ph. 3, 9, 11)*, 137 *(ph. 13, 14)*, 146–151, 152-153, 158-159, 166–169, 170–173, 180-181, 182–185, 186-187 *(ph. 1, 8)*, 195 *(ph. 5)*, 218–221, 222-223, 260–263, back cover *(2nd row ph. 3, 3rd row ph. 1)*.
Hervé Lorgeré: p. 41 *(ph. 12)*.
Nicolas Mathéus: pp. 28 *(top and bottom)*, 28-29, 30-31, 188–191, 232–235.
Philippe Matsas: p. 195 *(ph. 7)*.
Éric Morin: pp. 142-143.
François Mouriès: pp. 26 *(ph. 2, 3, 12)*, 40-41 *(ph. 2,4, 7, 8)*, 60 *(ph. 4, 5, 7, 9, 10)*, 96-97 *(ph. 9, 10, 11)*, 131 *(ph. 5)*, 136-137 *(ph. 3, 4, 7, 11)*, 145 *(ph. 3, 12)*.
Jacques Primois: pp. 176–179.
M. Renaudeau/Hoaqui: p. 252 *(top left)*.
Ivan Terestchenko: pp. 264–269.
Vincent Thibert: p. 97 *(ph. 12)*.
Gilles Trillard: pp. 62-63, 100-101, 108 *(ph. 3)*, 122-123, 144 *(ph. 9)*.
Buss Wojtek/Hoaqui: pp. 252 *(center)*, 253 *(top right)*.
A. Wolf/Explorer: p. 250.
All rights reserved: pp. 144 *(ph. 5)*, 165 *(ph. 2)*.

Style by:

Alexandra d'Arnoux: pp. 224-225.
François Baudot: pp. 36–39, 58-59, 84–87, 98-99, 102-103, 108 *(ph. 1)*, 174-175, 187 *(ph. 3, 4, 5)*.
Marie-Claire Blanckaert: cover, pp. 20–25, 32-35, 46-47, 54-57, 62-63, 64-67, 68–71, 72–75, 100, 101, 104-107, 108 *(ph. 1, 3, 4)*, 110 111, 112 113, 122-123, 128-129, 138–141, 144 *(ph. 9)*, 146–151, 158-159, 160-161, 162-163, 164 *(ph. 1, 9)*, 176–179, 180-181, 182-185, 187 *(ph. 8)*, 196–199, 214–217, 222-223, 226–229, 238–241, 242-245, 254-259, 264–269, 270-275, back cover *(1st row ph. 2, 2nd row ph. 1)*.
Marie-Claire Blanckaert and Gérard Pussey: pp. 182-185.
Barbara Bourgois: pp. 132-135, 144 *(ph. 5)*.
Jérôme Coignard: pp. 166–169.
Françoise Delbecq: pp. 152-153.
Françoise Delbecq and Françoise Labro: pp. 118-119.
Laurence Dougier: pp. 188–191, 232–235.
Marie-Claude Dumoulin: pp. 88–91.
Armel Ferroudj-Begou: pp. 116-117.
Inès Heugel: pp. 164 *(ph. 5)*, 165 *(ph. 3, 8)*, back cover *(4th row ph. 2)*.
Marie Kalt: pp. 92–95, 164-165 *(ph. 2, 4, 6, 7, 10, 11, 12)*, 192, 194, 200-201 *(ph. 1–6 and 8–12)*, 208-209.
Marie Kalt and Gérard Pussey: pp. 260–263.
Françoise Labro: pp. 154–157.
Marie-Maud Levron: pp. 136 *(ph. 5)*, 144-145 *(ph. 4 and 6)*.
Catherine Mamet: pp. 210–213, back cover *(1st row ph. 3)*.
Franck Maubert: p. 195 *(ph. 6)*.
Nathalie Nort: pp. 186-187 *(ph. 2, 6, 7)*.
Marie-Paule Pellé: pp. 52-53.
Olivia Phélip-and Monique Duveau: pp. 80-81.
Marie-France Pochna: pp. 218–221.
Misha de Potestad: pp. 108-109 *(ph. 6, 7, 8)*, 136 *(ph. 6)*.
Gérard Pussey: pp. 114-115.
Isabelle Rosanis: pp. 246–249, back cover *(3rd row ph. 2)*.
Catherine Scotto: pp. 97 *(ph. 8, 12)*, 142-143, 202–207, 236-237.
Elsa Simon: pp. 48–51, back cover *(4th row ph. 4)*.
Sylvie Tardrew: pp. 18 *(ph. 2)*, 27 *(ph. 13)*, 40 *(ph. 1)*, 41 *(ph. 11)*, 61 *(ph. 6, 11, 12)*, 96-97 *(ph. 1, 3, 6, 7, 13)*, 124, 131 *(ph. 3 and 11)*, 137 *(ph. 13 and 14)*.
Laure Verchère and Olivier Chapel-Stick: p. 200 *(ph. 7)*.
Laure Verchère: pp. 26–27 *(ph. 2–12)*, 28–31, 40–41 *(ph. 2–10, 12)*, 42 à 45, 60-61 *(ph. 2, 4, 5, 7, 9, 10)*, 76 *(top right and bottom left)*, 82-83, 96-97 *(ph. 2, 4, 5, 9, 10, 11)*, 109 *(ph. 9)*, 130-131 *(ph. 1, 2, 4–8, 10, 13)*, 136-137 *(ph. 1, 3, 4, 7, 8, 11, 12)*, 144-145 *(ph. 1–3, 7, 10–12)*, 170–173, 186 *(ph. 9)*, 230-231, 252 *(bottom left)*, 253 *(top left and bottom right)*, back cover *(2nd row ph. 2 and 3, 4th row ph. 1)*.
Olivier de Vleeschouwer: p. 61 *(ph. 8)*.
Agnès Waendendries: p. 137 *(ph. 2)*.
Clara Whitebus: p. 186 *(ph. 1)*.
Claire Wilson: pp. 120-121.

The content of this book was taken solely from *Elle Décoration* and appeared only in France.